Proceedings of a
Symposium on

Numerical and Physical Aspects of Aerodynamic Flows

Numerical and Physical Aspects of Aerodynamic Flows

Edited by Tuncer Cebeci

With 302 Illustrations

Springer-Verlag
New York Heidelberg Berlin

Tuncer Cebeci
California State University
Long Beach, CA 90840
U.S.A.

Library of Congress Cataloging in Publication Data
Symposium on Numerical and Physical Aspects of Aerodynamic Flows (1981: California State University, Long Beach)
Numerical and Physical Aspects of Aerodynamic Flows.
1. Aerodynamics—Congresses. 2. Boundary layer—Congresses. I. Cebeci, Tuncer. II. Title.
TL574.F5S95 1981 629.132′32 81-18433
AACR2

Typeset by Composition House Limited, Salisbury, Wilts, England.
Printed and bound by R. R. Donnelley & Sons, Harrisonburg, VA.
Printed in the United States of America

9 8 7 6 5 4 3 2 1

ISBN 0-387-11044-5 Springer-Verlag New York Heidelberg Berlin
ISBN 3-540-11044-5 Springer-Verlag Berlin Heidelberg New York

Preface

This volume contains revised and edited forms of papers presented at the Symposium on Numerical and Physical Aspects of Aerodynamic Flows, held at the California State University from 19 to 21 January 1981. The Symposium was organized to bring together leading research workers in those aspects of aerodynamic flows represented by the five parts and to fulfill the following purposes: first, to allow the presentation of technical papers which provide a basis for research workers to assess the present status of the subject and to formulate priorities for the future; and second, to promote informal discussion and thereby to assist the communication and development of novel concepts.

The format of the content of the volume is similar to that of the Symposium and addresses, in separate parts: Numerical Fluid Dynamics, Interactive Steady Boundary Layers, Singularities in Unsteady Boundary Layers, Transonic Flows, and Experimental Fluid Dynamics. The motivation for most of the work described relates to the internal and external aerodynamics of aircraft and to the development and appraisal of design methods based on numerical solutions to conservation equations in differential forms, for corresponding components. The chapters concerned with numerical fluid dynamics can, perhaps, be interpreted in a more general context, but the emphasis on boundary-layer flows and the special consideration of transonic flows reflects the interest in external flows and the recent advances which have allowed the calculation methods to encompass transonic regions. The desire for numerical methods to aid the design process is prompted by the need to reduce the number and cost of experimental tests, particularly those with large or full-scale equipment. It must be recognized that numerical methods, and usually the equations which they are used to solve, introduce uncertainties, and experiments remain essential, though (as indicated in Part 5: Experimental Fluid Dynamics) these can be of reasonably small scale and cost.

The papers presented here have been modified and improved as a consequence of suggestions made by the session chairmen and the authors themselves. It is a pleasure to record the willingness with which authors set out to make their contributions as useful and stimulating as possible, and thereby helping to ensure that this volume has archival value and at the same time provides an overview of current knowledge. The first chapter in each part introduces material which provides an overview of its topic and makes specific reference to the papers which follow. Thus in Parts 1, 4, and 5, Dr. Keller, Dr. Ballhaus and his coworkers, and Dr. Roshko have modified the material which they presented at the symposium to fulfill this purpose: the introductions to Parts 2 and 3 have been prepared to provide the same service.

The Symposium was made possible partly by financial support provided to the California State University by the Office of Naval Research and also by the cooperation of authors, session chairmen, participants, and colleagues at the University. Particular thanks are due to Dr. R. E. Whitehead of the Office of Naval Research, Mr. Ralph Cooper formerly of the same office, and Professor Hillar Unt of the University. The editing process benefited considerably from the efforts of Nancy O'Barr and Sue Schimke, and it is a pleasure to acknowledge their help.

It is appropriate to recognize, at this time, the service given to the Office of Naval Research and to the research community by Mr. Morton Cooper, who retired after 18 years of service as an Engineering Science Advisor. Many of us have enjoyed long associations with him through the Fluid Dynamics Program of the Office of Naval Research: this volume is dedicated to him in recognition of the many successful research programs which he organized and supported in an efficient and understanding manner.

Long Beach, California
April 1982

TUNCER CEBECI

Contents

PART 2. Interactive Steady Boundary Layers

PART 3. Singularities in Unsteady Boundary Layers

PART 4. Transonic Flows

PART 5. Experimental Fluid Dynamics

PART 1

NUMERICAL FLUID DYNAMICS

CHAPTER 1

Continuation Methods in Computational Fluid Dynamics[1]

Herbert B. Keller*

1 Introduction

Continuation methods are extremely powerful techniques that aid in the numerical solution of nonlinear problems. Their use in computational fluid dynamics has, until very recently, been minimal. This is somewhat surprising, since they are rather well known in solid mechanics. Furthermore, their mathematical foundations—homotopy methods—were laid by mathematicians very much concerned with fluid-dynamical problems and in particular with the construction of existence proofs for the Navier–Stokes equations. We shall attempt here to recall or expose some of these ideas and to indicate some of their current uses in computational fluid dynamics. Brief remarks on their relevance to the papers on numerical fluid dynamics in this meeting are made in Section 2.

Continuation methods are based on the assumption that when a solution of a given problem, say $P(\lambda)$, is known, then it is easy to solve a "nearby" problem, say $P(\lambda + \delta\lambda)$. This is all quite vague and not always correct, but it does give the basic idea. To present useful methods of this type we will be more specific later on.

A key feature in all continuation methods is that some quantity (e.g. λ in the above description) will vary during the solution procedure. The quantity λ which varies may be a steady physical parameter of the problem, for

[1] This research was supported by the DOE under contract EX-76-S-03-0767, Project Agreement No. 12, and by the USARO under contract No. DAAG 29-78-C-0011.

* Applied Mathematics, California Institute of Technology, Pasadena, CA 91125.

example, a Reynold's number, the entrance velocity in a duct, the amplitude of a periodic disturbance, or the length of a body. In all of these cases each problem that is solved in the continuation procedure, $P(\lambda_1), P(\lambda_2), P(\lambda_3), \ldots$, is a physically meaningful problem whose solution may be of independent interest. Indeed, in design calculations such a sequence of solutions is frequently the desired goal.

The variable quantity λ may be an artificially introduced parameter designed to aid in solving a very difficult problem. This is the mathematical idea of homotopy—a continuous change from an easy problem to a hard one. More recently I have introduced a different type of continuation parameter, arc length or pseudo-arc length, for a deeper purpose [1.1], to detect and circumvent limit points and bifurcation points which invalidate normal continuation procedures. We shall illustrate these ideas later.

Finally, the quantity could be the time ($\lambda \equiv t$), an intrinsic variable not appearing explicitly in the problem. Of course this is a very old and frequently used idea to get steady states. It is also a continuation method. There are many reasons both for and against using this method to get steady states. I am usually against its use—if other methods are available and if the steady states are the primary goals. But the problems are so varied that no such blanket choice can reasonably be made.

Another benefit in developing efficient techniques for applying continuation methods in fluid dynamics is their use in bifurcation phenomena. Bifurcation is of course quite well known in theoretical studies of fluid flows. The engineering importance and significance of this phenomenon is not yet fully appreciated. The time will surely come, in the not too distant future, when the computation of such flows will be one of the main concerns in applied programs. But somehow in this area the fluid engineers remain in the dark ages, unlike their colleagues in solid mechanics. Perhaps that is because buckling is such an obvious and important phenomenon for structures, and the corresponding fluid behavior is not nearly so obvious (though it is equally important).

Of course, as the miniaturization of computers continues and their tremendous potential for distributed control is realized, the practical importance of bifurcating flows will also be realized. Then the need for the ability to compute unstable (as well as stable) bifurcating flows will be clear to all. It would be more comforting, however, if the tooling up for this future need were more actively pursued in the present. Independently of any agreement on the importance of bifurcation in practical fluid dynamics, there can be no disagreement on the fact that all practical methods for computationally determining the conditions of bifurcation use some form of continuation.

Finally, another basic use for continuation in computational fluid dynamics is just beginning to be realized. This concerns the crucial question of assuring that the results of any computation have some reasonable validity as approximations to some theoretically possible flow. It would be ridiculous to require instead approximations to observable (physical) flows, as that would assume we were sure of the relevant equations of motion. So all we can reasonably require is that the results of the computations approximate some appropriate solutions of the equations we are using to model the flows of interest. This problem is hardly ever mentioned in the computational fluid-dynamics literature. However, recent computations on the driven-cavity problem [1.2, 1.3] have revealed, via continuation methods, how spurious solutions of the numerical problem develop on a fixed grid as the Reynolds number is varied. On a much simpler example, using Burger's equation, deeper insights into this problem are obtained [1.4, 1.5], and a mesh Reynolds-number concept, first introduced in [1.6] (pp. 125–126), is clearly related to some of the spurious solutions. From a purely mathematical point of view the idea of continuation in the mesh spacing, $h \to 0$, is related to spurious solutions of very special difference equations [1.7]. This subject is at present much too difficult and ill developed for any assurances of validity for any single set of computations. But we should now be aware that spurious solutions, looking very much like the "real thing", can be present in many numerical schemes. Then we should be prepared to be able to rule them out, perhaps by additional numerical evidence.

2 Continuation and the Current Papers

Rather than review all the papers in the current sessions devoted to numerical fluid dynamics, I will confine my comments, for the most part, to their relation to continuation methods.

Professor Moretti's interesting experimental calculations on the response of different methods for treating the flow over a bump in channel flow contain numerous possibilities for the productive use of continuation methods. Some have been used, others have not. Two continuations of interest include variations in the bump amplitude b or the Mach number, for example. The former has been done, but it is not indicated that the methods of continuation theory have been used. Rather the implication is that for each amplitude value b, a completely new calculation was done—an inefficient procedure. But, of course, the purpose of this work was not to develop efficient codes but rather to explore the response of different numerical methods to various difficulties—an admirable goal.

The paper of Prof. Wu is a rough survey of hybrid methods he has used on a variety of problems in fluid dynamics. Insufficient detail is given to allow us to determine if these methods use any form of continuation. However, one important idea he stresses—to use different formulations and numerical methods in different regions, depending upon the local length scale—is very much related to some uses of continuation. Indeed recent work on adaptive methods for boundary-layer calculations proceeds by continuation in the "thickness" of the boundary layer.

Professor Ferziger's paper surveys some current methods for subgrid-scale turbulent modeling. It seems surprising that the numerical methods employed are never described and hence appear to play no role in the results of the computations. This is at great variance with simpler (i.e. laminar) fluid-flow problems, where the difference schemes and mesh Reynolds numbers are crucial. It suggests that subgrid-scale modeling is somehow more physical than the other approximations entering into numerical fluid dynamics. My own naive view would suggest just the opposite and that, in fact, the discrete approximations must somehow be "tuned" to the small-scale motions to get accurate approximations. Continuation procedures are remote from the considerations of this paper.

A tour de force of computation coupled with inspired interpretation of their results led Orszag and Patera to very important conjectures regarding transition in plane channel flows. The numerical methods use Fourier and Chebyshev expansions. The solution techniques very clearly employ arc-length continuation [1.1, 1.8], and with such extensive parameter searches additional continuation procedures could no doubt be (or perhaps were) extremely useful. Unfortunately the unsteady curves used to give important insight into the possibly occurring phenomena are not clearly enough described to allow others to check the computations or to do similar calculations on related problems. This defect could be lodged against 95% of the numerical fluid-dynamics papers; but it only matters in those of some significance, such as the present one.

The contribution by Roache solves the steady two-dimensional Navier–Stokes equations in a slightly curved channel for six values of the Reynolds number, $\mathrm{Re} = 10^n$, $n = 2, 3, 4, 5, 6, 7$. The curvature is adjusted so that a weak separation region is formed and a similarity law for the solutions in the variable $\tilde{x} = x/\mathrm{Re}$ as $\mathrm{Re} \to \infty$ is observed. Continuation in Re could have been extremely profitable in this example. Indeed, most of the computations could have been eliminated; only the solution at $\mathrm{Re} = 10^2$ need involve more than one or two iterations. Since 32 and 60 iterations were used on the coarse and fine grids, respectively, for $\mathrm{Re} = 10^n$ with each $n = 3, 4, 5, 6, 7$, we can safely estimate that at least 90% of the iterations could have been eliminated by using continuation in Re to supply the initial guesses. The gain will not always be so dramatic, but in the present case, where a similarity law is being verified, it seems quite natural.

McDonald and Briley sketch some very important considerations for three-dimensional Navier–Stokes problems. Only the unsteady approach, $t \to \infty$, is considered, and the problem of mesh selection in stressed and discussed. Again great reductions in computer time could have been obtained by using continuation of the initial data. Of course, some interpolation would have to be done on previously computed solutions to employ them on the altered nets, but this is a trivial price to pay. The major problems attacked in this paper, of devising efficient splitting algorithms and mesh-refinement methods for three-dimensional flows, will be with us for many years. It would enhance their development if continuation were used to make tests more efficient.

3 Brief Sketch of Continuation

We sketch here a few of the basic ideas in some current continuation studies. Suppose we have two problems of the same "type": an easy one, say E, and a hard one, say H. Then we can consider a one-parameter family of problems

$$P(\lambda) \equiv \lambda H + (1 - \lambda)E. \tag{1}$$

For $\lambda = 0$, we have $P(0) = E$, the easy problem, and for $\lambda = 1$ we have $P(1) = H$, the hard problem. If $U(\lambda)$ represents the solution of $P(\lambda)$, then we can find $U(0)$ (since E is easy), and using it to start, we may be able to generate the path of solutions $\{U(\lambda) \text{ for } 0 \leq \lambda \leq 1\}$, to get the desired hard solution, $U(1)$. This is the basic "homotopy" method invented by mathematicians to prove existence theorems (in fluid dynamics, no less). It can frequently be shown that no solutions are on the boundary $\partial\Omega$ of some set Ω in solution space. Then the path stays in some cylinder as shown in Fig. 1. This situation can be made the basis for practical computational methods. Specifically, the solution $U(\lambda)$ for $P(\lambda)$ will in general be a very close approximation to the solution for problem $P(\lambda + \Delta\lambda)$, if $\Delta\lambda$ is not too large, and so a good starting value for iterations is available. We need simply continue from $\lambda = 0$ to $\lambda = 1$ in appropriate steps. Just how to do this for large classes of problems is currently a very active area of study by numerical analysts.

Two sketches of how the solution paths may look are given in Fig. 1. In case A the path goes monotonely from $U(0)$ to $U(1)$ as λ goes from 0 to 1. This is the classical behavior that was first proposed and analyzed. Deeper studies, using so-called degree theory, allowed paths as sketched in case B.

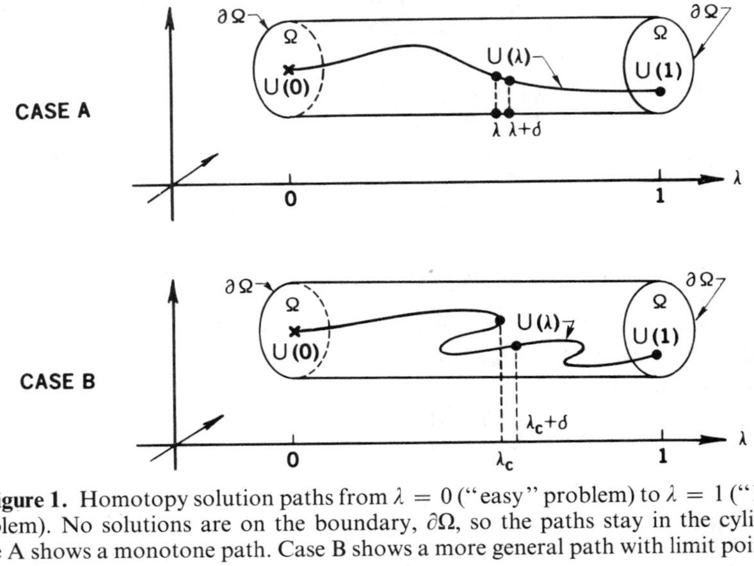

Figure 1. Homotopy solution paths from $\lambda = 0$ ("easy" problem) to $\lambda = 1$ ("hard" problem). No solutions are on the boundary, $\partial\Omega$, so the paths stay in the cylinders. Case A shows a monotone path. Case B shows a more general path with limit points, as at $\lambda = \lambda_c$.

For mere existence proofs, both cases are equally useful. However, the usual computing applications for those paths in case B must lead to failure. At the point λ_c the solution path $U(\lambda)$ turns back—we call $U(\lambda_c)$ a limit point solution—so that $U(\lambda_c)$ is not close to $U(\lambda_c + \Delta\lambda)$. The standard continuation methods thus fail near such limit points. Note that there are four such limit points on the solution path in Fig. 1, case B.

Numerous ways have now been devised to circumvent the indicated difficulty posed by limit points [1.1]. With a little thought most experienced numerical fluid-dynamicists should have no trouble in rediscovering these methods or modifications of them and endowing them with suitable names and acronyms. It is important to recall, however, that the phenomenon may also occur when λ is a physical parameter of the problem, not an artificially introduced homotopy parameter. Fluid-dynamical examples of precisely this type have been discussed in [1.2, 1.8, 1.9, 1.10]. Indeed, in the problem in [1.2]—driven cavity flows—the continuation procedure around limit points has exposed spurious numerical solutions. Unfortunately, these continuation processes are not capable, in themselves, of determining which solutions are physical or spurious.

The simplest way to continue about limit points can be sketched by writing the problem in the form

$$P(U, \lambda) = 0. \tag{2}$$

Then if $U^0 = U(\lambda_0)$ is the solution of (2) at $\lambda = \lambda_0$, we can seek the solution $U(\lambda)$ at $\lambda = \lambda_0 + \Delta\lambda$ by Newton's method:

$$P_U(U^\nu, \lambda)(U^{\nu+1} - U^\nu) = -P(U^\nu, \lambda), \qquad \nu = 0, 1, 2, \ldots. \tag{3}$$

This usually fails if λ is too close to a limit point, as there $P_U(U, \lambda)$ is singular (or near-singular). When such behavior is detected or suspected, we give up direct continuation in λ and continue in a new parameter, say s, that is an approximation to the arc length on the path $\{U(\lambda), \lambda\}$.

If this new representation of the solution path is denoted by $\{U(s), \lambda(s)\}$ with $\{U(s_0), \lambda(s_0)\} = \{U^0, \lambda_0\}$ say, then

$$P[U(s), \lambda(s)] = 0. \tag{4}$$

Thus the path must satisfy the differential equation

$$P_U \dot{U} + P_\lambda \dot{\lambda} = 0, \qquad U(s_0) = U^0, \quad \lambda(s_0) = \lambda_0. \tag{5}$$

Numerical approximations to the solution of this initial-value problem is one of the current powerful methods for continuation. Since (5) is an under-determined system, one scalar constraint can be added, and it serves to fix the parameter s. For example, if the arc length is desired, we adjoin the condition

$$\|\dot{U}(s)\|_2^2 + \dot{\lambda}^2(s) = 1. \tag{6}$$

The system (5), (6) can be written together as

$$\begin{pmatrix} P_U[U(s), \lambda(s)] & P_\lambda[U(s), \lambda(s)] \\ \dot{U}(s)^T & \dot{\lambda}(s) \end{pmatrix} \begin{pmatrix} \dot{U}(s) \\ \dot{\lambda}(s) \end{pmatrix} = \begin{pmatrix} 0 \\ 1 \end{pmatrix}. \tag{7}$$

The crucial fact is that the coefficient matrix in (7) is *nonsingular* at the simple limit points indicated above. Thus approximations to this matrix are also nonsingular, and so it is not difficult to accurately approximate the solution of (5), (6).

Another approach, which we have used in the fluid-dynamical problems of [1.2, 1.8, 1.9, 1.10], is to approximate (6) by

$$N(U, \lambda) \equiv \dot{U}(s_0)^T[U(s) - U^0] + \dot{\lambda}(s_0)[\lambda(s) - \lambda_0] - \Delta s = 0. \quad (8)$$

Then we solve (2) and (8) simultaneously. If Newton's method is used for this purpose, we get

$$\begin{pmatrix} P_U(U^\nu, \lambda^\nu) & P_\lambda(U^\nu, \lambda^\nu) \\ \dot{U}(s_0)^T & \dot{\lambda}(s_0) \end{pmatrix} \begin{pmatrix} \Delta U^\nu \\ \Delta\lambda^\nu \end{pmatrix} = -\begin{pmatrix} P(U^\nu, \lambda^\nu) \\ N(U^\nu, \lambda^\nu) \end{pmatrix}. \quad (9)$$

Note that the coefficient matrix in (9) is "close" to that in (7) for Δs not too large. Thus simple limit points cause no difficulty in this pseudo-arc-length continuation method.

All of the above ideas can be made quite rigorous and shown to be valid for an astonishingly large class of problems. Some specific applications are illustrated in Section 4.

4 Fluid-Dynamical Applications

The continuation methods indicated in Section 3, and in particular those based on (2), (8), and (9), have been employed in many fluids calculations. We indicate a few such applications here.

4.1 Driven Cavity Flows[2]

Using the stream function formulation

$$u = \psi_y, \qquad v = -\psi_x, \quad (10)$$

the Navier–Stokes equations can be replaced by the fourth-order scalar equation:

$$\frac{1}{\text{Re}} \nabla^4\psi + \frac{\partial(\psi, \nabla^2\psi)}{\partial(x, y)} = 0. \quad (11)$$

The boundary conditions specify $v = 0$ on four sides of the square cavity, $u = 0$ on three of the sides, and $u = 1$ on, say, the top side. With a uniform net of spacing $\Delta x = \Delta y = h$ on the unit square, we approximate (11) by

$$\frac{1}{\text{Re}} \Delta_h^2\psi - (D_y^0\psi D_x^0\Delta_h\psi - D_x^0\psi D_y^0\Delta_h\psi) = 0. \quad (12)$$

[2] With R. Schreiber [1.2, 1.3].

Here Δ_h is the usual five-point difference approximation for the Laplacian, while D_y^0 and D_x^0 are the centered y- and x-difference quotients, respectively. The boundary conditions are approximated by centered differences *on* the boundaries, and (12) is imposed at all *interior* netpoints of the unit square. Thus the first set of netpoints exterior to the square can be eliminated using the boundary conditions, and as many equations as netpoints *on* the unit square result. These nonlinear difference equations are solved using a form of Euler–Newton continuation in Re, with pseudo-arc-length continuation

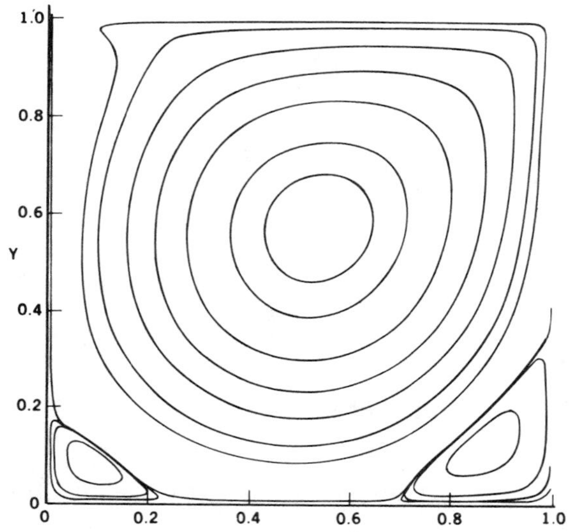

Figure 2. Streamlines for the driven-cavity problem at Re = 1000.

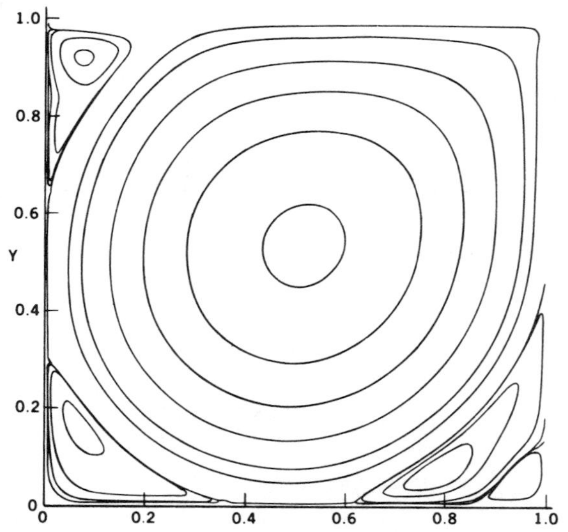

Figure 3. Streamlines for the driven-cavity problem at Re = 10,000.

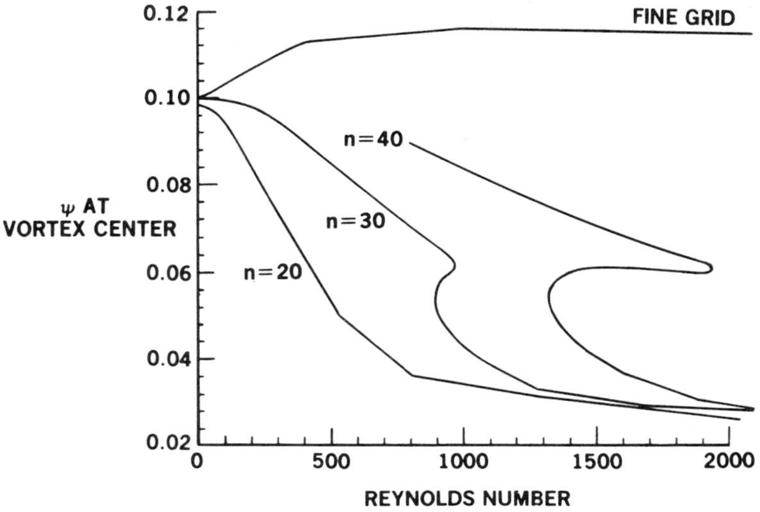

Figure 4. Stream-function values at the center of the large eddy in the driven-cavity problem. As the net spacing is refined ($n \to N$), the curves lose the limit points, shown for $n = 30$ and $n = 40$. The bottom branch contains spurious solutions with two large eddies.

invoked when the *LU* factorization of the Jacobian matrix shows signs of approaching singularity (as detected by rapid decrease in the value of the minimum pivot element in *U*). In fact, we used a combination of chord and Newton iterates, dictated by which was most cost-effective for reducing the error, but we do not wish to digress into these very important matters here.

Solutions for $0 \leq \text{Re} \leq 10^4$ were obtained on nets as fine as $h = \frac{1}{180}$. No bifurcations or limit points were detected on the path of most accurate solutions; we show some typical solutions in Figs. 2 and 3. However, we did detect limit points when we intentionally retained a crude mesh while increasing the Reynold's number. Refining the mesh moved the limit points to higher Re-values; see Fig. 4, where $h = 1/n$ on the various grids. The obviously spurious solutions, on the lower branch of these paths, have been reported as the "correct" solutions in several papers in the literature. Indeed, upwind differencing has been promoted, in part, because it so readily generates these (spurious) solutions at high Re-values with little difficulty. More good evidence that something for nothing is worthless.

4.2 Flows between Rotating Disks[3]

These flows have been studied numerically and analytically for many years. Using continuation methods, we have been able to correlate a large amount, perhaps all, of this previous work. The "answer" is that as a function of

[3] With R. Szeto [1.9, 1.11].

$R = \Omega_0 d^2/\nu$ (a Reynold's number) and $\gamma = \Omega_1/\Omega_0$ (the ratio of the disk rotation speeds), the sheets of solutions contain many folds and a few bifurcations. A good idea of how these sheets of solutions are connected is obtained, but the full picture is not yet completely clear.

In the dimensionless Karman similarity variables, with cylindrical velocity components

$$w = f(z), \qquad u = -rf'(z)/2, \qquad b = rg(z), \tag{13}$$

the Navier–Stokes equations reduce to

$$f'''' = R[ff''' + 4gg'], \tag{14a}$$

$$g'' = R[fg' - gf']. \tag{14b}$$

Here $z \in [0, 1]$ is the axial variable, and the boundary conditions are (since the fluid moves with the disks) are

$$f(0) = f'(0) = 0, \qquad g(0) = 1, \tag{15a}$$

$$f(1) = f'(1) = 0, \qquad g'(1) = \gamma. \tag{15b}$$

These equations are replaced by a system of six first-order equations which are then approximated by the box scheme on a nonuniform (adaptive) grid. The resulting system of nonlinear difference equations are solved using Euler–Newton continuation, pseudo-arc-length continuation, and bifurca-

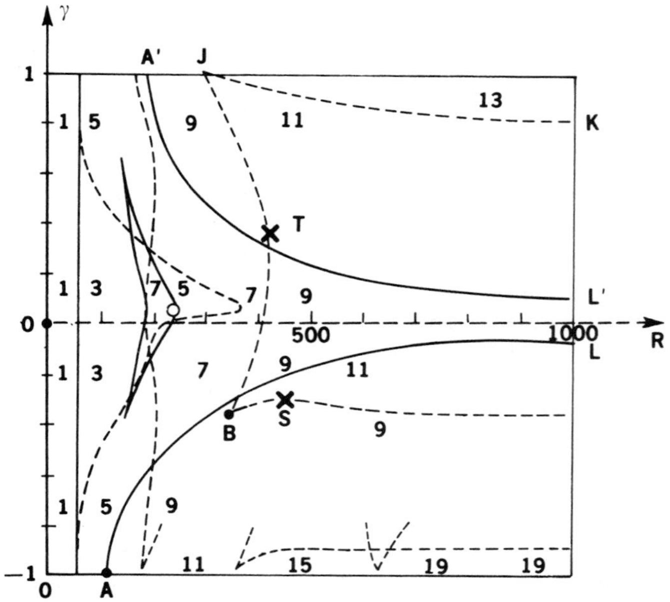

Figure 5. Folds and special critical points on the sheets of solutions for the flows between rotating disks. The integers tell how many sheets or solutions cover a given region. Bifurcations occur at A, B, S, and T, while an ISOLA center is at O.

tion techniques [1.9, 1.11]. A summary of some of the results is shown in Fig. 5. The curves represent folds in the sheets of solutions, and the integers tell how many sheets (i.e. solutions) cover each area. More details of what the solutions look like in different regions of the (R, γ) plane will be given elsewhere. All currently reported analytical and numerical solutions for this problem have been located on our various sheets [1.11]. Current work on this problem continues, and in particular we have developed new continuation techniques which are capable of following the folds in the solution sheets. This work will also be reported elsewhere.

CHAPTER 2

Experiments on Initial and Boundary Conditions[1]

Gino Moretti *

1 Introduction

Experimentalists have learnt to appreciate the importance of a proper installation, a careful calibration of instruments, and an analysis of environmental interferences. Data obtained by the Wright brothers in their historic wind tunnel enabled them to evaluate the possibility of flying; by current standards they would probably not even fit in what engineers call the "ballpark" range. The designer's problem is efficiency; errors larger than 1% in basic information should not be tolerated.

Surprisingly, a large amount of numerical work is far from reaching such standards. More surprisingly, very little seems to be being done to improve the situation, and this is all the more disturbing in that claims are occasionally made that numerical analysis should replace experiments.

The present paper is a modest attempt to show a possible approach to understanding the reliability of numerical analysis. It is shaped as a series of numerical experiments. Empiricism, however, is not suggested. Quite to the contrary, the work described in this paper is inspired by the idea that a numerical procedure describes, more or less accurately, a physical model and that the understanding of such a model will lead us to judge whether or not our calculation makes the physical sense which it should.

[1] Part of this work was sponsored by the NASA Langley Research Center under Grant No. NSG 1248 and performed during the author's visit at ICASE in August 1980.

* Polytechnic Institute of New York.

In preparing the present paper, a very large number of cases were computed, expressing different lines of thought, and for each one of them, tests of different parameters were made. A detailed discussion transcends the limits of this presentation. Therefore, I will limit myself to showing the guidelines of the investigation, and some of its highlights, without attempting to be exhaustive or even to draw conclusions, which (as will appear) could at this stage still be hasty and inappropriate.

2 A Channel Flow

On 18–19 September 1979, a workshop was held in Stockholm, the object of which was the comparison of results obtained by using different numerical methods on two assigned problems, the second of which was formulated as follows.

> Internal two-dimensional flow through a parallel channel having a 4.2% thick circular arc "bump" on the lower wall. The ratio of static downstream pressure to total upstream pressure is 0.623512 (corresponding to $M = 0.85$ in isentropic flow), and the distance between the walls is 2.073 times the chord length of the bump.

Obviously, the emphasis of the assignment was on steady solutions and transonic flow with an imbedded shock. The latter requirement adds a number of complications to the problem of a subsonic, steady, isentropic flow in a channel. The assigned data were so close to producing a choked flow that some of the methods generated a choked flow (all potential fully conservative methods) and others did not (all potential nonconservative methods and Euler solvers, and Hafez's artificial compressibility method). Scatter of results and conflicts between conclusions are not new in our short history of numerical analysis. As I recall, the first numerical contest was inspired by Morton Cooper in 1965, a calculation of blunt-body shock layers for an ellipsoid of revolution with a 2 : 1 axis ratio, at a freestream Mach number of 3 [1.12]. Techniques ranging from truncated series expansions to integral relations to inverse methods offered a variety of results. Comparing them with what has now been accepted as standard, that is, a second-order finite-difference calculation with bow-shock fitting, we see that methods focused on the stagnation line gave good results near the stagnation line and poor results away from it, whereas methods focused on the sonic line had the opposite behavior [1.13]. The contest clearly showed a need for a different numerical approach, more general and powerful.

The object of a workshop is, indeed, to promote healthy competition and unrestrained debate, not to solve problems or to render verdicts; a workshop can be considered successful if it inspires new, and deeper, work. In studying the results of the Stockholm workshop, I decided to take a closer look at the channel flow, at least for a certain brand of Euler solvers.

Obvious questions to be answered were:

1. Is a steady state reached?
2. Do results depend on the type and size of the computational mesh?

3. How do different treatments of the left and right boundary of the computed region affect the results?
4. Can any detail be provided of the flow near the leading and trailing edges of the bump?

Note that I abstain from mentioning integration schemes. The virtues and shortcomings of such schemes, including their ability to capture shocks, their numerical diffusion and dispersion, etc., are out of context. They cannot be tested in the channel-flow problem unless the questions above have been exhaustively answered. On the other hand, there are general features of the flow which should be revealed and which should provide clues to the questions, regardless of the integration scheme used, at least so long as the flow is well within the subsonic range.

I decided, thus, to limit a preliminary investigation to subsonic, isentropic flows, and I adopted the MacCormack predictor–corrector scheme to the equations of motion in the form

$$\begin{aligned} P_t + \mathbf{V}\cdot\nabla P + \gamma\nabla\cdot\mathbf{V} &= 0, \\ \mathbf{V}_t + (\mathbf{V}\cdot\nabla)\mathbf{V} + T\nabla P &= 0, \end{aligned} \tag{1}$$

where P is the logarithm of the pressure, $\mathbf{V}$ the velocity, T the temperature, and γ the ratio of specific heats. Pressure and temperature are related by

$$P = \frac{\gamma}{\gamma - 1}\ln T. \tag{2}$$

For subsonic flows, the MacCormack scheme is safely applicable to (1) and has the advantage of great simplicity. To maintain second-order accuracy at the boundaries, where the MacCormack scheme can be applied only at the predictor (or the corrector) level for want of external data on the other level, any derivative at such a level is discretized by differences of the type

$$2f_1 - 3f_2 + f_3,$$

where values at three adjacent points, from the boundary in, are denoted by 1, 2, and 3 sequentially.

3 Computational Grid

The grid suggested for the Stockholm workshop was a Cartesian grid, normalized between upper and lower wall of the channel, and stretching from $-\infty$ to $+\infty$ with a strong accumulation of grid lines over the bump. I used this type of grid, forcing two grid lines to originate exactly at the leading and trailing edges (Fig. 1).

I also adopted a different grid (shown in Fig. 2) which is obtained using a conformal mapping of the Kármán–Trefftz type:

$$\frac{z - 1}{z + 1} = \left(\frac{\zeta - 1}{\zeta + 1}\right)^{\delta}, \tag{3}$$

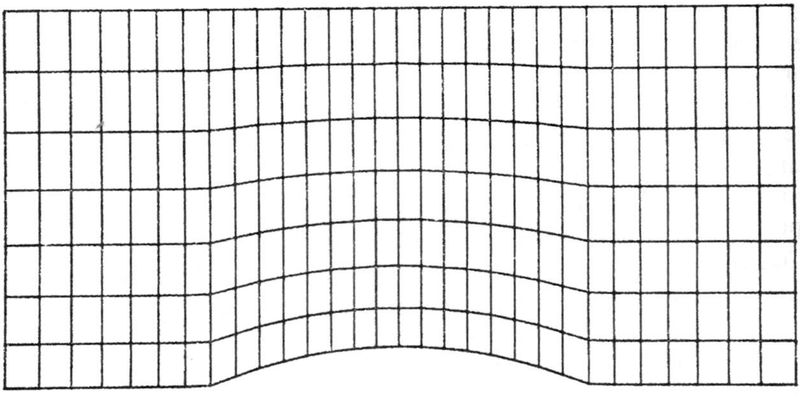

Figure 1. "Cartesian" computational grid.

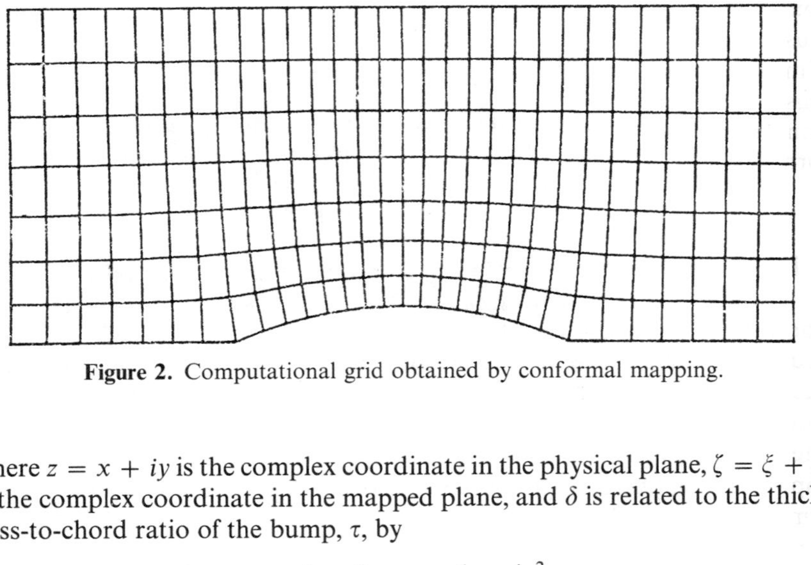

Figure 2. Computational grid obtained by conformal mapping.

where $z = x + iy$ is the complex coordinate in the physical plane, $\zeta = \xi + i\eta$ is the complex coordinate in the mapped plane, and δ is related to the thickness-to-chord ratio of the bump, τ, by

$$\delta = \frac{1}{2} + \frac{1}{\pi} \arctan \frac{1 - 4\tau^2}{4\tau}. \tag{4}$$

The η-coordinates are normalized between the upper and lower walls, and the ξ-coordinates are stretched in the horizontal direction as in the Cartesian grid. Denoting by u and v the velocity components in the directions of the Cartesian axes for the first grid, and in the directions of the ξ = constant, η = constant axes for the second grid, it turns out that v vanishes exactly along the upper wall when the first grid is used, and along the lower wall when the second grid is used. Consequently, one may expect the calculation to be easier and, perhaps, the results to be better along the wall where v vanishes. In any event, the boundary conditions at rigid walls are enforced by first integrating the Euler equations as at interior points, and then by correcting the pressure to satisfy the vanishing of the velocity component normal to the wall [1.14, 1.15]. Along walls where v is not identically zero, the u-momentum equation is replaced by an equation along the tangent to the wall.

Calculations are actually performed in a computational plane, (X, Y), where the grid is evenly spaced in both directions. If the first grid is used, additional coefficients appear, containing dX/dt, $\partial Y/\partial x$ and $\partial Y/\partial y$. If the second grid is used, two independent sets of additional coefficients and terms appear: the first due to normalization and stretching, and containing $dX/d\xi$, $\partial Y/\partial \xi$, $\partial Y/\partial \eta$, and the second due to the mapping, and containing

$$g = Ge^{i\omega} = \frac{d\zeta}{dz}, \qquad \phi = \phi_1 + i\phi_2 = \frac{d \log g}{d\zeta}. \tag{5}$$

The major physical difficulty is offered by the leading- and trailing-edge corners, where the flow stagnates. In using the first grid, the difficulty is reflected in the discontinuity in the slope of the lower wall (which affects all points on the vertical grid lines issuing from the corners). If the second grid is used, then a mapping singularity appears at the corners, and the equations of motion, expressed in terms of ξ and η, become indeterminate. In both cases, thus, some special treatment must be given to the corner points and their immediate neighbors. If the grids are laid to avoid passing through the corners, the effect of neglecting them has to be evaluated.

4 Inlet and Outlet Boundary Conditions

Another critical issue regards the treatment of the arbitrary computational boundaries which delimit an inlet and an outlet to the region of interest. Such boundaries cross regions of subsonic flow, and some physical model is required to supply the information from outside which is necessary. New interest in this problem seems to have arisen in recent times, but the physical implications of modeling a subsonic boundary seem not to have been grasped firmly yet. The problem of subsonic boundaries cannot be disassociated from the problem of choosing initial conditions [1.16]. In internal flows, several simple physical models can be adopted, of which here is a sample:

1. The region of interest is a channel of finite length, connecting two infinite cavities; the gas is at rest everywhere; at $t = 0$, the stagnation pressure is increased in the cavity at the left, until a given value is reached, and then kept constant.
2. The same setting is used, but at $t = 0$ the pressure in the cavity at right is decreased until a given value is reached, and then kept constant.
3. The channel is infinitely long, and it contains a gas at rest; at $t = 0$, the channel is accelerated towards the left, until a cruising speed is reached.

In the first two cases, two models of transitions from the interior to infinite cavities are adopted at each boundary point on the left and on the right. As explained in [1.16], one can stipulate that the fictitious flows in the transitions are quasisteady (the length of the transition being assumed vanishingly small), so that, for the purpose of closing the boundary data sets, steady equations of motion can be differentiated in time. On the left, the total pressure and the slope of the velocity vector at each entry point are assigned. The latter condition brings in the largest arbitrariness in the model. Physically, one can always justify a choice of slopes by assuming that the inlet is equipped with a series of guiding vanes. In the present case, for example, one can assume that all velocity vectors are parallel to the rigid walls; this is obviously not the case for an infinitely long channel, and the effect of such a restrictive assumption on the rest of the flow has to be evaluated.

The equations used at the inlet are:

1. the definition of total pressure, differentiated in time under the assumption that the total pressure itself may be a function of time:

$$TP_t + u(1 + \sigma^2)u_t = T_0 P_{0t}, \tag{6}$$

   where the index 0 denotes stagnation values in the infinite cavity, and $\sigma = v/u$ is a prescribed value;
2. a left-running characteristic equation:

$$aP_t - \gamma u_t = R, \tag{7}$$

   where R is the left-hand side of (7) as computed by the standard integration routine.

The outlet model is simpler, since the v-component of the velocity is determined on the basis of internal information only [1.17]; in the present case, it is sufficient to prescribe the exit pressure as the pressure in the infinite cavity, and compute the u-component of the velocity accordingly. The equations are thus

1. the continuity equation:

$$\rho u P_t + \gamma \rho u_t = \rho_\infty u_\infty P_{\infty t} + \gamma \rho_\infty u_{\infty t}, \tag{8}$$

2. the definition of total pressure:

$$TP_t + uu_t + vv_t = T_\infty P_{\infty t} + u_\infty(1 + \sigma_\infty^2)u_{\infty t}, \tag{9}$$

3. a right-running characteristic equation:

$$aP_t + \gamma u_t = R, \tag{10}$$

   where R has the same meaning as in (7).

In (8), (9), and (10), v_t is computed by the standard routine; $u_{\infty t}$ is unknown, but it can be eliminated easily. Naturally, here too there is an element of arbitrariness, whose effects have to be checked. For example, in an infinitely long channel, the pressure across the channel is not exactly constant at a finite distance from the bump.

The inlet and outlet boundaries just described allow perturbations proceeding from the interior to interact with the conditions in the infinite cavities. For each perturbation reaching the boundary, a new perturbation is generated and transmitted in the opposite direction. The process will eventually reach an asymptotic steady state, but the number of waves of sizable amplitude moving back and forth can be very large.

The third model relies on a simple idea: if the motion were one-dimensional, all perturbations would travel outwards as simple waves, at the end of the acceleration phase. A simple wave is easy to describe using information from the interior and the constancy of one Riemann invariant from the exterior. In a two-dimensional problem of internal flow, the waves cannot be exactly simple waves, but no major errors are expected if the velocity vector is forced to be parallel to the rigid walls at the inlet. The simple wave equations at the inlet are modified as follows:

$$\begin{aligned} aP_t - \gamma(1 + \sigma^2)^{1/2}u_t &= R, \\ aP_t + \gamma(1 + \sigma^2)^{1/2}u_t &= 0, \end{aligned} \tag{11}$$

Similarly, at the outlet,

$$\begin{aligned} aP_t + \gamma u \frac{u_t}{q} &= R, \\ aP_t - \gamma u \frac{u_t}{q} &= \gamma v \frac{v_t}{q}, \end{aligned} \tag{12}$$

where v_t is computed by standard routines and q is the modulus of the velocity.

5 Two-Dimensional Calculations with Models 1 and 2

We describe now the general features of calculations made using the first two models mentioned in Section 4. One of the problems presented by a study like the present one is the large number of data produced by a single run and the necessity for organizing them in a series of simple plots, easy to interpret. I decided to store the following information:

1. at every step, P and u on the lower wall, at the inlet, at two selected points, and at the highest point on the bump,
2. at every step, the location of selected isobars on the lower wall [to build an isobar pattern on an (x, t) plane], and
3. at selected steps, P, u, and v at all the grid points; this information can be easily processed to provide Mach numbers and total pressures.

The basic geometry has been defined as a channel with width equal to 2, containing a bump which extends from $x = -1$ to $x = 1$ and which has a maximum thickness of 0.2. This defines a corner angle of 157.38°, and $\delta = 0.8743$. To avoid initial complications at the corners, so that our attention can be focused on the wave propagation and the effects of boundaries, we use a smooth lower wall, which can be easily obtained from the mapping function by defining the wall as the image, in the z-plane, of a line $\eta = b$, where b is a constant greater than 0. The same definition can be transferred

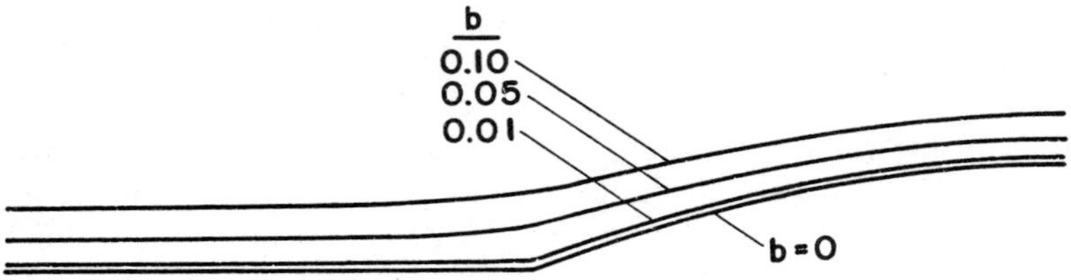

Figure 3. Shapes of lower wall for different values of b.

to the code which uses the Cartesian grid. In Fig. 3 are shown some shapes of the lower wall for different values of b; one can observe that, for b less than 0.01, there is no practical difference between the wall so defined and the wall corresponding to $b = 0$.

In this paper, we will limit our analysis to a channel with a very smooth wall, defined by $b = 0.1$. The computational mesh has 7 intervals between the rigid walls and 30 intervals in the x-direction, stretched between $x = -2.345$ and $x = 2.345$; 16 intervals cover the bump region. The stagnation pressure is raised (in the first model) or the exit pressure is lowered (in the second model) to produce final values of the Mach number "at infinity" of the order of 0.1. A plot of P vs. time at the 6th node on the lower wall is shown in Fig. 4, for the case where the first model is applied. The oscillations are obviously produced by waves going back and forth along the channel (at such low Mach numbers, the speed of propagation is practically the same in both directions, and the phenomenon shows a well-defined

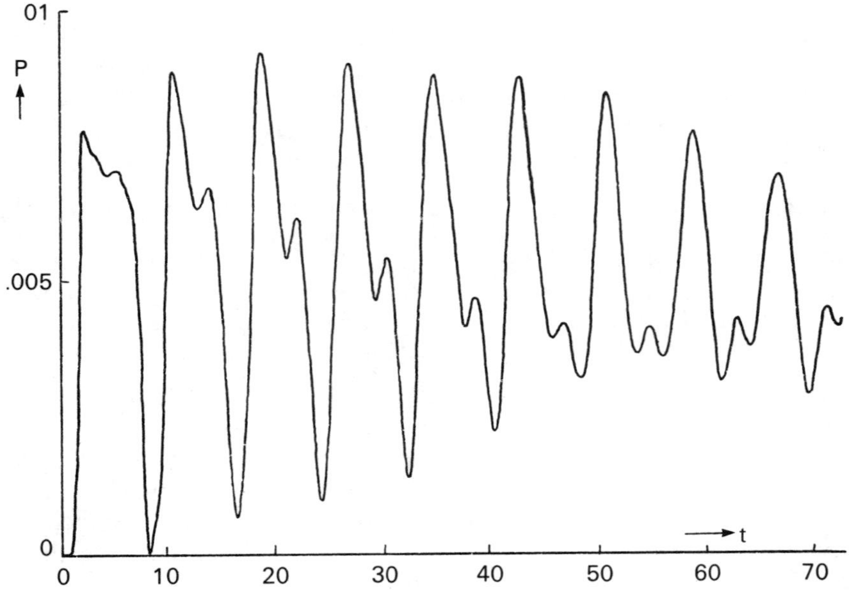

Figure 4. $P(t)$ at sixth node on lower wall, $M = 0.1$, model 1.

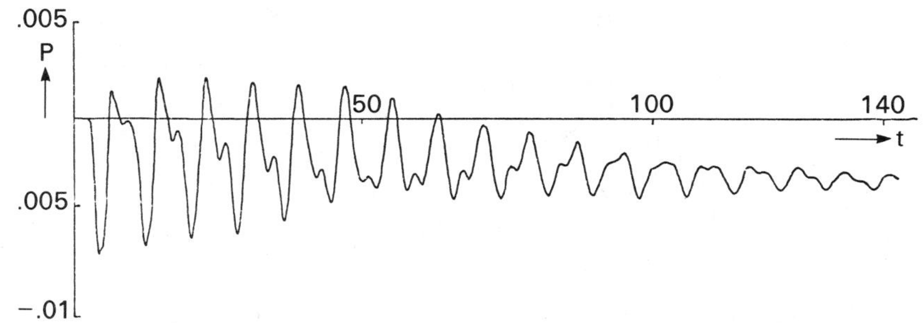

Figure 5. $P(t)$ at sixth node on lower wall, $M = 0.1$, model 2.

frequency); one disturbing feature of the model is the smallness of the damping factor.

A similar pattern appears (Fig. 5) using the second model; the oscillations are smaller in amplitude but still damped very slowly. Figure 6 shows $u(t)$ at the same node, and from it we see that the velocity presents smaller oscillations, but that a steady state is far from having been reached after 2000 computational steps ($t = 146$).

Such details are hard to detect from plots of level lines at a given step. For example, isobars and isomachs at step 1000 (Figs. 7 and 8) look very reasonable, although the isobars would not pass a closer scrutiny, due to a clear lack of symmetry. As a matter of fact, if we plot P at the 6th node vs. P at the 24th node as they evolve in time, we see that, after 2000 steps, the plotted line still oscillates between -0.00283 and -0.00353, whereas at both points P should be about -0.0032.

An analysis of these oscillations should take at least two elements into consideration: the first is the Mach-number effect, and the second is the influence of geometry. To have an idea of the Mach-number effect, let us rerun the above cases for a Mach-number of 0.5. Plots of P and u vs. time are shown in Figs. 9 and 10 for the first model; Fig. 9 should be compared with Fig. 4. Oscillations still appear, but they seem to be damped much more quickly. Similar behavior is seen in Figs. 11 and 12, which refer to the second

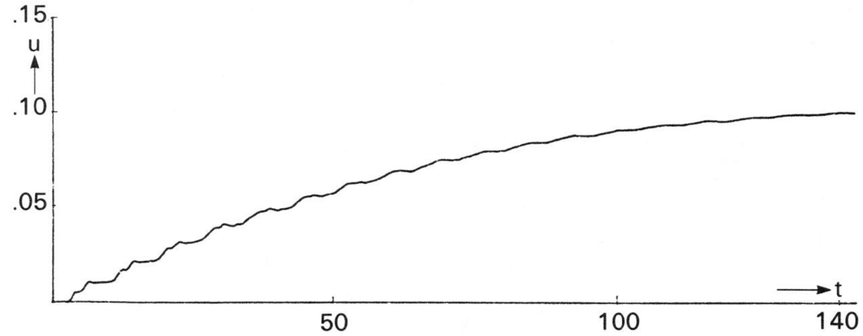

Figure 6. $u(t)$ at sixth node on lower wall, $M = 0.1$, model 2.

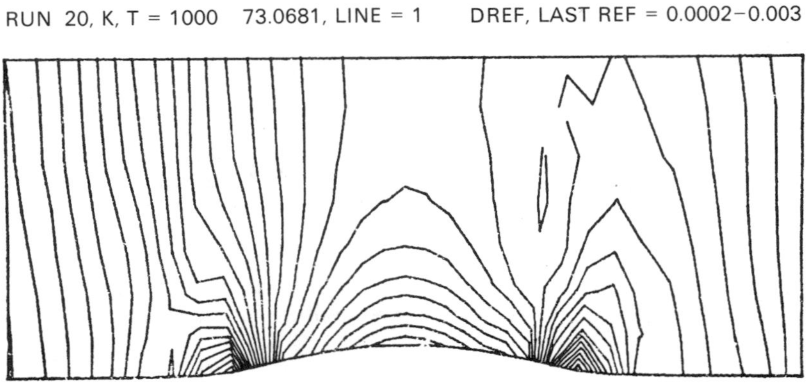

Figure 7. Isobars at step 1000, $M = 0.1$.

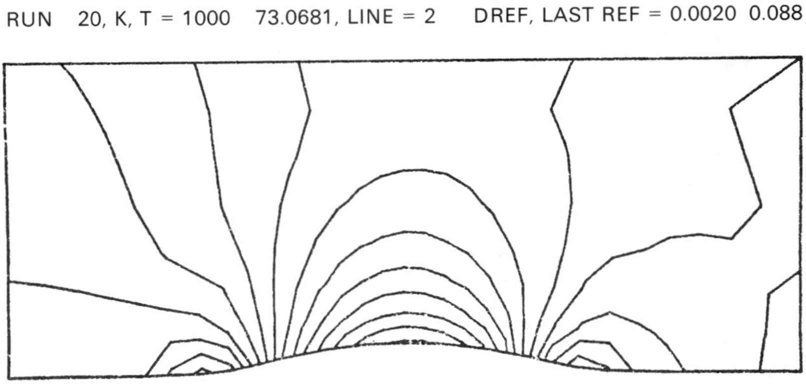

Figure 8. Isomachs at step 1000, $M = 0.1$.

model. One should, however, take care not to draw hasty conclusions in comparing Figs. 4 and 9, or Figs. 5 and 11. The scale of P in Fig. 9 is 20 times smaller than in Fig. 4, and in Fig. 11 is 40 times smaller than in Fig. 5; the correct conclusion is that very small pressure waves take a long time to be eliminated. The patterns of level lines (isobars in Fig. 13, isomachs in Fig. 14) are much better than their counterparts for $M = 0.1$ (Figs. 7 and 8). Even the $v =$ constant lines, which are very critical, look good (Fig. 15). At this stage, it pays to take a look at lines of constant stagnation pressure (Fig. 16); here a new element appears. In fact, the stagnation pressure is practically constant everywhere, but it drifts away in the vicinity of the "corners" (or whatever remains of them in the smoothed wall). The stagnation pressure is a very sensitive parameter indeed, and it is the proper indicator of local inaccuracies, when a steady state is apparently reached numerically. In this case, it is obvious that inaccuracies should be attributed to the vicinity of a singular point of the mapping and to the consequent worsening of the metric.

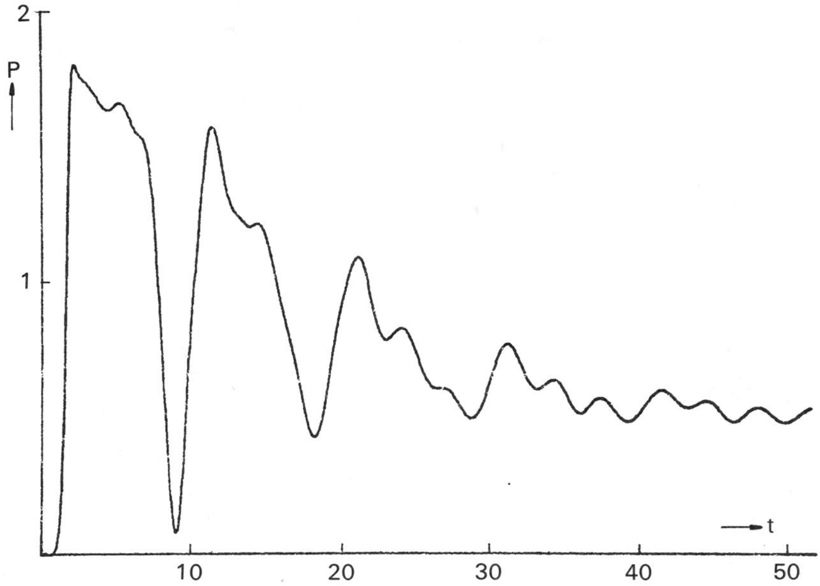

Figure 9. $P(t)$ at sixth node on lower wall, $M = 0.5$, model 1.

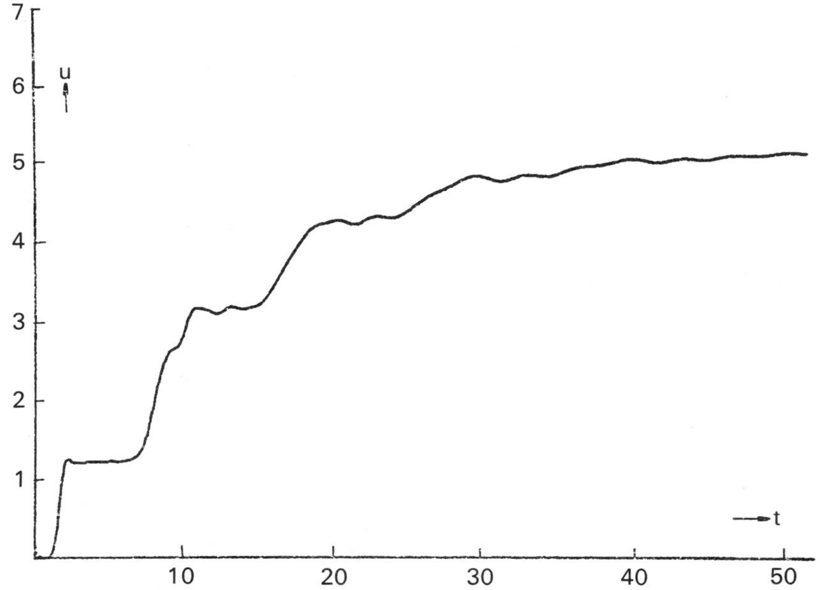

Figure 10. $u(t)$ at sixth node on lower wall, $M = 0.5$, model 1.

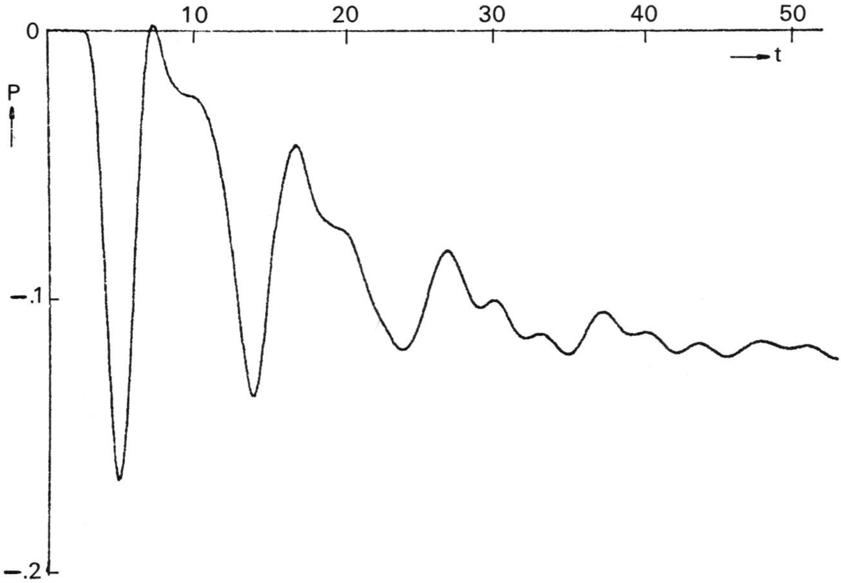

Figure 11. $P(t)$ at sixth node on lower wall, $M = 0.5$, model 2.

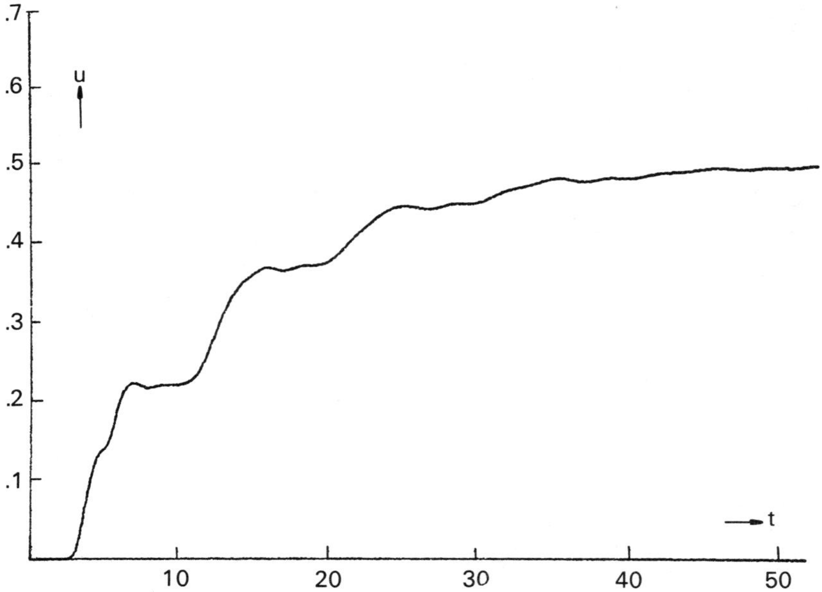

Figure 12. $u(t)$ at sixth node on lower wall, $M = 0.5$, model 2.

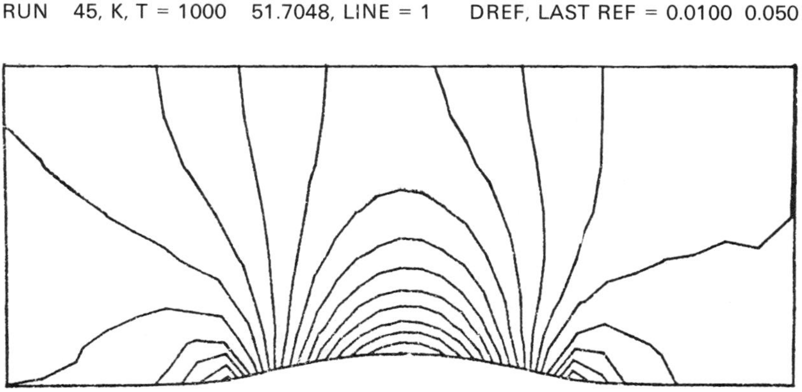

Figure 13. Isobars at step 1000, $M = 0.5$.

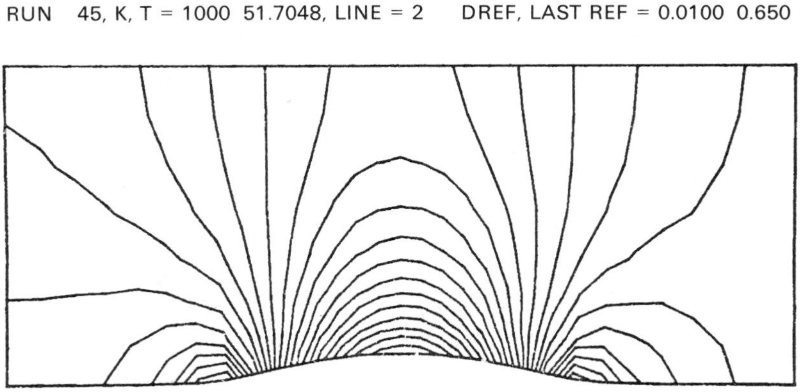

Figure 14. Isomachs at step 1000, $M = 0.5$.

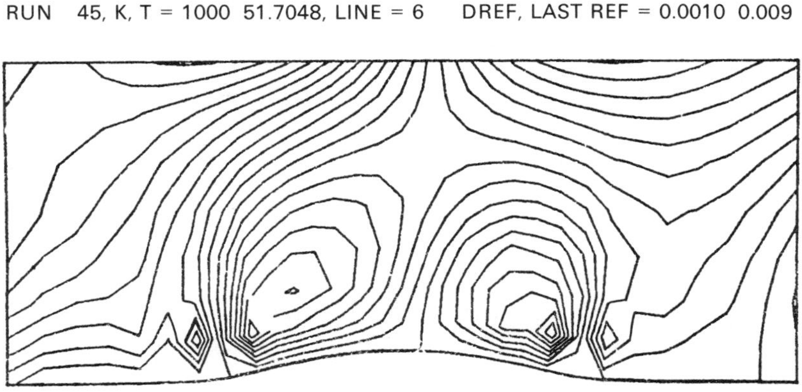

Figure 15. v = constant lines at step 1000, $M = 0.5$.

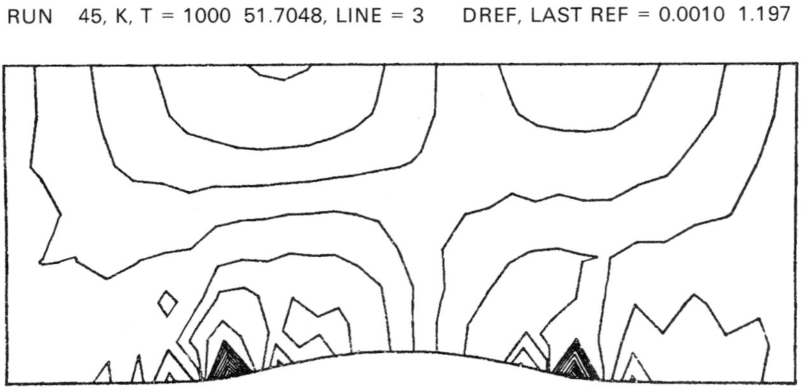

Figure 16. P = constant lines at step 1000, M = 0.5.

Nevertheless, it would be advisable to undertake the delicate analysis of the corner singularity only after having acquired more familiarity with the wave propagation pattern for models 1 and 2, and their possible relationship with the existence of a bump. We have seen, so far, that waves tend to continue swaying back and forth, with very little damping, at low Mach numbers. To judge whether the geometry, and particularly the presence of a bump, has anything to do with the wave behavior, we repeat the analysis for a straight channel (which can be easily obtained from either the code using the Cartesian grid or the code using the mapping by setting the thickness of the bump equal to zero).

Here are some results, for comparison with the previous cases. The first (Figs. 17 and 18) uses model 2, with a Mach number of 0.1. Compare Fig. 17

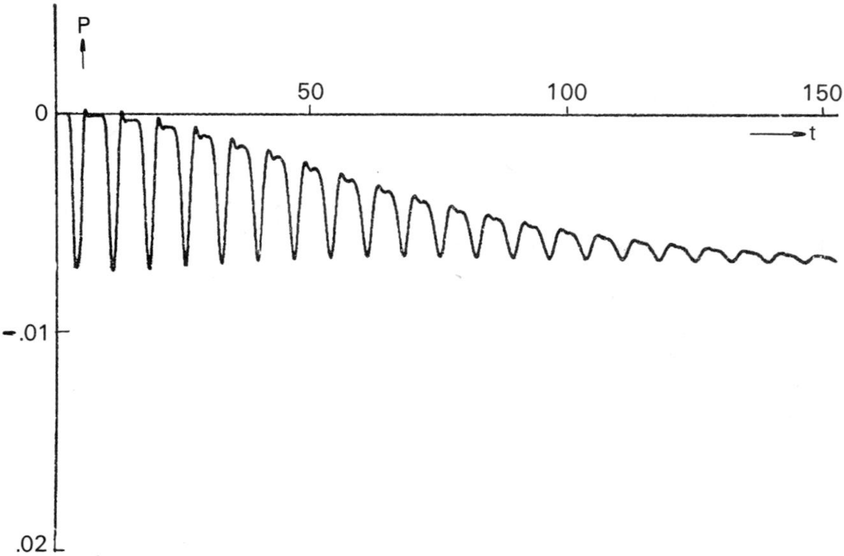

Figure 17. $P(t)$ at sixth node on straight channel, M = 0.1, model 2.

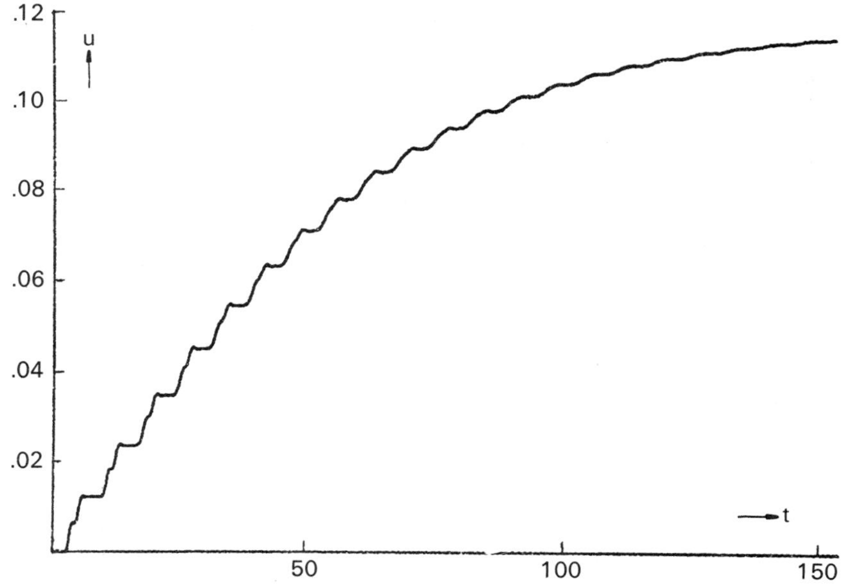

Figure 18. $u(t)$ at sixth node on straight channel, $M = 0.1$, model 2.

with Fig. 5, and Fig. 18 with Fig. 6. The second uses model 2 again, but with a Mach number equal to 0.5 (Figs. 19 and 20). Compare Fig. 19 with Fig. 11, and Fig. 20 with Fig. 12. Note that in this case the steady flow in the channel is uniform, with a pressure equal to the exit pressure; the transition from stagnation pressure ($P = 0$) to the channel pressure takes place in the

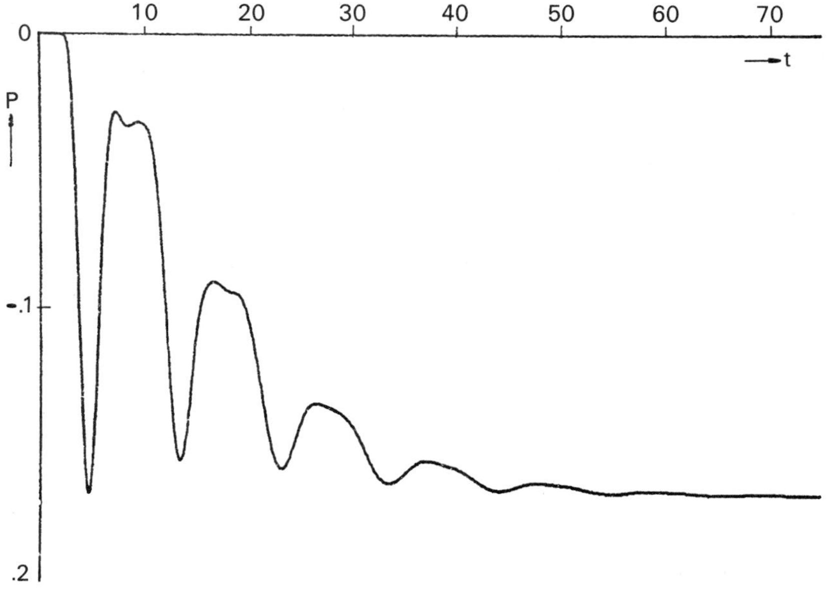

Figure 19. $P(t)$ at sixth nodel on straight channel, $M = 0.5$, model 2.

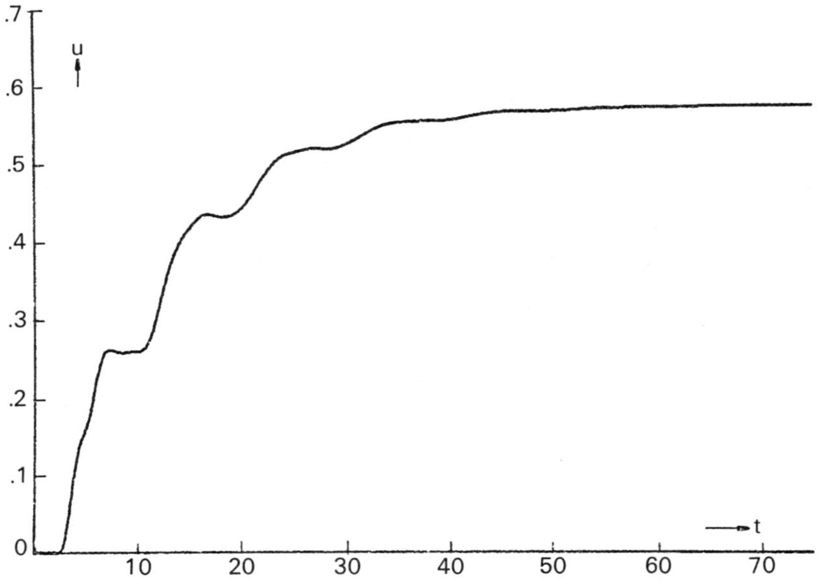

Figure 20. $u(t)$ at sixth node on straight channel, $M = 0.5$, model 2.

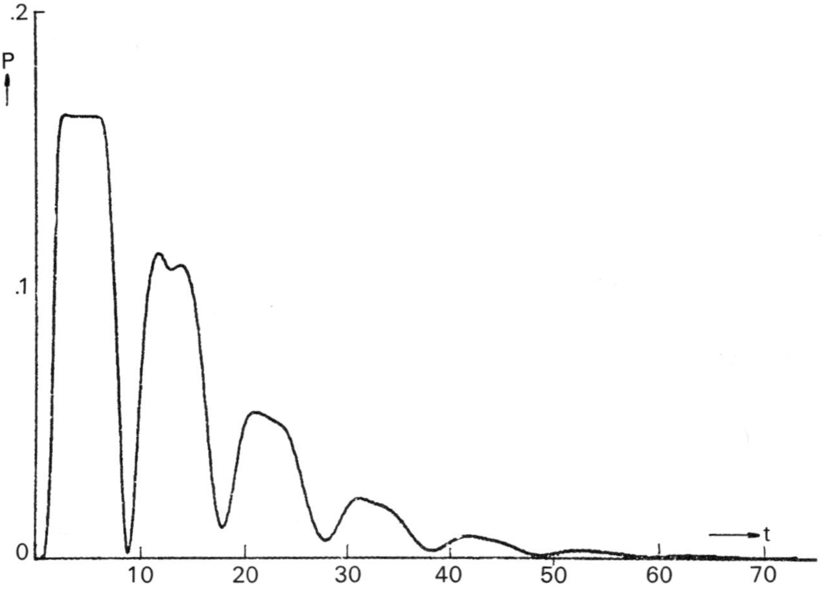

Figure 21. $P(t)$ at sixth node on straight channel, $M = 0.5$, model 1.

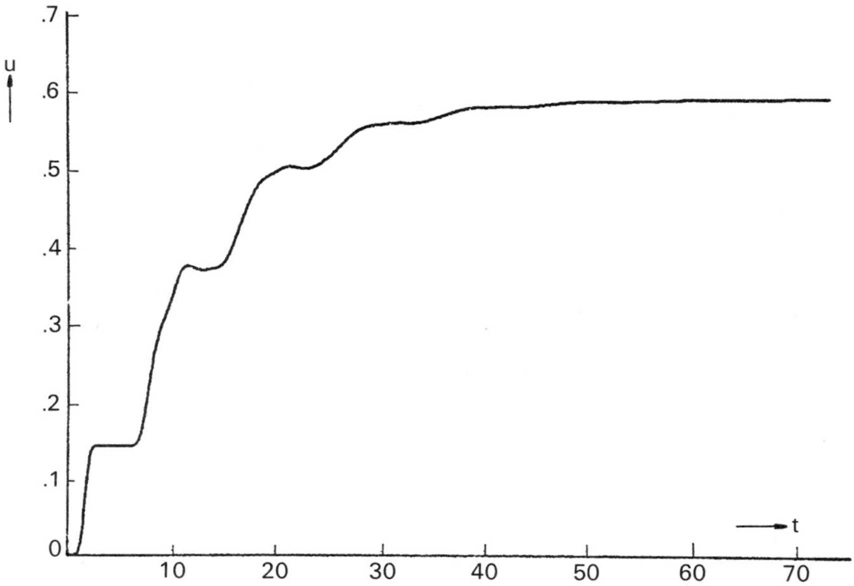

Figure 22. $u(t)$ at sixth node on straight channel, $M = 0.5$, model 1.

fictitious transition which has been modeled at the inlet. In Figs. 21 (to be compared with Fig. 9) and 22 (to be compared with Fig. 10) the first model is used. Here the stagnation P is raised to a positive value and the exit P remains equal to zero; the latter is thus the asymptotic value of P in the whole channel.

All these patterns are similar to the ones with the bump; we can conclude that the oscillations are produced by the models of the boundaries and that they have a physical interpretation of their own, unrelated to the geometrical complications of the channel.

6 Two-Dimensional Calculations with Model 3

We expect calculations made using the third model of Section 4 to converge to a steady state much faster than the previous ones, since the initial perturbation affects the entire flow field and whatever is not pertinent to the final state is promptly eliminated through the boundaries, which in this model are not reflective. The expectations are confirmed by Figs. 23 and 24 (which should be compared with Figs. 4, 5, and 6). Note also, in Figs. 25 and 26 (isobars and isomachs, respectively) how close the pattern is to the symmetric pattern of a steady state; compare these figures with Figs. 7 and 8. We omit results for $M = 0.5$; they are equally good and not dissimilar from the ones obtained using the second model, although a close inspection may reveal some advantage in using the third model (for example, the v = constant lines appear more symmetrical than in Fig. 15).

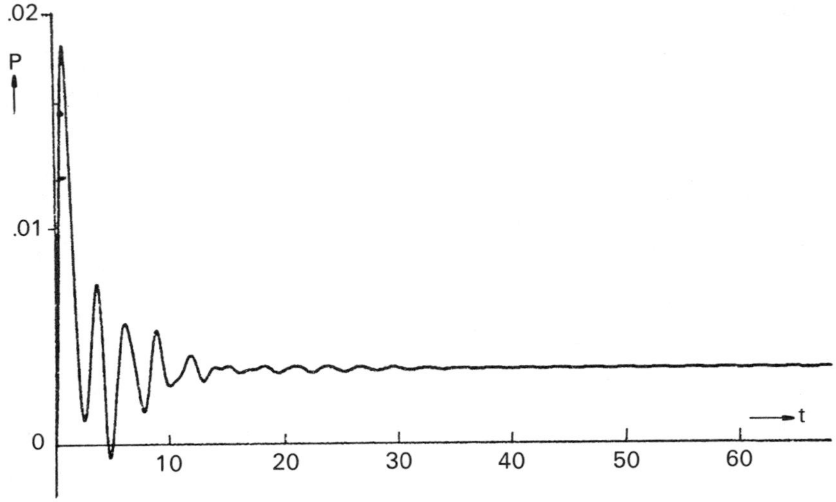

Figure 23. $P(t)$ at sixth node on lower wall, $M = 0.1$, model 3.

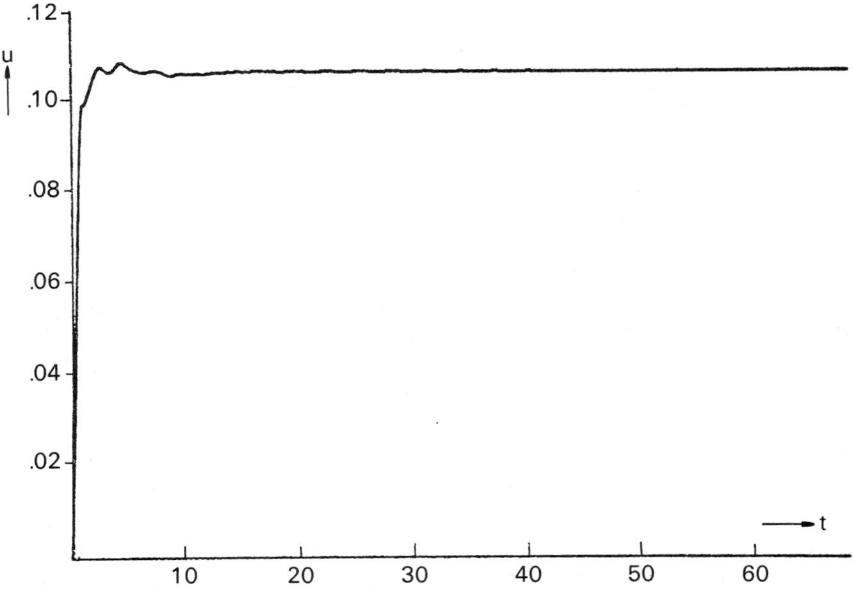

Figure 24. $u(t)$ at sixth node on lower wall, $M = 0.1$, model 3.

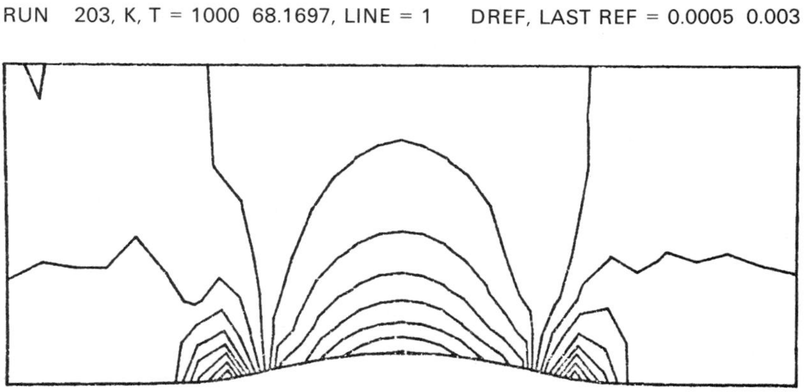

Figure 25. Isobars at step 1000, $M = 0.1$, model 3.

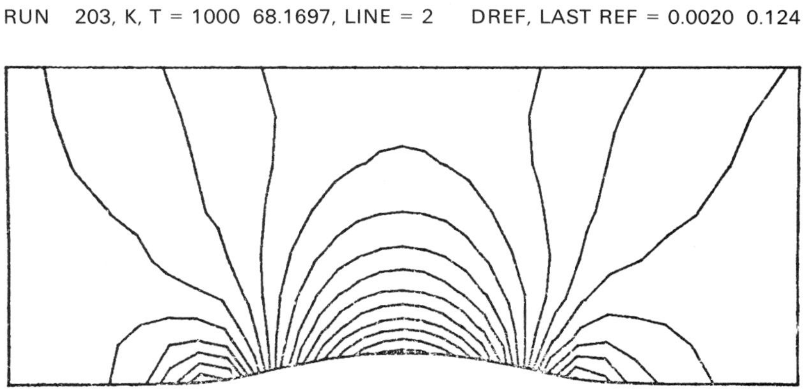

Figure 26. Isomachs at step 1000, $M = 0.1$, model 3.

7 Conclusions

At this stage, we conclude that:

1. all three models are acceptable as the description of a physical evolution;
2. all three models are acceptable for the evaluation of a subsonic steady state;
3. in all three models the computational region can be limited to a small portion of the channel, bracketing the bump;
4. the third model, however, provides faster convergence to a steady state, particularly for low Mach numbers;
5. in the first two models, small perturbations are eliminated slowly, and plots of typical parameters, such as $P(t)$, are necessary to detect the presence of unwanted oscillations;
6. a perfect symmetry in the steady state patterns cannot be provided by the first two models, due to the finite distance between the two infinite

capacities; the third model provides a better description of an infinitely long duct, for which we expect the steady-state solution to be symmetric.

Finally, let us make clear that in this analysis our notion of speed is merely related to the number of computational steps necessary to achieve a steady state. The physical phenomena, as described by our calculations, are indeed very fast. Take for example the case of a duct, 1 meter in width, at sea level; according to our conventions, the reference time is 0.00164 sec. The plots shown in Figs. 5 and 6 are interrupted at $t = 146$, that is, after only 0.24 sec from the beginning. Therefore, all further investigations will be conducted using the third model.

CHAPTER 3

Hybrid Procedures for Computing General Viscous Flows

J. C. Wu*

1 Introduction

In the computational fluid-dynamics community, the general viscous flow, i.e., a flow in which viscous separation is an important feature, is receiving a great deal of current attention. Because of its fundamental nature and its practical importance, the problem of finding predictive methods for general viscous flows has been traditionally a focal point of research in fluid dynamics. It has long been recognized that the mathematical difficulties attendant on the analytical solution of the Navier–Stokes equations describing the general viscous flows are formidable. The numerical approach offers the only promise, for the foreseeable future, of accurate quantitative solutions under reasonably general circumstances [1.18, 1.19].

At present, useful information for two-dimensional general viscous flows at moderate flow Reynolds numbers can be generated on modern computers using the state-of-the-art numerical procedures. The cost of computing such flows is large, but not prohibitively so. It is known, however, that the computational efforts required increase rapidly with increasing flow Reynolds number. In consequence, very few numerical solutions of high-Reynolds-number general viscous flows are presently available. In engineering applications, high-Reynolds-number flows are of much greater importance than are low-Reynolds-number flows. The "Reynolds-number limit" of the computational approach is therefore of great concern to the designers and researchers in fluid dynamics.

* School of Aerospace Engineering, Georgia Institute of Technology, Atlanta, GA 30332.

For engineering design applications, the ultimate goal of computational fluid dynamics is to be able to simulate three-dimensional high-Reynolds-number flows routinely. Because of the drastically larger operation and data-storage requirements, even the next generation of general-purpose supercomputers are not expected to be adequate for the routine computation of high-Reynolds-number three-dimensional general viscous flows. For this reason, some researchers have advocated the development of special-purpose computers designed specifically for the solution of viscous flow equations [1.20, 1.21]. Obviously, the severity of new computer requirements may be moderated, or even eliminated, by the development of drastically more efficient solution algorithms. In this paper, the use of several hybrid procedures that offer superior solution efficiency in the computation of general viscous flows is considered. For brevity, the present discussions are centered upon laminar incompressible flows past the exteriors of solid bodies. This type of flow possesses the essential features of interest and serves to bring into focus the most important concepts associated with the hybrid approach. It has been found that this approach is also well suited for turbulent flows [1.22, 1.23] compressible flows [1.24, 1.25], and internal flows [1.24].

2 Vorticity Dynamics

In this section, several prominent features of the general viscous flow are reviewed. The necessary interplay between the physical and numerical aspects of the general viscous flow is then brought into focus.

The incompressible flow of a viscous fluid is familiarly formulated mathematically as differential continuity and the Navier–Stokes equations. It is convenient [1.26] to introduce the concept of vorticity, $\boldsymbol{\omega}$, defined by

$$\nabla \times \mathbf{v} = \boldsymbol{\omega}, \tag{1}$$

and partition the overall problem into its kinematic and kinetic aspects. The kinematic aspect of the problem is described by Eq. (1) and the continuity equation:

$$\nabla \cdot \mathbf{v} = 0. \tag{2}$$

This aspect expresses the relationship between the vorticity field at any given instant of time and the velocity field at the same instant. The kinetic aspect of the problem is described by the vorticity transport equation. For a time-dependent flow, this equation is

$$\frac{\partial \boldsymbol{\omega}}{\partial t} = \nabla \times (\mathbf{v} \times \boldsymbol{\omega}) + \nu \nabla^2 \boldsymbol{\omega}, \tag{3}$$

where t is the time and ν is the kinematic viscosity, considered to be uniform for simplicity.

Equation (3) is obtained from the Navier–Stokes equations by taking the curl of each term in the latter equation and using Eqs. (1) and (2). It is known that the net viscous force acting on a fluid element is zero wherever the vorticity field and its first derivative are zero. Also, vorticity cannot be generated or destroyed in the interior of a viscous fluid [1.27]. The "no-slip" condition on the solid–fluid interface, however, provides a mechanism for the generation of vorticity. This vorticity spreads into the interior of the fluid domain through the kinetic processes of convection and diffusion. If the flow Reynolds number is not very small, then the convective speed is much greater than the effective speed of diffusion. The vorticity field is then nonzero only in a small region surrounding and trailing the solid body. This small vortical region is surrounded by a much larger, in fact infinitely large, region which is irrotational and therefore inviscid.

In the solution of a time-dependent general viscous-flow problem, it is convenient to follow the kinetic development of the vorticity field in the fluid. A numerical procedure therefore can be established in which the solution is advanced from an old time level to a new time level through a computation loop consisting of the following three steps:

1. With known vorticity and velocity values at the old level, Eq. (3) is solved to obtain vorticity values in the interior of the fluid domain at the new level.
2. New boundary values of the vorticity are computed using the "no-slip" condition.
3. New velocity values corresponding to the new vorticity values are obtained by solving Eqs. (1) and (2).

Step (1) is the kinetic part of the computation loop. Since Eq. (3) is parabolic in its time–space relation, step (1) requires the solution of an initial-value–boundary-value problem. The process of vorticity generation on the solid boundary is not described by the kinetic process of vorticity diffusion and convection [1.28]. Boundary values of vorticity, however, are necessary to advance the solution further in time. This step is critical to the accuracy of the time-dependent solution. It has been shown [1.28] that step (2), where the boundary values of vorticity are computed, is a kinematic step. Step (2) completes the computation of vorticity values at the new time level. Step (3) utilizes the just-computed vorticity values to establish a set of new velocity values. This step requires the solution of a boundary-value problem, since Eqs. (1) and (2) constitute an elliptic system of differential equations.

The general procedure just outlined for the time-dependent flow can be utilized to obtain steady-state solutions asymptotically at large times. Alternative, steady-state equations may be solved directly.

It is clear that a general viscous flow field is composed of a relatively small vortical (viscous) region surrounded by a large potential (irrotational) region. The viscous region is further composed of three distinct flow components: the boundary layer, the recirculating flow, and the wake. These components are clearly seen in numerical results for flows about various solid bodies. In Fig. 1 are shown computed mean streamlines and equi-vorticity contours [1.23] about a 12%-thick Joukowski airfoil at a 15° angle

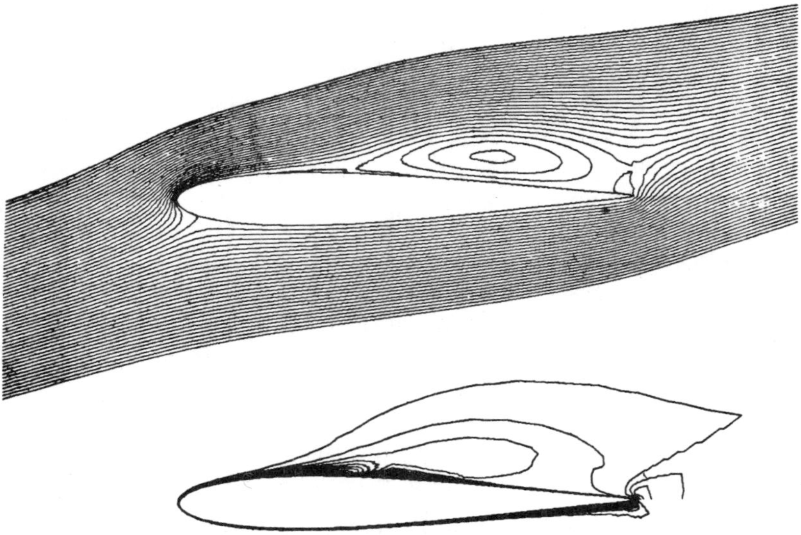

Figure 1. General viscous flow pattern.

of attack and a flow Reynolds number of $3.\dot{6} \times 10^6$. This flow is turbulent. For laminar flows, similar features are present.

The thickness of the boundary layer is of the order $Re^{-1/2} L$, where Re is a flow Reynolds number based on the freestream velocity and the characteristic length L of the solid body. The length scale of the detached-viscous-flow components are obviously of order L. Since the gradient of the vorticity field is zero in the potential region, the length scale there is infinite. There are therefore present in a high-Reynolds-number general viscous flow field diverse length scales for different components of the flow. The simultaneous accommodation of these diverse length scales presents serious difficulties in the numerical solution of general viscous-flow problems. In particular, the design of efficient grid systems for high-Reynolds-number flows has been a focal research problem for many years. The hybrid procedure offers a possibility of adopting different grid systems in different regions of the flow to accommodate the diverse length scales. The hybrid procedure therefore offers important advantages in computing high-Reynolds-number general viscous flows.

The kinetic processes of convection and diffusion, which redistribute the vorticity in the fluid, generally proceed at vastly different speeds in a high-Reynolds-number flow. Only in the vicinity of a solid wall, provided that the flow is attached, is the convection speed comparable to the diffusion speed (in the direction normal to the wall). Many of the current difficulties in solution accuracy for general viscous flows can be traced to the fact that the two kinetic processes are different in character. It is appropriate to consider the possible use of different numerical procedures in the simulation of these two processes. This possibility has not received adequate attention in the past. It is obvious, however, that in order to treat these two processes separately, a hybrid approach which takes into account the interplay between the physical and the numerical aspects of the general viscous flow is needed.

3 Prevailing Procedures

In computing general viscous flows, the differential equations describing the flows are first approximated by sets of algebraic equations, which are then solved using computers. With respect to the method of setting up the algebraic equations, the numerical procedures currently in use can be grouped into three major classes, each of which possesses certain desirable attributes.

Finite-Difference Procedures

Until recently, computational fluid dynamicists have emphasized the development of the finite-difference procedures for computing general viscous flows. For problems involving complex boundary shapes, the solution field is often transformed into one with simple boundary shapes. The transformed equations are then replaced by their finite-difference analogs. The major advantages of the finite-difference procedures are the simplicity in the mathematical concepts used and the simplicity of the coefficient matrices of the resulting algebraic equations.

Because of these advantages and of the extensive research efforts carried out in the past two decades, the finite-difference approach represents today the most popular method for computing general viscous flows. The error analysis for the finite-difference approach is at a rather advanced, though by no means conclusive, stage of development. The principal limitations of the finite-difference procedures are the very large amounts of computation required for high-Reynolds-number flows; the presence of numerical viscosity and spurious results which, though often appearing small, cast doubt on the accuracy of the solution for high-Reynolds-number flows; and the difficulties associated with the numerical treatment of certain boundary conditions [1.28]. These limitations can be traced to the prominent features of the general viscous flow described in Section 2. In particular, with a finite-difference procedure, the solution field must contain the potential-flow region as well as the viscous region. The currently used procedure for accommodating the diverse length scales of these regions is the use of expanding grids, i.e., grids with increasing spacing as the distance from the solid (and from the viscous region) increases. It is known, however, that expanding grids give rise to grid-associated errors which can be large for the rapidly expanding grids used for high-Reynolds-number flows. Furthermore, the finite-difference representation of the convective and diffusive processes presents substantial difficulties. For high-Reynolds-number flows, it is usually necessary to use upwind differencing for the convective term. The diffusion term is generally represented by central differencing. Such a practice tolerates the approximation of different terms in the same differential equation by difference expressions with different orders of accuracy. It admits into the solution a numerical, or artificial, viscosity which is misleading for high-Reynolds-number flows with small real viscosity.

Finite-Element Procedures

During the past five years, there has been a rapid growth in the application of the finite-element procedures to flow problems. With prevailing finite-element procedures, the integral relations are obtained through the concept of variational principle or that of residuals (e.g. the Galerkin process). The resulting algebraic equations for the kinematics of the general viscous flow, like those obtained through finite-differencing, are implicit. That is, each equation contains more than one unknown value of the field variable. The implicitness of the algebraic equations makes it necessary to compute both the potential-region and the viscous parts of the flow. In consequence, the limitations experienced in the application of the finite-difference procedures have reappeared in connection with the use of finite-element procedures.

With a finite-element procedure, the differential equations describing the flow are first recast as integral relations. The solution field is then mapped into elements with associated nodes. It is worthy of note that, if the elements and their associated nodes are arranged in a regular and uniformly spaced rectangular array, the algebraic equations obtained through the variational principle or through residuals are often identical to those obtained through a finite-difference procedure using the same grid system. In consequence, the familiar finite-element approach is not as fundamentally different from the finite-difference approach as some researchers have claimed it to be. The principal advantages of the finite-element approach are its inherent ability to accommodate complex boundary shapes and its flexibility in the selection of data-node locations. These advantages result from the fact that integrals, rather than derivatives, are being approximated. When the data nodes are arranged irregularly, however, the coefficient matrix of the finite-element equations are more complex than that of the finite-difference equations. The algorithms for solving the algebraic equations and the associated error analyses are more difficult to establish.

Integral-Representation Procedures

The integral-representation procedure utilizes the finite-element methodology. Unlike prevailing finite-element procedures based on the variational principle and on residuals, however, integral representations are obtained through the use of the principal, or fundamental, solutions. Many researchers have referred to the integral-representation procedures as "boundary-integral" or "boundary-element" methods [1.29]. The integral representation, however, in general contains integrals over a solution field as well as over the boundary. The integral representation reduces to a boundary integral only in the special case of potential flows, where the vorticity field is zero everywhere except on the boundary.

An integral representation for the kinematics of the incompressible flow is [1.26]

$$\mathbf{v}(\mathbf{r}) = \int_R \boldsymbol{\omega}_0 \times \nabla_0 P \, dR_0 + \oint_B (\mathbf{v}_0 \cdot \mathbf{n}_0 + \mathbf{v}_0 \times \mathbf{n}_0 \cdot \mathbf{x}) \nabla_0 P \, dB_0 \tag{4}$$

where B is the boundary of R; $\mathbf{n}$ is the outward unit normal vector on B; the subscript 0 indicates that the variable, the differentiation, or the integration is in the $\mathbf{r}_0$-space; and P is the principal solution of the scalar Poisson's equation, defined by

$$P(\mathbf{r}, \mathbf{r}_0) = \begin{cases} -\dfrac{1}{4\pi|\mathbf{r} - \mathbf{r}_0|} & \text{for three-dimensional problems, (5)} \\[2ex] -\dfrac{1}{2\pi} \ln \dfrac{1}{|\mathbf{r} - \mathbf{r}_0|} & \text{for two-dimensional problems. (6)} \end{cases}$$

The integral representation (4) is completely equivalent to the differential equations (1) and (2) and is valid for both the time-dependent and the steady flows.

For a two-dimensional time-dependent flows, the kinetics of the problem, described by Eq. (3), can be recast into the form [1.30, 1.31]

$$\begin{aligned} \boldsymbol{\omega} = & \int_0^t dt_0 \oint_B (Q \nabla_0 \omega_0 - \omega_0 \nabla_0 Q) \cdot \mathbf{n}_0 \, dR_0 \\ & + \int_R (Q\omega_0)_{t_0 = 0} \, dR_0 + \int_0^t dt_0 \int_R Q(\mathbf{v}_0 \cdot \nabla_0) \omega_0 \, dR_0, \end{aligned} \tag{7}$$

where Q is the principal solution of the parabolic differential equation and is

$$Q = \frac{1}{[4\pi\nu(t_0 - t)]^{d/2}} \exp\left(-\frac{|\mathbf{r}_0 - \mathbf{r}|^2}{4\nu(t_0 - t)}\right). \tag{8}$$

Equation (7) can be generalized for three-dimensional flows. For steady flows, an integral representation for the kinetics of the flow is presented in [1.24] and utilized in computation in [1.32] and [1.33].

It is obvious from Eq. (3) that in the kinetic part of the computation, i.e., in step (a) of the computation loop, velocity values need to be known only in the viscous region of the flow. The integral representation (4) permits the computation of velocity values explicitly, point by point. It is therefore possible to compute velocity values only in the viscous region. Since the right-hand side of Eq. (3) vanishes in the potential region where the vorticity and its spatial derivatives vanish, in the kinetic part of the computation it is only necessary to compute new vorticity values in the viscous region. The entirety of the computation procedure can thus be confined to the viscous region [1.26]. As discussed earlier, the viscous region usually occupies only a small portion of the total flow field. The use of the integral-representation approach therefore requires a drastically smaller number of data nodes than that required by prevailing finite-difference and finite-element methods. Computational difficulties resulting from the incompatibility of the length scales of the potential and the viscous region are eliminated by the use of Eq. (4). It should be emphasized, however, that within the viscous region there exists another incompatibility of the length scales, between the boundary-layer component and the detached-flow component.

Previous methods for computing the boundary vorticity values, based upon extrapolation procedures, often experience stability and accuracy difficulties. The integral-representation approach for computing boundary vorticity values has been shown to produce stable and accurate results. This approach simulates the physical process of vorticity generation on solid boundaries accurately and is discussed fully in [1.28].

The integral representation (4) represents an elliptic differential equation. The contributions of the boundary conditions and of the inhomogeneous term of the elliptic equation are expressed separately, respectively as a boundary integral and an integral over the solution field. The ability of Eq. (4) to simulate the process of vorticity generation on solid boundaries is a consequence of the fact just stated. It is noted in passing that the boundary integrals often can be evaluated explicitly. For example, the boundary integral in Eq. (4) gives simply $\mathbf{v}_\infty$, the freestream velocity, for external flows.

The integral representation (7) represents a parabolic differential equation. It describes an initial–boundary-value problem. With Eq. (7), the contributions of the boundary condition, of the initial condition, and of the inhomogeneous term of the parabolic equation are all expressed separately. They are, respectively, the first, second, and third terms on the right-hand side of Eq. (7). The inhomogeneous term in Eq. (3) represent the process of convection. The fact that its contribution to the vorticity distribution is expressed separately suggests that with an integral-representation formulation, a special numerical procedure can be developed to compute the contributions of the convective process.

The major advantages of the integral-representation procedures are:

1. its ability to confine the solution field to the viscous region of the flow,
2. its ability to accurately simulate the process of vorticity generation on solid boundaries,
3. the possibility of treating different physical process using different numerical procedures,
4. its inherent flexibility in locating the nodes and in accommodating complex boundary shapes.

4 Hybrid Procedures

Several hybrid procedures utilized by this author and his coworkers in the solution of incompressible general viscous flows are reviewed in this section. These hybrid procedures reflect the research emphasis of this author during the past decade in the development of the integral-representation procedures for general viscous flows. The several hybrid procedures are only outlined here, and a brief account of important features is provided for each procedure. Suitable references containing detailed analyses, descriptions of computational techniques, and extensive numerical illustrations are given for each of the procedures reviewed.

Integrodifferential Procedure [1.26, 1.34]

With an integrodifferential procedure, the kinematic aspect of the problem is formulated as an integral representation such as Eq. (4). The kinetic aspect is kept in its familiar differential form, such as Eq. (3), and is solved using a finite-difference procedure. The integrodifferential procedure therefore is a hybrid procedure. In the initial stage of development of the integrodifferential approach, the integral representation for the velocity vector is approximated by a set of algebraic equations in the following form:

$$v_i = \sum_{y=1}^{J} G_{ij}\omega_j + F_i, \tag{9}$$

where v_i is the value of a velocity component at the node i, ω_j is the vorticity value at the node j, G_{ij} is the geometric coefficient, dependent only upon the relative locations of the nodes i and j, and F_i is the contribution of the velocity boundary condition to the velocity component v_i.

Using Eq. (9), with the geometric coefficients G_{ij} and the terms F_i precomputed, the computation of each velocity-component value at each node requires J multiplications. The number J is drastically smaller than the number, say N, of grid points involved in a prevailing finite-difference or finite-element method, since with the integral-representation approach the solution field is confined to the viscous region. Nevertheless, the number J is not small for high-Reynolds-number flows. To compute the values of each velocity component at all the J points using the integral-representation procedure, J^2 multiplications are required. Also, the number of geometric coefficients G_{ij} needed is, in the general case, J^2. The storage of all the geometric coefficients may exceed the central core capability and may lead to additional computation cost. In comparison, a finite-difference procedure for the kinematics of the problem requires only several times N data storage, which is usually not excessive. The operation count required for a finite-difference procedure, depending on the specific technique used, ranges from $N \ln N$ to $N^{3/2}$.

Several two- and three-dimensional general viscous-flow problems were solved using the basic integrodifferential procedure in the initial stage of its development. Because of the large computer-time and data-storage requirements, only flows at low Reynolds numbers and involving simple boundary geometries were computed. For two-dimensional flows, the computer time used was actually greater than required by the then prevailing finite-difference procedures. For three-dimensional flows, the ability of the integral-representation procedure to confine the solution field to the viscous region leads to a much greater factor of reduction of grid points needed. (It is obvious that if the factor of reduction of grid points is $J/N = A$ for a two-dimensional problem, then it is approximately A^2 for a three-dimensional problem.) In consequence, even without the more sophisticated devices incorporated later, the integrodifferential procedure produced superior solution efficiency for three-dimensional flows.

Hybrid Procedure for Kinematics [1.35, 1.36]

Equation (9) can be utilized to compute only the velocity-component values at grid points surrounding the viscous region. Once this is accomplished, a finite-difference or an efficient finite-element procedure can be utilized for computing velocity values at grid points within the viscous region. In this manner, a substantial reduction in computational effort is realizable, even if J is not very much smaller than N. This hybridization of the integral-representation and the finite-difference (or the finite-element) procedures for the kinematics of the problem indeed ensures a superior solution efficiency, since the most efficient procedure, including those that may be developed in the future, can be used, and the number of grid points J is smaller than that needed without hybridizing.

The accuracy and the practicality of the hybrid procedure for the kinematics of the general viscous flow problem were conclusively demonstrated by solving several test problems as well as a time-dependent two-dimensional flow past an airfoil at a high angle of attack and a moderate Reynolds number of 1000 [1.35, 1.36]. The actual computations were carried out using the stream function in place of the velocity components. In addition to the hybrid procedure just described, the solution field was segmented and the stream-function counterpart of Eq. (9) is used to compute the stream-function values on the boundary of each segment. The stream-function values at the interior points of each segment are then computed using a finite-difference procedure. Since the computations within each segment are performed independently of that within other segments, the average operation count required for computing each stream-function value is further reduced by this flow-field segmentation technique.

The results for the airfoil problem obtained using this procedure were sufficiently encouraging so that, recently, the computer code developed for this problem was revised and a manual [1.37] for this code was prepared. This code is being used by researchers elsewhere to treat problems of interest to them.

Hybrid Procedure for Kinetics [1.38, 1.39]

The use of finite-element procedures enables complex geometries of the flow boundary to be accommodated satisfactorily. However, finite-element equations are more complex, more difficult to analyze, and more time-consuming to solve than finite-difference equations. A hybrid procedure using a finite-element method near the solid boundary and a finite-difference method far from the solid boundary offers the advantages of both methods. Such a hybrid procedure was developed for the solution of the kinetic aspect of the time-dependent general viscous flow problem.

A small portion of the solution field enveloping the airfoil is mapped into triangular elements. The number of nodes in this inner region is kept small, and the nodes are spaced closely. Algebraic equations for this region are obtained through the use of Galerkin's procedure. These algebraic equations possess rather complicated coefficient matrices. The size of the matrices is small, since the number of the nodes in the finite-element region is small. It is therefore not time-consuming to handle these matrices.

The outermost layer of elements of the finite-element region is made to overlap with a finite-difference grid for the outer region. The finite-difference grid lines are constructed using a Cartesian coordinate system. The grid-line spacing is coarser than the finite-element node spacing. The resulting finite-difference equations possess matrices that are relatively large but are simple in form and much easier to handle.

Several test problems as well as several cases for an oscillating airfoil problems, for a moderate Reynolds number of 1000 and covering a range of amplitudes and frequencies, were computed using the hybrid procedure for the kinetics and the integral-representation procedure for the kinematics of the problem. It was found that no computational errors were introduced by the hybridization of the finite-element and finite-difference procedures. For the test problems solved, the results were found to be in excellent agreement with available experimental data in the steady-state limit.

Boundary-Layer and Detached Regions [1.40]

The use of the integral-representation approach permits the solution field to be confined to the viscous region of a general viscous flow. As discussed earlier, however, the viscous region is composed of attached and detached flow components. The incompatibility of length scales of these two components makes it impractical to devise a common numerical grid that provides sufficient solution resolution for the attached flow and yet does not contain an excessively large number of grid points in the detached part of the viscous region. Furthermore, the retention of terms that are in reality negligible in the boundary layer places an excessive demand on the numerical procedure. This demand can lead to inaccuracy as well as inefficiency.

An optimal way to eliminate the difficulties caused by the incompatibility of length scales within the viscous region is the use of a hybrid procedure treating the attached and the detached flow components differently. Specifically, the Navier–Stokes equations are solved only in the detached part of the viscous region, and the boundary-layer equations are solved in the attached part, wherever the boundary-layer simplifications are justified. With such a hybrid procedure, the computer-time requirement becomes relatively insensitive to the flow Reynolds number.

To treat the attached viscous flow component using the boundary-layer equations, the mathematical description of the flow must include a specification of the velocity at the outer edge of the boundary layer or, equivalently, the pressure gradient along the boundary layer. In a general viscous flow, the presence of "strong viscous–inviscid interaction" makes it unacceptable to determine the boundary-layer-edge velocity through a potential-flow calculation based on the solid-body shape. With a finite-difference or finite-element procedure that computes the entire flow field, it is necessary to carry out repeatedly individual computations of the attached flow component, the detached flow component, and the potential flow. The boundary-layer-edge velocity is then determined through an iterative matching procedure. Such a procedure demands very large amounts of computation. In particular, for time-dependent flows, iterative computations need to be performed for each time step and the amount of computation can be excessive.

If the integral-representation procedure is used in conjunction with the hybrid "boundary-layer Navier–Stokes" procedure, then the potential part of the flow is excluded from the computation. There is therefore no viscous–inviscid interaction involved in the procedure, and the amount of computation required is drastically reduced. For time-dependent flows, separate computations of the attached and detached viscous flow components can be carried out without an iterative matching process.

A hybrid procedure has been developed for the time-dependent general viscous flow. With this procedure, the attached and detached viscous flow components are treated separately. This procedure, described fully in [1.40], is based on the integrodifferential approach [1.35]. Several test problems have been solved using this procedure. Computations of a general viscous flow past an airfoil have also been carried out.

5 Sample Results and Discussions

A variety of general viscous-flow problems have been treated using the hybrid procedures described in this paper. Selected results are given here to demonstrate the application of the hybrid approach under a range of circumstances.

With each of the hybrid procedures, a solution for a general viscous flow past an airfoil has been obtained. Shown here are some typical results obtained for a flow past 9%-thick symmetric Joukowski airfoil. The airfoil is initially at rest in a fluid also at rest. At a given time level, $T = 0$, the airfoil is set in motion. Thereafter, it is kept moving at a uniform speed with an angle of attack of 15° and a Reynolds number based on the chord length and the freestream velocity of 1000.

The numerical results show that, although the motion of the airfoil after the impulsive start is time-dependent, a steady-state solution is not reached asymptotically at large time levels. Rather, a periodic shedding of vortices occurs at large time levels. In [1.35] the computation was carried to a dimensionless time level of 7.75, the reference time being the chord length divided by the freestream velocity. At that time level, the flow field surrounding the airfoil has apparently gone through a little more than one cycle of shedding of vortices. The computation of [1.35] was recently extended to a time level of 12.19, covering a second cycle of shedding of vortices. In Fig. 2 are shown constant stream function contours for three different time levels. The contours at the time levels 3.19 and 8.64 are strikingly similar. Furthermore, they are drastically different from the contours at the time level 12.19.

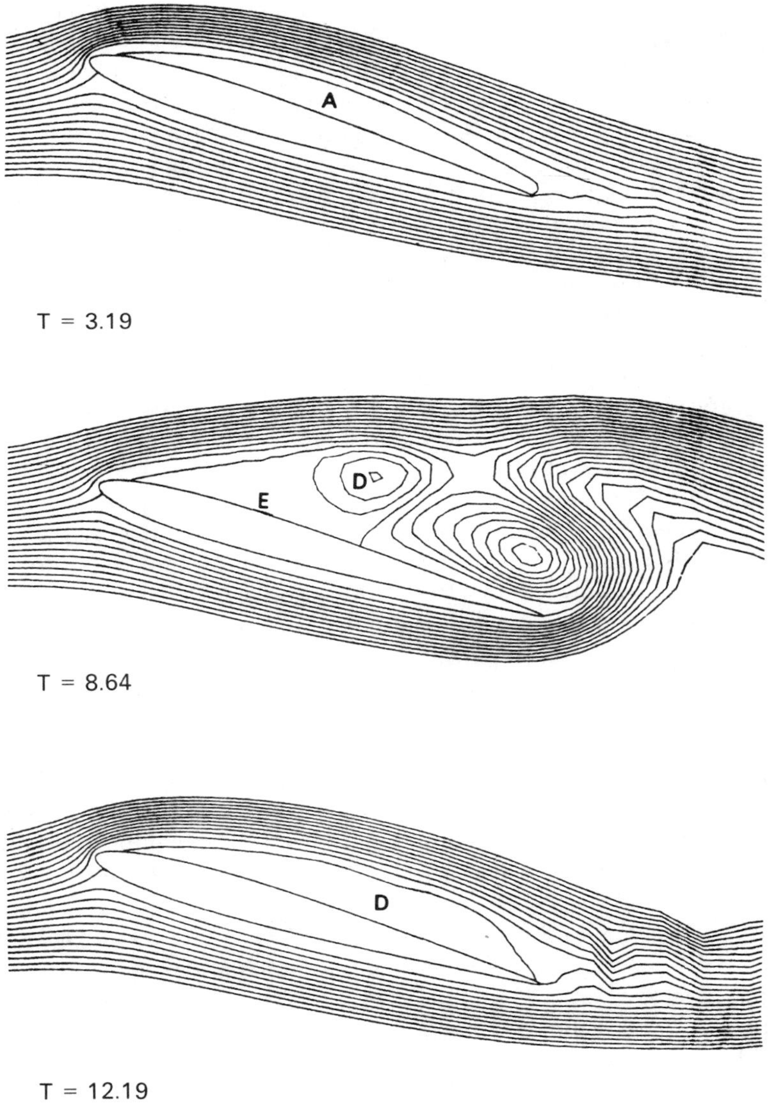

Figure 2. Flow around an airfoil.

Comparisons of numerical results show some differences between the first and the second cycles. There also exist some quantative differences between the present results and that obtained using other numerical procedures. There is no doubt, however, that the phenomenon of periodic shedding of vortices occurs regularly. The numerical solution clearly simulates the important flow features occurring in the fluid, including the development and shedding of a starting vortex, the transport of this starting vortex with the fluid and its viscous dispersion in the fluid, the formation of separation bubbles on the upper surface of the airfoil, their subsequent growth and bursting, and the accompanying shedding of vortices. Additional descriptions of the flow details are presented in [1.36].

As discussed earlier, the kinetic processes of convection and diffusion generally proceed at vastly different speeds in a high-Reynolds-number flow. Many of the current difficulties in solution accuracy can be traced to the fact that the two kinetic processes are different in character. It is noted that the integral representation (7) expresses the diffusion and convection contributions as separate integrals. This suggests that a hybrid procedure can be developed and these two processes can be simulated by different algorithms. In a recent study [1.31], the usefulness of the integral representation (7) in computing general viscous flows was established. Several test problems as well as the airfoil problem just described were solved using Eq. (7). The possibility of using different algorithms to simulate the processes of diffusion and convection, however, was not explored in [1.31]. This possibility will be examined in the near future.

The use of different algorithms to simulate the flows in the boundary layers and in the detached viscous components has been explored [1.40], as discussed briefly in Section 4. The results obtained using this procedure show that the computer time required to solve a given general viscous-flow problem is insensitive to the flow Reynolds number. This hybrid procedure therefore removes the "Reynolds-number limitation" experienced in the use of prevailing methods. Also, no spurious vorticities or noises are observed with the use of this hybrid procedure.

In Fig. 3 are shown three sets of steady-state pressure distributions on a circular cylinder at a Reynolds number of 40,000 based on the cylinder diameter and the freestream velocity. The distributions are (a) numerical results obtained using the hybrid approach just mentioned, (b) finite-difference results obtained by others [1.41] treating the entire flow, including the potential and the viscous regions, using the Navier–Stokes equations, and (c) experimental results reported in [1.42]. The numerical results were obtained by solving the time-dependent equations and carrying out the computation to a large-time-limit steady solution. The agreement between the present results and the experimental results is remarkably good. The small differences near the rear stagnation point are to be expected, since in the present computation no provision is made to model flow turbulence occurring in the recirculating flow and in the wake downstream from the cylinder. In contrast with the remarkably good agreement just described, the finite-

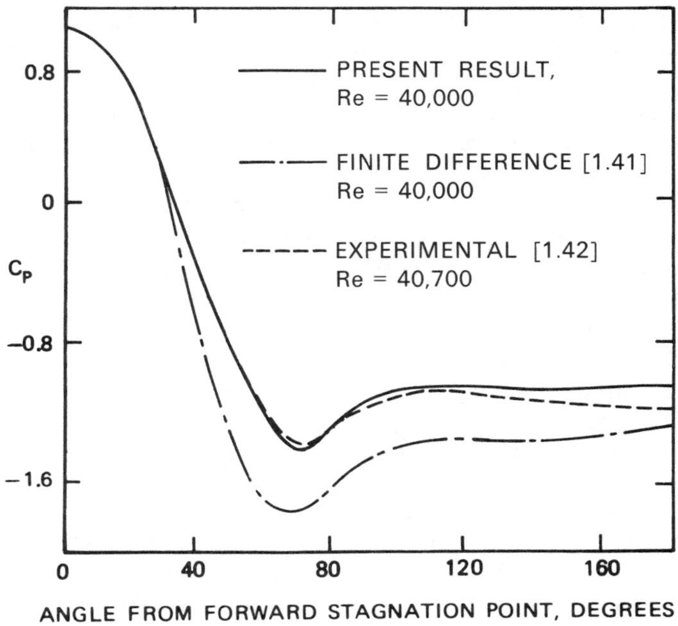

Figure 3. Pressure coefficient on a circular cylinder.

difference results obtained treating the entire flow on the basis of the Navier–Stokes equations differ substantially from the experimental results.

In Fig. 4 are shown a computed steady-state pressure distribution on a circular cylinder at a Reynolds number of 3.6×10^6 compared with experimental data presented in [1.43]. The computation was performed using a procedure similar to that described in [1.36] in conjunction with a

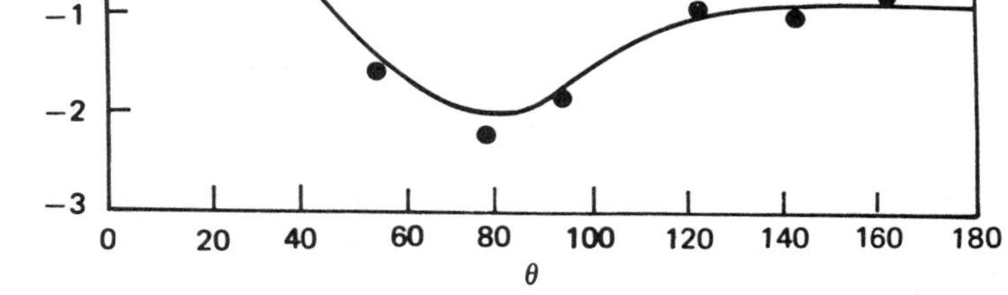

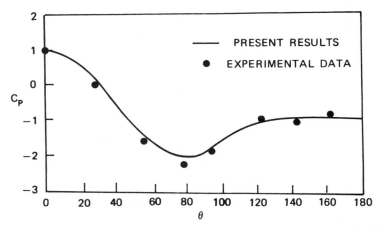

Figure 4. Turbulent flow past a circular cylinder: surface pressure distribution ($Re = 3.6 \times 10^6$).

two-equation differential model for turbulence. The agreement between the computed and experimental results is satisfactory. The computed skin-friction distribution differs substantially from the measured distribution due to the assumption of fully turbulent flow in the entire computation. (The experimentally observed transition point is about 65° from the front stagnation point.)

In Fig. 5 are shown computed drag coefficients for the circular cylinder problem at various Reynolds numbers. Values of the drag coefficient computed using several different hybrid procedures [1.22, 1.30, 1.35, 1.39] for Reynolds number 40 are in excellent agreement with each other and with the experimental data. For Reynolds numbers of 1000 and 40,000, the boundary-layer and Navier–Stokes hybridization procedure gives results in excellent agreement with the experimental data.

The most encouraging aspect of the hybrid approach is the continual improvement in computational efficiency it offers. In Table 1 are summarized the computer-time requirements for computing the steady flow around a circular cylinder at various Reynolds numbers. A peculiar feature appearing in this table is that, at the present, the computation of the flow at the higher Reynolds numbers of 1000 and 40,000 actually requires less computer time than at Reynolds number 40. This is because at the low Reynolds number of 40, boundary-layer simplifications are not justifiable in the attached region and the hybridization of the boundary-layer and Navier–Stokes approaches is not useful.

The development of a routine computational capability for complex viscous flows is not yet near completion. In particular, it is anticipated that the difficulties associated with the accuracy of time-dependent solutions, with the efficiency of three-dimensional computations, and with the reliability of turbulence models in recirculating flows will require extensive and persist-

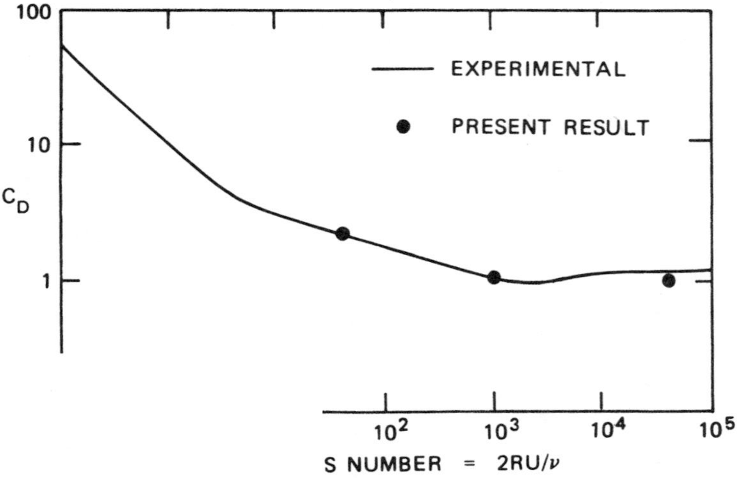

Figure 5. Drag coefficient for a circular cylinder.

ent efforts over a number of years. In this regard, the hybrid approach offers some encouragement because of its inherent efficiency and its ability to accommodate the complex and diverse physical features present in a general viscous flow.

Table 1 Computer-time requirement for the circular-cylinder problem.

Reynolds number	CDC-6600 equivalent time, min.			
	1970	1975	1980	1985
40	95	40	25	(20)[a]
1,000	(300)	80	20	(6)
40,000	(1200)	(250)	20	(6)
3,600,000	—	—	80	(15)

[a] Numbers in parentheses are estimated values.

Acknowledgment

This research is supported by the Office of Naval Research under contract No. N00014-75-C-0249. The author expresses his sincere appreciation to Morton Cooper for his continual interest in this research and for many stimulating suggestions.

CHAPTER 4

State of the Art in Subgrid-Scale Modeling

Joel H. Ferziger *

1 Introduction

The need for turbulence models is well known. Only in the last few years has it become possible to compute any turbulent flow without models. We can now simulate homogeneous turbulent flows at moderate Reynolds numbers and are beginning to simulate simple inhomogeneous low-Reynolds-number flows. Although the range of flows accessible to direct or exact simulation will increase as further improvements are made in computer hardware, there is no possibility of this approach becoming an engineering tool in the foreseeable future. Even if large enough computers were available, it is unlikely that an engineer would want to generate the enormous amount of data exact simulation would produce in order to get a few pieces of significant information. What is desired, obviously, is a method which produces the information needed for design purposes (pressure distribution, skin friction, etc.) and as little else as possible.

Experiments are becoming more expensive as hardware becomes more sophisticated. Likewise, the cut-and-try approach to design is becoming increasingly costly. The downward trend of computer costs makes it attractive to substitute computational simulation for hardware wherever possible. Thus the current interest in turbulent flow simulation.

It has proved difficult to construct turbulence models of great generality. We have special-purpose models that do an excellent job on particular classes of flows; the various boundary-layer models are probably the best examples.

* Department of Mechanical Engineering, Stanford University, Stanford, CA 94305.

We would like to simulate more complex turbulent flows without requiring extensive laboratory work to tune the model; this means that we shall have to find more general models or another way to determine the model constants for particular flows.

It is possible to use exact simulations to help develop models. This approach is attractive both because computations can often be done at smaller expense than experiments and because they can produce results (such as fluctuating pressures) which need to be modeled but are not currently amenable to laboratory measurement. A significant disadvantage is the limitations to relatively low Reynolds numbers and simple geometry. Thus, laboratory experiments remain essential, but the burden on them can be reduced; the two approaches are, for the most part, complementary.

The value of any increase in the range of flows which can be simulated accurately is obvious. The limited range of flows amenable to exact simulation on the one hand and the lack of models of general applicability on the other suggest that a hybrid approach might have significant value.

This has led to the development of large-eddy simulation (LES). LES mimics exact simulation with respect to the large eddies; they are treated as three-dimensional, time-dependent structures which are simulated as accurately as possible. The small-scale structures must be modeled; much of what is done in this area can be copied from time-average modeling ideas. LES increases the number of flows which can be simulated. It also offers the more distant possibility of becoming an engineering method in the future. For these reasons, it has been the center of the author's attention and that of his colleagues and students for the past eight years or so. For reviews on this subject, see [1.44], [1.45], and [1.46].

In this paper, we shall compare and contrast subgrid-scale (SGS) modeling used in LES with the time-average modeling that is in more widespread use. Obviously many ideas are applicable in both areas. We will first briefly review the theory and practice of LES and will then proceed to consideration of the equations for and modeling of the subgrid scale component of turbulence. The similarities and differences between the behavior of the quantities modeled in the two cases will be examined with an eye to seeing which ideas from time-average modeling can be borrowed for use in subgrid scale modeling. We will also look at the application of direct simulation to the testing of both kinds of models and the use of LES for testing time-averaged models and at ideas which are applicable only to SGS modeling. Although there is considerable interest in turbulent mixing of scalar quantities, we shall consider only models for the hydrodynamics and mention mixing only in passing.

2 Approaches to LES

Large-eddy simulation had its origins in the attempts of meteorologists to predict the global weather. Many of the ideas were developed in that context, and the first paper [1.47] on engineering applications of LES was written by a meteorologist, Deardorff, in 1970. For the earlier work see Smagorinsky [1.48] and Lilly [1.49, 1.51].

All of the approaches in use today begin by averaging the Navier–Stokes equations. The formalisms differ somewhat, but they all amount to averaging the Navier–Stokes equations over a small control volume, and the resulting equations are

$$\frac{\partial \bar{u}_i}{\partial t} + \frac{\partial}{\partial x_j}\overline{u_i u_j} = -\frac{\partial \bar{p}/\rho}{\partial x_i} + \nu \frac{\partial^2 \bar{u}_i}{\partial x_j \partial x_j} \tag{1}$$

which, for incompressible flow, must be solved together with the continuity equation

$$\frac{\partial \bar{u}_i}{\partial x_i} = 0. \tag{2}$$

As always in turbulence, the trouble comes from the nonlinear term in Eq. (1). It is common to write

$$u_i = \bar{u}_i + u_i' \tag{3}$$

so that the nonlinear terms become

$$\overline{u_i u_j} = \overline{\bar{u}_i \bar{u}_j} + \overline{\bar{u}_i u_j'} + \overline{u_i' \bar{u}_j} + \overline{u_i' u_j'}. \tag{4}$$

This is the point at which the differences of approach show up. These differences are more of opinion and style than of content. Deardorff, Schumann, and their coworkers prefer to introduce the grid system to be used in the numerical solution immediately, and the averaging in the above equations is interpreted as an average over a grid volume. The averaging process is thus closely tied to the numerical method, and the small-scale field u_i' is properly called the subgrid-scale (SGS) component of the turbulence. In this approach, $\bar{u}_i$ is taken to be constant within each grid volume, as shown in Fig. 1. However, this means that $\bar{u}_i$ is discontinuous. A consequence of this definition is that u_i' is also discontinuous, but $\overline{u_i'} = 0$ and the central two terms in Eq. (4) are identically zero. Furthermore, if the averages of each velocity component are defined on the same grid, we have

$$\overline{\bar{u}_i \bar{u}_j} = \bar{u}_i \bar{u}_j. \tag{5}$$

However, most of the calculations done using this approach have used the Harlow–Welsh staggered grid, which uses different averaging volumes for the various equations. Equation (5) is then regarded as an approximation, but it has been widely used. This leaves just the term $\overline{u_i' u_j'}$ to be modeled. Schumann's approach is a slight modification of the one just described. In the averaging procedure, he evaluates the integral in the direction of the derivative analytically and then has to deal with averages over the faces of the finite volumes. One then has three types of surface averages as well as volume averages to deal with; these must be related to each other before one can proceed further. Although this method has some appeal, we believe that the extra effort involved in relating the planar and volume averages does not

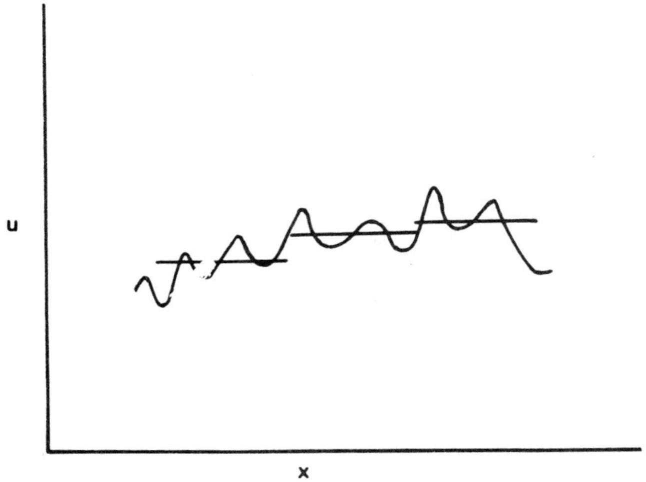

Figure 1. The grid average velocity according to the Deardorff–Schumann prescription. The curve is $u(x)$; the horizontal lines represent u, and the difference is u'.

yield a sufficient dividend and prefer the method described above. An important advantage of the Deardorff–Schumann approach over the one described below is that it is "cleaner"; that is, the process of going from the basic equations to a computer program is simpler.

The Stanford group, of which the author is a member, has preferred to look at the fundamentals of LES and has spent effort on trying to define terms carefully and to understand the basis of the models. This requires that the definition of the subgrid-scale component of the turbulence be uncoupled from the numerical method insofar as possible; in particular, it should be independent of the mesh used in the numerical calculation. It is useful to allow $\bar{u}_i$ and u_i' to be continuous functions. This led to decoupling the process of filtering from that of numerical solution, and we shall describe it briefly here. The chief disadvantage of this approach is that it is more cumbersome than those described above.

We prefer to define the large-scale field, i.e., the component which will be simulated, by means of a filtering operation [1.50]:

$$\bar{u}_i(\mathbf{r}) = \int u_i(\mathbf{r}')G(\mathbf{r} - \mathbf{r}')\, d^3r'. \tag{6}$$

This definition allows a lot of flexibility. It also defines the large-scale field everywhere in the flow, as desired, and makes it and the subgrid-scale field variables continuous functions of the coordinates, as shown in Fig. 2. The kernel of the filter (G) can be any reasonably smooth function; we prefer the Gaussian for its nice mathematical properties. The characteristic width, Δ, of the kernel is arbitrary, and different widths can be used in each direction. From a practical point of view, it is best if the width is approximately twice the computational grid size, but this is not absolutely required. A significant

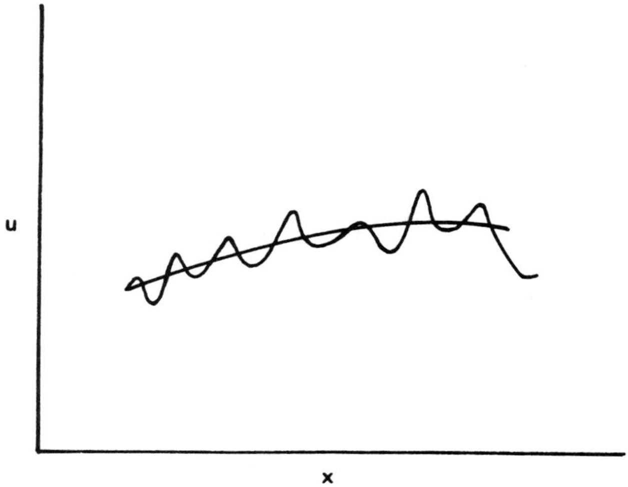

Figure 2. The filtered velocity according to Eq. (6). The velocity $u(x)$ is the same as in Fig. 1; the smooth curve represents $\bar{u}(x)$, and the difference is u'.

disadvantage to this method is that none of the simplifications of the Deardorff method applies here; this has some advantages, as it permits one to sort out various effects. The difficulties with this method are traceable to $\overline{u_i'}$ not being zero according to these definitions. This leads to the appearance of terms that would be zero in the time-average approach. We shall say more about this later.

3 SGS Stress Equations

Essentially all models, whether for time-average calculations or for LES, can be derived by making approximations to the exact equations describing the Reynolds stresses in question. The terms which need modeling are most conveniently represented as

$$\tau_{ij} = \overline{\bar{u}_i \bar{u}_j} - \overline{\bar{u}_i \bar{u}_j} = \overline{\bar{u}_i \bar{u}_j'} + \overline{u_i' \bar{u}_j} + \overline{u_i' u_j'}. \tag{7}$$

which is the subgrid-scale Reynolds stress.

It is also worth mentioning the Leonard term [1.50],

$$\lambda_{ij} = \overline{\bar{u}_i \bar{u}_j} - \bar{u}_i \bar{u}_j. \tag{8}$$

In the Deardorff approach, this term is zero. When it is nonzero, it can be treated by any of several methods. We have used three different ones: (a) A Taylor series expansion which approximates λ_{ij} in terms of derivatives of the large scale field, (b) explicit calculation via fast Fourier transform (this is possible only in homogeneous flows), and (c) allowing the numerical truncation errors to approximately represent this term.

The equations for the subgrid-scale stresses can be derived by the usual procedure and can be written in a number of ways. One, which is similar to the standard time-average form, is

$$\underset{\text{(convection)}}{\frac{\partial \tau_{ij}}{\partial t} + \bar{u}_k \frac{\partial}{\partial x_k} \tau_{ij}} = - \underset{\text{(production)}}{\left(\overline{(\tau_{ik} + \lambda_{ik}) \frac{\partial \bar{u}_j}{\partial x_k}} + \overline{(\tau_{jk} + \lambda_{jk}) \frac{\partial \bar{u}_i}{\partial x_k}} \right)}$$
$$+ \underset{\text{(redistribution)}}{\overline{\frac{p}{\rho} \left(\frac{\partial u_i}{\partial x_j} + \frac{\partial u_i}{\partial x_j} \right)} - \frac{\bar{p}}{\rho} \left(\frac{\partial \bar{u}_i}{\partial x_j} + \frac{\partial \bar{u}_j}{\partial x_i} \right)}$$
$$- \underset{\text{(dissipation)}}{2\nu \left(\overline{\frac{\partial u_i}{\partial x_k} \frac{\partial u_j}{\partial x_k}} - \overline{\frac{\partial \bar{u}_i}{\partial x_k} \frac{\partial \bar{u}_j}{\partial x_k}} \right)}. \qquad (9)$$
$$+ \text{DIFFUSION}$$

There are many diffusion terms, and for that reason they are not written explicitly.

All of the terms in Eq. (9) are analogous to terms in the more familiar equations of time-average modeling. The interpretations are also similar. However, the differences are also quite important. The equations (9) contain more terms than the equations for the time-average Reynolds stresses because some items that are zero in time-average approach are not zero when filtering is used. In particular, we note the appearance of the Leonard term in the production term and, more importantly, the fact that the production term is filtered. The various terms in these equations can be computed by the methods described in the next section, but this has not yet been done.

The most common assumption in turbulence modeling is that production and dissipation terms dominate the turbulence budget and, as a first approximation, we can equate them and ignore the other terms. For the time-average equations, this approximation is reasonable when applied to the budget for the turbulent kinetic energy away from solid boundaries, but it is less valid for the component equations because the redistribution term is quite large. Near walls, the diffusion terms become quite important and the approximation is even more questionable; the low Reynolds numbers also mean that viscous effects may need to be considered. Nevertheless, it is frequently applied.

For LES, the situation is somewhat different. It is important to note that the model is assumed to represent a kind of spatial average of the *instantaneous* small-scale turbulence *at a given locale*. This is quite different from what is modeled in time- or ensemble-average modeling, and its nature is such that our understanding of it (and consequently, our ability to model it) is more limited. The compensating feature is that, because we are modeling just the small scales of the turbulence and explicitly computing a large part of the flow field, the overall results are much less sensitive to the accuracy of the modeling approximations. However, we can look at the magnitudes of some of the effects.

In particular, because the small scales of turbulence are highly intermittent, we should expect that the gradient of subgrid-scale quantities might be relatively large. This being the case, it is probable that the convection and diffusion terms, which are ignored in many conventional models, are more important in SGS modeling. On the other hand, we have recently found evidence that the pressure fluctuations and, more particularly, the pressure-strain correlations reside mainly in the large scales and may perhaps be less important in SGS modeling than they are in conventional modeling. Despite these differences, most SGS models to date have relied on ideas developed for time-average models.

4 Computational-Model Validation

Two approaches are commonly used for developing and testing time-average models. One method, favored by Lumley, Reynolds, and others, uses simple turbulent flows (usually the homogeneous flows) to test the validity of the models and to determine the adjustable parameters in the model. The major objection to this approach is that the homogeneous flows may differ considerably from engineering flows. The other method, used by Spalding, Launder, and others, adjusts the parameters to fit flows similar to the ones that one wishes to calculate.

It is more difficult to select and test models for LES. Information about the small scales of most turbulent flows is scarce; consequently, the constants have to be chosen in some indirect manner. At high Reynolds numbers, one cannot use enough grid points to assure that the turbulence at the cutoff between the large and small scales is in the inertial subrange. Despite this, the properties of the inertial subrange have been used to choose both the models and the parameters in them (e.g., [1.51]. Of course, the general properties of the small-scale field (positivity of the energy, the behavior of the SGS stresses as functions of the large-scale stress, etc.) can be used as well. To a large extent, the choice of the parameters has relied largely on results from time-average modeling. As we saw in the previous section, the two kinds of models are quite different, and it is hard to recommend this approach.

There is another approach that is quite useful. With the current generation of computers it is possible to compute the homogeneous turbulent flows with no approximations other than the well-defined ones present in any numerical simulation. At present, it is possible to do such calculations with grids as large as $64 \times 64 \times 64$ and, in a limited number of cases, $128 \times 128 \times 128$. This allows simulation at Reynolds numbers based on Taylor microscale up to 40 (80 with the larger grids). These simulations can be regarded as realizations of physical flow fields and are an interesting complement to laboratory results in that the kinds of data are very different.

Having a realization of one of these flows, we proceed in much the same manner an experimentalist would. The computed field can be filtered to give the large-scale component of the flow field; the small-scale component is obtained by difference. We can then compute the terms which need to be modeled, and from the large-scale field we can also compute what the model predicts these terms ought to be. Direct comparison between the model and the exact value is then possible. This can be done in a couple of ways. This method has been used to evaluate a number of models which will be described later.

The results of the simulations can also be used to validate time-average models. However, the means of doing this are somewhat different because, in a homogeneous flow, we obtain only a single value of each time-average variable, and we cannot verify directly whether an assumption of proportionality is valid or not. What we can do is assume the validity of the model, find the parameter, and then determine whether it is independent of the independent variables of the problem such as the Reynolds number, mean strain (or shear) rate, and other parameters peculiar to the flow. For tensor quantities, the value of the parameter obtained for different tensor components may be different, and this provides a direct test of the validity of the model. A number of interesting, but as yet unpublished, results of this type (cf. [1.52] and [1.53]) have been recently obtained; a preview of some of these will be given later.

The final way in which computed results can be used to check models is to use LES to check the validity of time-average models. This has the disadvantage, relative to the method described in the last paragraph, that the quantities of interest are not computed exactly in this method. The large-scale components are computed by LES, but the small-scale contributions are estimated from the SGS model; if the latter contribution is a large part of the total, this method can be inaccurate. However, at the present time, this is the only method which is applicable to inhomogeneous flows. Relatively little work has been done in this direction to date, but some is now in progress.

5 Eddy-Viscosity Models

Eddy-viscosity models can be "derived" from the "production equals dissipation" argument discussed earlier. This is done in a number of places and need not be repeated here. For subgrid-scale turbulence, the eddy-viscosity model amounts to assuming that the subgrid-scale Reynolds stress is proportional to the strain in the large-scale flow:

$$\tau_{ij} = 2\nu_T \bar{S}_{ij} = \nu_T\left(\frac{\partial \bar{u}_i}{\partial x_j} + \frac{\partial \bar{u}_j}{\partial x_i}\right). \tag{10}$$

The eddy viscosity ν_T has the dimensions of a kinematic viscosity. Most of the early work on the subject was based on the assumption that the eddy viscosity could be represented by

$$\nu_T = (C\Delta)^2 |\bar{S}|, \tag{11}$$

where Δ is the width associated with the filter and $|\bar{S}| = (\bar{S}_{ij}\bar{S}_{ij})^{1/2}$. More recently, it has been shown by a number of authors that this is correct only if the integral scale of the turbulence is smaller than Δ. Since LES is designed for this not to be the case, it is better to assume that

$$\nu_\tau = C\Delta^{4/3}L^{2/3}|\bar{S}|, \tag{12}$$

where L is the integral scale of the turbulence. Usually L is estimated from $L = q^3/\varepsilon$, where ε is the dissipation. However, most work done to date has used Eq. (11) rather than Eq. (12). A further modification to the eddy-viscosity models that is necessary for some homogeneous flows was suggested by Bardina, Ferziger, and Reynolds [1.54].

Eddy-viscosity models have a long record of reasonable success in time-average modeling of simple shear flows, and one might expect them to do well as SGS models. In fact, they have been found to do an adequate job in the homogeneous flows. Using them, one is able to predict energies, spectra, and a few higher-order quantities. Most higher-order quantities are not well predicted; this is not unexpected, as the higher-order quantities depend more strongly on the small scales than do the lower-order quantities and probably cannot be predicted with any model.

However, these models are clearly incapable of handling other types of flows. For example, in transitional flows, we must expect that most of the energy will be in the large scales, i.e., the small scales are not in equilibrium with the large scales and the "production equals dissipation" argument is incorrect. Furthermore, although Moin et al. [1.55] had reasonable success in simulating channel flow with these models, later extensions by Moin and Kim [1.56] clearly showed the deficiencies of the model. They found that eddy-viscosity models (several were tried) were unable to maintain the energy of the turbulence. The problem is only partially due to the model, as the turbulence tends to decay even when the model is eliminated. The problem is due to the fact that, near a solid boundary, the structures are very small, and it is impossible to use a grid small enough to resolve them without also missing the large-scale structures of the flow. The simulation then produces fewer, but larger, structures than are seen in the laboratory; this results in the production being too small to maintain the turbulence.

Clark et al. [1.57], McMillan and Ferziger [1.58], and McMillan et al. [1.59] have applied the model-testing method described in the previous section to eddy-viscosity SGS models. A typical result is shown in Fig. 3, in which the exact subgrid-scale stress is plotted against the model value. It can be seen that there is a little correlation between the two data sets (the correlation coefficient is approximately 0.4 for the case shown), but it is even more clear that this is far from an adequate model. This result is fairly typical, although there are variations in the correlation coefficient with many of the significant parameters.

This result shows that eddy-viscosity models are rather poor and, in fact, become even poorer when there is mean strain and/or shear in the flow. However, it is not easy to find more accurate models (we shall look at this in later sections of this paper), and we may be forced to use eddy-viscosity models until something better is developed. Furthermore, as McMillan and

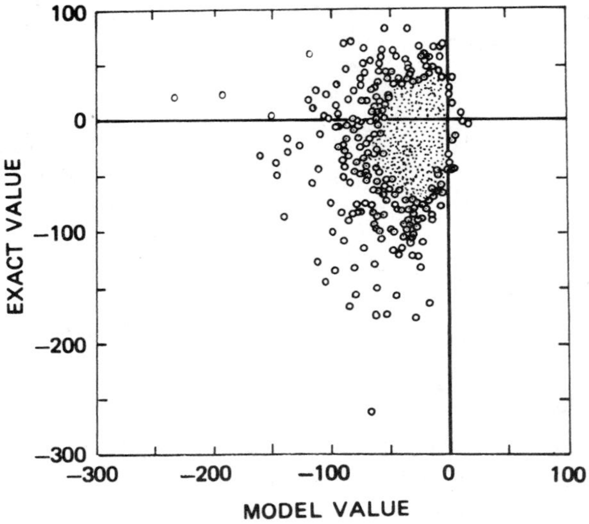

Figure 3. A "scatter" plot of the exact and model values of the local SGS Reynolds stress (this plot is for a homogeneous strain flow).

Ferziger have shown, the method can be used to predict the effect of Reynolds number on the model parameter. Their results are shown as Fig. 4. However, when these results were applied to channel flow by Kim and Moin (unpublished), they did not produce the desired effects, probably for the reasons given above.

Most of the successful simulations of channel flow to date have not used the LES procedures laid out above. Schumann ([1.60] and in later papers) has followed Deardorff's method [1.61] of simulating only the central region of the flow; the logarithmic and viscous layers are treated by means of an artificial boundary condition. This boundary condition serves to produce the energy that the turbulence needs to survive.

Kim and Moin [1.62] have recently succeeded in simulating the channel flow including the wall region. Their approach includes a two-part model

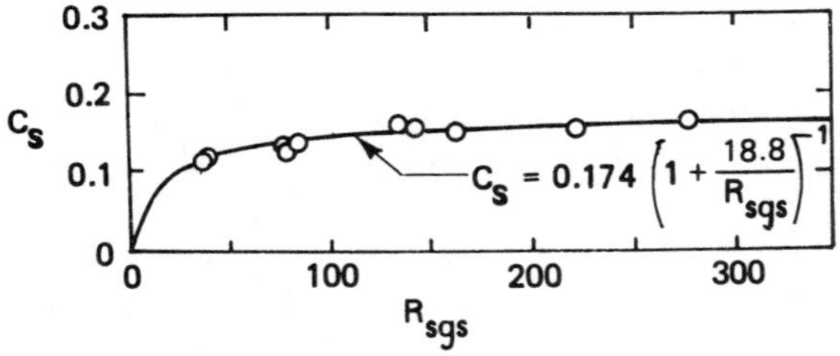

Figure 4. The variation of the constant in the Smagorinsky model as a function of a subgrid-scale Reynolds number. From McMillan and Ferziger [1.58].

similar to the one used by Schumann. The SGS Reynolds stress is divided into two parts: a mean component and a local component:

$$\tau_{ij} = \langle\tau_{ij}\rangle + \tilde{\tau}_{ij}, \tag{13}$$

where $\langle\ \rangle$ denotes an average over a plane parallel to the wall. The first term is modeled in terms of the mean flow; this modification differs from the LES strategy outlined above but seems to be necessary for maintaining the energy of the turbulence. The second part is treated by means of the kind of eddy-viscosity model suggested above. Their results are very encouraging and show many of the features of the laboratory flow.

The development of LES given above suggests that the SGS Reynolds stress is local in both space and time. Use of spatial averaging can be justified, but it is difficult to see why the average used should be one over a plane parallel to the wall. There is clearly more work to be done here before a completely satisfactory model is available.

6 The Role of Theory

Of course, theoretical insight plays a considerable role in understanding the physics of turbulence and contributes considerably to modeling it. Turbulence is, however, a problem of such complexity that the role of theory in our present state of knowledge is smaller than in most areas of physics or engineering. Progress has been frustratingly slow. A review of recent theories is given by Leslie [1.63].

Most theories provide limited information about turbulence. These theories were developed for homogeneous turbulence and have proved difficult to generalize.

The theories which have attracted the most attention recently are Kraichnan's direct-interaction approximation and others related to it. These theories are statistical in nature, i.e., they attempt to make statements about averages of turbulence quantities rather than the detailed dynamics. The question of whether this theory could be extended so as to yield information about the small scales of turbulence and thus to provide a SGS model has been investigated by Leslie and his coworkers. Some interesting results have been produced.

The theory necessarily deals with statistically averaged SGS turbulence. Thus we might imagine an ensemble of flows which have the same large-scale motions but different small-scale motions and ask for the average behavior of the small-scale motions. Whether this is adequate for modeling purposes is an important question, but whatever information can be generated is more than welcome. This theory, like many others, is capable of predicting the existance of an inertial subrange, and it can predict the Kolomogorov constant as well.

Love and Leslie [1.64] extended the theory and were able to show that a form of the eddy-viscosity model could be deduced from the theory. In particular, they were able to predict the constant in the model and showed that the mean strain rate that appears in the eddy-viscosity model ought to be not the local one but a spatial average. The constant predicted in this way is in good agreement with that obtained by other arguments and with the constants obtained from empirical fits to experimental data.

With respect to the spatial averaging of the strain rate in the eddy viscosity, the evidence is mixed. Love and Leslie [1.64] found that it was important in the solution of Burgers' equation, but Mansour et al. [1.65] found that it did not matter much either way whether averaging was used or not.

A number of other issues were investigated by Leslie and Quarini [1.66]. In particular, they divided the SGS terms into "outscatter" and "backscatter" terms representing, respectively, the energy flows to and from the subgrid scale. They find that eddy-viscosity models appear to represent the outscatter fairly well but that not much can be said about the backscatter.

Although limited, these theories are proving useful in choosing and validating models.

7 A Scale-Similarity Model

All models, by definition, relate the SGS Reynolds stress to the large-scale flow field. Eddy-viscosity models view τ_{ij} as a stress and make an analogy between it and the viscous stress. These models have the important virtue that they are guaranteed to extract energy from the large-scale field (i.e., they are dissipative). It is difficult to construct other models with this property.

It is important to observe that the interaction between the large- and small-scale components of the flow field takes place mainly between the segment of each that is most like the other. The major interaction is thus between the smallest scales of the large-scale field and the largest scales of the small-scale field. This is what the SGS term in the filtered equations represents (a mathematical demonstration of this can be given). Since the interacting components are very much alike, it seems natural to have the model reflect this. To do this requires that we find some way of defining the smallest-scale component of the large-scale field $\bar{u}_i$. One way to do this was suggested by Bardina et al. [1.54]. Since $\bar{u}_i$ represents the large-scale component of the field, filtering $\bar{u}_i$ again produces a field ($\bar{\bar{u}}_i$) whose content is still richer in the largest scales. Then

$$\tilde{u}_i = \bar{u}_i - \bar{\bar{u}}_i \tag{14}$$

is a field which contains the smallest scales of the large-scale component. This suggests that a reasonable model might be

$$\tau_{ij} = c\tilde{u}_i\tilde{u}_j, \tag{15}$$

or, better yet,

$$\tau_{ij} = c(\bar{u}_i\bar{u}_j - \bar{\bar{u}}_i\bar{\bar{u}}_j). \tag{16}$$

Preliminary tests have shown that this model is not dissipative, but it does correlate very well with the exact stress. This suggests that a combination of the two models might be better yet. The correlation is largely due to the fact that, with a Gaussian filter, the two fields in question contain much the same structures. With other filters, particularly one which is a sharp cutoff in Fourier space, the correlation is smaller. These models are currently being investigated.

8 Length Scales

In all of the above, we have implicitly used the fact that the natural length scale of the SGS eddies is the width Δ associated with the filter. By definition, this is the scale that defines whether an eddy is large or small, and there is little doubt that this is a correct choice.

However, when the filter is anisotropic, as it should be in computing shear flows, it is not quite so clear what is the correct length scale. Almost everyone has used the cube root of the filter volume:

$$\Delta = (\Delta_1 \Delta_2 \Delta_3)^{1/3}. \tag{17}$$

However, Bardina et al. [1.54] shows that a better choice might be

$$\Delta = (\Delta_1^2 + \Delta_2^2 + \Delta_3^2)^{1/2}, \tag{18}$$

and Kim and Moin have recently used this with success in the channel-flow simulation described above. It is recommended that Eq. (18) be adopted for general use.

9 Higher-Order Models

The inadequacies of eddy-viscosity models in time-average modeling have been known for a long time. It wasn't until large computers were available, however, that improvements could be put into practice. A number of types of models have been used, and since they have analogs in SGS modeling, a brief review of this topic is in order.

Many of the improvements are based on the notion that proportionality between Reynolds stress and mean strain rate is valid, but the eddy-viscosity formulation needs improvement. In these models one writes

$$\nu_T = C_1 q l, \tag{19}$$

where q and l are, respectively, velocity and length scales of the turbulence. In the simplest such models, the length scale is prescribed and a partial differential equation for the turbulence kinetic energy ($q^2/2$) is solved, along with the equations for the mean flow field. These are known as one-equation models; their record has not been particularly good, and most people now use more complex models. In particular, the assumption of a prescribed length scale was questioned, and methods of predicting the length scale were proposed. Of these, the most widely accepted models are those in which an equation for the dissipation of turbulent kinetic energy (actually, the energy transferred to the small scales) is added to the equations used in one-equation models. The length scale is related to the dissipation ε by

$$\varepsilon = C_2 q^3/l, \tag{20}$$

and we have the so-called two-equation models. This is the most popular method of computing time-average flow fields at present.

Finally, the most recent development has been the use of the full Reynolds stress equations. In two dimensions, three PDEs are needed to define the Reynolds stress, while in three dimensions, six are required. Clearly, this is a rather expensive approach.

A way of avoiding the computational cost of full Reynolds-stress methods is obtained by noting that the convective and diffusive terms can frequently be neglected. If they are, and approximations are made to the redistribution terms, the equations reduce to algebraic ones. Algebraic models have been popular in recent years. However, there is doubt as to whether the neglect of diffusion is correct near the wall.

All of these models have analogs in SGS modeling, and a number of them have been used. Let us consider them in the order in which they were introduced above.

First, consider one- and two-equation models. They have as their fundamental basis the proportionality of the SGS Reynolds stress and the large-scale stress. We saw earlier that the Smagorinsky model (an algebraic eddy-viscosity model) correlates poorly with the exact SGS Reynolds stress. Clark et al. [1.57] also looked at the behavior of one-equation models as well as an "optimized" eddy-viscosity model. In the latter, the eddy viscosity was chosen, at every point, to give the best local correlation between the SGS Reynolds stress and the large-scale strain. By definition, no eddy-viscosity model can do better than this. It was found that the correlation coefficient improved somewhat relative to the Smagorinsky model (from approximately 0.35 to 0.50 in a typical case), but this still leaves the model far short of what we would like to have. The lack of correlation seems to be due to the difference between the principal axes of the two tensors. The modeling assumptions need to be changed if further improvement is to be obtained (cf. McMillan and Ferziger [1.58]). It is clear that more complex models are required. Whether this result has implications for time-average modeling is unclear. Schumann has used one-equation models without finding improvement over algebraic eddy-viscosity models.

Next, recall the earlier remark that convection and diffusion are likely to be more important in SGS modeling than they are in time-average modeling. This means that the approximations needed to reduce the full Reynolds-stress equations to a set of algebraic model equations are less likely to be valid in the SGS case. However, several authors have used algebraic models. The applications have been almost exclusively to meteorological and environmental flows in which stratification and buoyancy effects are important. These flows tend to be sensitive to small variations in both properties and model, making it difficult to assess the accuracy of a model with precision. To our knowledge, no applications of these models to engineering flows have yet been made.

It is probable that to obtain a significant improvement over the Smagorinsky eddy-viscosity model, we shall need to go to full Reynolds-stress models. This, of course, is not something to be looked forward to. The only use of these equations to date was in the meteorological flows mentioned above and was made by Deardorff [1.61]. He reported that the computer time increased by a factor of approximately 2.5. Furthermore, the results were not improved to the degree that he had hoped for. Although this is discouraging, Deardorff's simulation was considerably ahead of its time and had the additional difficulties associated with buoyancy, so it is hard to reach definitive conclusions. Thus, we cannot conclude that these models are not worthwhile, but it is clear that quite a bit of work needs to be done in this area before these methods become useful tools of the trade.

10 Other Effects

The author's group has been doing computations involving the effects of compressibility on turbulence and the mixing of passive scalars in turbulent flows. To date, the work has concentrated on evaluating time-average models, because it was felt that this is the area in which the work will have the most immediate impact. Other groups have been actively developing methods of simulating flows in which buoyancy is important.

The effects of compressibility on SGS turbulence are probably quite small. The effect on the turbulence as a whole has been found to be fairly weak, except for effects due to the propagation of acoustic pressure waves. Since the latter are large-scale phenomena and the Mach number of the SGS turbulence is small, we expect that compressibility will have only a weak effect on SGS modeling.

On the other hand, SGS modeling of turbulent mixing is quite important. If we are to simulate combusting flows, it will be necessary to treat the small scales quite accurately, since that is where the action is in these flows. Although the effect of the Prandtl–Schmidt number on time-average models is fairly small, we expect that the effects on the SGS models may be quite profound. Furthermore, the specific effects due to combustion are also likely to be important on the small scales. We intend to look at SGS modeling of mixing flows in the near future.

Another effect of considerable importance in application is buoyancy, which was mentioned earlier in connection with the meteorological simulations. These, too, are particularly difficult flows, and a great deal of work will need to be done on engineering flows in this area. Important work in this area has been done by the Karlsruhe group [1.67], and further work is under way in London [1.68].

Finally, we should state that meteorologists and environmental engineers have a great interest in both mixing and buoyancy effects, and considerable effort has been made by these people. In particular, we note again the work of Deardorff cited above and that of Sommeria [1.69], Schemm and Lipps [1.70] and Findikakis [1.71]. One of the principal difficulties of these flows is that the scales are so large that eddies of length scale equal to the grid size are still quite important. Consequently, the SGS eddies do not behave entirely like "small eddies"; they carry a significant fraction of the total energy and are therefore hard to model.

11 Conclusions

From the arguments given above, we can reach the following conclusions about the current state of the art in SGS modeling.

1. Although they are inadequate in detail, eddy-viscosity models can be used in simulating homogeneous turbulent flows. However, they seem to be inadequate for most inhomogeneous flows.
2. For models in which the length scale is prescribed, the length scale of Eq. (18) is preferred to that of Eq. (17).
3. One- and two-equation turbulence models are unlikely to provide significant improvement relative to algebraic eddy-viscosity models. An exception to this might be transitional flows.
4. Full Reynolds-stress models offer the most promise as future SGS models. However, the modeling assumptions probably need to be different from those used in time-average modeling.
5. The scale-similarity model is promising, but only when used in conjunction with other, dissipative, models.
6. Exact simulations seem to be the best way available at present for testing SGS models and determining the parameters in them. Turbulence theories can also be profitably used in this regard.
7. Exact simulations and SGS can both be used in time-averaged model building. This is the area in which both types of simulations will make their greatest impact in practical engineering calculation in the near future.

Acknowledgments

The work reported herein was supported in part by NASA—Ames Research Center, the Office of Naval Research, and the National Science Foundation. The author also wishes to thank a number of colleagues, without whom this work would have been impossible. In particular, the contributions of Professor W. C. Reynolds, P. Moin, J. Kim, and D. C. Leslie, of Drs. O. J. McMillan, R. S. Rogallo, A. Leonard, and of Mr. M. Rubesin must be mentioned. Finally, the author wishes to thank several students who contributed to this work. Included among these are Drs. W. J. Feiereisen, A. B. Cain, E. Shirani, and J. Bardina.

CHAPTER 5

Three-Dimensional Instability of Plane Channel Flows at Subcritical Reynolds Numbers[1]

Steven A. Orszag and Anthony T. Patera*

1 Introduction

There is a large discrepancy between the results of parallel-flow linear stability theory and experimental observations of transition in plane channel flows. Experiments by Davies and White [1.72], Kao and Park [1.73], and Patel and Head [1.74] indicate that plane Poiseuille flow can undergo transition at Reynolds numbers R of roughly 1000 when the background disturbance level is about 10%. The strongly nonlinear nature of the problem is reflected in the fact that Nishioka et al. [1.75] have maintained laminar flow to $R = 8000$ by keeping the background disturbance level sufficiently low. While there have been fewer experimental investigations of plane Couette flow, available data [1.76] suggest a transitional Reynolds number as low as or lower than that for plane Poiseuille flow.

The disparity between the experimental evidence and the results of parallel-flow linear theory (which predicts a critical Reynolds number of 5772 for plane Poiseuille flow and ∞ for plane Couette flow) requires consideration of finite-amplitude effects. Nonlinear stability analysis typically involves the search for finite-amplitude equilibria, the stability of the flow being inferred by assuming the equilibria to be critical points in a one-dimensional phase

[1] This work was supported by the Office of Naval Research under Contracts No. N00014-77-C-0138 and No. N00014-79-C-0478. Development of the stability codes was supported by NASA Langley Research Center under Contract No. NAS1-15894. The computations were performed at the Computing Facility of the National Center for Atmospheric Research, which is supported by the National Science Foundation.

* Massachusetts Institute of Technology, Cambridge, MA 02139.

space for the amplitude of the (primary) disturbance. A large number of these analyses (following the seminal work of Meksyn and Stuart [1.77] and Stuart [1.78]) rely on amplitude expansions that have very small [1.79] or unknown radii of convergence. In plane Poiseuille flow (where a finite linear neutral curve exists) these techniques can successfully predict small-amplitude equilibria close to the neutral curve. However, upon extension to situations where a finite neutral curve does not exist (e.g. plane Couette flow, Hagen–Poiseuille flow), these techniques yield spurious equilibria.

In contrast to the amplitude expansion techniques, the iterative method due to Zahn et al. [1.80] and Herbert [1.81] is able to reliably calculate large-amplitude as well as small-amplitude equilibria. Here, the nonlinear eigenvalue problem for traveling-wave solutions to the Navier–Stokes equations (rendered steady by an appropriate Galilean transformation) is solved numerically using a spectral expansion and Newton iteration. It is on the basis of results obtained using this method [1.82] and full simulations that we claim plane Poiseuille flow has both subcritical and supercritical finite-amplitude equilibria whereas plane Couette flow [1.83, 1.84] and Hagen–Poiseuille [1.85] have none.

Most nonlinear stability results obtained to date are two-dimensional. There have been investigations of low-order two- and three-dimensional interacting-wave systems that exhibit resonance-induced instabilities [1.86, 1.87, 1.88]; however, it is not yet clear whether these instabilities lead directly to transition. For instance, in Hagen–Poiseuille flow, near-synchronization between the primary and harmonic occurs for all (a, R), but slow secular growth is maintained for such a short time that the perturbation remains linear [1.85].

It is generally believed that the instability leading to transition in plane channel flows is a "secondary instability" [1.89], i.e. an instability which acts on the combined flow consisting of the basic parallel flow and a finite-amplitude two-dimensional cellular motion. Furthermore, the three-dimensional nature of shear-flow turbulence suggests that the secondary instability should be three-dimensional. That such an instability would lead to aperiodic, apparently random behavior upon attaining finite amplitude rather than undergoing nonlinear saturation has been demonstrated numerically by Orszag and Kells [1.83] and Patera and Orszag [1.84]. Indeed, the transition process is inherently three-dimensional, as shown by the fact that even initially "turbulent" two-dimensional fields relax to a laminar state. In this paper we present a three-dimensional secondary instability that predicts transitional Reynolds numbers in good agreement with experiment.

In Section 2 we describe the two-dimensional steady and time-dependent properties of plane Poiseuille and plane Couette flows, using the iterative technique of Zahn et al. [1.80] and Herbert [1.81] and full numerical simulation of the Navier–Stokes equations. Two spectral codes (both implemented on the Cray-1 computer) have been used to verify the behavior reported here. The first employs a splitting technique similar to that described in

[1.83], while the second involves a full-step second-order (in time) method. Details of the latter are given in the Appendix.

In Section 3 it is shown that the finite-amplitude two-dimensional states investigated in Section 2 are strongly unstable to very small three-dimensional perturbations. Our aim here is to show (by full numerical simulation) that this explosive secondary instability can explain the subcritical transitions that often occur in real flows.

Finally, in Section 4, we show that the three-dimensional instability described in Section 3 can be analyzed by a linear stability analysis of a two-dimensional flow consisting of the basic parallel flow and a steady (or quasi-steady) finite-amplitude two-dimensional cellular motion.

A brief presentation of the results of the present paper was given by Orszag and Patera [1.90].

2 Two-Dimensional Quasi-equilibria

The instability to be described in Section 3 involves the rapid exponential growth of three-dimensional perturbations superposed on two-dimensional finite-amplitude flows. As the existence of the instability depends critically on the time scales associated with the evolution of the two-dimensional finite-amplitude motions, it is necessary to understand two-dimensional behavior before discussing three-dimensional effects. Just as in the linear regime plane Poiseuille flow and plane Couette flow differ in that the former has a finite neutral curve whereas the latter does not, so in the nonlinear regime they differ in that plane Poiseuille flow admits finite-amplitude equilibria while plane Couette flow apparently does not. We begin our discussion by investigating equilibrium solutions in plane Poiseuille flow.

In plane Poiseuille flow two-dimensional finite-amplitude solutions to the Navier–Stokes equations are sought in the form of traveling waves [1.80, 1.81].

$$\mathbf{v}(x, z, t) = \mathbf{F}(x - ct, z) + (1 - z^2)\hat{\mathbf{x}} \tag{1}$$

where $\hat{\mathbf{x}}$ is a unit vector in the streamwise direction x, z is the cross-stream direction, and c is a real wave speed. No-slip boundary conditions hold at the walls, $z = \pm 1$, and periodicity with wavelength $\lambda = 2\pi/\alpha$ is assumed in the streamwise direction.

The locus of points in (E, R, α) space for which a solution of the form (1) exists is called the neutral surface, where E is the energy of the disturbance relative to the basic flow $\mathbf{U} = (1 - z^2)\hat{\mathbf{x}}$ and R is the Reynolds number based on the centerline velocity and channel half width, $h = 1$. Such a neutral surface (as calculated by Herbert) is plotted in Fig. 1.

The stability analysis given later involves linearization about the neutral states (1), and so we summarize here our numerical methods (similar to those of [1.81]) for finding these solutions of the Navier–Stokes equations. If the

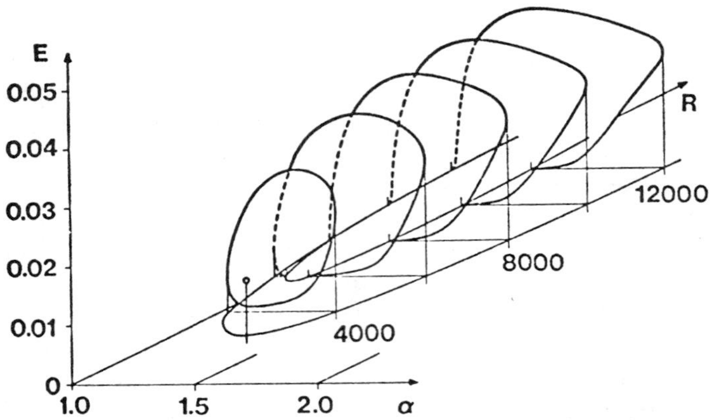

Figure 1. The neutral surface in (E, R, α) space for plane Poiseuille flow. Finite-amplitude neutral disturbances exist down to a Reynolds number $R \approx 2900$. The critical wavenumber, α_c, is shifted from the linear value of 1.02 at $R \approx 5772$ up to 1.32 at $R \approx 2900$. (Courtesy of Dr. Th. Herbert.)

streamwise and cross-stream velocities are written in terms of a stream-function as $u = \partial\psi/\partial z$ and $\omega = -\partial\psi/\partial x$, respectively, the Navier–Stokes equations become

$$\frac{\partial}{\partial t}\nabla^2\psi + \frac{\partial(\psi, \nabla^2\psi)}{\partial(z, x)} = \nu\nabla^4\psi, \tag{2}$$

where $\nu = 1/R$. We look for solutions periodic in x and steady in a frame moving with speed c relative to the laboratory frame,

$$\psi = \sum_{n=-\infty}^{\infty} \tilde{\psi}_n(z)e^{in\alpha(x-ct)}.$$

The nonlinear eigenvalue problem for $(\tilde{\psi}_n, c)$ is then given by

$$\nu(D^2 - n^2\alpha^2)^2\tilde{\psi}_n - ian([\overline{U} - c](D^2 - n^2\alpha^2)\tilde{\psi}_n - (D^2\overline{U})\tilde{\psi}_n)$$
$$+ i\alpha \sum_{m=-\infty}^{\infty} [(n-m)\tilde{\psi}_{n-m}(D^3 - m^2\alpha^2 D)\tilde{\psi}_m - mD\tilde{\psi}_{n-m}(D^2 - m^2\alpha^2)\tilde{\psi}_m] = 0, \tag{3}$$

where $D = \partial/\partial z$ and $\overline{U}$ is the basic (parabolic) profile. Reality of ψ requires

$$\tilde{\psi}_n = \tilde{\psi}^{\dagger}_{-n}, \tag{4}$$

where superscript † denotes complex conjugation. We also impose the symmetry requirement that $\psi(x, z) = \psi(x + \pi/\alpha, -z)$ or

$$\tilde{\psi}_n(z) = (-1)^{n+1}\tilde{\psi}_n(-z) \tag{5}$$

(which is consistent with (3)), in order to reduce the computational complexity of the problem.

Two boundary conditions must be provided at each wall for each Fourier mode. Noting the symmetry condition (5), it suffices to impose

$$\tilde{\psi}_n(1) = D\tilde{\psi}_n(1) = 0, \qquad n \neq 0, \tag{6}$$

$$D\tilde{\psi}_n(1) = D^2\tilde{\psi}_n(1) = 0, \qquad n = 0, \tag{7}$$

where (6) follows from the no-slip condition while (7) derives from the additional requirement of no mean disturbance stress at the wall (i.e. the mean pressure gradient remains $2/R$ as in the basic undisturbed flow).

To solve the equations (3) numerically, a Galerkin approximation in terms of Fourier modes in the x-direction is used (i.e. (3) is truncated at $m = N - 1$), and a Chebyshev pseudospectral method [1.91] is used in the z-direction. The resulting nonlinear algebraic eigenvalue problem is solved using a Newton iteration in conjunction with an arc-length continuation method [1.8] to avoid problems at limit points. Note that only one of the boundary conditions for each Fourier mode can be handled by replacing the dynamical equation at the wall. The second boundary condition results in an additional equation and a corresponding τ-factor [1.91]. The solution to (3) is unique only to within an arbitrary phase, and upon specifying this the numerical problem is completely determined.

Laminar equilibria of the form (1) exist down to Reynolds number $R \approx 2900$ (see Fig. 1). However, transition typically occurs (and the three-dimensional instability described in Section 3 obtains) down to $R = O(1000)$. In order to understand how an instability mechanism which requires a two-dimensional secondary flow can persist down to a Reynolds number at which all two-dimensional flows decay, one must investigate the time scales associated with approach to equilibrium.

A subcritical ($2900 < R < 5772$) slice of the neutral surface (Fig. 1) is given at $R = 4000$ in Fig. 2, the upper part of the oval being termed the upper

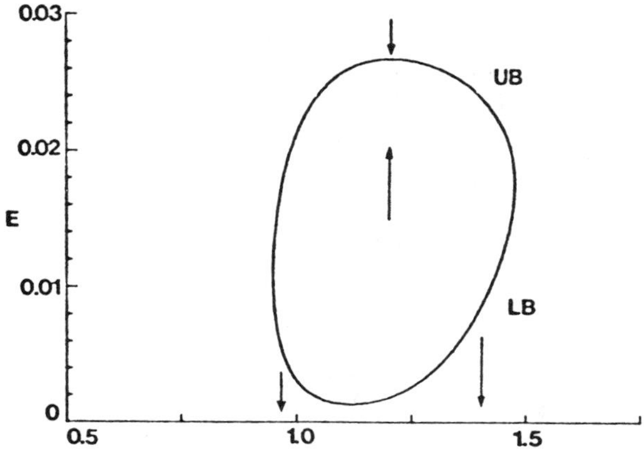

Figure 2. A subcritical (E, α) slice of the neutral surface for plane Poiseuille flow at $R = 4000$. The stability of solutions is indicated by the arrows. The behavior shown in this plot is typical for $2900 < R < 5772$.

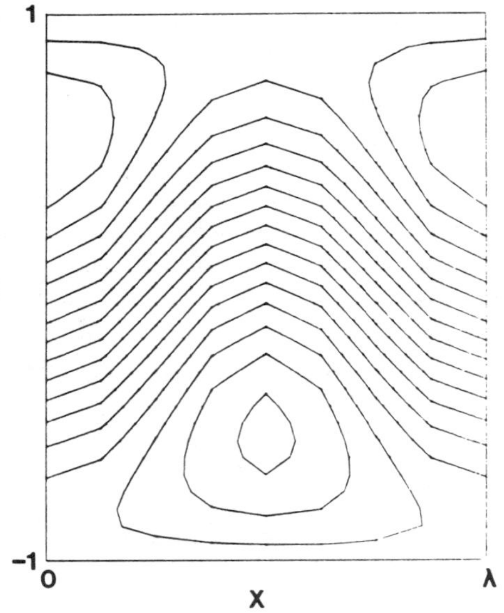

Figure 3. Streamlines of the steady (stable) finite-amplitude two-dimensional plane Poiseuille flow of the form (1) at $R = 4000$, $\alpha = 1.25$. The secondary motion appears as counter-rotating eddies. Here λ $(=2\pi/\alpha)$ is the wavelength of the primary.

branch (UB), and the lower part the lower branch (LB). An oversimplified argument based on a one-dimensional phase-space representation for E correctly predicts the UB solutions to be stable and the LB solutions to be unstable (since in a one-dimensional phase space the stability of critical points typically alternates).

We can determine the time scale on which flows approach UB solutions by noting that in a periodic steady flow, vorticity varies by at most $O(1/R)$ along an *interior* streamline. Indeed, when $R = \infty$, the Jacobian in (2) must vanish in a reference frame moving with speed c. This result is illustrated in Figs. 3 and 4, in which streamlines of the steady flow (1) and vorticity contours, respectively, are plotted. Note the similarity of the plots in the interior of the flow. The evolution of a typical flow to the equilibrium state (1) may then be described as follows. First, on a (fast) convective time scale, the flow achieves a state in which vorticity is a function of the stream function to $O(1/R)$. Final approach to equilibrium then occurs on a purely diffusive time scale, of order R. (Note this suggests that if a strong secondary instability exists, it is not two-dimensional).

This prediction of diffusive approach to equilibrium is validated in Fig. 5 as a plot of the projection of numerical solutions to the two-dimensional Navier–Stokes equations on the two-dimensional phase space $(\sqrt{E_1}, \sqrt{E_2})$. Here E_n is the energy of that part of the flow which depends on x like $e^{i\alpha nx}$. In Fig. 5, the dots, equally spaced in time, indicate the actual evolution of trajectories emanating from different initial conditions. The arrows show

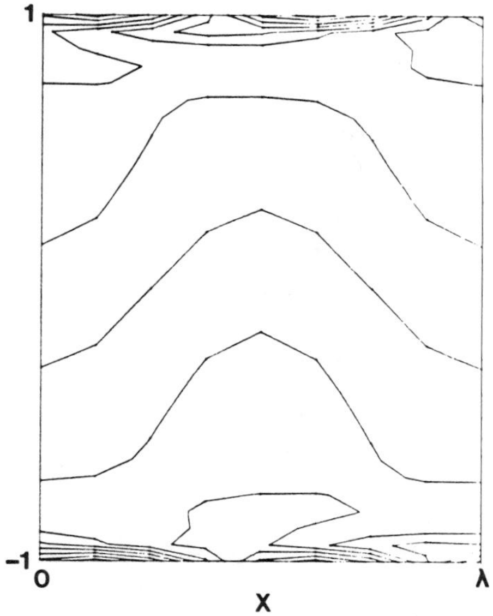

Figure 4. Vorticity contours of the steady (stable) finite-amplitude two-dimensional plane Poiseuille flow of the form (1) at $R = 4000$, $\alpha = 1.25$. Note that in the interior of the flow (where viscosity is unimportant), the vorticity contours are very similar to the streamlines in Fig. 3. This implies that the nonlinear interaction is small away from the boundaries.

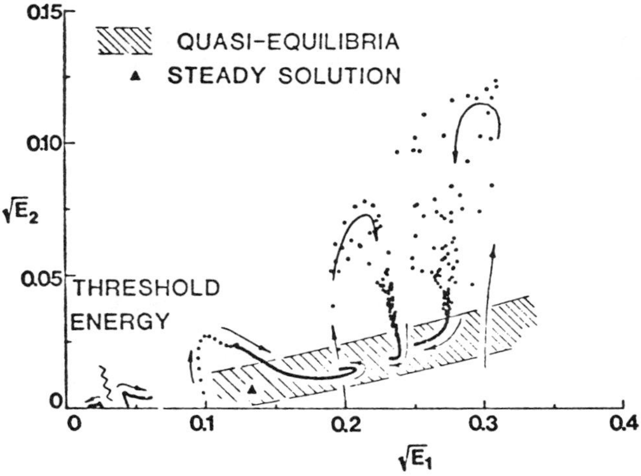

Figure 5. A phase portrait of disturbances to laminar parallel plane Poiseuille flow in $(\sqrt{E_1}, \sqrt{E_2})$ space at $R = 4000$, $\alpha = 1.25$. The dots, equally spaced by 1.25 in time, indicate the trajectories of flows evolving from different initial conditions proportional to the least stable Orr–Sommerfeld mode at this (α, R). Following an initial transient, flows evolve to a state within a band of quasi-equilibria and reach the steady solution only on times of the order of R.

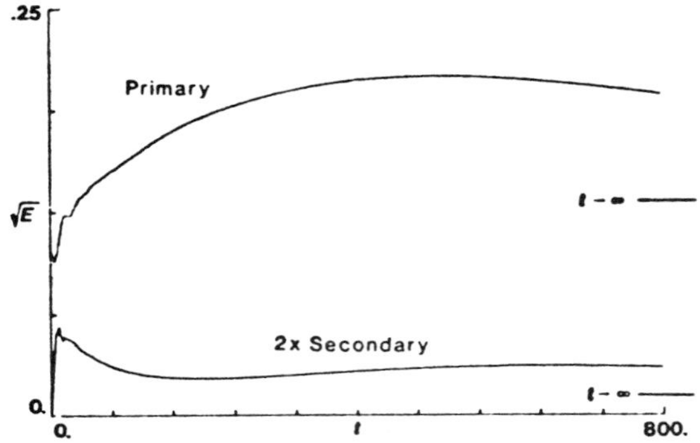

Figure 6. The evolution of a primary disturbance and its harmonic in plane Poiseuille flow at $R = 4000$, $\alpha = 1.25$. Significant deviations from the steady solution are only eliminated via cross-stream diffusion on a time scale of order R.

schematically the orbits followed by the different flows. With regard to time scales, the important point to note is that each initial condition above the threshold energy quickly evolves on a time scale of order 10 to a state within a band of quasiequilibria (illustrated by the shaded region in Fig. 5), and then only very slowly (indicated by the clustering of dots) approaches the steady solution on a time scale of order R. The integration times in Fig. 5 are as large as 800. A plot of E_1 and E_2 vs. t for one initial condition is given in Fig. 6, showing that appreciable deviations from the equilibrium solution only decay by cross-streamline diffusion on a time scale on the order of R. The fact that the vorticity-diffusion argument given above applies to such a wide band of phase space around the equilibrium suggests that it may also apply for nearby flows with $R < 2900$ (i.e. at Reynolds numbers where there are no finite-

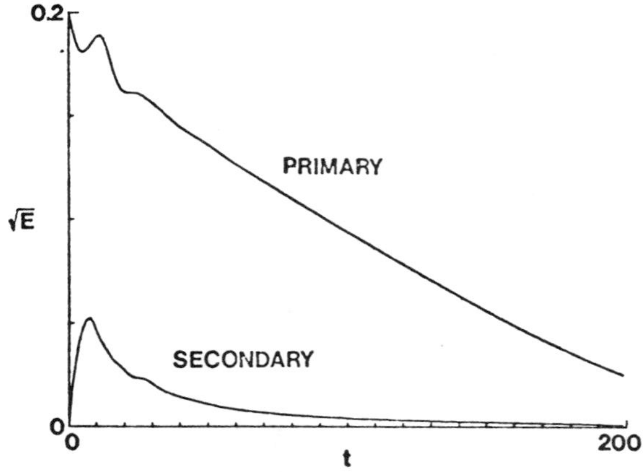

Figure 7. The decay of a disturbance in plane Poiseuille flow at $R = 1500$, $\alpha = 1.32$. The slow decay at finite amplitude (four times slower than the linear decay rate) reflects the existence of equilibria at higher Reynolds numbers.

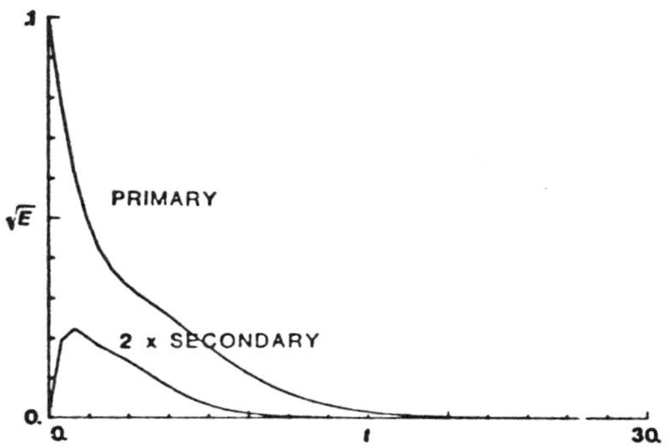

Figure 8. Decay of a finite-amplitude disturbance in plane Couette flow at $\alpha = 3.0$, $R = 500$. This (α, R) is predicted to be dangerous by amplitude-expansion techniques.

amplitude equilibria). This is indeed the case, as shown in Fig. 7 by the slow decay of a disturbance at $R = 1500$, $\alpha = 1.32$.

To date no equilibria have been found in plane Couette flow. There is also no evidence of quasiequilibria, i.e. all perturbations decay on a time scale short compared to the diffusive scale. In Fig. 8 and Fig. 9 we show the rapid decay of two disturbances. The parameters ($\alpha = 3.0$, $R = 500$) of the flow shown in Fig. 8 are in that class predicted to allow equilibria according to amplitude-expansion nonlinear techniques [1.92].

3 Three-Dimensional Instability

Numerical solution of the Navier–Stokes equations for plane channel flows shows that a very small three-dimensional disturbance superposed on a basic (parallel) flow and a two-dimensional secondary flow undergoes rapid

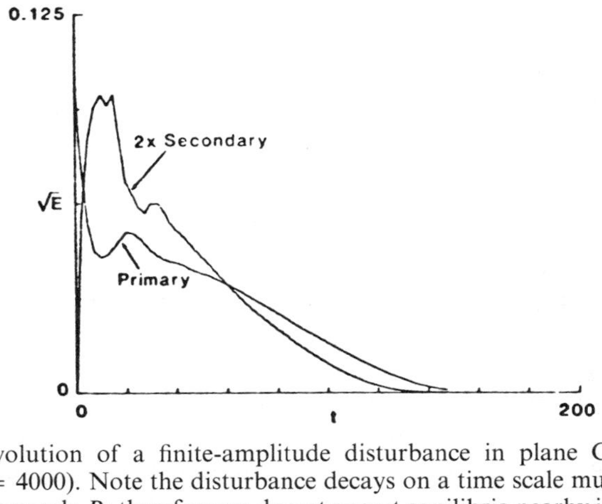

Figure 9. Evolution of a finite-amplitude disturbance in plane Couette flow at ($\alpha = 1.25$, $R = 4000$). Note the disturbance decays on a time scale much shorter than the diffusive time scale R; therefore we do not expect equilibria nearby in the parameter space.

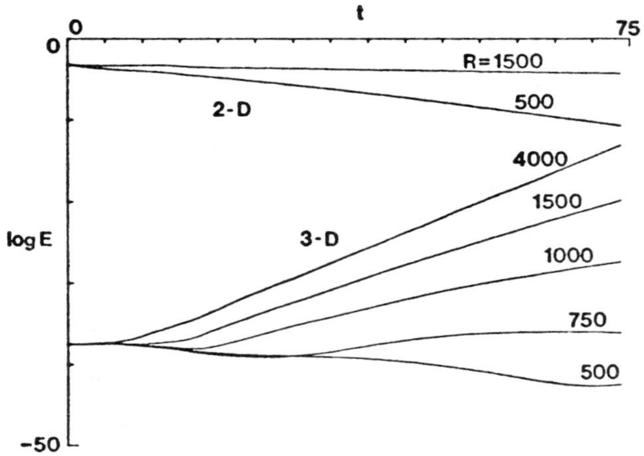

Figure 10. A plot of the growth of three-dimensional perturbations on finite-amplitude two-dimensional states in plane Poiseuille flow at $(\alpha, \beta) = (1.32, 1.32)$. Here $E_{2\text{-D}}$ is the total energy (relative to the basic laminar flow) in wavenumbers of the form $(n\alpha, 0)$, while $E_{3\text{-D}}$ is the total energy in wavenumbers $(n\alpha, \beta)$. For $R > 1000$ we obtain growth, whereas at $R = 500$ the three-dimensional perturbations ultimately decay. The growth rate of the three-dimensional disturbance amplitude at $R = 4000$ is about 0.18, and depends only weakly on R as $R \to \infty$.

exponential growth. We illustrate this in Figs. 10 and 11 for plane Poiseuille and plane Couette flows, respectively, in which we plot the logarithm of three-dimensional energy, $E_{3-\mathrm{D}}$, vs. time for flows resulting from the initial conditions

$$\mathbf{v}(x, y, z, t = 0) = \bar{U}(z)\hat{\mathbf{x}} + A\mathbf{v}_{2-\mathrm{D}}(x, z) + \varepsilon\mathbf{v}_{3-\mathrm{D}}(x, y, z),$$

where $\bar{U}(z) = (1 - z^2)$ for plane Poiseuille flow and $\bar{U}(z) = z$ for plane Couette flow. The initial two-dimensional disturbance $\mathbf{v}_{2-\mathrm{D}}$ is an Orr–Sommerfeld mode with wave vector $(\alpha, 0)$ and amplitude, A, corresponding to an energy $E_{2-\mathrm{D}}$ of 0.04. The initial three-dimensional disturbance $\varepsilon\mathbf{v}_{3-\mathrm{D}}$ is an infinitesimal $(\varepsilon = 10^{-8})$ Orr–Sommerfeld mode with wave vector $(0, \beta)$.

Observe from Figs. 10 and 11 that this instability singles out a critical Reynolds number of about 1000, in good agreement with transitional Reynolds numbers observed experimentally. The instability is only effective in forcing transition if the three-dimensional perturbation growth rate (and/or initial energy) is sufficiently large *and* the decay rate of the secondary two-dimensional flow is sufficiently small to allow three-dimensional non-linear effects to develop. Therefore, from the results of Section 2 we would expect three-dimensional threshold energies in plane Couette flow to be greater than in plane Poiseuille flow due to the larger two-dimensional decay rates associated with the former flow. In fact, it will be shown in the next section that for $R > 2900$ in plane Poiseuille flow (i.e. where equilibria do exist) *infinitesimal* three-dimensional perturbations are sufficient to trigger and maintain the instability.

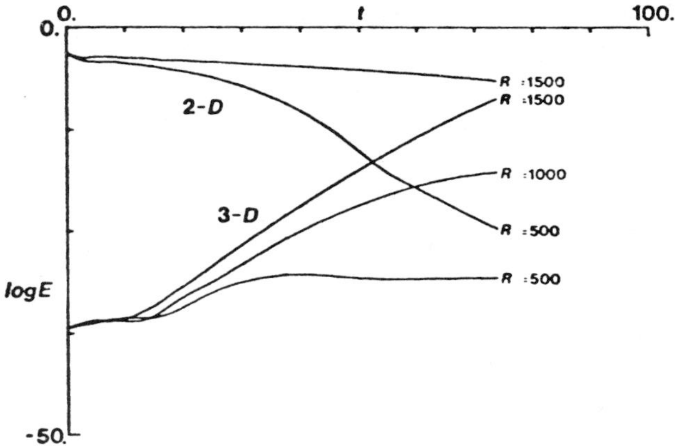

Figure 11. A plot of the growth of three-dimensional perturbations on finite-amplitude two-dimensional states in plane Couette flow at $(\alpha, \beta) = (1.0, 1.0)$. $E_{2\text{-D}}$ and $E_{3\text{-D}}$ are defined as in Fig. 10. The instability singles out a critical Reynolds number on the order of 1000 in accordance with experiment. The large decay rates of the two-dimensional states in plane Couette flow imply that larger threshold three-dimensional energies are required to force transition in this flow than in plane Poiseuille flow.

Note the time scale (typically of order 10 for amplitude growth by a factor of 10) on which the three-dimensional perturbation grows is one to two orders of magnitude smaller than the viscous time scales associated with the growth of supercritical (e.g. $R = 10^4$) linear disturbances. This leads us to consider the three-dimensional instability as "inviscid" in the sense that the growth rate σ becomes independent of R as $R \to \infty$ (see Fig. 10).

The time scales relevant to the three-dimensional growth are the mean flow and primary convective scales, $h/\hat{U}$ and $h/\sqrt{E_{2-\text{D}}}$, respectively, where $\hat{U}$ characterizes the mean flow velocity. Although it is easily verified that σ scales with $\hat{U}/h$, the dependence on $\sqrt{E_{2-\text{D}}}/h$ is more complicated. There does appear to be a region of simple proportionality; however, threshold and saturation phenomena at large amplitude are also present. For instance, at $R = 1500$, an initial two-dimensional amplitude of approximately 0.1–0.2 gives strong three-dimensional growth, whereas a two-dimensional amplitude of 0.01 results in linear decay of both the two-dimensional and three-dimensional disturbances.

The effect of the streamwise wavenumber α reflects the tradeoff between three-dimensional growth and two-dimensional decay. (The effect of the spanwise wavenumber β is briefly described in Section 4). For instance, in plane Poiseuille flow $\alpha = 1.32$ does not result in the largest three-dimensional growth rate for given two-dimensional energy, although it is close to that wavenumber giving the least stable two-dimensional flow. This is illustrated in Fig. 12. The disturbances followed in Fig. 10 are considered "most dangerous" in the sense that two-dimensional decay appears more significant in determining the low-Reynolds-number cutoff than three-dimensional perturbation behavior.

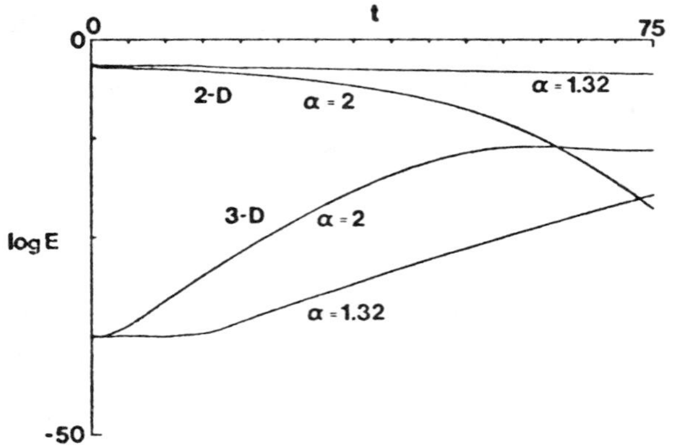

Figure 12. The growth of three-dimensional disturbances on finite-amplitude states in plane Poiseuille flow at $R = 1500$. At $\alpha = 2$ the three-dimensional growth rate is larger than at $\alpha = 1.32$; however, strong two-dimensional decay limits the effectiveness of the instability in forcing transition. At large times the three-dimensional instability is turned off. At $\alpha = 1.32$ the relatively steady two-dimensional flow permits persistent growth, ultimately resulting in nonlinear three-dimensional interactions.

Varying the spatial resolution of our calculations indicates that the instability reported here is not peculiar to any low-order model but is in fact a property of the Navier–Stokes equations. The large parameter space (e.g. α, R, $E_{2-\mathrm{D}}$) which describes the instability precludes an exhaustive study of the characteristics of the phenomenon, so we restrict ourselves here to demonstrating that the mechanism predicts transition in accordance with experiment. The large growth rates, small threshold energies, and experimentally reasonable dependence on Reynolds number indicate that this instability leads to transition in plane channel flows. The strength of the instability explains why the finite-amplitude two-dimensional states (1) have never been observed.

4 Linear Perturbation Analysis

The existence of equilibria for plane Poiseuille flow allows us to study the mechanism of three-dimensional growth in the relatively simple case in which the time dependent behavior of the two-dimensional component need not be considered. In particular, the exponential growth of three-dimensional perturbation energy in Figs. 10 and 11 indicates that a linear mechanism is involved. In this section we derive and solve the stability equations resulting from linearization of the Navier–Stokes equations around the finite-amplitude (nonparallel) flow (1). (The procedure of finding nonlinear two-dimensional states and performing a linear stability analysis about them has been used previously by Clever and Busse [1.93] for the case of Benard convection.) In the rest frame (moving with speed c relative to the laboratory frame) the stability problem is separable in time and periodic in x, and so we assume a solution of the form

$$\mathbf{v}(x, y, z, t) = (1 - z^2)\hat{\mathbf{x}} + \mathbf{F}(x, z) + \varepsilon \operatorname{Re}\left\{e^{\sigma t} \sum_{m=-1}^{1} \sum_{n=-\infty}^{\infty} \mathbf{u}_{nm}(z) e^{i\alpha n x} e^{i\beta m y}\right\} \tag{8}$$

where σ is a complex frequency and β is the spanwise wave number.

Upon inserting (8) into the Navier–Stokes equations and linearizing with respect to ε, the following equations result:

$$\begin{aligned}&\{\sigma(D^2 - k_{n,m}^2) - \nu(D^2 - k_{n,m}^2)^2\} w_{n,m} \\ &\quad - im\beta D\{(\overline{U} * v_x)_{n,m} + (\overline{W} * v_z)_{n,m}\} \\ &\quad - in\alpha D\{(\overline{U} * u_x)_{n,m} + (u * \overline{U}_x)_{n,m} + (\overline{W} * u_z)_{n,m} + (w * \overline{U}_z)_{n,m}\} \\ &\quad - k_{n,m}^2\{(\overline{U} * w_x)_{n,m} + (u * \overline{W}_x)_{n,m} + (\overline{W} * w_z)_{n,m} \\ &\quad + (w * \overline{W}_z)_{n,m}\} = 0,\end{aligned} \tag{9}$$

$$\begin{aligned}&\{\sigma - \nu(D^2 - k_{n,m}^2)\}\zeta_{n,m} - i\alpha n\{(\overline{U} * v_x)_{n,m} + (\overline{W} * v_z)_{n,m}\} \\ &\quad + im\beta\{(\overline{U} * u_x)_{n,m} + (u * \overline{U}_x)_{n,m} + (\overline{W} * u_z)_{n,m} \\ &\quad + (w * \overline{U}_z)_{n,m}\} = 0,\end{aligned} \tag{10}$$

$$i\alpha n u_{n,m} + im\beta v_{n,m} + D w_{n,m} = 0, \tag{11}$$

$$im\beta u_{n,m} - i\alpha n v_{n,m} = \zeta_{n,m}, \tag{12}$$

where $D = \partial/\partial z$, $(\overline{U}, \overline{W})$ is the two-dimensional flow (1), $k_{n,m}^2 = \alpha^2 n^2 + \beta^2 m^2$, and the convolution operator $*$ is defined by

$$(h * g)_{n,m} = \sum_{p+q=n} h_{p,m} g_{q,m}.$$

Here u, v, and w are the streamwise (x), spanwise (y), and cross-stream (z) velocities, respectively, and ζ is the cross-stream vorticity. Elimination of $\zeta_{n,m}$ and $v_{n,m}$ in (9) and (10) results in one fourth-order equation for $w_{n,m}$ and one second-order equation for $u_{n,m}$. At the wall the perturbation velocities must satisfy the no-slip condition

$$w_{n,m}, Dw_{n,m}, u_{n,m} = 0 \qquad (z = \pm 1).$$

At this point we invoke a result from our full numerical simulations of the Navier–Stokes equations in order to reduce the number of unknowns and, hence, simplify the problem. The solutions to the Navier–Stokes equations indicate that the three-dimensional perturbation travels at the same speed as the two-dimensional finite-amplitude wave $\mathbf{F}$, and so $\operatorname{Im} \sigma = 0$ in (8). (This assumption is validated by the solution of the eigenvalue problem (9)–(12).) The spatial dependence of the three-dimensional perturbation in (8) must therefore be real:

$$\mathbf{u}_{n,m} = \mathbf{u}_{-n,-m}^{\dagger}.$$

In addition we note the following symmetries consistent with (9)–(12):

$$\{u_{n,m}(z), v_{n,m}(z), w_{n,m}(z)\} = \pm(-1)^{n+1}\{u_{n,m}(-z), v_{n,m}(-z), -w_{n,m}(-z)\} \tag{13}$$

and

$$\{u_{n,m}(z), v_{n,m}(z), w_{n,m}(z)\} = \pm\{u_{n,-m}(z), -v_{n,-m}(z), w_{n,-m}(z)\}. \tag{14}$$

In this section we restrict ourselves to the upper signs in (13) and (14) exclusively. The three symmetries above allow us to solve for only the modes $m = 1, n \geq 0$ on half the channel.

The numerical procedure used to solve the linear eigenvalue problem (9)–(12) is similar to that used to solve the nonlinear equations (3). A Galerkin procedure is used in x (truncated at $n = N - 1$), and Chebyshev collocation is used in z. The dynamical equation at the wall is dropped in favor of one boundary condition. Additional boundary conditions (e.g. $Dw_{n,m} = 0$) result in additional equations and corresponding τ-factors. If K is the number of collocation points in the half channel, the number of (real) unknowns is $K(4N - 2) + N$. The algebraic eigenvalue problem is solved locally with a Newton iteration or globally with the QR algorithm. When a local algorithm is used, an arbitrary normalization is required, so the matrix is of rank $K(4N - 2) + N + 2$.

The (real) growth rate predicted by the linear problem (9)–(12), σ, is plotted vs. β at $R = 4000$, $\alpha = 1.25$ with $N = 2$, $K = 17$ in Fig. 13. The good agreement between these growth rates and those obtained from the full simulations indicates that the linear mechanism isolated here is indeed responsible for the growth of three-dimensional perturbations described in

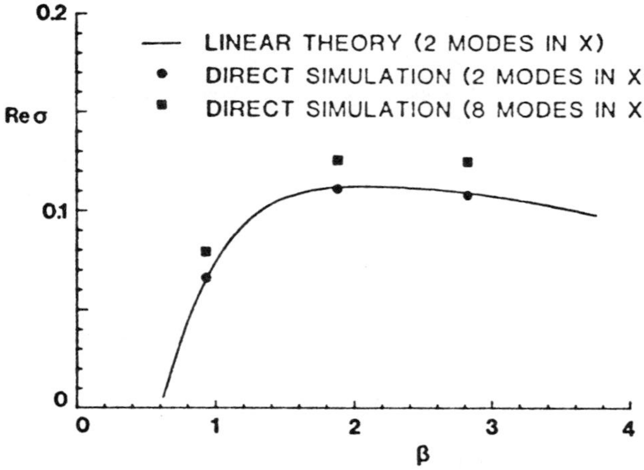

Figure 13. A plot of the growth rate of three-dimensional perturbations, σ, as a function of β at $R = 4000$, $\alpha = 1.25$. Note the good agreement between the linear calculation and the full simulation. Increasing the number of nodes in the x-direction increases the growth rate, but the error in the $N = 2$-mode model is not large.

the previous section. The effect of increasing N (which is more easily done in the full simulation than in the linear problem) is to *increase* σ; however, the error in the $N = 2$-mode model is not large. Increasing K indicates that the solution has converged in z. From Fig. 12 we also see that the growth rate is fairly insensitive to β once a certain "threshold" three-dimensionality is achieved.

Although the linear analysis presented here strictly applies only when equilibria exist, separation of time scales allows it to be used where quasi-equilibria exist as well by freezing the two-dimensional motion.

5 Discussion

The "inviscid" nature of the instability described here might suggest an explanation based on the instability of instantaneously inflectional velocity profiles created by the two-dimensional eddy motions. However, such an explanation could not adequately describe the most distinctive characteristic of the instability, namely its three-dimensionality. Indeed, the classical results of Squire, Fjortoft, and Rayleigh show that three-dimensional inflectional inviscid instability implies two-dimensional inflectional inviscid instability, which is not true here.

The complexity of the three-dimensional instability can best be described by vorticity dynamics, i.e. the stretching and tilting of vortex lines. In particular, we can write the equation for the perturbation vorticity as

$$\frac{\partial \omega_1}{\partial t} + \{(\mathbf{v}_0 \cdot \nabla)\boldsymbol{\omega}_1 - (\boldsymbol{\omega}_1 \cdot \nabla)\mathbf{v}_0\} + \{(\mathbf{v}_1 \cdot \nabla)\boldsymbol{\omega}_0 - (\boldsymbol{\omega}_0 \cdot \nabla)\mathbf{v}_1\} = 0, \quad (15)$$

where subscript 0 refers to a two-dimensional equilibrium (independent of y) and subscript 1 denotes the three-dimensional perturbation. We neglect the effect of viscosity and assume periodic boundary conditions. Note that conserved "inviscid" integrals of motion will in fact decay in the presence of finite dissipation.

The first bracketed term in (15) represents the action of the two-dimensional flow on the three-dimensional perturbation (primarily stretching), while the second term represents the action of the perturbation on the two-dimensional flow (primarily tilting). We now show (within the periodic restriction outlined above) that any mechanism proposed to explain the three-dimensional instability must include both these effects. Thus, a simple explanation of the phenomenon based on stretching of perturbation vortex filaments in the plane of the two-dimensional flow, i.e. a "vortex" dynamo, cannot be valid, just as two-dimensional magnetic dynamos do not exist [1.94].

Including only the effects of tilting, (15) becomes

$$\frac{\partial(\nabla \times \mathbf{v}_1)}{\partial t} = (\boldsymbol{\omega}_0 \cdot \nabla)\mathbf{v}_1 - (\mathbf{v}_1 \cdot \nabla)\boldsymbol{\omega}_0 = \nabla \times (\mathbf{v}_1 \times \boldsymbol{\omega}_0)$$

where $\boldsymbol{\omega}_1 = \nabla \times \mathbf{v}_1$. Upon integrating out the curl and forming the energy integral, we have

$$\frac{1}{2}\frac{\partial}{\partial t}\int (\mathbf{v}_1 \cdot \mathbf{v}_1)\, d\mathbf{x} = \int \mathbf{v}_1 \cdot (\mathbf{v}_1 \times \omega_0)\, d\mathbf{x} - \int (\mathbf{v}_1 \cdot \nabla\phi)\, d\mathbf{x},$$

where ϕ is a potential. The integrand of the first term on the right-hand side vanishes identically. The second term can be converted to a surface integral which vanishes due to periodicity. Thus we find that tilting alone cannot result in rapid growth.

On the other hand, including only the stretching effects, (15) becomes

$$\frac{\partial \omega_1}{\partial t} = (\boldsymbol{\omega}_1 \cdot \nabla)\mathbf{v}_0 - (\mathbf{v}_0 \cdot \nabla)\boldsymbol{\omega}_1 = \nabla \times (\mathbf{v}_0 \times \boldsymbol{\omega}_1).$$

It follows that the convective derivative of $\boldsymbol{\omega}_1 \cdot \hat{\mathbf{y}}$ vanishes, so that we can restrict attention to a two-dimensional solenoidal vorticity field, $\boldsymbol{\omega}_1 = \nabla \times \psi\hat{\mathbf{y}}$. Upon integrating out the curl we get

$$\frac{\partial\psi(x, y, z, t)}{\partial t} + (\mathbf{v}_0(x, z) \cdot \nabla)\psi(x, y, z, t) = -\frac{\partial\phi(y, t)}{\partial t}. \tag{16}$$

Including the potential ϕ in the convective derivative (noting $\mathbf{v}_0 \cdot \hat{\mathbf{y}} = 0$), we obtain

$$\frac{\partial(\psi + \phi)}{\partial t} + (\mathbf{v}_0 \cdot \nabla)(\psi + \phi) = 0.$$

Thus it follows that $\omega_1 \to 0$ and $\frac{1}{2}\int (\mathbf{v}_1 \cdot \mathbf{v}_1)\, d\mathbf{x} \to 0$ (since the only potential flow in a periodic domain is $\mathbf{v} = 0$). We conclude that the physics of the present instability involves a delicate balance between vortex tilting and vortex stretching.

Appendix: Numerical Methods

As the treatment of the nonlinear terms in the Navier–Stokes equations is the same using splitting methods [1.83] and using full-step techniques, we will not discuss them here. The differences between the two methods is best illustrated in the context of the Stokes equations

$$\begin{aligned} \frac{\partial \mathbf{v}}{\partial t} &= -\nabla\Pi + \nu\nabla^2\mathbf{v}, \\ \nabla \cdot \mathbf{v} &= 0, \\ \mathbf{v} &= 0, \qquad z = \pm 1. \end{aligned} \tag{A1}$$

The splitting method factors these equations into a pressure operator (to impose incompressibility) and a viscous operator. As these operators do not commute when no-slip boundary conditions are imposed, one must incur a first-order error in time (though in practice this is not found to be restrictive for shear flows). Using the splitting technique each time step requires the solution of four Poisson equations.

In the full-step method we reduce the Stokes equations to a fourth-order equation for the cross-stream velocity, w.

Using the (second-order) Crank–Nicolson method to advance the solution in time, the equation for w at time t is then given by

$$\{D^2 - k^2\}w = \zeta,$$

$$\left\{D^2 - \left(k^2 + \frac{2}{\nu\Delta t}\right)\right\}\zeta = f, \tag{A2}$$

$$w = Dw = 0, \qquad z = \pm 1,$$

where $D = \partial/\partial z$, k^2 represents a Fourier wave number, and Δt is the time step. Here f represents a forcing term (or the results of the nonlinear step). This coupled system is solved using a Chebyshev-tau Green's-function technique, in which we write the solution as

$$\zeta_m = \sum_{i=1}^{5} \tau^i \zeta_m^i,$$

$$w_m = \sum_{i=1}^{5} \tau^i w_m^i,$$

where, in general, h_m is the coefficient of the Chebyshev polynomial of degree m in a Pth-order spectral expansion of $h(z)$. The equations for the w^i and the ζ^i are then given by

$$\{D^2 - k^2\}w_m^i = \zeta_m^i, \qquad 0 \le m \le P - 2,$$

$$\sum_{m=0}^{P} w_m^i(\pm 1)^m = 0,$$

and

$$\left\{D^2 - \left(k^2 + \frac{2}{\nu\Delta t}\right)\right\}\zeta_m^i = e_m^i, \qquad 0 \le m \le P - 2,$$

$$\sum_{m=0}^{P} \zeta_m^i(\pm 1)^m = e_n^i, \qquad P - 1 \le n \le P,$$

respectively. The τ^i ($i = 1, \ldots, 4$) are determined by the four equations

$$\{D^2 - k^2\}w_m = \zeta_m, \qquad P - 1 \le m \le P,$$

$$\sum_{p=0}^{P} (\pm 1)^p p^2 w_p = 0,$$

and $\tau^5 = 1$. The e^i_m are given by

$$\begin{aligned} e^1_m &= \delta_{m,P-1}, \\ e^2_m &= \delta_{m,P}, \\ e^3_m &= D(T_P)_m, \\ e^4_m &= D(T_{P-1})_m, \\ e^5_m &= f_m, \end{aligned}$$

where T_P is the Chebyshev polynomial of degree P, and δ is the Kronecker delta function. Thus w^5_m, ζ^5_m is a particular solution of (A2), while w^i_m, ζ^i_m $(i \le 4)$ are used to enforce the boundary conditions.

The solution of the fourth-order equation is thus reduced to the solution of four Poisson equations per time step per Fourier mode. In order to complete the time step an additional Poisson equation (for the cross-stream vorticity) must also be solved. Each of the Poisson equations can be represented as an essentially tridiagonal diagonally dominant system [1.91] which can be efficiently inverted. A detailed comparison of the accuracy of these methods will be given elsewhere.

CHAPTER 6

Scaling of High-Reynolds-Number Weakly Separated Channel Flows[1]

Patrick J. Roache*

1 Introduction

This paper presents evidence for scaling behavior of high-Reynolds-number weakly separated flows. The scaling behavior is of interest in itself, and for the interpretation of accuracy limits on high-Reynolds-number Navier–Stokes computational solutions.

It is shown analytically and verified numerically that weakly separated laminar two-dimensional incompressible channel flows can display a self-similar solution. If the channel length x is increased proportional to the Reynolds number Re, the solutions become self-similar in the scaled longitudinal variable $\tilde{x} = x/\mathrm{Re}$.

The first indication of this result for the full Navier–Stokes equations was presented by the author in [1.95], which also contains the limit analysis given below. However, [1.95] contained only a coarse mesh (11×11 grid) solution. In the present work, we perform convergence testing for $\Delta x \to 0$ as well as $\mathrm{Re} \to \infty$. The results of [1.95] are verified. Rubin et al. [1.96] have pointed out that the 3-D boundary layer equations for channel flows contain this scaling behavior.

[1] This work was partially supported by the Fluid Dynamics Program of the Office of Naval Research under contract N000 14-76-C-0636, and by the U.S. Army Research Office under contract DAAG 29-79-C-0214.

* Ecodynamics Research Associates, Inc., P.O. Box 8172, Albuquerque, NM 87198.

2 Basic Numerical Method

The numerical method used herein, and in [1.95], is the split NOS method, one of a class of "semidirect methods" which are very efficient for steady-state and slowly time-varying solutions [1.97].

The basic concept of semidirect methods is to use the recently developed fast (direct, or noniterative) linear solvers to solve linearized equations, which are then iterated to solve the nonlinearity.

The governing equations are the vorticity transport equation

$$\zeta_t = -\mathrm{Re}\, \nabla \cdot \mathbf{V}\zeta + \nabla^2 \zeta \tag{1}$$

and the Poisson equation for the stream function,

$$\nabla^2 \psi = \zeta, \tag{2}$$

where ζ is the vorticity, Re is the Reynolds number, $\mathbf{V}$ is the vector velocity, and ψ is the stream function defined by the relation

$$\mathbf{V} = (\psi_y, -\psi_x). \tag{3}$$

Time is t, and subscripts indicate partial differentiation.

In the split NOS method, a steady-state solution is assumed, so $\zeta_t = 0$. The velocity $\mathbf{V}^k$ at the kth iteration is split into an initial guess $\mathbf{V}^0$ and a perturbation $\mathbf{V}'$, not necessarily small. That is,

$$\mathbf{V}^k = \mathbf{V}^0 + \mathbf{V}'. \tag{4}$$

Then the steady-state form of the vorticity transport equation (1) is solved for the kth iteration using standard $O(\Delta x^2)$ centered differences, as

$$L(\zeta^k) = \mathrm{Re}\, \nabla \cdot (\mathbf{V}'\zeta)^{k-1} \tag{5}$$

where the linear operator L is defined by

$$L(\zeta) \equiv \nabla^2 \zeta - \mathrm{Re}\, \nabla \cdot (\mathbf{V}^0 \zeta). \tag{6}$$

The solution of the linear equation (5) is accomplished noniteratively with a direct marching method [1.98, 1.99]. With the new kth value for the vorticity ζ at interior points, the Poisson equation (2) is solved for the new kth value of the stream function ψ, also using a marching method. With these new kth values for ζ and ψ at interior points, new kth values for ζ at the boundaries are calculated.

3 Coordinate Transformation

The origin of the physical coordinates (x, y) is on the lower wall at the inflow boundary, where the channel half height has been normalized to $y = 1$. The length of the channel is L_c. The lower wall coordinates are described by a function $y_l(x)$, which is arbitrary but single-valued. The transformation used is the simple linear stretch in y to map (x, y) into the rectangular region (X, Y). The transformation equations are

$$X = x, \tag{7}$$

$$Y = \frac{y - y_l}{y_u - y_l}. \tag{8}$$

The nonconservation form of the vorticity transport equation, obtained by replacing $\nabla \cdot \mathbf{V}\zeta$ by $\mathbf{V} \cdot \nabla\zeta$ in Eqs. (1), (5), and (6), was used because it was easier to program in the transformed coordinates and gave some minor savings in operations and in storage for the right-hand side of (5). Using the transformation (8) and the chain rule, the split NOS method of Eqs. (5) and (6) then becomes

$$\tilde{L}(\zeta^k) = \mathrm{RHS}(\zeta, \mathbf{V}')^{k-1} \tag{9a}$$

where

$$\tilde{L}(\zeta) \equiv \zeta_{XX} + p\zeta_{YY} + 2p_{21}\zeta_{XY} + p_{23}\zeta_Y - \mathrm{Re} \cdot u^0(\zeta_X + p_{21}\zeta_Y) - \mathrm{Re} \cdot v^0 p_{22}\zeta_Y \tag{9b}$$

and

$$\mathrm{RHS}(\zeta, \mathbf{V}') \equiv \mathrm{Re} \cdot u'(\zeta_X + p_{21}\zeta_Y) + \mathrm{Re} \cdot v' p_{22}\zeta_Y \tag{9c}$$

The velocity components are evaluated from the stream function as

$$u = \psi_y = p_{22}\psi_Y, \tag{10a}$$

$$v = -\psi_x = -(\psi_X + p_{21}\psi_Y). \tag{10b}$$

The Poisson equation (2) for stream function becomes

$$\psi_{XX} + p\psi_{YY} + 2p_{21}\psi_{XY} + p_{23}\psi_Y = \zeta. \tag{11}$$

The transformation parameters in the above equations are defined as follows (primes here indicate total differentiation with respect to x):

$$H(x) = y_u(x) - y_l(x), \tag{12a}$$

$$p_{21} = -\frac{y_l' + YH'}{H}, \tag{12b}$$

$$p_{22} = 1/H, \tag{12c}$$

$$p = p_{21}^2 + p_{22}^2, \tag{12d}$$

$$p_{23} = -p_{22}(y_l'' 2H'(y_l' - Y)p_{22} - H''Y)). \tag{12e}$$

(The p_{23} term was erroneously omitted in [1.95].)

It is noteworthy that the linear operator $\tilde{L}$ in Eq. (9), which is to be solved noniteratively, contains variable-coefficient first derivatives and cross-derivatives.

4 Boundary Conditions at Outflow

The computational boundary conditions used in the present study are simple adaptations of those commonly used in time-dependent methods. At $X = L_c$ or $i = \mathrm{IL}$, outflow conditions are set from finite-difference analogs of

$$\left.\frac{\partial f}{\partial X}\right|_{\mathrm{IL}} = 0, \qquad f = \zeta \text{ or } \psi. \tag{13}$$

Since the channel shape used (see below) has a height that is constant to five figures over the last 20% of the channel, the conditions (13) are virtually the same as those commonly used in straight-channel calculations. However, the commonly used implementation of (13), setting

$$f^k_{\mathrm{IL},j} = f^k_{\mathrm{IL}-1,j} \tag{14}$$

is not adequate for the present study. Equation (14) can be interpreted as a second-order approximation to (13) where the outflow has been moved to the cell midpoint, $X = X_{\mathrm{IL}} - \Delta X/2$, which is ordinarily acceptable, since (13) is itself heuristic. However, this confuses the convergence study, since the position of the outflow boundary now is a function of ΔX. For the convergence study, we must view (14) as a first-order approximation to (13) at $X = X_{\mathrm{IL}}$, which makes convergence difficult to judge, and which invalidates the use of Richardson extrapolation (see below).

Consequently, it was necessary to use a second-order approximation to (13), even though the formal accuracy of outflow boundary conditions is usually of little interest. The usual 3-point form used was

$$\frac{f_{\mathrm{IL}-2} - 4f_{\mathrm{IL}-1} + 3f_{\mathrm{IL}}}{2\,\Delta x} = 0. \tag{15}$$

The marching method for the 9-point operator [1.98] involves an implicit march procedure in the j-direction. At each (i, j) location, this 9-point stencil is solved for a linear combination of the three values $\zeta(i, j+1)$ and $\zeta(i \pm 1, j+1)$. The march in j then proceeds similar to a line SOR iteration by implicit solution of the new line of values at $j+1$ using a tridiagonal algorithm. The form of (15) extends the bandwidth of the tridiagonal matrix at $i = \mathrm{IL}$, but this can be readily cured by applying Gauss elimination just at $i = \mathrm{IL}$, so that (15) is incorporated directly into the marching solution. This works well for the stream-function equation (11), but fails for the vorticity transport equation (9) at high Re. As $\mathrm{Re} \to \infty$, the corner terms of the 9-point operator drop out of (9), so that the operator reverts to a 5-point form. The combination of high Re with (15) and the implicit march at $j+1$ deteriorates the error-propagation characteristics of the marching method, and large roundoff errors result at high Re and fine mesh resolution.

The solution to this annoying problem is obtained by applying (15) as an iterative correction to (14). That is, the direct linear solution at iteration level k is obtained using the first-order form of (14) with a nonzero gradient g, which is underrelaxed by the factor r:

$$\frac{f^k_{\mathrm{IL},j} - f^k_{\mathrm{IL}-1,j}}{\Delta x} = rg_j^{k-1} + (1-r)g_j^{k-2}, \tag{16}$$

where

$$g_j = \frac{1}{3}\frac{f_{\mathrm{IL}-2,j} - f_{\mathrm{IL}-3,j}}{\Delta x}. \tag{17}$$

The results are not very sensitive to r, although $r < 1$ is required for stability. The study reported herein used $r = \frac{1}{2}$ for all cases. Computational experience shows that the error of this lagging correction is at least two orders of magnitude less than the convergence tolerance of the overall solution, so it does not affect the convergence time or the roundoff accuracy, while improving the formal truncation-error accuracy.

5 Boundary Conditions at Inflow and Upper Boundary

The inflow boundary conditions used specify both ψ and ζ as fully developed Poiseuille flow. The normalizing system used defines Re based on the bulk mixing velocity of the inflow, with $\psi(0, 1) = 1$. This gives inflow conditions of

$$\psi(0, Y) = \tfrac{3}{2}Y^2 - \tfrac{1}{2}Y^3, \tag{18a}$$

$$\zeta(0, Y) = 3(1 - Y), \tag{18b}$$

$$u(0, Y) = 3Y - \tfrac{3}{2}Y^2, \tag{18c}$$

with the maximum inflow u-velocity $= u(0, 1) = \frac{3}{2}$.

The upper boundary is a symmetry condition, with

$$\psi(X, 1) = 1, \qquad \zeta(X, 1) = 0 \tag{19}$$

The velocity $u(X, 1)$ develops as part of the solution, decreasing from the inflow value $u(0, 1) = \frac{3}{2}$ as the channel expands. It is evaluated by 3-point one-sided differences to maintain $O(\Delta^2)$ accuracy.

6 Boundary Conditions On the No-Slip Wall

As pointed out in [1.95], conventional boundary conditions for the wall vorticity give a null iteration as $\Delta \to 0$, and confuse the interpretation of iterative convergence. In this work, we use an implementation of Israeli's method [1.100] which is more straightforward than that of [1.95].

Israeli's method directly drives a finite-difference approximation to the no-slip condition, $u_w = 0$, by the following iteration, wherein r is a relaxation parameter:

$$\zeta_w^k = r \cdot u_w + \zeta_w^{k-1}. \tag{20}$$

Clearly, when $u_w = 0$, convergence is reached. The value of u_w may be determined from (10a) using any consistent difference for ψ_Y. In the present study, we use $O(\Delta Y^4)$ one-sided differences. (At least third-order accuracy is required here so that the trivial solution of a straight channel may be compatible with the inflow Poiseuille solution, Eq. (18).) An analysis shows that, for the simple case of one-dimensional Poiseuille flow, the optimum value for r is

$$r_{\text{opt}} = \frac{3\psi_{\max}}{Y_{\max}^2} = 3. \tag{21}$$

This simple optimal value was used in all calculations in this study, even though the deviation of the solution from Poiseuille flow was significant.

7 Initial Conditions and Channel Shape

The initial conditions used were those of fully developed Poiseuille flow at each X-position, allowing for the expanding channel. This gives, for $i > 1$,

$$\psi_{i,j}^0 = \psi_{1,j}, \tag{22a}$$

$$\zeta_{i,j}^0 = \zeta_{1,j}\left(\frac{H_1}{H_i}\right)^2. \tag{22b}$$

It was also verified that the solutions were path-independent by starting the iteration from different initial conditions.

The channel shape for this study was chosen after some experimentation. It was desirable to have separated flow with a nearly constant channel height H at outflow, and a shape defined by smooth analytic function. The shape of the lower wall was defined by a shifted hyperbolic tangent function, given by

$$y_l(x) = \frac{1}{2}\left\{1 - \tanh\left(\text{sc}\,\frac{x - x_F}{L_c}\right)\right\} + \delta, \tag{23}$$

where L_c = length of the channel, x_F = location of the inflection point in the tanh profile, and sc is a scale factor. The small adjustment δ is set to give $y_l(0) = 0$. For the parametric cases presented here, L_c is scaled with Re as

$$L_c = \text{Re}/C_{L_c} \tag{24}$$

This scaling is necessary to keep the separated flow region within the computational mesh, and results in the self-similarity of the solutions at high Re. The particular channel parameters used in the present study were

$$\text{sc} = 10, \qquad X_F/L_c = 0.2, \qquad C_{L_c} = 3. \tag{25}$$

8 Iterative Convergence Criterion and Performance

It is known by computational experience that the iterative convergence can be judged by examining only the vorticity at the wall. Convergence was satisfied when

$$DQ_w = \max_i |\zeta_{i,1}^k - \zeta_{i,1}^{k-1}| < \varepsilon. \tag{26}$$

Experience with this semidirect method has also shown that this ε is a faithful indicator of ultimate convergence; that is, if (26) is satisfied, then

$$\max_i |\zeta_{i,1}^{\infty} - \zeta_{i,1}^k| < \varepsilon \tag{27}$$

is also satisfied. (This is not always the case with time-dependent explicit methods, especially if weak convergence criteria such as $\varepsilon = 10^{-2}$ or 10^{-3} are used in a fine grid.)

If a "reasonable" value of ε is used, the number of iterations NIT to reach convergence is roughly independent of grid size and (in the present study) of Re. For example, for grids from 11 × 11 to 81 × 81 nodes, and Re from 10^2 to 10^7, the minimum NIT = 13 and most cases had NIT = 15, for $\varepsilon = 10^{-4}$. However, for the present study we used an unreasonable value of

$$\varepsilon = 10^{-8}. \tag{28}$$

This removed any doubt about iterative convergence, and allowed the use of Richardson extrapolation with confidence (see below). This value of ε is of the same order of magnitude as the roundoff error in the solution of the marching method at high resolution (81 × 81 grid). The interaction of this round-off with the iteration error slows convergence at this tolerance, so that NIT for $\varepsilon = 10^{-8}$ increased from NIT = 32 in the 11 × 11 grid to NIT = 60 in the 81 × 81 grid, for Re from 10^3 to 10^7. However, this does not represent a proportional increase in computer time, since the relatively expensive initiation of the marching method [1.98] is done only once for each case.

The CPU times on the CDC 6600 were relatively insensitive to Re. For Re = 10^3 to 10^7, the CPU times ranged from 1.5 seconds in a 11 × 11 grid to 218 seconds in the 81 × 81 grid, using a "cold start" and $\varepsilon = 10^{-8}$. (The Re = 100 cases were about 5% higher.) These correspond to the range of 15 to 34 milliseconds/cell. Considering the algebraic complexity of the transformed equations, these CPU times to obtain unequivocally converged steady-state solutions compare well with the CPU time for just a few time steps using conventional time-dependent methods.

9 Richardson Extrapolation

Solutions accurate to $O(\Delta^2)$ were obtained on four grids, each doubling the previous resolution: 11 × 11, 21 × 21, 41 × 41, and 81 × 81. Richardson extrapolation was then applied to the 41 × 41 and 81 × 81 grid solutions to give $O(\Delta^4)$ solutions on the 41 × 41 grid.

Richardson extrapolation can be very delicate; if misused, it can give deterioration of accuracy rather than improvement. First of all, the solution on either grid cannot exhibit oscillations ("wiggles" due to cell Re effects, for example.) This condition is satisfied by the present solutions. Second, the solutions must be globally $O(\Delta^2)$. This is the reason for the attention described above to the formal truncation error of the outflow condition, which is ordinarily of little interest. Third, the two solutions must be obtained with tight iterative convergence criteria. Otherwise, the iteration errors will be magnified by the extrapolation procedure. This is the reason for the tight tolerance $\varepsilon = 10^{-8}$ described above.

If these conditions are met, as in the present study, Richardson extrapolation can be quite useful. In simple coordinate systems (say, Cartesian coordinates with at most independent stretching in x and y) it is not too difficult to obtain $O(\Delta^4)$ solutions using conventional high-order finite-difference or spline methods. In the present case of a nonconformal transformation, this would be quite difficult because of the presence of the cross-derivative terms, especially at near-boundary points. For these problems, Richardson extrapolation is well suited. Also, the method gives a ready indication of truncation-error convergence by providing a comparison between the $O(\Delta^2)$ solution in the 81×81 grid and the $O(\Delta^4)$ solution in the 41×41 grid.

The extrapolated solution f_{ext} in the 41×41 grid is obtained from the $O(\Delta^2)$ solutions in the 41×41 and 81×81 grids, denoted by f_{41} and f_{81}, as

$$f_{ext} = \tfrac{4}{3} f_{81} - \tfrac{1}{3} f_{41} \tag{29}$$

with $f = \psi$, ζ, or u.

10 Truncation-Error Convergence

Because of the scaling behavior of the solutions, the truncation-error convergence is very regular and changes little with Re over the range covered, from Re $= 10^2$ to 10^7. For each case, we considered the wall vorticity $\zeta_w(X)$ and calculated the maximum deviation between the $O(\Delta^2)$ solution in the 81×81 grid and the extrapolated $O(\Delta^4)$ solution. Likewise, we considered the u-velocity profile at the station $X = Lc/4$, denoted by $U_{1/4}(Y)$, which traverses the separation bubble. The ranges of the maximum deviations are

Table 1 Maximum deviation for truncation error

	Maximum function value	Range of maximum deviation for truncation error
$\zeta_w(X)$	3	.005142–.005302
$U_{1/4}(Y)$	1.25–1.354	.001064–.001065

shown in Table 1. Referred to the maximum function values, these represent deviations of approximately 0.17% for ζ_w and 0.13% for $U_{1/4}$. On this basis, we judge the $O(\Delta^4)$ solution to be converged with respect to truncation error.

11 Re Scaling Behavior

Having established the iterative convergence and the truncation-error convergence, it is relatively easy to establish the Re scaling behavior. If, as claimed, the solutions become self-similar in X/Re as $\text{Re} \to \infty$, the solutions at the same finite-difference grid points should become identical as Re increases. We ran cases for $\text{Re} = 10^2, 10^3, \ldots, 10^7$, obtaining the $O(\Delta^4)$ solutions in the 41×41 grid. We again considered $\zeta_w(X)$ and $U_{1/4}(Y)$ and considered the maximum deviation of each solution at a given Re from that solution at $\text{Re} = 10^7$. The results are given in Table 2. Even at $\text{Re} = 10^2$, the solution is a fair approximation to the asymptotic values.

We conclude that the $\text{Re} = 10^7$ solution adequately represents the $\text{Re} = \infty$ solution. The $\text{Re} = 10^7$ solutions for wall vorticity $\zeta_w(X)$ and u-velocity $U_{1/4}(Y)$ from the $O(\Delta^4)$ extrapolations are shown in Figs. 1 and 2.

Table 2 Maximum deviation for high-Re scaling

	Maximum function value	Maximum deviation from $\text{Re} = 10^7$ value				
		$\text{Re} = 10^2$	10^3	10^4	10^5	10^6
$\zeta_w(X)$	3	6.83×10^{-2}	1.01×10^{-2}	1.06×10^{-3}	1.06×10^{-4}	9.70×10^{-6}
$U_{1/4}(Y)$	1.25–1.354	1.02×10^{-1}	5.86×10^{-3}	5.43×10^{-4}	5.43×10^{-5}	4.80×10^{-6}

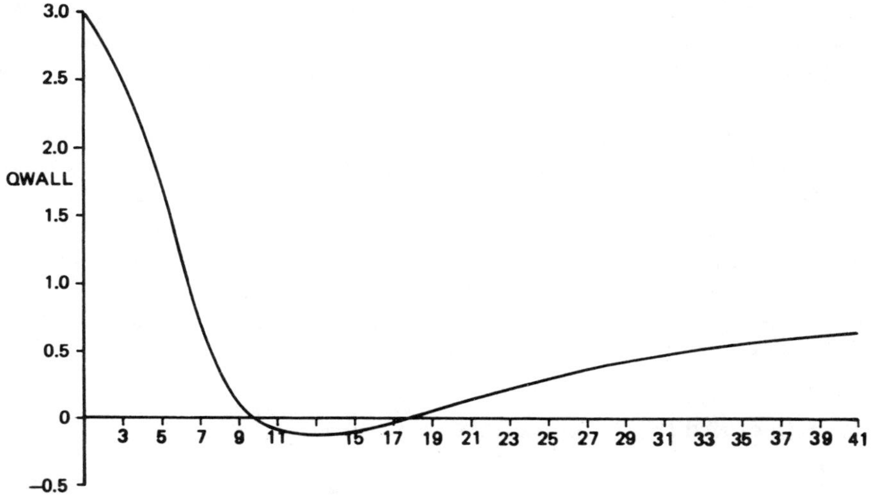

Figure 1. Wall vorticity at $\text{Re} = 10^7$.

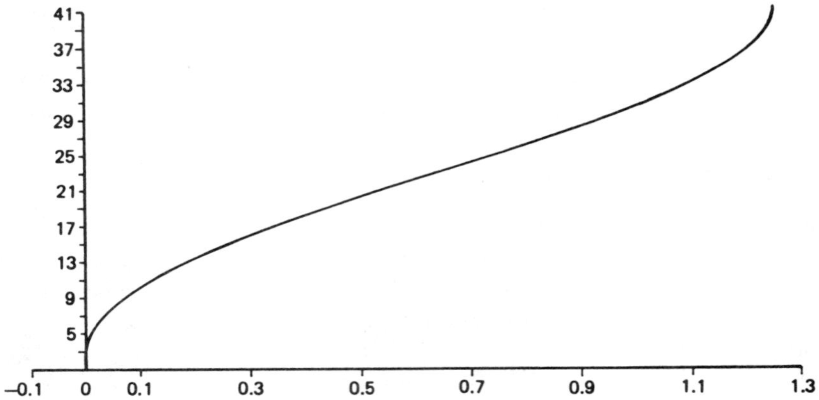

Figure 2. u-velocity at $X/L_c = \frac{1}{4}$, Re $= 10^7$.

12 Limit Analysis

This Re scaling can be rationalized by writing the continuum equations in the scaled variable $\tilde{x} = x/\text{Re}$. Equations (1), in nonconservation form, and (2) become

$$\zeta_t = -u\frac{\partial \zeta}{\partial \tilde{x}} - \text{Re}\cdot v\frac{\partial \zeta}{\partial y} + \frac{1}{\text{Re}^2}\frac{\partial^2 \zeta}{\partial \tilde{x}^2} + \frac{\partial^2 \zeta}{\partial y^2}, \tag{30}$$

$$\frac{1}{\text{Re}^2}\frac{\partial^2 \psi}{\partial \tilde{x}^2} + \frac{\partial^2 \psi}{\partial y^2} = \zeta. \tag{31}$$

In the usual boundary-layer scalings, the thickness of the boundary layer at some x-position scales as $1/\sqrt{\text{Rx}}$, where Rx is a length Re based on the distance from a leading edge. In the present geometry, with a fully developed inflow profile independent of Re, no such boundary-layer scaling applies. For boundary-layer solutions at a fixed thickness, $v \propto 1/\text{Re}$. Using this, Eqs. (30) and (31) become independent of Re as Re $\to \infty$, giving self-similarity in $(\tilde{x}, y)$.

The streamwise diffusion terms become negligible as Re and L_c increase in these problems, so that those terms do not contribute significantly to the solution at large Re. Thus, although the full Navier–Stokes equations are calculated, we are really approaching a kind of strongly coupled interacting boundary-layer calculation at large Re. Also, it is clear that these high-Re separated flows physically would not be stable (although the counterparts in hypersonic flow might be) but, since the scaling is a fair approximation even at Re $= 100$, they can be of practical interest.

13 Effect of Other Boundary Conditions

The solutions presented herein use a longitudinal coordinate that is stretched only by a scaling parameter, with heuristic outflow boundary conditions (13) which are a reasonable approximation for some channel problems but which neither strictly apply to any real problem of interest, nor have the advantage of analytic simplicity. Likewise, the fully developed Poiseuille inflow condition is heuristic, since conditions at $x = 0$ would be affected by the channel expansion. Two "neater" analytic problems could be solved. The first problem would use a mapping function (such as tanh) to map the region $(-\infty, +\infty)$ onto $(-1, +1)$ so that asymptotic upstream and downstream boundary conditions could be used. This was not done in the present study, simply because the transformation introduces additional terms (including additional cross-derivative contributions) which were not included in the code. (The code was developed originally to calculate flow in flush inlets of ships powered by water pumps, so that asymptotic boundary conditions were not of interest.) The second problem would use periodic boundary conditions in X and a moving wall at $Y = 1$, modeling a high-Re lubrication problem. The marching method has been recently extended to handle periodic boundary conditions with a 9-point operator, and user-oriented software is available [1.99].

It is doubtful that either of these problem solutions would change the general conclusion of the present study.

14 Conclusion

Steady-state separated flow solutions have been obtained quite efficiently with semidirect methods using tight iterative convergence criteria. Fine-mesh solutions and Richardson extrapolation have been used to achieve truncation-error convergence. The resulting solutions have been examined as Re increases with channel length $\propto$ Re.

It has been verified numerically that weakly separated laminar two-dimensional incompressible channel flows can display a self-similar solution. If the channel length x is increased proportionally to the Reynolds number Re, the solutions become self-similar in the scaled longitudinal variable $\tilde{x} = x/\text{Re}$.

This scaling behavior has implications for the interpretation of accuracy limits on high-Re Navier–Stokes calculations. A perpetual question in computational fluid dynamics is the Re limitation on accuracy. It is generally recognized that boundary-layer-type flows may be solved to high accuracy, since the directional cell Reynolds number (based on velocity component and mesh spacing) is small in the direction in which diffusion is important. However, more general flows—"strongly separated" in some sense—are severely limited by the cell Re limitation, which is ultimately related to the Nyquist frequency limitation (see e.g. [1.101]).

The present results may help to quantify the concepts of "strongly separated" and "weakly separated" flows in the present context. Since the Re = 100 solution is easily calculated with accuracy and since this solution agrees qualitatively with the high-Re solutions in the scaled coordinate, it is clear that these high-Re solutions are calculated accurately. This is indeed verified by the truncation convergence testing. We conclude that these flows in a scaled channel with $L_c \propto$ Re are then "weakly separated" even though Re is unbounded.

It is not known if these results would apply to a step channel, which has no length scale. It seems that the same results would exist (certainly the limit analysis of Section 12 applies), but the verification would be tricky because of the vorticity singularity at the sharp corner. On the other hand, it is doubtful that turbulent base flows at high Re could be "weakly separated" in the present context, since it is known experimentally that the length of the separation regions in these cases is a very weak function of Re. Certainly the cell Re limitations are very pronounced in calculating stagnation flows, as is well known.

CHAPTER 7

Some Observations on Numerical Solutions of the Three-Dimensional Navier–Stokes Equations[1]

H. McDonald and W. R. Briley*

Introduction

A number of important flow problems are inherently spatially three-dimensional in nature and feature significant diffusive effects in all three directions. These flows may be either laminar or turbulent or both, and for laminar flow the use of the three-dimensional Navier–Stokes equations is required to adequately represent the motion. For turbulent flows, these same equations are usually ensemble-averaged, and with the introduction of current-generation turbulence models to represent the turbulent correlations, the resulting equations bear a formal similarity to the unaveraged Navier–Stokes equations. Thus, except where noted, no formal distinction is drawn here between the laminar and the turbulent equation systems.

In view of the importance of some of these flows there have been a number of efforts devoted to developing, improving, and implementing numerical techniques to obtain solutions to the three-dimensional Navier–Stokes equations for a variety of flows. The high computational cost of obtaining such solutions has resulted in a desire for algorithm efficiency. However, the realities of the state of the art of solving the Navier–Stokes equations often results in prohibitively long run times even with what can only be charac-

[1] This research was supported by the Office of Naval Research under contracts N00014-72-C-0183, N00014-77-C-0075, and N00014-79-C-0713 with Morton Cooper as Contract Technical Monitor.

* Scientific Research Associates, Inc., Glastonbury, CT 06033.

terized as questionable mesh densities, affordable only by the computer-rich of the computational community. Doubtless, more efficient algorithms will be developed and less costly computers will become available, but even with this envisaged progress there still seem to be grounds for attempting to minimize the cost of this type of calculation.

In the present paper some observations are made on a solution methodology which, for a particular nontrivial class of both laminar and turbulent flows, has resulted in what is believed to be numerically reasonably accurate solutions of the compressible three-dimensional Navier–Stokes equations.

Flow Problems Considered and Their Impact on the Algorithm

The driving problems for the particular approach adopted here can be thought of as "difficult" three-dimensional boundary or shear-layer-type flows—difficult in the sense of not permitting typical boundary-layer simplifications. Here the interest lies in making flow-field predictions at high Reynolds numbers where the boundary or shear layers might initially be thin, and possibly turbulent. As a result of the geometry or boundary conditions (including the initial conditions), multiple length scales are present within the flow. Interaction between these various scales of the flow is considered of primary importance. The flows considered are not explicitly restricted to subsonic Mach numbers, but for the present shock waves are not considered to be present in the flow.

As examples of this category of flow, two specific problems are considered. The first concerns developing or fully developed laminar or turbulent flow in a duct whose centerline turns rapidly through ninety degrees, with the results taken from the report by Buggeln, Briley, and McDonald [1.102]. The second problem concerns a boundary layer growing on a surface encountering an elliptical strut mounted normal to the surface, and the results are taken from the paper by Briley and McDonald [1.103]. The major axis of the ellipse is set at a small incidence angle to the approach stream. In both problems, a core-flow length scale and boundary-layer thickness scale normal to the no-slip surfaces are augumented by a cross-flow thickness scale which is initiated upstream of the turn or the strut leading edge via the pressure field. In addition, in both problems there are two length scales associated with the throughflow direction. The first is associated with the definition of the pressure field, and the second is associated with the possible separation region. This latter length scale poses a particular problem which will be elaborated on subsequently, inasmuch as the location and extent of the separation region is not known *a priori*. For the results shown subsequently, the streamwise mesh was determined largely to define the streamwise development expected of the pressure field, with the resulting possible penalty that small regions of separated flow might be poorly resolved. In the strut problem, an additional length scale is involved due to the boundary layers growing on the strut surface. Each of the scales must be considered in turn from the point of view of the required mesh resolution and the resulting possible impact on the algorithm.

In some problems extreme resolution near the wall may not be required and one may be able to assume some form for the unresolved dependent-variable profile beneath the first grid point. In many two-dimensional turbulent flows, the remarkable range of validity of the so-called "law of the wall" makes this wall-function approach feasible and probably realistic, particularly in flows where the wall flow is well behaved and the real interest is in the core region far from the bounding walls. In the problems considered here, the near-wall flows, particularly the near-wall cross flows, are critical to the fluid mechanics in view of their impact on the core-flow distortion. Although three-dimensional velocity profile families such as that due to Johnson [1.104] are available and of sufficient accuracy for use in estimating scales, a general profile family of sufficient accuracy to serve as a crossflow wall-function boundary condition in the envisaged problems—encompassing such phenemona as the presence of corners, strong skew with perhaps changing sign of the transverse pressure field—does not appear to exist (see discussion, [1.105]). Thus the wall-function approach is not expected to be useful for the problems considered here, and it is anticipated that all the relevant scales normal to the wall in the turbulent boundary layer will have to be defined.

The basic philosophy adopted here is to use a nonuniform mesh to resolve the various scales, since this provides for economic utilization of a limited number of grid points. In three space dimensions, the current state of the art is such that using a mesh distributed nonuniformly to resolve the appropriate scales is mandatory to obtain solutions in any reasonable amount of run time in a multiscale flow problem. Dynamic rezoning of the mesh as the solution develops is potentially a very attractive way to treat this problem, but until it is developed for the very complicated circumstances of the problems considered here, some means of estimating scales are very useful in reducing the overall labor to achieve a resolved flow. In attempting to make estimates for the length scales involved in these very complex flows, appeal can be made to experiment, approximate or related analysis, or existing solutions. Ultimately, mesh-refinement studies should be performed to demonstrate that the relevant scales have been resolved. However, for the complex flows considered, mesh-refinement studies are very expensive, sometimes prohibitively so, and consequently only a very limited amount of three-dimensional mesh refinement can be performed. Asymptotic analyses such as triple deck (Stewartson [1.106]) are very useful for estimating scales on certain classes of very high-Reynolds-number boundary-layer problems, but for the somewhat different, complex three-dimensional, even turbulent problems considered here, these analyses are of limited utility. Experimental and/or analytical evidence is most useful when it has been collapsed into the form of similarity "laws", and consequently here much use will be made of this approach for estimating length scales.

Boundary Conditions

In order to determine the number of boundary conditions required on each surface by the governing equations, a study of the characteristics is required. Note that although this information is vital, it is incomplete in that only the number and not the specific boundary conditions emerge from the analysis. Following Courant and Hilbert [1.107], the equations are written in first-order form. The case of boundary conditions for the impermeable wall is considered first, and here the two-dimensional Navier–Stokes equations are written for Stokes flow, where the convective terms may be neglected in comparison with the viscous terms. This reduction is valid very close to the wall, where the boundary conditions are being applied, and since the sublayer is in fact resolved, no loss in generality is expected from the approximation.

The reduced governing equations are therefore parabolic in nature and require all of the dependent variables be specified throughout space at time t_0 together with (in principle) a boundary condition for each dependent variable. If, however, one examines the governing equations and integrates the interior points, say explicitly, from time t_0 till $t_1 = t_0 + \Delta t$, it is clear that only two function values at the boundary may be specified independently. For example, if one specifies the velocity components on the boundary at time t_1, the known interior solution at time t_1 combines with the boundary values to prescribe the density evolution from the continuity equation. Consequently, the density evolution at the wall cannot be arbitrarily assigned. The extension of these arguments to three dimensions is immediate.

Consider now the specific wall boundary conditions in laminar and turbulent flow. According to the preceding argument, no further boundary conditions at the wall are required by the governing equations. Any additional boundary conditions which might be imposed result in an overspecification of the problem, although computationally this in itself need not be disastrous. A problem arises with the discrete form of the equations written in implicit form, as this usually introduces the values of all k dependent variables at the boundary. If n boundary conditions are all that the governing boundary equation requires, some way has to be found to eliminate the $k - n$ dependent variables at the wall from the system without imposing inaccurate overspecified boundary conditions. In the particular cases under consideration the wall density is the variable of concern. Note here that if on the basis of observation or order-of-magnitude analysis it is believed that a valid flow approximation can be made which could be imposed as an additional boundary condition (e.g. $\partial p/\partial n = 0$), the ability to obtain a solution with this overspecification would be entirely based on the stability and convergence properties of the numerical scheme. If obtainable, the resulting solution would of course reflect the approximation inherent in the additional boundary condition, and in general this may or may not be acceptable on physical grounds.

An increasingly popular way to avoid overspecification, and one which is adopted here, has been used by a number of investigators, e.g. Rubin and

Lin [1.108] and Briley and McDonald [1.109]. Here the wall density is the problem, and the governing equations can be written in one-sided difference form, so that the density at the wall does not appear. For the present problems, at a no-slip wall, the density derivative normal to the wall only appears in the continuity and the normal momentum equation, making either of these equations candidates for the elimination process. Since the one-sided form of the equations is not a boundary condition, the problem of overconstraint is avoided. In the present problems, the normal momentum equation is approximated by $\partial p/\partial n = 0$ and utilized to remove the wall density; possibly aided by the well-resolved wall layer, it functioned very well. The approximate form of the normal momentum equation was compared in one flow with use of the full equation, and no significant difference was observed.

Turning now to the inflow and outflow boundaries, here the required number and type of boundary conditions for the compressible Navier–Stokes equations have not been rigorously derived to date. We will not go through the labor of deriving the appropriate characteristics, since eventually a numerical experimental demonstration of the validity of the hypothesized boundary conditions is required. Appeal is made to a characteristics analysis for an inviscid flow, however, since over a large part of the inflow and outflow boundaries for the flow problems considered here, the core flow behaves in a rotational but inviscid manner. In these flows, the inflow and outflow planes are placed sufficiently far upstream and downstream that they are roughly normal to the core flow streamlines, and the flow at the boundary is essentially down the duct (no recirculation at the boundary). With these caveats, a governing equation system is considered, comprising the Euler equations together with either an inviscid entropy or a stagnation energy equation and the gas law added. With all the streamlines (actually particle lines) pointing down the duct and nearly normal to the bounding planes, and with subsonic core flow, it is clear from the characteristic analyses of such inviscid systems given by Von Mises [1.110], for instance, that the characteristics are all real with two positive and one negative. Using the counting principle of one boundary condition per entering characteristic [1.107], on the inflow this requires two boundary conditions to be specified, and on the outflow only one.

In spite of the physical plausibility of approximating the inflow and outflow core regions as rotational inviscid regions, the governing equations solved by the numerical algorithm in this region remain the full Navier–Stokes equations. Consequently, there is no guarantee that the physically negligible small stress terms will not disrupt this boundary-condition count. Hence, once the specific boundary conditions are introduced, the resulting system is verified by numerical experimentation and the characteristics analysis is not developed further.

Insofar as the specific boundary conditions used are concerned, appeal is made to physics. In a typical duct flow the usual putative flow proceeds from some very large reservoir through the duct to exhaust into a plenum. The reservoir conditions of the flow at rest are known. The plenum (uniform) static pressure is also known. The flow rate through the duct is thus determined by these conditions together with the losses created by the precise duct configuration. This putative flow model leads immediately to the proposed computational boundary conditions: prescribing the reservoir conditions of the gas at rest (stagnation conditions), together with a plenum static pressure. The specified stagnation temperature and pressure constitute the required two inflow boundary conditions for the rotational inviscid flow, with the specified static pressure constituting the one outflow condition. Unfortunately, it is not computationally convenient to proceed upstream into the reservoir or downstream into the plenum. Instead, these same boundary conditions are imposed within the duct just upstream and downstream of the regions of interest. This poses two problems. The first is that the upstream duct generates a nonuniform total pressure and temperature in the duct, and this has to be taken into account in setting the inflow conditions. The second problem is that the static pressure on the outflow boundary may not be uniform (or unknown precisely) over the duct cross-section. Both of these problems reflect the effect of the upstream and downstream flow development from the reservoir and into the plenum, and both effects must be taken into account.

At the inflow plane sufficient information must be available to estimate the total pressure and temperature distribution. For the problem considered here this was reduced to specifying the core-flow total conditions together with a boundary-layer thickness and profile shape. Using this profile shape, given in terms of velocity and stagnation temperature, the inflow total pressure distribution within the boundary layer (and only within the boundary layer) was updated at each time step to maintain the required velocity and temperature profile shape.

The outflow plane in the turning duct problems was situated far enough downstream of the corners that uniform static pressure in the outflow plane was a reasonable approximation. In the strut–endwall problem, the outflow boundary was placed at the strut midchord and in a region of very obvious nonconstant static pressure. In this problem the static pressures on the strut surface in the outflow plane were set and the distribution normal to the strut surface found by integrating the inviscid normal-momentum equation $\partial p/\partial n = q^2/R_I$. Here q is the local computed velocity magnitude and R_I is the local potential-flow streamline radius of curvature. The resulting static pressure distribution is imposed on the outflow plane, assuming no variation normal to the endwall. The flow velocity used in this approximation is updated after each time step. The justification for this approximation is twofold. First, the prime interest is in the immediate vicinity of the leading edge, and it is simply desired to truncate the domain of computation in a graceful manner. The second justification is that away from the endwall and the immediate strut surface, the flow does behave inviscidly at these Reynolds numbers, and that using the inviscid-flow streamline curvature is a reasonable first approximation.

The foregoing deals with the boundary conditions required by the partial differential equations. The appearance of all the dependent variables on the inflow and outflow boundaries due to the implicit formulation parallels that already touched on for the wall. The problem on flow boundaries is not to eliminate dependent variables but to ensure that they satisfy a nonlinear relationship such as a prescribed total pressure. Instead of using one-sided differencing, the previously mentioned strategy is used of developing physically plausible flow approximations which lead to additional boundary conditions and an overconstrained system. The hypothesis is that if the overconstrained system permits a solution, the solution is only defective in that it contains the flow approximations used to generate the additional extraneous boundary conditions. For the flows considered the extraneous conditions were as follows. Consider Cartesian velocity components u, v, w, where u is the streamwise velocity down the duct or parallel to the ellipse major axis, v is normal to the duct wall or the ellipse major axis, and w is normal to the endwall or to the duct wall. It was first of all supposed on inlet that v was proportional to u so that the flow could be at some incidence angle to the ellipse. For ducts v was set to zero. It was then supposed that the second derivative normal to the inflow boundary of the cross-flow velocity w was zero. In this same normal direction the second derivative of the pressure was set to its potential-flow value (corrected for blockage) for the ellipse, and set to zero for the duct. For both flows the outflow static pressure was specified as remarked on earlier, and the outflow extraneous boundary conditions used were that the second derivatives in the streamwise direction were zero. The physical plausibility of these extraneous boundary conditions is evident, at least for the flows considered. Based on the numerical evidence subsequently obtained, little difficulty was observed with the particular boundary conditions used. Note that these boundary conditions are all reasonable in the transient as well as at steady state. Also, the mass flow in the duct is not set *a priori*, and since the inflow static pressure is not specified, pressure waves can escape upstream, thus avoiding the problem of standing waves, discussed in [1.111].

Initial Conditions

In order to advance the solution from time $t = t_0$ to time $t = t_0 + \Delta t$, the solution at time $t = t_0$ must be known. Thus, values for all of the dependent variables must be either given or able to be deduced from given information at time $t = t_0$. In very general terms, the velocity and static pressure fields are first constructed. The density field is then deduced from the gas law and the condition of specified stagnation temperature. Other strategies are possible, such as specifying only two components of the velocity and deducing the third component from the continuity equation to give zero time acceleration of the density. This latter approach was not investigated, since with the given approximations the time evolution was quite well controlled.

It is often suggested, and certainly true in the limit, that the better the initial guess, the less computer time will be used to achieve convergence. On the other hand, for these complex three-dimensional flows only very crude approximations to the flow structure can be made *a priori*, without a very major effort. What is clear from the results obtained subsequently is that using comparatively crude initial conditions, rapid convergence to the steady state is still obtainable. Minor improvements to the initial flow field did not result in significant improvement in the convergence rate. Conversely, it was not necessary to input a close approximation to the final solution in order to obtain the rapid convergence observed in the results.

Governing Equations, Coordinates, and Differencing Procedures

Various forms of the compressible time-dependent Navier–Stokes equations are available in the open literature and will not be reproduced here. The precise form used was that given by Hughes and Gaylord [1.112] for general orthogonal curvilinear coordinates with the density and the velocity components aligned with the coordinates as dependent variables. For the present work, the first-derivative flux terms were put in conservative form. The usual perfect-gas law was adopted, and although various forms of the energy equation are available within the code, the energy equation was deleted and replaced by the assumption that the stagnation temperature was constant.

An equation governing the balance of turbulence kinetic energy, $\overline{q^2}$, in curvilinear orthogonal coordinates was derived from the Cartesian tensor form of this equation given by Launder and Spalding [1.113]. An equation governing the dissipation of turbulence, ε, is often used to generate a turbulent-length-scale equation, but earlier numerical difficulties were experienced with this equation due to extreme stiffness, and so for the present series of calculations this equation was not used. Instead a very simple three-layer length scale was constructed, with the outer or wake scale determined by a one-dimensional estimate of the growth of this wake length scale from its value on inlet. This growth rate was essentially obtained from the von Karman momentum integral equation, assuming the wake length scale would grow roughly as the boundary-layer thickness. Near the nearest wall the length scale was assumed to vary in accordance with von Karman's linear relationship $l = \kappa y$. In the viscous sublayer the length scale was damped by viscous effects according to Van Driest's formulation. Finally, the turbulent effective viscosity μ_t was obtained from the Prandtl–Kolomogorov constitutive relationship $\mu_t = l(\overline{q^2})^{1/2}$. The turbulent viscosity was then supposed isotropic and the stress tensor in the ensemble averaged equations determined by adding the turbulent viscosity to the kinetic viscosity. This latter course is quite approximate, and future work will involve an improved formulation. The turbulent kinetic energy near the wall was damped out according to the suggestion of Shamroth and Gibeling [1.114]. For the ducts, the compressible Navier–Stokes equations in general orthogonal coordinates are solved using analytical coordinate data for a system of coordinates aligned with the duct geometry. The coordinate system is shown in Fig. 1.

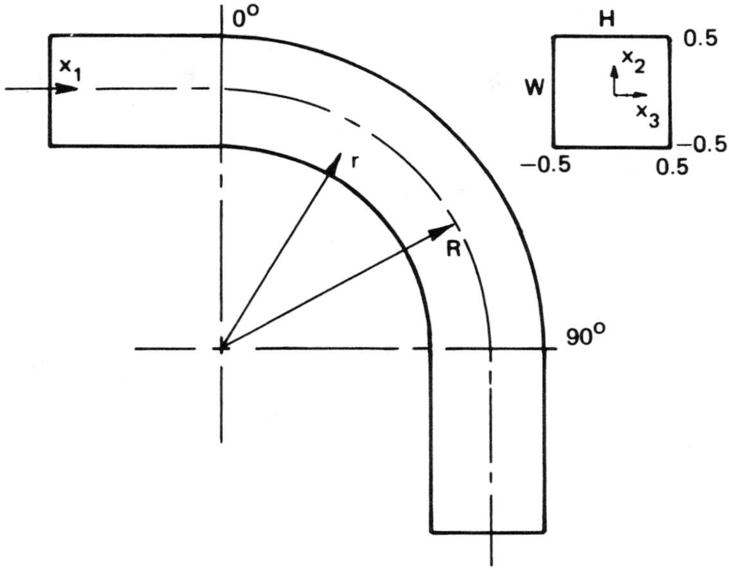

Figure 1. Geometry of square duct (to scale): $R/W = 2.3$, $H/W = 1.0$.

For the elliptical strut endwall problem the compressible Navier–Stokes equations in general orthogonal coordinates are solved using analytical coordinate data for an elliptic-cylindrical coordinate system which fits all solid surfaces within the computational domain but is not aligned with the direction of the freestream flow. In selecting the computational domain, a "zone embedding" approach is adopted whereby attention is focused on a subregion of the overall flow field in the immediate vicinity of the leading-edge horseshoe vortex flow. A perspective view of the geometry, the coordinate system (ξ, η, z), and a representative computational grid is shown in Fig. 2.

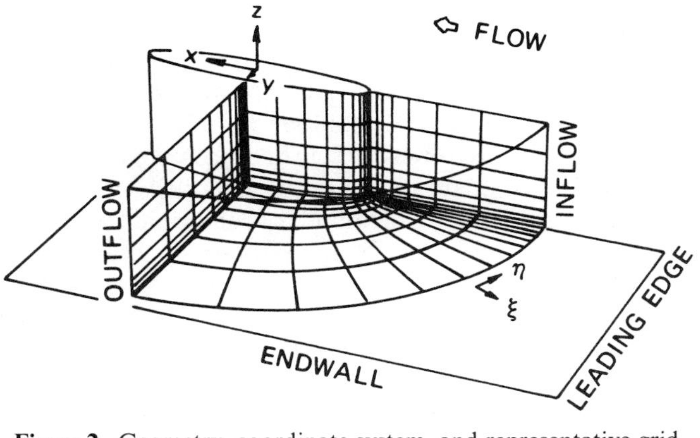

Figure 2. Geometry, coordinate system, and representative grid.

The differencing procedures used are a straightforward adaptation of those used by the authors [1.109] in Cartesian coordinates for flow in a straight duct. The definition of stagnation enthalpy and the equation of state for a perfect gas can be used to eliminate pressure and temperature as dependent variables, and solution of the energy equation is unnecessary. The continuity and three momentum equations are solved with the density and the velocity components as dependent variables. Three-point central differences were used for spatial derivatives, and second-order artificial dissipation terms are added as in [1.109] to prevent spatial oscillations at high cell Reynolds number. This treatment lowers the formal accuracy to first order but does not seriously degrade accuracy in representing viscous terms in thin shear layers. The analytical coordinate transformation of Roberts [1.115] was used to redistribute grid points and thus improve resolution in the boundary layers near the leading edge. Derivatives of geometric data were determined analytically for use in the difference equations.

Derivation of Split LBI Schemes by Approximate Factorization

The governing equations are written as a general system of time-dependent nonlinear partial differential equations (PDEs). As mentioned earlier, an implicit formulation is favored because of the desired sublayer resolution. To treat the nonlinearity, Taylor-series expansion in time is used to linearize the system to the same order accuracy as the temporal discretization. To efficiently eliminate the large block banded matrix, a splitting of the matrix into a sequence of more easily treated component matrices is performed. In view of this construction, the resulting algorithm is termed a split linearized block implicit (LBI) scheme.

The numerical algorithm used is developed in detail by the present authors in [1.109, 1.116, 1.117]. In this development, the authors split their LBI scheme by following the procedure described by Douglas and Gunn [1.118] in their generalization and unification of scalar ADI schemes. Recently, a derivation of this scheme (and other related schemes), based on the approximate factorization approach of Yanenko and D'Yakonov [1.119], has been given by Warming and Beam [1.120]. Here, a brief account is given of the derivation by approximate factorization, which also includes consideration of the intermediate boundary conditions required by split schemes. By considering these intermediate boundary conditions a commutative condition on the boundary conditions is identified, and it is shown that the treatment of these intermediate boundary conditions can affect transient accuracy, and even cause error in steady solutions.

To illustrate this derivation, let

$$\frac{\phi^{n+1} - \phi^n}{\Delta t} = \beta D(\phi^{n+1}) + (1 - \beta)D(\phi^n) + O[\Delta t^2, (\beta - \tfrac{1}{2})\,\Delta t] \qquad (2)$$

approximate $\partial\phi/\partial t = D(\phi)$, a system of time-dependent nonlinear PDEs for the vector ϕ of dependent variables, where D is a multidimensional vector spatial differential operator, and t is a time variable with discretization $\Delta t = t^{n+1} - t^n$. A local time linearization (Taylor expansion about ϕ^n) of requisite formal accuracy is introduced, and this serves to define a *linear* differential operator L such that

$$D(\phi^{n+1}) = D(\phi^n) + L^n(\phi^{n+1} - \phi^n) + O(\Delta t^2) \tag{3}$$

Equation (2) can thus be written as the linear system

$$(I - \beta \, \Delta t \, L^n)(\bar{\phi}^{n+1} - \phi^n) = \Delta t \, D(\phi^n), \tag{4a}$$

$$\bar{\phi}^{n+1} - \phi^n = \phi^{n+1} - \phi^n + O[\Delta t^3, (\beta - \tfrac{1}{2}) \, \Delta t^2], \tag{4b}$$

where $\bar{\phi}^{n+1}$ differs from ϕ^{n+1} by the linearization error.

To obtain a split scheme, the multidimensional operation L is divided into m "one-dimensional" suboperators $L = L_1 + L_2 + \cdots + L_m$ which are usually associated with coordinate directions. For simplicity, only two suboperators are considered here; inclusion of additional suboperators is straightforward. Equation (4) can then be replaced by the approximate factorization

$$(I - \beta \, \Delta t \, L_1^n)(I - \beta \, \Delta t \, L_2^n)(\phi^* - \phi^n) = \Delta t \, D(\phi^n), \tag{5a}$$

$$\phi^* - \phi^n = \phi^{n+1} - \phi^n + O[\Delta t^3, (\beta - \tfrac{1}{2}) \, \Delta t^2], \tag{5b}$$

where ϕ^* differs from $\bar{\phi}^{n+1}$ by the factorization error. Although ϕ^*, $\bar{\phi}^{n+1}$, and ϕ^{n+1} are interchangeable without formal loss of accuracy, the distinction is worthwhile, since they entail different types of error which in practice may differ widely in magnitude. Note that the approximate factorization (5a) does not represent a simplification of Eq. (4) until it is written in a split form.

The most obvious splitting of Eq. (5a) is

$$(I - \beta \, \Delta t \, L_1^n)\psi = \Delta t \, D(\phi^n), \tag{6a}$$

$$(I - \beta \, \Delta t \, L_2^n)(\phi^* - \phi^n) = \psi, \tag{6b}$$

where ψ is an intermediate quantity whose physical significance follows from its definition (6b), which implies that ψ approximates the time increment $\phi - \phi^n$ to order Δt. If spatial derivatives appearing in L_1 and L_2 are replaced by three-point difference formulas, then each step in Eqs. (6a, b) can be solved by a block-tridiagonal elimination.

The derivation of the algorithm is incomplete, however, since Eq. (6a) cannot be solved for ψ without deriving boundary conditions for ψ from those given for ϕ. If function values of ϕ are given, then boundary values of ψ can always be derived from the definition (6b) of ψ for use in Eq. (6a). In practice, however, more complex boundary conditions such as normal derivatives may be specified, and so here we consider a much more general nonlinear boundary condition which, after linearization as in Eq. (3), can be written in the form $B^n(\phi^{n+1}) = g(t, \phi^n)$. Here, B^n is a linearized operator which may include the same derivatives as L_2^n, and g is given. Applying the operator B to Eq. (6b) gives

$$B^n(\psi) = B^n(\phi^* - \phi^n) - \beta \, \Delta t \, B^n[L_2^n(\phi^* - \phi^n)]. \tag{7}$$

Since $B^n(\phi^* - \phi^n) \approx g^{n+1} - g^n$, without formal loss of accuracy Eq. (7) can be used to derive exact boundary conditions for ψ from the given boundary conditions *provided* B^n and L_2^n commute, which is unfortunately often not the case. The need for commutativity occurs because $L_2^n \phi^*$ cannot be computed at this step in the algorithm, whereas $B^n(\phi^*)$ can be replaced by the given g^{n+1}.

When B^n and L_2^n do not commute, an approximate solution $\tilde{\psi}$ can be computed instead of ψ, where $\tilde{\psi}$ satisfies

$$B^n(\tilde{\psi}) = g^{n+1} - g^n, \tag{8a}$$

$$\tilde{\psi} = \psi + O[\Delta t\,(\phi^* - \phi^n)]. \tag{8b}$$

instead of Eq. (7). This is of course equivalent to using uncorrected (i.e. "physical") boundary conditions for $\phi^* - \phi^n$ as boundary conditions for ψ. Substituting Eq. (8b) into Eqs. (6) shows that an additional error of order $\Delta t\,(\phi^* - \phi^n)$ is introduced by the use of "uncorrected" boundary conditions as in Eq. (8a). The overall accuracy in terms of approximating Eq. (2) is $O[\Delta t^2, (\beta - \frac{1}{2})\,\Delta t, \phi^* - \phi^n]$, where in turn $\phi^* - \phi^n$ is of order Δt. Note, however, that in the steady state $\phi^* - \phi^n \equiv 0$, and Eq. (8a) becomes an exact boundary condition. Consequently, leaving boundary conditions uncorrected as in Eq. (8a) introduces no error in steady solutions, which in turn satisfy $D(\phi^*) = 0$. It is worth noting that unless the solution ϕ of $D(\phi) = 0$ is unique, the steady solution ϕ^* obtained need not be the same as would be obtained by repeating Eq. (2) until $\phi^{n+1} - \phi^n = 0$. This completes the derivation of the scheme (6a, b) and its boundary conditions.

It is of interest to develop an alternative splitting of Eq. (5), based on a dependent variable ϕ instead of the time increment $\phi - \phi^n$. Moving all terms involving ϕ^n to the right-hand side, the following D'Yakonov-type of splitting can be obtained:

$$(I - \beta\,\Delta t\,L_1^n)\xi = (I - \beta\,\Delta t\,L_1^n)(I - \beta\,\Delta t\,L_2^n)\phi^n + \Delta t\,D(\phi^n), \tag{9a}$$

$$(I - \beta\,\Delta t\,L_2^n)\phi^* = \xi. \tag{9b}$$

The intermediate boundary condition analogous to Eq. (7) is given by

$$B^n(\xi) = B^n(\phi^*) - \beta\,\Delta t\,B^n[L_2^n \phi^*]. \tag{10}$$

Again, Eq. (10) cannot be used unless B^n and L_2^n commute. The approximate solution $\tilde{\xi}$ obtained with uncorrected boundary conditions analogous to Eq. (8) satisfies

$$B^n(\tilde{\xi}) = g^{n+1}, \tag{11a}$$

$$\xi = \tilde{\xi} + O(\Delta t) \tag{11b}$$

In this instance, however, the error is of order Δt instead of $\Delta t\,(\phi^* - \phi^n)$ as in Eq. (8b), and the overall error in terms of approximating Eq. (2) increases to $O(1)$. Furthermore, the $O(\Delta t)$ error in the boundary condition (11) persists in the steady state, and the steady solution, therefore, depends on the time step. One remedy for this problem is to compute a different solution $\hat{\xi}$, which satisfies the following approximation:

$$B^n(\hat{\xi}) = g^{n+1} + \beta \, \Delta t \, B^n[L_2^n \phi^n], \tag{12a}$$

$$\xi = \hat{\xi} + O[\Delta t \, (\phi^* - \phi^n)]. \tag{12b}$$

Equation (12a) is an implicit boundary condition derived by making explicit corrections (setting $L_2^n \phi^n \approx L_2^n \phi^*$ in the correction term) to the physical boundary conditions.. Although less convenient, the accuracy of Eqs. (12) is the same as Eqs. (8), and the steady solution ϕ^* is now independent of Δt.

In summary, the recommended splitting is Eqs. (6a, b), since both $O(\Delta t)$ accuracy in the transient and a steady solution independent of Δt can be obtained without the inconvenience of computing boundary corrections. Use of the splitting (9a, b) requires boundary corrections such as Eq. (12a), since uncorrected boundary conditions lead to $O(1)$ transient accuracy and a steady solution which depends on Δt. If B^n and L_2^n commute and exact boundary corrections are made, the results from the splittings (6) and (9) are identical. Warming and Beam [1.120] have observed that an advantage of the "delta form" factored scheme over the nondelta form (equivalent to Eqs. (6) and (9), respectively) is a steady solution independent of Δt. This is true only if no boundary corrections are made to the nondelta form; otherwise, the two splittings give identical steady solutions. It is also noted that the treatment of boundary conditions depends on the splitting itself and not on the choice of dependent variable. The splitting (6) can be coded as written for the time increments ψ and $\phi^* - \phi^n$, or rearranged and coded for the dependent variables $\psi + \phi^n$ and ϕ^*. Finally, in the Douglas–Gunn [1.118] generalization of scalar ADI schemes used by the authors [1.109, 1.116] to derive the split LBI scheme (6a, b), the development ensures that each intermediate step of the scheme is a consistent approximation to the unsplit equation. This development leads to the splitting (6a, b) but excludes (9a, b), whose intermediate steps are inconsistent.

Results

Mesh refinement studies represent a critical test in the evaluation of the numerical results. In view of the very high computational cost of performing mesh-refinement studies with three space dimensions, many of the required studies were performed in two space dimensions, although this demands care in selecting those results which should carry over into three dimensions. Since the three-dimensional computations are usually performed with the coarsest mesh required to satisfy some (subjective) criteria of accuracy, a valid necessary precondition on the mesh would be that in an appropriate two-dimensional version of the problem, the same type of coarse mesh would give acceptable results. On the other hand, too much cannot be read into the two-dimensional results, since a two-dimensional mesh-refinement study cannot establish the accuracy of resolution of three-dimensional phenomena which do not even exist in two space dimensions.

The ducts presented here are all of square cross-section, and the computational results are taken from the report by Buggeln, Briley, and McDonald [1.102]. The strut–endwall results are taken from the paper by Briley and McDonald [1.103]. A schematic of the duct and the strut is shown in Figs. 1 and 2. In two dimensions a channel is considered to represent the duct, and for the strut-endwall flow a two-dimensional elliptical cylinder is considered. The channel-flow results are examined first, and the turning radius of the channel centerline was the same as for the duct, i.e. 2.3 times the width of the duct or channel. The entrance Reynolds number based on bulk velocity was 7.9×10^2 for laminar flow and 4×10^4 for turbulent flow. The bends turned through 90°, and the inlet boundary-layer thickness was set at 0.2 of a channel width. These conditions were chosen to replicate the experiment of Taylor, Whitelaw, and Yianneskis [1.121]. The experiments were performed in water, and the calculations performed for air at an inlet Mach number of

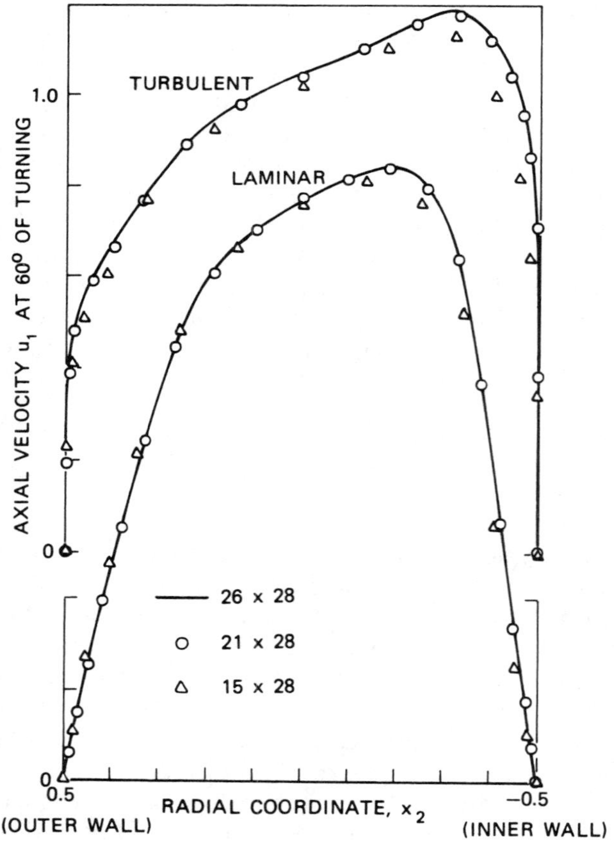

Figure 3. Radial mesh refinement for two-dimensional channel.

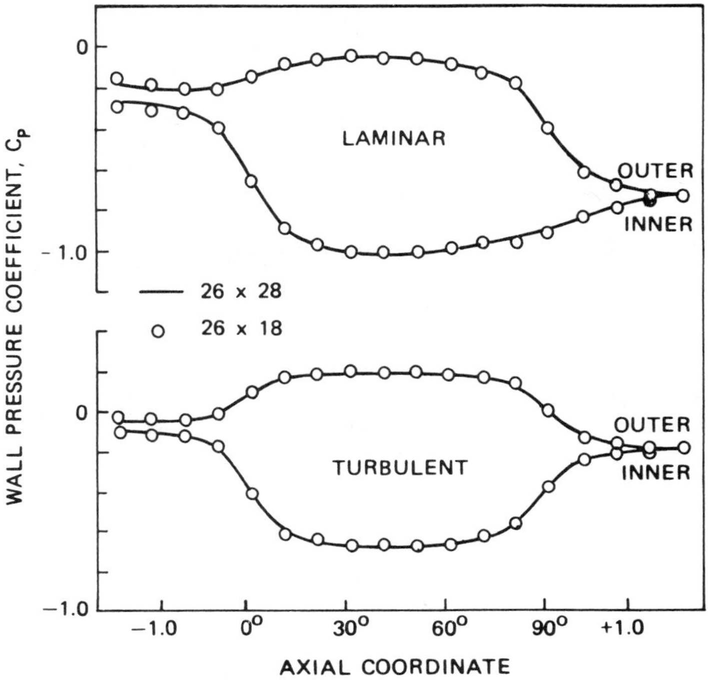

Figure 4. Axial mesh refinement for two-dimensional channel.

0.05. In one instance the channel inlet Mach number was doubled to 0.1, with no discernable influence on the results. The inlet and exit planes were located approximately 1.5 duct or channel widths upstream of the bend.

First of all, the mesh density and distribution in the two-dimensional channel was established in accordance with earlier scale estimates. The effect of varying the mesh density in the radial direction with a fixed streamwise mesh of 28 grid points is shown in Fig. 3. Note that in these and similar plots, in the straight sections the axial distance is presented normalized by duct widths, with origin at the start or finish of the turn. Within the turn, the axial location is given in degrees of turn. The streamwise velocity for both laminar and a resolved-sublayer turbulent flow after 60° of turning are shown. It is clear that mesh independence of this flow prediction is obtained when twenty or more radial mesh points are used with an appropriate stretch. Next an equally spaced axial mesh was examined with a fixed stretched radial mesh of 26 grid points, and this is shown for a laminar and a resolved-sublayer turbulent flow in Fig. 4. The pressure distributions on the inner and outer wall are shown for axial meshes of 18 and 28 equally spaced points. Not shown are the results with 23 axial mesh points, since these were indistinguishable from the results at 28 points. The effect of the location of the inflow and outflow boundaries for both the laminar and turbulent flows is shown in Fig. 5 using the 26-point stretched radial mesh. From this figure it is clear

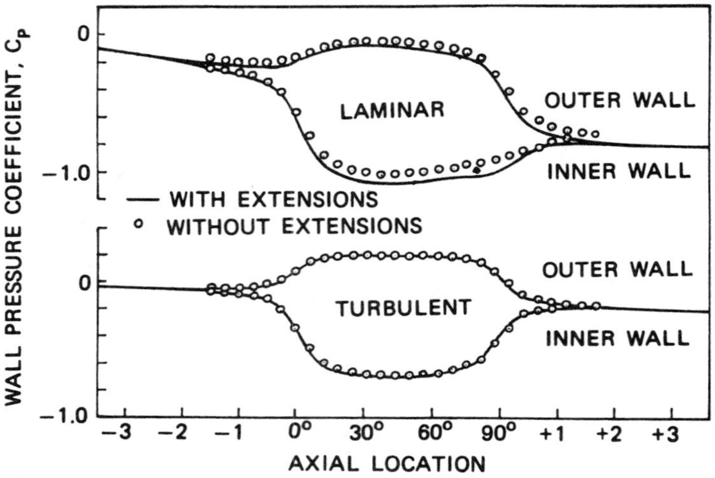

Figure 5. Effect of extending upstream and downstream segments of two-dimensional channel.

that the upstream movement of the inflow and outflow boundaries did not significantly affect the turbulent results.

As a final observation on the channel-flow results, in Fig. 6 the maximum change of streamwise velocity in the computational zone is shown plotted against nondimensional time. Here 6.75 units of nondimensional time represents the time of residence within the computational domain of a particle traveling with the reference velocity. Also shown on the plot is the time-step number. In all the calculations performed, initially a limited number of constant time steps were taken to eliminate the major initial transients. Thereafter a cyclical sequence of time steps was taken ranging over a factor of about 100, up to a maximum of 0.5 nondimensional time units. The changes in velocities shown in Fig. 6 are taken at this maximum, comparatively large time step. Note that Fig. 6 would not be meaningful unless the change were taken at a very large time step, since any temporal change could always be made small by taking a very small time step. With 10^{-4} as the criterion, convergence to the steady state was usually obtained at around 80 time steps in both two- and three-dimensional flow. Residuals in the governing equations were also examined at convergence and noted to be small in comparison to the other terms in the governing equations. As a final observation, with 10^{-4} as the criterion, the plotted results exhibited little or no discernable change with further iteration.

Attention is now turned to the three-dimensional ducts. Here, based on the two-dimensional results, 28 equally spaced mesh points were used with 26 radial points distributed nonuniformly to resolve the inner- and outer-wall boundary layers, including sublayers. Since only the symmetric half of the duct cross section was considered, only one endwall boundary layer was present, and hence 13 points distributed to resolve the sublayer were used,

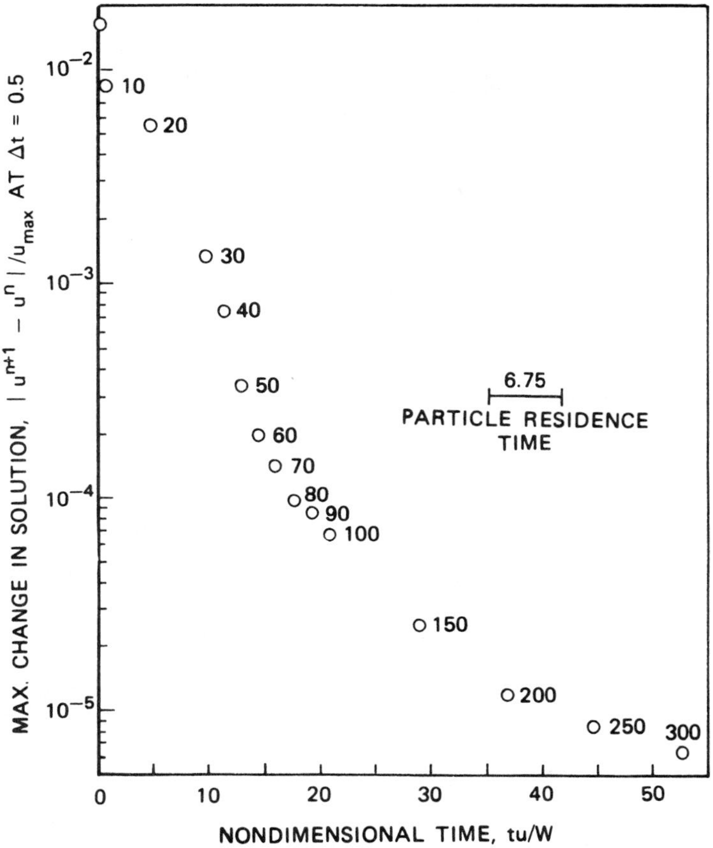

Figure 6. Convergence rate for two-dimensional turbulent channel flow.

by analogy with the radial mesh. A laminar and a turbulent square-duct calculation were performed, followed by a circular-cross-section pipe, also turning through 90°. Only the turbulent square-duct results are presented here, since mesh-refinement and mesh-distribution studies have not been completed for all the cases at present. The preliminary indications are that with the distribution of inlet vorticity present in the laminar flow, much stronger secondary flows are created. These strong secondary flows require careful resolution, as they are formed over a short period of axial development. It would appear that a nonuniform distribution of the existing axial mesh, together with perhaps an increase in resolution normal to the endwall, will be required for accurate resolution of such flows with large inlet vorticity. A limited mesh-refinement study normal to the endwall was performed for the turbulent case, which has considerably less inlet vorticity than the laminar case. Although not yet complete, the turbulent results appear free from significant numerical error. A comparison between the predictions of the computed results and the recent laser Doppler velocimeter results of Taylor,

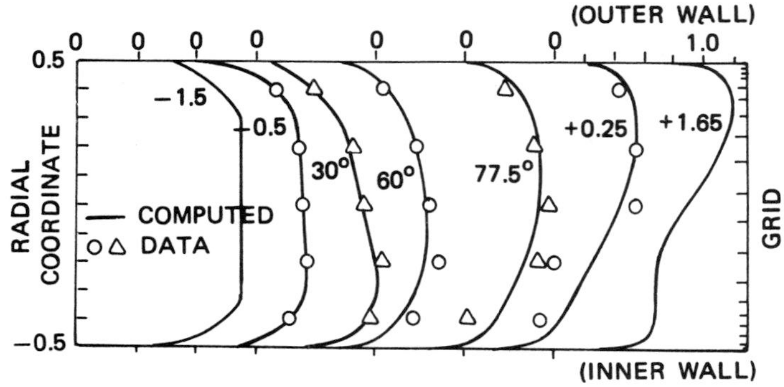

Figure 7. Axial velocity profiles in a three-dimensional turbulent duct with Re = 4×10^4 at various axial locations on the midplane.

Whitelaw, and Yianneskis [1.121] are given in Figs. 7 and 8, after 77.5° of turning. The station chosen for comparison represents the most severe test of the computation available from the measured data. The general level of agreement with the data at this worst station is very encouraging. As mentioned earlier, the peak secondary velocities are large and occur near the edge of the viscous sublayer, thus demonstrating the need for sublayer resolution. The resulting distortion of the streamwise flow by these very large secondary flows is quite remarkable, the peak velocity being shifted from the suction (inner) surface, where it occurs in two-dimensional flow, all the way over to the pressure (outer) surface. This clearly demonstrates the importance of accurately defining the three-dimensional nature of the flow and the limited relevance of two-dimensional channel flow to this problem.

Turning now to the strut-endwall problem, here accurate solutions to the two-dimensional Navier–Stokes equations for laminar flow over a 10:1 ellipse at zero incidence and a chord Reynolds number of 200 have been

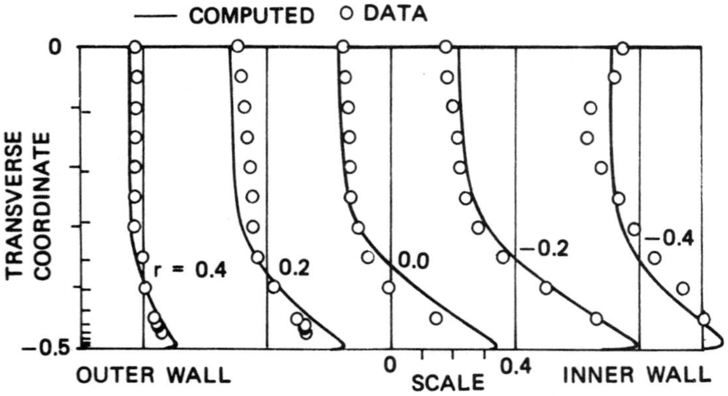

Figure 8. Radial velocity profile after $77\frac{1}{2}°$ of turning in three-dimensional turbulent duct at Re = 4×10^4.

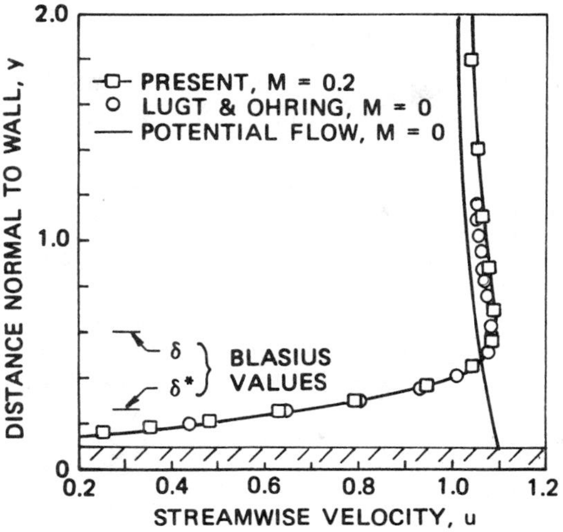

Figure 9. Velocity profiles at outflow boundary for 10 : 1 ellipse, $\alpha = 0°$, Re = 200.

obtained by Lugt and Ohring [1.122]. In two dimensions the present calculations for the half chord of the ellipse were examined both by mesh refinement and comparison with the prior Lugt–Ohring solutions. This latter comparison for the streamwise velocity at the present outflow boundary ($\eta = \pi/2$) is shown in Fig. 9. The agreement is very reassuring in the light of the comparatively coarse mesh used in the present calculation and of the treatment of outflow boundaries. Although not shown here, good agreement was also obtained for surface pressure distributions. Using the present scheme, two-dimensional solutions were computed for the forward half of a 5 : 1 ellipse at zero incidence and a chord Reynolds number of 200. Symmetry about the chord line, $\eta = \pi$, was assumed, and a mesh of 14×14 and 28×28 was used. The results of this study are shown in Fig. 10 and are given

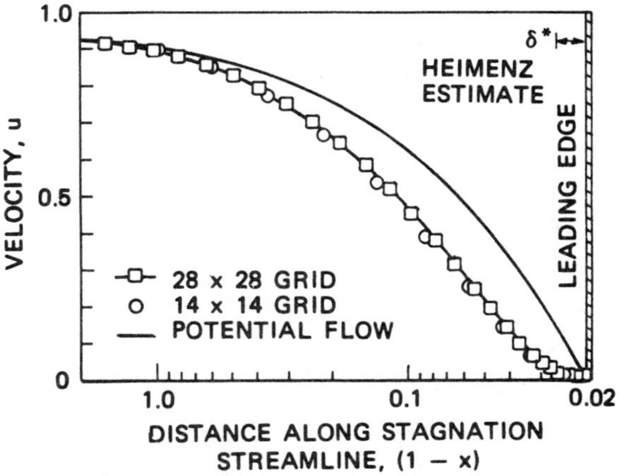

Figure 10. Velocity along stagnation streamline for 5 : 1 ellipse, $\alpha = 0°$, Re = 400.

in the form of a distribution of the velocity along the stagnation streamline, since this is a very sensitive indicator of solution accuracy. The small amount of mesh dependence shown in Fig. 10 indicates that without assuming symmetry about the chord, a 14×28 mesh in planes parallel to the endwall should not introduce major error.

With the favorable two-dimensional results as background, the resolution normal to the wall was examined by performing two three-dimensional calculations at 5° incidence using a mesh of $14 \times 28 \times 10$ and a mesh of $14 \times 28 \times 14$. The chordwise profile of velocity at $x = -1.06$, $y = 0$ near separation proved to be a sensitive indicator of the mesh-induced changes, and this is shown in Fig. 11. Here it can be seen that the less-coarse mesh improves the resolution considerably very near the endwall in the region of the flow reversal, but makes little difference elsewhere. This result is very encouraging, since if substantiated by further study, it implies that small unresolved separation zones do not destroy the accuracy of the flow prediction in resolved regions removed from the local underresolved region. Further, the typical manifestion of underresolution in these calculations seems to be a local smoothing of the solution, in itself an acceptable "failure" mode.

Further, results from the three-dimensional flow predictions at 5° incidence are shown in Fig. 12. These solutions converged in 80 time steps, as before, and shown in Fig. 12 is a vector plot of the velocity in a plane one grid point away from the no-slip endwall surface. Here a saddle-point type of flow is observed on the high-pressure side of the ellipse. Also evident is the overturning of the wall streamlines towards the ellipse after the leading-edge region has been negotiated. Both of these very interesting effects have been observed in various flow visualization studies. Thus, the flow predictions

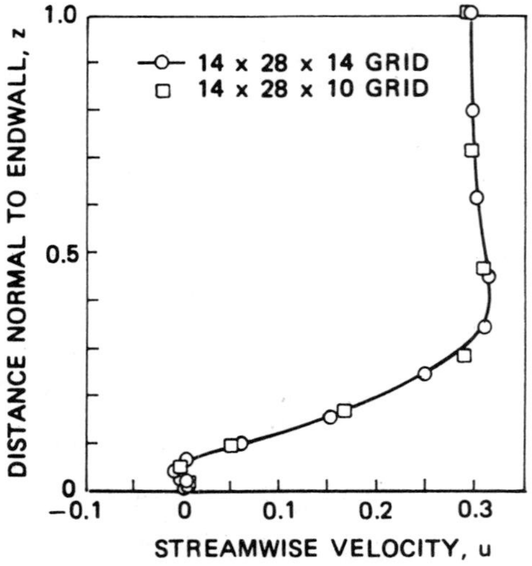

Figure 11. Streamwise velocity profiles near separation for 5 : 1 ellipse, $\alpha = 5°$, Re = 400.

appear qualitatively correct, and in the near future a quantitative comparison with data will be performed on a turbulent-boundary-layer–elliptical-strut interaction.

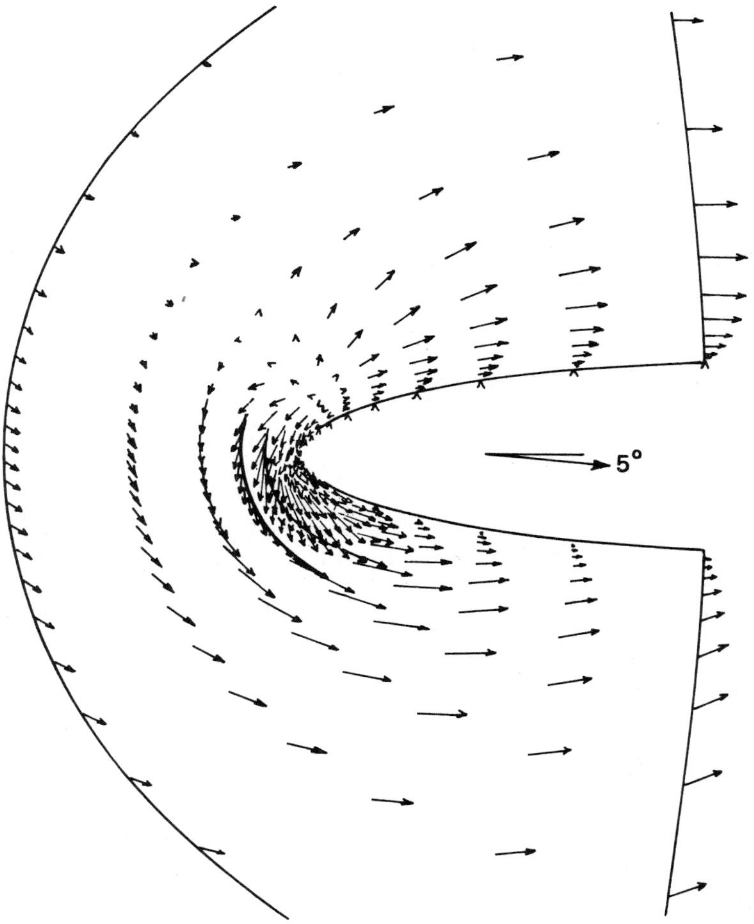

Figure 12. Computed velocity in the plane of grid points adjacent to endwall surface, 5 : 1 ellipse, $\alpha = 5°$, Re = 400, $M = 0.2$.

Concluding Remarks

It was noted that most of the runs in both two and three space dimensions achieved the desired convergence tolerance in approximately 80 time steps. The computed results were obtained on two different CDC 7600 computers, and the computer time required to achieve the requisite number of time steps varied greatly. Two computer codes were employed, and although the computational algorithm was the same in both cases, the code constructed by Buggeln [1.102] was considerably more general and much more flexible than the other, at the expense of a great deal more overhead computation. As a consequence, computer time for the eighty time steps ranged from a low of 20 minutes for 5.5×10^3 grid points on the faster CDC machine using the less general deck to a high of 180 minutes for 9.5×10^3 grid points on the slower CDC machine using the much more general computer code. As a result, comparison of computer times with other Navier–Stokes solvers should be performed with caution, and code generality must be factored in. The eighty time steps to achieve the stated convergence criterion is independent of the code complexity, and this in conjunction with an operation count would seem to be a more reasonable basis for algorithm comparison.

The experience gained here from attempting to obtain accurate three-dimensional flow predictions leads to the observation that while the grid resolution problems are certainly not insignificant, for the class of problems considered one can do surprisingly well with a comparatively coarse mesh. Key requirements for this to be true are that the available mesh be distributed in a nonuniform manner to give at least modest resolution to the scales present and, as a corollary, that only a limited number of scales be present. It also follows that since resolved multiple length scales are present, the numerical algorithm must maintain its efficiency in this environment, and in addition should not be limited in some global manner by criteria associated with the minimum mesh size present in the domain. As a consequence of resolving the scales involved, albeit in perhaps a rather limited manner, the effect of the nonisotropic artificial viscosity, which is independently set for each coordinate direction, is minimized, since the artificial viscosity must act on a gradient to produce an artificial diffusion before the solution is degraded. With the important scales resolved, the artificial viscosity is reduced in the critical regions where, if significant, it could produce a large artificial diffusion by acting on a large gradient. The converse is also true, in that the mesh is opened up where the gradients are small, and hence although the cell Reynolds number and hence artificial viscosity in this particular coordinate direction might become high, the artificial diffusion remains negligible and solution integrity is maintained. These observations go some way to explain the present somewhat surprisingly good results of the mesh-refinement studies.

In distributing the mesh *a priori* to resolve the multiple scales present, much use was made of existing fluid-mechanics information. However, it is clear that an adaptive mesh which would adjust to resolve the developing flow gradients would be of great value. Although still in the early phase of development, the results obtained in simple flows by Dwyer, Kee, and Sanders [1.123] are impressive. More recently, within the context of the present algorithm, Shamroth [1.124] has constructed a time-dependent mesh with a multidimensional adaptive mesh clustering capability. Thus, as much fluid-mechanics information on scales as is known for the given problem should be used where possible, and adaptive meshes should be developed for use in defining scales whose *a priori* location is not known.

References for Part 1

[1.1] H. B. Keller: In: *Applications of Bifurcation Theory*, P. Rabinowitz (ed.), 359 (Academic Press, Inc., New York 1977).

[1.2] R. S. Schreiber, H. B. Keller: *SIAM J. Sci. Stat. Comp.* (1982) in press.

[1.3] R. S. Schreiber, H. B. Keller: *J. Comp. Phys.* (1982) in press.

[1.4] R. B. Kellogg, G. R. Schubin, A. B. Stephens: SIAM J. Num. Anal., **17**, 733 (1980).

[1.5] G. R. Shubin, A. B. Stephens: To be published.

[1.6] H. B. Keller, H. Takami: In: *Numerical Solutions of Nonlinear Differential Equations*, D. Greenspan (ed.) (J. Wiley & Sons, Inc., New York 1966).

[1.7] E. Allgower: In: *Proc. Conf. Roy. Irish Acad.* (Academic Press 1974).

[1.8] M. Lentini, H. B. Keller: SIAM J. Appl. Math. **38**, 52 (1980).

[1.9] H. B. Keller, R. Szeto: In: *Computing Methods in Applied Sciences and Engineering*, R. Glowinski and J. L. Lions (eds.), 51 (North-Holland, New York 1980).

[1.10] R. Meyer-Spasche, H. B. Keller: J. Comp. Phys. **35**, 79 (1980).

[1.11] R. Szeto: Ph.D. Thesis, California Institute of Technology, Pasadena (1977).

[1.12] J. Perry: AIAA J. **4**, 1425 (1966).

[1.13] G. Moretti, M. Abbett: AIAA J. **4**, 2136 (1966).

[1.14] M. J. Abbett: In: *Proc. AIAA Comp. Fl. Dyn. Conf.*, Palm Springs, 153 (1973).

[1.15] T. de Neef: Rept. LR-262 (Dept. of Aerospace Engineering), Delft University of Technology (1978).

[1.16] G. Moretti, M. Pandolfi: AIAA J. **19**, 449 (1981).

[1.17] M. Pandolfi, L. Zannetti: VII Int. Conf. on Numerical Methods in Fluid Dynamics (1980).

[1.18] D. R. Chapman et al.: Astro & Aero **13**, 22 (1975).

[1.19] A. Gessow, D. J. Morris: NASA SP-394, Washington, D.C. (1977).

[1.20] Anon.: NASA CP-2032, Moffett Field, CA (1978).
[1.21] E. C. Gritton et al.: Rept. R-2183-RC, Rand Corp. (1977).
[1.22] J. C. Wu, A. Sugavanam: AIAA J. **16**, 948 (1978).
[1.23] A. Sugavanam: Ph.D. Thesis, Georgia Institute of Technology, Atlanta (1978).
[1.24] J. C. Wu: In: *International Conference on Finite-Element Methods in Engineering Proceedings*, 827 (Clarendon Press, Sydney 1974).
[1.25] J. C. Wu, M. M. El Refaee, S. G. Lekoudis: AIAA Paper No. 81-0046 (1981).
[1.26] J. C. Wu, J. F. Thompson: J. Comp. and Flu. **1**, 197 (1973).
[1.27] J. C. Wu, N. L. Sankar: AIAA Paper No. 80-0011 (1980).
[1.28] J. C. Wu: AIAA J. **14**, 1042 (1976).
[1.29] P. K. Banerjee, R. Butterfield (eds.): *Developments in Boundary Element Methods I* (Applied Science, London 1979).
[1.30] J. C. Wu, Y. M. Rizk: In: *Lecture Notes in Physics* **80**, 558 (Springer-Verlag 1978).
[1.31] Y. M. Rizk: Ph.D. Thesis, Georgia Institute of Technology, Atlanta (1980).
[1.32] J. C. Wu, M. Wahbah: In: *Lecture Notes in Physics* **59**, 448 (Springer-Verlag, New York 1976).
[1.33] M. M. Wahbah: Rept., Georgia Institute of Technology, Atlanta (1978).
[1.34] J. F. Thompson: Ph.D. Thesis, Georgia Institute of Technology, Atlanta (1971).
[1.35] J. C. Wu, S. Sampath: AIAA Paper 76-337 (1976).
[1.36] S. Sampath: Ph.D. Thesis, Georgia Institute of Technology, Atlanta (1977).
[1.37] A. Sugavanam, S. Sampath, J. C. Wu: Rept., Georgia Institute of Technology, Atlanta (1980).
[1.38] J. C. Wu, N. L. Sankar: AIAA Paper No. 78-1225 (1976).
[1.39] N. L. Sankar: Ph.D. Thesis, Georgia Institute of Technology, Atlanta (1977).
[1.40] J. C. Wu, U. Gulcat: AIAA J. **19**, 20 (1981).
[1.41] O. M. Thoman, A. A. Szewcyzk: Phys. Fluids **12**, 76 (1969).
[1.42] S. Goldstein (ed.): *Modern Developments in Fluid Dynamics* **2**, 424 (Dover Publications, New York 1965).
[1.43] E. Achenbach: J. Fluid Dyn. **34**, 625 (1968).
[1.44] J. H. Ferziger: AIAA J. **15**, 1261 (1977).
[1.45] U. Schumann, G. Grotzbach, L. Kleiser: In: *Von Karman Institute Lecture Series 1979–2*, Brussels (1979).
[1.46] J. H. Ferziger, D. C. Leslie: AIAA Paper 79–1441 (1979).
[1.47] J. W. Deardorff: J. Fluid Mech. **41**, 452 (1970).
[1.48] J. Smagorinsky: Mon. Wea. Rev. **91**, 99 (1963).
[1.49] D. K. Lilly: Mon. Wea. Rev. **93**, 11 (1965).
[1.50] A. Leonard: Adv. in Geophysics **18A**, 237 (1974).
[1.51] D. K. Lilly: Proc. IBM Sci. Comp. Symp. on Env. Sci. (1967).
[1.52] W. J. Feiereisen: Dissertation, Stanford University (1981).
[1.53] E. Shirani: Dissertation, Stanford University (1981).
[1.54] J. Bardina, J. H. Ferziger, W. C. Reynolds: AIAA paper 80-1357 (1980).
[1.55] P. Moin, W. C. Reynolds, J. H. Ferziger: Rept. TF-12 (Dept. of Mech. Engrg.), Stanford University (1978).
[1.56] P. Moin, J. Kim: Private communication (1980).
[1.57] R. A. Clark, J. H. Ferziger, W. C. Reynolds: J. Fluid Mech. **91**, 92 (1979).
[1.58] O. J. McMillan, J. H. Ferziger: AIAA J. **17**, 1340 (1979).
[1.59] O. J. McMillan, J. H. Ferziger, R. S. Rogallo: AIAA Paper 80-1339 (1980).
[1.60] U. Schumann: J. Comp. Phys. **18**, 376 (1975).

[1.61] J. W. Deardorff: J. Fluids Engrg. **95**, 429 (1973).
[1.62] J. Kim, P. Moin: AGARD Symp. on Turbulent Boundary Layers, The Hague (1979).
[1.63] D. C. Leslie: *Developments in the Theory of Turbulence* (Oxford University Press 1973).
[1.64] M. D. Love, D. C. Leslie: In: *Proc. Symp. on Turbulent Shear Flows*, Penn. State University (1976).
[1.65] N. N. Mansour, P. Moin, W. C. Reynolds, J. H. Ferziger: In: *Proc. First Int. Symp. on Turbulent Shear Flows*, Penn. State University (1977).
[1.66] D. C. Leslie, G. L. Quarini: J. Fluid Mech. **91**, 65 (1979).
[1.67] G. Grotzbach: Dissertation, University of Karlsruhe (1977).
[1.68] D. C. Leslie: Private communication (1980).
[1.69] G. Sommeria: J. Atmos. Sci. **33**, 216 (1976).
[1.70] C. E. Schemm, F. B. Lipps: J. Atmos. Sci. **33**, 102 (1976).
[1.71] A. Findikakis: Dissertation (Dept. of Civil Engrg.), Stanford University (1980).
[1.72] S. J. Davies, C. M. White: Proc. Roy. Soc. A **119**, 92 (1928).
[1.73] T. W. Kao, C. Park: J. Fluid Mech. **43**, 145 (1970).
[1.74] V. Patel, M. R. Head: J. Fluid Mech. **38**, 181 (1969).
[1.75] M. Nishioka, S. Iida, Y. Ichikawa: J. Fluid Mech. **72**, 731 (1975).
[1.76] H. Reichardt: Mitt. Max-Planck-Inst. Stromforsch. **22**, Göttingen (1959).
[1.77] M. Meksyn, J. T. Stuart: Proc. Roy. Soc. A **208**, 517 (1951).
[1.78] J. T. Stuart: J. Fluid Mech. **9**, 353 (1960).
[1.79] T. Herbert: AIAA J. **18**, 243 (1980).
[1.80] J. P. Zahn, J. Toomre, E. A. Spiegel, D. O. Gough: J. Fluid Mech. **64**, 319 (1974).
[1.81] T. Herbert: In: *Proc. 5th Int. Conf. on Numerical Methods in Fluid Dynamics*, A. I. van de Vooren, P. J. Zandbergen (eds.), 235 (Springer 1976).
[1.82] T. Herbert: In: *Laminar-Turbulent Transition*, AGARD Conf. Proc. **224**, 3-1 (1977).
[1.83] S. A. Orszag, L. C. Kells: J. Fluid Mech. **96**, 159 (1980).
[1.84] A. T. Patera, S. A. Orszag: In: *Proc. 7th Int. Conf. on Numerical Methods in Fluid Dynamics*, R. W. MacCormack, W. C. Reynolds (eds.) (Springer 1980).
[1.85] A. T. Patera, S. A. Orszag: *J. Fluid Mech.* **112**, 467 (1981).
[1.86] D. J. Benney, C. C. Lin: Phys. Fluids **3**, 656 (1960).
[1.87] A. D. D. Craik: J. Fluid Mech. **99**, 247 (1980).
[1.88] L. H. Gustavsson, L. S. Hultgren: J. Fluid Mech. **98**, 149 (1980).
[1.89] T. Herbert, M. V. Morkovin: In: *Laminar-Turbulent Transition*, IUTAM Conf. Proc., R. Eppler, H. Fasel (eds.), 47 (Springer 1980).
[1.90] S. A. Orszag, A. T. Patera: Phys. Rev. Lett. **45**, 989 (1980).
[1.91] D. Gottlieb, S. A. Orszag: *1974 Numerical Analysis of Spectral Methods: Theory and Applications*, NSF-CBMS Monograph 26 (Soc. Ind. App. Math., Philadelphia 1974).
[1.92] A. Davey, H. P. F. Nguyen: J. Fluid Mech. **45**, 701 (1971).
[1.93] R. M. Clever, F. H. Busse: J. Fluid Mech. **65**, 625 (1974).
[1.94] H. K. Moffatt: *Magnetic Field Generation in Electrically Conducting Fluids*, 121 (Cambridge University Press, Cambridge 1978).
[1.95] P. J. Roache: AIAA Paper 77-647 (1977).
[1.96] S. G. Rubin, P. K. Khosla, S. Saari: Comp. and Flu. **5**, 151 (1977).
[1.97] P. J. Roache: Semidirect Methods in Fluid Dynamics (to appear).
[1.98] P. J. Roache: Num. Heat Trans. **1**, 1, 163, 183 (1978).

[1.99] P. J. Roache: In: *Proc. LASL Conf. on Elliptic Solvers*, M. Schultz (ed.), 399 (Academic Press 1981). Also, *Num. Heat Trans.* **4**, 395 (1981).
[1.100] M. Israeli: Studies in Appl. Math. **39**, 327 (1970).
[1.101] P. J. Roache: J. Comp. Phys. **27**, 204 (1978).
[1.102] R. C. Buggeln, W. R. Briley, H. McDonald: Final Rept. Contract 00014-79-C-0713 (1980).
[1.103] W. R. Briley, H. McDonald: In: *7th Int. Conf. on Numerical Methods in Fluid Dynamics*, Stanford University and NASA Ames (1980).
[1.104] J. P. Johnston: J. Basic Engrg. **82**, 233 (1960).
[1.105] G. G. Sovran (ed.): *Proc. of Symp. on Fluid Mechanics of Internal Flows*, General Motors Research Labs, Warren, Mich. (Elsevier, Amsterdam 1967).
[1.106] K. Stewartson: *Advances in Applied Mechanics* **14**, 145 (Academic Press 1974).
[1.107] R. Courant, D. Hilbert: *Partial Differential Equations II* (Interscience Publishers 1962).
[1.108] S. G. Rubin, T. C. Lin: Rept. 71-8, PIBAL (1971).
[1.109] W. R. Briley, H. McDonald: In: *Proc. 4th Int. Conf. on Numerical Methods in Fluid Dynamics* (Springer-Verlag, New York/Berlin 1975); also Rept. M911363-6, United Aircraft Research Labs (1973).
[1.110] R. von Mises: *Mathematical Theory of Compressible Flow* (Academic Press 1958).
[1.111] D. H. Rudy, J. C. Strikwerda: J. Comp. Phys. **36**, 55 (1980).
[1.112] Hughes, Gaylord: *Basic Equations of Engineering Science* (Schaum Publishing Co. 1964).
[1.113] B. E. Launder, D. B. Spalding: *Comp. Methods Appl. Mech. and Engrg.* **3**, 269 (1974).
[1.114] S. J. Shamroth, H. J. Gibeling: NASA CR3183 (1979).
[1.115] G. O. Roberts: In: *Proc. 2nd Int. Conf. Numerical Methods in Fluid Dynamics*, 171 (Springer-Verlag 1971).
[1.116] W. R. Briley, H. McDonald: J. Comp. Phys. **24**, 372 (1977).
[1.117] W. R. Briley, H. McDonald: J. Comp. Phys. **34**, 54 (1980).
[1.118] J. Douglas, J. E. Gunn: Numerische Math. **6**, 428 (1964).
[1.119] N. N. Yanenko: *The Method of Fractional Steps* (Springer-Verlag, New York 1971).
[1.120] R. F. Warming, R. M. Beam: SIAM-AMS Proc. **11**, 85 (1978).
[1.121] A. M. K. P. Taylor, J. H. Whitelaw, M. Yianneskis: Rept. NASW-3258 (1980).
[1.122] H. J. Lugt, S. Ohring: J. Fluid Mech. **79**, 127 (1977).
[1.123] H. A. Dwyer, R. J. Kee, B. R. Sanders: AIAA Paper 79-1464 (1979).
[1.124] S. J. Shamroth: Private communication (1980).

PART 2

INTERACTIVE STEADY BOUNDARY LAYERS

Introduction

Tuncer Cebeci

This part begins with a chapter by K. Stewartson, who reviews some of the recent advances in the asymptotic theory of high-Reynolds-number flows. There are two broad classes of such flows: attached, in which the boundary layer does not separate, and detached, in which there is an extensive eddy of slowly moving fluid behind the body. The first class is well understood, but there are still some controversies associated with the second class. The chapter discusses the progress made towards understanding the transition between these classes; it is conjectured that it is possible for such a transition to occur unless there is circulation, in which case there is still a definite gap between them. The relevance of these studies to practical calculations for both laminar and turbulent flows is discussed and the need is stressed for for finding an improved model for describing the behavior of the inner boundary layer near separation and under a strong adverse pressure gradient.

The contents of the chapter by Burggraf and Duck may be summarized as follows: The computation of interactive boundary-layer flows according to triple-deck theory requires the determination of the pressure by a Hilbert integral if the flow is subsonic. In their paper the authors propose a pseudo-spectral method in which the nonlinear terms in the equations are regarded as perturbations on the linear terms, so that problems may be solved by a sequence of Fourier transforms. The fast-Fourier-transform method may now be used with advantage, and the authors find that convergence is achieved more rapidly than using conventional methods even when flow reversal occurs. A parallel procedure may be used in supersonic flows.

The triple-deck theory is an asymptotic theory for laminar flows and is not directly applicable to many flows of practical interest. In his paper, Inger suggests that nevertheless the crucial concepts of this theory may be used to consider turbulent-boundary-layer–shock-wave interactions. He generalizes the original approach of Lighthill [2.1] by accounting for the wall turbulence in the inner deck, which is otherwise assumed to be of laminar form. The model with the main deck is made along the line, whereas the velocity normal to the surface is zero according to the inviscid theory of that deck. The conclusions of this theory lead to some favorable comparisons with experiments on the interaction of shock waves with transonic boundary layers.

An approximate way of solving the Navier–Stokes equations at high Reynolds number is to parabolize them by neglecting the streamwise diffusion terms. This approach has the advantage over classical boundary-layer theory that the pressure gradient is not completely fixed at the outset of the calculations, so that a self-adjusting mechanism operates which prevents the occurrence of a singularity at separation. However, the pressure gradient is prescribed at the outer edge of the boundary layer, and in this regard the scheme differs conceptually from triple-deck theory. Chapter 11, by Rubin, describes a stable marching procedure and its application to two flows with separation and reattachment. For problems in which the subsonic layers are not thin, a global relaxation of the pressure interaction is required, and effective ways of carrying this out have been found. An application has also been made to the calculation of boundary layers on sharp corners at large incidences to an oncoming supersonic stream.

Recently Veldman [2.2] has proposed a new method of integrating the boundary-layer equations when the pressure gradient is partly controlled by a Hilbert integral depending on the displacement thickness. By also making use of earlier ideas, Davis and Werle have developed this method further and applied it to the study of interacting laminar and turbulent boundary layers. A feature of the method is that the coordinate system is not based on the body, but one of the axes is placed so that it is aligned as nearly as possible to the shear flow. The authors find it to be effective in a number of test cases, such as the flat-plate trailing edge and the Carter–Wornom trough [2.3]. Some new applications are made to blunt trailing edges, and it is possible to account for limited regions of separated flow.

Another application of high-Reynolds-number theories to flows of practical interest, when the boundary layer is turbulent, is described by Melnik and his associates, who have developed a successful theory for cusped trailing edges [2.4–2.7]; the present contribution applies these ideas to wedges. A weak-interaction theory in which the external velocity is prescribed must fail at the trailing edge, since it is a stagnation point. Two procedures are investigated to overcome this difficulty. The first is an interactive boundary-layer theory in which the displacement thickness of the boundary layer is allowed to influence the external flow. The second method explicitly takes account of the pressure variation across the boundary layer induced by the variation of flow properties near the trailing edge. Both methods are consistent, but the second has more physical content. The pressure-drag contribution from the trailing edge computed by the second method is twice that of the first method.

CHAPTER 8

Some Recent Studies in Triple-Deck Theory

K. Stewartson*

1 Introduction

The elucidation of the phenomenon of the breakaway of fluid moving past a rigid surface, when the characteristic Reynolds number R is large, has been a continual challenge to theoreticians for many years. The usual approach, and one which I favor, is to regard it as controlled by an interaction between an external inviscid and a viscous boundary layer. For some time the discovery [2.8] that a laminar boundary layer invariably appears to become singular at separation when the external velocity is prescribed proved a severe obstacle to progress; but recently the development of the triple-deck theory of mutually interacting inviscid flows and boundary layers has enabled solutions to be computed which pass smoothly through this point. I shall not review this theory here and assume instead that the reader is broadly familiar with the concept. For further details reference may be made to a number of comprehensive reviews [1.106, 2.9, 2.10, 2.11, 2.12]. It might have been thought that once the separation hurdle had been cleared, then the way would be open to a rapid and successful attack on the fundamental problems associated with flows at $R \gg 1$, but while considerable progress has been made, some puzzling questions remain.

For example, I have been very impressed by Sychev's assertion [2.13] that "the scheme of flow, assuming that the separation of the stream on the smooth surface occurs as a result of an adverse pressure gradient distributed over a finite segment of the body . . . apparently does not have any place in actuality." Of course we know that in laminar flow this is not correct at the finite values

* Department of Mathematics, University College London.

of R at which the experiments or numerical calculations were carried out, but this is not what Sychev has in mind. He is concerned with the role of boundary-layer theory, which is strictly valid only in the limit $R \to \infty$, and he claims that in this limit the segment of the body subject to an adverse pressure gradient must be zero. When the available studies are considered with this limit process in mind, they tend to support Sychev's view. Now we can also construct inviscid flows past bodies which include regions of adverse pressure gradient, but the corresponding boundary layer only separates if some parameter λ, describing the geometrical properties of the flow, exceeds a critical value λ_c. One example is provided by the flow near the trailing edge of a finite plate with a suitably rounded leading edge, and another by the flow near the leading edge of a thin aerofoil or equivalently a parabolic body. The angle α^* of incidence of the oncoming stream to the airfoil plays the role of λ in these examples. We shall discuss these properties of the flows below and consider the implications for the limit solution as $R \to \infty$.

In well-separated flows such as occur over bluff bodies, the most promising candidate for the limit solution is the Kirchhoff free-streamline flow with separation occurring at the pressure minimum on the body. In this case, as Sychev conjectures, the adverse pressure gradient ahead of separation is confined to a vanishingly short segment, and the fluid velocity on the free streamline is the same as it is at infinite distances upstream. The theoretical description of this candidate is incomplete, because we do not yet fully understand the asymptotic structure of the closed eddy that must form behind the body. Nor do we understand the nature of the change from fully attached flow to fully separated flow as the relevant geometrical parameter is varied. It seems that the limit solutions change from the one to the other discontinuously at a critical value of this parameter.

It is natural to wonder how this asymptotic theory of two-dimensional flows may be generalized to include three-dimensional and unsteady effects. There are tremendous numerical problems involved because of the need to solve the three-dimensional boundary-layer equations coupled with a relation connecting the pressure and displacement thickness which may be written most simply as a double integral. A number of exploratory studies have been carrried out on the interaction problem which have revealed interesting new features of the flow field, and in addition an efficient numerical scheme for integrating the basic boundary-layer equations has recently been devised. Important new progress in this area may well be made in the near future. Similar remarks apply to unsteady flows.

Much of the work described so far is of a fundamental nature and may be thought of as aiming to set up mathematical conjectures, i.e. to make a catalog of flow properties which are self-consistent and therefore have a reasonable chance of being proved to be correct by rigorous argument. Such aims are in my view quite defensible and even laudable, but they are not enough if the theoretician wishes to make a contribution to the fluid mechanics of the real world. To be sure, some of the conclusions drawn from the theory have already been shown to be in quantitative accord with experiment even at moderate values of R, but to be relevant to the majority of practical problems, the boundary layer must be assumed turbulent. Even using simple models of turbulence, the interactive characteristics of the flows are quite different from the laminar case, but nevertheless the attitudes of mind which permeate the laminar theory have proved very helpful here too.

2 Attached Trailing-Edge Flows

Consider the motion of a viscous fluid past a finite flat plate of zero thickness and length l, the characteristic Reynolds number R of the flow being large. If the angle of attack, α^*, of the plate is zero, a uniformly valid first approximation to the asymptotic expansion of the solution in descending powers of R may, to all intents and purposes, be written down [1.106, 2.12]. Its structure over the majority of the flow field is formally simple, but it is very complicated near the trailing edge, a dominant role being played by a triple deck. For explicit calculations it is fortunately only necessary to integrate the boundary-layer equations for an incompressible fluid subject to slightly unusual boundary conditions and including an interactive condition on the pressure. In the usual notation, after appropriate choice of scalings, we have to solve

$$u\frac{\partial u}{\partial x} + v\frac{\partial u}{\partial y} = -\frac{dp}{dx} + \frac{\partial^2 u}{\partial y^2}, \qquad \frac{\partial u}{\partial x} + \frac{\partial v}{\partial y} = 0,$$

with

$$u \to y \qquad \text{as } x \to -\infty, \tag{2.1}$$

$$u - y \to A(x) \quad \text{as } y \to \infty$$

$$u = v = 0 \quad \text{if } x < 0,\ y = 0, \tag{2.2a}$$

and

$$v = \frac{\partial u}{\partial y} = 0 \quad \text{if } x > 0,\ y = 0. \tag{2.2b}$$

In addition,

$$p = \begin{cases} -A'(x) & \text{if } M_\infty > 1, \\ \dfrac{1}{\pi}\displaystyle\int_{-\infty}^{\infty} \frac{A'(x_1)\,dx_1}{x - x_1} & \text{if } M_\infty < 1, \end{cases} \tag{2.3}$$

where M_∞ is the Mach number of the undisturbed stream. The triple-deck region extends a distance $O(R^{-3/8}l)$ from the trailing edge, and the induced pressure coefficient is $O(R^{-1/4})$. Thus the pressure gradient is large but localized.

One immediate consequence of this precise theory is to provide further encouragement in interactive problems to the classical technique of using standard boundary-layer methods together with an appropriate equation connecting the pressure and displacement thickness, the latter being equivalent to A. The main reservation about this procedure is that when the interaction is significant, special attention should be paid to the bottom of the boundary layer, where rapid changes tend to occur.

Explicit numerical solutions of (2.1) subject to (2.2) and (2.3) have been found both when $M_\infty < 1$ [2.14, 2.15, 2.4] and when $M_\infty > 1$ [2.16, 2.17]. Comparisons with experimental studies and the solution of the full Navier–Stokes equations [2.14] are very favorable as far as drag is concerned. However at the lowest values of R (~ 10) this is certainly fortuitous, since the triple-deck region extends beyond the plate.

If the plate is at incidence, there is an immediate danger of separation at the leading edge (see Section 3 below), which can be avoided if it is suitably rounded. For example, if the radius of curvature of the leading edge is small but larger than any negative power of R, a meaningful discussion can be carried out of the flow near its trailing edge provided that separation does not occur there when $\alpha^* = 0$. For subsonic flow and a thermally insulated plate with a sharp trailing edge, the critical angle α^* of attack is of order $R^{-1/16}$, and then the flow, near but upstream of the trailing edge, can be considered to be dominated by two triple decks, one above and one below the plate. These join downstream of the leading edge to form a smooth wake. The null condition on p as $x \to -\infty$, implied by (2.2), is replaced by

$$p_\pm(x) \pm \alpha(-x)^{1/2} \to 0. \tag{2.4}$$

according as $y > 0$ or $y < 0$, where

$$\alpha^* = R^{-1/16}\lambda_B^{9/8}(1 - M_\infty^2)^{7/16}\alpha, \qquad \lambda_B = 0.332 \tag{2.5}$$

The appropriate equations were integrated by Chow and Melnik [2.5], who found that the skin friction on the upper side of the plate develops a minimum which deepens as α increases and simultaneously moves towards the trailing edge. Representative graphs of the skin friction are displayed in Fig. 1. On

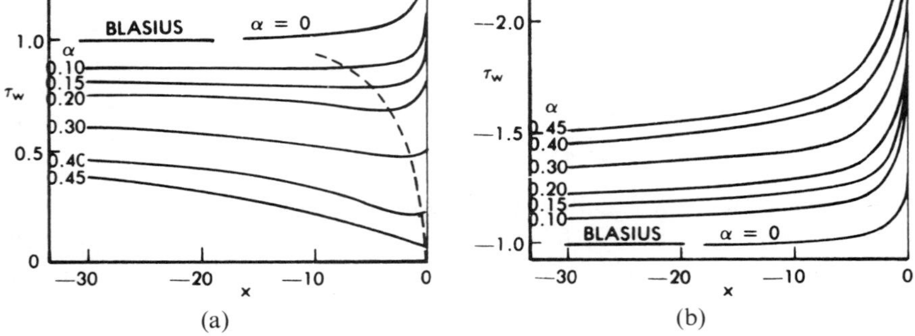

Figure 1. Skin-friction distribution in triple-deck region of a sharp trailing edge [2.5]: (a) left, (b) right.

the upper side of the plate separation occurs when $\alpha \approx 0.47$, precisely at the trailing edge, and for larger values of α no solutions could be found. Thus it appears that the concept of attached flow is valid for a thin airfoil at zero incidence and remains so as α increases from zero provided $\alpha < 0.47$, but not thereafter. The nature of the limit solution for $\alpha > 0.47$ is likely to be quite different, being no longer attached, and discontinuous with the limit solution for $\alpha < 0.47$. One possibility that might be envisaged is that separation then occurs near the trailing edge but outside its triple-deck region. If so, it must still occur through a triple-deck phenomenon, the governing equations being the same as (2.1), but now

$$\begin{aligned} p + \alpha(-x)^{1/2} &\to 0 \qquad \text{as } x \to -\infty, \\ p &\to \text{const} \quad \text{as } x \to +\infty. \end{aligned} \tag{2.6}$$

The numerical problem posed by (2.1), (2.6), and the no-slip condition at $y = 0$ has been solved by Smith [2.18], who finds that $\alpha \approx 0.44$.

Thus we may have a hysteresis phenomenon in which the evolution of the limit solution as α increases follows a different path from that as α decreases in the range $0.44 < \alpha < 0.47$. As α decreases from large values, the limit solution could be a Kirchhoff free-streamline flow, and conceivably the last member of this family would separate near but outside the triple deck of the trailing edge at $\alpha \approx 0.44$. On the other hand as α increases from zero the limit solutions are attached until $\alpha \approx 0.47$. In each case there is a discontinuity in the limit solution at the critical value of α with an abrupt transition to the other family.

A similar situation occurs in supersonic problems [2.16, 2.17]. Here α^* is smaller, and we define α by

$$\alpha^* = (M_\infty^2 - 1)^{1/4} \lambda_B^{1/2} R^{-1/4} \alpha. \tag{2.7}$$

The conditions on p corresponding to (2.4) are that

$$\begin{aligned} p_\pm \pm \alpha &\to 0 \quad \text{as } x \to -\infty, \\ p &\to 0 \quad \text{as } x \to \infty, \end{aligned} \tag{2.8}$$

according as $y > 0$ or $y < 0$. Again solutions come to an end as α increases, when $\alpha = 2.05$, by the occurrence of separation on the upper side of the plate at the trailing edge. It is known that a self-induced separation can only occur near the trailing edge, but outside the trailing-edge triple deck, if $\alpha \approx 1.80$ [1.106], this being the plateau pressure in a supersonic free interaction. Again therefore a hysteresis phenomenon may well occur in the evolution of the limit solution as α varies in the range $1.8 < \alpha < 2.05$. Also the limit solution apparently has to change discontinuously as α increases through 2.05.

The occurrence of separation is not necessarily fatal to the notion of attached flow. So long as it occurs simultaneously on both sides, the triple deck can tolerate it. Thus for an airfoil with a wedgelike trailing edge of vertex angle $2\beta^*$ with $\alpha^* = 0$ and $M_\infty < 1$, the relevant equations are still (2.1), the chief difference being that we must add a term $\beta \log |x|$ to the right-hand side of (2.3), where

$$\beta^* = |M_\infty^2 - 1|^{1/4} \lambda_B^{1/2} R^{-1/4} \beta. \tag{2.9}$$

Solutions of these equations have been found with separation when $\beta > 2.3$ [2.19, 2.20]. The behavior of these solutions as β increases indefinitely has not yet been clarified.

3 Leading-Edge Separation

It is well known that separation can occur near the leading edge of thin airfoils at rather small angles α^* of attack, and that further downstream a very complicated flow field can develop which includes short and long bubbles, transition, and bursting. Extensive accounts of the phenomena have been given by Tani [2.21] and Chang [2.22]. The flow properties near the critical angle of attack which just provokes separation may be studied within the purview of triple-deck theory. Suppose, to fix matters, the fluid is incompressible and the airfoil is sharp-tailed and thin with a thickness ratio $t \ll R^{-1/16}$ to prevent separation at the trailing edge. Alternatively we may suppose it to be a thin parabola ($t \ll 1$) and to extend to infinity downstream. Then at sufficiently small angles of attack, the flow is attached everywhere and the uniformly valid first approximation to the flow as $R \to \infty$ can be found by standard methods. However, at a finite value of α^*/t the boundary layer separates near the leading edge, and this terminates the hierarchical approach. The underlying reason why the boundary layer is liable to separate is that there is an adverse pressure gradient just downstream of the velocity maximum. It is known that the boundary layer develops a catastrophic singularity when the external velocity is prescribed [2.8], and that it cannot be removed by a local triple-deck theory [2.23]. On the present instance, when separation first occurs as α^* is increased, the strength of the singularity is weak and we may expect that an interaction theory can smooth it out. This possibility is under study at the moment [2.24], and the principal results are as follows.

Suppose marginal separation occurs at a distance x_s^* from stagnation when $\alpha^* = \alpha_s^*$. Then if

$$\begin{aligned} \alpha^* - \alpha_s^* &= \beta R^{-2/5}, \quad x^* - x_s^* = X R^{-1/5}, \\ c_f R^{1/2} &= A(x) R^{-1/5} \end{aligned} \tag{3.1}$$

where c_f is the local skin-friction coefficient, the triple-deck theory may be applied provided an A can be found to satisfy the integral equation

$$A^2 - c_1(X^2 - \beta) = c_2 \int_x^\infty \frac{A''(X_1)\, dx_1}{\sqrt{X_1 - X}} \tag{3.2}$$

where c_1, c_2 are positive constants. The boundary conditions on A are that

$$A \to +c_1^{1/2}|x| \quad \text{as } |x| \to \infty,$$

so that the separated region is small and the solutions with separation are smooth continuations of the unseparated solutions as β increases from large negative values. The numerical studies are not complete, but taking $c_1 = c_2 = 1$, it appears that $A > 0$ so long as $\beta < 2.4$ and that no solution of (3.2) can be found if $\beta > 2.75$.

The most likely conclusion is that as with trailing-edge flows the notion of attached flow as a limit of the solution of the Navier–Stokes equation is valid only in sharply defined ranges of the geometrical parameters. Once such a range is exceeded the limit solution must change abruptly to a new form, which we shall discuss further in the next section. The most likely candidate now is the Kirchhoff free-streamline solution, in which separation occurs at the pressure minimum. The extent of the discontinuity can be gauged by taking the example of the thin parabola at incidence. We define a typical point on it in Cartesian coordinates by $x = \frac{1}{2}\xi^2 t^2 l$, $y = \xi t^2 l$, where l is a length scale and t is small, and then the slip velocity can be written as

$$\frac{\xi + \xi_0}{\sqrt{1 + \xi^2}},$$

ξ_0 being related to the angle of attack. Marginal separation occurs when $\xi_0 \approx 1.16$ [2.25, 2.26], and then $\xi_s \approx 3.4$. If a Kirchhoff free streamline is a valid limit, separation must occur at the pressure minimum, namely $\xi = \xi_0^{-1} \approx 0.86$, and quite different from ξ_s.

From a practical point of view a more satisfactory procedure of treating leading-edge separation problems is to use the displacement thickness computed from the conventional boundary-layer formulation to modify the external velocity, and this procedure is well adapted for extension to turbulent flows. The triple-deck theory justifies this approach in the limit $R \to \infty$, but the numerical work is more complicated, since formally the whole boundary layer must be computed at the same time, unless further approximations are made, and there is a very small numerical factor in the contribution from the boundary layer to the external stream. These aspects of the interactive theory have recently been studied [2.27], and results obtained in accord with the triple-deck theory. In particular we demonstrate that the attached solution comes to an end shortly after reversed flow occurs in the boundary layer. In Fig. 2 we display a set of graphs of the reduced skin friction at different values of ξ_0 for an airfoil in which $t = 0.1$ and $R = 10^6$, and in Fig. 3 we demonstrate the failure of the iteration sequence at a slightly larger value of ξ_0. It is interesting to note that observations provide support to the notion of a critical value of ξ_0 beyond which attached flow is impossible [2.28], but it should be remembered that transition is likely to occur in flows which have separated and this may well lead to reattachment.

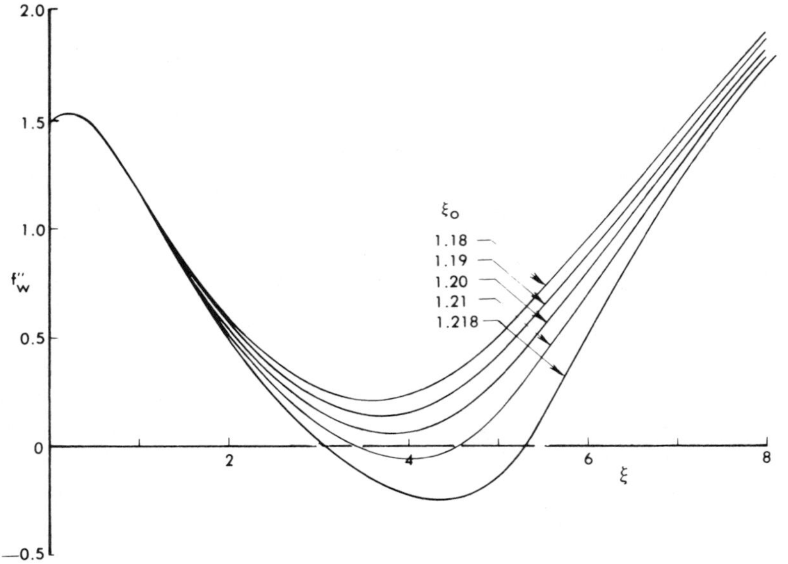

Figure 2. Variation of laminar wall shear parameter f''_w with ξ for various reduced angles of attack ξ_0.

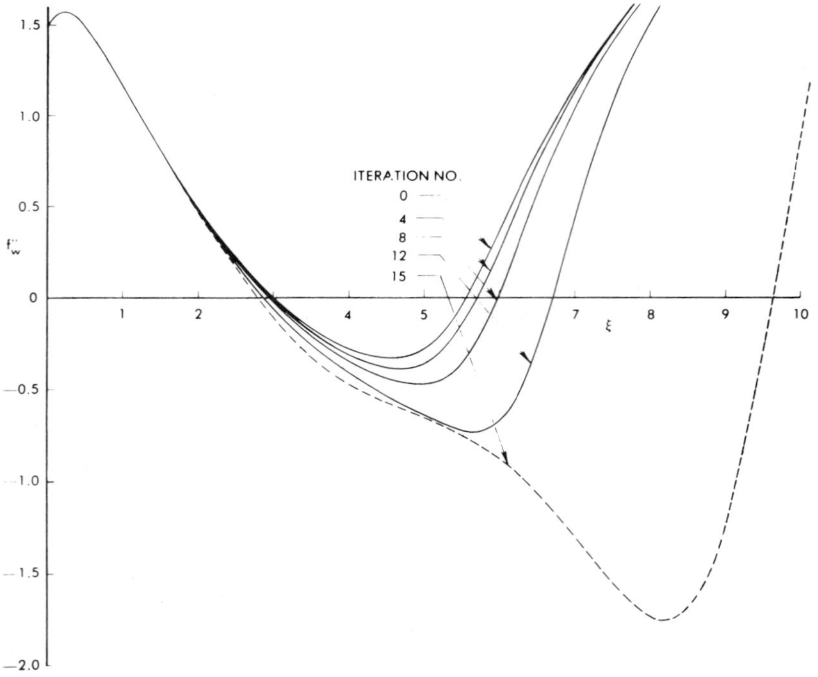

Figure 3. Variation of laminar wall shear parameter in various iterations for $\xi_0 = 1.220$. In the dashed curve, the interaction region extended downstream beyond the range of the integration.

4 Detached Flows

Once the "bootstrap" approach of determining the limit solution as $R \to \infty$, by the smooth variation of the geometrical properties of the body from the finite flat plate, has come to an end, the correct alternative form is not immediately obvious. As has already been pointed out, the most likely candidate is Kirchhoff flow, in which the oncoming fluid detaches from the surface of the body at a pressure minimum, setting up a free-streamline flow, behind which the fluid in the wake is at rest. This flow is quite different from attached flow and is not continuous with it in any sense as the geometrical parameters of the body vary. Furthermore, although the theory of the flow is self-consistent as far as it goes [2.29], it is not complete, the recirculating flow in the wake bubble at high Reynolds number proving to be too recondite at present to be elucidated or for a flaw in the logic of its assumed properties to be revealed. An account of the theory of Kirchhoff flows regarded as limits of viscous flows has recently been given by Stewartson [2.12], and it will not be repeated in detail here. Instead, some remarks are made on a very recent and extensive numerical integration of the Navier–Stokes equations for flow past a circular cylinder at values of R up to 150 (based on radius) by Fornberg [2.30], which in the present context has some puzzling features.

The principal results are that while the computed drag on the cylinder agrees very well with the asymptotic formulas computed by Smith [2.18, 2.31] on the assumption that Kirchhoff theory is correct, the length L of the wake bubble diminishes for $R > 130$, whereas the limit theory requires $L \sim 0.34R$ as $R \to \infty$. Since little notice, and that misleading, is taken of the asymptotic theory in [2.30], and yet the results raise many questions in the reader's mind, it is perhaps appropriate to make some comments here. First, the flow properties on the cylinder, i.e. pressure and skin friction, are well in line with the Kirchhoff theory. If the pressure minimum is p_M occurring at x_M, then $p_M < p_\infty$ and is increasing with R, while x_M is decreasing, with $x_M > 1.2$. Similarly x_s, the separation point, is a decreasing function of R_s with $x_s > 1.6$. According to Kirchhoff theory $x_M = x_s = 0.96$, $p_M = p_\infty$. In addition p_B, the pressure at the rear stagnation point, is less than p_∞ and increasing with R as required by the theory. Even at $R = 150$, $2(p_\infty - p_B)/pU_\infty^2 = +0.09$ as against the asymptotic value of 0.112. Thus if we compared only flow properties at the body, the partisans of the Kirchhoff flow as the limit solution would be well satisfied with the agreement, in trends and even quantitatively. Moreover this part of the flow field is the easiest to compute accurately.

When we compare the far field, however, the situation is quite different. The incompleteness of the theory cannot be gainsaid, but it does agree well with numerical data and experiment at lower values of R, as Fig. 4 shows. At $R = 100$ two estimates of L are given: one using a "coarse" mesh gives $L \approx 33$, one using a "fine" mesh gives $L \approx 28$, and these straddle the straight line in Fig. 4. Moreover if the drag on the body is finite at large R, a long bubble is necessary in order that the momentum defects of the boundary layers from the two sides of the cylinder increase from value $O(R^{-1/2})$ at

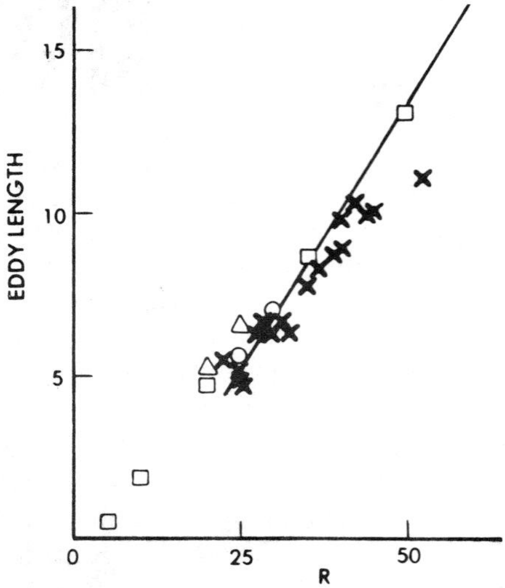

Figure 4. Eddy length as a function of Reynolds number. The solid line is the asymptotic result apart from an origin shift. The crosses are experimental and the other symbols are numerical values.

separation to be $O(1)$ when they finally unite. For, once they unite behind the wake bubble, this defect remains constant and may be directly related to the drag. Indeed, this line of thought leads directly to the view that $L \sim R$. Only if the limit of the drag as $R \to \infty$ is zero can a finite bubble be contemplated at the present time, and the close agreement between the asymptotic and the numerical solutions on this specific feature would seem to preclude this possibility except in bizarre circumstances. It is clear that while more study would be useful on the asymptotic aspects of the solution, a detailed examination of the numerical results does raise doubts about their reliability at the highest values of R.

5 Three-Dimensional Flows

Some useful studies on the extension of the ideas of the triple deck to three dimensions have been made. If z^* denotes the third direction (in the body surface and perpendicular to the external velocity) and w^* the corresponding component in the lower deck, the length scale in the z^*-direction is $O(R^{-3/8})$, as in the x^*-direction, and the governing equations are

$$\frac{\partial u}{\partial x} + \frac{\partial v}{\partial y} + \frac{\partial w}{\partial z} = 0,$$

$$u\frac{\partial u}{\partial x} + v\frac{\partial u}{\partial y} + w\frac{\partial u}{\partial z} = -\frac{\partial p}{\partial x} + \frac{\partial^2 u}{\partial y^2}, \tag{5.1}$$

$$u\frac{\partial w}{\partial x} + v\frac{\partial w}{\partial y} + w\frac{\partial w}{\partial z} = -\frac{\partial p}{\partial z} + \frac{\partial^2 w}{\partial y^2},$$

after appropriate scaling.

The boundary conditions are

$$u - y \to A(x, z), \quad yw \to D(x, z) \quad \text{as } y \to \infty$$

where $\partial D/\partial x = -\partial p/\partial z$ [2.31]. If the fluid is incompressible, we obtain from the governing equations of the upper deck the relation

$$p = -\frac{1}{2\pi}\iint_{-\infty}^{\infty} \frac{(\partial^2 A/\partial \xi^2)\, d\xi\, d\eta}{\sqrt{(x-\xi)^2 + (y-\eta)^2}} \tag{5.2}$$

connecting A and p. If the flow is supersonic, the appropriate relation is

$$p = -\frac{1}{\pi}\iint \frac{(\partial^2 A/\partial \xi^2)\, d\xi\, d\eta}{\sqrt{(x-\xi)^2 - (y-\eta)^2}}, \tag{5.3}$$

the domain of integration being confined to a wedge in the ξ-η plane, $\xi < x$, $(x-\xi)^2 > (y-\eta)^2$ [2.12]. In most studies attention has been concentrated on incompressible flow, and we shall discuss some of the results obtained here.

An interesting feature of the three-dimensional triple deck is that it is possible to write down a self-generating solution which can lead to separation. This is not the case in two dimensions, where it is necessary to assume that $p \propto (-x)^{1/2}$ as $x \to -\infty$, but by allowing a sinusoidal variation with z, a solution can be written down in which $p \to 0$ as $x \to -\infty$ [2.31, 2.32]. Smith [2.33] has studied the evolution of this solution and concludes that it has some unusual features. With the sinusoidal variation of p postulated, the pressure over part of the range of z begins by rising as x increases, and eventually flow reversal occurs, just as in two-dimensional compressive *supersonic* free interactions. However, whereas in that case the pressure approached a constant value as $x \to \infty$, now this cannot occur, and the most likely conclusion from his numerical study appears to be that a singularity develops, terminating the integration.

In order that such a numerical study may be embedded in a three-dimensional flow past a body, it is necessary to consider how such a free interaction might be provoked by some feature on the surface, such as a hump. Now not only is the origin of the interaction uncertain, as it is in supersonic flow, but so also is the appropriate dependence of p on z as $x \to -\infty$; Smith [2.33] took it to be $\propto \sin z$. These aspects of three-dimensional triple decks have been examined by Sykes [2.32], taking the hump to be of height

$$h(x, z) = h_0 \cos^2 \pi \times \cos \frac{\pi y}{\delta} \tag{5.4}$$

if $|x| < \frac{1}{2}$ and zero elsewhere, where h_0 is typically 2 and δ typically 1. As an illustration of the results he obtained, the skin-friction lines computed in the case $h_0 = 2$, $\delta = 1.2$ are displayed in Fig. 5.

The computational effort in solving (5.1), (5.2) is bound to be very large in view of the interactive nature of the problem and the presence of three independent variables. The method adopted by both Smith and Sykes is to make a Fourier analysis of the solution in the z-direction, retaining only a

Figure 5. Surface stress patterns for the profile (5.4) with $h_0 = 2$ and $\delta = 1.2$[2.23].

finite number of modes. Although this is a useful start on a very difficult problem, it can be argued that it masks a special difficulty of integrating the three-dimensional boundary-layer equations. This arises from the fact, apparent from (5.1), that the disturbances are convected downstream along the local streamline and diffused normal to the surface. Hence the domain of influence on any point P of the boundary layer is a curvilinear wedge (Raetz wedge) formed by the envelope of the streamlines, through the normal to the surface at P, together with *their* normals. Once the vertex angle of this wedge becomes large, as can happen near separation, when the surface streamline and the external streamlines are often pointing in markedly different directions, numerical analysts have had difficulty in advancing the solution further.

Recently however the Keller box scheme [2.34] has been successfully adapted to integrate the equations in such circumstances. An important feature of this scheme is that the equations are converted into five first-order equations, and so both the value of Δy and the direction of integration in the x-z plane can be varied as y increases. The *characteristic-box* method [2.35] integrates the momentum equations along the local streamline in the x-z plane, its position being found by iteration, and so long as it does not turn through an obtuse angle between the wall and the external flow, no insuperable difficulty about advancing the integration should arise. A number of successes have already been reported in both laminar and turbulent flows [2.35–2.37]. It is expected that the use of this method in conjunction with the triple-deck concept will facilitate the determination of the properties of the flow field.

6 Unsteady Trailing Edges

One aim of the steady triple-deck theory has been to justify the Kutta condition at sharp trailing edges. The generalization of this condition to include unsteady flows has also attracted some interest, partly in connection with noise radiation. With the definition $S = \omega^* l/U_\infty$, where ω^* is the frequency of the imposed oscillation, Brown and Daniels [2.38] showed that the triple-deck formulation is unaltered if $S \ll R^{1/4}$, but in view of the relation connecting the external pressure p_e^* with the external velocity U_e^* in classical boundary-layer theory, namely

$$-\frac{1}{\rho_e}\frac{\partial p_e^*}{\partial x^*} = u_e^* \frac{\partial u_e^*}{\partial x^*} + \frac{\partial u_e^*}{\partial t^*}, \tag{6.1}$$

care must be taken upstream of the triple deck to ensure a proper match between it and the oncoming boundary layer. Over the majority of this region only the last term of (6.1) is important, but there is a primary foredeck region of length lS^{-1} where both terms are significant. Further, if $1 \ll S \ll R^{1/4}$, Brown and Cheng [2.39] have found the necessity for yet another precursor region or foredeck of thickness $\sim lS^{-3/2}$, wherein the lower deck adjusts to achieve the quasisteady form required in the triple deck.

The viscous connection to the Kutta condition has been studied in detail by Daniels [2.40] when $S \sim R^{1/4}$ and the lower-deck equations have to be modified by the addition of the unsteady term $\partial u/\partial t$ to the momentum equation of (2.1). There is some uncertainty about the correct conditions to apply at the trailing edge if the oscillation is weak but, if the amplitude $\geq O(R^{-7/16}l)$ in magnitude, only the application of the full Kutta condition leads to consistent results. A further discussion of these studies has been given by Rienstra [2.41].

7 Turbulence

In practical problems the boundary layer is often turbulent, and it would be desirable to extend the theory of laminar interactions to include such flows. A major difference between the types of flow was pointed out by Lighthill [2.1] in his seminal paper from which the triple-deck theory originated. Whereas the lateral extent of a weak laminar interaction is formally large in comparison with the boundary-layer thickness, this is not the case with turbulence. Moreover the oncoming turbulent boundary layer already has a double structure, with the outer part governed by the law of the wake and the inner part by the law of the wall. In the interactive region the behavior of the inner part of the boundary layer is uncertain. The situation is exacerbated if separation occurs, because even under moderate pressure gradient there is a marked unsteadiness in its position [2.42]. Doubts have in fact been expressed about the value of attempting to set up a model for turbulent separated flow with the present state of knowledge about its properties [2.43].

Nevertheless, the principles underlying the concept of the triple deck have proved useful to students of interactions when the boundary layer is turbulent, provoked either by shock waves or by trailing edges [2.6, 2.44]. The crucial assumptions made are that in the outer part of the boundary layer, where the law of the wake normally applies, the turbulence is frozen, the length scale of the interaction is the same as the boundary-layer thickness, and the perturbations in the flow are small compared with the oncoming velocity. Such a hypothesis was made by Lighthill [2.1] many years ago, but he also assumed that the oncoming velocity $U_0(y)$ in this part of the boundary layer approached zero at the lower end, which meant that the induced pressure variation on the wall must vanish. Modern theories avoid this difficulty by taking

$$U_0(y) = u_e\left(1 + \varepsilon u_0\left(\frac{y}{\delta}\right)\right) \tag{7.1}$$

where u_e is the external velocity $\varepsilon = (\log R)^{-1}$, $\delta = O(\varepsilon)$, and u_0 has a logarithmic singularity as $y/\delta \to 0$. The inner part of the boundary layer occupies a region of thickness $o(R\varepsilon)^{-1}$ below it. Typically $\varepsilon = 0.05$, and so an asymptotic theory can be envisaged in which the solution is expanded in powers of ε. In this way the interactive problem is often reduced to the determination of a rotational inviscid flow. Alternatively $U_0(y)$ may be assumed to vary significantly across the outer part of the boundary layer, either with an *ad hoc* slip velocity at the wall or a power-law profile $U_0^n \propto y$ with $n > 2$, but the conclusion is the same.

Once this conclusion is reached, and especially when $U_0(y)$ is given by (6.1), an extensive theory of interactions between turbulent boundary layers and external inviscid flows can be built up. The reader is referred to the excellent reviews by Melnik [2.6] and by Adamson and Messiter [2.44] for accounts of the progress made in studying trailing-edge flows and shock-wave interactions when the boundary layer is assumed to remain unseparated.

It appears to be an exceptionally formidable task at the present time to develop a theory of separated turbulent flows. Some of the more basic problems about the inner layer structure have already been mentioned, but on the other hand the spectacular properties that can arise must be generated from this region. There does not seem to be any way in which an inviscid theory of the outer region can (for example) give rise to stall. Several authors have attempted to use Lighthill's ideas about the inner region. Most notably, Inger [2.45] develops a quasilaminar theory in which the kinematic viscosity is allowed to vary in the same way as in the undisturbed flow ahead of the interaction. Attention is primarily concentrated on a linearized form of the equation which enables him to obtain a description of the skin friction which compares reasonably well with experiment. He specifically excludes the possibility of separation, but it may be included formally without disturbing his basic assumptions apart from the choice of eddy viscosity. This approach has been criticized by Melnik [2.6] as arbitrary and unnecessary, but it *does* open possibilities of predicting stall, albeit perhaps not very well. It is interesting to note that in the Melnik theory of trailing-edge flows, uniqueness of solutions is achieved by applying the Kutta condition at the bottom of the rotational region. Our experience of laminar flows tells us that the inner boundary layer, if of an equivalent form, must then develop a Goldstein singularity with catastrophic effects. This result must be regarded as another argument against such a simple view of the inner layer, but it does show at least that there is a way for this layer to exert a significant influence on the rest of the flow field.

In Section 3 we showed that when the boundary layer is allowed to interact with the external stream near the leading edge of a thin airfoil, the notion of smooth attached flow fails, both on the exact triple-deck theory or on an approximate interacting-boundary-layer theory, unless the separated region is small or nonexistent. An exploratory study [2.27] was also carried out, assuming the transition to turbulence occurred near separation and taking a very crude model for the eddy viscosity. The breakdown of attached flow was also found to occur at a slightly larger angle of attack. To be sure, it was still considerably smaller than the stall angles observed in practice, but the fact that an estimate is possible does give some cause for optimism. Veldman [2.46] has also shown that an approximate interacting-boundary-layer theory of trailing-edge flows at finite R leads to breakdown of the solution at a certain critical angle of attack, just as in the Chow–Melnik triple-deck theory [2.4], but the extension to turbulent flow, for any model of the inner region, has not been carried out.

8 Discussion

In this review an account has been given of some of the recent developments in the theory of high-Reynolds-number flows. Of the questions left unanswered at the present time three may be mentioned specifically here. The first and the most important from a practical point of view is to develop a more satisfactory theory of the wall region of a turbulent boundary layer, suitable for predicting separation. The second is to improve the range of applications of numerical schemes to boundary-layer solutions which include regions of reverse flow. The method recently proposed by Veldman [2.46] is satisfactory for interactive problems at high or infinite values of the Reynolds number provided the region of reversed flow is not too extensive, and has been used with success in a number of situations [2.26]. When the reversed-flow region is thick, however, the method proposed by Williams [2.47, 2.48], involving upstream–downstream iterations, seems to be the most suitable, but it has not been extended beyond flows in which the thickness of the reversed-flow region is about eight times that of the displacement thickness of the undisturbed boundary layer. Our appraisal of the Sychev–Smith [2.13, 2.28] theory of detached flows would be sharper if we understood the structure of finite eddies, and a computational study of large eddies in boundary layers may be a useful way of approaching this problem.

The third question concerns the transition from attached flow to detached flow. For each type the interaction between the boundary layer and the inviscid external flow is of significance in controlling the structure. However, each type is disjoint from the other, and in the intermediate ranges of parameters when the flow is adjusting from one to another, it is not at all clear how its structure is to be elucidated.

CHAPTER 9

Spectral Computation of Triple-Deck Flows

Odus R. Burggraf* and P. W. Duck†

1 Introduction

In recent times the most common means of solving (numerically) Prandtl's boundary-layer equations has been finite-difference marching methods. Despite the nonlinearity of these equations, these techniques (incorporating Newton iteration) can provide rapid accurate solutions on modern computers. The marching method was applied successfully, for example, by Jobe and Burggraf [2.14] in their solution of the triple-deck problem of interaction between boundary layer and freestream near a trailing edge. Nevertheless, there are several types of flow for which these marching techniques either fail or become difficult to implement. Of these difficult types, perhaps separated flow is the most obvious. The physical cause of the difficulty is the region of reversed flow, in which the fluid travels in the direction opposite to that of the main body of fluid. As a result there is a change of character of the governing boundary-layer equations. Usually these are of parabolic type for which marching techniques are valid; however, the flow reversal due to separation changes the boundary-layer equations to quasielliptic type, with information being propagated both upstream and downstream. Any attempt to obtain an *accurate* solution by marching through such reversed-flow regions must fail due to the improper direction of information flow. In

* The Ohio State University, Columbus, OH 43210.

† University of Manchester, England. Formerly, Research Associate, The Ohio State University. This work was sponsored by Office of Naval Research under Contract N00014-76-C-0333.

finite-difference procedures, the failure is usually exhibited either by lack of convergence of the iterative procedure or by severe oscillations in the "solution".

Reyhner and Flügge-Lotz [2.49] have demonstrated a simple, though approximate, remedy to this problem of treating reversed flows. Their approximation was to neglect the product of the streamwise velocity component u and its streamwise derivative u_x in the governing equations wherever u became negative. This technique has been used by a number of authors [2.50, 2.51] with good results for the computed pressure distribution. However, much of the success of this technique owes to the fact that the reversed flow is very slow in many situations, and so the exclusion of this component of the inertia terms is of little consequence in those cases.

Williams [2.47] has suggested a more rational (and more complicated) procedure for treating reversed flows based on bidirectional marching techniques. The Reyhner–Flügge-Lotz (FLARE) procedure is used as a first approximation, followed by a backwards sweep in the reversed-flow region only. The previously neglected uu_x term in the reversed-flow region is now accounted for in the next forward sweep by treating it as a known quantity evaluated from the preceding backwards sweep. This downstream-upstream iteration (DUIT) is then continued until convergence is achieved. Williams usually finds only five to ten of these sweeps to be necessary. Obviously this method is considerably less straightforward to program than simple marching procedures, and also takes rather longer to compute.

Another class of problems for which the standard marching techniques require iterative application is that of free-interaction flows. Originally postulated by Lighthill [2.1], these flows are essentially eigensolutions of the boundary-layer equations that provide the means of upstream influence in the flow, despite the parabolic nature of the governing equations. Stewartson and Williams [2.52] and later Smith and Stewartson [2.53] used a shooting approach to generate their free-interaction solutions. Here the pressure is perturbed by a small jump and the resulting solution is obtained by marching downstream; the correct amount of perturbation is determined so that the free-interaction solution is suppressed far downstream. As usual with shooting methods, this approach proves more and more difficult as the range of integration is extended downstream, owing to the ultimate exponential growth of the free-interaction solutions.

The above methods all solve the steady-flow equations by marching in the flow direction. An alternative is time marching, with which the unsteady equations are integrated forward in time until a steady-state solution is obtained. This approach permits retention of all the inertia terms in the equations of motion, even in reversed-flow regions. Jenson et al. [2.54] and later Rizzetta et al. [2.55] applied the time-marching procedure to the triple-deck formulation for supersonic separated flows over ramp configurations. Their method was based on shear stress as the primary dependent variable. Upstream influence was then accounted for through an interaction condition applied as a wall boundary condition on the stress: namely, the boundary-layer compatibility condition with the pressure gradient evaluated from the displacement function. This condition is an exact requirement of the interacting flow, and permits downstream conditions to influence the upstream flow at each time step of the computation. The method works well, but flows with extensive separation are rather expensive to compute due to their slow development in time. An interacting boundary-layer program based on similar ideas has been presented by Werle and Vatsa [2.56] and has been shown to give results that asymptote the triple-deck results for very large Reynolds number [2.57].

Various alternatives to these finite-difference approaches have been proposed. The spectral method seemed to us to have characteristics advantageous for computing separated flows, and we have applied it to the triple-deck problem in this study. The (nonlinear) governing equations are transformed from physical to spectral variables using the Fourier integral transform in the main-flow direction, together with finite differences in the transverse direction. The solution is computed (iteratively) in spectral space, and then the inverse transform applied to obtain the solution in physical variables. A major advantage of the method is that reversed flows present no difficulty; each point in spectral space relates to all points on the path of integration in physical space. Thus the reversed and forward-flow information is diffused together in transform variables. This interconnection between the physical grid points via each spectral grid point endows the spectral method with a physically implicit nature, suggesting an accelerated convergence of the iterations. As will be seen, the spectral computations exhibit rapid convergence, even for flows exhibiting separated regions large enough that other (finite-difference) schemes fail.

Details of the method as well as results of computations for both incompressible and supersonic flows are given below. In both cases, the spectral results are shown to compare well with those of finite-difference computations.

2 The Mathematical Problem

The problem considered is the flow disturbance produced by a shallow perturbation of the surface height at a distance L from the leading edge of an otherwise plane wall. Let u_∞ be the flow speed and ν_∞ the kinematic viscosity of the fluid, the subscript (∞) referring to conditions in the undisturbed freestream. The Reynolds number $R = u_\infty L/\nu_\infty$ is regarded as large. As is common in triple-deck analyses, it is convenient to define the small parameter ε as $R^{-1/8}$. Then if the surface height perturbation is of order $\varepsilon^5 L$ and extends over a length of order $\varepsilon^3 L$, the flow disturbance is contained in the triple-deck structure originally deduced by Messiter [2.58], Neiland [2.59], and Stewartson [2.52]. Since this asymptotic flow structure has been described by many authors, we merely state here that the problem reduces to solving the conventional incompressible boundary-layer equations (governing the lower-deck flow) subject to an unconventional outer boundary condition whose form depends on whether the outer flow is incompressible or supersonic. (A hypersonic version exists as well, but is not considered in this work; see [2.60]).

It is convenient to describe the triple-deck disturbance using variables with magnitude of order one. These are denoted by uppercase letters, while lower case letters designate the corresponding physical variable. Thus, the coordinate representing distance parallel to the surface is $x = \varepsilon^3 aLX$, while the normal coordinate is $y = \varepsilon^5 bLY$. The origin is taken at a convenient point on the disturbed surface. The corresponding velocity components are $u = \varepsilon(d/b)u_\infty U$ and $v = \varepsilon^3(d/a)u_\infty V$, while the pressure $p = p_\infty(1 + \varepsilon^2 cP)$. Here a, b, c, and d are constant scale factors depending on the Mach number, the wall-temperature ratio, and the surface stress of the undisturbed flow. Their values are given, for example, in [2.52] and [2.55] for supersonic flow, and in [2.14] and [2.61] for incompressible flow. Following Jenson et al. [2.54, 2.55], we introduce the shear $\tau = U_Y$ as basic dependent variable, where subscripts denote partial derivatives. Then the equation to be solved in the lower deck is the shear transport equation,

$$U\tau_X + V\tau_Y = \tau_{YY}, \tag{1}$$

together with the continuity equation

$$U_X + V_Y = 0. \tag{2}$$

The conventional no-slip conditions apply:

$$U = V = 0 \quad \text{on } Y = 0. \tag{3}$$

The remaining boundary conditions take a form governed by the triple-deck structure. Upstream the shear must match that of the undisturbed flow near the wall; thus,

$$\tau \to 1 \quad \text{for } X \to -\infty \tag{4}$$

Moreover, the triple-deck structure requires that the main-deck solution correspond to a simple vertical displacement of the original undisturbed boundary layer by the lower deck. Consequently, matching of main and lower deck solutions requires

$$\tau \to 1 \quad \text{as } Y \to \infty. \tag{5}$$

The corresponding condition for U is

$$U \sim Y + A(X) \quad \text{as } Y \to \infty,$$

where $A(X)$ is the (negative) displacement of the main deck. For supersonic flow, the pressure is given by linearized potential theory as

$$P = H'(X) - A'(X),$$

where $H(X)$ is the height of the surface perturbation in lower-deck scaling The boundary-layer compatibility condition states that the pressure gradient balances the normal gradient of the shear at the wall. Expressing A in the form

$$A(X) = \lim_{Y \to \infty} (U - Y) = \int_0^\infty (\tau - 1)\, dY,$$

the compatibility condition for supersonic flow can then be expressed as

$$\tau_Y|_{Y=0} = H''(X) - \int_0^\infty \tau_{XX}\, dY. \tag{6}$$

On the other hand, the elliptic nature of incompressible flow gives rise to a Hilbert integral for the pressure:

$$P(X) = \frac{1}{\pi} \int_{-\infty}^{\infty} \frac{H'(S) - A'(S)}{S - X}\, dS. \tag{7}$$

Thus, an interaction condition very similar to Eq. (6) can be given for incompressible flow. This will be postponed, however, as the similarity is much more remarkable in spectral variables.

The Fourier transform is now applied to the governing equations. Using an overbar to denote transformed variables, the transform $\bar{\phi}(\omega)$ of any real function $\phi(X)$ is given as

$$\bar{\phi}(\omega) \equiv F\{\phi(X)\} \equiv \int_{-\infty}^{\infty} \phi(X) e^{-i\omega X}\, dX.$$

Actually, to permit convergence of the integrals, the transformation is performed on the perturbation flow quantities. Define

$$\begin{aligned} \tilde{\tau} &= \tau - 1, \\ \tilde{U} &= U - Y, \\ \bar{\tau}(\omega, Y) &= F\{\tilde{\tau}(X, Y)\}, \\ \bar{U}(\omega, Y) &= F\{\tilde{U}(X, Y)\}, \\ \bar{V}(\omega, Y) &= F\{V(X, Y)\}. \end{aligned}$$

In terms of these perturbation variables, the shear transport equation (1) becomes

$$Y\tau_X - \tau_{YY} = -(\tilde{U}\tau_X + V\tau_Y) \equiv R, \tag{8}$$

where the second-order perturbation terms have been placed on the right side of the equation. The spectral form of this equation is

$$i\omega Y\bar{\tau} - \bar{\tau}_{YY} = \bar{R} \equiv \overline{-(\tilde{U}\tau_X + V\tau_Y)}. \tag{9}$$

The right-side function $\bar{R}$ has the form of a complicated convolution integral. However, it need not be given here, since the term will be evaluated by another method (pseudospectrally). The continuity equation (2) is expressed spectrally as

$$\bar{V} = -i\omega\bar{U} = -i\omega \int_0^Y \bar{\tau}\, dY. \tag{10}$$

The boundary conditions (3)–(6) also must be expressed in spectral variables. Thus,

$$\bar{U} = \bar{V} = 0 \quad \text{on } Y = 0, \tag{11}$$

$$\bar{\tau} \to 0 \quad \text{as } X \to -\infty, \tag{12}$$

$$\bar{\tau} \to 0 \quad \text{as } Y \to \infty, \tag{13}$$

and the interaction condition (6) for supersonic flow becomes

$$\bar{\tau}_Y|_{Y=0} = -\omega^2\left\{\bar{H}(\omega) - \int_0^\infty \bar{\tau}(\omega, Y)\, dY\right\} \tag{14}$$

For incompressible flow, the pressure is given by Eq. (7), so that the compatibility condition becomes

$$\bar{\tau}_Y|_{Y=0} = i\omega\bar{P}.$$

Applying the Fourier transform to the Hilbert integral in Eq. (7) thus yields the incompressible interaction condition

$$\bar{\tau}_Y|_{Y=0} = -i\omega|\omega|\left\{\bar{H}(\omega) - \int_0^\infty \bar{\tau}(\omega, Y)\, dY\right\}, \tag{15}$$

which is very similar in form to the supersonic interaction condition. In fact, this slight difference between Eqs. (14) and (15) is the only difference in the triple-deck theories for supersonic and incompressible flow. Yet, as will be demonstrated, large apparent differences in the physical flow properties result. The significance with regard to computer programming is that the change of a single FORTRAN statement permits computing either supersonic or incompressible flows with the same computer code.

3 The Solution Procedure

The spectral shear-transport equation (9) has been written in the form of a linear ordinary differential operator on the left side, with the nonlinear inertia terms collected together in the right-side function $\bar{R}$. The solution is obtained by iteration, solving first with $\bar{R}$ set identically to zero; then in later iterations $\bar{R}$ is evaluated from the subsequent solution, as

$$\begin{aligned} i\omega\bar{\tau}^{(n)} - \bar{\tau}_{YY}^{(n)} &= \bar{R}^{(n-1)} \\ &= -\overline{(\tilde{U}^{(n-1)}\tau_X^{(n-1)} + V^{(n-1)}\tau_Y^{(n-1)})}. \end{aligned} \tag{16}$$

Central differences in Y are applied to Eq. (16), as well as to the interaction condition, either (14) or (15). The result of the first iteration is the linearized theory of Stewartson [2.62], which provides a reasonably accurate prediction for small disturbances. For disturbances large enough to produce flow separation, further iteration is necessary to bring in the effects of the nonlinear inertia terms.

The right-side function $\bar{R}$ in Eq. (16) can be expressed as a convolution of the velocity referred to two different points in spectral space. However, evaluating this convolution integral is a very inefficient way of determining $\bar{R}$, requiring of order $N_Y N_\omega^2$ multiplications, for each complete iteration, where N_Y and N_ω are the number of grid points in the Y and ω (spectral) directions, respectively.

A much more efficient method is to use the so-called fast-Fourier-transform (FFT) algorithm of Cooley and Tukey [2.63]. The procedure used here is to invert the transforms $\bar{\tau}$ and $i\omega\bar{\tau}$ to obtain the physical variables τ and τ_X. From these $\tilde{U}$, U_X, and V are determined, permitting the inertia function R of Eq. (8) to be evaluated. Applying the direct transform to R then results in the desired inertia transform term $\bar{R}$. Equation (16) is then solved for the next approximation, and the above procedure is iterated until the solution repeats to the desired number of decimal places. The advantage of this method arises from the efficiency of the FFT algorithm; the scheme outlined here requires only of the order of $N_Y N_\omega \log_2 N_\omega$ multiplications for each complete iteration, compared with $N_Y N_\omega^2$ for the direct method based on the convolution integral. An order-of-magnitude reduction in computing time is achieved for quite reasonable grid resolution. It may be noted that the method used here is the reverse of the pseudospectral method described by Orszag [2.64] and Roache [1.101], who solve the equations of motion in the physical plane and use the Fourier transform to evaluate derivatives.

The discretization of the Fourier transforms was carried out as follows. Define the discrete physical and spectral variables X_j and ω_k as

$$X_j = (j - 1 - M)\,\Delta X \text{ for } j = 1, 2, \ldots, N, \tag{17a}$$

$$\omega_k = (k - 1 - M)\,\Delta\omega \text{ for } k = 1, 2, \ldots, N, \tag{17b}$$

where for convenience we take $N = 2M$. Denote any physical function $\phi(X_j)$ by ϕ_j, and the corresponding spectral function $\bar{\phi}(\omega_k)$ by $\bar{\phi}_k$. Choose the range of variables such that $\phi(X)$ is negligible for $X < X_1$, $X > X_{N+1}$, and $\bar{\phi}(\omega)$ is negligible for $\omega < \omega_1$, $\omega > \omega_{N+1}$. Then the Fourier integral transform $\bar{\phi}(\omega)$, defined in the last section, can be approximated by the finite sum

$$\bar{\phi}_k = \Delta X \sum_{j=1}^{N} \phi_j e^{-i\omega_k X_j} \tag{18}$$

The range of integration is slightly uncentered, from $X = -(M + \frac{1}{2})\,\Delta X$ to $X = (M - \frac{1}{2})\,\Delta X$. Since $\phi(X)$ is real, it follows from the form of the transform that $\bar{\phi}(-\omega) = \bar{\phi}^*(\omega)$, where the asterisk denotes the complex conjugate. Because of this property the transform variables need not be stored for, say, positive values of ω_k. The midrange parameter M in Eqs. (17) is now required to be an even integer, and the grid spacings ΔX and $\Delta\omega$ satisfy the relation

$$\Delta X\,\Delta\omega = \frac{2\pi}{N}. \tag{19}$$

Then, Eq. (18) can be reduced to the finite Fourier transform

$$\bar{\phi}_k = (-1)^k \, \Delta X \sum_{j=1}^{N} (-1)^j \phi_j e^{-i\,2\pi(k-1)(j-1)/N} \tag{20}$$

For the inverse Fourier integral transform, the centered range of integration $\omega_1 = -M\,\Delta\omega$ to $\omega_{N+1} = M\,\Delta\omega$ is taken, and, noting that $\bar{\phi}_{N+1} = \bar{\phi}_1^*$, the trapezoidal rule results in the finite sum

$$\begin{aligned}\phi_j &= \frac{\Delta\omega}{2\pi}\left[\tfrac{1}{2}(\bar{\phi}_1 + \bar{\phi}_1^*) + \sum_{k=2}^{N} \bar{\phi}_k e^{i\omega_k X_j}\right] \\ &= \frac{\Delta\omega}{2\pi}\left[\tfrac{1}{2}(\bar{\phi}_1^* - \bar{\phi}_1) + \sum_{k=1}^{N} \bar{\phi}_k e^{i\omega_k X_j}\right]\end{aligned}$$

Since ϕ_1 is real and $(\bar{\phi}_1^* - \bar{\phi}_1)$ is imaginary, the result is

$$\phi_j = \mathrm{Re}\left[\frac{\Delta\omega}{2\pi} \sum_{k=1}^{N} \bar{\phi}_k e^{i\omega_k X_j}\right]$$

Substituting for ω_k and X_j from Eq. (17), and again recalling that M is even, the inverse finite Fourier transform results:

$$\phi_j = \mathrm{Re}\left[(-1)^j \frac{\Delta\omega}{2\pi} \sum_{k=1}^{N} (-1)^k \bar{\phi}_k e^{i\,2\pi(k-1)(j-1)/N}\right] \tag{21}$$

Equations (20) and (21) are best evaluated using the Cooley–Tukey FFT algorithm [2.63], for the reasons discussed above. In that case, M must be a power of 2.

The range of integration was approximately centered about the origin in both physical and spectral variables because of the symmetry property $\bar{\phi}(-\omega) = \bar{\phi}^*(\omega)$. The real variable $\phi(X)$ may be quite unsymmetrical, however, and aliasing errors may be observed. This effect arises because the nonperiodic function $\phi(X)$ is represented by the finite Fourier transform as the periodic function $\Phi(X)$, with period $N\,\Delta X$. Consequently, if an improperly aligned interval of length $N\,\Delta X$ is sampled from $\Phi(X)$, aliasing may occur on either the left or right side of the sample interval. If on the left, for example, the aliased values actually would correspond to the correct values taken from $\Phi(X)$ on the right of the interval. No such errors exist in $\bar{\phi}(\omega)$ as computed here, since the inertia terms are evaluated at each X-grid point independently. Aliasing does occur in the physical results, but is easily recognized, since the flow properties decay to the undisturbed Blasius values at infinity. The effect can be avoided either by restricting the X-range of the data presented, or by use of a special inversion formula based on a properly uncentered integration interval. Both methods were used for the results presented in the following section. A better method of suppressing aliasing errors would be to evaluate the truncated tails of the Fourier integral by asymptotic analysis and add their contributions to the FFT result. This procedure is cumbersome and was not necessary for the cases dealt with here.

4 Results

Two surface-height configurations, indicated by $H_1(X)$ and $H_2(X)$, have been chosen to illustrate the spectral triple-deck method. The shapes are shown in Fig. 1, and are defined by the relations

$$H_1(X) = \frac{\alpha}{1 + X^2}$$

$$H_2(X) = \begin{cases} 0 & \text{for } |X| > 1, \\ \alpha(1 - X^2)^2 & \text{for } |X| < 1. \end{cases}$$

The first of these shapes exhibits a long gentle variation of height, but has a Fourier transform that is much sharper, being significant only over a narrow range of ω:

$$\overline{H}_1(\omega) = \alpha\pi e^{-|\omega|}.$$

The second shape has the opposite behavior: it differs from zero only over a narrow range, but has a slowly varying Fourier transform:

$$\overline{H}_2(\omega) = -16\,\frac{3\omega \cos\omega + (\omega^2 - 3)\sin\omega}{\omega^5}$$

with

$$\overline{H}_2(0) = \tfrac{16}{15}.$$

The opposing behavior of these two shapes in physical and spectral coordinates makes them useful for case studies of the spectral method. The second shape $H_2(X)$ was used earlier by Sykes [2.65], and his finite-difference results will serve as a check case for incompressible flow. For supersonic flow, an independent check case was run for $H_1(X)$ using the time-marching finite-difference program of Jenson [2.54] and Rizzetta [2.55], modified to incorporate second-order-accurate Crank–Nicolson differencing in X.

Figure 2 displays results for supersonic flow over the surface contour $H_1(X)$ for values of amplitude α from -3 to $+3$. (Positive α corresponds to a

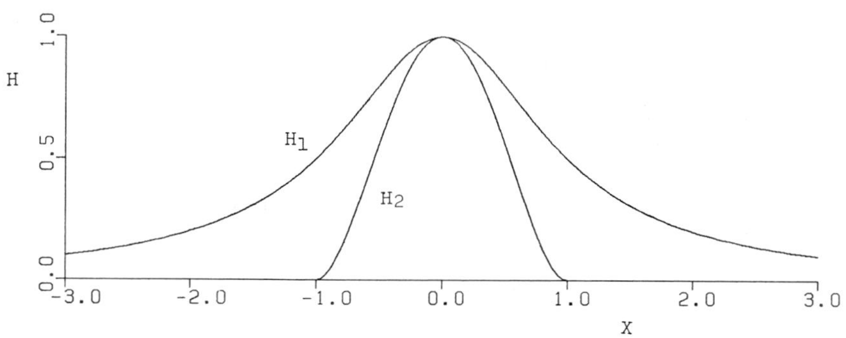

Figure 1. Surface height contours.

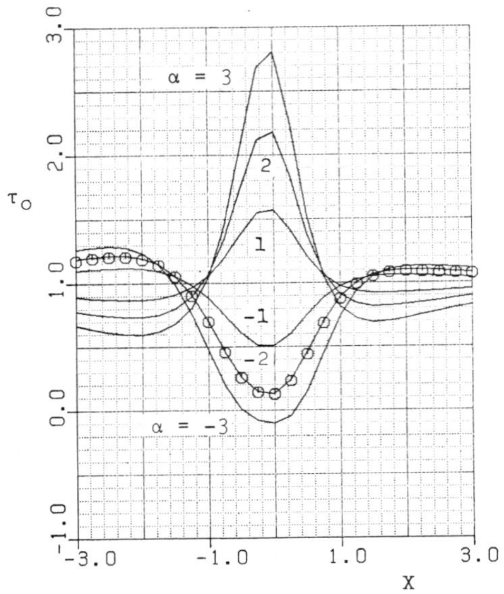

Figure 2. Surface stress for supersonic flow past contour $H_1(X)$.

hump, negative α to a hollow in the surface.) The curves represent surface stress computed by the spectral program for $N_\omega = 128$, $\Delta X = 0.25$, $N_Y = 25$, $\Delta Y = 0.5$. The grid interval is obvious in the figure, since values at grid points are connected by straight lines. The symbols are for the check-case results from the finite-difference program for $\alpha = -2$, with $N_X = 81$, $\Delta X = 0.25$, $N_Y = 37$, $\Delta Y = 0.25$. Obviously, the results of the two methods are in excellent agreement. The spectral method was much faster, however, requiring only 10 iterations for $\alpha = -2$, while the time-marching finite-difference program required 300 time steps to converge to the steady state. In terms of processing time, the spectral method is an order of magnitude faster, requiring for these cases about 10 seconds CPU time on an Amdahl 470/V6 computer, versus about 81 seconds for the finite-difference program.

For the humps ($\alpha = 1, 2, 3$), the flow coming from the left at first decelerates, reducing the shear below the Blasius value ($\tau_0 = 1$ in lower-deck scaling). For sufficiently tall humps ($\alpha > 7$, approximately) the flow would be expected to separate here, as well as on the lee side of the hump. Near the top of the hump the lower-deck flow is squeezed by the rapidly increasing surface height, and the surface stress peaks to very high values. For the hollows ($\alpha = -1, -2, -3$) the reverse of these effects occurs. In this case, a flow-separation bubble occurs at the bottom of the hollow for $\alpha < -2.6$ (approximately). The nonlinear effects can be seen in Fig. 2, but are more clear in the table of values below, where the difference between the center-point shear and the Blasius value of unity is normalized with respect to α. The effect of nonlinearity is demonstrated by the deviation from the value 0.537 for $\alpha = 0$. Clearly nonlinearity is more important for the hollows than for the humps, accounting for a change of about 30% from the linear-theory value for $\alpha = -3$.

Table 1 Surface stress at $X = 0$: $H(X) = \alpha/(1 + X^2)$

α	$\tau_0(0)$	$[\tau_0(0) - 1]/\alpha$
3	2.807	0.602
2	2.179	0.589
1	1.568	0.568
0	0.0	0.537
−1	0.508	0.492
−2	0.132	0.434
−3	−0.100	0.369

The effect of spatial resolution on accuracy is indicated in Fig. 3 for $\alpha = -2$. The symbols represent results for a coarse grid having $N_\omega = 32$, $\Delta X = 1.0$, and the curve represents fine-grid results with $N_\omega = 128$, $\Delta X = 0.25$. (The fine-grid results are the same as those of Fig. 2.) For both grids $N_Y = 25$ and $\Delta Y = 0.5$. Both cases used the same X-range ($-16 \le X \le 16$, approximately), which is clearly large enough to yield accurate results. As N_ω is reduced below 32, with the same X-range, the accuracy degrades quickly owing to poor approximation to the surface shape for $\Delta X > 1$. Also, reducing the X-range below about $-8 \le X \le 8$ severely degrades the accuracy, most likely due to truncation of the wakelike flow decay downstream.

A comparison of separated supersonic-flow results with those for incompressible flow past the surface contour $H_1(X)$ is given in Fig. 4. The amplitude $\alpha = -5$ was chosen for the comparison, since the incompressible flow does not separate for $\alpha < -4$. The incompressible flow (upper curve) converged in 30 iterations, but the supersonic flow (lower curve) did not converge to the same accuracy until 80 iterations. In the supersonic case, the residuals did not decay monotonically, and an underrelaxation factor of 0.5

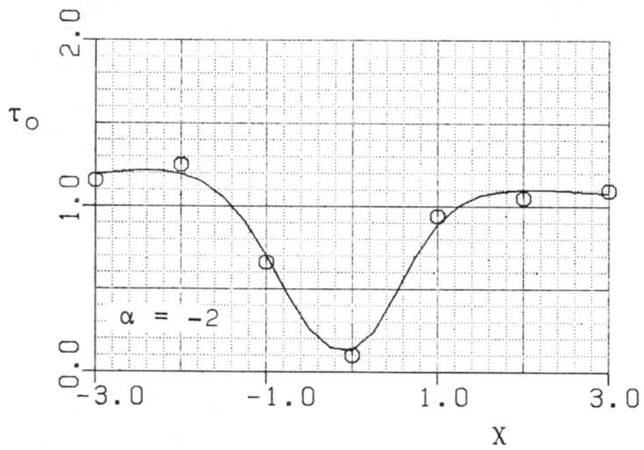

Figure 3. Effect of grid size on accuracy: supersonic flow past contour $H_1(X)$.

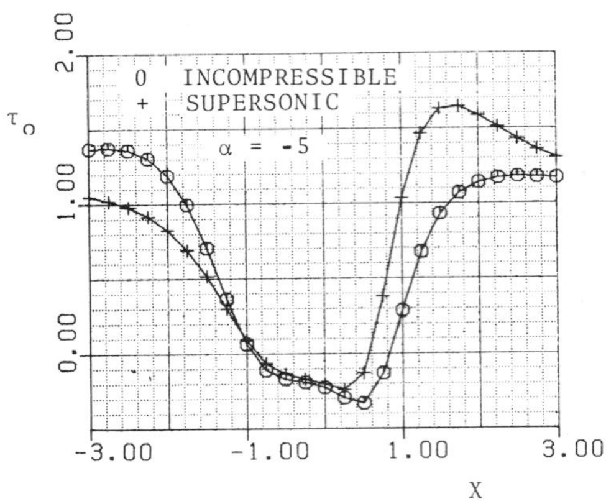

Figure 4. Comparison of supersonic and incompressible flows past contour $H_1(X)$.

was applied to obtain convergence. This case was the only one of those analyzed in this study that required underrelaxation. In general, the supersonic and incompressible flows show quite similar surface shear-stress patterns, especially in the separated region. It may be noted that the separated $\tau_0(X)$ distribution is very much like that computed for other configurations such as in the ramp study of Rizzetta et al. [2.55].

Our last comparison is for the quartic hump $H_2(X)$, which provides a severe test case for triple-deck computations. Sykes [2.65] considered incompressible flow past this shape as an example of topographic effects on the Earth's boundary layer; his results are compared here with results obtained spectrally. Ragab and Nayfeh [2.66] also obtained solutions for this shape with moderate amplitude ($\alpha \leq 2.4$), but reported that their finite-difference program would not converge for $\alpha = 3$, even with underrelaxation factors as small as 0.1. A similar failure was reported for Napolitano's [2.67] finite-difference scheme. The spectral method described here treated the problem with no difficulty, converging in 30 iterations without needing underrelaxation. The spectral results are shown in Fig. 5 by the solid curve, and Sykes's results are indicated by the symbols. The grid parameters are $N_\omega = 256$, $\Delta X = 0.125$, $N_Y = 31$, $\Delta Y = 0.5$ for the spectral results, and $N_x = 256$, $\Delta X = 0.08$, $N_Y = 60$, $\Delta Y = 0.25$, for the finite-difference results of Sykes. The two numerical solutions compare fairly well, with the spectral results displaying a smoother variation in the reversed flow region. There appears to be a slight oscillation in the finite-difference results, suggesting the need for a finer grid, already finer than that used in the spectral computations. It is interesting that both results show a sharp break in the slope of the shear-stress curve just downstream of the leading edge of the hump. These results for surface stress in incompressible flow appear qualitatively quite similar to those for supersonic flow past the $H_1(X)$ hump. The main difference is that the supersonic flow showed a stronger tendency to separate upstream of the hump, while the incompressible flow generates a rather large separation

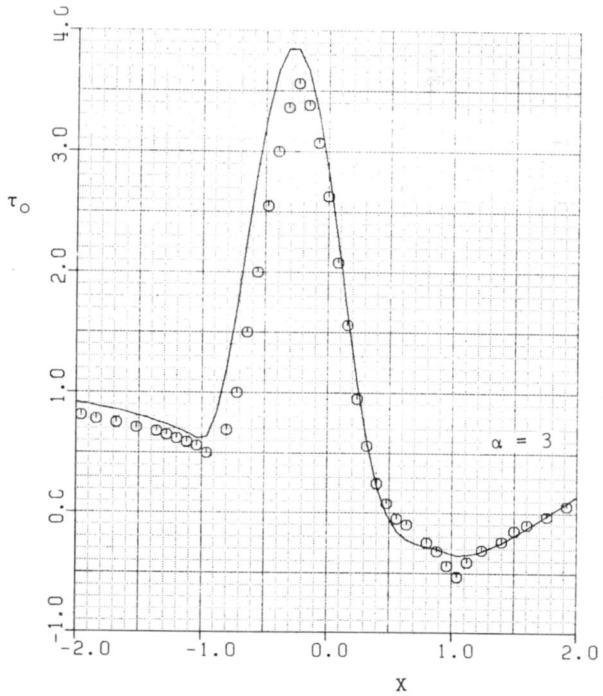

Figure 5. Surface stress for incompressible flow past contour $H_2(X)$.

bubble in the lee of the $\alpha = -3$ hump. These trends would not appear to contradict physical insight regarding the behavior of supersonic and incompressible flows.

5 Concluding Remarks

This study has demonstrated that the spectral method is fast and effective for solving triple-deck problems, with the ability to handle easily problems that some recent finite-difference methods are unable to solve. Moreover, the accuracy of the spectral method is compatible with that of the better finite-difference schemes. In this regard, it is often stated that the spectral method is of infinite-order accuracy, converging faster than any power of N. This statement would be true for finite domains if the boundary conditions did not introduce boundary discontinuities into the function represented by the Fourier series. Chebyshev polynomial representation is preferred for this reason [2.64]. For infinite domains, as in the present application, the Fourier integrals themselves lead to algebraic-order accuracy when discretized (second-order when using the trapezoidal rule). In addition, truncating the infinite range of integration leads to additional error depending on the asymptotic behavior of the physical and spectral functions involved. The use of too restricted a range of integration can lead to quite large errors. As can be seen from Eq. (19), the finite-range ($N\ \Delta X$) error in physical variables is equivalent to grid-size ($\Delta\omega$) error in spectral variables, and vice versa. Consequently, it is necessary to use care in choosing the grid size in both physical and spectral variables to ensure an accurate solution. Of course, similar considerations hold for finite-difference solutions.

CHAPTER 10

Nonasymptotic Theory of Unseparated Turbulent-Boundary-Layer – Shock-Wave Interactions with Application to Transonic Flows

G. R. Inger*

Nomenclature

C_f	Skin-friction coefficient, $2\tau_w/\rho_{e0} U_{e0}^2$
C_p	Pressure coefficient, $2p'/\rho_{e0} U_{e0}^2$
H_i	Incompressible shape factor, δ_i^*/θ_i^*
M	Mach number
p	Static pressure
p'	Interactive pressure perturbation, $p - p_1$
Δp	Pressure jump across incident shock
Re_l	Reynolds number based on length l
T	Absolute temperature
Υ	Basic interactive wall-turbulence parameter
u', v'	Streamwise and normal interactive-disturbance-velocity components, respectively
U_0	Undisturbed incoming boundary-layer velocity in x-direction
x, y	Streamwise and normal distance coordinates (origin at the inviscid shock intersection with the wall)
$y_{w\,\mathrm{eff}}$	Effective wall shift seen by interactive inviscid flow
β	$\sqrt{M_1^2 - 1}$
γ	Specific-heat ratio
δ	Boundary-layer thickness
δ^*	Boundary-layer displacement thickness

* Professor and Chairman, Department of Aerospace Engineering Sciences, University of Colorado, Boulder, CO 80309.

ε_T Kinematic turbulent eddy viscosity
μ Ordinary coefficient of viscosity
ν μ/ρ
ω Viscosity temperature-dependence exponent, $\mu \propto T^{\omega}$
ρ Density
θ^* Boundary-layer momentum thickness
τ Total shear stress

Subscripts

1 Undisturbed inviscid values ahead of incident shock
e Conditions at the boundary-layer edge
inv Inviscid-disturbance solution value
0 Undisturbed incoming-boundary-layer properties

1 Introduction

Shock–turbulent-boundary-layer interactions are important in the aerodynamic design of high-speed aircraft wings, and of turbine and cascade blades in turbomachinery and air-breathing-engine inlets and diffusors. Of particular importance are the features of upstream influence, boundary-layer displacement, skin friction, and incipient separation dominated by the thin interactive shear-stress disturbance layer very close to the surface. Lighthill's pioneering study [2.1] of this region, however, takes into account only the laminar portion of the incoming turbulent-boundary-layer profile, which is inaccurate for the higher Reynolds numbers pertaining to full-scale aircraft. On the other hand, more recent work on an improved theory either has been confined to the treatment of the transonic regime by asymptotic methods [2.68, 2.69] that entail a severe limiting model of the interactive physics as $\mathrm{Re}_l \to \infty$, or has involved approximate double-layered models for supersonic flow [2.70–2.72] with insufficient consideration of the basic flow structure in the shear-disturbance sublayer [2.73]. Consequently, there is a need for a more general theory at ordinary practical Reynolds numbers, applicable to both transonic and supersonic flow. The present paper addresses this objective for the case of 2-D steady adiabatic nonseparating flows up to Mach numbers that are not "hypersonic"; we describe a unified triple-deck theory for practical Reynolds numbers, in which strong emphasis is placed on the law-of-the-wall turbulence effects that extend the Lighthill theory to overlap the extremely high-Reynolds-number trends implied by asymptotic theory.

We consider small disturbances of an incoming turbulent boundary layer due to a weak external shock (see Fig. 1) and examine the detailed perturbation field with a nonasymptotic triple-deck flow model patterned in many ways after Lighthill's approach because of its essential soundness, its similarity

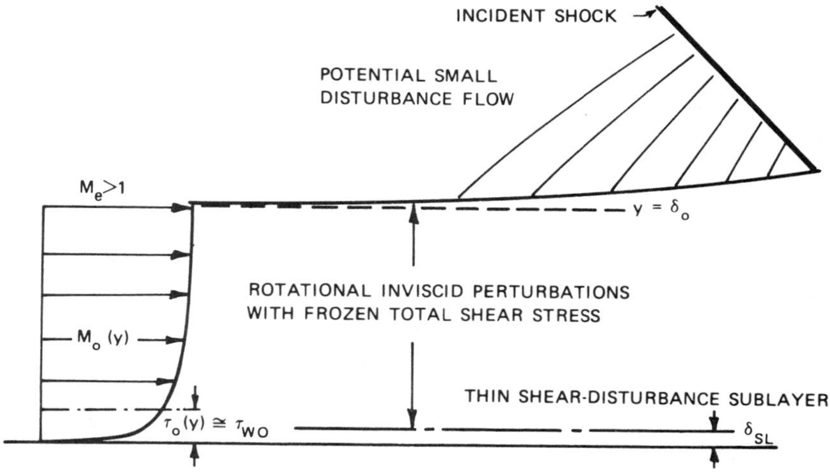

Figure 1. Triple-deck structure of the interaction (schematic).

to related multiple-deck theories that have proven highly successful in treating turbulent boundary layer response to strong adverse pressure gradients [2.45], and the large body of data that supports the predicted results in a variety of specific problems [2.45]. In many results the present work is the turbulent counterpart of the laminar nonasymptotic theory by Tu and Weinbaum [2.74] and a supplement to Rosen et al. [2.72] that supplies their missing consideration of the skin-friction disturbances.

2 Outline of the Nonasymptotic Small-Disturbance Interaction Theory

2.1 The Triple-Deck Structure

At high Reynolds numbers it has been established that the interaction field in the neighborhood of the shock organizes itself into three basic layers or "decks" (Fig. 1): (a) an outer region of potential inviscid flow above the boundary layer, which contains the incident shock and interactive wave systems; (b) an intermediate deck of frozen shear-stress–rotational inviscid disturbance flow, occupying the outer 90 % or more of the incoming boundary-layer thickness; (c) an inner shear-disturbance sublayer adjacent to the wall, which accounts for the interactive skin-friction perturbations (and hence any possible incipient separation) plus most of the upstream influence of the interaction. The "forcing function" of the problem here is thus impressed by the outer deck upon the boundary layer; the middle deck couples this to the response of the inner deck, but in so doing can itself modify the disturbance field, while the slow viscous flow in the thin inner deck reacts strongly to the pressure-gradient disturbances imposed by these overlying decks.

2.2 Formulation of the Disturbance Problem in Each Deck

Outer Potential-Flow Deck. If the incident shock and its reflection system are weak with isentropic nonhypersonic flow, we have a small disturbance potential inviscid motion in the undisturbed uniform flow U_{0e}, M_{0e}:

$$\frac{\partial^2 p'}{\partial y^2} + \left[1 - M_{0e}^2 - 2\frac{u' M_{0e}^2}{U_{0e}}\right]\frac{\partial^2 p'}{\partial x^2} = 0, \tag{1}$$

where the third term within the square brackets is significant in the transonic case $1 < M_{0e} < 1.2$–1.3 and includes shock jump conditions to this order of approximation. Since various solution methods are available in either transonic or purely supersonic flow (in which case Eq. (1) reduces to an Ackert problem), we assume that such a solution may be carried out for all x on $y \geq \delta_0$ subject to the usual far-field conditions as $y \to \infty$. The remaining disturbance boundary condition along $y = \delta_0$ then couples this solution to the underlying double deck: it requires that both v'/U_{01} and p' be continuous there.

Middle Rotational-Disturbance Flow Deck. This layer contributes to and transmits the displacement effect, contains the boundary-layer lateral pressure gradient due to the interaction, and carries the influence of the incoming boundary-layer profile shape. Our analysis rests on the key simplifying assumption that for nonseparating interactions the changes in the turbulent Reynolds shear stress are small enough to have a negligible back effect on the mean flow properties along the interaction zone; hence, they are "frozen" along each streamline at the appropriate value in the undisturbed incoming boundary layer. This approximation is well supported by Rose's detailed experimental studies [2.75, 2.76] of a nonseparating-shock–turbulent-boundary-layer interaction, which showed that over such short-range interactions the pressure gradient and inertial forces outside a thin layer near the wall are at least an order of magnitude larger than the corresponding changes in Reynolds stress. Thus the disturbance field is one of small rotational inviscid perturbation of the incoming nonuniform turbulent boundary-layer profile, governed by

$$\frac{\partial}{\partial y}\left[\frac{v'(x, y)}{U_0(y)}\right] = \frac{1 - M_0^2(y)}{\gamma M_0^2(y)} \cdot \frac{\partial(p'/p_0)}{\partial x} \tag{2}$$

$$\frac{\partial^2 p'}{\partial y^2} - \frac{2}{M_0} \cdot \frac{dM_0}{dy} \cdot \frac{\partial p'}{\partial y} + \left(1 - M_0^2 - \frac{2u' M_0^2}{U_0}\right)\frac{\partial^2 p'}{\partial x^2} = 0, \tag{3}$$

where $U_0(y)$, $\rho_0(y)$ are arbitrary functions of y with δ_0, δ_0^*, and τ_{w0} as constants. Now Eq. (3) is a slight generalization of Lighthill's well-known pressure perturbation equation for nonuniform flows, which includes a nonlinear correction term for transonic effects including the diffracted impinging shock above the sonic level of the incoming boundary-layer profile. Excluding the hypersonic regime, Eqs. (2) and (3) therefore apply to a wide range of initially supersonic external flow conditions and across the boundary layer except at the singular point $M_0 \to 0$ (which we avoid by consideration of the inner deck as shown below). Whatever the method used to solve this middle-deck disturbance problem, we imagine that it provides the disturbance pressure distribution $p'(x, y)$; then y-integration of Eq. (2) gives the disturbance streamline slope as

$$\frac{v'(x, y)}{U_0(y)} = \underbrace{\left[\frac{v'}{U_0}\right](x, y_{w\,\text{eff}})}_{0} + \frac{\partial}{\partial x}\left\{\int_{y_{w\,\text{eff}}}^{y} \frac{p'}{\gamma p_{e_1}}\left[\frac{1 - M_0^2(\bar{y})}{M_0^2(\bar{y})}\right] d\bar{y}\right\}, \tag{4}$$

where $y_{w\,\text{eff}} > 0$ is the effective wall height of the inner deck defined such that the inviscid $v'(x, y_{w\,\text{eff}})$ and hence $\partial p'/\partial y(x, y_{w\,\text{eff}})$ both vanish (see below).

The corresponding displacement thickness growth along the interaction is then given by streamwise quadrature of the continuity-equation integral as

$$\Delta\delta^*(x) \approx \int_{y_{w\,\text{eff}}}^{\delta_0} \frac{p'}{p_{e_1}}\left[\frac{1 - M_0^2(\bar{y})}{M_0^2}\right] d\bar{y} + (\delta_0 - \delta_0^*)\left[\frac{M_{e_1}^2 - 1}{M_{e_1}^2 p_{e_1}}\right] p_w'(x) \tag{5}$$

Inner Shear-Disturbance Deck. This thin inner deck contains the significant viscous and turbulent shear-stress disturbances due to the interaction. It lies well within the law-of-the-wall region of the incoming boundary layer and below the sonic level of the profile [2.45]. The original work of Lighthill [2.1] and others further neglected the turbulent stresses and considered only the laminar sublayer effect; while this yields an elegant analytical solution, the results are in error at high Reynolds numbers and cannot explain the ultimate asymptotic behavior pertaining to the $\mathrm{Re}_l \to \infty$ limit. The present theory remedies this by including the effect of turbulent stress in the entire law-of-the-wall region. Note that our consideration of the entire law-of-the-wall region, combined with the application of the effective-inviscid-wall concept to the inner-deck displacement effect, eliminates the "blending layer" [2.69] that is otherwise required to match the *disturbance* field in the laminar sublayer region with the middle inviscid deck; our inner solution effectively includes this blending function, since it imposes a boundary condition of vanishing *total* shear disturbance at the outer edge of the deck. In addition, retention of the disturbance pressure-gradient term provides the correct physics at practical Re_l.

To facilitate a tractable theory, we retain only the main physical effects by introducing the following simplifying assumptions:

1. The incoming-boundary-layer law-of-the-wall region is characterized by a constant total (laminar plus turbulent eddy) shear stress and a Van Driest–Cebeci type of damped-eddy-viscosity model.
2. For weak incident shocks, the sublayer disturbances are small perturbations upon the incoming boundary layer; however, *all* the physically important effects of streamwise pressure gradient, streamwise and vertical acceleration, and both laminar and turbulent disturbance

stresses are retained. Moreover, since the *form* of the resulting set of linear equations is in fact unaltered by nonlinear effects, the quantitative accuracy is expected to be good until close to separation.

3. For adiabatic flows at low to moderate external Mach numbers, the undisturbed and perturbation flow Mach numbers are both quite small within the shear-disturbance sublayer; consequently, the influence of the density perturbations may be neglected, while the corresponding compressibility effect on the undisturbed profile is treated by using incompressible relations based on wall-recovery-temperature properties.
4. The turbulent fluctuations and the small interaction disturbances are uncorrelated.
5. The thinness of the inner deck allows neglect of its lateral pressure gradient, $p' \approx p'_w(x)$.

Under these assumptions, the disturbance field is governed by the continuity and momentum equations

$$\frac{\partial u'}{\partial x} + \frac{\partial v'}{\partial y} = 0, \tag{6}$$

$$U_0 \frac{\partial u'}{\partial x} + v' \frac{\partial U_0}{dy} + (\rho_{w_0}^{-1}) \frac{dp'_w}{dx} = \frac{\partial}{\partial y}\left(v_{wo} \frac{\partial u'}{\partial y} + \varepsilon'_{T_0} \frac{\partial u'}{\partial y} + \varepsilon_T \frac{\partial U_0}{dy}\right), \tag{7}$$

where ε_T is the kinematic eddy-viscosity perturbation. The corresponding undisturbed turbulent-boundary-layer law-of-the-wall profile $U_0(y)$ is governed by

$$\tau_0(y) = \text{const} = \tau_{wo} = [\mu_{wo} + \rho_{wo}\varepsilon_{TO}(y)]\frac{dU_0}{dy}, \tag{8}$$

where the Van Driest–Cebeci eddy-viscosity model with $y^* = y\sqrt{\tau_{wo}/\rho_{wo}}/v_{wo}$ yields for nonseparating flow that

$$\varepsilon_{TO} = [0.41y(1 - e^{-y^*/A})]^2 \frac{dU_0}{dy}, \tag{9a}$$

$$\varepsilon'_T \approx \left(\frac{\partial u'/\partial y}{dU_0/dy}\right)\varepsilon_{TO}, \tag{9b}$$

where we take the commonly accepted value $A = 26$.

We seek to solve Eqs. (6)–(9) subject to the impermeable-wall no-slip conditions $U_0(0) = u'(x, 0) = v'(x, 0) = 0$ plus an initial condition requiring all interactive disturbances to vanish far upstream. Furthermore, sufficiently far from the wall, u' must pass over to the inviscid solution u^i_{inv} along the bottom of the middle deck governed by Eq. (6) plus

$$U_0 \frac{\partial u_{\text{inv}}}{\partial x} + v'_{\text{inv}} \frac{dU_0}{dy} + (\rho_{wo})^{-1} \frac{\partial p_w}{\partial x} \approx 0, \tag{10}$$

while the corresponding total shear disturbance ($\approx \partial u'/\partial y$) vanishes to the desired accuracy.

Following Lighthill [2.1], we now differentiate Eq. (7) with respect to x, substituting Eq. (6) so as to eliminate u' and then differentiating the result with respect to y so as to eliminate p'_w; one thus obtains the following fourth-order equation for v':

$$\frac{\partial}{\partial x}\left(U_0 \frac{\partial^2 v'}{\partial y^2} - \frac{d^2 U_0}{dy^2} v'\right) = \frac{\partial^2}{\partial y^2}\left[(v_{0w} + 2\varepsilon_{T0}) \frac{\partial^2 v'}{\partial y^2}\right]. \tag{11}$$

This equation contains a threefold influence of the turbulent flow: the profile $U_0(y)$, its curvature d^2U_0/dy^2 (nonzero outside the laminar sublayer), and a new eddy-disturbance stress term $2\varepsilon_{T0}$.

Equation (11) is to be solved together with Eqs. (8), (9) and the wall boundary conditions $v'(x, 0) = \partial v'/\partial y(x, 0) = 0$. A third condition involving v' is obtained by satisfying the x-momentum equation right at the wall:

$$\frac{\partial^3 v'}{\partial y^3}(x, 0) = -(2\mu_{0w})^{-1} \frac{d^2 p'_w}{dx^2}. \tag{12}$$

The fourth boundary condition is the outer inviscid matching from Eq. (10):

$$\frac{\partial}{\partial x}\left(U_0 \frac{\partial^2 v'_{\text{inv}}}{\partial y^2} - v_{\text{inv}} \frac{d^2 U_0}{dy^2}\right) = 0 \tag{13}$$

along with $\partial^2 v'/\partial y^2 \approx 0$ (i.e., vanishing total disturbance shear). Once this $v'(x, y)$ field is obtained, the attendant streamwise velocity and hence the disturbance shear stress) may then be found from Eq. (6).

An important feature here is the "effective inviscid wall" position (or displacement thickness) that emerges from the asymptotic behavior of v' far from the wall [2.1] (Fig. 2). This is defined by the value $y_{w\,\text{eff}}$ where the "back projection" of the v_{inv} solution vanishes; physically, $y_{w\,\text{eff}}$ thus represents the total mass-defect height due to the shear-stress perturbation field and hence the effective wall position seen by the overlying inviscid middle-deck disturbance flow. As indicated in Fig. 2(b), this serves to couple the

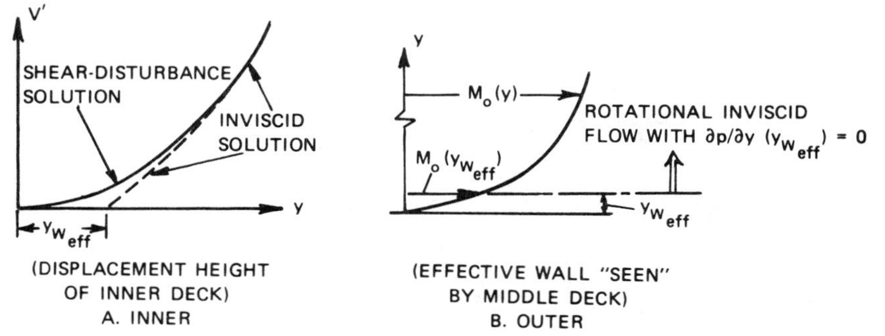

Figure 2. Middle–inner-deck matching by the effective-wall concept: (A) inner; (B) outer.

inner- and middle-deck solutions by providing the nonsingular inner equivalent slip-flow boundary conditions

$$\frac{\partial p'}{\partial y}(y_{\text{eff}}) = v'_{\text{inv}}(y_{w\,\text{eff}}) = 0 \quad \text{at } U_0(y_{w\,\text{eff}}) > 0$$

for the middle-deck solution.

2.3 Approximate Solution by Operational Methods

An analytical solution is further achieved (as described in detail in [2.45]) by Fourier-transform methods, yielding the interactive pressure rise and displacement thickness growth as inputs to the above extended theory of the inner-disturbance sublayer. In particular, the matching of the outer two decks with the inner shear-disturbance deck in connection with the Fourier inversion process yields the determination of the upstream influence distance, the inner-deck displacement thickness $y_{w\,\text{eff}}$, and the interactive skin-friction relation

$$\frac{\tau'_w(x)}{\tau_{w0}} = -\left(\frac{\kappa_{\min}}{\lambda}\right)^{2/3} S(\Upsilon)\sqrt{\frac{\beta}{C_{f0}}}\, C'_{pw}\left[\frac{2(C'_{pw})^{3/2}}{3\kappa_{\min}\int_{-\infty}^{x}(C'_{pw})^{3/2}\,dx}\right]^{2/3}. \tag{14a}$$

with

$$\Upsilon = (0.41)^2\left[\frac{C_{f0}}{2}\frac{\text{Re}_{\delta_0}}{(T_w/T_e)^{1+2\omega}}\right]^{1/3}(\kappa_{\min}\delta_0)^{-2/3}, \tag{14b}$$

$$\lambda\delta_0 = \frac{0.744\beta^{3/4}(C_{f0}/2)^{5/4}\text{Re}_{\delta_0}}{(T_{w0}/T_{e0})^{\omega+1/2}}, \tag{14c}$$

where $\kappa_{\min} = l_u^{-1}$ is given in [2.45], while the functions $H(\Upsilon)$ and $S(\Upsilon)$, given in Fig. 3, represent the wall-turbulence effects on the interactive displacement effect and skin friction, respectively, of the inner deck. Figure 3 is a central result, providing a unified account of the entire Reynolds-number range in terms of the single new turbulent interaction parameter Υ, from the limiting behavior for negligible wall-turbulence effect (pertaining to Lighthill's

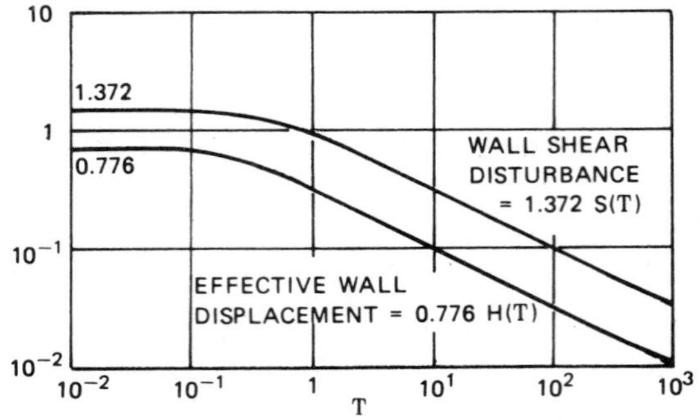

Figure 3. Wall-turbulence effect on inner-deck displacement thickness and disturbance skin friction.

theory at $\Upsilon \to 0$) to the opposite extreme of wall-turbulence-dominated behavior at $\Upsilon \gg 1$ (pertaining to an asymptotic type of theory at very large Reynolds numbers, where the inner-deck thickness and disturbance field become vanishingly small). It is also seen that the asymptotic trends at very large Re_l cannot be extrapolated down to ordinary Re_l values; by the same token, the extreme approximation involved in the $\Upsilon = 0$ limit breaks down at larger Re_l's pertaining to $\Upsilon \gg 1$.

A computer program has been constructed to carry out the foregoing solution method. The incoming turbulent boundary layer is treated by the compressible version of an analytical universal composite law-of-the-wall–law-of-the-wake model due to Walz [2.77] that is characterized by *three* arbitrary parameters: the preshock Mach number, the boundary-layer displacement-thickness Reynolds number, and the incompressible shape factor H_{1i}.

3 Application to Transonic Normal-Shock–Turbulent-Boundary-Layer Interactions

We illustrate the above theory for the problem of weak normal-shock–turbulent-boundary-layer interaction because of its importance in transonic aerodynamics, the availability of a proven pressure-field solution method for the two outer decks in the mixed transonic flow regime involved [2.78], and the existence of experimental data to check the theory. Nonseparating nearly normal-shock-boundary-layer interactions up to $M_1 \approx 1.3$ involve a much simpler type of interaction pattern [2.79], amenable to analytical treatment. Such a solution is achieved by assuming small linearized disturbances ahead of, behind, and below the *nonlinear* shock jump plus an approximate treatment of the shock structure within the boundary layer [2.78], which gives reasonably accurate predictions for all the properties of engineering interest when $M_1 \gtrsim 1.05$. The resulting solution contains all the essential physics of the mixed transonic viscous interaction field for non-separating flows; numerous comparisons with experiment have shown that it gives a good engineering account of the interaction over a wide range of Mach–Reynolds-number conditions. This theory has further proven successful as a local interactive module imbedded within global viscous–inviscid prediction programs for supercritical airfoil [2.80] and transonic projectile flow fields [2.81].

A sample comparison with the transonic experimental data of Ackeret, Feldman, and Rott [2.79] is presented in Fig. 4 to illustrate the typical behavior and good agreement of the predicted interactive pressure and displacement-thickness distributions; also shown are fairly good results for the skin-friction variation. A second comparison with more recent data is illustrated in Fig. 5 for the nonseparating interaction zones on two supercritical airfoils in the DFVLR–AVA (Gö) transonic wind tunnel [2.80]; again, the theory is well supported by this detailed experimental study.

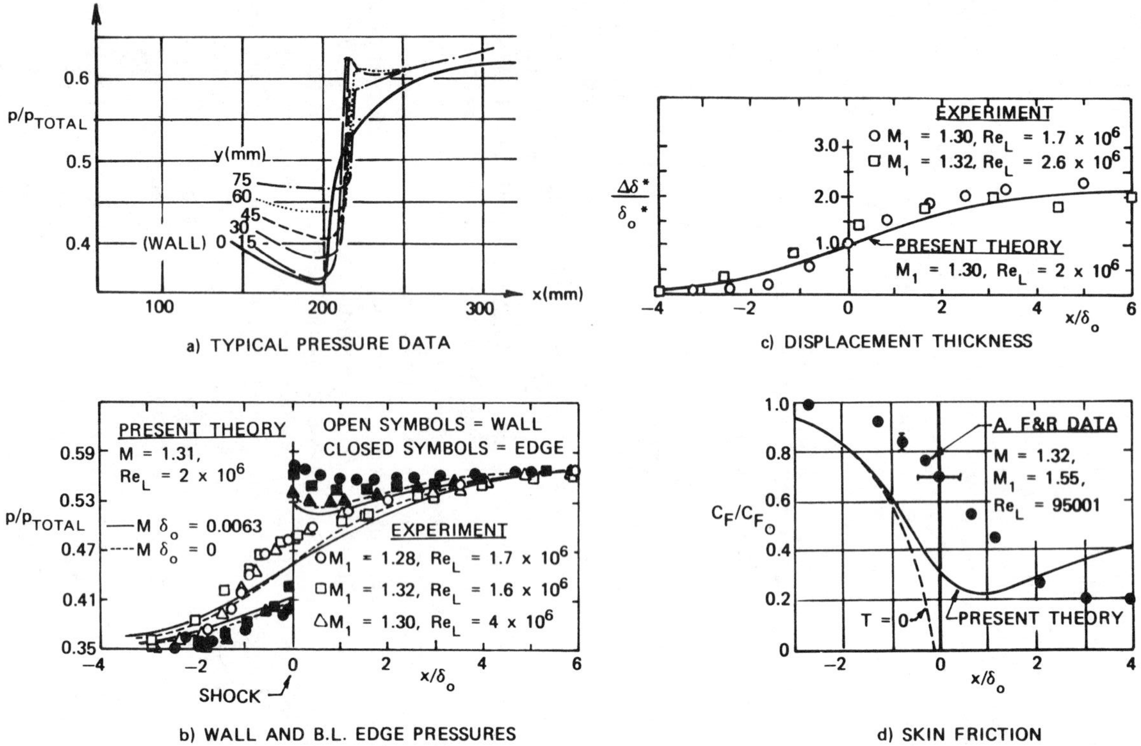

Figure 4. Theoretical comparisons with data of Ackeret, Feldman, and Rott [2.79]: (a) typical pressure data, (b) wall and boundary-layer edge pressures, (c) displacement thickness, (d) skin friction.

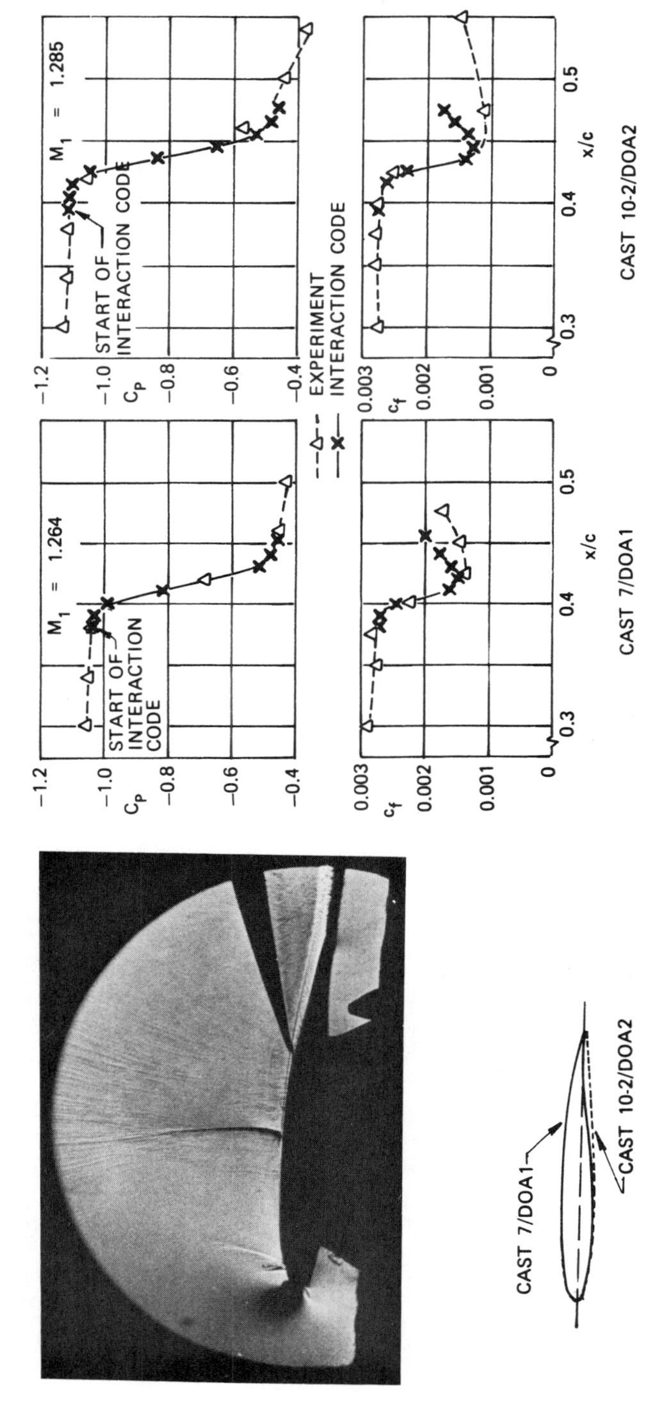

Figure 5. Comparisons of present interaction theory with experimental results of Nandenan, Stanewsky, and Inger [2.80].

CHAPTER 11

A Review of Marching Procedures for Parabolized Navier–Stokes Equations

S. G. Rubin*

1 Introduction

It has now been generally accepted that boundary-layer methodology can be extended to the so-called parabolized Navier–Stokes (PNS) equations for a significant variety of flow problems. In a recent paper, Davis and Rubin [2.82] have reviewed several viscous flow computations in which parabolized or thin-layer techniques have been applied in order to accurately determine the flow characteristics. This publication also reviews some of the early history of the PNS development.

The purpose of the present paper is to discuss some recent investigations using the PNS equations. In particular, we are concerned here with efficient three-dimensional algorithms, a clearer understanding of the limits of applicability of PNS marching techniques, and pressure interaction relaxation for separated flows and other problems where upstream influence is of importance. In this regard, three solution procedures are considered: (a) single-sweep "boundary-layer-like" marching for two- and three-dimensional flows; (b) multiple-sweep iteration or global pressure relaxation, where upstream influence and possibly axial flow separation are important, but regions of subsonic flow are small; and (c) global relaxation where subsonic flow domains are large. For the last two classes of problems, the analysis draws heavily on that of interacting boundary layers and inviscid subsonic relaxation methods where applicable.

* Department of Aerospace Engineering and Applied Mechanics, University of Cincinnati, Cincinnati, OH 45221.

In the course of proceeding to specific examples, a brief review of the limitations associated with PNS marching or relaxation is necessary. The PNS equations, which for simplicity are given here only for "two-dimensional incompressible flow", are as follows:

$$u_t + uu_x + vu_y = -p_x + \frac{1}{R} u_{yy}, \tag{1a}$$

$$v_t + uv_x + vv_y = -p_y + \left(\frac{1}{R} v_{yy}\right), \tag{1b}$$

$$u_x + v_y = 0. \tag{1c}$$

R is the Reynolds number; x is defined as the axial flow direction, and y as the normal direction. The system (1) differs from the complete Navier–Stokes equations only by the omission of the axial diffusion terms. Strictly speaking the inclusion of the v_{yy} term in (1b) is inconsistent with the omission of u_{xx} in (1a). Either the former should be neglected (this is probably a more appropriate definition of the PNS equations) or the latter retained. In fact, these terms have little effect on any of the results presented here. The mathematical character of (1) is controlled by the p_x term in (1a). When p_x is prescribed (assumed known), the system is parabolic. This was the case in the original merged-layer analysis of Rudman and Rubin [2.83], where for hypersonic cold-wall flow p_x can, in fact, be neglected in (1a). It should be emphasized, however, that axial pressure gradients are still present and are evaluated through the momentum equation (1b) and the energy equation in the compressible case.

When the p_x term in (1a) is retained implicitly (not prescribed), the system (1) is no longer parabolic, as an elliptic pressure or acoustic interaction occurs in regions of subsonic flow [2.82]. The "parabolic" form for the velocities has led to the expression "parabolized Navier–Stokes equations". This pressure interaction also appears for boundary-layer equations, when p_x is not prescribed. The resulting upstream interaction has been analyzed by Lighthill [2.1], who demonstrated the existence of exponential growing solutions in the interaction zone. Similar behavior has been encountered with marching procedures for the PNS equations. The primary difference between the boundary-layer and the PNS equations is that for the latter the pressure interaction is manifested through both the outer pressure boundary formulation and the normal-momentum equation (1b).

2 Single-Sweep Marching

For problems where upstream influence and axial flow separation are not significant, it is natural to consider the system (1) by boundary-layer marching techniques, i.e., backward differences are applied for all x-gradients. If p_x is prescribed, this approach is quite acceptable, as the equations are in fact parabolic. For implicit numerical schemes, the marching calculation should be unconditionally stable for all Δx marching steps; see Davis and Rubin [2.82] for additional references. On the other hand, if p_x is assumed unknown, then the "elliptic" pressure interaction of Lighthill is introduced, and therefore the exponential growth, representative of upstream influence, can be anticipated. Lubard and Helliwell [2.84] have examined the stability of the backward-difference approximation for p_x in (1a), and they have shown that for $\Delta x < (\Delta x)_{\min}$ instability or departure solutions will occur. Similar results were found earlier by Lin and Rubin [2.85]. For $\Delta x > (\Delta x)_{\min}$ the marching scheme is stable. Therefore $(\Delta x)_{\min}$ would appear to represent a measure of the upstream elliptic interaction. For $\Delta x < (\Delta x)_{\min}$ the marching scheme attempts to represent this interaction, and therefore the Lighthill behavior should be recovered.

Since the backward difference formula for p_x does not provide any upstream contribution, it does not properly represent the differential form for $\Delta x < (\Delta x)_{\min}$. When the backward-difference approximation is less representative of p_x, the error introduced serves to reduce $(\Delta x)_{\min}$. For example, at $x = x_i$, $(p_{i-1} - p_{i-2})/\Delta x$ is less severe than $(p_i - p_{i-1})/\Delta x$, and for p_x prescribed, $(\Delta x)_{\min} = 0$. In view of this behavior, several investigators (Vigneron et al. [2.86], Yanenko et al. [2.87], and Lin and Rubin [2.88]) have attempted to eliminate the pressure interaction by incorporating "small" inconsistencies into the difference approximation. They have assumed (a) a variable Δx in the difference form for p_x such that $\Delta x > (\Delta x)_{\min}$ locally [2.86], (b) "regularization" functions of the type $(1/\sqrt{R})\, f(u_x, v_x, p_x)$ as modified coefficients for the uu_x and p_x terms in (1a) [2.87], and (c) the use of finite temporal iteration to modify the convection velocity in each step of the marching procedure [2.88]. Each of these techniques introduces some inconsistency into the difference equations in subsonic zones; reasonable results have been obtained with these methods for certain problems. In order for these techniques to be effective, the inconsistency must be large enough to suppress the elliptic character, yet small enough to maintain an acceptable order of accuracy.

The PNS model has been considered in some detail by Rubin and Lin [2.89]. For the system (1) with a backward difference for $\partial/\partial x$ and central y-differences, the linear von Neumann stability analysis leads to the following condition for the eigenvalues λ [2.89]:

$$\frac{u(\lambda - 1) + 4b\lambda \sin^2(\beta/2) + Ic\,\lambda \sin\beta}{u(\lambda - 1)\cos^2(\beta/2) + 4b\lambda \sin^2(\beta/2) + Ic\,\lambda \sin\beta} = \frac{(\lambda - 1)F(\lambda)}{4a^2\lambda^2 \sin^2(\beta/2)}, \tag{2}$$

where $a = \Delta x/\Delta y$, $b = \Delta x/(R\,\Delta y^2)$, $c = av$, $I = \sqrt{-1}$, $F = \lambda - 1$.

It can be shown that the value of $\lambda_{\max}$ is closely related to the highest-frequency mode, so that when the number of grid points across the layer $N \gg 1$, $\beta \approx \pi/(N - 1)$. Equation (2) then takes the simplified form

$$\frac{(\lambda - 1)F(\lambda)}{A^2\lambda^2} = 1, \tag{3}$$

where $A = \pi \, \Delta x/y_M$ and y_M is the layer thickness; $y_M = (N) \, \Delta y$. The condition (3) indicates that $\Delta x/y_M$ is the relevant stability parameter. From (3), the marching procedure will be stable for

$$A = \pi \, \Delta x/y_M > 2.$$

Therefore

$$(\Delta x)_{\min} \approx y_M. \tag{4}$$

The complete numerical solution of (2), for all β, has been obtained, and the analytic result $(\Delta x)_{\min} \leq (2/\pi) y_M$ is confirmed. The extent of the elliptic numerical interaction is of the order of the thickness of the total layer. If the system (1) is used to solve boundary-layer problems, then $y_M = O(R^{-1/2})$ and therefore $(\Delta x)_{\min} = O(R^{-1/2})$. For interaction regions where the triple-deck [2.2] structure is applicable, $(\Delta x)_{\min} = (2/\pi) y_M = O(R^{-3/8})$, which is the extent of the Lighthill [2.1] upstream influence. This would tend to confirm the idea of a limited elliptic zone contained in the PNS formulation. For $\Delta x > (\Delta x)_{\min}$, this elliptic effect is suppressed. When the upstream influence is negligible, this inconsistency should have little effect on the solution. For truly interactive flows the ellipticity must be retained and the global forward-difference concept discussed in the next section is required.

The results for the PNS and other "transonic" equations clearly indicate that step sizes of the order of the subsonic region, which for supersonic mainstreams is $O(R^{-1/2})$ or $O(\mathrm{Re}^{-3/8})$ in a triple-deck region, will provide stable and accurate solutions for flows in which upstream effects are not dominant. In a later section, where a strongly implicit algorithm is introduced to obtain marching solutions for the supersonic flow over a cone at incidence, this $(\Delta x)_{\min}$ dependence on y_M will be shown for the compressible PNS equations.

3 Multiple-Sweep Marching—Global Iteration

If consistency, for $\Delta x \to 0$, of the difference formulation is to be achieved, or if upstream influence is important and/or separation occurs, then backward differencing of the p_x term should be rejected. With any form of forward or central differencing for p_x, relaxation (multiple marching sweeps or global iteration) is required. Three possibilities will be discussed: (a) forward differencing, (b) central differencing, and (c) use of the Poisson equation for pressure.

Forward differencing for p_x introduces the upstream value p_{i+1}, which is prescribed in each marching sweep, and also has the advantage of including the local pressure value p_i. This provides for coupling of the pressure–velocity system (1) and allows for a free surface pressure interaction. This is important for problems where axial flow separation occurs.

At this point, a few remarks concerning the role of the p_x term for separated flow are relevant. In recent years there has been considerable analysis of free pressure–boundary-layer interactions for separated flows [2.56]. There is general agreement that, for limited separation bubbles, boundary-layer equations can be used to calculate such flows. Moreover, the singularity at separation does not appear if the $p'(x)$ term is not prescribed but allowed to develop. Inverse methods, in which the displacement thickness or shear stress is prescribed during the marching step and then updated in subsequent relaxation sweeps, have been considered, as have procedures in which temporal terms are retained in order to introduce the pressure or displacement thickness implicitly. In these procedures, the $p'(x)$ term is replaced with an interaction expression (displacement slope for supersonic flow [2.56] or Cauchy integral for subsonic flow [2.2]). The local pressure or displacement thickness then appears implicitly and is coupled with the velocity evaluation. In all of these interaction analyses, those components of the p_x approximation that introduce upstream terms are updated during the relaxation sweeps.

From interacting-boundary-layer analysis we can then conclude that if the elliptic character is to be modeled consistently, the p_x representation for the PNS equations should introduce downstream contributions. If the separation singularity is to be circumvented in any relaxation sweep, a free pressure interaction through the outer boundary condition or through the y-momentum equation (1b) must be introduced. Forward differencing would appear to satisfy both of these constraints. The stability analysis for Eqs. (1) has been extended in [2.89] to a variety of p_x approximations. For forward differencing of p_x, the stability condition (4) is modified solely by the factor $F(\lambda)$, such that $F(\lambda) = -\lambda$. From (4) it can be inferred that for $\beta = \pi/N - 1$, this procedure is unconditionally stable. The stability curves for general β-values are given in [2.89]. Forward differencing is unconditionally stable for all β at all R. It is significant, however, that as the convective velocity v increases, both eigenvalues asymptote to one. So stability is marginal when the subsonic region is large.

During the first sweep of the global iteration procedure the value p_{i+1} must be prescribed by an initial guess. p_{i+1} can be chosen as a constant equal to the boundary condition at the outer boundary, or the surface, or some combination thereof. Since the variation of p across the subsonic layer is small, any of these values will generally suffice.

In order to test the applicability of forward differencing for p_x, two boundary-layer problems have been considered with the full PNS system (1):

1. Flat plate:

$$u = p = 1 \quad \text{at} \quad y = y_M;$$

$$u = v = 0 \quad \text{at} \quad y = 0.$$

2. Separation bubble:

$$\left.\begin{array}{ll} u = p = 1 & \text{if } x < 0 \\ u = 1 - x,\ p_x = uu_x & \text{if } 0 \le x \le 0.25 \\ u = 0.75,\ p_x = 0 & \text{if } x \ge 0.25 \end{array}\right\} \text{ at } y = y_M$$

$$u = v = 0 \qquad \text{at } y = 0.$$

The pressure gradient p_x is forward-differenced. All other x-derivatives are backward differenced. These are neglected in regions of reverse flow. All y-derivatives are central-differenced, except for the continuity equation (1c) and the momentum equation (1b), where the trapezoidal rule is used. Multiple sweeps or global iteration was stable and converged for $(\Delta x)_{\min} \approx y_M/6$; the value of $(\Delta x)_{\min} = y_M/60$ was also tested and with forward differencing was stable. It is significant that *in this relaxation procedure it is necessary to store only the pressure field for each successive iteration level.* The velocities are reevaluated during each marching sweep. The results are in excellent agreement with published results for both problems. The free surface pressure interaction introduced by the y-momentum equation has eliminated the separation singularity for the bubble problem. A value of Δx equal to one-sixth the boundary layer thickness $y_M = O(R^{-1/2})$ appears to be adequate. For the smaller value of $(\Delta x)_{\min} = y_M/60$ or with $y_M = O(R^{-3/8})$

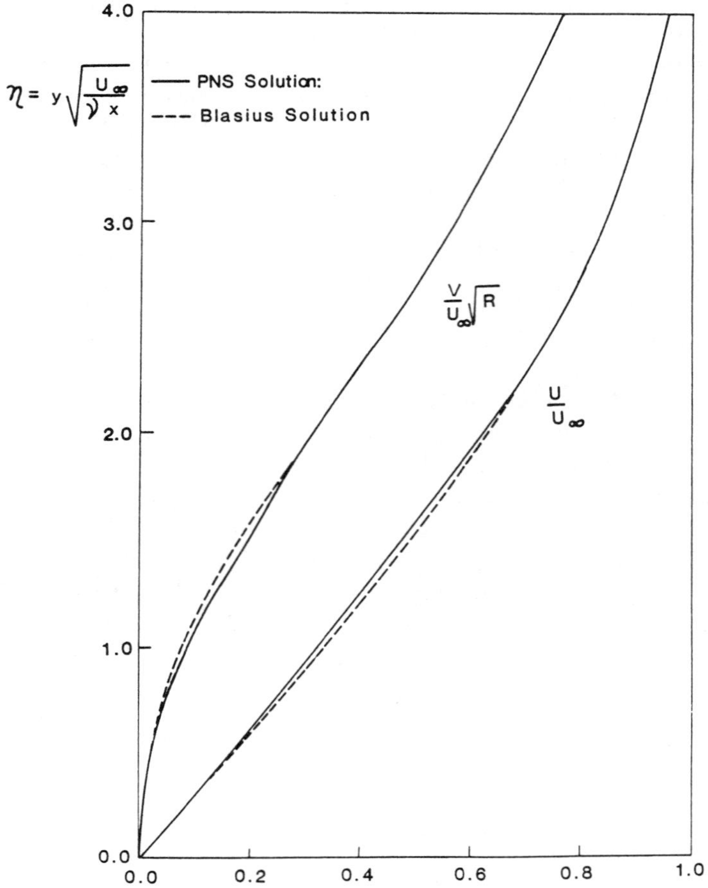

Figure 1. Comparison of the PNS profiles with the Blasius solutions: $R = 10^3, 10^4$, $N = 101$.

(i.e., the triple-deck interaction length) rather than $R^{-1/2}$ (the boundary-layer length scale), some variations in the solution were obtained. These are not given here. From the stability results, we note that the value of Δx can be made arbitrarily small; however, for the present examples this is unnecessary. For the cone geometry, to be considered in a following section, considerably smaller values of Δx are used. Some typical results for the boundary layer examples are shown in Figs. (1) to (4).

For the flat-plate case, comparisons between the PNS and the Blasius solution are shown, for $R = 10^3$ and 10^7, in Fig. 1 for the velocity components, and in Fig. 2 for the surface skin-friction coefficient C_f. The agreement is quite reasonable. The maximum error occurs at the surface, and this can be seen from the figures. Additional results for the pressure variation across the boundary layer are given in [2.89]. The pressure p_{i+1} is updated during each sweep of the global iteration procedure. For the initial iteration $p_{i+1}(x, y)$ was taken equal to the prescribed edge value; i.e., $p(x_{i+1}, y) = p(x_{i+1}, y_M)$. During the relaxation process small pressure variations are calculated across the boundary layer. The qualitative agreement with the third-order Blasius pressure distribution is good [2.89]. Solutions for the separation bubble case are given in Fig. 3 for typical isovels, and in Fig. 4 for the surface pressure variation. The predicted separation point value of $x_{sep} = 0.1180$ is close to the boundary-layer value of 0.1198. Both separation and reattachment points exhibited smooth transitions and convergence. Since the outer boundary conditions were fixed and the second-order Cauchy-integral displacement

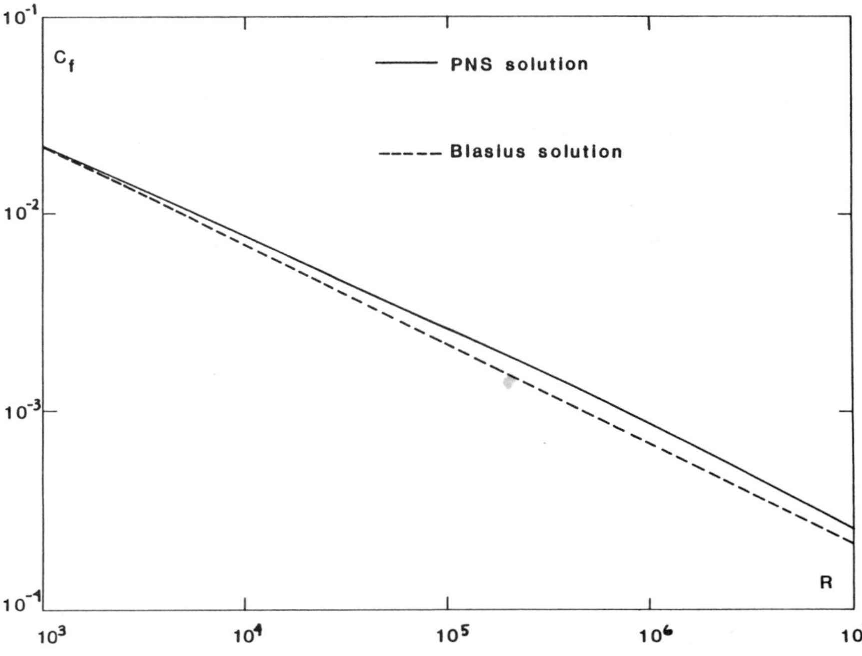

Figure 2. Comparison of the local skin friction coefficient for a flat plate.

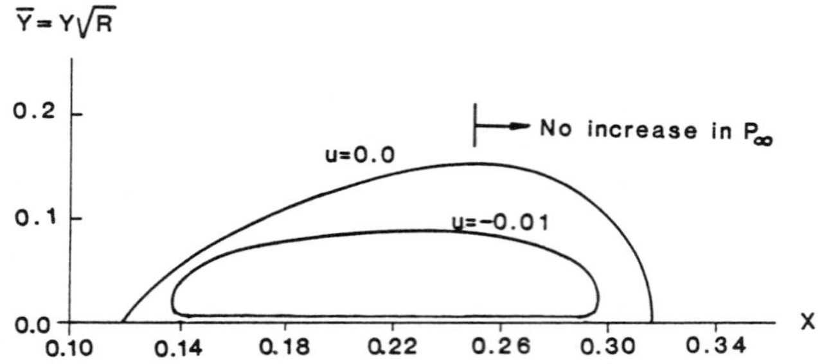

Figure 3. Isovels for separation bubble: PNS equations.

condition was not imposed, the free interaction was manifested solely through the y-momentum equation (1b). Inclusion of the displacement boundary condition should have a slight effect on the solutions. Convergence of the global relaxation procedure is quite rapid. Only five to ten iterations are required [2.98]. Of course, for the problems considered here the pressure variation across the layer is small, so that the initial guess is quite good. In view of the stability analysis previously discussed, and since the PNS system includes all of the elements of both boundary-layer and triple-deck equations, the present solutions with $y_M = O(R^{-3/8})$ should reproduce the results obtained with these approximations. Detailed comparisons will be the subject of future studies.

If central differencing is used for p_x, the downstream point p_{i+1} is introduced once again; however, the value p_i no longer appears and the y-momentum equation will be uncoupled from the velocities unless p_i is re-introduced. Several possibilities exist: (a) p_i appears directly through the outer boundary condition, as in interacting-boundary-layer theory [2.2, 2.56]; (b) a temporal relaxation term p_τ is introduced in (1a) (this has been used in ADI solutions

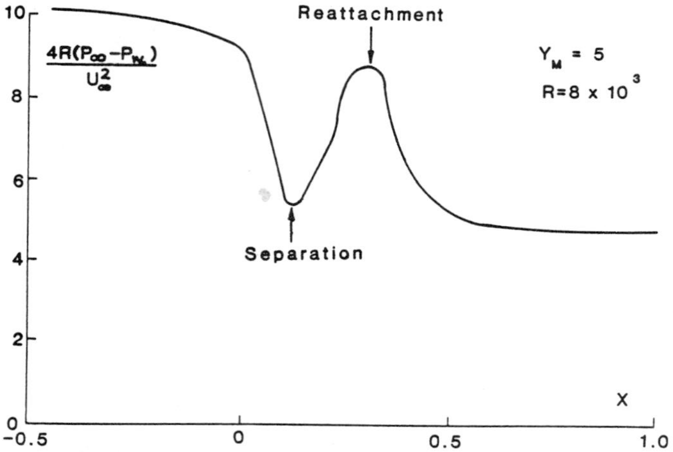

Figure 4. Wall pressure variation through the separation zone.

for interacting boundary layers) [2.56]; (c) a K–R [2.90] approximation, where forward differencing is corrected during the relaxation sweeps, is applied.

From the stability analysis for each sweep of the marching procedure, it is seen that with central differencing, the function $F(\lambda)$ in (2) is equal to -1. This is an unconditionally unstable condition and therefore further reinforces the need for a p_i contribution. With an appropriate p_i contribution $F(\lambda) = \sigma\lambda - 1$, where σ reflects the p_i term. From (2), the marching procedure is stabilized conditionally for all $\sigma \neq 0$ (recall the earlier $(\Delta x)_{\min}$ condition for $\sigma = 1$); however, for $\sigma \leq -1$, unconditional stability results. Actual experience with the various p_x approximations for the compressible PNS system is discussed in a following section on flow over a cone at incidence.

Finally, a stability analysis for convergence of the global iteration procedure has also been considered. When p_x is treated implicitly, to some degree, preliminary conclusions are that convergence is assured. On the other hand, if p_x is treated explicitly, i.e., from a previous marching solution, it would appear that the global iteration procedure will diverge. The pressure results solely from the uncoupled normal momentum equation (1b).

4 Global Iteration for Subsonic Flow

For problems where subsonic regions are not confined to thin layers, single-sweep backward differencing for p_x will generally not be reliable, since the stability condition requires that $\Delta x \geq (\Delta x)_{\min} = O(y_M) = O(1)$. Therefore, global relaxation for the pressure interaction is required in all cases.

Two procedures for calculating such flows have been considered. In the first, the pressure interaction is evaluated with the Poisson form of the pressure equation. In the second, which is currently under development, the pressure is coupled directly to the elliptic velocity solver by a splitting procedure in the spirit of Dodge [2.91]. There are a number of significant differences, however, that allow for a completely consistent global relaxation formulation. For both of these formulations a Poisson operator appears explicitly in the equations, and therefore the elliptic character of the equations is further strengthened. Only the former approach will be described here in any detail. Solutions are presented for an asymmetric channel having a moderate constriction. The complete analysis is given in [2.92]. A short review is presented here.

Equations (1a) and (1b) are rewritten in the form

$$p_x = F_1 - u_t,$$

$$p_y = F_2 - v_t,$$

so that differentiating and adding, we find

$$\nabla^2 p = F_{1x} + F_{2y} - (u_x + v_y)_t. \tag{5}$$

The continuity equation (1c) is replaced with (5). In the global iteration procedure the condition (1c) is applied in obtaining appropriate boundary values for p, and this equation is satisfied through the convergence of the pressure solver.

The primary differences between this method or the velocity-split approach for subsonic flows, and the global pressure relaxation procedure described previously for thin subsonic regions, are twofold: (a) the pressure gradients in the momentum equations are treated as given from the previous iteration, and therefore the pressure calculation is uncoupled from that for the velocities as well. In view of the temporal gradients appearing in (1a), (1b), and (5), all variables are required from previous iterations. Solutions cannot be obtained in the steady-state mode ($\Delta t = \infty$) discussed in the previous section. A line relaxation procedure coupling (5) with (1a), (1b) was not tested. The pressure equations were solved independently with an ADI or SOR technique. The temporal gradients in (1a) and (1b) allow for smooth passage through separated-flow regions.

A typical solution is shown in Fig. 5. These PNS solutions are compared with full Navier–Stokes results in [2.92]. The agreement is good, as the axial diffusion effects are quite small even in the regions of separated flow. This provides another example where steady-state marching procedures, with global pressure relaxation, can be used effectively to solve the PNS equations or Navier–Stokes equations where the influence of axial diffusion is small. A similar approach for unbounded subsonic regions is being tested with the velocity-split method mentioned previously.

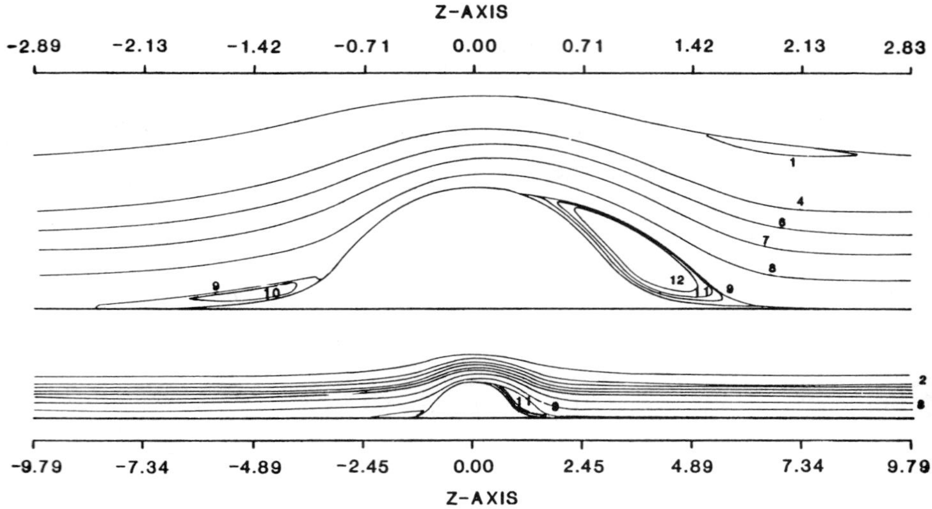

Figure 5. Streamline contours for flow in channel: $R = 10^3$ (from [2.92]).

5 Example: Cone at Incidence—Supersonic Flow

In order to test the global pressure relaxation procedure for the complete compressible PNS equations, the supersonic flow over a sharp cone at incidence has been considered [2.93]. The pressure gradient p_x in (1a) has been approximated with forward, central, and backward (single-sweep) differences. Lubard and Helliwell [2.84] and Lin and Rubin [2.85, 2.88], among others, have already investigated the latter case. Our primary interest in this regard is to evaluate the limiting value of $(\Delta x)_{\text{min}}$ in each of the cases and to determine whether the forward-difference approximation retains the effective unconditional stability predicted in the previous analysis.

In a recent study, Schiff and Steger [2.94] have presented a global iteration method, but not with the intent of treating problems where upstream interaction or axial flow separation is important. An estimate for p_x is obtained by an initial backward sweep with $\Delta x \geq (\Delta x)_{\text{min}}$. Subsequently, p_x is treated as known from the previous iterative value of p. In addition, the sublayer approximation presented by Lin and Rubin (see [2.85]) is applied in order to reduce $(\Delta x)_{\text{min}}$; i.e., p_x is assumed constant across the subsonic portion of the boundary layer. As noted earlier, the convergence analysis for procedures in which p_x is treated explicitly, i.e., from the previous sweep, would indicate that these global relaxation methods are unstable. In [2.94] the appearance of oscillations is noted after four global iterations.

A second important feature of the present analysis, which is described in detail in [2.93], is the application of the coupled strongly implicit procedure of Rubin and Khosla [2.95] for the cross-plane (normal to the axial flow or x direction) solution. This is considered to be an improvement over ADI or SOR or other splitting techniques, as there is an immediate coupling of all the boundaries, i.e., shock, body, lee, and wind planes. The strongly implicit character of the algorithm also appears desirable for capturing imbedded shock waves and for evaluating secondary flow separation at larger angles of incidence. In addition, in comparison with earlier studies [2.95] convergence rates are improved, a direct steady-state solution is possible, and artificial dissipation has not been required for iterative convergence.

The compressible PNS equations are given by an expanded form of (1) with appropriate energy and state equations. All x-derivatives are backward or K-R [2.90] differenced except for p_x, which is forward or central differenced in certain cases. A sublayer approximation is not assumed. All y-derivatives are central differenced. The trapezoidal rule is used for the continuity and y-momentum equations. Boundary-layer-like marching is applied in the x-direction; the normal velocity v is prescribed only at the surface. The outer value of v is coupled with and determined by the Rankine–Hugoniot conditions at the shock. This condition also provides for mass conservation in the shock layer.

The strongly implicit algorithm [2.95] is used to couple the velocities u, v, w for the calculation in the (y, ζ) cross plane. The temperature is obtained independently from a similar algorithm, and the pressure is updated from the normal (y) momentum equation. The strongly implicit algorithm is of the form

$$\mathbf{V}_{ij} = \mathbf{GM}_{ij} + T_{1\,ij}\mathbf{V}_{i-1,j} + T_{2\,ij}\mathbf{V}_{i,j+1}, \tag{6}$$

where

$$\mathbf{V}_{ij} = \begin{pmatrix} u \\ v \\ w \end{pmatrix}_{ij}.$$

The algorithm (6) is scalarized to improve computational efficiency. The coefficients $\mathbf{GM}_{ij}$, $(\mathbf{T}_{1,2})_{ij}$ are determined from recursive formulas whereby

$$\mathbf{GM}_{ij} = f(\mathbf{GM}_{i+1}, \mathbf{GM}_{i,j-1}, \ldots) \tag{7}$$

etc. [2.95]. The boundary conditions on all surfaces are imposed either in (6) or in (7). Symmetry conditions are used for the wind and lee planes, and the Rankine–Hugoniot relations are imposed at the outer shock wave.

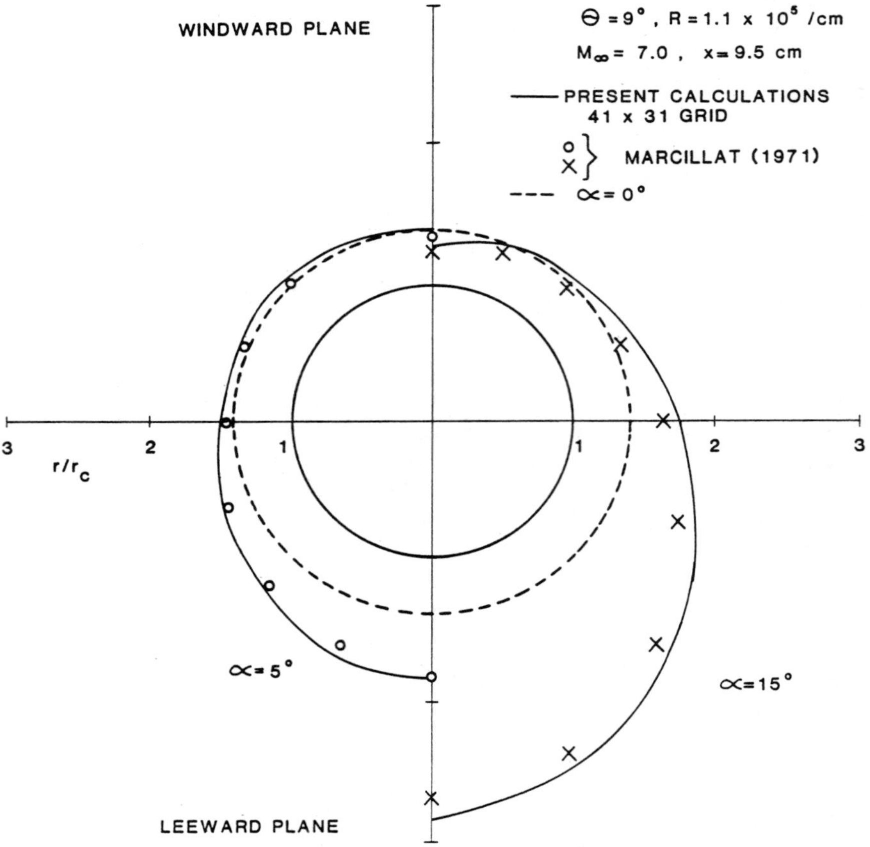

Figure 6. Shock location around cone.

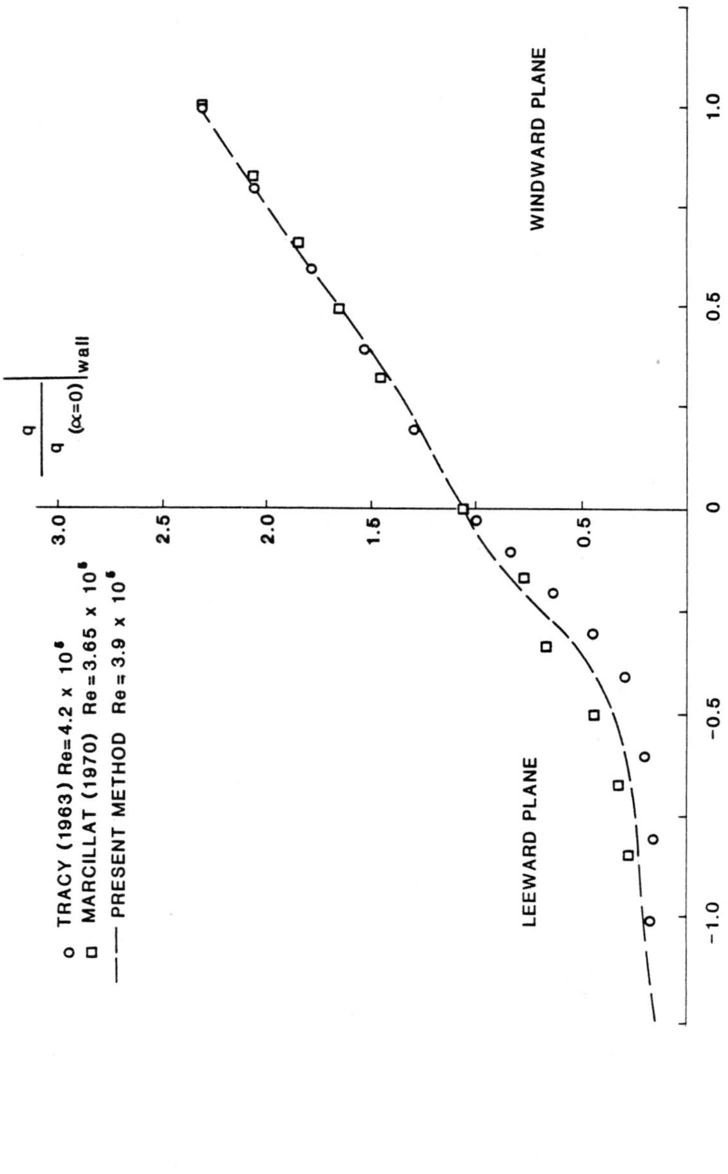

Figure 7. Symmetry plane heat transfer.

Typical solutions are given in Figs. 6 and 7 for the shock-layer thickness and the heat-transfer coefficient. The convergence criteria are 10^{-5} for $\mathbf{V}$ and 10^{-7} for p and T. The difference grid was quite coarse in the cross plane, 41×31 points. With finer grids the accuracy can be improved [2.93]. For $x/\delta(x) \geq 2$, where $\delta(x)$ is the shock-layer thickness, the subsonic layer thickness $y_M \leq 0.2\delta$. Several numerical experiments were made in order to estimate the values of $(\Delta x)_{\min}$. The results are given in Table 1. Converged results were obtained with central differencing for several marching steps; however, the solutions were oscillatory and extremely inaccurate. When these calculations were continued further, the instability predicted by the relation (4) soon leads to divergence of the solution. With the backward and forward differences, behavior similar to that found for the incompressible equation (1) was recovered here. For the backward p_x calculations, $(\Delta x)_{\min}$ appears to be approaching an asymptotic limit as R increases. For the incompressible problem, at $R = 10^4$, $(\Delta x)_{\min}/y_M \approx 0.22$, as opposed to the value 0.25 obtained for the cone flow. For the forward p_x calculations, the present results confirm those for the incompressible equation—i.e., unconditional stability. Finally, a very weak stability condition on Δx is obtained when the modified "forward-difference" condition is applied (see Table 1). As seen from the solutions presented here, this approximation is quite good for the cone geometry. Solutions for angles of incidence up to 45° are presented in [2.93].

Table 1 Estimated values of $(\Delta x)_{\min}/y_M$.[a]

	$R = 10^3$	10^4
Backward p_x	0.2	0.25
Modified "forward" p_x	10^{-3}	2×10^{-4}
Forward p_x	0	0
Central p_x	Unstable for $O(1)$	Unstable for $O(1)$

[a] $M_\infty = 7.95$; $\alpha = 18°$.

6 Summary

The parabolized Navier–Stokes equations have been considered for subsonic and supersonic flows. It has been shown that with single-sweep marching and backward differencing for all axial derivatives, including the pressure, the elliptic influence is numerically suppressed for marching steps $\Delta x \geq (\Delta x)_{\min} \approx y_M$, where y_M is the thickness of the subsonic zone. For subsonic boundary layers or the PNS equations with supersonic outer flow conditions, $y_M = O(R^{-1/2})$ and therefore $(\Delta x)_{\min} = O(R^{-1/2})$. For triple-deck regions, $y_M = O(R^{-3/8})$, so $(\Delta x)_{\min} = O(R^{-3/8})$. For problems with large regions of subsonic flow, $(\Delta x)_{\min} = O(1)$ and large truncation errors can be expected.

If global relaxation is considered, i.e., multiple marching sweeps, then central differencing for p_x is unstable, but forward p_x differencing is unconditionally stable. Forward differencing also has the desirable property of allowing for a complete pressure coupling with the velocities and a free surface pressure interaction. Therefore, separation regions can be evaluated with this global relaxation method. Examples for a flat-plate boundary layer and for a separation bubble have been presented.

For supersonic outer flows the global relaxation method requires only that the pressure be retained from the previous iteration. Therefore, computer storage is minimal. With large regions of subsonic flow, a global iteration method is also presented; however, pressure and velocity data are required in this procedure, and therefore computer storage will be increased. Solutions are presented for separated channel flow.

Numerical experiments have been conducted for the supersonic flow over a cone at incidence. A coupled strongly implicit numerical algorithm was applied with backward, central, and forward differencing for p_x. The solutions confirm the analytic stability results for the incompressible equations. Backward differencing is conditionally stable ($\Delta x \geq (\Delta x)_{\min} \approx y_M$); central differencing is unstable, and forward differencing is unconditionally stable. In view of the results obtained herein, we conclude that for flows with thin subsonic layers, forward differencing for the p_x term leads to an optimal global pressure relaxation procedure, with free pressure interaction and minimum stability limitations. Global relaxation solutions have been obtained for subsonic flows; however, optimization of such techniques requires further study.

Acknowledgment

The author would like to thank Professors P. Khosla and A. Lin for many useful conversations and for their help in preparing this paper.

The research was supported in part by the Office of Naval Research under Contract N00014-78-0849, Task No. NR 061-258, and in part by the Air Force Office of Scientific Research under Grant No. AFOSR 80-0047.

CHAPTER 12

Progress on Interacting Boundary-Layer Computations at High Reynolds Number[1]

R. T. Davis* and M. J. Werle†

1 Introduction

It is becoming increasingly clear that many viscous-flow problems involving separation can be handled with Prandtl boundary-layer equations which are allowed to interact through the displacement thickness with an outer inviscid flow. We will call this model the interacting-boundary-layer (IBL) model. The triple-deck (TD) theory of Stewartson [1.106], Sychev [2.13], and Messiter [2.96], governing the small-separation[2] problem, contains no terms which are not included in the IBL model, and therefore the model is correct for at least the small-separation problem in a composite sense. In addition, the Sychev [2.13] and Smith [2.18, 2.97] theories of incompressible massive separation indicate that up to and through separation, the triple-deck theory (and therefore IBL theory) holds if one has properly taken care of the downstream wake in the interaction model.

The purpose of this paper is to examine some of the existing techniques for solving the subsonic laminar and turbulent interacting-boundary-layer equations and to propose and demonstrate some improvements. The first

[1] This research was supported (for R.T.D.) by NASA under Consortium Agreement NCA2-OR130-901 and ONR under N00014-76-C-0364, and (for M.J.W.) by Naval Air System Command Contract N00019-80-C-0057.

* Professor of Aerospace Engineering and Applied Mechanics, University of Cincinnati, Cincinatti, OH 45221.

† Manager of Gas Dynamics and Thermophysics, United Technologies Research Center, East Hartford, CT 06108.

[2] By small separations we mean those totally contained within the triple-deck structure.

improvement consists of developing a method for incorporating the influence of the upstream and downstream (wake or nonwake) displacement-thickness effect on a strong (nonlinear) interaction region. The other improvements are related to the manner in which the IBL equations are handled in the strong-interaction region. Two methods are examined for coupling the Hilbert integrals (airfoil integrals) to the boundary-layer equations. The first is due to Carter [2.98], and the second is a modification of the method proposed by Veldman [2.2]. In addition, methods for the numerical evaluation of the Hilbert integrals themselves are discussed. Also, an improved method is presented, for the inversion of the coupled set of equations (difference equations for continuity and momentum plus boundary conditions) which result at a given solution station. Finally, a new generalized coordinate system is demonstrated for separated flow calculations where classical surface-oriented coordinates are inappropriate.

Several model problems are considered. The first set are laminar flows past a finite flat plate, an infinite flat plate with a depression-induced (Carter–Wornom trough [2.3]) separation, and a flow past a finite-thickness trailing edge (Werle–Verdon trailing edge [2.20]). The second set involve turbulent flows past the same finite flat plate and finite-thickness trailing edge.

The laminar finite-flat-plate problem is used to demonstrate and assess the IBL concept through a detailed comparison with the triple-deck solution as the Reynolds number approaches infinity. The other problems are chosen to demonstrate the ability of the IBL model to predict flow properties, including those involving separation and turbulence.

While the present methods have not yet been used to solve problems involving massive separation, it is felt that some of the concepts presented provide a basis for developing a method capable of handling this situation if an appropriate wake model is incorporated.

2 Governing Equations and Boundary Conditions

The incompressible interacting-boundary-layer equations used in this study are the usual ones except that they are written in a coordinate system which may be situated a small distance (i.e., within the boundary-layer) from the body surface or wake centerline; see Fig. 1. This coordinate system, introduced by Werle and Verdon [2.20], is especially useful for trailing-edge flows where the coordinate base curve is chosen to be aligned as nearly as possible with the shear layer. An obvious choice for the coordinate base curve would be the displacement thickness; however, that has not been done here, since that would require updating the coordinate system as the displacement thickness changes with each iteration of the solution process. The reason for choosing such a coordinate system is that it is more physically reasonable than the original body-oriented one. Viscous effects smooth the body surface, and the outer flow sees a body consisting of the original body thickened by displacement thickness, which is smoother than the original body itself. In

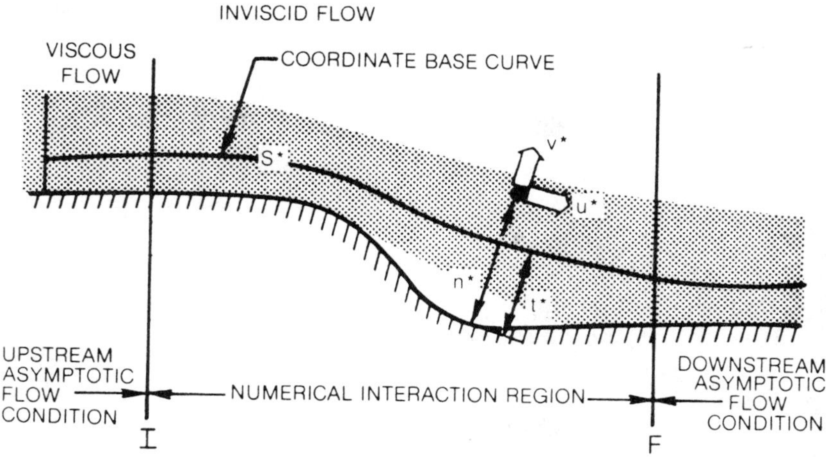

Figure 1. Flow–field nomenclature.

addition, the governing boundary-layer equations written in the base-curve coordinate system may be physically more accurate, since neglecting streamwise diffusion along a properly chosen base curve should be more accurate than along a tangent to the original body surface, especially for separated flows. The only difficulty left with the base-line coordinate system is that the body surface no longer lies along a coordinate line. However, following the lead of Jenson [2.99], this is overcome by the use of Prandtl's transposition theorem.

The dependent and independent boundary-layer flow variables are nondimensionalized and the Prandtl transposition theorem is applied simultaneously as follows:

$$u = \frac{u^*}{u_r^*}, \qquad v = \left(w^* + \frac{dt^*}{ds^*} u^*\right) \frac{\sqrt{\text{Re}}}{u_r^*}, \tag{2.1}$$

$$p = \frac{p^* - p_\infty^*}{\rho_\infty^* u_r^{*2}}, \tag{2.2}$$

$$s = \frac{s^*}{L_r^*}, \qquad n = y + t = (y^* + t^*) \frac{\sqrt{\text{Re}}}{L_r^*}, \tag{2.3}$$

where n^* is measured from the body surface, y^* is measured from the base line, t^* is the normal distance from the base curve to the body surface, and the Reynolds number Re is defined by

$$\text{Re} = \frac{\rho_\infty^* u_r^* L_r^*}{\mu^*}.$$

Note that the transposition theorem appears in Eqs. (2.1) and (2.3), where w is the physical normal velocity.

In these variables the incompressible boundary-layer equations become

$$\frac{\partial u}{\partial s} + \frac{\partial v}{\partial n} = 0 \tag{2.4}$$

and

$$u\frac{\partial u}{\partial s} + v\frac{\partial u}{\partial n} - U_e\frac{dU_e}{ds} = \frac{\partial}{\partial n}\left(\mu_T\frac{\partial u}{\partial n}\right) \tag{2.5}$$

where μ_T is given by $\mu_T = 1 + e\gamma_T$, e being the eddy-viscosity term and γ_T a longitudinal intermittency factor. Most of the examples presented here will be laminar flow cases, and therefore e is taken to be zero. For the turbulent cases the model used is given in Werle and Verdon [2.20] and will not be repeated here.

In addition to the governing equation we must prescribe boundary conditions. On a solid surface with no ejection, the no-slip condition implies $u(s, 0) = v(s, 0) = 0$.

For a wake centerline, the conditions are more complicated. However, in the examples we will consider, the angle between a tangent to the baseline coordinate surface and the wake centerline is small. Since, for the present cases, the wake centerline is a plane of symmetry, $w_n^* = 0$, where w_n^* is the physical velocity component normal to the centerline. Note that for small slopes $w_n^* \approx +(dt^*/ds^*)u^*$, where w^* and u^* are the physical velocity components in the base-line coordinate system. Therefore from Eq. (2.1) the nondimensional transposed velocity component v is zero. The proper boundary condition on v is therefore $v(s, 0) = 0$ on the wake centerline. A similar argument will show that under the same conditions, due to symmetry and small angles,

$$\frac{\partial u}{\partial n}(s, 0) = 0 \tag{2.6}$$

on the wake centerline (see Werle and Verdon [2.20] for more details).

In addition to conditions on the body or wake centerline surface, we must prescribe an outer-edge condition. The usual one is to say that $u(s, n) \to U_e$ as $n \to \infty$, where U_e is in the present case the potential-flow velocity on the body surface thickened by the displacement thickness. However, from the matching conditions at the outer edge (see Van Dyke [2.100]) it can be shown that

$$v - nv_n \to \frac{d}{ds}(U_e\delta) \quad \text{as } n \to \infty, \tag{2.7}$$

where δ is the classical displacement thickness. This condition turns out to be a more convenient outer-edge expression to use. It should be noted that in the above expression v is the Prandtl transposed velocity, i.e. $v = \overline{w} + (dt/ds)u$, where $\overline{w}$ is the physical dimensionless normal boundary-layer velocity component.

The matching conditions further provide the dimensionless inviscid normal surface injection velocity $W(s, 0)_{\text{inv}}$ to be applied along the base line (see Fig. 1) as

$$W(s, 0)_{\text{inv}} = \frac{1}{\sqrt{\text{Re}}} \frac{d}{ds} [U_e(\delta - t)]. \tag{2.8}$$

Note that if we have chosen t to be δ, then $W(s, 0)_{\text{inv}} = 0$ as it should.

The condition (2.8) provides the surface condition on the baseline for an inviscid-flow calculation. If, for example, the inviscid flow field is symmetric with the freestream velocity vector $U_\infty \mathbf{i}$ parallel to the x-axis and governed by linear theory, the base line might be chosen to be the x-axis and therefore

$$t = -f, \tag{2.9}$$

where f is half the airfoil thickness function. Incompressible linear airfoil theory therefore relates $W(s, 0)_{\text{inv}}$ to $U_e(s, 0)$ by

$$U_e(s, 0) = 1 + \frac{1}{\pi\sqrt{\text{Re}}} \oint_{-\infty}^{\infty} \frac{1}{s - \bar{s}} \frac{d}{d\bar{s}} U_e(\delta + f)\, d\bar{s} \tag{2.10}$$

and this condition along with Eq. (2.7) provides the outer-edge condition on the interacting-boundary-layer calculation.

In order to provide the proper scaling in a triple-deck region it is useful to use variables which are scaled according to the lower deck of triple-deck theory rather than according to Prandtl boundary-layer theory as has been given in Eqs. (2.1)–(2.3). These lower-deck variables provide the proper resolution for the smallest scales existing in a problem. The lower-deck variables are given by Stewartson [1.106] and are repeated here for the incompressible case for convenience. Let s_1 be a Prandtl variable measuring distance along the base line from a triple-deck-type disturbance which is located at s_0, i.e.,

$$s_1 = s - s_0. \tag{2.11}$$

The quantity s_0 could be, for example, the location of the trailing edge of a finite flat plate. The lower-deck variables are then related to the Prandtl variables by

$$U = \frac{u}{\varepsilon\lambda^{1/4}}, \qquad V = \frac{v\varepsilon}{\lambda^{3/4}}, \qquad P = \frac{p}{\varepsilon^2\lambda^{1/2}}, \tag{2.12}$$

$$S = \frac{s_1\lambda^{5/4}}{\varepsilon^3}, \qquad N = \frac{n\lambda^{3/4}}{\varepsilon}, \tag{2.13}$$

where $\varepsilon = \text{Re}^{-1/8}$, and $\lambda = 0.332$ is the Blasius skin-friction constant which is appropriate for flat-plate triple decks.

The equations (2.13) set the mesh sizes for numerical calculations in triple-deck regions. Computations should be done in these regions with small step sizes ΔS and ΔN rather than Δs_1 and Δn. Some of the results to be given later will show that if this is done, IBL results will approach triple-deck results as $\mathrm{Re} \to \infty$.

Rather than solve the governing equations in physical variables, it is possible to obtain more accurate solutions with a given number of grid points by using similarity-type variables. This is true for two reasons. First, similarity variables take advantage of the locally similar nature of the flow and minimize streamwise gradients, and second, similarity variables more properly account for the growth of the boundary layer and allow one to use a fixed number of mesh points in the normal direction for the entire flow field.

We will use a modified form of Görtler variables in the present study. Let $U_{e0}(s)$ be the inviscid surface speed for flow past a representative surface. Therefore coordinates ξ, η are defined by

$$\xi = \int_0^s U_{e0}(s)\tilde{g}(s)\, ds, \qquad \eta = \frac{U_{e0}(s)n}{\sqrt{2\xi}}. \tag{2.14}$$

The quantity $\tilde{g}$ is a function defined by Werle and Verdon [2.20] in order to control the growth of a turbulent boundary layer. For the laminar case $\tilde{g}$ is equal to one. The dependent variable u is replaced by

$$F = \frac{u}{U_{e0}}. \tag{2.15}$$

Using these variables and defining a similarity type V function as usual we obtain after transformation the similarity form of Eqs. (2.4) and (2.5) as

$$\frac{\partial V}{\partial \eta} + F + 2\xi \frac{\partial F}{\partial \xi} = 0, \tag{2.16}$$

$$\frac{\partial}{\partial \eta}\left(l \frac{\partial F}{\partial \eta}\right) - V \frac{\partial F}{\partial \eta} + \beta_1 - \beta_0 F^2 - 2\xi F \frac{\partial F}{\partial \xi} = 0, \tag{2.17}$$

where

$$\beta_0 = \frac{2\xi}{U_{e0}} \frac{dU_{e0}}{d\xi}, \qquad \beta_1 = \frac{2\xi}{U_{e0}^2} U_e \frac{dU_e}{d\xi}. \tag{2.18}$$

Here β_0 is the pressure-gradient parameter for the representative base flow, and β_1 is the pressure-gradient parameter for the interacted flow, i.e., U_e is the surface inviscid speed for flow past the body thickened by the displacement thickness. Finally, the quantity l is given by $l = (1 + e\gamma_T)/\tilde{g}$.

If we choose the representative base flow to be the inviscid flow past the displacement body, then $\beta_0 = \beta_1$, both of which must be determined from the solution. This is the approach used by Werle and Verdon [2.20] and Vatsa, Werle and Verdon [2.101]. However, an alternate approach could be to take U_{e0} to be that of the flow over the base-line surface of Fig. 1, thereby isolating the unknown pressure-gradient effect into the single linear term β_1. This is the approach used by Davis [2.102].

The solid-surface no-slip boundary conditions are replaced by $F(\xi, 0) = V(\xi, 0) = 0$, and the wake centerline conditions are replaced by $V(\xi, 0) = 0$ and

$$\frac{\partial F}{\partial \eta}(\xi, 0) = 0. \tag{2.19}$$

The edge conditions are more complicated, but after some algebra they become

$$F \to \frac{U_e}{U_{e0}} = 1 + \frac{\bar{U}_e}{U_{e0}} \quad \text{as } n \to \infty \tag{2.20}$$

and

$$V - nV_n \to \sqrt{2\xi}\,\frac{d}{d\xi}\,[U_e\delta] \quad \text{as } n \to \infty. \tag{2.21}$$

Finally, the surface injection condition (2.8) becomes

$$W_{\text{inv}}(\xi, 0) = \frac{U_{e0}\tilde{g}}{\sqrt{\text{Re}}}\frac{d}{d\xi}\,[U_e(\delta - t)]. \tag{2.22}$$

3 Numerical Methods

The first issue to be resolved in attempting to solve the interacting boundary-layer equations numerically is to develop a method for the solution of the inviscid-flow problem. Next we need to couple this inviscid solution method to a boundary-layer solution method and through some global iterative process produce the solution which satisfies the full set of interacting equations. The manner in which this is done is crucial to producing a stable and accurate method which converges in a reasonable number of global iterations. Early methods for solving the triple-deck or interacting boundary-layer equations for incompressible and subsonic flow tended to be slowly converging. More recently, methods have been developed by Carter [2.98] and Veldman [2.2] which show much higher convergence rates. We will use ideas from both the methods of Carter and Veldman in our present study.

Let us assume that the outer inviscid flow can be computed using linear airfoil theory. In Fig. 1 we can take the base-line curve to be the x-axis. The total thickness function T then becomes the body shape $f(x)$ plus the displacement thickness δ.

Therefore, if we let

$$T(x) = U_e\,\frac{f(x) + \delta(x)}{\sqrt{\text{Re}}}, \tag{3.1}$$

Eq. (2.10) becomes

$$U_e(x) = 1 + \frac{1}{\pi}\oint_{-\infty}^{+\infty}\frac{T'(\xi)}{x - \xi}\,d\xi, \tag{3.2}$$

where $'$ means differentiation with respect to the given argument.

A numerical method can be developed directly from (3.2) by assuming particular forms for $T(\xi)$ on an element and developing element expressions; see Napolitano, Werle, and Davis [2.103]. Alternatively, if care is taken with the singular point at $x - \xi = 0$, we can also integrate (3.2) by parts to produce expressions of the type [3]

$$U_e(x) = 1 + \frac{1}{\pi} \int_{-\infty}^{+\infty} T''(\xi) \ln |x - \xi| \, d\xi \tag{3.3}$$

or

$$U_e(x) = 1 - \frac{1}{\pi} \int_{-\infty}^{+\infty} \frac{T(\xi)}{(x - \xi)^2} \, d\xi. \tag{3.4}$$

Veldman [2.2] and Veldman and Dijkstra [2.104] have used the form (3.3) along with the assumption that $T''(\xi)$ is a constant on an element. In addition, they have replaced $T''(\xi)$ with the three-point central-difference expression. We have found that the expression (3.4) is more convenient for several reasons to be discussed below.

First, we subdivide the thickness function $T(x)$ into elements as shown in Fig. 2 in the interaction region. The x_i locations are the boundary-layer solution points, and the points $x_i - h/2$ and $x_i + h/2$ are the midpoints between solution points. The thickness functions upstream and downstream of the interaction region are replaced with expressions which properly model the asymptotic nature of the flow. For example, if the upstream flow is a laminar flat-plate one, $T(x)$ can be replaced with the flat-plate displacement thickness $C\sqrt{x}$, where C is determined by matching with the computed function $T(x)$ at the first solution point. Similar types of expressions can be used for laminar flat-plate flow beyond the interaction region, with matching provided between these downstream conditions and the final solution point. For this case these upstream and downstream forms for $T(x)$ can be integrated analytically, and therefore the numerical procedure need only focus attention on the region between some initial point I and final point F of the interaction region. The general numerical expression for Eq. (3.4) is replaced by

$$U_{ei} = -\frac{1}{\pi} \int_I^F \frac{T(\xi)}{(x_i - \xi)^2} \, d\xi + g(x_i), \tag{3.5}$$

where

$$g(x_i) = 1 - \frac{1}{\pi} \int_{-\infty}^{I} \frac{T(\xi)\, d\xi}{(x_i - \xi)^2} - \frac{1}{\pi} \int_F^{\infty} \frac{T(\xi)}{(x_i - \xi)^2} \, d\xi, \tag{3.6}$$

and the last two integrals in (3.6) are replaced with the analytical expressions mentioned previously. For more complicated flows, numerical procedures can be straightforwardly developed for evaluating the integrals of Eq. (3.6), as will be discussed in a later section.

[3] These integrals and later ones are considered to be written in symbolic form since care must be taken in evaluating them around the point $x - \xi = 0$.

$$U_{e_i} = -\frac{1}{\pi}\int_I^F \frac{T}{(X_i-\xi)^2}\,d\xi + g(X_i) = \frac{1}{\pi h}\sum_{j=1}^{N} T_j\,D(j-i) + g(X_i)$$

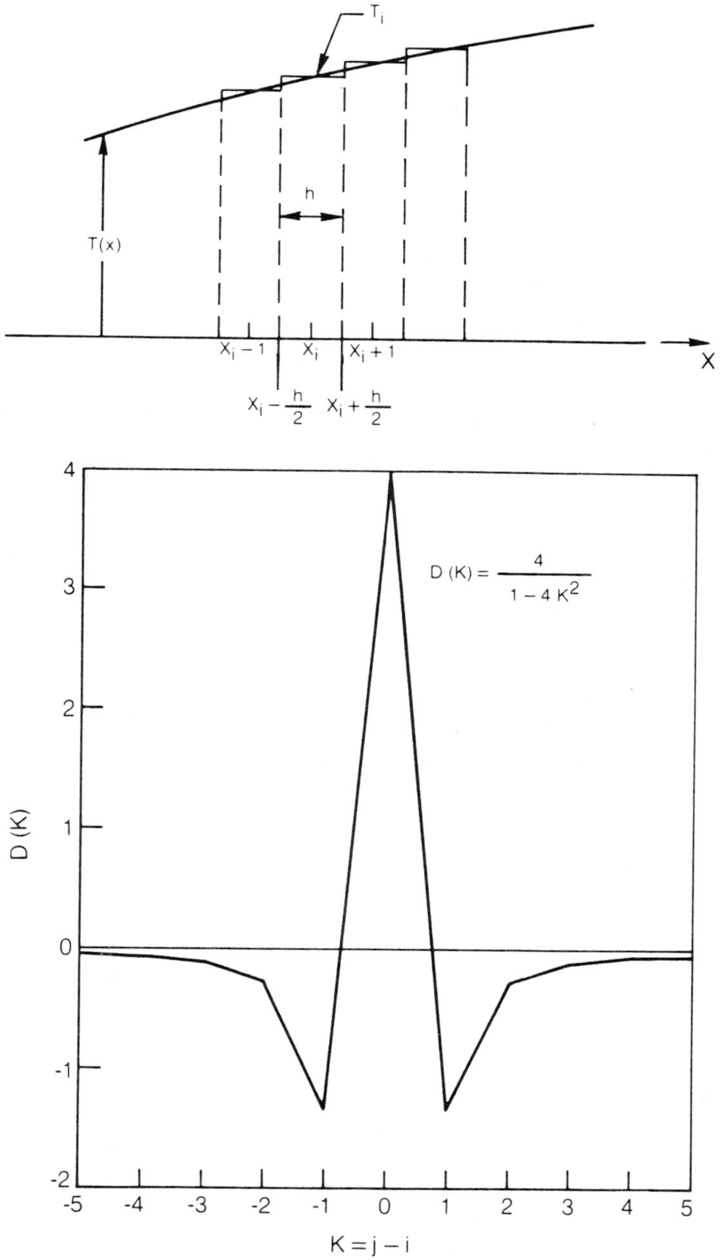

Figure 2. Inviscid-flow solver.

Finally, the integral in (3.5) is evaluated by assuming $T(\xi)$ is a constant on an element (see Fig. 2(a)). This results in

$$U_{ei} = \frac{1}{\pi h} \sum_{j=1}^{N} T_j D(j-1) + g(x_i), \tag{3.7}$$

where

$$D(j-i) = \frac{4}{1-4(j-i)^2} \tag{3.8}$$

and N is the number of the last element.

It is interesting to observe the behavior of the function $D(j-i)$. If we focus attention on the point $K = j - i = 0$ and examine the values of D at this and neighboring locations, we find that $D(0) = 4$, $D(\pm 1) = -\frac{4}{3}$, and D drops off rapidly as K increases; see Fig. 2(b). Therefore, the ratio

$$\frac{D(0)}{D(\pm 1)} = -3 \tag{3.9}$$

indicates that Eq. (3.7) has excellent properties for the development of an iterative solution technique. Veldman [2.2] reached the same conclusion by using Eq. (3.3) rather than (3.4) to develop element expressions. The advantage here is that Eq. (3.8) is simpler than Veldman's expression in that the present expression involves no log-type functions. Since the only index appearing is $j - i$, Eq. (3.8) can be stored as a one-dimensional array. Napolitano, Werle, and Davis [2.103], as well as Veldman [2.2] and Veldman and Dijkstra [2.104], observed this feature in evaluating similar-type integral expressions. If one develops expressions for variable mesh spacings, this property is lost in all methods.

In the present study, two methods have been used for solving the interacting boundary-layer equations. The first method (method A) is due to Werle and Verdon [2.20] and Vatsa, Werle, and Verdon [2.101]. The manner in which the solution is globally iterated is due to Carter [2.98]. The second method (method B) is due to Davis [2.102], and the manner in which the solution is globally iterated is due to Veldman [2.2].

3.1 Method A

First the boundary-layer equations (2.16) and (2.17) are written in difference form by using two-point differences in the η-direction. The resulting difference scheme is second-order accurate in the η-direction and may be first- or second-order accurate in the ξ-direction, depending upon the manner in which averaging and quasilinearization of the nonlinear terms are handled. The pressure-gradient parameter β_1 ($\beta_0 = \beta_1$ in this method) is treated as a parameter in the problem, and both homogeneous and particular solutions to linearized difference equations are found by inverting the difference equations twice using the appropriate surface or wake boundary conditions and outer-edge conditions. This produces a solution of the form

$$F = \beta_1 F_a + F_b, \qquad V = \beta_1 V_a + V_b, \tag{3.10}$$

where F_a, F_b, V_a, and V_b are now known. With the definition of displacement thickness and using Eq. (3.10), we obtain

$$\delta = \frac{\sqrt{2\xi}}{U_{e0}} \int_0^\infty (1 - F_b)\, d\eta + \beta_1 \frac{\sqrt{2\xi}}{U_{e0}} \int_0^\infty F_a\, d\eta, \tag{3.11}$$

and therefore β_1 can be determined if δ is given. Therefore a $\delta(x)$ distribution function is assumed (in the spirit of Carter's [2.98] inverse method), and β_1 is determined from the above equation. The solution is continued downstream until the final streamwise location is reached.

The calculated β_1 distribution is now integrated with respect to x to produce a U_e distribution. Let us call this value U_{ev}. Next, the assumed $\delta(x)$ distribution is used to form the $T(x)$ expression given by Eq. (3.1), and a numerical solver for the inviscid flow like Eq. (3.7) is used to determine a U_e, which we will call U_{ei}. Normally U_{ev} and U_{ei} will be different at each streamwise station. Carter's [2.98] technique,[4]

$$\delta^{n+1} = \delta^n[1 + \omega(U_{ei} - U_{ev})], \tag{3.12}$$

is used to produce a new δ and the whole process is repeated iteratively until convergence is achieved. The reader is referred to Werle and Verdon [2.20] details of this method.

3.2 Method B

The continuity and momentum equations are differenced in much the same was as in Method A, except that the $2\xi F_\xi$ term in the momentum equation (2.17) is replaced by $-F - V_\eta$ from the continuity equation (2.16). Again the resulting difference equations may be first- or second-order accurate in the ξ-direction, depending upon the manner in which the averaging and quasi-linearization are carried out. Rather than use a superposition technique on β_1, an additional differential equation is added of the form

$$\frac{d\beta_1}{d\eta} = 0, \tag{3.13}$$

which is also written in finite-difference form. This results in a coupled set of three difference equations which can be inverted directly once the boundary conditions are prescribed.

On the surface or wake centerline the usual boundary conditions, as in Method A, are used. However, the outer-edge conditions are different.

[4] The actual solution algorithm employs the local pressure levels instead of the velocity levels. Since that is equivalent to this expression, details are not provided here.

The outer-edge condition on β_1 is prescribed by evaluating the difference form of the momentum equation (2.17) as $\eta \to \infty$. Next the $T_j D(j-i)$ term at $j = i$ in Eq. (3.7) is isolated, and the remaining terms are assumed to be known, i.e., from Eq. (3.7),

$$U_{ei} = \bar{A}_i + \frac{4}{\pi h} T_i, \tag{3.14}$$

where $\bar{A}_i$ is given by

$$\bar{A}_i = g(x_i) + \sum_{\substack{j=1 \\ j \neq i}}^{N} T_j D(j-i). \tag{3.15}$$

Finally we substitute Eq. (3.1) into (3.14), isolate $(U_e\delta)_i$, and rewrite Eq. (3.14) as

$$U_{ei} = \bar{C} + \bar{D}(U_e\delta)_i \tag{3.16}$$

Assuming $\bar{C}$ and $\bar{D}$ contain the most recent values of $U_{e\delta}$, Eq. (3.16) is an excellent candidate for an iterative technique for the reason stated earlier; see Fig. 2(b). This step is similar to the procedure used by Veldman [2.2].

Finally, $(U_{e\delta})_i$ is determined from Eq. (3.16) as

$$(U_e\delta)_i = \frac{U_{ei}}{\bar{D}} - \frac{\bar{C}}{\bar{D}} \tag{3.17}$$

and substituted into Eq. (2.21) to produce an outer-edge condition written partially in finite-difference form, i.e.

$$V - \eta V_\eta \to \frac{\sqrt{2\xi}}{\Delta\xi}\left[\frac{U_{ei}}{\bar{D}} - \frac{\bar{C}}{\bar{D}} - (U_e\delta)_{i-1}\right] \quad \text{as } \eta \to \infty. \tag{3.18}$$

The η-derivatives are also written in finite-difference form, and therefore Eq. (3.18) serves as an outer-edge boundary condition on the coupled set of difference equations for continuity, momentum, and normal pressure gradient.

An algorithm has been developed for the direct inversion of this set of equations and boundary conditions without superposition or iteration.

The problem is therefore solved by repeated passes in the ξ-direction until convergence is achieved. This method will be reported on in detail in a fourthcoming paper by Davis [2.102].

4 Results

We have chosen several example cases of laminar and turbulent flow in order to assess and demonstrate the ability of the interacting boundary-layer solution techniques (methods A and B) to produce meaningful results. The first example case is that of laminar flow past a finite flat plate.

4.1 Laminar Flow past a Finite Flat Plate

This case has been chosen in order to demonstrate the approach of the interacting boundary-layer solutions to the infinite-Reynolds-number triple-deck limit for the trailing-edge flow. In addition, it has been used to determine the influence of the assumed forms of the upstream and downstream displacement-thickness distributions on the solution.

The Prandtl mesh spacings Δx and $\Delta\eta$ have been chosen according to the lower-deck scaling laws. For example, for flow at Reynolds number 10^5, based on the plate width, a longitudinal lower-deck step size of 0.2 translates to a Prandtl step size of 0.0106, whereas a lower-deck normal step size of 0.282 translates to a Prandtl normal step size of 0.153 (see Eq. (2.13)). As the Reynolds number increases, the situation becomes even more severe. For example, at a Reynolds number of 10^7 the same lower-deck step sizes translate to Prandtl step sizes of $\Delta x = 0.00188$ and $\Delta\eta = 0.086$ respectively.

The numerical calculations have therefore been carried out by fixing the mesh spacings according to the lower-deck scaling. The coordinate system is oriented at the trailing edge, and calculations have been performed with the above lower-deck mesh spacings over a range of Reynolds numbers. The same calculations have also been repeated with half the original spacings and again with one-quarter the original longitudinal step size. The upstream boundary condition is applied as a Blasius profile at a triple-deck location of $X = -8.0$, and the calculations terminate downstream at a value of $X = 16.0$ (for the Prandtl values see Eq. (2.13)). These locations are sufficiently far from the trailing edge to capture the triple-deck structure. In the region upstream of the initial solution station, it is assumed that the boundary-layer displacement thickness is of the form $C_1\sqrt{x}$, where x is the Prandtl variable measured from the leading edge and C_1 is determined by matching with the first solution point. For the downstream wake beyond the last solution point, we assume

$$\delta = 0.664 + \frac{C_2}{(x - C_3)^{1/2}}, \tag{4.1}$$

after Goldstein. The constants C_2 and C_3 are determined by matching the slopes and magnitudes of the displacement thickness at the last solution stations.

The calculations have been found to be insensitive to the locations of the initial and final solution stations as long as they are outside the triple-deck region. It has been found that the solutions are influenced significantly by the inclusion of the leading-edge and wake displacement-thickness functions. Calculations performed at finite Reynolds number with and without these functions produce significantly different results (particularly for the pressure distribution). The differences are larger at the lower values of the Reynolds number, as they should be.

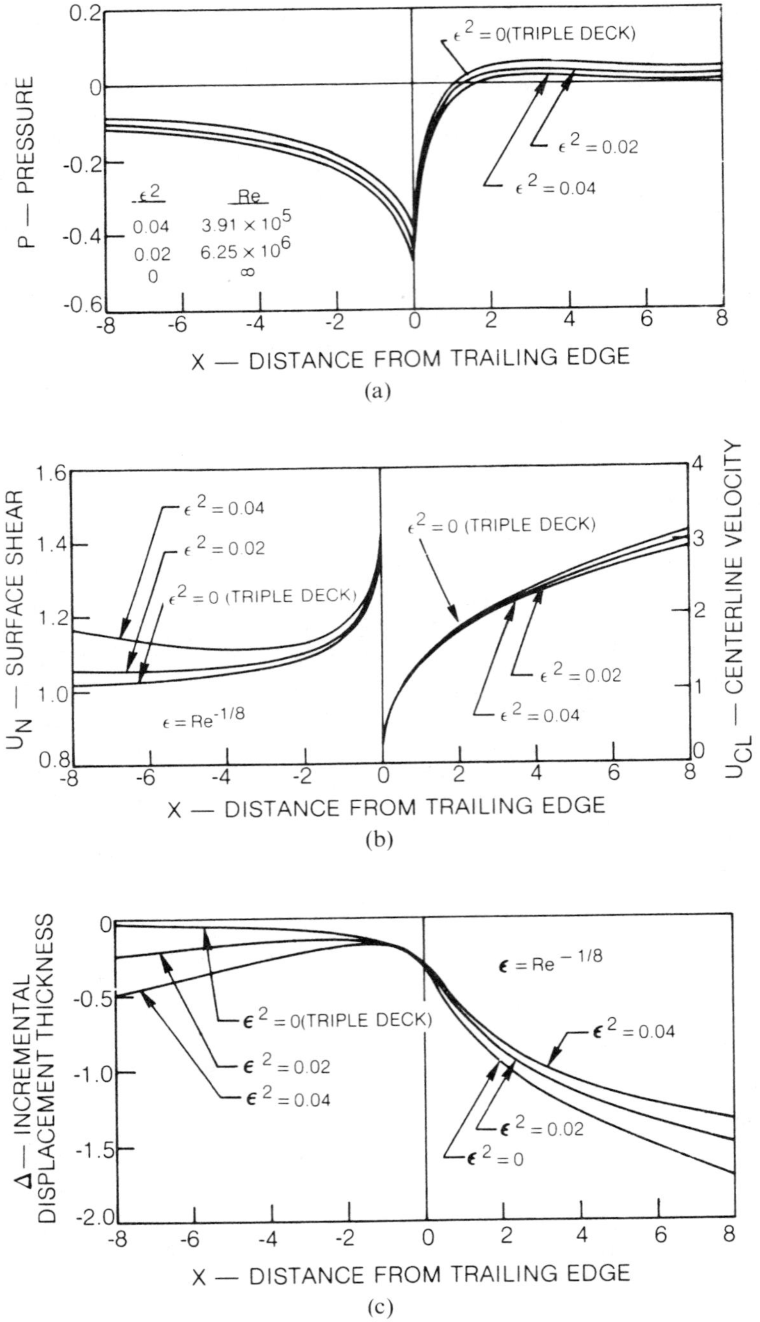

Figure 3. Flat-plate interacting-boundary-layer solutions: (a) surface pressure; (b) surface shear stress and wake centerline velocity; (c) incremental displacement thickness.

The calculations have not been found to be insensitive to mesh spacings as long as the lower-deck scale laws are honored. From an engineering viewpoint, the coarser mesh sizes mentioned previously would be adequate.

Figures 3(a), (b), (c) present results obtained from Method B for high Reynolds number to show the approach to the triple-deck limit. The results are for the fine mesh spacings. The triple-deck results are from Jobe and Burggraf [2.14], Melnik and Chow [2.4], and Veldman [2.105]. All of these results are essentially the same to the scale of these plots. All results are presented in the triple-deck variables (2.12), (2.13). In all results given in Figs. 3(a), (b), (c) we can visualize a smooth approach to the triple-deck limit, indicating that the interacting-boundary-layer model is correctly capturing the triple-deck limit. The incremental displacement thickness in Fig. 3(c) is defined by

$$\Delta = (\delta - 1.7208)\frac{\lambda^{3/4}}{\varepsilon}, \tag{4.2}$$

where 1.7208 is the Blasuis value of the displacement thickness at the trailing edge.

Figures 4(a), (b), (c), (d) present results in Prandtl variables for the particular case of $\mathrm{Re} = 10^5$. Both methods A and B produce essentially the same results, and the results have been found to approach one another as mesh sizes are refined. The results of method B were done on a finer mesh and are therefore believed to be more accurate.

The most important result is shown in Fig. 4(a), where it is found that the pressure distribution from the present results (methods A and B) differs significantly from previous results due to Veldman [2.2]. This is primarily due to the inclusion of the leading-edge and wake displacement functions (which were in effect ignored by Veldman [2.2]) in the present analysis. When these functions are ignored in the present analysis, the pressure distribution is in good agreement with Veldman's results. The effect on the remaining quantities of neglecting this effect is less severe: see Fig. 4(b) for wall shear and wake centerline velocity.

Finally, a convergence-rate study is presented in Fig. 4(d) for the finite-flat-plate case at Reynolds number 10^5. The results are for method B. Method A shows an order-of-magnitude slower rate of convergence. This is primarily due to the manner in which the airfoil integral is coupled to the boundary-layer equations in method B, and verifies the conclusion reached by Veldman [2.2] that the semidirect coupling proposed by him (see Eq. (3.16)) leads to fast convergence.

4.2 Laminar Blunt-Trailing-Edge Flow

Method A outlined in the previous section has been applied by Vatsa et al. [2.101] to the case of blunt-trailing-edge flows typical of that encountered on axial compressor blades. The actual geometry used to model such flows

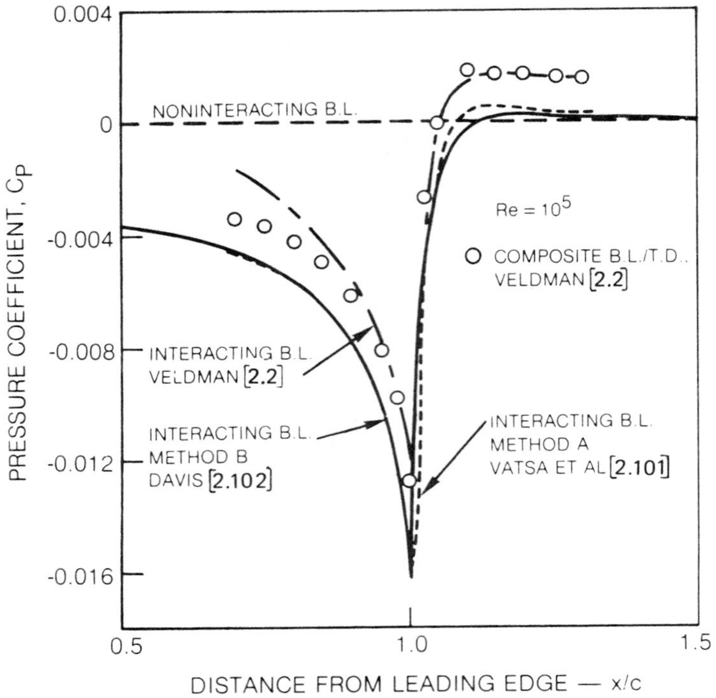

(a)

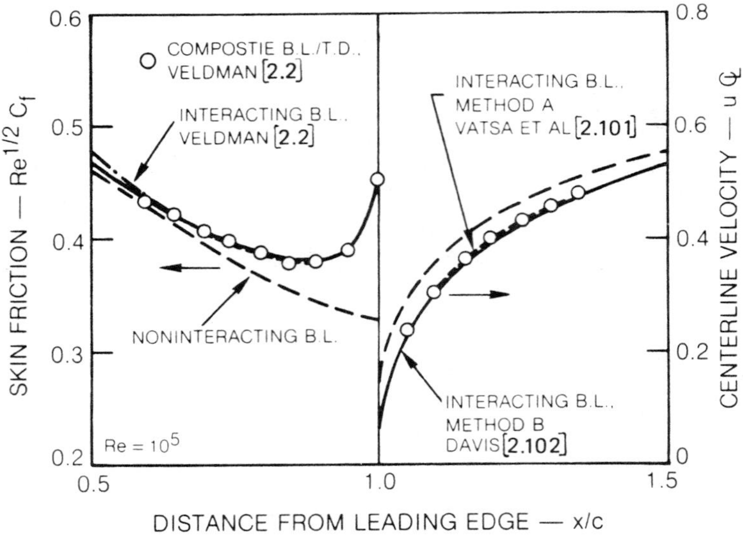

(b)

Figure 4. Comparison with Veldman's flat-plate results: (a) surface pressure; (b) skin friction and centerline velocity.

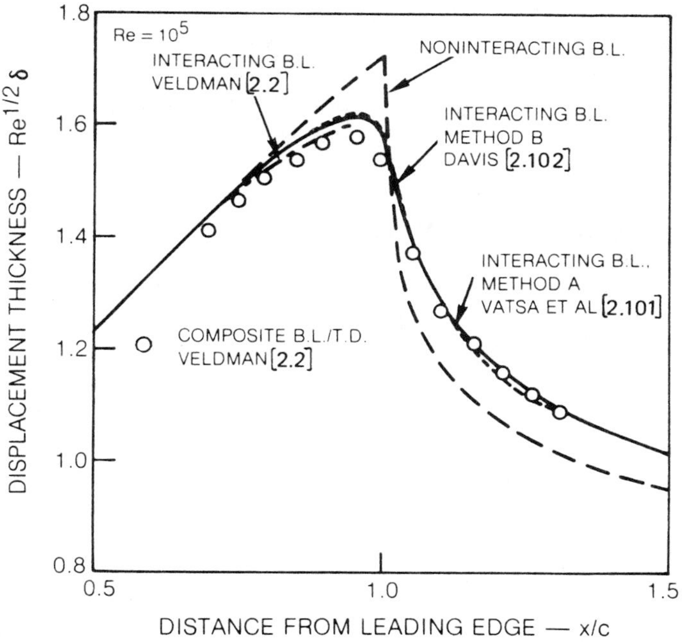

(c)

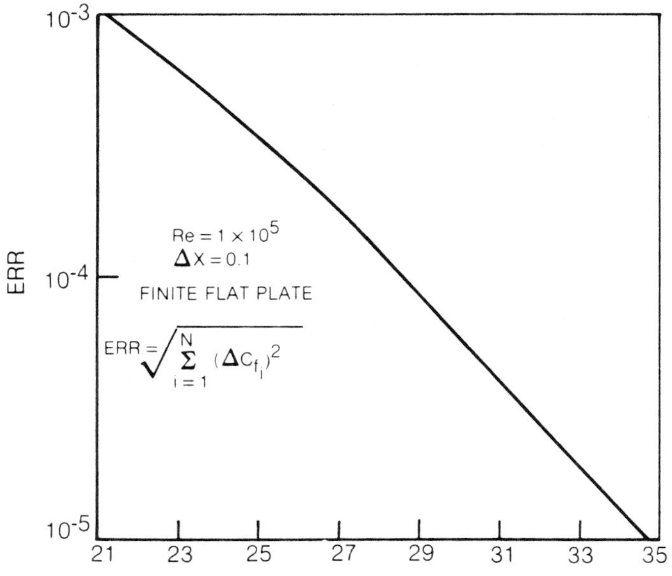

(d)

Figure 4. Comparison with Veldman's flat-plate results: (c) displacement thickness distributions; (d) numerical scheme convergence (Method B).

was flat plates with elliptical trailing-edge geometries as depicted in Fig. 5(a). Solutions for this geometric family for asymptotically large Reynolds numbers (the triple-deck limit) had already been presented by Werle and Verdon [2.20], and thus were available for assessment of the interacting-boundary-layer solutions. For a freestream Reynolds number (based on plate length) of 10^5 and a Mach number of 0.1, elliptical conic-section geometries (see Fig. 5(a)) were studied for a length $l = 0.2$, and heights $h_0 = 0.070$ and 0.14. These parameters give inner-deck scalings of $L = 4$, $H_0 = 4$ and 8, and $\gamma = 2$, corresponding exactly to two of the triple-deck solutions given in [2.20]. Solutions were obtained in the IBL case with the strong-interaction zone set from 40 to 160% chord and with a classical Blasius-type square-root boundary-layer growth in the leading-edge region and a Goldstein-like wake

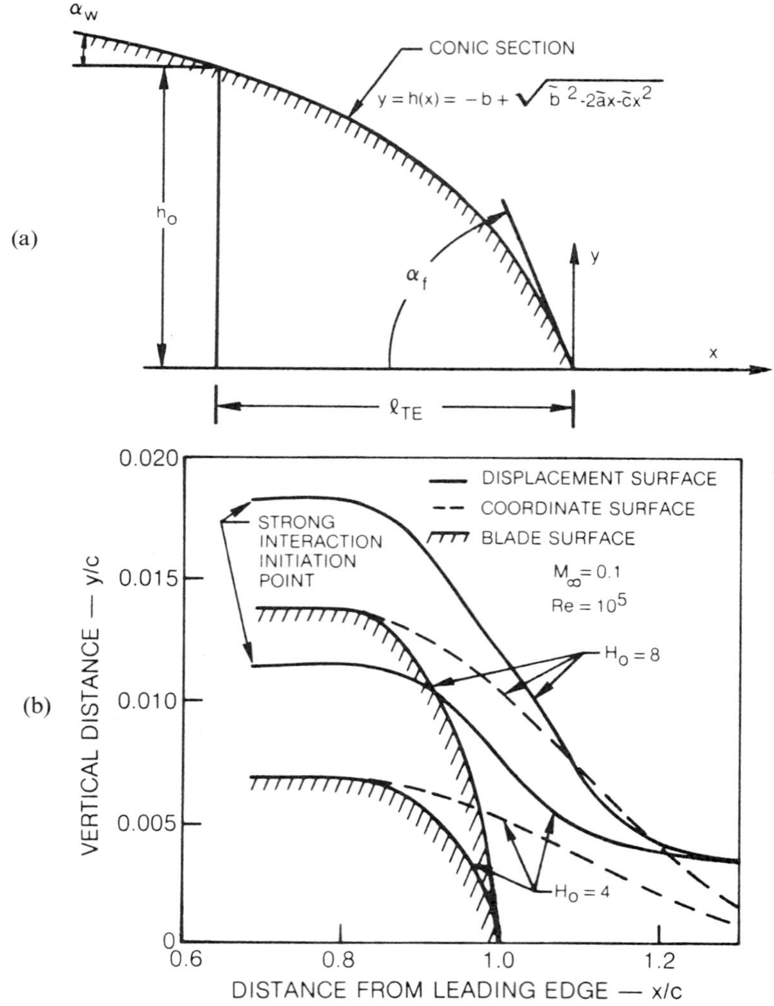

Figure 5. Laminar blunt-trailing-edge solutions: (a) trailing edge geometry; (b) displacement surface distributions.

(c)

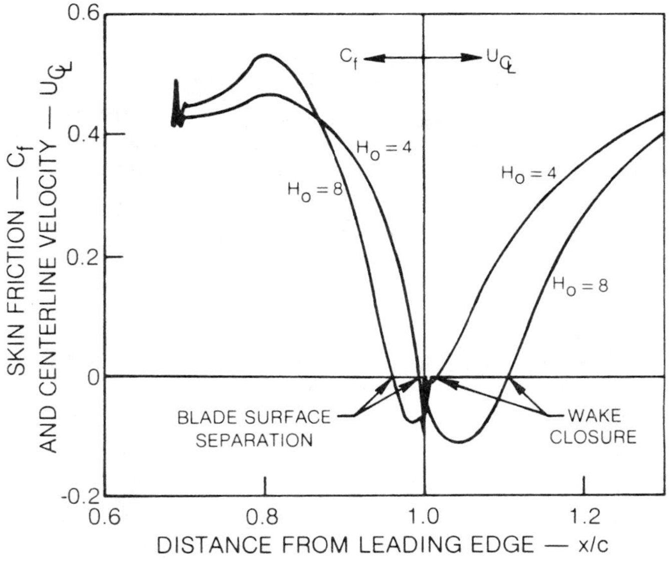

(d)

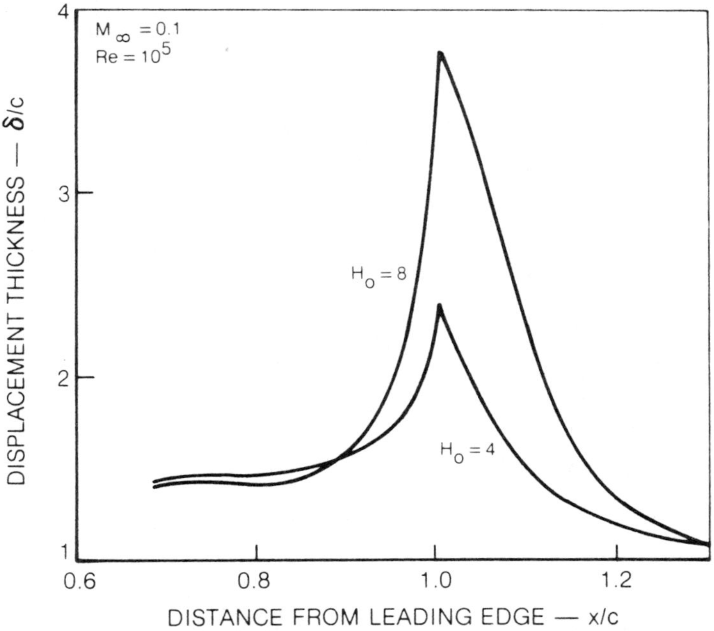

Figure 5. Laminar blunt-trailing-edge solutions: (c) skin friction and wake velocities; (d) displacement thickness distributions.

decay downstream. For this case, the coordinate base curve was taken as a simple cubic (see Fig. 5(a)), tangent at the trailing edge juncture point and at 120% chord on the wake centerline. The resulting interaction solutions are shown in Figs. 5(b)–(d) where the displacement surfaces show steady growth up to near the juncture point followed by a rapid smooth decay in the near wake down toward constant wake valves. The accompanying surface skin-friction and wake centerline velocity distributions are shown where the thick-blade case, $H_0 = 8$, is seen to experience considerable separation off the blunt base. The oscillations in the skin friction at the initialization point represent the rapid adjustment of the assumed initial self-similar velocity profile to a compatible nonsimilar state as it enters the interaction zone. The thick-plate case experiences a skin-friction growth immediately downstream of the initialization point which is due to the boundary-layer thinning induced by a flow overspeed as it accelerates around the juncture corner.

4.3 Laminar Flow past the Carter–Wornom Trough

The so-called Carter–Wornom trough [2.3] has become a test case for interacting-boundary-layer and triple-deck solvers; see Veldman [2.2] and Veldman and Dijkstra [2.104]. This problem has been solved by method B in order to test the method for a separated-flow case.

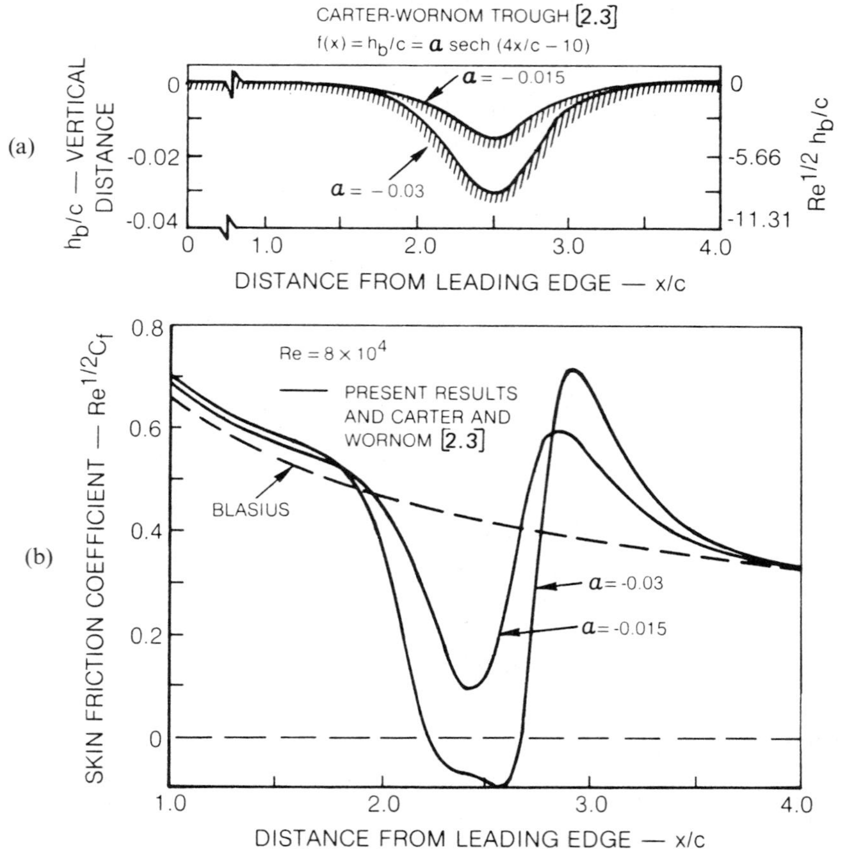

Figure 6. Solutions for the Cater–Wornom trough: (a) geometry description; (b) surface skin friction.

Figure 6(a) provides the geometry for the problem, which is a flat plate with a hyperbolic-secant depression centered at $x = 2.5$. The values of the thickness parameter α have been chosen so that both cases shown would show separation according to noninteracting-boundary-layer theory. The upstream and downstream assumed displacement thickness distributions have been taken to be Blasius flat-plate distributions ($C\sqrt{x}$) and matched with the first and last solution-point values in the same manner as was discussed previously in the finite-flat-plate case. Carter and Wornom [2.3] used a similar approach.

Figure 6(b) presents results for skin friction for both cases, $\alpha = -0.015$ and $\alpha = -0.03$. The upstream flow at $x = 1.0$ was assumed to be a Blasius profile, and the interacting calculation was terminated downstream at $x = 4.0$. The calculations were performed with a Blasius step size Δx of 0.025 and step size $\Delta\eta$ of 0.33.

The results are identical, on the scale of the plot, to those of Carter and Wornom [2.3]. Shown in Fig. 6(b) are surface skin-friction results; the results for other quantities show the same type of agreement.

The case of $\alpha = -0.03$ shows a significant separated-flow region. It was found that no special treatment of the convective terms in the boundary-layer equations was needed in this region. Rapid convergence was obtained in both cases. For larger separated regions and or smaller mesh spacings in the streamwise direction, upwind differencing of the convective terms may be required.

4.4 Turbulent Trailing-Edge Flows

Solution method A has been applied by Vatsa et al. [2.101] to the turbulent-base-flow cases corresponding to the laminar cases discussed above. The turbulence model used was essentially that presented by Cebeci, et al. [2.106] for flat-plate wakes, generalized slightly in [2.101] to accommodate finite thickened blades.

For these turbulent cases, the pressure contribution from the region ahead of the strong-interaction zone in Eq. (3.5) cannot be evaluated analytically. Thus the upstream displacement-thickness distribution was taken directly from classical noninteracting boundary-layer calculations with instantaneous transition immediately aft of the leading edge. The appropriate far-wake displacement-thickness distribution is of the Goldstein square-root decay type in the current variables, and thus the procedure employed for the laminar case carries over directly.

Solutions for flat-plate flow have been obtained in [2.101] for a case corresponding to the experimental study of Chevray and Kovaznay and the noninteracting-boundary-layer solution of Cebeci et al. The IBL pressure distribution obtained in [2.101] for this case is shown in Fig. 7(a) along with a rather enlightening comparison with a "second-order" boundary-layer solution. In this latter case a noninteracting-boundary-layer solution was first obtained over the length of the plate and into the wake region to obtain

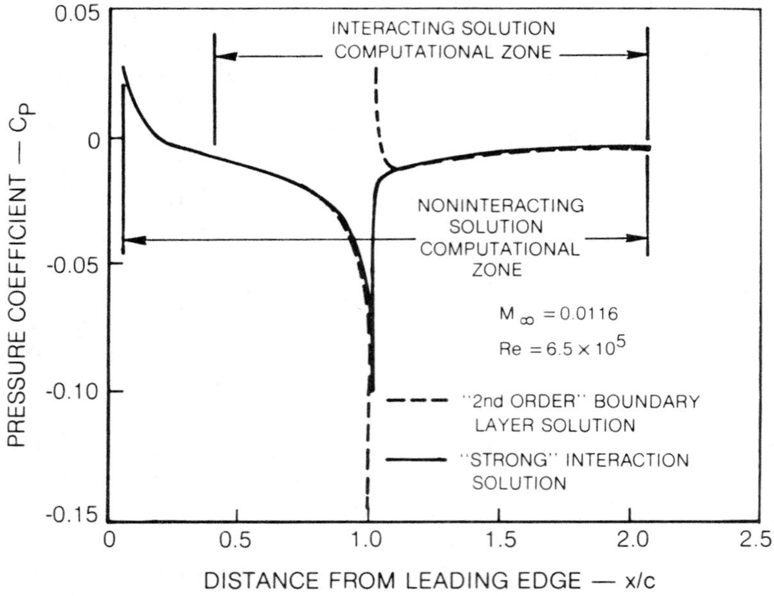

(a)

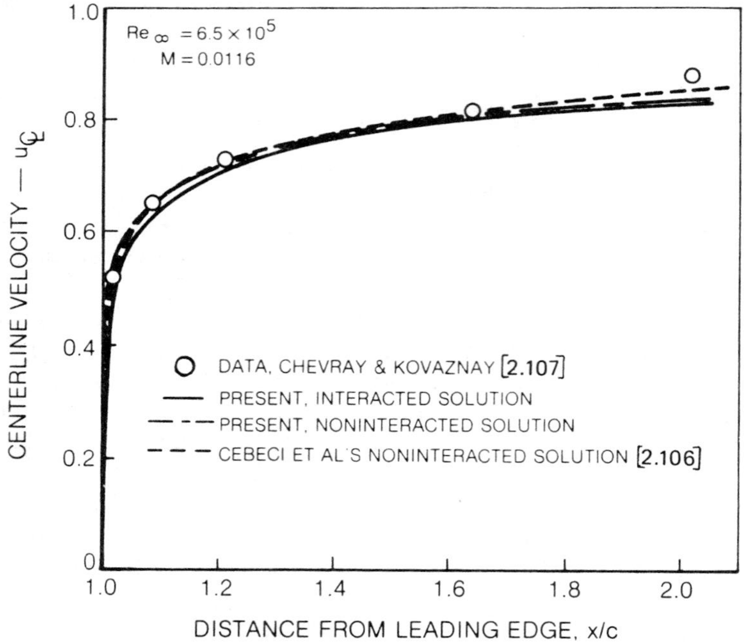

(b)

Figure 7. Turbulent flat-plate results: (a) pressure coefficients; (b) wake centerline velocity.

a "first-order" displacement-thickness distribution. In the spirit of higher-order boundary-layer corrections, the pressure distribution was then calculated using this displacement in the Cauchy solver of Eqs. (3.4) or (3.2). The resulting pressure distribution is shown in Fig. 7(a), where note is made

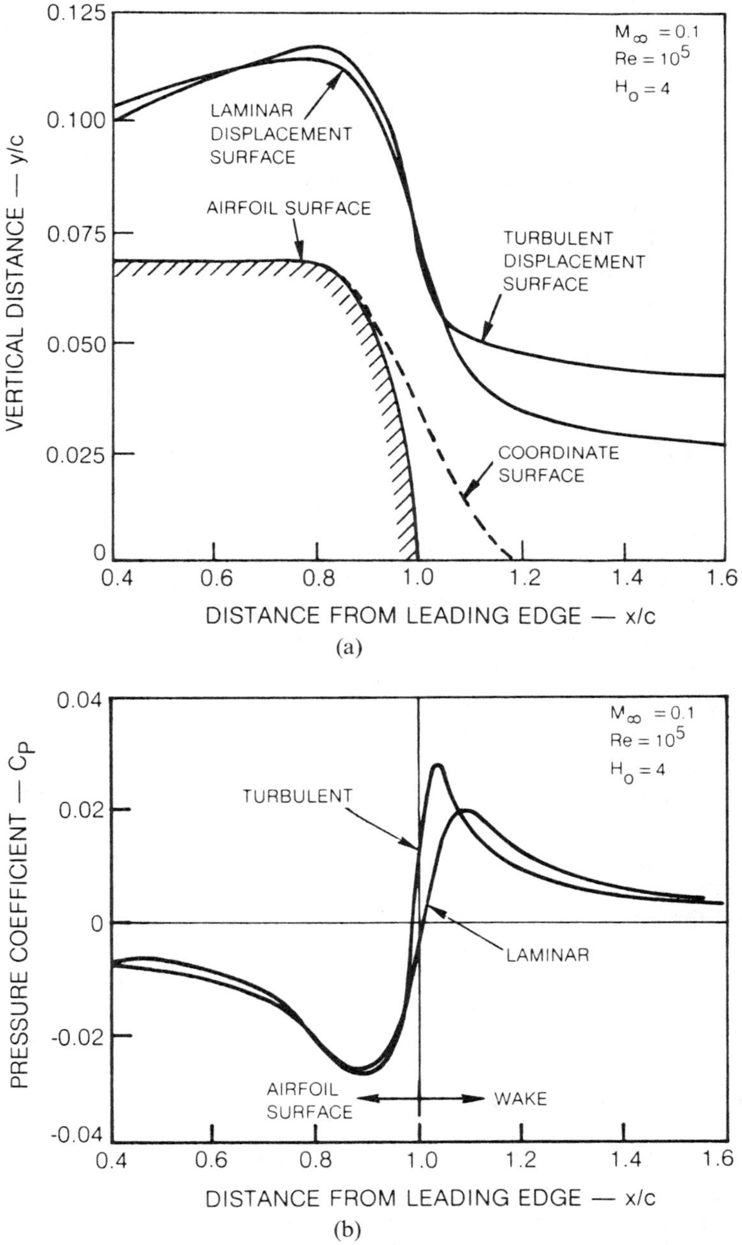

Figure 8. Turbulent blunt-trailing-edge solutions: (a) displacement surface; (b) pressure coefficient.

of the expected singular behavior across the trailing-edge point. Interaction solutions are also shown in Fig. 7(a), where it is seen that, as anticipated by Melnik [2.7], the displacement interaction effect is highly confined to the near trailing-edge point and serves principally to eliminate the pressure singularity in this region of the flow. It is most important to take note that the most dominant displacement effect is produced by the weak interaction or global effect caused by the wake suckdown of the displacement thickness in the near-wake region. For the turbulent case this is seen to set the major portion of the upstream and downstream pressure field with very little "correction" due to strong-interaction effects at the trailing edge. The conclusion drawn here is that while strong-interaction effects are important, they do not dominate weak- or global-interaction effects, at least not for the Reynolds-number range considered here.

A comparison of the predicted wake centerline velocity with experimental data is presented in Fig. 7(b). The important feature here is that the interacting solutions give a welcome slight improvement over the noninteracting results in the near wake and that both slightly underpredict the far-wake case. Such differences are well within the uncertainty surrounding the turbulence model for such flows and are not of concern here. Also shown in Fig. 7(b) is a favorable comparison with the noninteracting results of Cebeci and Stewartson for the same case.

The final case presented here is that of turbulent flow off the trailing edge of one of the finite thick blunted plates discussed previously (see Fig. 5). As shown in Fig. 8(a), the geometry, Reynolds number, and Mach number are identical to those employed in the laminar case for $H_0 = 4$. The resulting displacement body and surface pressure distributions are compared with their laminar counterparts in Fig. 8(a) and 8(b). Not surprisingly, the turbulent results show steeper gradients of displacement surface and pressure across the trailing-edge region than those encountered for the laminar case. This is seen to result in a slightly higher peak pressure in the near wake. Also note that the wake asymptotic displacement thickness appears to be larger for the turbulent case than for the laminar, apparently directly due to the increased drag induced on the plate surface by the turbulent flow. Again it is interesting to note the degree to which the global-interaction effects influence the pressure distribution.

The combination of geometric and displacement neckdown effects so dominates the problem that the difference between laminar and turbulent flow can only be detected over a small region of the near wake.

Acknowledgment

The authors wish to thank Dr. Veer N. Vatsa for his extensive contributions to the results presented in this paper. We greatly appreciate his most accommodating efforts in the running and rerunning of many test cases in a timely fashion to allow us to meet deadlines.

CHAPTER 13

On the Turbulent Viscid–Inviscid Interaction at a Wedge-Shaped Trailing Edge[1]

R. E. Melnik and B. Grossman*

Introduction

It is generally recognized that zonal-type methods, based on boundary-layer concepts, lead to very effective procedures for computing viscous flows at high Reynolds numbers. To achieve their full potential, these methods should account for strong-interaction effects that arise at trailing edges. The trailing-edge interaction is of particular importance because of its influence on the Kutta condition, leading to large global effects on lift and drag. The present paper deals with the theoretical description of strong viscid–inviscid interaction at the trailing edge of an airfoil, when the boundary layer approaching the trailing edge is a fully developed turbulent flow.

In an earlier study [2.7] the first author, R. Chow, and H. R. Mead developed a rational asymptotic theory for turbulent interactions at the trailing edge of a lifting airfoil with a cusped trailing edge. That paper clearly demonstrated the importance of normal pressure-gradient effects in the trailing-edge region, when the boundary layer is turbulent. The present study generalizes the earlier Melnik–Chow theory to airfoils with nonzero trailing-edge angles. In order to focus on the key aspects of the theory that are related to the wedge-shaped trailing edge, the present paper will deal only with the basic problem of incompressible flow over a nonlifting, symmetric profile. Furthermore, in

[1] This work was partially supported by the NASA Langley Research Center under Contract NAS 1-12426.

* Grumman Aerospace Corporation, Bethpage, New York 11714. Present address: Polytechnic Institute of New York, Department of Mechanical and Aerospace Engineering, Farmingdale, NY 11735.

order to make use of thin-airfoil approximations, we assume the airfoil is thin and can be characterized by a single small parameter, the thickness ratio t. The present formulation is then based on the double limit of thickness ratio $t \to 0$ and Reynolds number $\mathrm{Re} \to \infty$.

The high-Reynolds-number limit leads to a standard inviscid–boundary-layer description in the central part of the flow away from the trailing edge. The formal approach reduces the problem to that of solving a hierarchy of inviscid and boundary-layer equations in sequence, starting from the solution of the inviscid equation for the pressure distribution. The hierarchical approach leads to a valid description in the central part of the flow, but fails near the trailing edge, due to appearance of nonuniformities in the formal solutions. The nonuniformities arise from two distinct sources, one from the discontinuous change in the no-slip condition across the trailing edge, and the other from the stagnation-point singularity arising in the inviscid solution at a wedge-shaped trailing edge.

In its simplest form, the first nonuniformity arises on a flat plate at zero incidence and is associated with the sudden transition between a boundary layer and wake behavior. Although this is a basic problem in turbulent interactions that has received some attention in the past [2.108, 2.109], definitive results have not been obtained. Simple consideration of the two-layer structure of turbulent boundary layers suggests that the interaction associated with this effect occurs on a length scale that is much smaller than a boundary-layer thickness (for some experimental evidence, see [2.110]) and can be expected to have only a slight effect on the pressure distribution on the plate. This aspect of the trailing-edge flow will not be considered in this paper. Instead we concentrate on the strong-interaction effect produced by the singularity in the inviscid pressure distribution. Results of the present study will demonstrate that interactions associated with the stagnation-point singularity produces an $O(1)$ change to the pressure distribution (or more properly, $O(t)$ for the thin-airfoil case considered herein).

The present paper describes two theoretical approaches for dealing with the trailing-edge interaction, one based on interacting boundary-layer theory (IBLT) and the other on a formal, large-Reynolds-number, asymptotic analysis including normal pressure-gradient effects. We recall that IBLT employs the usual viscous matching condition, but couples the inviscid and boundary-layer equations to obtain self-consistent solutions for the pressure and displacement thickness. Interacting boundary-layer theory has been widely employed to make engineering predictions of viscous effects in many problems. In this approach, it is implicitly assumed that the singularities appearing in the hierarchy approach are due to the use of the inviscid pressure distribution in the leading term. It is assumed (or hoped), without justification, that coupling of the inviscid and boundary-layer solutions eliminates the singularities that would otherwise arise from the use of the inviscid solution for the pressure. The triple-deck theories of Stewartson [1.106] and others demonstrate this to be true for laminar flow at the trailing edge of cusped airfoils. It is also likely true for laminar flow at wedge-shaped trailing edges, although solutions for this problem have not yet been found.

The situation in this regard is less clear when the boundary layer is turbulent. The evidence from the many previously published numerical solutions suggests that the coupling of the inviscid and boundary-layer solutions completely eliminates the stagnation-point singularity. This seems to occur through a mechanism in which the computed displacement thickness develops a corner at the trailing edge that exactly cancels the geometric corner in the airfoil, leading to a smooth equivalent body and no stagnation point. Although the evidence for this behavior in IBLT seems strong, it has never been analytically demonstrated that solutions of IBLT have this property.

In the present study, we use the method of matched asymptotic expansions to develop a local solution of IBLT that is valid near the trailing edge in the double limit of $t \to 0$, $\mathrm{Re} \to \infty$. This results in a closed-form analytic solution that clearly indicates the manner in which the coupling eliminates the stagna-ation-point singularity. The theory also indicates that the streamwise length scale associated with the local inner solution is the order of the displacement thickness, which in turbulent flow is too small for the theory to be rationally consistent. Simple scaling analysis, which includes the normal momentum equation, indicates that for this small a length scale, normal pressure gradients across the boundary layer will dominate the displacement effect of IBLT. Thus, although IBLT eliminates the nonuniformity associated with the stagnation-point singularity, it does not provide a consistent solution to the interaction at the trailing edge in turbulent flow.

The second approach to the problem discussed in the paper is based on a more complete asymptotic analysis. This new formulation is also based on a two-parameter asymptotic expansion for $t \to 0$ and $\mathrm{Re} \to \infty$, but it now also includes an accounting of the normal momentum equation. A more complete scaling analysis indicates that normal pressure-gradient effects appear in the lowest-order description of the flow near the trailing edge and that the interaction is spread upstream over a distance the order of a boundary-layer thickness by inviscid flow mechanisms. In the present study, we are also able to obtain a closed-form solution for this strong-interaction problem. The analytical solution clearly indicates the manner in which normal pressure gradients affect the local trailing-edge flow. The paper will compare the local solutions obtained by the interacting boundary-layer and strong-interaction theories.

The only other analytical treatment of the turbulent strong interaction at wedge-shaped trailing edges is Küchemann's [2.111] analysis (see also Smith [2.112]). Küchemann considered a more general problem in which the airfoil was not regarded as thin or the trailing-edge angle as small, as in the present study. On the basis of physical arguments, he reasoned that the shear stresses could be neglected and the flow could be treated as an inviscid, rotational flow near the trailing edge. He then developed special solutions of the exact, inviscid rotational-flow equations for linear initial velocity profiles. Although Küchemann's assumption of inviscid rotational flow is proper and is consistent with the more detailed asymptotic description of the present study, the theory is not completely satisfactory. In order to avoid a trivial solution of constant pressure on the airfoil surface, Küchemann was forced to introduce an arbitrary cutoff of the initial velocity profile to a nonzero "slip" velocity at the airfoil surface. Unfortunately, the solution is strongly influenced by the choice of the cutoff, and no indication is offered on how the slip velocity is to be determined, even in principle.

The new strong-interaction (S-I) theory discussed in the present paper can be viewed as an approximation to the Küchemann theory for small trailing-edge angles and also as an improvement that enables one to avoid the use of an arbitrary cutoff to the initial velocity profile. The new theory shows that a cutoff can be avoided by properly accounting for the two-layer structure and logarithmic behavior of the initial profile for $y \to 0$ that is appropriate for fully-developed turbulent flows. The theory, however, is in itself not complete, in that it also leads to a logarithmic singularity at the trailing edge. The implication is that a further inner solution on an even smaller scale than the boundary-layer thickness is required to complete the solution. Analysis indicates that the solution in this region is governed by the full Reynolds equations on a length scale of the order of the thickness of the turbulent wall layer and that no further simplification is possible.

Formulation

We consider the flow of a viscous, incompressible stream at zero incidence to a nonlifting, symmetric profile of thickness ratio t, with a trailing-edge (included) half-angle of $t\theta_{te}$. Without loss of generality we assume the density, freestream speed, and airfoil chord are unity. We employ a Cartesian coordinate system x, y with origin at the trailing edge and velocity components u, v as sketched in Fig. 1. Further, we assume the airfoil is thin and the Reynolds number Re is large, so that formal asymptotic solutions can be developed on the basis of the double limit $t \to 0$ and $\mathrm{Re} \to \infty$.

The boundary layer is assumed to experience transition to turbulent flow well upstream of the trailing edge. For turbulent flow, the Reynolds-number effects are most conveniently scaled in terms of the dimensionless friction velocity $u^* \equiv \sqrt{\tau_{wr}}$, where τ_{wr} is a representative value of the skin friction. Accordingly, to indicate the magnitude of the Reynolds-number effects, we introduce a small parameter ε defined by

$$\varepsilon = O(u^*) = O[(\ln \mathrm{Re})^{-1}]. \tag{1}$$

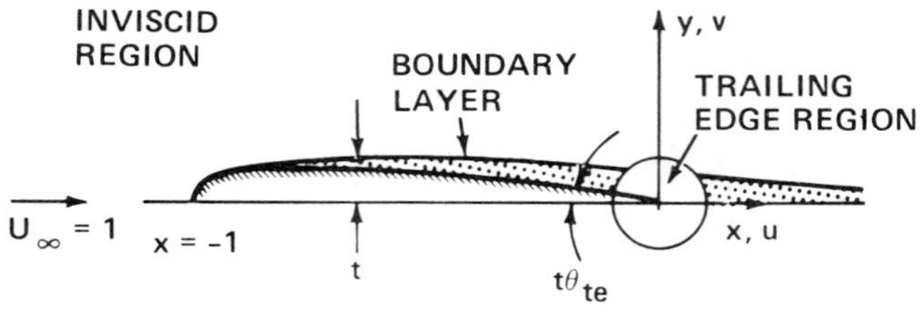

Figure 1. Flow-field schematic.

The parameter ε is employed in the analysis only to indicate the relative magnitude of the various terms appearing in the expansions, but should be set equal to unity in all final formulas. For future reference, we note the magnitude of the boundary-layer thickness δ and displacement and momentum thickness δ^* and θ, given in terms of ε by

$$\delta = O(\varepsilon), \qquad \delta^*, \theta = O(\varepsilon^2), \tag{2}$$

and by definition, the skin friction τ_w by

$$\tau_w = O(\varepsilon^2) \tag{3}$$

For large Reynolds number ($\varepsilon \to 0$), the flow field develops the familiar multilayer structure sketched in Fig. 1 consisting of (a) an outer inviscid flow, (b) an inner boundary layer, and (c) a localized strong-interaction zone at the trailing edge. It is well known from simple scaling arguments, and supported by experimental evidence, that the turbulent boundary layer itself develops a two-layer structure consisting of an outer "defect" layer and an inner "law of the wall" region [2.113]. Mellor [2.114] developed a formal asymptotic description of turbulent boundary layers, including this feature, using the method of matched asymptotic expansions.

The formal theory indicated how the two-layer structure, the logarithmic behavior of the velocity profiles near the wall, and the logarithmic skin-friction law can be deduced from first principles. Mellor's theory is valid for the weak interaction in the central part of the airfoil, but fails at the trailing edge because of the appearance of nonuniformities in the formal solution, as discussed previously. The present work and Melnik and Chow's earlier contribution for cusped airfoils represent a generalization of Mellor's theory to the strong interaction at trailing edges. In the following section, we review some of the details of the weak-interaction theory, including, in particular, a discussion of the nonuniformity that arises at a wedge-shaped trailing edge.

Weak-Interaction Theory

Weak-interaction theory is based on a large-Reynolds-number ($\varepsilon \to 0$) asymptotic expansion of the full viscous flow equations. In the present application, the theory provides a uniformly valid description of the flow, except in the vicinity of the trailing edge. Using the method of matched asymptotic expansions, Mellor [2.114] presented a formal development of the theory for incompressible turbulent flow. In this section, we briefly outline those parts of the theory that are needed to elucidate the nature of the breakdown at the trailing edge.

As indicated in Fig. 1, the solution in the central portion of the flow field involves the usual division of the flow into inviscid and boundary-layer regions. The solution in the outer inviscid region is represented in the following two-parameter asymptotic expansions for the tangential and normal components of velocity:

$$u(x, y) = [1 + tu_{01} + \cdots] + \varepsilon^2[u_{20} + tu_{21} + \cdots] + \cdots, \tag{4}$$

$$v(x, y) = [tv_{01} + \cdots] + \varepsilon^2[v_{20} + tv_{21} + \cdots] + \cdots. \tag{5}$$

The first group of terms represent the thin-airfoil expansion of the outer inviscid solution, while the second group represent corrections for the boundary-layer displacement effect (we recall $\delta^* = O(\varepsilon^2)$). Of the latter, the first arises from the basic flat-plate boundary layer, while the second stems from the effect of the pressure gradient on the boundary-layer displacement thickness. Since the effect of the boundary layer on the outer inviscid flow enters only through the boundary conditions, the complex velocity, $u - iv$, is an analytic function of the complex variable $z \equiv x + iy$. It follows that the outer solution is completely determined from specified values of the normal velocity component on the real axis ($y = 0$) by the Hilbert integral representation

$$u(x, y) = \frac{1}{\pi} \int \frac{v(s, 0)}{z - s}\, ds, \tag{6}$$

where the limits of integration extend over the region of nonvanishing $v(s, 0)$. The individual terms of the series solution also satisfy the same relation. The first inviscid term, $v_{01}(x, 0)$, is determined from the profile shape, $y = ty_b(x)$, by

$$v_{01}(x, 0) = \frac{dy_b}{dx}. \tag{7}$$

The corresponding terms for the second-order displacement effect are determined from an expansion, in powers of t, of the transpiration velocity v_σ defined in the usual way by

$$v_\sigma \equiv \frac{d}{dx}(u_e \delta^*), \tag{8}$$

where u_e is the surface value of the inviscid solution,

$$u_e \equiv 1 + tu_{e1} + \cdots, \tag{9}$$

and $u_{e1} \equiv u_{01}(x, 0)$ and δ^* is the displacement thickness. Thus if the expansion for v_σ is given by

$$v_\sigma = \varepsilon^2[v_{\sigma 0}(x) + tv_{\sigma 1}(x) + \cdots], \tag{10}$$

then

$$v_{20}(x, 0) = v_{\sigma 0}(x), \qquad v_{21}(x, 0) = v_{\sigma 1}(x). \tag{11}$$

The displacement thickness is determined from the solution of the boundary-layer equations, with the velocity at the boundary-layer edge prescribed equal to the inviscid surface velocity, $u_e(x)$.

In turbulent flow, the solution in the boundary layer also develops a two-layer structure, consisting of an outer or defect layer and a much thinner wall layer. Mellor [2.114] provided a complete derivation of the equations governing the first few terms of the solution, including the matching conditions coupling the outer inviscid and boundary-layer solutions. Although the analysis is a straightforward application of the method of matched asymptotic expansions, the details are lengthy and will not be repeated here. Instead we indicate how the required transpiration velocity can be simply determined from the von Kármán momentum integral equation. From this equation and the definition of the shape parameter, $H \equiv \delta^*/\theta$, we can arrive at the following expression for v_σ:

$$v_\sigma = -\delta^*\left[(H+1)\frac{1}{u_e}\frac{du_e}{dx} - \frac{1}{H}\frac{dH}{dx}\right] + \frac{H\tau_w}{u_e^2}, \tag{12}$$

where τ_w is the surface shear stress. Expressions for the perturbation quantities $v_{\sigma 0}$ and $v_{\sigma 1}$ can be developed by expanding the boundary-layer quantities and edge velocity appearing in Eq. (12) about their flat-plate value ($t = 0$). The expansion for the edge velocity is given by Eq. (9), and those for the boundary-layer parameters can be represented as follows:

$$\delta^* = \varepsilon^2[\delta_0^*(x) + O(t)], \tag{13}$$

$$H = H_0[1 + O(\varepsilon t)], \tag{14}$$

$$\tau_w/u_e^2 = \varepsilon^2\tau_{w0}[1 + O(\varepsilon t)], \tag{15}$$

where the subscript 0 on δ_0^*, H_0, and τ_{w0} denotes flat-plate solutions. We note that there is no $O(t)$ term in the expressions for the shape factor and skin friction. The absence of this term in the expression for H follows from the fact that $H \to 1 + O(\varepsilon)$ for $\text{Re} \to \infty$, so that pressure-gradient effects enter only through the $O(\varepsilon)$ variable part of H. The justification for Eq. (15) is somewhat more complicated but follows directly from an analysis of small pressure-gradient effects on the logarithmic skin-friction law (which is a direct result of Mellor's formal asymptotic solution). Substitution of the above expansions into Eq. (12) yields the following expressions for the transpiration velocity terms:

$$v_{\sigma 0} = \tau_{w0}, \tag{16}$$

$$v_{\sigma 1} = -\delta_0^*(H_0 + 1)\frac{du_{e1}}{dx}. \tag{17}$$

Somewhat surprisingly, we see that the thin-airfoil expansion has resulted in very simple algebraic expressions for the transpiration velocity, involving only boundary-layer information from flat-plate (constant-pressure) solutions. To be consistent, the flat-plate shape factor H_0 appearing in Eq. (17) should be set equal to its limiting value ($H_0 \to 1$) for $\varepsilon \to 0$. However, since $H_0 \approx 1.3$ for most Reynolds numbers of interest in aerodynamics, we prefer to retain the full flat-plate value, H_0, in the hope of obtaining improved accuracy.

The expressions given in Eq. (16) and (17) enable us to complete the second-order outer solution with little difficulty, once the basic flat-plate boundary-layer solution is determined. Since a complete solution for the flat plate including the wake effect is not yet available (because of theoretical difficulties in describing the boundary-layer–wake transition), the second-order solution involving the pressure-gradient effect, u_{21}, cannot now be fully determined. However, the above expressions enable us to determine the local behavior of this term at the trailing edge.

Since the shear stress τ_{w0} drops discontinuously to zero across the trailing edge, the transpiration velocity term $v_{\sigma 0}$ must also be discontinuous at the trailing edge. From complex-variable theory it follows that the displacement effect due to the flat-plate boundary-layer–wake transition must have a logarithmic singularity at the trailing edge. Thus

$$u_{20}(x, 0) = -\frac{\tau_{w0}}{\pi} \ln |x|, \qquad |x| \to 0. \tag{18}$$

This effect on the outer inviscid flow is equivalent to a small expansion corner at the trailing edge with the corner angle equal to the value of the skin friction at the trailing edge.

The nonuniformity of the weak-interaction theory at a wedge trailing edge is due to the singularity in the inviscid solution. Thus for a small wedge angle equal to $t\theta_{te}$, the inviscid surface speed is given by

$$u_{e1} = \frac{\theta_{te}}{\pi} (\ln |x| + a_1), \qquad |x| \to 0 \tag{19}$$

where a_1 is a parameter determined by the overall profile shape. Equations (17) and (19) imply a corresponding singularity in the transpiration velocity given by

$$v_{\sigma 1} = -\frac{c\theta_{te}}{\pi} x^{-1} + O(1), \qquad |x| \to 0, \tag{20}$$

where

$$c = \delta^*_{te}(H_{te} + 1). \tag{21}$$

Here δ^*_{te} and H_{te} are the flat-plate values of δ^* and H evaluated at the trailing edge.

The pole singularity in the normal-velocity component in Eq. (20) gives rise to a point-vortex behavior in the outer inviscid flow, whereby

$$u_{21}(x, 0) = \frac{c\theta_{te}}{\pi} \hat{\delta}(x) + O(1), \tag{22}$$

with the notation $\hat{\delta}(x)$ for the delta function. Equation (22) indicates that the displacement effect at a wedge-shaped trailing edge leads to a highly-localized negative spike in the second-order pressure distribution. It is of

An unedited soft-bound volume of camera-ready papers presented at the symposium can be obtained from Dr. H. Unt, Mechanical Engineering Department, California State University, Long Beach, CA 90840. This volume contains papers on three-dimensional flows and complements the present book.

interest to note that Eq. (22) also implies a concentrated force that contributes to the profile drag an increment ΔC_d equal to

$$\Delta C_d = \varepsilon^2 c(t\theta_{te})^2/\pi, \tag{23}$$

which is analogous to the leading-edge suction force in thin-airfoil theory.

The surface-pressure distribution can be determined from the above expressions by noting that the perturbation in pressure is equal to the negative of the streamwise velocity perturbation. There is an additional contribution to the surface pressures arising from the pressure drop across the boundary layer. Although this term makes only a small, $O(\varepsilon^3 t)$ contribution to the surface pressure over the main part of the airfoil, it will turn out to be a dominant viscous effect near the trailing edge. The pressure drop arises from the turning of the low-momentum fluid in the boundary layer along the curved streamlines of the inviscid flow. This term can be computed from an integration of the normal-momentum equation using the velocity profiles from the first term of the boundary-layer solution. The behavior near the trailing edge can be deduced from a straightforward, but lengthy analysis which will not be repeated here. Because of its importance in understanding later parts of the paper, we provide the final result for this term, which is

$$\Delta p = -2\varepsilon^3 k_{\text{te}} \frac{t\theta_{\text{te}}\gamma_{\text{te}}}{\pi} \delta_{\text{te}}^2 x^{-2} \tag{24}$$

where $\gamma_{\text{te}} = \sqrt{\tau_{w\,\text{te}}}/k$ $(k = 0.41)$, $\tau_{w\,\text{te}}$ and δ_{te} are the flat-plate values of the skin friction and boundary-layer thickness at the trailing edge, and k_{te} is a parameter that depends on an integral of the flat-plate velocity profile at the trailing edge. If the velocity profile is written in defect form as

$$u_0(0, y) = 1 + \gamma_{\text{te}} f\left(\frac{y}{\delta_{\text{te}}}\right),$$

Then k_{te} is equal to

$$k_{\text{te}} = 4 \int_0^1 \eta f(\eta)\, d\eta$$

The behavior of the pressure distribution near the trailing edge is summarized in the following formula:

$$\begin{aligned} p(x, 0) \to & \left[p_\infty - \frac{t\theta_{\text{te}}}{\pi}(\ln|x| + a_1) + \cdots\right] \\ & + \varepsilon^2\left[\tau_{w\,\text{te}} \ln|x| - \frac{ct\theta_{\text{te}}}{\pi}\hat{\delta}(x) + \cdots\right] \\ & - 2\varepsilon^3\left[\frac{\gamma_{\text{te}} t\theta_{\text{te}}}{\pi} k_{\text{te}} \delta_{\text{te}}^2 x^{-2} + \cdots\right] \quad \text{for } |x| \to 0. \end{aligned} \tag{25}$$

This result clearly illustrates the significant effect of the boundary layer on the pressure distribution near the trailing edge. There are three principal viscous effects evident in the above relation. Within the weak-interaction theory, they are all unbounded and they all act to reduce the pressure at the trailing edge. The first two viscous terms in Eq. (25) are due to the displacement effect of the boundary layer arising from the discontinuity in the no-slip condition at the trailing edge and from the stagnation-point singularity in the inviscid pressure distribution. The former leads to a weak, logarithmic singularity in the pressure, while the latter produces a compact delta-function behavior and a concentrated drag force at the trailing edge. Equation (25) indicates that the most important viscous effects near the trailing edge is due to the delta function and the normal pressure-gradient term. The pressure perturbations from the normal-pressure-gradient term are formally an order of magnitude smaller than those induced by the displacement effect in the central part of the profile. Nevertheless, Eq. (25) indicates that normal-pressure-gradient and displacement effects increase to equal magnitudes in the immediate vicinity of the trailing edge, when $x = O(\delta) = O(\varepsilon)$.

Interacting Boundary-Layer Theory (IBLT)

The analysis of the previous section demonstrated that weak-interaction theory fails at trailing edges, because of the occurrence of singularities associated with both displacement and normal pressure-gradient effects. Moreover, the results summarized in Eq. (25) suggest that the normal pressure-gradient term contributes a dominant viscous effect on the pressure distribution near trailing edges. Nevertheless, IBLT, which completely ignores the normal-pressure-gradient effect, has been one of the standard approaches for treating viscous effects in high-Reynolds-number flows. In this section we seek to determine if the strong coupling inherent in IBLT is by itself sufficient to eliminate the singularities associated with the displacement effect. We consider only the nonuniformity due to the stagnation-point singularity of the inviscid solution. The nonuniformity associated with the boundary-layer–wake transition produces a comparatively weak effect on the pressure distribution (provided $O(t) > O(\varepsilon^2)$) and hence will not be considered further in this paper.

In our approach, we develop a local analytic solution valid near the trailing edge by applying the method of matched asymptotic expansions to the coupled inviscid–viscid equations of IBLT. The analysis of the last section clearly displayed the rapid growth of displacement thickness induced by the stagnation-point pressure distribution. In order for IBLT to eliminate the nonuniformity associated with the stagnation point, the displacement effect must generate a pressure distribution that exactly cancels the logarithmic singularity in the inviscid solution. This can only occur if the flow deflections generated by the displacement thickness are the same order as those produced by the profile thickness. From Eqs. (9) and (12) we see that for this to occur, the streamwise length scale, Δx, of the local inner solution must be

$$\Delta x = O(\delta^*) = O(\varepsilon^2). \tag{26}$$

Using the above length scale, we seek a local solution in which the disturbance velocities induced by the profile and displacement thickness distributions are the same order of magnitude, equal to $O(t)$. Thus we introduce stretched coordinates $\bar{x}$, $\bar{y}$ defined by

$$x = \varepsilon^2 \bar{x}, \qquad y = \varepsilon^2 \bar{y} \tag{27}$$

and seek a series representation of the outer inviscid solution in the form

$$u = \left[1 + 2t \frac{\theta_{\text{te}}}{\pi} \ln \varepsilon + t\bar{u}(\bar{x}, \bar{y}) + \cdots \right] + O(\varepsilon^2 \ln \varepsilon), \tag{28}$$

$$v = [t\bar{v}(x, y) + \cdots] + O(\varepsilon^2 \ln \varepsilon). \tag{29}$$

The logarithmic term appearing in the first bracket in Eq. (28) is included in order to allow a matching of the above solution to the weak interaction in an overlap domain defined by $\bar{x} \to \infty$, $x \to 0$. The $(\varepsilon^2 \ln \varepsilon)$ error term arises from the nonuniformity associated with the boundary-layer wake transition in the basic flat-plate solution.

The perturbation velocities $\bar{u}$, $\bar{v}$ satisfy the exact Cauchy–Riemann equations governing the outer inviscid flow and hence must also satisfy the Hilbert integral identities relating the real and imaginary parts of the analytic function $\bar{u} - i\bar{v}$. Thus on the real axis we have

$$\bar{v}(\bar{x}, 0) = \frac{1}{\pi} \int_{-\infty}^{\infty} \frac{\bar{u}(s, 0)}{s - \bar{x}} \, ds. \tag{30}$$

Equation (30) is taken as the basic equation governing the inviscid solution. The problem formulation is completed by specifying the normal component of velocity, $\bar{v}(x, 0)$ on the real axis. There are two contributions to this factor: one due to the profile thickness, and a second due to the viscous matching conditions. Near the trailing edge, the profile geometry is represented, to lowest order, by a straight wedge with corner angle $t\theta_{\text{te}}$. Thus

$$\bar{v}(x, 0) = \theta_{\text{te}} \hat{H}(-\bar{x}) + \bar{v}_\sigma(\bar{x}), \tag{31}$$

where $\hat{H}(x)$ is the unit step function and $\bar{v}_\sigma$ is the $O(t)$ contribution of the boundary-layer transpiration velocity defined by

$$v_\sigma \equiv \frac{d}{dx}(u_e \delta^*) = t\bar{v}_\sigma(\bar{x}) + \cdots. \tag{32}$$

Expressions relating the transpiration $\bar{v}_\sigma$ to the inviscid surface velocity $\bar{u}(\bar{x}, 0)$ are determined from series solutions of the turbulent-boundary-layer equation expanded in terms of the two small parameters t and ε. As in the weak-interaction solution, discussed previously, the required expressions can be obtained most simply from the exact relation for v_σ given in Eq. (12). Expansions for the boundary-layer parameters are assumed in the form

$$\delta^*(\bar{x}) = \delta^*_{\text{te}}[1 + O(t \ln \varepsilon)], \tag{33}$$

$$H(\bar{x}) = H_{\text{te}}[1 + O(t\varepsilon)], \tag{34}$$

$$\tau_w/u_e^2 = \tau_{w\text{te}}[1 + O(t\varepsilon)]. \tag{35}$$

These expressions are similar to those employed in the weak-interaction solution, except that here the boundary-layer parameters δ^*_{te}, H_{te}, and $\tau_{w\,te}$ are equal to the trailing-edge values of the basic flat-plate solution. As in the expansion for the inviscid solution, the logarithmic term appearing in Eq. (33) arises from the matching of the above solution to the outer weak-interaction solution. Substitution of Eqs. (32)–(35) into Eq. (12) yields the following expression for $\bar{v}_\sigma$

$$\bar{v}_\sigma = -c\,\frac{d\bar{u}(\bar{x}, 0)}{d\bar{x}}, \tag{36}$$

where c is an $O(1)$ constant defined by Eq. (21). Using Eqs. (36) and (31) to evaluate the left side of Eq. (30) leads to the following equation governing the local trailing-edge solution:

$$c\,\frac{d\bar{u}}{d\bar{x}} + \frac{1}{\pi}\int_{-\infty}^{\infty} \frac{\bar{u}(s, 0)}{s - \bar{x}}\,ds = \theta_{te}\hat{H}(-\bar{x}). \tag{37}$$

For $c = 0$, the displacement effect vanishes and the solution to Eq. (37) can be shown to reduce to the incompressible wedge solution,

$$\bar{u}_i(\bar{x}, 0) = \frac{\theta_{te}}{\pi}(\ln|\bar{x}| + b), \tag{38}$$

where b is an arbitrary constant. (It should be noted at this point that a constant is an arbitrary homogeneous solution of Eq. (37).) With the choice $b = a_1$ the above expression, when combined with the logarithmic term in Eq. (28), exactly matches the outer inviscid solution.

In order to scale out the parameters c and θ_{te} and to obtain an equation better suited for solution by Fourier transforms, we introduce the following change of variables:

$$\bar{u} = \theta_{te}\left[\left(\frac{1}{\pi}\ln|\bar{x}| + a_1 + f(\xi)\right)\right] \tag{39}$$

and $\xi = \bar{x}/c$.

The function $f(\xi)$ can be shown to satisfy the following singular integro-differential equation:

$$\frac{df}{d\xi} + \frac{1}{\pi}\int_{-\infty}^{\infty} \frac{f(s)}{s - \xi}\,ds = -\frac{1}{\pi\xi}. \tag{40}$$

We define the Fourier transform of $f(\xi)$ by

$$F(\alpha) = \int_{-\infty}^{\infty} f(\xi)e^{i\alpha\xi}\,d\xi. \tag{41}$$

Its inverse is then given by

$$f(\xi) = \frac{1}{2\pi}\int_{-\infty}^{\infty} F(\alpha)e^{-i\alpha\xi}\,d\alpha \tag{42}$$

Application of the Fourier transform to Eq. (40) results in the following algebraic relation:

$$-i\alpha F(\alpha) - \frac{1}{\pi}[i\pi \operatorname{sgn}(\alpha) F(\alpha)] = -\frac{1}{\pi}[i\pi \operatorname{sgn}(\alpha)],$$

which can be solved for $F(\alpha)$. Thus

$$F(\alpha) = \frac{\operatorname{sgn}(\alpha)}{\alpha + \operatorname{sgn}(\alpha)} = \frac{1}{1 + |\alpha|}. \tag{43}$$

Inversion of $F(\alpha)$ by Eq. (42) yields the solution

$$f(\xi) = \frac{1}{2\pi}\int_{-\infty}^{\infty} \frac{e^{-i\alpha\xi}}{1 + |\alpha|}\, d\alpha, \tag{44}$$

which can be manipulated into the form

$$f(\xi) = \frac{1}{\pi}\int_{0}^{\infty} \frac{\cos \alpha\xi}{\xi + \alpha\xi}\, d(\alpha\xi). \tag{45}$$

The integral appearing in Eq. (45) can be related to the classical sine and cosine integral functions defined, for example in [2.115]. Thus for, $\xi > 0$,

$$\pi f(\xi) = g(\xi) \equiv -\left\{\mathscr{C}_i(\xi)\cos\xi + \left[\mathscr{S}_i(\xi) - \frac{\pi}{2}\right]\sin\xi\right\}, \tag{46}$$

where $g(\xi)$ is the function defined in [2.115]. The function $g(\xi)$ is symmetric in ξ, so that the above relation is sufficient to define the solution on the entire real axis.

An asymptotic expansion of $g(\xi)$ for large ξ is obtained by expanding the above integral representation in a power series in α and evaluating the individual terms from a standard table of Fourier transforms. Thus for $\xi \to \infty$

$$g(\xi) = \pi\hat{\delta}(\xi) + \sum_{m=1}^{\infty} \frac{(-1)^{m+1}(2m-1)!}{\xi^{2m}}$$

$$= \pi\hat{\delta}(\xi) + \frac{1}{\xi^2} + \cdots. \tag{47}$$

Curiously, the delta-function contribution was missed in the results presented in [2.115].

The behavior near the origin can be deduced by expanding the integral representation of $f(\xi)$ for $\alpha \to \infty$. The resulting series is somewhat more complicated, so we give only the first few terms. Thus for $\xi \to 0$,

$$g(\xi) \to -\left(\ln|\xi| + \gamma - \frac{\pi}{2}|\xi| + \cdots\right), \tag{48}$$

where Euler's constant $\gamma = 0.577216\ldots$. A plot of the function $g(\xi)$ is given in Fig. 2, showing the smooth transition between the limiting behavior at $\xi = 0$ and $\xi \to -\infty$.

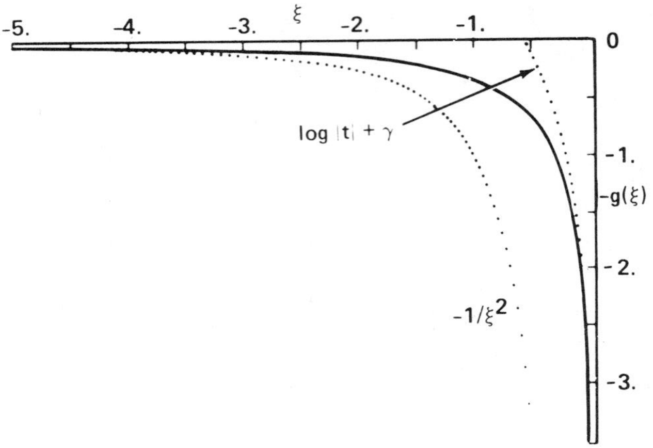

Figure 2. Pressure distribution—interacting-boundary-layer theory.

The solution for the chordwise velocity component u takes the form

$$u = 1 + \frac{t\theta_{\text{te}}}{\pi}(\ln|x| + a_1) + g\left(\frac{x}{\varepsilon^2 c}\right). \tag{49}$$

Since $g(\xi) \to \pi\hat{\delta}(\xi)$ for $|\xi| \to \infty$, the above solution clearly matches the weak-interaction solution for $x \to 0$.

The solution for the pressure distribution can be written in the form

$$p = p_{\text{inv}}(x) - \frac{t\theta_{\text{te}}}{\pi} g\left(\frac{x}{\varepsilon^2 c}\right), \tag{50}$$

where $p_{\text{inv}}(x)$ is the inner limit of the inviscid solution for $x \to 0$ and is given by

$$p_{\text{inv}}(x) = p_\infty - \frac{t\theta_{\text{te}}}{\pi}(\ln|x| + a_1) \tag{51}$$

A plot of the pressure distribution illustrating the effect of Reynolds number is provided in Fig. 3. The flat-plate boundary-layer parameters required in the solution were determined from a standard, equilibrium flat-plate boundary-layer solution given in [2.34] assuming a transition point at 10% chord.

The behavior of the pressure distribution near the trailing edge follows from Eq. (48); thus for $x \to 0$,

$$p \to p_\infty + \frac{t\theta_{\text{te}}}{\pi}\left\{\left[\ln\left(\frac{1}{\varepsilon^2 c}\right) + \gamma - a_1\right] - \frac{\pi}{2}\varepsilon^2 c\,|x|\right\}. \tag{52}$$

This result indicates that the displacement effect of the boundary layer generates a logarithmic singularity which just cancels the corresponding singularity in the inviscid solution. The pressure in the interacting-boundary-layer solution is bounded at the trailing edge for finite Reynolds number ($\varepsilon \neq 0$) and experiences a finite jump in pressure gradient across the trailing edge. We also note that the pressure at the trailing edge becomes unbounded

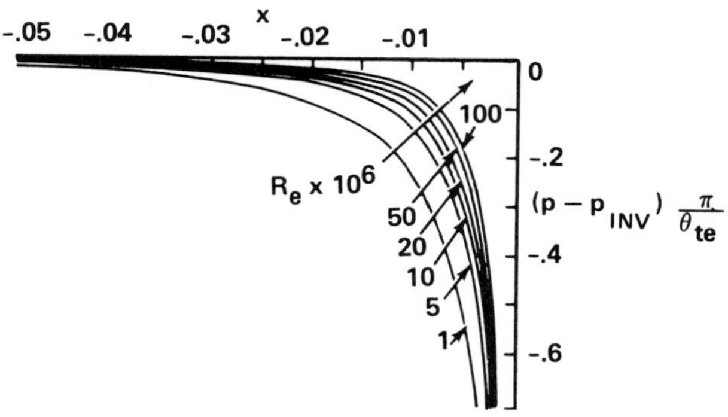

Figure 3. Reynolds-number effect on pressure distribution—interacting-boundary-layer theory.

like $(\ln \varepsilon)^{-1}$ as the inviscid limit is approached ($\varepsilon \to 0$) keeping the trailing-edge angle $t\theta_{te}$ fixed.

The contribution of the local trailing-edge solution to the profile drag can be determined from an integration of the quantity, $p - p_{inv}$, given by Eq. (50) over the interval $(-\infty, 0)$. This integration can be evaluated exactly, to yield the value

$$\Delta C_d = 2(t\theta_{te})^2 c = 2(t\theta_{te})^2(H_{te} + 1)\delta^*_{te}. \tag{53}$$

In concluding this section, we note that the question raised in the beginning of this study has been answered in the affirmative—the strong coupling employed in IBLT completely eliminates the stagnation-point singularity occurring in the inviscid solution. Were it not for normal pressure-gradient effects, IBLT would provide perfectly acceptable solutions for wedge-shaped trailing edges.

Strong-Interaction Theory (S-I)

The analysis of the previous section indicated that IBLT is inadequate for turbulent interactions at trailing edges because these interactions are too compact—occurring over a small streamwise length scale that is the order of the displacement thickness, $\Delta x = O(\delta^*) = O(\varepsilon^2)$. Such narrow interaction regions imply large streamline curvature and normal-pressure-gradient effects across the boundary layer that should be included in the leading terms of the solution.

In this section, we present a systematic approach to the trailing-edge interaction problem that properly accounts for normal pressure gradients in the boundary-layer solution. As in the theories discussed previously, the strong-interaction theory (S-I) developed here is also based on formal asymptotic procedures employing the double limit $t \to 0$, $\varepsilon \to 0$. However, as distinct from IBLT, the new theory employs a more complete formulation that includes the normal-momentum equation. The theory to be described represents an extension of the Melnik–Chow [2.7] (MC) asymptotic theory of turbulent strong interactions at trailing edges.

The MC theory applies to lifting airfoils with cusped trailing edges. The present study extends the MC theory to airfoils with small but nonzero trailing-edge angles, but is limited to symmetric profiles at zero incidence and to incompressible flow. However, the new solution developed in this paper can be superposed on the MC solutions to obtain more general solutions for lifting profiles with nonzero trailing-edge angles. The restriction to incompressible flow is not significant, as compressibility effects can be incorporated into the present theory using the same scale transformations developed in MC.

The flow near the trailing edge develops the same multilayer structure at high Reynolds number as in MC. Referring to Fig. 4, the flow upstream of the trailing-edge interaction develops a conventional two-layer structure, consisting of an outer defect layer and inner wall layer. The flow near the trailing edge develops a three-layer structure spread over a streamwise distance that is of the order of the boundary-layer thickness, $\delta = O(\varepsilon)$. Because of the small length scale, the shear-stress terms can be ignored in comparison with the inertia terms in both momentum equations. Thus, the leading terms in the outer solution can be obtained from the inviscid rotational flow equations. The inner shear layers are required to satisfy the no-slip condition at the wall. An extra blending layer between the outer (inviscid) and inner wall layer is required to account for differences in the growth rate of the Reynolds shear stresses in the outer and wall layers. A much smaller, inner wall layer (not mentioned in MC) is required near the trailing edge to account for the boundary-layer–wake transition across the trailing edge as well as to resolve a weak nonuniformity that develops in the present solution at the trailing

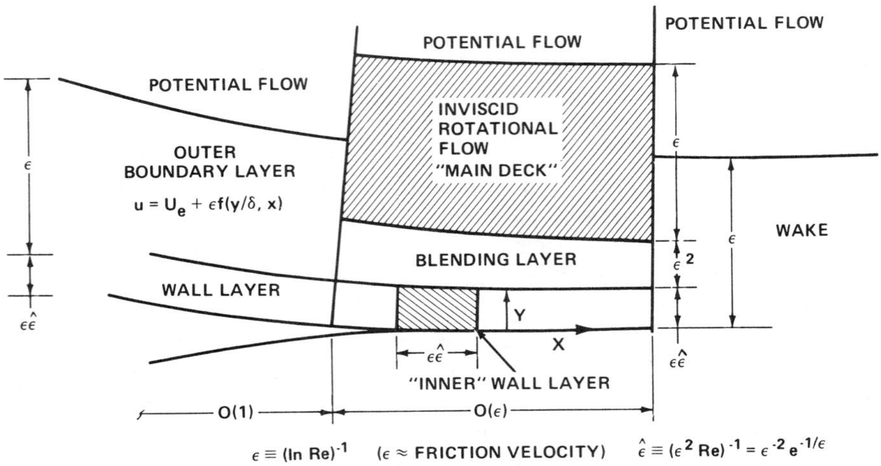

Figure 4. Asymptotic flow-field structure in Melnik–Chow theory [2.7].

edge. This region occurs over a length scale of the order of the wall-layer thickness in both directions. It can be shown that the flow in this region is governed by the full, turbulent Reynolds equations with no approximation possible, except for those associated with the use of the local profile geometry.

Following MC, we assume the velocity profile in the upstream boundary layer approaches a limit at the trailing edge (for $t \to 0$, $\varepsilon \to 0$) that can be represented in a Coles law-of-the-wake–law-of-the-wall form, $U_0(y) = 1 + \varepsilon f(y/\delta_{te})$, where

$$f(y/\delta_{te}) = \gamma_{te}[\ln(y/\delta_{te}) - \tilde{\pi} W(y/\delta_{te})], \tag{54}$$

where δ_{te} is the flat-plate boundary-layer thickness at the trailing edge, $\gamma_{te} = \sqrt{\tau_{w\,te}}/k$, and where $\tilde{\pi}$, $W(y/\delta_{te})$ are Coles's wake parameter and wake function. We employ the polynomial representation for the wake function used by MC,

$$W(r) = 1 - \frac{C_1 r}{8}\left[\left(1 - \frac{r^2}{3}\right) + \frac{C_2}{2}\left(1 - \frac{r^4}{5}\right)\right],$$
$$r = \frac{1 + y/\delta_{te}}{2}, \tag{55}$$

where $C_1 = 15.39616$ and $C_2 = -0.36536$.

The present work will be concerned only with the solution in the outer, inviscid-rotational-flow region (see Fig. 4). Because the inner shear layers are thin, it can be shown that the flow deflections produced by these layers are an order of magnitude smaller than the disturbance in the outer layer. Thus the solution in the outer layer can be completed without consideration of the inner solutions. Furthermore, it can be shown that the pressure variation across the inner layers can be ignored in the evaluation of the surface pressure on the profile.

Following MC, we introduce stretched inner variables ξ, η defined by

$$x \equiv \varepsilon\delta_{te}\xi, \qquad y \equiv \varepsilon\delta_{te}\eta \tag{56}$$

and seek an asymptotic solution in the outer trailing-edge region in the form

$$u = \left[1 + \frac{t\theta_{te}}{\pi}\ln(\varepsilon\delta_{te}) + tU + \cdots\right] + \varepsilon f(\eta) + \varepsilon t\tilde{u}(\xi, \eta) + \cdots, \tag{57}$$

$$v = [tV + \cdots] + \varepsilon t\tilde{v}(\xi, \eta) + \cdots. \tag{58}$$

The first group of terms in Eqs. (57) and (58) are from contributions of the outer, inviscid-irrotational-flow solutions near the trailing edge, written in terms of inner variables (ξ, η). They are given in complex form by

$$U - iV = \frac{\theta_{te}}{\pi}(\ln z + a_1), \qquad z \equiv \xi + i\eta, \tag{59}$$

where a_1 is the global constant appearing in the earlier parts of the paper. The second term in Eq. (57) is a nonuniform shear-flow contribution arising from the upstream boundary layer and is assumed to have the form given in Eq. (54). The disturbance velocities $\tilde{u}$, $\tilde{v}$ then appear as perturbations to a weakly sheared ($\varepsilon \to 0$) nearly parallel flow ($t \to 0$).

The analysis is most conveniently carried out in terms of a disturbance pressure $\tilde{p}$ and flow deflection $\tilde{\omega}$ defined by

$$\tilde{u} = -\tilde{p} - Uf + \Psi f', \tag{60}$$

$$\tilde{v} \equiv \tilde{\omega} + Vf, \tag{61}$$

where Ψ is the real stream function corresponding to the complex velocity $U - iV$. Substituting the expansions given in Eqs. (57) and (58) into the exact inviscid-rotational-flow equations (since the shear stresses affect only the high-order terms of the solution), and using Eqs. (60) and (61) to convert to $\tilde{p}$, $\tilde{\omega}$ variables, we obtain the following equations governing the lowest-order solution in the outer trailing-edge region:

$$\frac{\partial \tilde{p}}{\partial \xi} - \frac{\partial \tilde{\omega}}{\partial \eta} = -2f(\eta)\frac{\partial U}{\partial \xi}, \tag{62}$$

$$\frac{\partial \tilde{p}}{\partial \eta} + \frac{\partial \tilde{\omega}}{\partial \xi} = -2f(\eta)\frac{\partial V}{\partial \xi}. \tag{63}$$

Since the right-hand sides involve only known functions, these equations can be identified as linear, inhomogeneous Cauchy–Riemann equations. They can be integrated using standard complex-function techniques. First, all parameters are removed from the basic equations by the following scale transformation:

$$\tilde{p} = \frac{\gamma_{\text{te}}\theta_{\text{te}}}{\pi}(\hat{p}_1 + \tilde{\pi}\hat{p}_2), \tag{64}$$

$$\omega = \frac{\gamma_{\text{te}}\theta_{\text{te}}}{\pi}(\hat{\omega}_1 + \tilde{\pi}\hat{\omega}_2). \tag{65}$$

Then a solution is sought as the sum of a particular solution plus a homogeneous solution. Since the homogeneous solution satisfies the Cauchy–Riemann equations, it can be represented as an arbitrary analytic function of the complex variable $z = \xi + i\eta$. A particular solution can be found as follows: Introduce a complex pressure $\lambda_{1,2}$ defined by

$$\lambda_{1,2} \equiv \hat{p}_{1,2} + i\hat{\omega}_{1,2}. \tag{66}$$

Then, considering $\lambda_{1,2}$ to be a function of two independent complex variables η, z, the inhomogeneous Cauchy–Riemann equations can be written in the following form (after using the scale transformation defined by Eqs. (64) and (65)):

$$\frac{\partial}{\partial \eta}\lambda_{1,2}(\eta, z) = \frac{2ih'_{1,2}(\eta)}{z - 2i\eta}, \tag{67}$$

where

$$h'_1(\eta) = \ln \eta, \qquad h'_2(\eta) = -W(\eta). \tag{68}$$

This equation must be solved subject to the boundary conditions $\hat{\omega}_{1,2}(\xi, \eta) = 0$. Particular integrals can be found by integrating Eq. (67) with respect to η. Thus on the axis ($\eta = 0$), we find for the first component of the particular solution, $\lambda_{1p}(\xi, 0)$,

$$\lambda_{1p}(\xi, 0) \equiv \beta_{1p}(\xi) + i\sigma_{1p}(\xi) = \int_0^1 \frac{\ln s}{s - \xi/2i}\, ds, \tag{69}$$

where β_{1p} and σ_{1p} are the real and imaginary parts of the complex function $\lambda_{1p}(\xi, 0)$. The above integral can be related to the various forms of the Euler dilogarithm, $\mathscr{L}_2$. Thus

$$\lambda_{1p} = \mathscr{L}_2(2i/\xi), \tag{70}$$

$$\beta_{1p}(\xi) = \tfrac{1}{4}\mathscr{L}_2(-4/\xi^2),$$

$$\sigma_{1p}(\xi) = \Lambda_2\left(\frac{2}{\xi}\right) \equiv \frac{1}{2i}\left[\mathscr{L}_2\left(\frac{2i}{\xi}\right) + \mathscr{L}_2\left(\frac{-2i}{\xi}\right)\right], \tag{71}$$

where the classical functions $\mathscr{L}_2$ and Λ_2 are defined in the integral tables in [2.116]. References [2.116] and [2.117] provide useful series representations of $\mathscr{L}_2$ which permit the numerical evaluation of the functions and which establish the behavior at the origin as well as in the far field.

The second component of the solution, associated with the law of the wake, can be written in the form

$$\begin{aligned}\lambda_{2p}(\xi, 0) &\equiv \beta_{2p}(\xi) + i\sigma_{2p}(\xi)\\ &= -\int_{-1}^{1} \frac{W(r)}{r + (1 + i\xi)}\, dr.\end{aligned} \tag{72}$$

The integral in Eq. (72) can be separated into its real and imaginary parts, and each part can be represented as the sum of several integrals of the form

$$I_n = \int_{-1}^{1} \frac{r^n\, dr}{r^2 + 2r + (1 + \xi^2)}, \tag{73}$$

all of which can be evaluated from standard integral tables.

The above expressions are solutions to the full inhomogeneous equation (67), but are not complete solutions to the problem, because the function $\sigma_2(\xi)$ is not zero as required by the surface boundary condition. The solution can be completed by adding to the above expressions for $\lambda_{1,2p}(\eta, z)$ an analytic function $\lambda_{1,2h}(z)$ that satisfies the boundary condition

$$\lambda_{1,2h}(z) = -\sigma_{1,2h}(\xi) \quad \text{for } \eta = 0 \tag{74}$$

An appropriate analytic function can be constructed by analytically continuing Eqs. (69) and (72) and then reflecting the resulting formula across the real axis. This gives

$$\lambda_{1,2h}(z) = \int_0^1 \frac{h'_{1,2}(s)}{s - iz/2}\, ds$$

From this it follows that on the real axis

$$\sigma_{1,2\,h}(\xi) = -\sigma_{1,2\,p}(\xi) \quad \text{and} \quad \beta_{1,2\,h}(\xi) = \beta_{1,2\,p}(\xi)$$

Hence the complete solution for $\beta_{1,2}(\xi)$ is given by

$$\beta_{1,2}(\xi) = \beta_{1,2\,p}(\xi) + \beta_{1,2\,h}(\xi) = 2\beta_{1,2\,p}(\xi). \tag{75}$$

It can be demonstrated that the solution for the pressure is symmetric about the origin, so that only negative values of the argument need be considered, with solutions for positive arguments computed from the relation $\beta_{1,2}(\xi) = \beta_{1,2}(-\xi)$. The limiting behavior in the far field is given by

$$\beta_1 \to 2\hat{\delta}(\xi) - 2\xi^{-2} + \cdots \quad \text{for } |\xi| \to \infty, \tag{76}$$

$$\beta_2 \to 2\hat{\delta}(\xi) - e_1\xi^{-2} + \cdots \quad \text{for } |\xi| \to \infty, \tag{77}$$

and near the origin by

$$\beta_1 \to -\ln^2|\xi/2| - \frac{\pi^2}{12} + \cdots \quad \text{for } |\xi| \to 0, \tag{78}$$

$$\beta_2 \to 4\ln|\xi/2| + e_2 + \cdots \quad \text{for } |\xi| \to 0, \tag{79}$$

where $e_{1,2}$ are positive constants given by

$$e_1 \equiv 2\left(2 - \frac{C_1}{15} - \frac{4C_1C_2}{105}\right) \approx 2.37861, \tag{80}$$

$$e_2 \equiv \frac{5C_1}{18} + \frac{13C_1C_2}{75} \approx 3.2959. \tag{81}$$

The functions $\beta_{1,2}(\xi)$ are plotted for negative arguments (airfoil surface) in Figs. 5 and 6 along with the first two terms of expansions valid in the far field and at the origin. The logarithmic behavior at the origin is clearly evident in

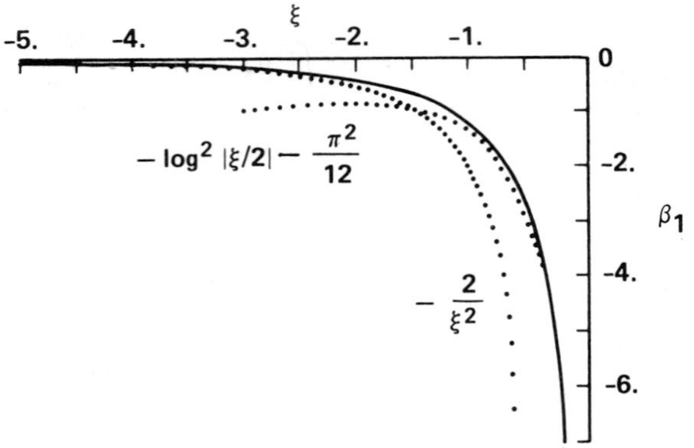

Figure 5. Pressure function β_1 (ξ) appearing in strong-interaction theory—law-of-the-wall contribution.

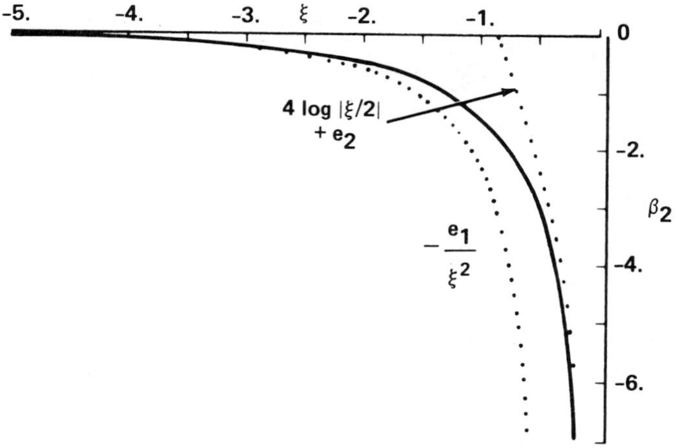

Figure 6. Pressure function β_2 (ξ) appearing in strong-interaction theory—law-of-the-wake contribution.

the plotted results. The solution, for the surface pressure distribution in terms of these functions, is given by

$$p = p_{te} - \frac{t\theta_{te}}{\pi} \ln |x| + \frac{\varepsilon\gamma_{te} t\theta_{te}}{\pi} \left[\beta_1\left(\frac{x}{\varepsilon\delta_{te}}\right) + \tilde{\pi}\beta_2\left(\frac{x}{\varepsilon\delta_{te}}\right) \right], \tag{82}$$

where the global constant a_1 has been absorbed into the parameter $p_{te} \equiv 1 - (t\theta_{te}/\pi)a_1$. Equation (82) is the basic result of this section. It expresses the solution for the pressure distribution in terms of two known universal functions, the trailing angle $t\theta_{te}$, and two viscous parameters $\varepsilon\gamma_{te}$ and $\varepsilon\delta_{te}$ given by the flat-plate values of the skin friction and boundary-layer thickness at the trailing edge.

To demonstrate that the inner solution given above matches the weak-interaction solution valid upstream of the trailing edge, we develop an outer limit of the inner solution (82) and compare it with the inner limit of the outer solution given in Eq. (25). Thus expanding Eq. (82) for $\varepsilon \to 0$ and fixed x using the asymptotic formulas given in Eqs. (76) and (77) leads to the desired result,

$$\begin{aligned} p \to & \left[p_\infty - \frac{t\theta_{te}}{\pi}(\ln |x| + a_1) + \cdots \right] \\ & + \varepsilon^2 \left[-\frac{2\delta^*_{te} t\theta_{te}}{\pi} \hat{\delta}(x) + \cdots \right] \\ & - \varepsilon^3 \left[\frac{2t\theta_{te}}{\pi} \gamma_{te}(1 + \tilde{\pi} e_1)\delta^2_{te} x^{-2} + \cdots \right], \end{aligned} \tag{83}$$

which should be compared with Eq. (25). Observing that the constant c in Eq. (25) approaches $2\delta^*_{te}$ as $\varepsilon \to 0$ and that the constant $k_{te} = 1 + \tilde{\pi} e_1$ for the velocity profile used in the inner solution (in Eq. (54)), we see that two

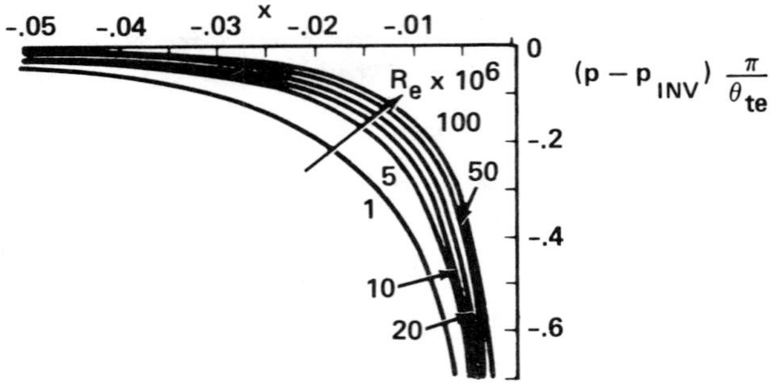

Figure 7. Reynolds-number effect on pressure distribution—strong-interaction theory.

expressions exactly match the terms displayed except for the $\varepsilon^2 \ln |x|$ term in Eq. (25). However, this is a small term in the overlap region that will be matched by higher-order terms (of order $\varepsilon^2 \ln \varepsilon$ and ε^2) of the inner solution, not included in Eq. (82). Thus it is evident that the inner solution matches all the relevant terms in the outer solution including those due to the displacement and normal-pressure-gradient effects. We note this is not the case for IBLT, which does not match the normal-pressure-gradient term in the outer, weak-interaction solution.

In Fig. 7, we present results that illustrate the Reynolds-number effect on the viscous correction to the surface pressure as computed from the last two terms in Eq. (82). Comparing these results with the interacting-boundary-layer solutions given in Fig. 3 indicates similar overall trends. In Fig. 8, we provide a direct comparison of the IBLT and strong-interaction solutions for the viscous pressure corrections for Re $= 10 \times 10^6$. We note the strong-interaction solution exhibits a significantly lower pressure over the entire trailing-edge region, and a somewhat slower decay in the far field. For reference, we have indicated the distance of one boundary-layer thickness on

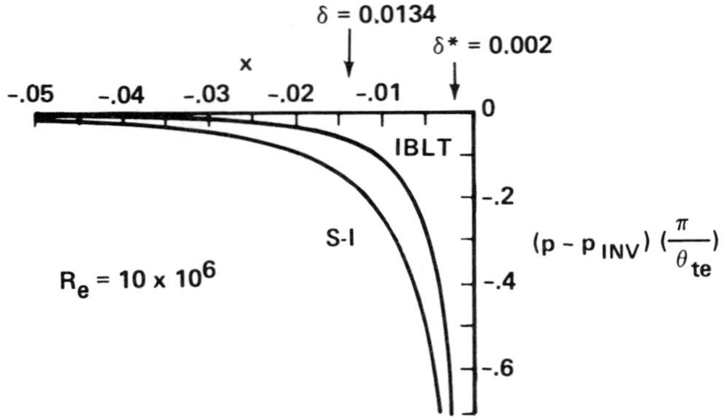

Figure 8. Viscous corrections to the pressure distribution—comparisons of strong-interaction and interacting-boundary-layer solutions.

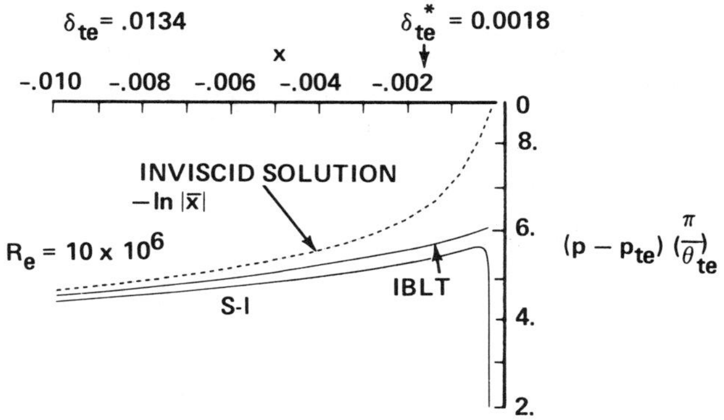

Figure 9. Pressure distribution—comparison of strong-interaction and interacting-boundary-layer solutions.

the axis. In Fig. 9, a similar comparison is presented, this time for the full pressure distribution, including the logarithmic contribution from the inviscid solution (note the changed scale). We also indicate the boundary-layer and displacement-thickness lengths in the figure. These represent the length scales associated with the strong-interaction and interacting-boundary-layer theories respectively. It is apparent from the results in this figure that the interaction occurs over a boundary-layer thickness rather than the much smaller displacement-thickness scale. The two viscous solutions are seen to be similar in character, except for the very noticeable difference in behavior at the trailing edge and the generally lower pressure predicted by the strong-interaction theory. The trailing-edge pressure is finite within IBLT, but is logarithmically divergent in the strong-interaction solution, which exhibits the following behavior near the origin:

$$p = p_{\mathrm{te}} - \frac{t\theta_{\mathrm{te}}}{\pi} \ln|x| - \frac{\varepsilon\gamma_{\mathrm{te}} t\theta_{\mathrm{te}}}{\pi} \times \left\{\left[\ln^2\left|\frac{x}{2\varepsilon\delta_{\mathrm{te}}}\right| + \frac{\pi^2}{12}\right] - \tilde{\pi}\left[4\ln\left|\frac{x}{2\varepsilon\delta_{\mathrm{te}}}\right| + 2e_2\right]\right\} + O(1) \quad \text{for } |x| \to 0. \tag{84}$$

The effect of the logarithmic singularities is evident only in the immediate vicinity of the trailing edge, in a region that is significantly smaller than the displacement thickness. The strong-interaction theory fails when the most singular viscous term in Eq. (84) grows to the same order as the leading inviscid term. On comparing the second-order logarithmic term in Eq. (84) with the inviscid term, we have the following estimate for the region of non-uniformity of the strong-interaction solution:

$$x_{\mathrm{sub}} = O(\varepsilon^{-1} e^{-1/\xi}) = \varepsilon\hat{\varepsilon}. \tag{85}$$

Equation (85) indicates the scale of this region to be the order of the inner wall-layer thickness (see Fig. 4), thus supporting our earlier discussion of the need for an inner wall layer near the trailing edge. It should be stressed, however, that the practical consequences of these logarithmic singularities is not great, since they imply a very small region of nonuniformity over which the pressure can be shown to vary by only $O(\varepsilon^2 \ln \varepsilon)$. A simple *ad hoc* cutoff of the logarithmic terms on a length scale of the order of x_{sub} (defined above) should be adequate for most purposes.

Since the singularities are integrable, the viscous pressure terms can be integrated to determine the contribution of the trailing-edge region to the pressure drag. The result is

$$\begin{aligned}(\Delta c_d)_{\text{S-I}} &= 8\varepsilon(t\theta_{\text{te}})^2 \delta_{\text{te}} \gamma_{\text{te}}(1 + \tilde{\pi}) \\ &= (8t\theta_{\text{te}})^2 \delta^*_{\text{te}}. \end{aligned} \tag{86}$$

This should be compared with the predictions of IBLT, which gave

$$(\Delta c_d)_{\text{IBLT}} = 2(H_{\text{te}} + 1)(t\theta_{\text{te}})^2 \delta^*_{\text{te}}. \tag{87}$$

We first note that both theories indicate the same dependence on the principle parameters of the problem, even though they are based on widely differing scales of pressure and length. The larger length scale of the strong-interaction theory is nearly compensated for by the much smaller levels of the pressure perturbations. Although both theories lead to the same functional dependence on the key parameters, the drag values are significantly different. The ratio of the drag from the two theories is a pure number in the limit $\text{Re} \to \infty$ (since $H_{\text{te}} \to 1$) and is given by

$$\frac{(\Delta c_d)_{\text{S-I}}}{(\Delta c_d)_{\text{IBLT}}} = \frac{4}{1 + H_{\text{te}}(\varepsilon)} \to 2. \tag{88}$$

Thus, although IBLT is nonsingular at the trailing edge, it underpredicts the pressure drag due to the trailing-edge interaction by a factor of two.

Concluding Remarks

In this paper we have presented two different asymptotic theories for the strong interaction at wedge-shaped trailing edges: one based on standard interacting-boundary-layer theory (IBLT), and the other on a new strong-interaction theory (S-I). The principle conclusions that may be drawn from the study are:

1. The strong coupling of the inviscid and boundary-layer solutions employed in IBLT clearly eliminates the singularities at the trailing edge that arise in weak-interaction theory. The pressure was shown to be bounded and the pressure gradient discontinuous at the trailing edge. This occurs because the displacement thickness generated by the self-consistent pressure distribution develops a corner that exactly cancels the geometric corner of the profile at the trailing edge.
2. The length scale of the interaction region associated with IBLT is $O(\delta^*) = O(\varepsilon^2)$, which is too small to be consistent with the normal

momentum equation. Thus, although IBLT leads to nonsingular solutions, it is not the proper large-Reynolds-number limit solution for turbulent interactions at trailing edges.

3. The inclusion of normal-pressure-gradient effects in the S-I theory of the present paper leads to the correct description of the flow near the trailing edge at large Reynolds numbers. The theory predicts an interaction length that is of the order of a boundary-layer thickness $\delta = O(\varepsilon)$ and significantly larger than the length scale indicated in IBLT. The systematic application of asymptotic methods accounting for the two-layer profile in the upstream boundary layer completely eliminated the need for the arbitrary cutoff required in Küchmann's theory.
4. The solutions presented in the paper indicate a nonuniformity in S-I theory in a very small region near the trailing edge. The behavior of the solution at the trailing edge suggests that the length scale of the region is the order of the thickness of the turbulent wall layer, $x_{\text{sub}} = O(\varepsilon\hat{\varepsilon})$. Further analysis indicates that the solution in this region is governed by the full Reynolds equation, with no further approximations possible. However, the small size of the region and the weak, $O(\varepsilon^2 \ln \varepsilon)$ pressure variations across it suggest that this nonuniformity should not be of practical significance. A simple cutoff of the logarithmic singularities at a length scale on the order of the wall-layer thickness should suffice for most applications of the theory.
5. The contribution of the trailing-edge region to the pressure drag is underpredicted by IBLT by a factor of two compared to S-I theory. This can be attributed to the neglect in IBLT of significant pressure variations across the boundary layer in this region.

Some interesting future directions that could be explored involve:

1. The completion of the inner shear layers to determine the skin friction.
2. The development of a suitable approximate solution of the full Reynolds equations for the inner wall-layer region at the trailing edge. Such a solution would be useful in clarifying the boundary-layer–wake transition.
3. The extension of the strong-interaction theory to airfoils with trailing-edge angles that are $O(1)$.
4. The incorporation of the local S-I solution into a viscous transonic airfoil computer code.
5. Comparison with experimental data.

Progress in these areas will go a long way toward the establishment of rational asymptotic methods for turbulent trailing-edge separation.

CHAPTER 14

A Comparison of the Second-Order Triple-Deck Theory with Interacting Boundary Layers[1]

Saad A. Ragab* and Ali H. Nayfeh†

1 Introduction

We consider viscous flows over a flat plate with a small hump situated downstream of the leading edge. The characteristic Reynolds number Re based on the freestream conditions (ρ_∞^*, U_∞^*, μ^*) and the distance L^* from the leading edge of the plate to the position of the hump is assumed to be large. The response of the boundary-layer flow to such a protuberance has been reviewed by Sedney [2.118]. Understanding this response is important in studying the effect of a roughness element on the transition of laminar flows to turbulent flows. Furthermore, in high-speed applications, the presence of such protrusions may contribute to the total drag of the vehicle and it may cause local high heating rates. The study is also relevant to atmospheric boundary-layer flows over hills and mountains [2.65, 2.119]; however, the two-dimensional and laminar-flow assumptions are over-simplifications for these applications.

The flow structure predicted by the triple-deck theory [2.52, 2.58, 2.120] can correctly describe the flow field in the region of such a small hump if the streamwise extent of the hump is $O(L^*\varepsilon^3)$ and its height is $O(L^*\varepsilon^5)$, where $\varepsilon = \mathrm{Re}^{-1/8}$. This is because for these scalings the important changes in the properties of the flow take place over a length $O(L^*\varepsilon^3)$, a basic assumption

[1] This work was supported by the Fluid Dynamics Program of the Office of Naval Research.

* Lockheed Georgia Co., Marietta, Georgia.

† Virginia Polytechnic Institute and State University, Blacksburg, Virginia 24061, and Yarmouk University, Irbid, Jordan.

in the triple-deck theory. Furthermore, the hump is imbedded in the lower-deck region, whose thickness is $O(L^*\varepsilon^5)$, where viscous terms are retained in order to satisfy the wall boundary conditions.

The problem was first analyzed in the context of the triple-deck theory by Smith [2.119]. He considered the first-order in the theory and linearized the lower-deck equations for very small heights of the hump in the lower-deck scaling. An analytic solution was obtained by using Fourier transforms. Numerical solutions of the nonlinear equations were obtained by Sykes [2.65] and Napolitano et al. [2.67]. The accuracy of the finite-difference scheme in the work of Sykes is second order in the mesh sizes Δx and Δz, where x and z are the coordinates along and normal to the wall, respectively. Napolitano et al. generalized the alternating-direction implicit-relaxation technique developed by Werle and Vatsa [2.56] for solving the supersonic interacting-boundary-layer equations to handle the first-order triple-deck equations for both supersonic and subsonic flows. Their finite-difference scheme is first-order accurate in the mesh size Δx. Therefore, extrapolation of the results to zero mesh size is necessary to obtain second-order accuracy in Δx.

From the previous paragraph we see that only the first-order triple-deck theory was considered for flows over a hump. Furthermore, no comparisons with solutions of the complete Navier–Stokes equations or interacting-boundary-layer equations were reported. In the present paper, we consider the second-order triple-deck theory for incompressible flows over a hump and provide comparisons with solutions of interacting-boundary-layer equations. The latter equations are the classical boundary-layer equations except that the inviscid flow is determined by the displacement body, which is composed of the given geometry and the viscous displacement thickness.

The work of Davis and Werle and their colleagues [2.51] on interacting-boundary-layer equations showed good agreement between the solutions of these equations and the Navier–Stokes equations for both supersonic and subsonic flows with small separation bubbles. Work on subsonic interaction problems was initiated by Carter and Wornom [2.3, 2.121], Klineberg and Steger [2.122], and Briley and McDonald [2.123]. Carter and Wornom [2.3] formulated the problem in terms of the vorticity and stream functions and used linearized potential-flow theory to predict the pressure over the displacement body. By using finite differences, they obtained solutions for flows over a dip with a small separation bubble. In the present work, the problem is formulated in terms of the Levy–Lees variables (ξ, η, F, V), while the interaction with the inviscid flow is accounted for by the Carter–Wornom procedure. Generalization to high subsonic flows with variable-fluid properties is straightforward when working with the Levy–Lees variables.

In Section 2, the second-order triple-deck equations with the proper matching and boundary conditions for incompressible flows over a hump are given. A second-order-accurate finite-difference scheme for solving these equations is presented in Section 3. In Section 4, the interacting-boundary-layer equations are considered along with the method of solution. The results and comparisons of the different theories for flows over a quartic hump are presented in Section 5.

2 The Triple-Deck Theory

Figure 1 shows a schematic of the triple-deck structure over a small hump in the physical coordinates x^* and y^*. The equation of the hump is assumed in the form

$$y = Hf(x) = F(x), \tag{1}$$

where x and y are the lower-deck transformed variables [1.106].

The numerical results reported in this paper are for the quartic hump

$$f(x) = \begin{cases} (1 - x^2)^2 & \text{if } |x| \le 1, \\ 0 & \text{if } |x| > 1. \end{cases} \tag{2}$$

This hump was considered by Sykes [2.65] and Napolitano et al. [2.67]. As shown by Stewartson [1.106], the flow in the lower deck is governed by the classical-boundary-layer equations

$$u\frac{\partial u}{\partial x} + v\frac{\partial u}{\partial y} = -\frac{dp}{dx} + \frac{\partial^2 u}{\partial y^2}, \tag{3}$$

$$\frac{\partial u}{\partial x} + \frac{\partial v}{\partial y} = 0. \tag{4}$$

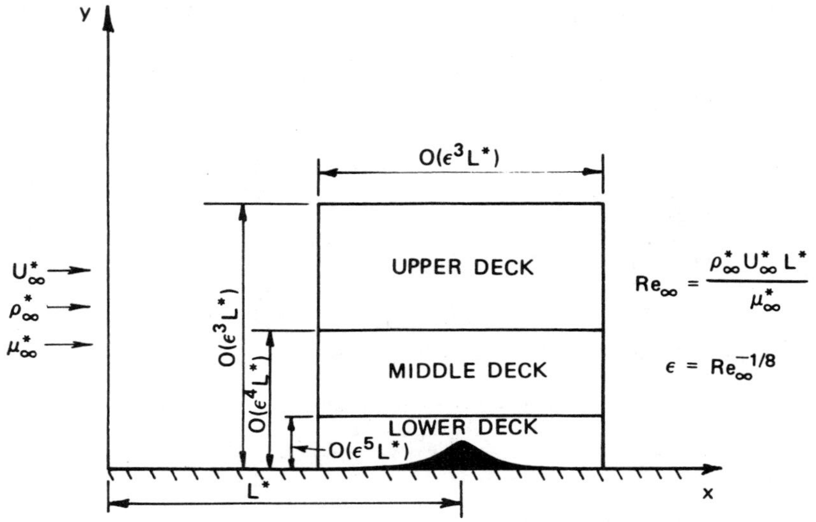

Figure 1. Schematic of the triple-deck structure over a small hump.

The boundary conditions on the wall are the usual no-slip and no-penetration conditions:

$$u = 0, \quad v = 0 \quad \text{at } y = F(x). \tag{5}$$

As $y \to \infty$, the solution in the lower deck should match with the solution in the main deck. This condition yields

$$u \to y + A + \varepsilon[B + \omega_1 p_2] + O(\varepsilon^2). \tag{6}$$

The interaction law which relates the pressure to the displacement function $A + \varepsilon B$ is

$$p = \frac{1}{\pi}\int_{-\infty}^{+\infty} \frac{\dfrac{dA}{d\xi} + \varepsilon\dfrac{dB}{d\xi}}{x - \xi}\, d\xi + \varepsilon\omega_2 \frac{d^2A}{dx^2} + O(\varepsilon^2), \tag{7}$$

where

$$p = p_2 + \varepsilon p_3, \tag{8}$$

where p_2, p_3, A, and B are unknown functions of x, and where

$$\omega_1 = -1.3058, \qquad \omega_2 = 0.60116. \tag{9}$$

Far upstream and downstream of the hump, the disturbance due the hump should decay and the Blasius flow should be recovered there. Therefore,

$$u \to y + O(\varepsilon^2) \quad \text{as } x \to -\infty \tag{10}$$

and

$$p_2, p_3, A, B \to 0 \quad \text{as } x \to \pm\infty. \tag{11}$$

Equations (4)–(11) are the governing equations of the lower deck correct to second order.

To simplify the boundary condition (6), we use the Prandtl transposition theorem [2.124]. We introduce new variables defined by

$$z = y - F, \qquad w = v - u\frac{dF}{dx}. \tag{12}$$

Under this transformation, the governing equations are unaltered except that y is replaced by z, and (6) becomes

$$u \to z + F + A + \varepsilon[B + \omega_1 p_2] + O(\varepsilon^2) \quad \text{as } z \to \infty. \tag{13}$$

Solutions of Eqs. (3) and (4) can be expanded in the form

$$u = u_1 + \varepsilon u_2 + O(\varepsilon^2), \tag{14}$$

$$w = w_1 + \varepsilon w_2 + O(\varepsilon^2), \tag{15}$$

$$p = p_2 + \varepsilon p_3 + O(\varepsilon^2). \tag{16}$$

Substituting these expansions into Eqs. (3) and (4) and equating coefficients of like powers of ε, one obtains equations for the first-order problem (u_1, w_1 p_2) and the second-order problem (u_2, w_2, p_3).

First, we consider the first-order problem. The differential equations governing u_1, w_1, and p_2 are again Eqs. (3) and (4). The matching condition (13) is

$$u_1 \to z + F + A \quad \text{as } z \to \infty. \tag{17}$$

The interaction law, Eq. (7), gives

$$p_2 = \frac{1}{\pi} \int_{-\infty}^{+\infty} \frac{\dfrac{dA}{d\xi}}{x - \xi} \, d\xi. \tag{18}$$

This relation gives the perturbation in pressure due the displacement surface $-A$. We identify Eq. (18) as the direct problem in thin-airfoil theory. In the indirect problem, p_2 is given and the displacement body $-A$ is sought. By using a vortex sheet along the x-axis of strength $\gamma = -2p_2$ per unit length, we obtain the following relation:

$$\frac{dA}{d\xi} = -\frac{1}{\pi} \int_{-\infty}^{+\infty} \frac{p_2}{x - \xi} \, d\xi. \tag{19}$$

Two functions, say $A(x)$ and $p_2(x)$, connected by Eqs. (18) and (19) are called Hilbert transforms (e.g., Titchmarsh [2.125, p. 120]). Differentiating Eq. (19) with respect to x and integrating the result by parts gives

$$\frac{d^2A}{dx^2} = -\frac{1}{\pi} \int_{-\infty}^{+\infty} \frac{\dfrac{dp_2}{d\xi}}{x - \xi} \, d\xi. \tag{20}$$

This equation is solved subject to the two boundary conditions of (11).

Now the pertinent equations to the first-order problem are Eqs. (3), (4), (17), (11), (10), and (20). A numerical procedure for solving these equations is presented in Section 3. Next, we show that the total second-order solution (i.e., $u_1 + \varepsilon u_2$, etc.) is governed by the same equations except that the shape of the hump is modified by a term proportional to p_2, which is already known from the solution of the first-order problem. From the point of view of computational work, this is an advantage over solving the two problems successively. Also the computer program used to solve the first-order problem can be used to determine the total solution with minor changes.

Let us define

$$\bar{A} = A + \varepsilon B - \varepsilon \omega_2 p_2, \qquad \bar{F} = F + \varepsilon(\omega_1 + \omega_2) p_2 \tag{21}$$

Now, the governing equations for the total solution are Eqs. (3), (4), (14), (17), and (20), with A, p_2 replaced by $\bar{A}$, p. They are identical to the governing equations of the first-order problem if the shape of the hump is modified as in Eq. (21).

3 Numerical Procedure for the Triple-Deck Equations

To integrate Eqs. (3) and (4), we use the procedure we developed for supersonic flows over compression ramps [2.126]. First, we differentiate Eq. (3) with respect to z to eliminate the unknown pressure and obtain an equation for the shear stress $\tau(x, z)$, where

$$\tau = \frac{\partial u}{\partial z}. \tag{22}$$

Initially, we guess a value for the shear distribution on the wall. Then, by a marching technique, we find a solution for the shear stress at all mesh points $\tau_{i,j}$, where i and j are indices defining the x and z coordinates of the mesh points; that is,

$$x_i = x_{i-1} + \Delta x, \qquad i = 2, M,$$

$$z_j = z_{j-1} + \Delta z, \qquad j = 2, N.$$

Here, x_1 is an assigned negative value, $z_1 = 0$, and M and N are the total numbers of points in the x and z directions, respectively. The pressure gradient is determined from

$$\frac{dp_2}{dx} = \left.\frac{\partial \tau}{\partial z}\right|_{z=0} \tag{23}$$

In finite-difference form, it is.

$$\left(\frac{dp_2}{dx}\right)_i = \frac{4\tau_{i,2} - 3\tau_{i,1} - \tau_{i,3}}{2\Delta z} + O(\Delta \bar{z}^2). \tag{24}$$

The second step is to solve Eq. (20) for the displacement function A subject to the boundary conditions (11). We use central differences for the second derivative d^2A/dx^2; that is,

$$-A_{i-1} + 2A_i - A_{i+1} = D_i, \qquad i = 2, M - 1, \tag{25}$$

where

$$D_i = \frac{\Delta \bar{x}^2}{\pi} \mathrm{CI}_i, \qquad \mathrm{CI}_i = \int_{-\infty}^{+\infty} \frac{dP_2/d\xi}{x_i - \xi}\, d\xi. \tag{26}$$

Here, CI_i denotes the principal value of the Cauchy integral at the ith nodal point. The method of evaluating Cauchy's integral is given in Appendix A. It is essentially the same as that presented by Napolitano et al. [2.67].

The tridiagonal system of equations (25) can be solved by using the Thomas algorithm. The asymptotic boundary condition $A \to 0$ as $x \to -\infty$ is applied at a finite negative value of $x = x_1$, where the integration of the momentum equation starts. For all calculations presented in this paper, $x_1 = -12$. This was numerically proven to be a good approximation: moving the range further upstream produced insignificant changes in the solution. Thus, one boundary condition for the tridiagonal system (25) is

$A_1 = 0$. The downstream boundary condition (that is, $A \to 0$ as $x \to +\infty$) could be applied at a finite value of x. But the linearized solution [2.119] for small heights of the hump indicates that A decays algebraically for large values of x; that is,

$$A \sim \frac{1}{x^{\gamma}} \text{ as } x \to \infty, \quad \text{where } \gamma = \tfrac{7}{3}. \tag{27}$$

It is better to make use of this asymptotic behavior in implementing the downstream boundary condition. By differentiating Eq. (27), we obtain

$$\frac{1}{A}\frac{dA}{dx} = -\frac{\gamma}{x}.$$

Thus at the final position $x = x_M$, we have

$$\frac{1}{A_M}\left(\frac{dA}{dx}\right)_M = -\frac{\gamma}{x_M},$$

or

$$\frac{3A_M - 4A_{M-1} + A_{M-2}}{2\,\Delta x} = -\frac{\gamma}{x_M}A_M. \tag{28}$$

It follows from (25) with $i = M - 1$ that

$$-A_{M-2} + 2A_{M-1} - A_M = D_{M-1}. \tag{29}$$

Eliminating A_{M-2} from Eqs. (28) and (29), we obtain

$$A_M = \frac{1}{1 + 1\gamma\,\Delta x/x_M}A_{M-1} + \frac{D_{M-1}}{2 + 2\gamma\,\Delta x/x_M}. \tag{30}$$

The recursion relation for inverting the tridiagonal system of equations is

$$A_i = E_i A_{i-1} + H_i,$$

where

$$E_M = \frac{1}{1 + \gamma\,\Delta x/x_M} \quad \text{and} \quad H_M = \frac{D_{M-1}}{2 + 2\gamma\,\Delta x/x_M}. \tag{31}$$

These relations complete the boundary conditions for the Thomas algorithm.

The third step in the numerical procedure is to use the calculated values of A to update the wall shear stress. This is realized by satisfying (17), which, from (22), is equivalent to

$$\int_0^z \tau\,dz \to z + F + A \quad \text{as } z \to \infty. \tag{32}$$

This asymptotic condition is also imposed at a finite value of z, say z_N. The integration in (32) is evaluated by using the trapezoidal rule, that is

$$\frac{\Delta z}{2}(\tau_{i,1} + \tau_{i,N}) + \Delta z \sum_{j=2}^{N-1} \tau_{i,j} = z_N + F_i + A_i,$$

or

$$\hat{\tau}_{i,1} = \frac{2}{\Delta z}(z_N + F_i + A_i) - 2 \sum_{j=2}^{N-1} \tau_{i,j} - \tau_{i,N}, \tag{33}$$

where ˆ denotes updated values of the wall shear stress. It was found that an underrelaxation factor R_f is necessary to avoid diverging solutions. Therefore, the new value of $\tau_{i,1}$ is taken to be

$$\tau_{i,1} \rightarrow R_f \hat{\tau}_{i,1} + (1 - R_f)\tau_{i,1}. \tag{34}$$

Values of R_f as low as 0.2 were necessary.

We repeat the basic three steps mentioned above until the solution does not change within a certain tolerance. The criterion for convergence was taken as

$$\frac{1}{M} \sum_{i=1}^{M} \left| \frac{dP_2}{dx} \right|_i < 10^{-7}. \tag{35}$$

That is, the average error over the nodal points is less than 10^{-7}.

To start the computation, we needed an initial guess for the wall shear stress. We could start with very small heights of the hump, for which the linearized solution given by Smith [2.119] can be used. Alternatively, we used an approximation based on (32). For small values of H, we expect that the shear stress will deviate little from unity except at the wall. Hence, we assume

$$\tau_{i,j} = \begin{cases} 1 & \text{for } j > J, \\ \tau_{i,1} & \text{for } j \leq J. \end{cases}$$

Thus, the integral in (32) becomes

$$\int_0^{z_N} \tau \, dz \approx \tau_{i,1}(J - 1)\,\Delta z + (N - J)\,\Delta z.$$

Now if we neglect A in (32), we obtain

$$\tau_{i,1} = 1 + \frac{1}{(J-1)\,\Delta z} F_i. \tag{36}$$

In the present calculations, we set $(J - 1)\,\Delta z = 1$. No convergence problems were observed due to this initial guess. After the solution was converged for this small value of H, we incremented H by ΔH and used the converged solution as an initial guess for the new value of H, and so on, until we reached the desired value of H.

4 The Interaction-Boundary-Layer Model

The governing equations for this model are the incompressible boundary-layer equations. In terms of the Levy–Lees variables, the problem is given by

$$2\xi F F_\xi + V F_\eta + \beta(F^2 - 1) - F_{\eta\eta} = 0, \tag{37}$$

$$2\xi F_\xi + V_\eta + F = 0, \tag{38}$$

subject to the boundary conditions

$$\begin{aligned} F &= 0,\ V = 0 && \text{at } \eta = 0, \\ F &\to 1 && \text{as } \eta \to \infty, \\ F &= F(\xi_0, \eta) && \text{at } \xi = \xi_0, \end{aligned} \tag{39}$$

where

$$\xi = \int_0^x U_e\, dx, \qquad \eta = \frac{U_e y \sqrt{\mathrm{Re}}}{\sqrt{2\xi}}, \tag{40}$$

$$F(\xi, \eta) = \frac{u}{U_e}, \qquad V(\xi, \eta) = \frac{2\xi}{U_e}\left[F\frac{\partial \eta}{\partial x} + \frac{v\sqrt{\mathrm{Re}}}{\sqrt{2\xi}}\right], \tag{41}$$

$$\beta = -\frac{2\xi}{U_e^3}\frac{dp}{dx}, \tag{42}$$

and x and y are the usual body-oriented axes, while U_e is the inviscid surface velocity including the displacement effect.

An expression for the displacement thickness is

$$\delta = \frac{\sqrt{2\xi}}{\sqrt{\mathrm{Re}}\, U_e}\int_0^\infty (1 - F)\, d\eta. \tag{43}$$

The friction coefficient defined by

$$C_f = \frac{\tau_w^*}{\rho^* U_e^{*2}} = C_{f0}\frac{u_e^2\sqrt{x}}{\alpha\sqrt{\xi}}\frac{\partial F}{\partial \eta}\bigg|_{\eta=0} \tag{44}$$

where $\alpha = 0.469600$ and C_{f0} is the friction coefficient for the Blasius flow (i.e., in the absence of the hump).

In the previous equations all variables were nondimensionalized with respect to the reference length L^*, the reference velocity U_∞^*, and the pressure $\rho^* U_\infty^{*2}$.

To solve (37)–(39), we define a viscous displacement body according to

$$(y_{\mathrm{DB}})_{\mathrm{vis}} = y_w + \frac{\delta}{\cos\theta}, \tag{45}$$

where y_w is the ordinate of the wall and θ its slope. Moreover, we define an inviscid displacement body based on the linearized potential-flow theory for a given pressure distribution according to

$$(y_{\mathrm{DB}})_{\mathrm{inv}} = \frac{1}{\pi} \int_{L.E.}^{\infty} \frac{p\, dx_1}{x - x_1}, \tag{46}$$

where p is the specified pressure.

Now, we need to find the pressure distribution such that

$$(y_{\mathrm{DB}})_{\mathrm{vis}} = (y_{\mathrm{DB}})_{\mathrm{inv}}. \tag{47}$$

To satisfy this condition, we use the following procedure. First, we prescribe a distribution for $(y_{\mathrm{DB}})_{\mathrm{inv}}$; then we determine β by solving Eqs. (37)–(39) so that the resulting profiles give a displacement body equal to the prescribed one. Thus (47) is satisfied for the prescribed $(y_{\mathrm{DB}})_{\mathrm{inv}}$. By integrating (42) we determine the pressure p. Next, we find the inviscid displacement body corresponding to this pressure by integrating (46). The inviscid displacement body thus obtained will be different from the assumed one. We update the assumed body by using the new calculated one and repeat the process until the change in the displacement body becomes negligible. The criterion for convergence is the same as (35) except that $|dp_2/dx|_i$ is replaced by $|\Delta y_i|$, where Δy_i is the change in the displacement body at the ith location.

It was found that underrelaxation of the pressure obtained in the first step and the inviscid displacement body calculated in the second step was necessary for convergence. Therefore, relations similar to (33) had to be used for the pressure and the displacement body.

To start the calculations, we needed an initial guess for the displacement body. We started with a hump of small height, for which the displacement body was taken to be

$$y_{\mathrm{DB}} = y_w + \frac{\delta_{\mathrm{B}}}{\cos\theta},$$

where δ_B is the displacement thickness of the Blasius flow. After achieving convergence for this hump, we increased the height of the hump by an increment and used the converged solution as an initial guess. The process was continued until we reached the desired hump height.

As shown by Van Dyke [2.127], the displacement effect for a semi-infinite flat plate vanishes to second order. Therefore the integral in (46) gives the difference between the actual displacement body and the displacement thickness of the Blasius flow. Thus to obtain $(y_{\mathrm{DB}})_{\mathrm{inv}}$, we added δ_B to the value given by (46).

The finite-difference scheme used to solve (37)–(39) utilized three-point backward differencing for the ξ-derivative and central differencing for the η-derivatives. The nonlinear terms in the momentum equation were quasi-linearized by using a Newton–Raphson method; hence the momentum and continuity equations were still coupled in their finite-difference forms. The system of equations was solved by using the scheme presented in Appendix A.

In the region of reversed flow (i.e., for $F < 0$), the convective term $2\xi F F_\xi$ was neglected to obtain stability for the numerical solution. This is known as the Reyhner–Flügge-Lotz approximation [2.49]. The distribution of the mesh points on the surface of the hump coincided with that used in solving the triple-deck equations. This ensured that the step size in the x-direction honored the triple-deck scalings. To be able to resolve the flow in the lower deck, we used a mesh with a variable step size in the η-direction, so that we were able to employ a larger number of points near the wall than in the rest of the boundary layer.

5 Results and Discussion

The following numerical values were used for all the results presented in this paper for the triple-deck equations: the region covered by the mesh points is $-12 \leq x \leq 20$, $0 \leq z \leq 12$; $\Delta z = 0.5$, $N = 25$, $R_f = 0.2$. We note that the value of γ in (27) controls the behavior of the pressure distribution near the downstream end of the region of computation. Its value was adjusted in the course of computation so that the pressure decreased monotonically at the far downstream end, as it should. It was found that the value of γ depends primarily on H, as

$$H = 0.1, \qquad \gamma \approx 0.52,$$

$$H = 2.4, \qquad \gamma \approx 0.70.$$

Changes in γ around the above values produced insignificant changes, $O(10^{-5})$, in the solution everywhere in the flow field except very near the downstream end of the computation region.

To confirm the second-order accuracy of the numerical scheme presented in Section 3, we conducted a step-size study on Δx for a hump with the maximum height, $H = 0.3$. The dependence of the pressure and wall shear stress at the three points $x = 1$, $x = 0$, and $x = +1$ on $\Delta \bar{x}^2$ is shown in Fig. 2(a), (b), and (c), respectively. From these graphs it is clear that the scheme is second-order accurate in the step size Δx.

The solution for the case $H = 0.1$ was calculated by Napolitano et al. [2.67]. Their results extrapolated to zero mesh size are in good agreement with the analytic solution of the linearized lower-deck equations. The same case was calculated by the present scheme with $\Delta x = 0.1$. The pressure and the wall shear stress distributions are shown in Figs. 3 and 4 respectively. The agreement with our scheme is also good within the plotting accuracy.

Now, we compare the triple-deck theory and interacting boundary layers. As mentioned in Section 4, the interacting-boundary-layer program utilizes a mesh with a variable step size in the η-direction. The variation of $\Delta\eta$ is given by

$$\Delta\eta_{j+1} = R\,\Delta\eta_j, \qquad j = 1, N - 1$$

where $\Delta\eta_1$ and R are specified values. The actual values used in this paper were $\Delta\eta_1 = 0.05$ and $R = 1.04$. The edge of the boundary layer was specified

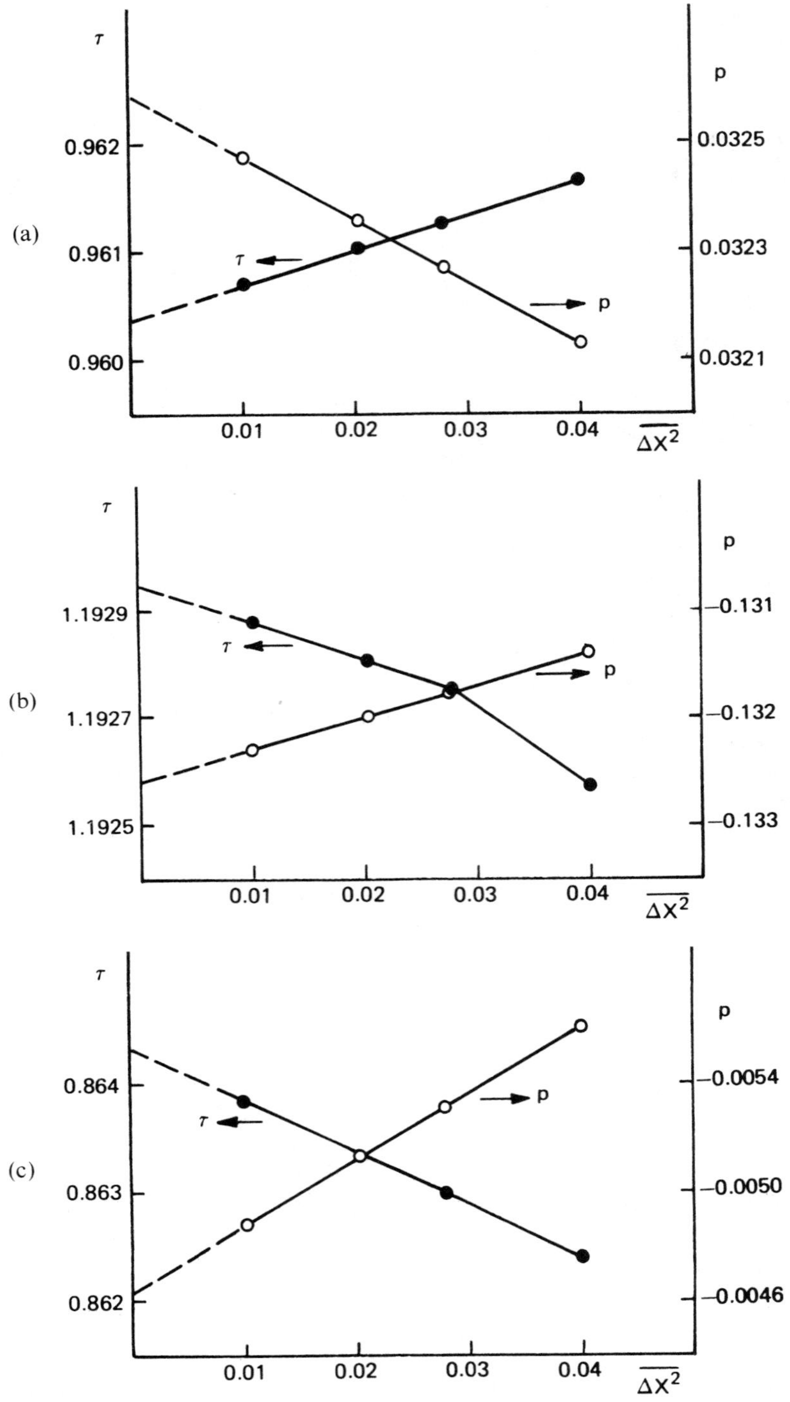

Figure 2. Longitudinal step size study for pressure and wall shear at (a) $x = -1$, (b) $x = 0$, (c) $x = -1$.

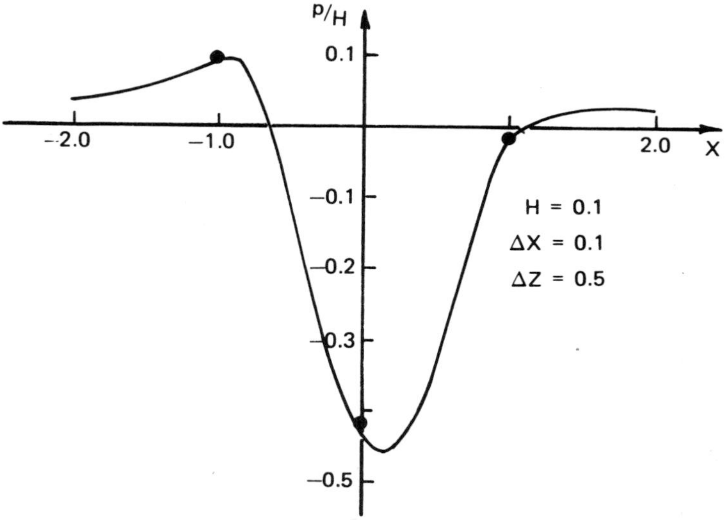

Figure 3. Pressure distribution for $H = 0.1$: •, extrapolated results of Napolitano et al. [2.67].

at $\eta_{\max} = 8.4$. In the lower-deck variables, the range of integration extended from $x = -14.8$ to $x = 25.2$. The distribution of the mesh points on the surface of the hump coincides with that used in solving the triple-deck equations. The underrelaxation factor for the pressure and the inviscid displacement body was 0.1 for all the results presented here. Finally, the Reynolds number Re used in the present calculation was 2.0×10^5, corresponding to $\varepsilon = 0.217$.

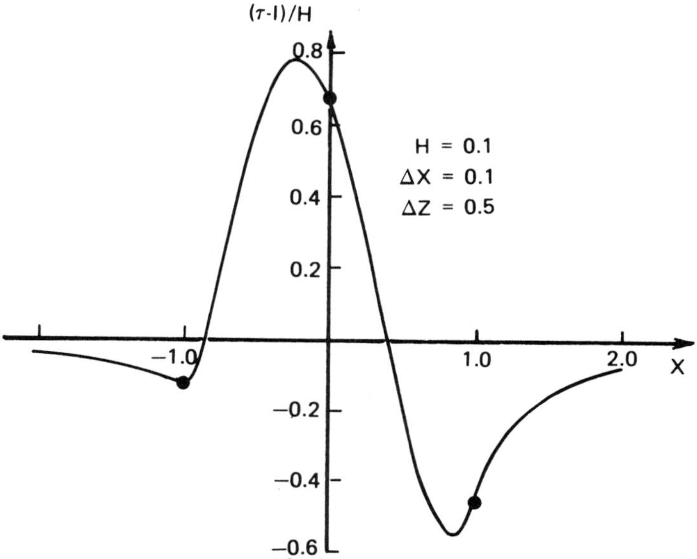

Figure 4. Wall shear-stress distribution for $H = 0.1$: •, extrapolated results of Napolitano et al. [2.67].

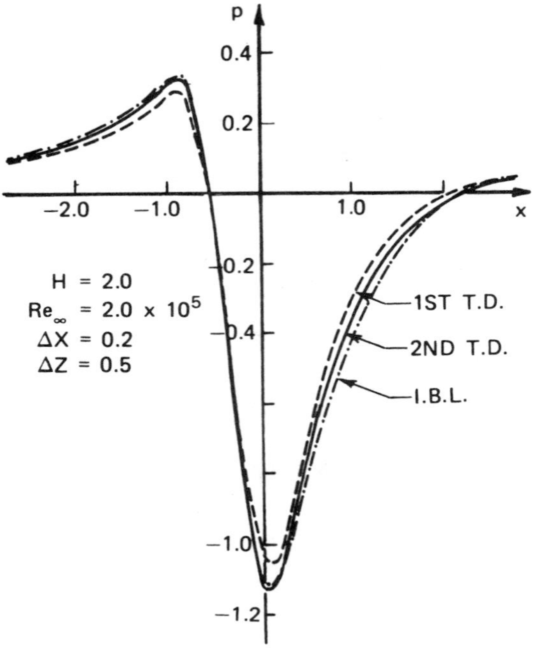

Figure 5. Pressure distribution for $H = 2.0$.

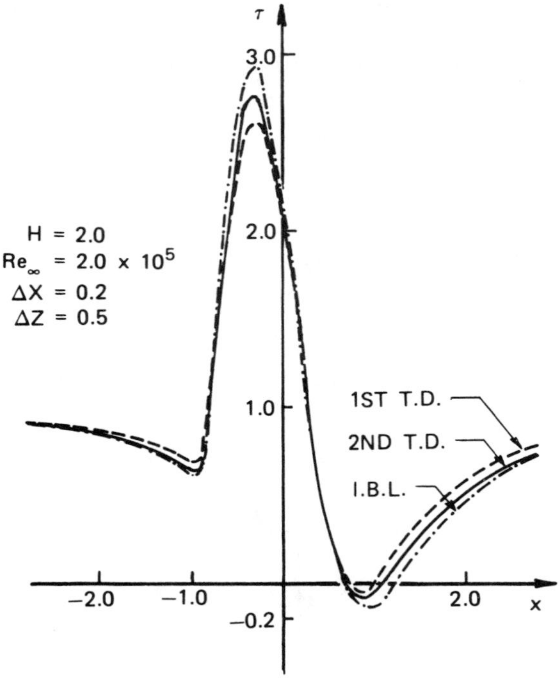

Figure 6. Wall shear-stress distribution for $H = 2.0$.

Solutions of the first- and second-order triple-deck equations were obtained for two values of H. For the first value $H = 2.0$, the pressure distribution and the wall shear stress are shown in Figs. 5 and 6, respectively. For this value of H, there is a small region of reversed flow on the downstream side of the hump. From these figures we see that the first-order triple-deck theory underpredicts the size of the separation region. However, its agreement with the solution of the interacting-boundary-layer equation is satisfactory. Including the second-order terms in the triple-deck equations yields better agreement, in contrast to the supersonic case [2.126].

The results for the second case, $H = 2.4$, are similar except that the region of separated flow is somewhat larger.

The last case considered here is $H = 3.0$. Numerical solutions of the first-order triple-deck theory for this case were obtained by Sykes [2.65]. The region of reversed flow predicted by Sykes is considerable. Our triple-deck program failed to converge for this case even when a relaxation factor of 0.1 was used. The same numerical trouble was encountered by Napolitano et al. [2.67], and they had to neglect both convective terms ($u\,\partial u/\partial x$ and $v\,\partial u/\partial y$) in the region of reversed flow in order to obtain convergence. However, the interacting-boundary-layer program converged for this case without any special adjustment. The results of the first-order triple-deck

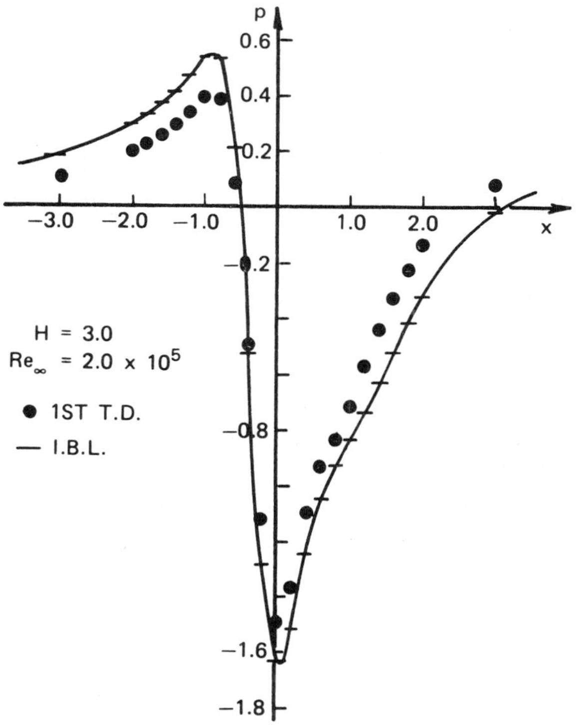

Figure 7. Pressure distribution for $H = 3.0$.

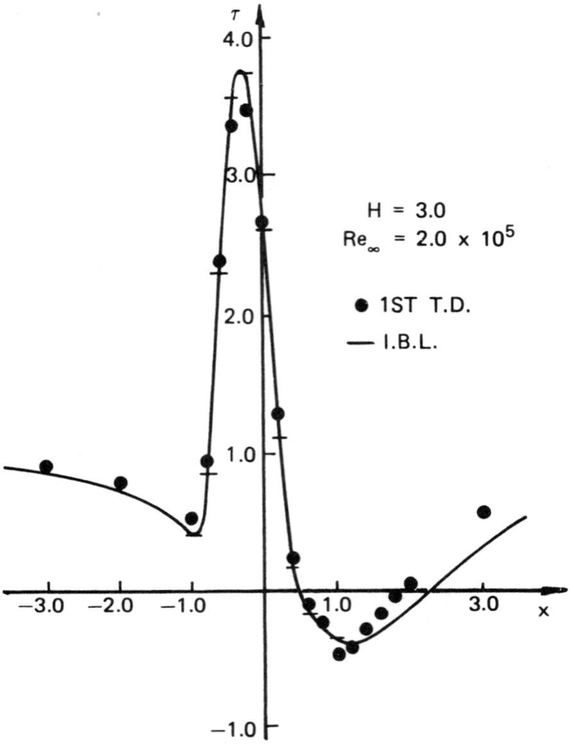

Figure 8. Wall shear-stress distribution for $H = 3.0$.

theory obtained by Sykes and those of the present interacting-boundary-layer equations are shown in Figs. 7 and 8. The shear stress predicted by Sykes shows oscillations in the regions of maximum reversed flow. The general trend of the first-order theory observed for the cases $H = 2.0$ and $H = 2.4$ is preserved for $H = 3.0$.

Appendix A. Numerical Evaluation of Cauchy's Integral

We are concerned here with the numerical evaluation of the principal value of the Cauchy integral

$$\mathrm{CI}(x) = \int_{-\infty}^{+\infty} \frac{f(\xi)}{x - \xi}\, d\xi. \tag{A1}$$

We divide this integral into three parts as

$$\mathrm{CI}(x) = C^{-} + C + C^{+}, \tag{A2}$$

where

$$C^- = \int_{-\infty}^{a} \frac{f(\xi)}{x-\xi}\,d\xi, \qquad C = \int_{b}^{\infty} \frac{f(\xi)}{x-\xi}\,d\xi, \qquad C^+ = \int_{a}^{b} \frac{f(\xi)}{x-\xi}\,d\xi, \quad \text{(A3)}$$

To evaluate the integral C, we divide the interval of integration into an arbitrary number of equally spaced grid points

$$x_i \quad (i = 1, 2, \ldots, M),$$

where $x_1 = a$ and $x_M = b$. The midpoint $\bar{x}_i$ of each interval is given by

$$\bar{x}_i = x_i - \tfrac{1}{2}\Delta x, \qquad i = 2, \ldots, M. \quad \text{(A4)}$$

The integral C at any grid point x_i can be written as

$$C(x_i) = \Delta_{i1} + \sum_{j=2}^{M-1} \Delta_{ij} + \Delta_{iM}, \quad \text{(A5)}$$

where

$$\Delta_{i1} = \int_{x_1}^{\bar{x}_2} \frac{f(\xi)}{x_i-\xi}\,d\xi, \qquad i \neq 1, \quad \text{(A6)}$$

$$\Delta_{ij} = \int_{\bar{x}_j}^{\bar{x}_{j+1}} \frac{f(\xi)}{x_i-\xi}\,d\xi, \quad \text{(A7)}$$

$$\Delta_{iM} = \int_{\bar{x}_M}^{x_M} \frac{f(\xi)}{x_i-\xi}\,d\xi, \qquad i \neq M. \quad \text{(A8)}$$

We approximate the function $f(x)$ over the interval $\bar{x}_j$ to $\bar{x}_{j+1}$ by a straight line. Then the principal value of Δ_{ij} in (A7) can be written as

$$\Delta_{ij} = f_j L_{ij} - S_j[1 - (i-j)L_{ij}], \quad \text{(A9)}$$

where

$$L_{ij} = \begin{cases} \log_2 \dfrac{2(i-j)+1}{(i-j)-1} & \text{if } i \neq j, \quad \text{(A10)} \\ 0 & \text{if } i = j, \quad \text{(A11)} \end{cases}$$

$$f_j = f(x_j), \qquad S_j = \tfrac{1}{2}(f_{j+1} - f_{j-1}). \quad \text{(A12)}$$

Similarly, by approximating $f(x)$ over the intervals x_1 to $\bar{x}_2$ and $\bar{x}_M$ to x_M by straight lines, we obtain

$$\Delta_{i1} = [f_1 + S_1(i-1)]L_{i1} - \tfrac{1}{2}S_1, \quad \text{(A13)}$$

$$\Delta_{iM} = [f_M - S_M(M-i)]L_{iM} - \tfrac{1}{2}S_M, \quad \text{(A14)}$$

where

$$f_1 = f(x_1), \qquad S_1 = f_2 - f_1,$$

$$f_M = f(x_M), \qquad S_M = f_M - f_{M-1}.$$

The value of $C(x_i)$ follows from Eqs. (A5), (A9), (A13), and (A14). The points

x_1 and x_M are excluded. However, we do not need the value of Cauchy's integral at these points in the numerical method described here.

The values of the integrals C^- and C^+ can be assessed if the asymptotic decay of $f(x)$ is known for large negative and large positive values of x. We recall that the function f stands for the pressure in the interacting-boundary-layer equations and for the pressure gradient in the triple-deck equations. For interacting boundary layers, the asymptotic decay of the pressure as $x \to \infty$ is assumed in the form

$$p = \frac{a_1}{x} + \frac{a_2}{x^2} + \frac{a_3}{x^3}. \tag{A15}$$

The factors a_1, a_2, and a_3 are determined by fitting the curve (A15) through the points (p_{M-3}, x_{M-3}), (p_{M-2}, x_{M-2}), and (p_{M-1}, x_{M-1}).

Now, the integral C^+ is obtained in the following form:

1. $x_i \neq 0$:

$$C^+(x_i) = \frac{a_1}{x_i} L + \frac{a_2}{x_i^2}\left(\frac{x_i}{x_M} + L\right) + \frac{a_3}{x_i^3}\left[\frac{1}{2}\left(\frac{x_i}{x_M}\right)^2 + \frac{x_i}{x_M} + L\right], \tag{A16}$$

where

$$L = \log\left(1 - \frac{x_i}{x_M}\right). \tag{A17}$$

2. $x_i = 0$:

$$C^+(0) = -\left[\frac{a_1}{x_M} + \frac{1}{2}\frac{1}{x_M^2} + \frac{1}{3}\frac{1}{x_M^3}\right]. \tag{A18}$$

The integral C^- does not exist for interacting boundary layers if the point x_1 is taken at the leading edge of the plate. However, in the actual calculations, the point x_1 is downstream of the leading edge of the plate. Nevertheless, we set $C^- = 0$, since the deviation of the flow from the Blasius flow is anticipated to be insignificant near the leading edge.

The asymptotic behavior of the pressure gradient in the triple-deck theory for large positive x is given [2.119] as

$$\frac{dp}{dx} = \frac{a_3}{x^3} \quad \text{as } x \to \infty. \tag{A19}$$

We use this result in evaluating the integral C^+. The constant a_3 is determined by patching at every iteration the numerical solution around x_M with the asymptotic expression (A19) for dp/dx.

Now, the result for C^+ is given by the third term in Eq. (A16) for $x_i \neq 0$ and (A18) for $x_i = 0$. We observe that, at the point x_1, the value of dp/dx is $O(10^{-4})$. Hence the contribution of C^- to Cauchy's integral can be neglected. However, the solutions obtained by neglecting C^- are not significantly different from those with C^- evaluated by assuming an asymptotic behavior of the form (A19) for large negative x; the discrepancy is less than 10^{-5}.

References for Part 2

[2.1] M. J. Lighthill: Proc. Roy. Soc. **A217**, 478 (1953).
[2.2] A. E. P. Veldman: Rept. NLR TR 79023 U, Nat. Aerospace Lab. (1979).
[2.3] J. E. Carter, S. F. Wornom: NASA SP-347, 125 (1975).
[2.4] R. E. Melnik, R. Chow: NASA SP-437, 177 (1975).
[2.5] R. Chow, R. E. Melnik: *Lecture Notes in Physics* **59**, 135 (Springer-Verlag, New York 1976).
[2.6] R. E. Melnik: Paper 10, AGARD Conf. Proc. No. 291 (1980).
[2.7] R. E. Melnik, R. Chow, H. R. Mead: AIAA Paper 77-680 (1977).
[2.8] S. Goldstein: Quart. J. Mech. & Appl. Math. **1**, 43 (1948).
[2.9] A. F. Messiter: 8th U.S. Natl. Cong. Appl. Mech., 157 (1978).
[2.10] V. Y. Neiland: No. 1529, Central Institute of Aero- and Hydro-dynamics, Moscow (in Russian) (1974).
[2.11] K. Stewartson: *Lecture Notes in Physics* **771**, 505 (Springer-Verlag, New York 1980).
[2.12] K. Stewartson: SIAM Review **23**, 308 (1981).
[2.13] V. V. Sychev: Izv. A. N. SSSR Mekh. Zhid. i Gaza **3**, 47 (1972).
[2.14] C. E. Jobe, O. R. Burggraf: Proc. Roy. Soc. **A340**, 91 (1974).
[2.15] A. E. P. Veldman, A. I. van de Vooren: *Lecture Notes in Physics* **35**, 422 (Springer-Verlag, New York 1974).
[2.16] P. G. Daniels: Quart. J. Mech. Appl. Math. **27**, 175 (1974).
[2.17] P. G. Daniels: J. Fluid Mech. **63**, 641 (1974).
[2.18] F. T. Smith: Proc. Roy. Soc. **A356**, 433 (1977).
[2.19] F. T. Smith: Private communication.
[2.20] M. J. Werle, J. M. Verdon: Rept. R79-914493-5, United Technologies Research Center, East Hartford, Conn. (1980).
[2.21] I. Tani: *Progress in Aeronautical Science* **5**, 70 (Pergamon Press, New York 1964).

[2.22] P. K. Chang: *Separation of Flow* (Pergamon Press, New York 1970).
[2.23] K. Stewartson: J. Fluid Mech. **44**, 347 (1970).
[2.24] K. Stewartson, F. T. Smith, K. Kaups: Studies in Appl. Math., to appear.
[2.25] M. J. Werle, R. T. Davis: J. Appl. Mech. **39**, 7 (1972).
[2.26] T. Cebeci, A. A. Khattab, K. Stewartson: J. Fluid Mech. **97**, 435 (1980).
[2.27] T. Cebeci, K. Stewartson, P. G. Williams: AGARD Conf. Proc. No. 291 (1980).
[2.28] K. Kraemer: Forsch a.d. Geb. d. Ing. **27**, 33 (1961).
[2.29] F. T. Smith: J. Fluid Mech. **92**, 171 (1977).
[2.30] B. Fornberg: J. Fluid Mech. **98**, 819 (1980).
[2.31] F. T. Smith, R. I. Sykes, P. W. M. Brighton: J. Fluid Mech. **83**, 163 (1977).
[2.32] R. I. Sykes: Proc. Roy Soc. **A373**, 311 (1980).
[2.33] F. T. Smith: J. Fluid Mech. **99**, 185 (1980).
[2.34] T. Cebeci, P. Bradshaw: *Momentum Transfer in Boundary Layers* (Hemisphere/McGraw-Hill, Washington, D.C. 1977).
[2.35] T. Cebeci, A. A. Khattab, K. Stewartson: J. Fluid Mech. **107**, 57 (1981).
[2.36] T. Cebeci, K. C. Chang, K. Kaups: J. Ocean Engrg. **7**, 229 (1980).
[2.37] T. Cebeci, A. A. Khattab, K. Stewartson: Proc. Second Int. Symp. on Turbulent Shear Flows (Springer-Verlag, New York 1981).
[2.38] S. N. Brown, P. G. Daniels: J. Fluid Mech. **67**, 743 (1975).
[2.39] S. N. Brown, H. K. Cheng: J. Fluid Mech. **108**, 171 (1981).
[2.40] P. G. Daniels: Quart. J. Mech. Appl. Math. **31**, 49 (1979).
[2.41] S. W. Rienstra: Ph.D. Thesis, Technische Hogeschool, Eindhoven (1979).
[2.42] R. L. Simpson, J. M. Strickland, P. W. Barr: J. Fluid Mech. **79**, 553 (1977).
[2.43] T. Cebeci, H. U. Meier: Paper 16, AGARD Conf. Proc. No. 271 (1979).
[2.44] T. C. Adamson, A. F. Messiter: Ann. Fluid Mech. **12**, 103, 138 (1980).
[2.45] G. R. Inger: AIAA Paper 80-1411 (1980).
[2.46] A. E. P. Veldman: AGARD Conf. Proc. No. 291 (1980).
[2.47] P. G. Williams: *Lecture Notes in Physics* **35** (Springer-Verlag, Berlin/Heidelberg/New York 1975).
[2.48] T. Cebeci, H. B. Keller, P. G. Williams: J. Comp. Phys. **31**, 373 (1979).
[2.49] T. A. Reyhner, I. Flügge-Lotz: Int. J. Nonlinear Mech., **3**, 173 (1968).
[2.50] F. T. Smith, K. Stewartson: J. Fluid Mech. **58**, 143 (1973).
[2.51] R. T. Davis, M. J. Werle: *Proc. 1976 Heat Transfer Fluid Mechanics Institute* (Stanford University Press 1976).
[2.52] K. Stewartson, P. G. Williams: Proc. Roy Soc. **A312**, 181 (1969).
[2.53] F. T. Smith, K. Stewartson: Proc. Roy. Soc. **A332**, 1 (1973).
[2.54] R. Jenson, O. R. Burggraf, D. P. Rizzetta: *Lecture Notes in Physics* **35** (Springer-Verlag, Berlin/Heidelberg/New York 1975).
[2.55] D. P. Rizzetta, O. R. Burggraf, R. Jenson: J. Fluid Mech. **89**, 535 (1978).
[2.56] M. J. Werle, V. N. Vatsa: AIAA J. **12**, 1491 (1974).
[2.57] O. R. Burggraf, D. P. Rizzetta, M. J. Werle, V. N. Vatsa: AIAA J. **17**, 336 (1979).
[2.58] A. F. Messiter: SIAM J. Appl. Math. **18**, 241 (1970).
[2.59] V. Y. Neiland: Akad. Nauk SSSR, Izv. Mekh. Zhidk. Gaza **3**, 19 (1970).
[2.60] S. N. Brown, K. Stewartson, P. G. Williams: Phys. Fluids **18**, 633 (1975).
[2.61] K. Stewartson: Mathematika **16**, 106 (1969).
[2.62] K. Stewartson: Quart. J. Mech. Appl. Math. **23**, 137 (1970); see also **24**, 387 (1971).
[2.63] J. W. Cooley, J. W. Tukey: Math. Comp. **19**, 297 (1965).
[2.64] S. Orszag: *Lecture Notes in Physics* **59** (Springer-Verlag, Berlin/Heidelberg/New York, 1976).

[2.65] R. I. Sykes: Proc. Roy. Soc. **A361**, 225 (1978).
[2.66] S. A. Ragab, A. H. Nayfeh: AIAA Paper 80-0072 (1980).
[2.67] M. Napolitano, M. J. Werle, R. T. Davis: AFL Rept. 78-6-42, Univ. of Cincinnati, Dept. of Aerospace Engng. (1978).
[2.68] T. C. Adamson, A. Feo: SIAM J. Appl. Math. **29**, 121 (1975).
[2.69] R. E. Melnik, B. Grossman: AIAA Paper 74-598 (1974).
[2.70] G. E. Gadd: J. Aero. Sci. **20**, 729 (1953).
[2.71] M. Honda: J. Aero/Space Sci. **25**, 667 (1958).
[2.72] R. Rosen, A. Roshko, D. L. Pavish: AIAA Paper 80-0135 (1980).
[2.73] M. S. Holden: AIAA Paper 77-45 (1977).
[2.74] K. M. Tu, S. Weinbaum: AIAA J. **14**, 767 (1976).
[2.75] W. C. Rose, D. A. Johnson: AIAA J. **13**, 884 (1975).
[2.76] W. C. Rose, M. E. Childs: J. Fluid Mech. **65**, 177 (1974).
[2.77] A. Walz: *Boundary Layers of Flow and Temperature*, 113 (Mass. Inst. Tech. Press, Cambridge 1969).
[2.78] G. R. Inger, W. H. Mason: AIAA Paper 75-831 (1975); also AIAA J. **14**, 1266 (1976).
[2.79] J. Ackeret, F. Feldman, N. Rott: NACA TM-1113 (1947).
[2.80] M. Nandanan, E. Stanewsky, G. R. Inger: AIAA Paper 80-1389 (1980).
[2.81] R. P. Reklis, J. E. Danberg, G. R. Inger: AIAA J. **19**, 1540 (1981).
[2.82] R. T. Davis, S. G. Rubin: Comp. & Fluids **8**, 101 (1980).
[2.83] S. Rudman, S. G. Rubin: AIAA J. **6**, 1883 (1968).
[2.84] S. C. Lubard, W. S. Helliwell: Comp. & Fluids **3**, 83 (1975).
[2.85] T. C. Lin, S. G. Rubin: Comp. & Fluids **1**, 37 (1973).
[2.86] Y. C. Vigneron, J. Rakich, T. C. Tannehill: AIAA Paper 78-1137 (1978).
[2.87] N. N. Yanenko, V. M. Konenya, G. A. Tarnavsky, S. G. Chernyi: 7th Int. Conf. on Numerical Methods in Fluid Dynamics, Stanford (1980).
[2.88] T. C. Lin, S. G. Rubin: AIAA Paper 79-0205 (1979).
[2.89] S. G. Rubin, A. Lin: Annual Conf. on Aviation & Astronautics, Tel Aviv, 60 (1980); also Israel J. Tech. **18** (1980).
[2.90] P. K. Khosla, S. G. Rubin: Comp. & Fluids **2**, 207 (1974).
[2.91] P. R. Dodge: Rept. 74-211196(6), Airesearch Mfg. Co. of Arizona (1976).
[2.92] U. Ghia, K. N. Ghia, S. G. Rubin, P. K. Khosla: Comp. & Fluids **9**, 123 (1981).
[2.93] A. Lin, S. G. Rubin: AIAA Paper 81-0192 (1981).
[2.94] L. B. Schiff, J. L. Steger: AIAA Paper 79-0130 (1979).
[2.95] S. G. Rubin, P. K. Khosla: AIAA Paper 79-0011 (1979).
[2.96] A. F. Messiter: AGARD Conf. Proc. No. 168 (1975).
[2.97] F. T. Smith: J. Fluid Mech. **92**, 171 (1979).
[2.98] J. E. Carter: AIAA Paper 79-1450 (1979).
[2.99] R. Jenson: Ph.D. Dissertation, Ohio State Univ., Columbus (1977).
[2.100] M. D. Van Dyke: J. Fluid Mech. **14**, 161 (1962).
[2.101] V. Vatsa, M. J. Werle, J. M. Verdon: Rept. R81-914986-5, United Technologies Research Center (1981).
[2.102] R. T. Davis: To appear.
[2.103] M. Napolitano, M. J. Werle, R. T. Davis: AIAA Paper 78-1133 (1978); also AIAA J. **17** (1979).
[2.104] A. E. P. Veldman, D. Dijkstra: 7th Int. Conf. on Numerical Methods in Fluid Dynamics, Stanford (1980).
[2.105] A. E. P. Veldman: Ph.D. Thesis, Mathematical Inst., Univ. of Groningen, The Netherlands (1976).

[2.106] T. Cebeci, F. Thiele, P. G. Williams, K. Stewartson: Num. Heat Trans. **2**, 35 (1979).
[2.107] R. Chevray, L. S. G. Kovasznay: AIAA J. **7**, 1641 (1969).
[2.108] J. L. Robinson: Rept. 1242, National Physical Lab. (1967).
[2.109] I. E. Alber: AIAA Paper 79-1545 (1979).
[2.110] J. Andreopoulos, P. Bradshaw: J. Fluid Mech. **100** (1980).
[2.111] K. Küchemann: Z. Flugwiss., Heft 8/9 (1967).
[2.112] P. D. Smith: RAE TM Aero. 1271 (1970).
[2.113] P. Bradshaw (ed.): *Topics in Applied Physics* **12**, 34 (Springer-Verlag 1978).
[2.114] G. L. Mellor: Int. J. Engr. Sci. **10**, 851 (1972).
[2.115] M. Abramowitz, I. A. Stegun: *Handbook of Mathematical Tables*, 231 (U.S. Government Printing Office 1964).
[2.116] W. Gröbner, N. Hofreiter: *Integraltafel, Zweiter Teil, Bestimmte Integrale*, 71 (Springer-Verlag 1958).
[2.117] K. Mitchell: Phil. Mag. **40** (1949).
[2.118] R. Sedney: AIAA J. **11**, 782 (1973).
[2.119] F. T. Smith: J. Fluid Mech. **57**, 803 (1973).
[2.120] V. Y. Neiland: Izv. Akad. Nauk. SSR, Rekh. Zhidk. Gaza **4**, 53 (1969).
[2.121] J. E. Carter, S. F. Wornom: AIAA J. **13**, 1101 (1975).
[2.122] J. M. Klineberg, J. L. Steger: AIAA Paper 74-94 (1974).
[2.123] W. R. Briley, H. McDonald: J. Fluid Mech. **69**, 631 (1975).
[2.124] L. Rosenhead (ed.): *Laminar Boundary Layers* (Clarendon Press, Oxford 1963).
[2.125] E. C. Titchmarsh: *Introduction to the Theory of Fourier Integrals*, 2nd ed., 120 (Clarendon Press, Oxford 1975).
[2.126] S. A. Ragab, A. H. Nayfeh: Phys. Fluids **23**, 1091 (1980).
[2.127] M. D. Van Dyke: *Perturbation Methods in Fluid Mechanics*, annot. ed. (Parabolic Press, Stanford 1964).

PART 3

SINGULARITIES IN UNSTEADY BOUNDARY LAYERS

Introduction

T. Cebeci

This part of the book is mainly concerned with the possibility that a singularity may spontaneously occur at a finite time at some position in an unsteady boundary layer, which is evolving under a prescribed pressure gradient. If it does, the consequences can be very important for the external inviscid flow, but the question has proved highly controversial in the past.

This part begins with a chapter by Cebeci, who reviews the history of the controversy. He points out that if the singularity occurs, it can be expected to be in one of two broad classes. First, it might be centered at S (which need not be at the wall) and be described by a form of the Goldstein asymptotic expansion. The requirement is that the MRS (Moore–Rott–Sears) criterion [3.1–3.3] should hold, i.e., that the stress should vanish at S and, viewed by an observer moving with the separation point, the fluid be at rest there. The second possibility is that the singularity is inviscid in character and is associated with an overall breakdown of the solution leading, for example, to infinite values of the displacement thickness. Illustrations of each class are given for laminar flow; in particular it is noted that for a boundary layer on a circular cylinder a singularity is highly likely and seems to have the characteristics of the second class. Some turbulent flows are also described, and one of them also seemed to be developing a singularity just downstream of the onset of reversed flow.

In Chapter 16, Wang examines the unsteady boundary layer on a circular cylinder in some detail using the Eulerian approach. He defines a family of pseudostreamlines, by analogy with the limiting streamlines in steady three-dimensional boundary layers, and fixes separation by the formation of an envelope of these streamlines. There are some doubts whether this phenomenon actually occurs in the evolution of the boundary layer.

The strongest evidence for the existence of the singularity in the unsteady boundary layer on a circular cylinder has been provided by the work of Van Dommelen and Shen. In their chapter, the authors present their method of calculating the flow, which is based on the Lagrangian method, and the singularity is recognized by the intersection of two particle paths at the same time. In view of the controversy about the existence of the singularity, the authors go to considerable length to establish the authenticity of their results, the finest mesh used being $289 \times 129 \times 960$. The unimportance of the viscous stresses at the onset of singularity is clearly brought out. It is also conjectured that in physically realistic flows at moderately high Reynolds number, the singularity manifests itself as a thin vorticity layer which rolls up into the trailing vortex. Further studies on this fascinating problem are eagerly awaited.

Somewhat in contrast to the previous study, Dwyer and Sherman present some calculations on unsteady boundary layers which show, downstream of the onset of reversed flow, no clear indication of a singularity. The cases studied include a generalization of Howarth's flow and a three-dimensional flow near a line of attachment (nodal saddle). In the first case, rapid growth of the boundary layer occurs once reversed flow is set up, and this appears to continue until the velocity profile occupies the whole range of integration in the direction normal to the surface. In the second case this rapid growth does not occur, although the approach to the steady state is slow.

The rapid growth of the boundary layer near the rear stagnation point presents severe problems for the computer, and in an earlier study [3.4] 10,000 steps were used in the grid normal to this point. In an attempt to reduce this computational effort, Williams presents a double-structured mesh scheme based on the asymptotic properties of the solution. The inner part of the grid consists of steps which remain constant for all time, while in the outer part, the step size increases exponentially with time. In this way the total number of steps is reduced to a few hundred. The matching of the two structures poses a problem, and various ways of overcoming it are described. The final results are quite satisfactory with respect to the skin friction, but less so with respect to the displacement thickness.

In their paper entitled "Form Factors near Separation", Barbi and Telionis examine an unsteady boundary layer approaching a singularity with the aim of providing a practical criterion for the onset of separation. They suggest that the form factor H might be useful in this context and that when it reaches a value ≈ 2.6, separation is very close at hand. Illustrations are provided from several sources, and in some of them the value of H appeared to be growing without limit at one station. The inference may be drawn that a singularity is indeed very close, and moreover that it is probably of the second class, in which viscous stresses are weak. For if a form of the Goldstein singularity is developing, the form factor remains finite upstream of separation for all time.

The unsteady evolution of similarity solutions of the steady boundary-layer equations is also investigated by Williams. In terms of a reduced time, defined by ξ, he finds that if the development of the solution terminates at a finite value ξ_F of ξ, the reason is either that disturbances from the leading edge can influence the similarity solution, or that the normal velocity at the edge of the boundary layer becomes infinite. In such cases the occurrence of the singularity is in agreement with the MRS criterion. Of special interest is the similarity solution corresponding to a trailing-edge wedge. It appears that if the wedge is blunted or sharp, no singularity is found, but there is a range of vertex angles for which the solution fails at a finite value of ξ.

CHAPTER 15

Unsteady Separation

Tuncer Cebeci

1 Introduction

Consider a steady boundary layer near a fixed wall and driven by a prescribed external velocity $u_e(x)$. It is well known that, in general, computations of the solution come to an end if and when the skin friction vanishes, due to the appearance of a catastrophic singularity centered at the wall [2.8]. The physical interpretation of this result is that $u_e(x)$ is incorrectly chosen and that in order to determine the properties of the fluid motion at high-Reynolds-number flow, the boundary layer must not be considered separately from the external inviscid flow: instead it is necessary to develop a theory which allows for the mutual interaction between the boundary layer and the inviscid flow.

The possibility of generalizing this result to other classes of flows and especially to unsteady flows has attracted many workers. It cannot be expected that the breakdown of the solution will always coincide with the vanishing of the skin friction, nor that it will be centered at the wall. We shall refer to the occurrence of the singularity as *separation* [3.5], and in this paper we shall examine the progress made in our understanding of the concept of unsteady separation both in laminar and in turbulent boundary layers.

2 Basic Equations

2.1 Laminar Boundary Layers

For two-dimensional incompressible unsteady laminar and turbulent shear layers, the continuity and momentum equations are well known and can be written as

$$\frac{\partial u}{\partial x} + \frac{\partial v}{\partial y} = 0, \tag{1}$$

$$\frac{\partial u}{\partial t} + u\frac{\partial u}{\partial x} + v\frac{\partial u}{\partial y} = \frac{\partial u_e}{\partial t} + u_e\frac{\partial u_e}{\partial x} + \frac{1}{\rho}\frac{\partial \tau}{\partial y}. \tag{2}$$

Here τ denotes the total shear stress given by

$$\tau = \mu\frac{\partial u}{\partial y} - \rho\overline{u'v'}. \tag{3}$$

These equations are subject to the usual boundary conditions, which in the case of wall boundary layers are

$$y = 0, \quad u = v = 0; \qquad y \to \infty, \quad u \to u_e(x, t). \tag{4}$$

A good general review of the properties of these equations was given recently by Telionis [3.5], who summarized the state of the art at that time and to whose paper reference may be made for further information about both laminar and turbulent flows.

It was suggested by Moore [3.1], Rott [3.2], and Sears [3.3] that for laminar flows, the solution of the system given by Eqs. (1) to (4) will indicate flow separation if and when there is a point $x_s(t)$, $y_s(t)$ in the flow field such that

$$u = \dot{x}_s, \qquad \frac{\partial u}{\partial y} = 0, \tag{5}$$

i.e., $u = \partial u/\partial y = 0$ at the separation point when viewed by an observer traveling with the speed of separation. This, the MRS criterion, is consistent with the Goldstein singularity for a steady flow past a fixed wall, but permits y_s to be nonzero. Its plausibility in more general flows was strengthened by a number of numerical investigations in which the unsteady problems can be reduced to steady problems with the moving walls [3.6, 3.7]. Moreover, it had been shown by Brown [3.8] that, if the flow is steady, an asymptotic expansion, similar to that obtained by Goldstein but with $y_s > 0$, can be written to describe the singular behavior of such boundary layers, and it appears that her solution can be matched to the terminal structure of the computations. It is, however, important that reversed flow has not occurred in the equivalent steady problem, for then Brown's arguments run into physical and mathematical difficulties. For she assumes $u > 0$ when $x < x_s$, and her expansion is valid only when $x_s - x$ is small and positive. In addition, the solution cannot be continued into $x > x_s$. If $u < 0$ when $x_s - x$ is small and positive, a Goldstein-type singularity is not possible. The reason is that if $u = \partial u/\partial y = 0$ at (x_s, y_s) and $u\, \partial u/\partial x$ is zero there, as this type of structure requires, then the properties of Eqs. (1) to (4) for laminar flow imply that $u \geq 0$ in that neighborhood of y ($\partial^2 u/\partial y^2 > 0$). A possible example of separation occurring in a region of reversed flow is provided by an upstream-moving wall in the presence of an adverse pressure gradient and has been studied by Tsahalis [3.9]. He finds strong evidence of some form of singularity, but fails to provide convincing proof that the MRS criterion is exactly satisfied there, since u is always a monotonic function of y in his computations.

An important feature of the studies on unsteady separation by Telionis and Werle [3.6] and by Williams and Johnson [3.7] is that the laminar-flow equations solved were a reduced form of Eqs. (1) to (4) obtained by letting $t \to \infty$ and assuming, in a sense, that $\partial/\partial t \to 0$. In taking this limit it is not necessarily assumed that x, y are held constant; instead two combinations of x, y, t are fixed. Thus the relevance of the MRS criterion was established by these computations in the limit $t \to \infty$.

A further generalization is to ask whether in any flow started from rest at $t = 0$ there exists a t^* such that for $t < t^*$ the solution is smooth throughout the entire domain of integration in the x, y plane but for $t > t^*$ it can only be found in a portion of the domain bounded in some way by a singularity in the solution. If so, one might ask whether the boundary is the line $y = y_s$ and whether the MRS criterion holds there. Were this the case, there would be profound implications for the theory of high-Reynolds-number flows. The classical view is that they evolve smoothly as t increases but the boundary layer in reversed-flow regions grows exponentially with time, leading eventually to significant changes in the inviscid flow. This view would need changing, and for $t \geq t^*$ the boundary layer would abruptly make a noticeable impact on the external flow in a way which is by no means understood, but might well be by initiating a jet of fluid into it, as has been suggested by Shen [3.10]. The questions we have raised are also important from a practical point of view in the problem of dynamic stall. Here Carr, McAlister, and McCroskey [3.11] have observed large eddies to break away from the boundary of slowly oscillating airfoils, but only after flow reversal has occurred in the boundary layer. These may be associated with the occurrence of a singularity in the solution of the unsteady equations.

These considerations have led to renewed interest in the laminar boundary layer on a circular cylinder moving with uniform speed after an impulsive start for which $u_e = \sin x$ in Eq. (4). For this problem the steady-state solution has a singularity at $x = x_s \approx 1.82 \approx 104°$, and for $x_s < x < \pi$ it does not exist. The previous views of the unsteady solution were that it rapidly approached the steady state if $x < x_s$, but for $x > x_s$ the boundary layer increased in thickness exponentially with time, the flow there being largely an inviscid eddy, but with a thin subboundary layer below it, moving fluid from the rear stagnation point at $x = \pi$ to $x = x_s$. Careful computations for $t \leq 1$ were carried out by a number of authors, most recently by Cebeci [3.12], and lent strong support to this description. Cebeci terminated his calculations partly because the rapid growth in the boundary-layer thickness made the computations tedious, but partly because no untoward feature was developing.

At about the same time van Dommelen and Shen [3.13] were carrying out an extensive numerical study of the laminar boundary-layer equations using a procedure new to boundary-layer theory, namely the Lagrangian method. Here x and y are taken as the dependent variables, being functions of ξ, η, t, where

$$x(\xi, \eta, 0) = \xi, \qquad y(\xi, \eta, 0) = \eta, \qquad \frac{\partial x}{\partial t} = u. \tag{6}$$

They confirmed Cebeci's results for $t \leq 1$, but found that for $t > 1$ a hump develops in the displacement thickness $\delta^*(x, t)$ in the neighborhood of $x = 2$, i.e., a little way into the reversed-flow region. This evolves into a very sharp singularity at $t = 1.502$, $x = 1.937$. This result is surprising, but it *was* obtained after a careful numerical study. Van Dommelen and Shen have looked at the analytic structure of the singularity, and conclude that the MRS criterion is satisfied but that it is not of the Goldstein type. Faced with this new information, Cebeci extended his computations to times greater than unity and confirmed the development of the hump in δ^* at $x = 2$ with increasing time. Figure 1 shows a comparison between the displacement-thickness values calculated by van Dommelen and Shen [3.14] and by Cebeci [3.15] for $t = 1.0$, 1.25, and 1.375. Figure 2 shows the variation of local skin friction with x for values of $t = 1.25$, 1.375. We note that while there is a hump or a rapid rise in δ^* at $x = 2$, the *c_f-values are smooth* and are free of any anomalies. Figure 3 shows the velocity profiles at different x-locations for three values of t. We note that while the velocity profiles for $t = 1$ have a conventional form, as t increases further they begin to develop the unusual feature of adhering to each other in the neighborhood of $x = 2$, $y = 1$, suggesting that $\partial v/\partial y$ is becoming small and even changing sign there. A comparison between the velocity profiles in Fig. 3 and those obtained by van Dommelen and Shen [3.14] shows reasonably good agreement for $t < 1.5$, but I do not consider them to provide strong evidence of the emergence of an MRS singularity. If the solution does develop a singularity, its structure is unlikely to be controlled by the viscous stresses in a significant way.

Figure 4 shows the results obtained by Cebeci [3.15] for values of t higher than those shown in Fig. 1. We see that for $t = 1.5$, the value of δ^*/L at $x = 2$ has increased to 12.5 from its value of 8 at $t = 1.375$. For $t = 1.55$, the value of δ^*/L at $x = 2$ increases further, but a second hump develops at $x = 2.2$. We feel that these calculations are not as reliable. We were able to perform these calculations by taking 1° intervals ($x = 0.0175$) after $x = 1.75$, and were able to essentially obtain the *same* results as those computed with 4.5°

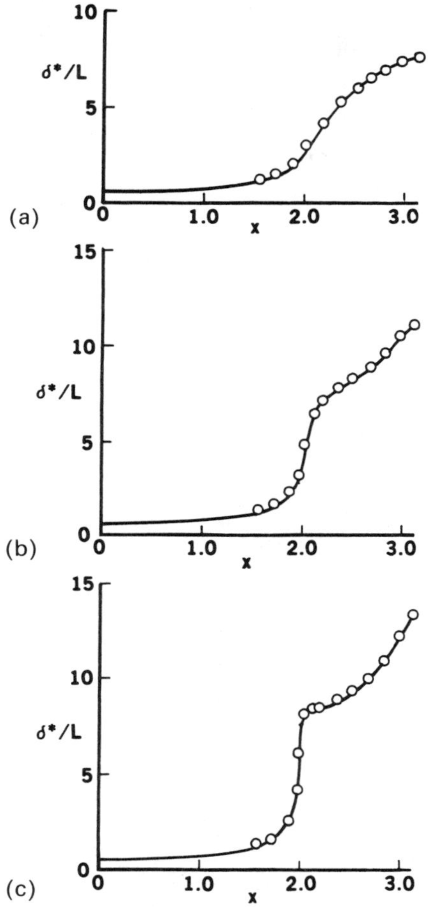

Figure 1. Comparison between the displacement-thickness values computed by van Dommelen and Shen [3.14] (circles) and by Cebeci [3.15] (solid line) for (a) $t = 1.0$, (b) $t = 1.25$, and (c) $t = 1.375$. x is in radians.

intervals ($x = 0.0785$) for $t \leq 1.5$. However, for $t = 1.55$, we were only able to do the calculations with 4.5° intervals.

The results in Fig. 3 also show that with increasing x and time, the boundary-layer thickness grows very rapidly. Computations involving such thick boundary layers as these are near the limit of feasible computations at the present time, and great care must be taken to obtain convincing evidence that any special features revealed are genuine properties of the solution and not just creatures of the numerical method. More will be said on this subject by Dwyer and Sherman in Chapter 18.

Before we leave the subject of laminar flows, it is worth noting that in more complicated situations singularities can develop after a finite time with a structure quite different from that envisaged by Goldstein and MRS. The

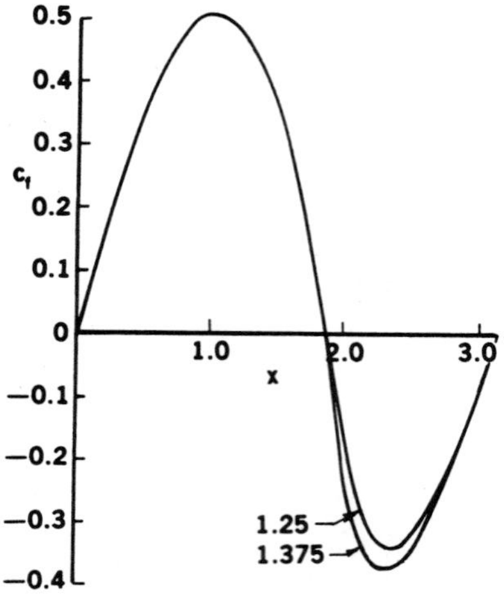

Figure 2. Variation of local skin-friction coefficient for $t = 1.25$ and 1.375. Calculations are due to Cebeci [3.15].

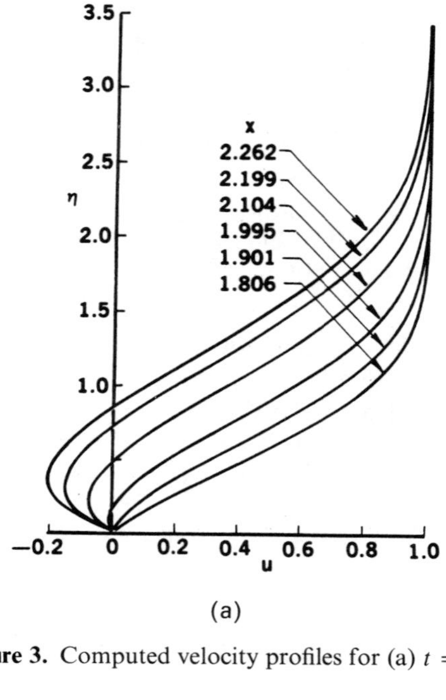

Figure 3. Computed velocity profiles for (a) $t = 1.0$.

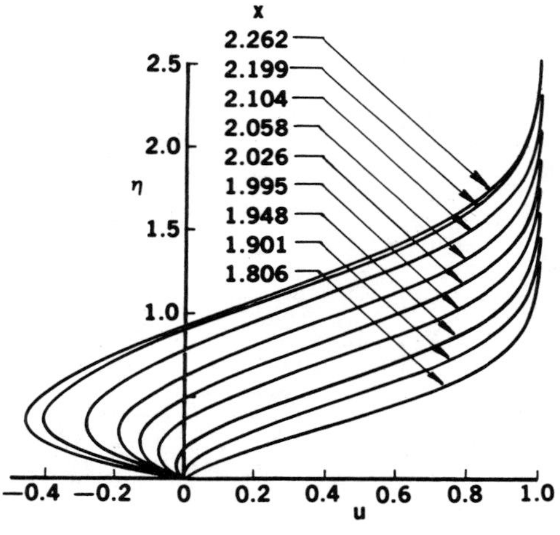

(b)

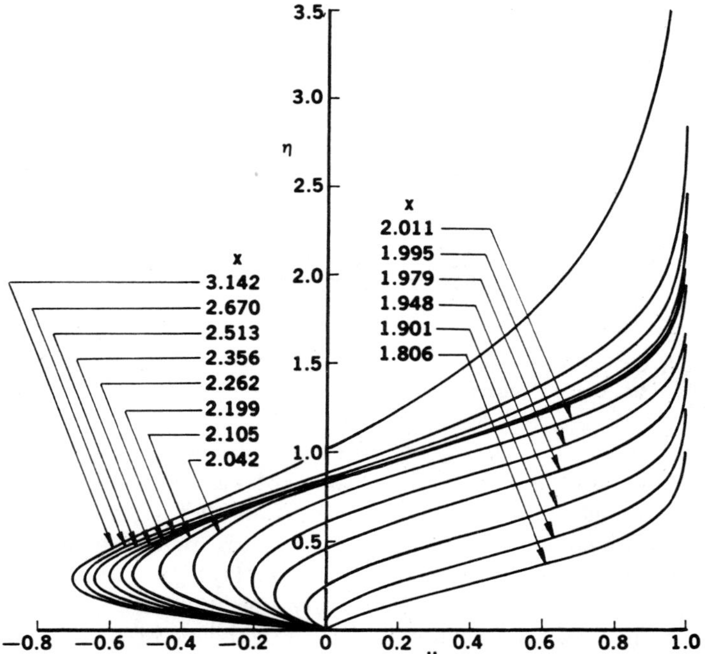

(c)

Figure 3. Computed velocity profiles for (b) $t = 1.25$, (c) $t = 1.375$.

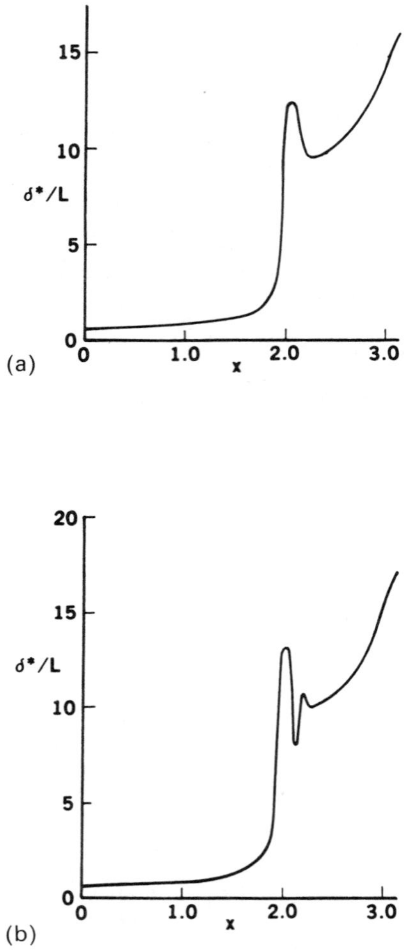

Figure 4. Computed displacement-thickness values for (a) $t = 1.5$ and (b) $t = 1.55$. Calculations are due to Cebeci [3.15].

first example arises in the boundary layer generated on a disk which, initially rotating with the same angular velocity as another, coaxial rotating disk, suddenly has its angular velocity reversed [3.16]. Before the disk has completed one further revolution, the boundary layer at the center becomes infinite in thickness, and simultaneously the angular velocity and axial velocity of the fluid become infinite there over the majority of the layer. The singularity may be associated with an eruption of the boundary layer and the initiation of an axial jet. Another, somewhat similar example is provided by the unsteady natural convection boundary layer on a heated horizontal circular cylinder [3.17], a singularity developing at the upper generator a finite time after the convection starts.

2.2 Turbulent Boundary Layers

The structure and prediction of unsteady separation in turbulent boundary layers have not received the same extensive and careful investigation as in laminar boundary layers. This is mostly due to the empirical nature of the closure assumptions for the Reynolds shear-stress term, $-\rho\overline{u'v'}$. In addition, depending on the turbulence model, the solution of the governing equations can be quite difficult. Except for the work of Patel and Nash [3.18], today most of the existing methods for unsteady boundary layers use the simple concepts of eddy viscosity and mixing length to satisfy the closure assumption. This choice, with Eq. (3) written as

$$\tau = (\mu + \rho\varepsilon_m)\frac{\partial u}{\partial y} \tag{7}$$

and with ε_m being a function of the velocity field, allows the laminar boundary-layer methods to be extended to turbulent flows. The work of Patel and Nash [3.18] uses a modified form of the Bradshaw–Ferriss–Atwell (BF) turbulence model [3.19]; unlike Bradshaw et al., who use a single first-order partial differential equation for the Reynolds shear-stress term, they use a second-order partial differential equation and solve it together with the continuity and momentum equations (1) and (2). Both methods neglect the contributions of laminar shear stress and apply the BF model *only* outside the viscous sublayer. Furthermore, the inner-wall boundary conditions given in Eq. (4) are applied outside the viscous sublayer, usually at $y = 50\,\nu(\rho/\tau_w)$, generally restricting the solution of the equations to problems with *positive wall shear*.

Recently the unsteady-laminar-boundary-layer method of Cebeci, used in performing the calculations for flow over a circular cylinder started impulsively from rest, has been extended to unsteady turbulent boundary layers with flow reversal [3.20]. Using the algebraic eddy-viscosity formulation of Cebeci and Smith [3.21], several test cases were computed to investigate the separation in unsteady turbulent boundary layers.

Figure 5 shows the variation of displacement thickness with x as a function of time for an external velocity distribution given by

$$u_e = u_\infty[1 - 40(x - x^2)(t^2 - t^3)], \qquad 0 < x < 1, \quad t > 0, \tag{8}$$

for a unit Reynolds number $u_\infty/\nu = 2.2 \times 10^6$/m. The results, as in laminar-flow calculations made earlier by Cebeci [3.22], exhibit no signs of singularity for all calculated values of t. This is in contrast to the findings of Patel and Nash, but is in agreement with the views recently expressed by Bradshaw [3.23]. Again as in laminar flows, we see the familiar rapid thickening of the boundary layer in the reversed flow region. If it had not been for this, the calculations would have been computed for greater values of t than those considered in [3.20].

Figure 6 shows the variation of the wall-shear parameter f''_w with distance as a function of time for an unsteady Howarth flow which starts from a

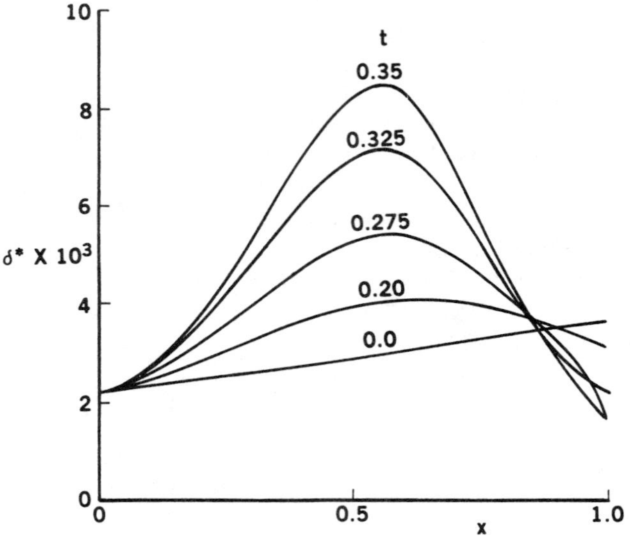

Figure 5. Variation of displacement thickness with x and t for the external velocity distribution given by Eq. (8).

well-established steady flat-plate flow, on which a linear deceleration u_e is imposed at $t = 0$. The external velocity for this flow is given by Carr [3.24]:

$$\frac{u_e}{u_\infty} = 1 - \bar{\alpha}(x - 1.24)t, \qquad 1.24 < x < 4.69, \tag{9}$$

where $\bar{\alpha}$ is a constant equal to 0.72 $\text{sec}^{-1}\ \text{m}^{-1}$. The flow was assumed to be steady up to $x = 1.24$ m; the velocity distribution (9) was then imposed as a function of x and t. This flow differs from that given by Eq. (8) in that, once the flow separates, it does not reattach. According to the calculations of [3.20], the wall shear vanishes first around $t \approx 0.22$, $x = 4.69$. Since the computation of boundary layers for values of x in the range $1.24 \leq x \leq 4.69$ for $t > 0.22$ depends on the specification of a velocity profile at $x = 4.69$, Cebeci and Carr [3.20] have generated such a profile by assuming it is given by the extrapolation of the two velocity profiles computed for $x < 4.69$. This procedure, in which the extrapolated station served as a downstream boundary condition, allowed the calculations to be continued in the negative-wall-shear region as shown in Fig. 6. Again, as in the results of Fig. 5, no signs of singularity were observed except for the rapid thickening of the boundary layer, which made the calculations difficult.

Figure 7 shows the variation of the wall shear parameter f''_w with distance as a function of time for an external velocity distribution given by

$$\frac{u_e}{u_\infty} = 1 + [A^2 + (Bt)^2(\xi - \xi_0)^2]^{1/2} - [A^2 + (B\xi_0 t)^2]^{1/2}, \tag{10}$$

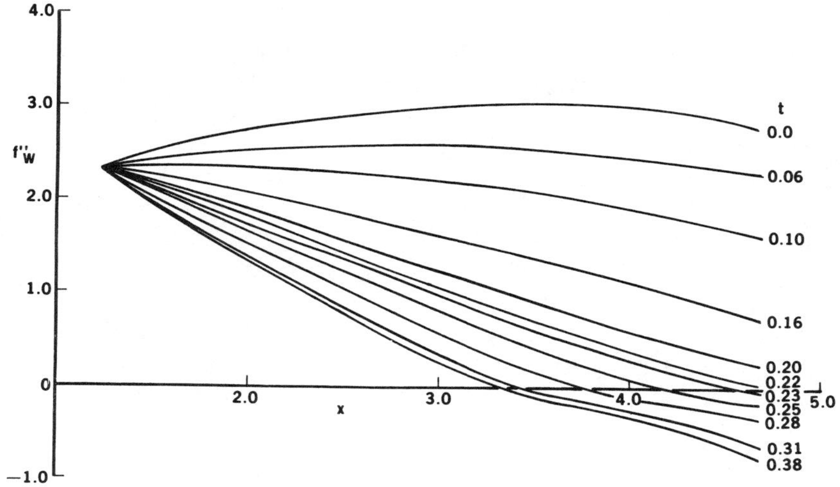

Figure 6. Variation of wall-shear parameter f''_w with distance as a function of time for the external velocity distribution given by Eq. (9).

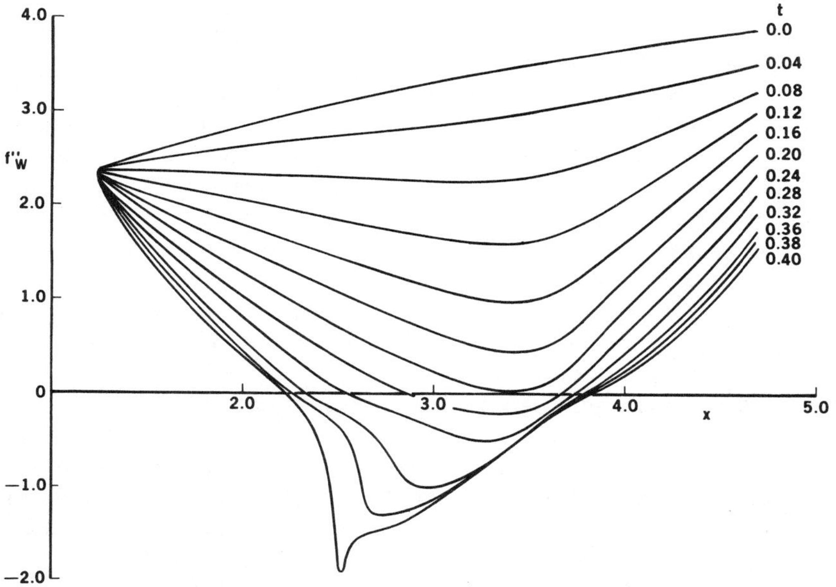

Figure 7. Variation of wall shear parameter f''_w with distance as a function of time for the external velocity distribution given by Eq. (10).

where $A = 0.05$, $B = 3.4\ \mathrm{sec}^{-1}$, $\xi = (x - 1.24)/3.45$, and the x-values are limited to the range $1.24 \leq x \leq 4.69$. The results in this case provide confirmation of the general trends in previous test cases, namely, that as in laminar flows, the unsteady turbulent boundary layers thicken rapidly with increasing flow reversal. A new feature, however, is the dip in the graphs of f''_w near $x = 2.5$ which develops as t increases towards 0.40. It is possible that a singularity occurs in the solution at a later time as in the case of unsteady laminar flow over an impulsively started circular cylinder, but we note that in that flow the most definite sign of its occurrence appeared in the displacement thickness, which showed spiky characteristics. Here the displacement thickness seems to be fairly smooth, but the skin friction becomes spiky. Further work is necessary to explore this puzzling situation.

3 Concluding Remarks

The purpose of most (if not all) of the studies conducted concerning the unsteady separation problem is to gain further insight into the structure of the boundary layers in regions where there is flow reversal across the layer. We know how important the singularity is in a steady flow, if only because in a practical problem the external velocity *must* be modified to remove it. We would like to know how far this is true for unsteady flows, since the presence of the singularity also has profound implications for the attainability of our goals in practical problems such as dynamic stall.

Until recently the present author did not believe there would be a singularity in an evolving boundary layer. There is now strong evidence, thanks to the work of van Dommelen and Shen [3.13, 3.14], in its favor. Although it could be that the features we have found are associated with our having reached the limits of our capability in computing for the time being, the results discussed in the previous sections for different flows *are* puzzling. We see, for example, that while the computed δ^*-distribution for the circular cylinder problem (Fig. 1) suggests a singularity is developing in δ^* at $x = 2$, the c_f-values are smooth and free from any anomalies. On the other hand, the turbulent flow results shown in Fig. 7 suggests that while a singularity is developing in c_f near $x = 2.5$, the δ^*-values are fairly smooth, a situation in contrast with that of circular cylinder. Obviously, further work is needed to explore this puzzling situation.

The closure assumption for the Reynolds shear stress can be satisfied by using several turbulence models. As long as the rate of change of Reynolds shear stress is not large, the ability of a turbulence model to predict unsteady flows can be gauged by its ability to predict steady flows. A comparison between the predictions of two turbulence models developed for steady flows—namely, the CS and BF models—when applied to unsteady flows using the same numerical method, shows that the two yield almost identical results, including the flows which are sufficiently strong in pressure gradient to cause flow reversal across the layer. It appears from the study of Cebeci and Carr [3.20] and Cebeci and Meier [2.43] that for wall boundary-layer flows in which the Reynolds shear stress and frequency do not change rapidly, the predictions for turbulent flows can be obtained satisfactorily by using simple algebraic eddy-viscosity expressions. On the other hand, when either the rate of change of shear stress or the frequency is large, it is not possible at present to calibrate appropriate turbulence models.

Although for a prescribed external velocity distribution the boundary-layer equations for steady flow are singular at separation, this is not the case for an interacting flow when either the displacement thickness or the wall shear is specified. This was demonstrated a long time ago by Catherall and Mangler [3.25], who solved the laminar boundary-layer equations in the usual way until separation was approached and then, by assuming a displacement-thickness distribution, calculated the external-velocity distribution and local flow properties through the recirculation region. Since that study, a large number of inverse boundary-layer procedures have been developed (see, for example, Bradshaw, Cebeci, and Whitelaw [3.26]), and laminar and turbulent boundary-layer equations have been solved for a wide range of problems. A procedure recently suggested by Veldman [2.2, 3.27] is especially promising and has already been applied to two problems [2.27]. Further work now should be advanced in that direction, and practical unsteady-flow problems should be attacked by using correct interactive boundary-layer procedures and especially inverse boundary-layer techniques. In my opinion the possible presence of a singularity will not then be a cause for concern.

CHAPTER 16

On the Current Controversy about Unsteady Separation

K. C. Wang*

1 Introduction

In the past few years, research on the problem of unsteady boundary-layer separation has become fairly active, but the results are contradictory and controversial. The present symposium provides a timely forum to sort out this problem.

During the investigation of three-dimensional steady, laminar boundary layers, it was noted that because of the similarity in structure between the respective governing systems of equations, important analogies exist between the three-dimensional steady boundary layer and the two-dimensional unsteady boundary layer.

One of the analogies is concerned with the concept of the zones of influence and dependence. The relevant dependence rule, first conceived by Raetz [3.28, 3.29] for steady three-dimensional boundary layers, was extended by the present author [3.30] to one for unsteady boundary layers. The idea of such an unsteady dependence rule has since been employed in other calculations.

Another analogy is the calculability of reversed flow. In the three-dimensional steady case, the author first demonstrated that, if the dependence rule is satisfied, vanishing of the circumferential component of skin friction does not prevent calculating the circumferential reversed flow. By analogy, it was argued [3.30] that two-dimensional unsteady boundary layers can

* Department of Aerospace Engineering and Engineering Mechanics, San Diego State University, San Diego, CA 92182.

also be calculated past the zero-skin-friction point if the corresponding dependence rule is satisfied. The same conclusion was also reached by other authors [3.31, 3.32] from calculations made largely on a trial basis without recognizing the above dependence rule.

At the same time, a third analogy was also conceived, i.e. the analogy in flow separation between the three-dimensional steady case and the two-dimensional unsteady case. However, reporting of this separation analogy was withheld until specific examples were calculated to check its validity [3.33].

We intended originally in [3.33] to present a new unsteady-separation criterion based on the aforementioned analogy in contrast to existing ones [3.34, 3.35], namely, the MRS (after Moore, Rott, and Sears) criterion and the unsteady Goldstein singularity criterion. Our calculation of the impulsively-started-cylinder problem was intended merely to provide an illustrating example. However, the conclusion of separation we reached there has led us into a debate with somewhat different emphasis, i.e., the question of singularity instead of the question of a specific separation criterion. These two questions are certainly related, but nevertheless not the same, as will be elaborated later.

In this paper, we discuss the current controversies surrounding the unsteady problem. There are two main sections: Section 2 is concerned with specific separation criteria, Section 3 with the particular cylinder problem.

2 Comparison of Separation Criteria

Prior to our proposal of the analogy separation criterion, the MRS criterion and the Goldstein singularity criterion were discussed for some years by Sears and Telionis. Recent work of Van Dommelen and Shen [3.13, 3.36] basically supports these two criteria. Their unprecedented Lagrangian calculation was undertaken to demonstrate clearly how a singularity develops in time, but nowhere, as far as this author knows, do they advocate outright that unsteady separation in general should be determined by their Lagrangian method of calculation. In other words, they did not propose a new separation criterion of their own. Likewise, Cebeci [3.12] concluded that singularity and separation do not occur for the cylinder problem at finite times, but he did not propose an unsteady criterion as such. Cebeci did, however, state [3.22] that "a singularity cannot develop in Eq. (1)—i.e. the standard unsteady boundary layer equations—at a finite time if the solution is free of singularities at earlier times." This is a rather general statement. It is not up to this author to speculate whether this statement implies that for this whole class of problem, there is no real need of a separation criterion. In any case, speaking of separation criteria, we shall confine ourselves to our analogy criterion along with the MRS criterion and the unsteady Goldstein singularity criterion.

Analogy Criterion

As mentioned in the Introduction, the analogy criterion was derived on the basis of the similarity of structure between the governing system of equations for the three-dimensional, steady boundary layer and that for the two dimensional unsteady boundary layer. In spite of the differences in the number of equations and velocity components, both systems are of the same type: multitime parabolic.

For steady, three-dimensional boundary layers, separation can be identified by the running-together of the limiting streamlines defined by

$$\frac{dy}{dx} = \frac{v_{z=0} + \left(\frac{\partial v}{\partial z}\right)_{z=0} \Delta z + \cdots}{u_{z=0} + \left(\frac{\partial u}{\partial z}\right)_{z=0} \Delta z + \cdots} \approx \frac{c_{fy}}{c_{fx}}, \tag{1}$$

where x, y are coordinates parallel and z normal to a body surface; u, v, w are the velocity components along x, y, z; and c_{fx} and c_{fy} are the components of the skin friction in the x- and y-directions. (In the literature, there is a dispute over whether the separation line is an envelope of the limiting streamline, or is itself also a limiting streamline. We are not concerned with this issue here.) Such a limiting flow pattern and separation line have been demonstrated by numerous experiments and a few careful calculations. Typically, over an inclined body [3.37] there is an open type and a closed type of separation (Fig. 1(a), (b)).

The present author contends that the well-established separation criterion for the three-dimensional, steady case can be analogously carried over to the two-dimensional, unsteady case. To do this, we define analogous limiting streamlines in the x, t plane,

$$\frac{h_x\, dx}{dt} = \left(\frac{\partial u}{\partial z}\right)_{z=0} \Delta z = c_f \sqrt{\mathrm{Re}}\, \Delta z, \tag{2}$$

where $\sqrt{\mathrm{Re}}\, \Delta z$ may be considered as a scale factor: it does not change the overall flow pattern, and its actual value need not be specified. To determine the analogous unsteady limiting flow pattern, one follows the same

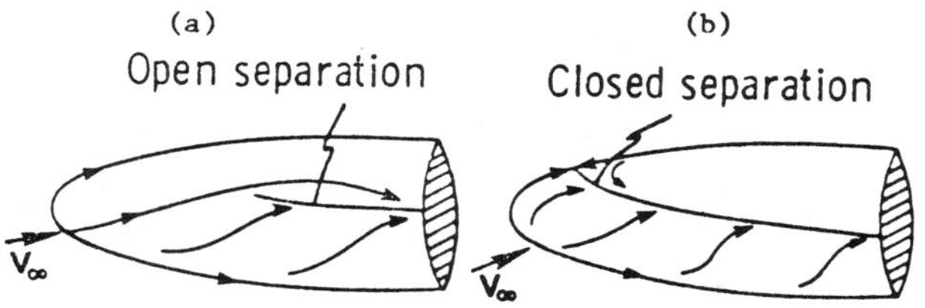

Figure 1. 3-D steady separation.

steps as in the three-dimensional steady case, i.e., one first solves the boundary-layer equations, then evaluates the skin friction $c_f(x, t)$, and then has the analogous limiting streamlines drawn according to Eq. (2). The crux of the matter then is to check whether those streamlines coalesce or not. It is proposed that unsteady separation is identified by such coalescence of streamlines. Again there is an open type and a closed type, as depicted in Fig. 2(a), (b).

An open type of unsteady separation corresponds to the case where separation does not occur at early times, but it gradually develops as time increases. An impulsively started cylinder provides an example of this type. Whether an open unsteady separation is possible is precisely the central point of current controversy. A closed unsteady separation occurs throughout the time under consideration. An unsteady flow resulting from a sudden change between two steady states falls into this type [3.33].

Just as in the three-dimensional steady separation, it should be noted that the analogous envelope of limiting streamlines represents an independent symptom of separation. It coincides with other common separation symptoms, such as the rapid growth of the boundary-layer thickness, the sharp increase in the w-velocity, and so on. Our calculated results bear out this contention, as will be seen later.

Because there lacks a precise criterion to follow, these common symptoms do not lead to reliable determination of separation. In contrast, our analogous limiting streamlines are precisely defined, and the steps to construct such a limiting flow pattern are straightforward to follow. Although the exact location of separation according to our proposed criterion is subject to the limit of accuracy in the graphical representation of the limiting flow pattern, yet general trends of the overall flow field are always clearly demonstrated, and the possible inaccuracy thus involved is minimal compared to the much greater errors resulting from uncertainties of other separation symptoms.

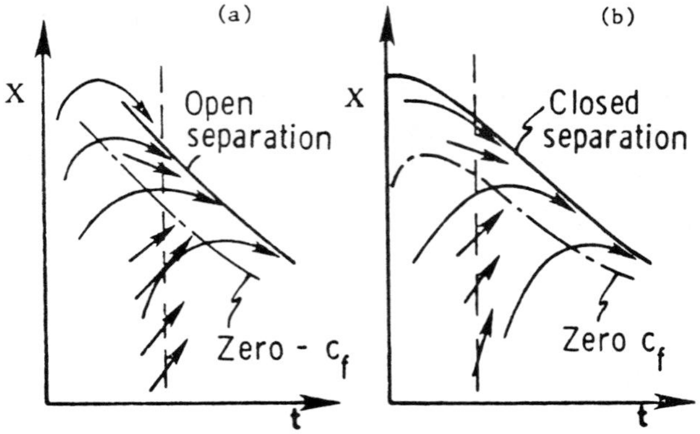

Figure 2. 2-D unsteady separation.

The MRS and Singularity Criteria

The question of singularity or no singularity is of fundamental importance to the mathematical understanding of boundary layer flow in general. To a lesser degree, the MRS model could also help to picture the separation structure. However, whether either one can be used as an implementable separation criterion is a different matter. In our opinion, the answer is, in fact, negative. The reason is that neither the MRS criterion nor the singularity criterion is convenient to implement. In the case of the MRS criterion, there is the need to attach a coordinate system moving with the velocity at separation, which in turn is not known until the location of the separation is determined. Thus one has at least to resort to some iteration scheme. Also, in the case of the singularity criterion, there is no simple way to know where the singularity is located short of an analytical solution. For numerical solutions, the separation therefore has still to be decided by the usual symptoms; thus the so-called criterion will be used only afterwards to help rationalize the interpretation. This procedure could be very unreliable and lead to an erroneous conclusion. In comparison, our proposed analogy criterion provides an independent judgement in identifying separation.

While the MRS and unsteady Goldstein singularity criteria were extended from that for two-dimensional, steady problems, the present criterion is patterned after that for three-dimensional steady cases. Because it does not specifically rely on the idea of singularity, our proposed criterion could also be applied to the Navier–Stokes solutions which are known to be free of Goldstein singularities. This possibility remains to be confirmed. Currently there is no definite method of determining separation in connection with Navier–Stokes solutions, other than those common separation symptoms.

3 Comparison of the Cylinder Problem

Agreements and Disagreements

Reference [3.33] compared our results with those due to Telionis and Tsahalis [3.31]. For this same cylinder problem, good agreement was found for the skin-friction distribution, but using their singularity criterion, they determined a much earlier separation than our results indicated. The same discrepancy was also found by others [3.12, 3.13]. This appears to confirm our view, expressed in the preceding section, that singularity is not an appropriate separation criterion because it cannot be located with sufficient certainty by numerical solutions. Otherwise, we found no basic contradictions between our work and Telionis and Tsahalis's for this problem.

After the completion of [3.33], the work of Cebeci [3.12] and of Van Dommelen and Shen [3.13] for the same impulsively-started-cylinder problem appeared in the literature. Van Dommelen and Shen calculated this problem by a Lagrangian method and found a singularity at time $t \approx 1.5$ and $\theta \approx 111°$. This finding lends support to the singularity hypothesis assumed by Sears and Telionis [3.34, 3.35]. Cebeci, [3.12] on the other hand, concluded that singularity and separation cannot develop at a finite time. He reported that rapid growth of the boundary-layer thickness did cause him to stop his calculation at $t = 1.4$, but his solution was smooth up to that time. He maintained that the thickness difficulty can be overcome in the future so that the calculation over the rear body can be continued to larger times. This claim, however, remains to be demonstrated.

In contrast, based on our calculations and our analogy separation criterion, we concluded that separation occurs approximately at $t = 1.4$ and $\theta = 110°$. Thus our conclusion contradicts Cebeci's but is fairly close to Van Dommelen and Shen's. The latter authors' method of approach is, however, entirely different from ours.

On the other hand, our difference with Cebeci is not really as bad as it sounds. In fact, reasonable agreement was indeed noted (as will be shown later) in the results for the skin friction and boundary-layer thickness. His thesis of no singularity and no separation was apparently influenced by Proudman and Johnson [3.38]. We shall comment on these questions after we first elaborate somewhat on our own results.

The Author's Results

We calculated the impulsively-started-cylinder problem by a Crank–Nicholson-type finite-difference scheme. A computer program originally developed for the steady three-dimensional boundary layer was employed for the present work. Reversed flow was calculated by a zigzag scheme to fulfill the rule of the dependence zone. In our first calculation presented in [3.33], the zigzag scheme necessitated the loss of one θ-station for each time-constant line. Subsequently, symmetry consideration on two sides of the rear stagnation point ($\theta = 180°$) enabled us to compute over the rear body until $t \approx 1.35$. Results from those repeated calculations are shown here.

Figure 3 shows the distribution of the skin friction c_f as a function of θ and t. Negative c_f implies the occurrence of reversed flow. The first time c_f becomes zero ahead of the rear stagnation point occurs at $t = 0.32$, in agreement with most results reported. Our area of calculation in the θ, t plane is also indicated in Fig. 3. For $t > 1.35$, only the front body was calculated.

Figure 4 shows our analogous limiting flow pattern in the θ, t plane. The arrows indicate the local directions in accordance with Eq. (2) which can be evaluated once the skin friction is known from the boundary-layer solution. The downward arrows correspond to negative skin friction due to flow

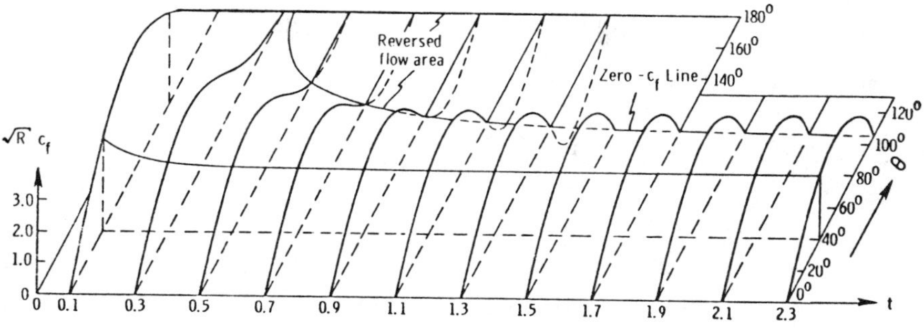

Figure 3. Distribution of skin friction.

reversal. The solid lines drawn parallel to those arrows are our analogous limiting streamlines. At larger times, the streamlines come close together to form a single line from both the lower and the upper side. This single line, is identified as the unsteady separation line in accordance with our proposed analogy criterion. Subject to the limitation of accuracy by graphical representation of numerical solutions, separation is seen to start around $t = 1.4$, and approaches asymptotically the steady value of Terrill [3.39].

The running-together of the streamlines is accompanied by a rapid increase in the displacement thickness Δ (Fig. 5). It is seen that Δ at the rear stagnation point is respectively 7.5, 5, 3.5, and 1.3 for $t = 1.3$, 1.0, 0.8, and 0.4. Compared to $\Delta \approx 0.15$ at the front stagnation point, the ratio is 15, 10, 7, and 2.6. Especially noteworthy is the growth rate of Δ for $t = 1.3$ at $110° \leq \theta \leq 120°$. The increase there is almost a sudden jump, which made computation very difficult to continue.

Comparison with Cebeci's Calculation

Comparisons of the skin friction and the displacement thickness are shown in Fig. 6(a), (b). Reasonable agreement is indeed evident.

Further comparisons of the computational details are given in the following table:

	Δt	Δθ	Max. z-steps	Method	Dependence rule
Wang	0.01	0.5°	70	Crank–Nicolson	Completely satisfied
Cebeci	0.025	4.5°	300	Box	Not completely satisfied

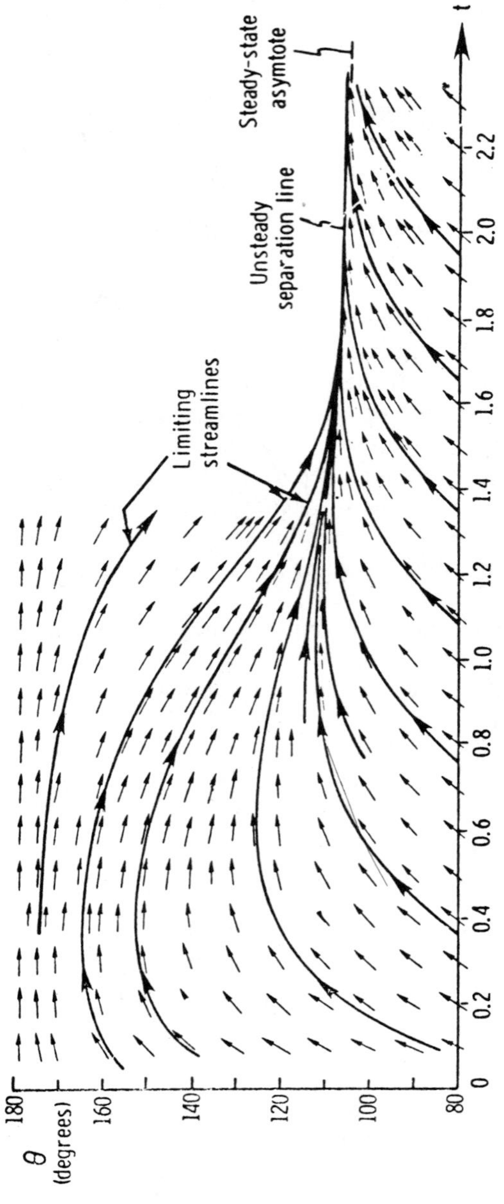

Figure 4. Unsteady limiting flow pattern and separation line in the θ, t plane.

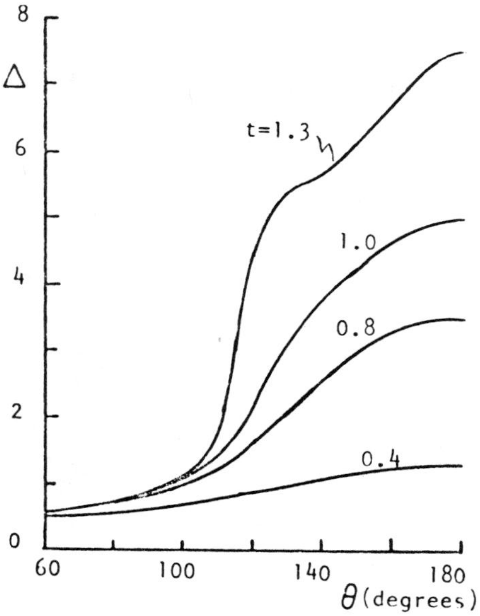

Figure 5. Displacement thickness.

His time step is larger than ours, and his θ-step is obviously too large, while 300 steps across the boundary layer are more than normally needed for the laminar case. Krause's zigzag scheme was employed in both cases to deal with reversed flow; however, Cebeci's treatment of this zigzag problem is faulty in two respects: (a) the dependence zone was not completely satisfied, and (b) the difference approximations were not centered on the same point. The failure to satisfy the dependence rule affects the calculation of reversed flow, which is precisely the source of controversy.

For a reversed profile across ABC (Fig. 7), Cebeci employed his standard box scheme for the part AB and his zigzag scheme for BC. If the zone of dependence is divided into four quarters, his procedure amounts to the quarter $ABA_1B_1A_2B_2$ being neglected. In other words, the rule of dependence was not completely satisfied in his calculation. For correctness, when reversed flow occurs, the zigzag scheme must be used across the entire layer AC instead of just across BC. The latter would have been correct if the problem were purely hyperbolic, but the diffusion process in the normal direction changes the picture for the present boundary-layer problems. Here the disturbances are not only propagated along the streamlines (or their projections) with finite speeds, but also spread instantaneously across the whole layer.

In short, these flaws in the details of Cebeci's computations may be responsible for the remaining differences between our results. This raises questions about his claims, including the smoothness of his solution and good agreement with others. The latter could happen for different reasons, such as the

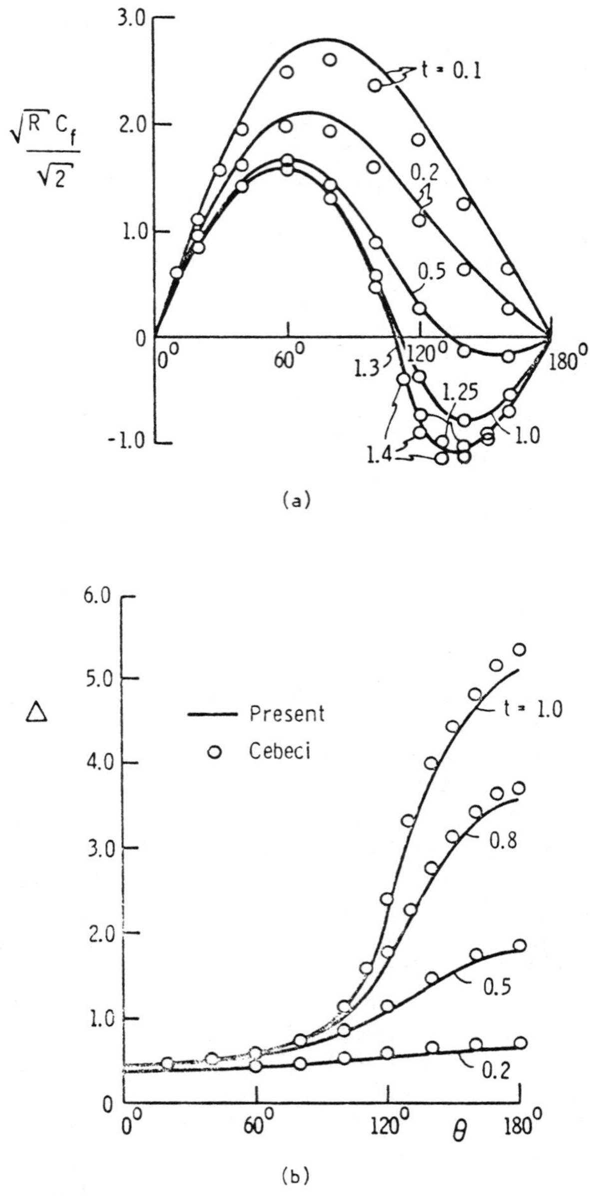

Figure 6. Comparison of (a) skin friction, (b) displacement thickness.

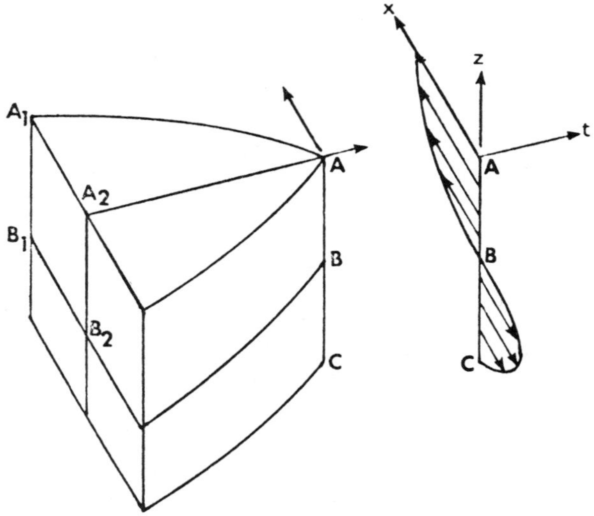

Figure 7. Dependence zone for reversed flow.

limitation of low Reynolds number in the Navier–Stokes solution and the inaccuracy of previous boundary-layer calculations.

Proudman and Johnson's Work

The work of Proudman and Johnson (P&G) plays an important role in the current controversy on unsteady separation. They originated the idea that no separation occurs at any finite time for this type of problem, and their analysis has been taken as the analytical basis for a similar claim. It is therefore important to ask whether P&G's work really applies to the cylinder problem as has been assumed so far.

The model studied by P&G dealt with a flat plate impulsively started from rest. The velocities were assumed to have the form

$$\begin{aligned} u &= -xF'(y, t), \qquad v = F(y, t), \\ x &= \frac{x^*}{\nu/\alpha}, \qquad y = \frac{y^*}{\nu/\alpha}, \qquad t = \frac{t^*}{\alpha}, \end{aligned} \tag{3}$$

where ν is the kinematic viscosity and α a rate of strain. Evidently, this flow was patterned after the well-known front stagnation flow with the velocity directions being reversed. Since t does not involve viscosity, P&G observed that for any fixed value of t, however large, the thickness of the viscous layer becomes arbitrarily small as $\nu \to 0$, and separation cannot begin at any finite time. Similarity solution of the inviscid approximation for the outer region yields

$$F(y, t) = e^{t}\left[\eta - \frac{2}{c}(1 - e^{-c\eta})\right] \tag{4}$$

with

$$\eta = ye^{-t}.$$

Equation (4) gives, for fixed y at large t,

$$F(y, t) \approx -y, \tag{5}$$

which is the often cited result that the outer flow for the rear stagnation point is the same for the front stagnation point. Robins and Howarth [3.4] later extended this work to include the inner solution for the viscous region close to the plate.

It seems clear that P&G's analysis was strictly for the idealized flat-plate problem. Nowhere does the cylinder geometry enter the picture. Thus the upstream curvature effects were completely neglected, whereas the curvature and its induced adverse pressure gradient are precisely the triggering mechanism for any flow separation. On this basis, we do not feel that application of P&G's work to the cylinder problem is justified.

P&G did stipulate that at large times (of the order of a multiple of a/V_∞, where a is the cylinder radius) separation would occur and eddies would develop in agreement with experimental observations. If we suppose this multiple to be about 1.4, then our conclusion seems in no way to contradict P&G's. However, we do not feel such a reconciliation is meaningful, since we are unconvinced that P&G's analysis holds for the cylinder problem.

From a somewhat different viewpoint, we may also argue that if we agree to disregard the upstream influence effects and if P&G's conclusion is valid for the circular cylinder problem, then the same conclusion should be equally valid for the impulsively started motion over any smooth two-dimensional symmetric body. The latter may include those with even greater adverse pressure gradients in the aft body, such as an elliptical cylinder with its major axis normal to the freestream direction. The consequence of this logic is evidently far-reaching, i.e., there is no separation whatsoever for any (impulsively started) two-dimensional problem at finite time. This is a situation very hard to imagine.

Cebeci stated that "a singularity cannot develop ... at a finite time if the solution is free from singularities at earlier times." He criticized others for lacking "any analytic argument in support", and seemed to imply that the opposite is true in his case. Admittedly, this author knows of no direct proof of the above statement even for the pure initial-value problem, let alone the present boundary-layer problem, which is of the mixed initial-value and boundary-value type.

In short, although P&G's work is interesting for the problem they posed, we are not yet convinced that their conclusion is directly applicable to the cylinder problem.

4 Conclusions

Current controversies surrounding the unsteady separation problem have been discussed. Compared to the MRS and Goldstein singularity criteria, our proposed analogy separation criterion is more specific in the detailed steps to identify separation, and those steps are straightforward to apply. The analogous limiting flow pattern we defined in the x, t plane provides an independent signal for locating separation, but it coincides with other symptoms of separation also. For the impulsively-started-cylinder problem, our prediction of the separation starting time is close to that of Van Dommelen and Shen, but fundamentally contradicts Cebeci's. Cebeci's is flawed in its computational methods, which may affect his reversed-flow calculation. It is argued that Proudman and Johnson's analysis applies strictly to the flat-plate case, but not to the cylinder problem.

Acknowledgment

This work was supported by the Office of Naval Research under Contract N00014-80-C-0307. The author wishes to express his deep gratitude to Morton Cooper for his support and discussions.

CHAPTER 17

The Genesis of Separation

L. L. Van Dommelen and S. F. Shen*

1 Introduction

For the classical boundary-layer problem of the impulsively started circular cylinder [3.40], both the Blasius small-time expansion [3.41] and the Proudman–Johnson solution for the rear stagnation point [3.4, 3.38] imply that the boundary layer remains, to its order of thickness, attached to the wall. However, experimental observations [1.42, 3.42] show "thin vortex layers" which "leave the surface of the body and curl round on themselves." Very recent experimental results of Bouard and Coutanceau [3.43], which became available after the original presentation of this paper, suggest even more precisely a "vorticity peak" which is "individualised into a 'vortex'" after roughly 0.75 diameter movement of the cylinder. These inconsistencies with experiment suggest that the Blasius and the Proudman–Johnson results, though locally correct, do not yet qualitatively describe the entire flow development.

Indeed, the present authors, Van Dommelen and Shen [3.13, 3.36], numerically discovered a singularity in the boundary-layer solution which arises after about 0.75-diameter movement through the fluid. This singularity is located at 111° from the forward stagnation point and moves upstream with velocity $0.52 V_\infty$, presumably [3.44] eventually reaching the 104.5° steady-state position calculated by Terrill [3.39].

* Sibley School of Mechanical and Aerospace Engineering, Cornell University, Ithaca, NY, 14853.

Since the discovery of the singularity by the authors in 1977, the computation has been repeated for a wide range of increasingly finer meshes. The results have further been verified in an independent calculation along quite different lines by Cebeci [3.12] and qualitatively by Wang [3.33]; these two authors discuss their work elsewhere in this volume.

The analytical structure of the singularity, as derived in the next sections, describes that the upper part of the boundary layer starts to "peel off" from the surface of the cylinder near the singularity. Thus a link is established toward the experimentally observed flow development [1.42, 3.42, 3.43]. The penetration of the separating layer becomes large on the $O(\mathrm{Re}^{-1/2})$ boundary-layer length scale in the transverse direction. Hence the approximation of boundary-layer theory that the Reynolds number is infinite, instead of finite, brings in the singularity.

Even though the singularity itself is not physical for finite Reynolds number [3.45], it should be expected that the real flow will follow the evolution until arbitrarily close to the singularity, provided only that the Reynolds number is sufficiently high. Indeed, during the evolution toward the singularity the boundary-layer solution is still smooth; thus the $O(\mathrm{Re}^{-1/2})$ displacement-thickness effect of the boundary layer on the attached irrotational outer flow will still be negligibly small [3.43] for high enough Reynolds numbers. This is supported by the experimental result of Schwabe that the prescribed pressure in the boundary layer is correct if the boundary is thin [3.46]. Also the neglected terms in the Navier–Stokes equations are $O(\mathrm{Re}^{-1/2})$ small according to the boundary-layer results. Thus Bouard and Coutanceau [3.43] find experimentally that the boundary-layer result for the time that reversed flow first appears is correct.

The occurrence of a singularity, signifying the need for local modifications in the boundary-layer problem, ties in with the concept for unsteady separation given by Sears and Telionis [3.35]. About 1955 Moore, Rott, and Sears already pointed out that in unsteady flow, or for moving walls, zero wall shear does not in general signify separation in the sense that the boundary layer leaves the wall. For steady separation over moving walls, Moore developed the picture of a singular bifurcation point of the streamlines inside the boundary layer. It should be located on the zero vorticity line and should move with the local flow speed in the downstream direction; these two conditions have become known as the Moore–Rott–Sears (MRS) conditions, as they were also formulated independently by Rott and Sears. Moore further reasoned that small unsteady perturbations on a singularly separating steady boundary layer would preserve the singular character [3.1]. Later, Sears and Telionis argued that the occurrence of such a singular MRS bifurcation point should be a defining feature of unsteady separation in general in the conventional boundary-layer approximation, and that the singularity is an exaggeration of the actual rapid thickening of the boundary layer [3.35]. The present results support this idea, even though there are many unexpected features in the solution.

Shen and Nenni [3.47] concluded on the basis of an approximate analysis that the singularity comes about as an intersection of characteristics, and Shen [3.10] developed this and the results of Van Dommelen and Shen [3.36] into a picture of the formation of a barrier inside the boundary layer by piling up of particles, analogous to wave steepening in compressible flow.

2 Description of the Singularity

The classical boundary-layer problem for the impulsively started circular cylinder is given by [3.40]

$$u_{,t} + uu_{,x} + vu_{,y} = UU'(x) + u_{,yy}, \tag{1}$$

$$u_{,x} + v_{,y} = 0, \tag{2}$$

with initial and boundary conditions

$$u(x, y, 0) = U(x) = \sin x \qquad (y \neq 0), \tag{3}$$

$$u(x, 0, t) = v(x, 0, t) = u(0, y, t) = u(\pi, y, t) = 0,$$

$$u(x, \infty, t) = U(x) = \sin x. \tag{4}$$

The choice of units is such that the x-coordinate along the surface is in radians, $x = 0$ corresponding to the forward stagnation point and $x = \pi$ to the rear one. The Reynolds number Re has been eliminated by means of the usual scalings of y and v [3.40].

By means of a small-time expansion of the solution, Blasius was able to show that after about 0.16-diameter movement, i.e. $t \approx 0.7$, the adverse pressure gradient near the rear stagnation point arrests the downstream motion of the boundary-layer particles, resulting in a region of reversed flow ($u < 0$). The minima of the velocity profiles in this region form a *zero vorticity line* of fundamental importance for the further evolution of the boundary layer [3.41].

Vorticity Dynamics of Separation

The boundary-layer problem as stated cannot be solved in closed form. Thus it is advantageous to consider the continuation problem for a "thick" initial boundary layer,

$$u(x, y, t_0) = u_0(x, \varepsilon y), \tag{5}$$

so that for $\varepsilon \to 0$ the characteristic boundary-layer thickness at the initial time t_0 becomes large on the $O(\text{Re}^{-1/2})$ transverse scale. The achieved simplification is then that the viscous $u_{,yy}$-term drops out of the boundary-layer equation. The resulting *vorticity-layer equation* becomes a simple ordinary differential equation when Lagrangian coordinates are used [3.48]:

$$x_{,tt} = UU'(x), \tag{6}$$

where x is now the streamwise particle position.

As usual [3.48], the Lagrangian coordinates ξ and η are defined as the particle position at the start of the motion:

$$(\xi, \eta) = (x, y) \quad \text{at time } t_0. \tag{7}$$

Continuity is equivalent to a mapping Jacobian equal to unity,

$$x_{,\xi} y_{,\eta} - x_{,\eta} y_{,\xi} = 1. \tag{8}$$

Further [3.10, 3.49],

$$\frac{\partial}{\partial x} = y_{,\eta} \frac{\partial}{\partial \xi} - y_{,\xi} \frac{\partial}{\partial \eta}, \tag{9a}$$

$$\frac{\partial}{\partial y} = x_{,\xi} \frac{\partial}{\partial \eta} - x_{,\eta} \frac{\partial}{\partial \xi}. \tag{9b}$$

By definition

$$u \equiv x_{,t}, \tag{10}$$

and using (9b), the vorticity ω becomes

$$\omega = x_{,\xi} x_{,\eta t} - x_{,\eta} x_{,\xi t}. \tag{11}$$

From the ordinary differential equation (6) the streamwise particle position x can be shown to be nonsingular for all time, but x does develop stationary points with respect to ξ and η [3.49]. These stationary points correspond to singularities in the transverse particle position y as a result of the unit Jacobian (8). In fact, the characteristics of this Jacobian,

$$d\xi : d\eta : dx : dy = -x_{,\eta} : x_{,\xi} : 0 : 1$$

combine into the *continuity integral*

$$y = \int_{\text{wall}}^{(\xi, \eta)} \frac{ds}{|\text{grad } x|} \tag{12}$$

if s is the arc length along the line of constant x through (ξ, η). Thus y is at least logarithmically infinite at stationary points in x.

It is not uncommon for singularities to be smoothed out by the presence of small nonzero viscosity, as in the case of the Burgers equation. However, for the present boundary-layer problem the viscous $u_{,yy}$-term does not remove the singularity. To verify this, x can be further expanded in powers of ε,

$$x \sim x_0 + \varepsilon^2 x_1 + \varepsilon^4 x_2 + \cdots. \tag{13}$$

By substitution in the full Lagrangian boundary-layer equation,

$$x_{,tt} = UU'(x) + \left[x_{,\xi}\frac{\partial}{\partial\eta} - x_{,\eta}\frac{\partial}{\partial\xi}\right]^2 x_{,t}, \tag{14}$$

it may now easily be shown [3.49] that the $x_1, x_2, \ldots$-terms are small smooth corrections which do not change the formation of the stationary points and hence the singularities in the y-particle position.

Numerical Solution

In order to verify the applicability of the results of the previous subsection to the full boundary-layer problem, the problem was integrated numerically in the Lagrangian form (14) [3.13, 3.36, 3.49]. The smoothness of x ensures good accuracy, as is evident from Table 1. Figure 1 shows that a stationary point first appears at time $t_S = 3.0045$, i.e. after 0.75-diameter movement. Located at $x_S = 1.9368$, 111° from the forward stagnation point, the point proves to move upstream with velocity $-u_S = 0.26$, or 0.52 times the undisturbed flow velocity.

The numerical results are in excellent agreement with an Eulerian calculation by Cebeci [3.12], who finds the same displacement thickness for times as close to the singularity as $t = 2.75$. We await a more detailed comparison of the velocity profiles, instead of only the displacement thickness and wall shear, in order to verify the detailed structure of the singularity. Qualitatively the same results were also obtained by Wang [3.33], but he observed some quantitative differences in wall shear and displacement thickness from both Cebeci's and our results, and finds separation at $t = 2.8$ instead of $t = 3.0045$. Wang's definition of separation as the formation of an envelope in the lines

Table 1 The error[a] $E(x)$ in the streamwise particle position at the instant of separation.

Mesh[b]	$E(x)$ Radians	$E(x)$ Degrees
19 × 9 × 60	0.0617	3° 32′
37 × 17 × 120	0.0218	1° 15′
73 × 33 × 240	0.0075	0° 25′
145 × 65 × 480	0.0026	0° 08′
289 × 129 × 960	0.0006	0° 02′

[a] Maximum error on the 19 × 9 mesh points of the coarsest mesh, using the Richardson extrapolate of the finest two meshes as the reference "exact" solution.

[b] Numbers of mesh lines in the ξ-, η-, and t-directions respectively.

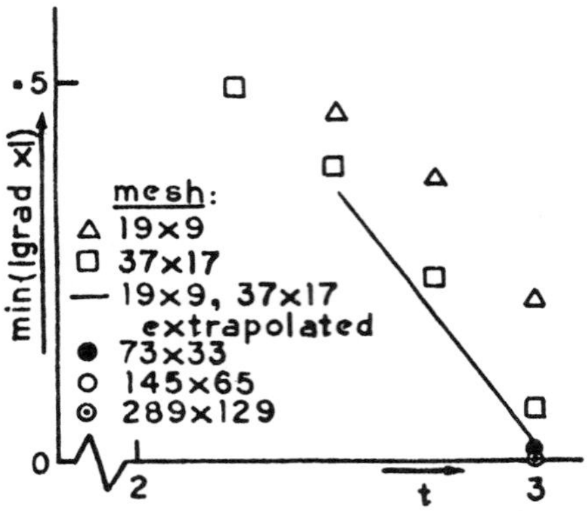

Figure 1. A stationary point in the streamwise particle position is formed in finite time.

$dx/dt = \tau_w$ (which requires, according to the well-known Lipschitz condition, infinite $\tau_{w,x}$) appears less precise than the present stationary-point criterion. In fact, our wall shear gradient is steep, but does not yet seem to be infinite at separation. Telionis and Tsahalis [3.31] experience numerical breakdown, suggesting a singularity appearing at time $t \approx 1.4$. They themselves admitted that the evidence was "of course not very strong," and since then there has been no additional support for a singularity at $t \approx 1.4$, at which time all other authors, including the present ones, find a smooth solution instead.

A final word in favor of the Lagrangian scheme as compared to the Eulerian ones near separation. The fact that x is smooth makes it possible to locate accurately the separation time t_S as in Fig. 1 and the streamwise location x_S (compare also Table 1). But in order to evaluate other quantities, the singular continuity integral still has to be integrated. To compare possible accuracy for this singular equation with the more usual Eulerian approach, in Fig. 2 the maximum value of the streamwise velocity gradient $-u_{,x} = v_{,y}$ over the entire boundary layer has been plotted. For an Eulerian scheme a computational bound on $u_{,x}$ can be formed with the undisturbed flow velocity V_∞ and the streamwise mesh spacing

$$[u_{,x}]_{\lim} = \frac{2V_\infty}{\Delta x} \approx \frac{2V_\infty}{\Delta t}; \tag{15}$$

clearly the computation can no longer be accurate when the velocity change between neighbouring mesh lines is as much as $2V_\infty$! For various Eulerian schemes [3.12, 3.33, 3.50] such limiting values have been plotted in Fig. 2; they prove an order smaller than what is possible in the Lagrangian computation.

Another advantage of the Lagrangian computation is a partial correction for the exponential Proudman–Johnson growth of the boundary-layer thickness near the rear stagnation point. A disadvantage is the region upstream of 104.5°, where the Eulerian solution tends to the steady Terrill

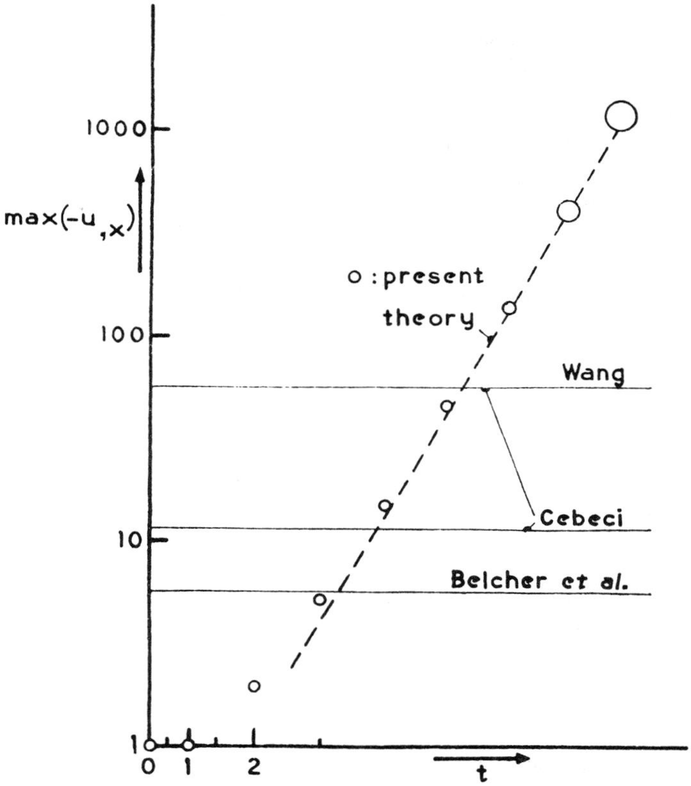

Figure 2. The streamwise velocity gradient blows up when the separation time $t = 3.0045$ is approached.

limit; in Lagrangian coordinates the solution remains unsteady, so that the numerical error keeps gradually increasing with time.

Asymptotic Expansion

Away from the separation point the streamwise particle position should be expected to tend toward a definite value when $t \to t_S$,

$$x \sim x(\xi, \eta, t_S), \qquad t \to t_S. \tag{16}$$

According to the Lagrangian boundary-layer equation (14), the acceptable gauge functions are integer powers of $t - t_S$; hence

$$x \sim \sum_{n=0}^{\infty} (t - t_S)^n \frac{1}{n!} \frac{\partial^n x}{\partial t^n} (\xi, \eta, t_S), \qquad t \to t_S, \tag{17}$$

which is no more than a Taylor series expansion.

Near the separation point the blowup of $u_{,x}$ shown in Fig. 2 suggests that the viscous and particle accelerations in the boundary-layer equation can be neglected locally. Thus the leading-order solution in Lagrangian coordinates is of the form

$$x \sim x(\xi, \eta, t_S) + (t - t_S)x_{,t}(\xi, \eta, t_S), \tag{18}$$

where in order for x to be smooth when $t < t_S$,

$$x(\xi, \eta, t_S),\ x_{,t}(\xi, \eta, t_S) \in C_\infty, \tag{19}$$

and where x must have a stationary point in order for a singularity to be present,

$$x_{,\xi}(\xi_S, \eta_S, t_S) = x_{,\eta}(\xi_S, \eta_S, t_S) = 0. \tag{20}$$

For this kind of asymptotic behaviour the Taylor series (2.17) for x remains uniformly valid in the separation region. In particular the Lagrangian boundary-layer equation shows that all x-derivatives are smooth at the stationary point:

$$x_{,tt},\ x_{,ttt},\ \ldots \in C_\infty. \tag{21}$$

Thus the indications of the previous subsections, that x is smooth but has stationary points corresponding to separation, are supported by this asymptotic expansion around $t = t_S$.

3 Physical Interpretation

In the previous section it was established that in Lagrangian coordinates (ξ, η) the x-position of the particles develops a stationary point, corresponding to a singularity in the y-position arising from the continuity integral

$$y = \int_{\text{wall}}^{(\xi,\eta)} \frac{ds}{|\text{grad } x|} \qquad (x = \text{constant}). \tag{22}$$

In order to assess the physical consequences of these results, in Fig. 3 the numerically found topology of the lines of constant x and vorticity is sketched for a time shortly before the time t_S when the stationary point in x is fully established at S. This stationary point satisfies the Moore–Rott–Sears (MRS) conditions mentioned in the introduction, as is immediately seen from the Lagrangian expressions (10) and (11) for the flow velocity and vorticity:

$$u_S = [x_{,t}]_S = \mathring{x}_S \qquad \text{(MRS I)}, \tag{23a}$$

$$\omega_S = [x_{,\xi}x_{,\eta t} - x_{,\eta}x_{,\xi t}]_S = 0 \qquad \text{(MRS II)}. \tag{23b}$$

As the numerically determined value of u_S is negative ($u_S = -0.26$), it follows that the separation point slips in the upstream direction.

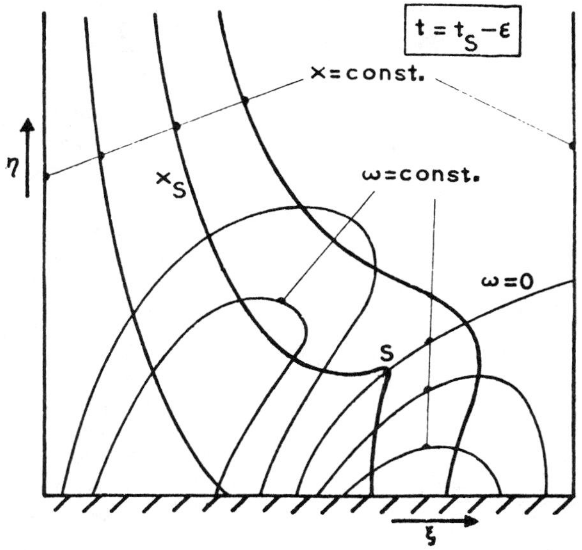

Figure 3. Lines of constant vorticity and streamwise particle position for a typical time shortly before separation. (Schematic).

Further, the boundary layer does indeed bifurcate at S, because when the continuity integral is evaluated along the line $x = x_S$ through S, strong growth in y occurs near S, where grad $x \approx 0$. The physical picture that results is one in which the boundary-layer vorticity divides into two layers as sketched in Fig. 4; in particular the lower part of $x = x_S$ in Fig. 3 corresponds to a vorticity layer near the wall, while the part of $x = x_S$ beyond the point S corresponds to a separating vorticity layer at large values of y. The vicinity of point S corresponds to a vorticity depleted *dead-water region* separating the two layers. This is illustrated by the computed vorticity profile (Fig. 5) shortly before separation; the wall and upper vorticity layers are fairly evident here, as is the flat minimum $u \approx u_S$ in the velocity profile.

The dominance of the part near the stationary point S in the continuity integral renders it possible to expand x locally in a Taylor series. In a convenient local Cartesian k, e-axis system centered at S and with k-axis aligned along the zero vorticity line, the Taylor series which describes the generation of a stationary point is of the form

$$x \sim x_S + \tfrac{1}{2}x_{,ee}e^2 + \tfrac{1}{6}x_{,kkk}k^3 + \cdots + (t - t_S)(u_S + x_{,kt}k + \cdots) + \cdots. \quad (24a)$$

Hence, from (10) and (11),

$$u \sim u_S + x_{,kt}k + \cdots, \quad (24b)$$

$$\omega \sim -x_{,ee}x_{,kt}e + \cdots. \quad (24c)$$

Numerical estimates for the coefficients are

$$x_{,ee} \approx 0.66, \qquad x_{,kkk} \approx 10.5, \qquad x_{,kt} \approx -0.86. \quad (25)$$

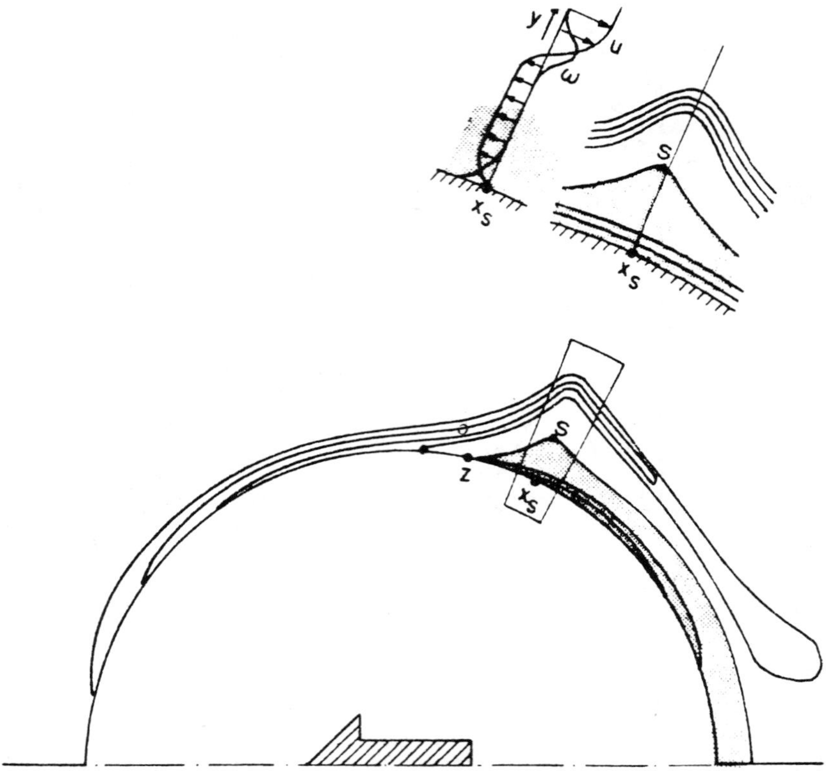

Figure 4. Physical geometry of the lines of constant vorticity corresponding to the Lagrangian plane of Figure 3. (Schematic).

Using the above Taylor series, the continuity integral can be evaluated as

$$
\begin{aligned}
y &= \int \frac{|d\Delta u|}{|\omega(\Delta u, \Delta x, t)|} \qquad (\Delta x,\ t = \text{constant}),\\
|\omega| &= \left\{2x_{,ee}x_{,kt}^2\left[\Delta x + \frac{-x_{,kkk}}{6x_{,kt}^3}\,\Delta u^3 + (t - t_S)\,\Delta u\right]\right\}^{1/2}, \qquad (26)\\
\Delta x &= x - [x_S + u_S(t - t_S)],\\
\Delta u &= u - u_S.
\end{aligned}
$$

In fact the integral expresses simply that $\omega = u_{,y}$. In order to find the y-position of the upper, separating vortex layer sketched in Fig. 4 and evident in the vorticity profile Fig. 5, the integration should be performed from the wall to the upper layer. But because the part of the integral near the stationary point dominates anyway, to a correct first approximation the integration can be extended towards infinity in both directions:

$$
y_u(\Delta x, t) = \int_{\infty}^{\infty} \frac{|d\Delta u|}{|\omega(\Delta u, \Delta x, t)|}. \qquad (27)
$$

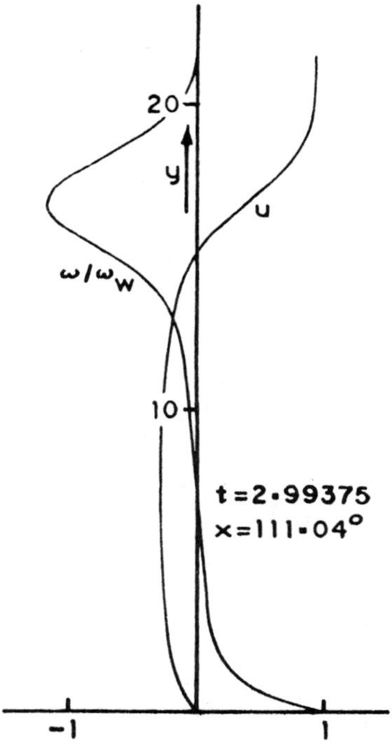

Figure 5. A typical computed vorticity profile and corresponding velocity profile close to separation. (mesh 289 × 129 × 960, $x_S \approx 110.96°$, $t_S \approx 3.006$).

This is a complete elliptic integral, of which the properties may fairly easily be assessed [3.49]. It proves that for $t \uparrow t_S$ and $\Delta x \to 0$, the value of y_u blows up—more precisely, that the scaled-down y-coordinate

$$Y = y|t - t_S|^{1/4} \tag{28a}$$

remains a well-behaved function of the stretched streamwise x-coordinate

$$X = \frac{\Delta x}{|t - t_S|^{3/2}} \tag{28b}$$

when $t \uparrow t_S$. This asymptotic position Y_U of the upper layer in the X, Y plane is shown in the bottom panel of Fig. 6. Thus when $t \uparrow t_S$ in the X, Y plane, the velocity and vorticity lines of the upper separating layer should collapse onto the curve shown, while those of the wall layer should collapse onto the X-axis.

In fact the Y-wise spacing between velocity and vorticity lines in both layers will vanish proportional to $|t - t_S|^{1/4}$, because the continuity integral ensures that the y-spacing between any two representative lines in either layer remains $O(1)$; in the X, Y plane this spacing is then scaled down by a factor $|t - t_S|^{1/4}$ according to (28a). Some of the computed results are shown in Figs. 6 and 7. Numerically it is very difficult to maintain accuracy so close to the singularity that $|t - t_S|^{1/4}$ is truly small. At $|t - t_S|^{1/4} = 0.38$ and 0.45

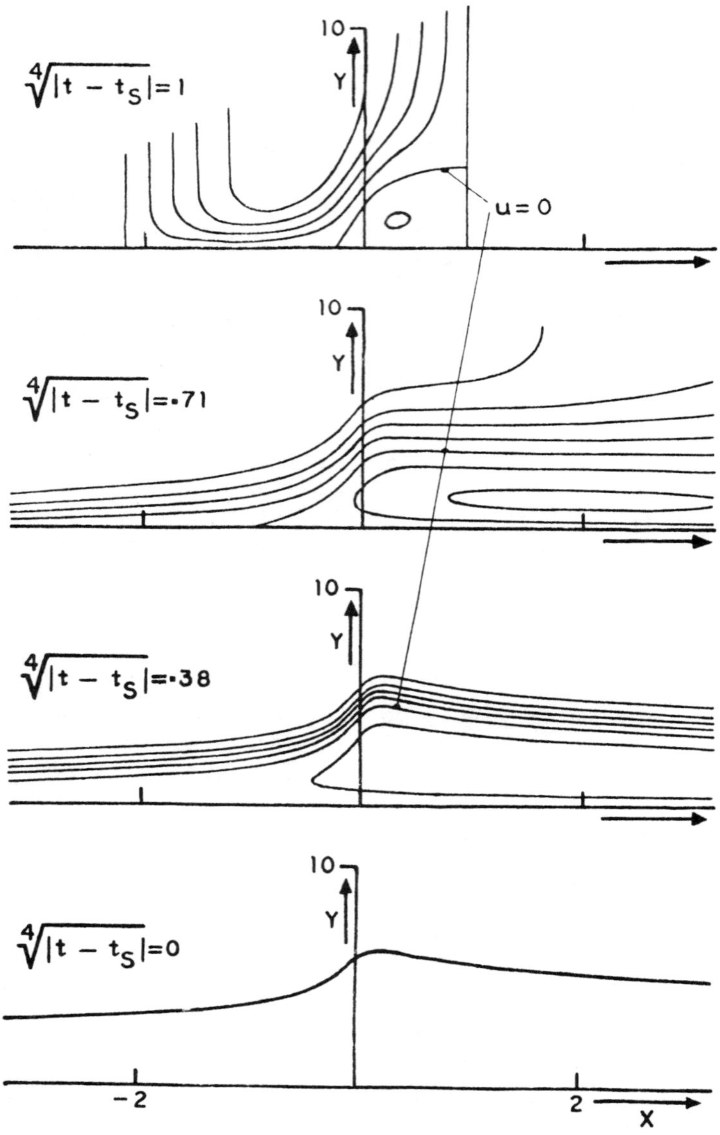

Figure 6. Lines of constant velocity $u = (-0.4)$, 0.2, 0, 0.2, 0.4, 0.6, 0.8, when separation is approached; see text. (mesh 289 × 129 × 960).

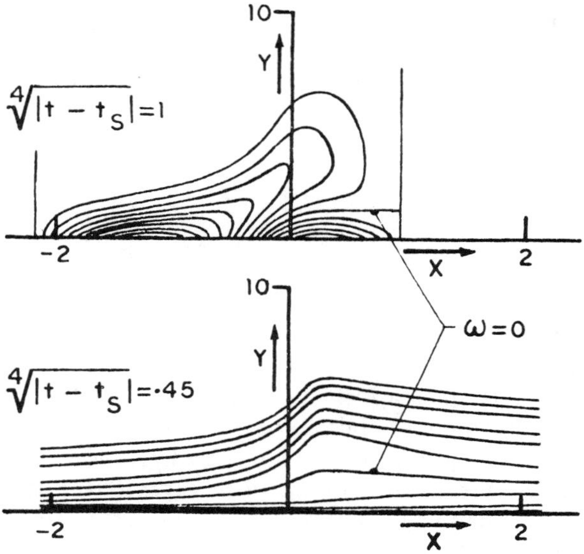

Figure 7. Lines of constant vorticity $\omega = 0, \pm 0.08, \pm 0.16, \ldots$ when separation is approached. (mesh $145 \times 65 \times 480$).

in Figs. 6 and 7 respectively, the lines of constant velocity and vorticity represent the limits where the present computation is still meaningful, and they have not yet collapsed.

The precise description of the wall layer below the stationary point is of course a Taylor series expansion,

$$u \sim \sum_{n=0}^{\infty} \sum_{m=0}^{\infty} (t - t_S)^n \, \Delta x^m f'_{mn}(y), \tag{29}$$

but for the separating vorticity layer above the stationary point the continuity integral implies that y is singular so that the above Taylor series cannot be used. Instead y can be related to any arbitrary reference line of particles in the upper layer,

$$\Delta y = y - y_{\mathrm{ref}} \qquad (y_{\mathrm{ref}} \sim y_u), \tag{30}$$

so that the continuity integral for Δy no longer passes the stationary point. Then Δy is regular and the description for the upper layer becomes again a Taylor series:

$$u \sim \sum_{n=0}^{\infty} \sum_{m=0}^{\infty} (t - t_S)^n \, \Delta x^m \, g'_{mn}(\Delta y). \tag{31}$$

From this result it may be shown [3.49] that asymptotically $-u_{,x}$ blows up proportional to $|t - t_S|^{7/4}$ at separation; according to Fig. 2 this prediction is in very good agreement with the numerical results.

The fact that the Navier–Stokes equations do not allow the singularity [3.45] suggests that it might be instructive to substitute the boundary-layer results in the full Navier–Stokes equations. Thus an estimate is obtained for the relative magnitude of the terms neglected in the boundary-layer approximation. It turns out here that when

$$|t - t_S| = O(\mathrm{Re}^{-2/7}), \tag{32}$$

$$\Delta x \approx y\,\mathrm{Re}^{-1/2} = O(\mathrm{Re}^{-3/7}), \tag{33}$$

then the neglected terms become fully as important as the leading ones and the obtained boundary-layer results cannot possibly remain meaningful.

Experimentally Tietjens [3.42] observed that the upper part of the boundary layer rolls up as a Kelvin–Helmholtz unstable free vortex layer. The present result simply establishes that the upper part of the boundary layer does indeed develop to such an unstable free vortex layer. According to our theory, however; in order for the upper layer to be really free from the wall (farther away than its thickness), $\mathrm{Re}^{-1/14}$ should be small in (33). This requirement is of course not met in the experiments. In the experiments of Bouard and Coutanceau [3.43] the center of rotation possibly corresponds to the core of the rolled-up sheet. The location appears to be significantly downstream of 111°, which may be a finite-Reynolds-number effect (Re = 9500).

For times beyond t_S, the singularity propagates upstream according to the MRS condition (23a) and will presumably tend to the steady position at 104.5° calculated by Terrill [3.39]. Outer-flow particles will propagate further downstream away from the singularity with the local outer flow velocity. Thus the influence of the singularity will be felt in a range of x-values, shaded in Fig. 8.

According to the Taylor series expansion (24a) for x in the Lagrangian plane, the singular line x_S describes a loop when $t > t_S$ and intersects itself at the stationary point S in Fig. 8. Assuming that for the upstream boundary layer the postulated outer flow still has meaning [3.49], this boundary layer will bifurcate when $x \uparrow x_S$, the particles beyond point S again penetrating toward large y-values on account of the continuity integral.

The analytical description of the singularity is obtainable, as before, from local Taylor series expansion of x and u near S, and integration of the continuity integral to find y [3.49, 3.51]. It is then found for the dead-water region around S that

$$\begin{aligned} \Delta u &= (\tfrac{1}{2}P_S^*|\Delta x|)^{1/2}\left(\sigma + \frac{1}{\sigma}\right), \\ \sigma &= (\tfrac{1}{2}P_S^*|\Delta x|)^{1/2} b \exp(y/a), \\ \Delta x &= x - x_S(t), \\ \Delta u &= u - u_S(t), \end{aligned} \tag{34}$$

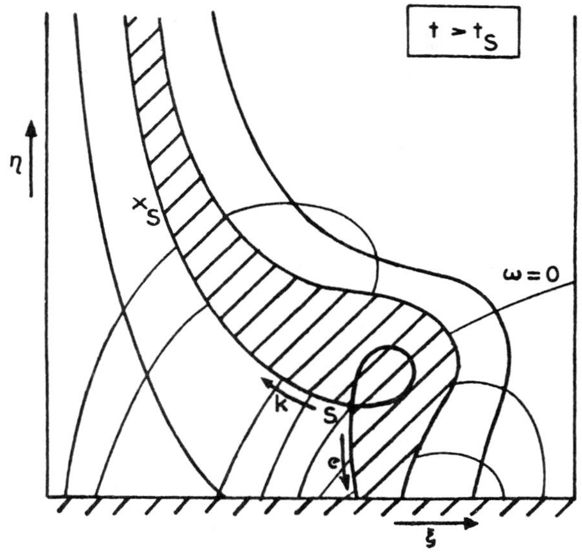

Figure 8. Lines of constant vorticity and streamwise particle position for times beyond t_S. (Schematic).

in which a and b are undetermined constants; a happens to be the determinant of second order x-derivatives at S, is infinite at time t_S, and presumably vanishes for infinite time [3.49]. The proper coordinate in the y-direction for the dead-water region is seen to be σ, with P_S^* being the pressure acceleration as seen in a system moving with separation

$$P_S^* = P_S + \mathring{u}_S. \tag{35}$$

The separating vorticity layer is described by large values of the transverse coordinate σ; hence

$$y_u \sim a \ln \frac{1}{P_S^* |\Delta x|} + O(1). \tag{36}$$

In the description (29), (31) of the upper and lower layers the Taylor series expansion should now of course be replaced by a continuous time dependence.

4 Discussion

According to the boundary-layer equations, it was found that after 0.75-diameter movement of the impulsively started cylinder, the upper part of the boundary layer starts to "peel off" from the surface at the rear half of the cylinder but upstream of the rear stagnation point. Because the penetration of the separating layer into the outer flow becomes large compared to $\mathrm{Re}^{-1/2}$, the conventional boundary-layer approximation can no longer describe the flow and responds with a singularity.

The experiments suggest that in the true development the separated layer proceeds to roll up into the shed vortices [1.42, 3.42, 3.43]. If so, the boundary layer equation will probably need to be replaced by the equations of inviscid flow for the description of the local evolution.

Various conjectures mentioned in the introduction, notably by Moore [3.1], Sears and Telionis [3.35], Shen and Nenni [3.47], and Shen [3.10], have been verified for the singularity. Some remarkable new results are that the upstream-slipping separation point is inviscid, that the wall shear is regular, and that the boundary-layer thickness blows up logarithmically at separation, except for the first appearance of the singularity, when the blowup is more severe [3.49]. In one form or the other, these features have all already been noted in literature, but remained unexplained. For example, Williams and Johnson [3.7] found a remarkably flat minimum in the velocity profile near separation, which is equivalent to inviscid separation at vanishing viscous particle acceleration $u_{,yy}$ and is in clear disagreement with the tentative viscous proposals advanced by Brown [3.8] and the more general one by Moore [3.1]. Also, Williams and Johnson's figures clearly show finite wall shear gradient at separation and a stronger growth of the boundary-layer thickness near separation than one would expect if it had been a Goldstein–Stewartson [2.8, 3.52, 3.53] square root; compare also their results and those of Barbi and Telionis elsewhere in this volume. Van Dommelen [3.49] further argues from an integral constraint that the viscous proposals of Moore and Brown cannot and do not describe bifurcation of the boundary layer. It should be noted, however, that the infinite boundary thickness also implies that the plots of various quantities at fixed y which appear in the literature have no clear meaning.

Thus the present upstream-slipping separation point is quite different from the nonslipping Goldstein–Stewartson point, which is viscous with wall shear tending to zero singularly and has $O(\mathrm{Re}^{-1/2})$ boundary-layer thickness at the separation point. Fortunately there is independent support for such differences from the description of nonslipping [2.13] and upstream-slipping [3.54] steady Helmholtz-type separation. Here Sychev arrives independently at the same conclusions that the upstream-slipping separation point is inviscid, has logarithmic blowup of the upstream boundary-layer thickness, and has nonzero, though logarithmically singular, wall shear. On the other hand the nonslipping bifurcation process is viscous at $O(\mathrm{Re}^{-1/2})$ boundary-layer thickness and around zero wall shear. Sychev's description of upstream-slipping separation is also very similar to the present one, where $P_S^*|\Delta x|$ in (34) and (36) should for this steady case be expressed in the equivalent form $(p_S - p)/\rho$ in order to accommodate the infinite pressure gradient of Helmholtz separation. Sychev's upsteam-slipping separation has been further developed than the present one, in that the singular upstream boundary layer has been matched to a separation region: because Helmholtz separation is tangential, simple viscous–inviscid interaction suffices here. However, Van Dommelen [3.49] has proposed a numerical solution of the interaction equations which is different from Sychev's.

Thus it appears that the upstream-slipping separation point is inviscid and quite generally described by the derived results. The nonslipping separation point is viscous. For the downstream-slipping separation point Moore [3.1] proposes that it should be viscous below and inviscid above, while Van Dommelen [3.49], reasoning from vorticity-dynamics arguments, instead arrives at the conclusion that it is inviscid below and viscous above.

Despite the wide differences in the actual bifurcation process, the mechanics of the upper, separating layer appears to be quite general. Invariably this layer has the essential character of a nonseparating layer which is swept upward by the rapid transverse expansion of the MRS-bifurcation region below the layer. In fact, this expansion simply adds a singular upflow $y_u(x, t)$ to the transverse particle position in the layer:

$$y \to y + y_u(x, t); \tag{37a}$$

hence,

$$v = \frac{Dy}{Dt} \to v + y_{u,t} + u y_{u,x}. \tag{37b}$$

If (37) is substituted in the boundary-layer equation, it is found that the velocity u remains unchanged under the transformation. Thus the boundary layer is invariant under the upflow.

Conversely, this *upflow invariance* of the boundary-layer equation also implies that any upflow y_u is allowable for the upper layer. Thus the precise form of y_u is solely determined by the nature of the MRS bifurcation process and was found to be (cf. (27) and (36))

$$y_u \sim \int_{\infty}^{\infty} \frac{|d\Delta u|}{|\omega(\Delta u, \Delta t, t)|}, \qquad t \uparrow t_S, \tag{38a}$$

$$y_u \sim a \ln \frac{1}{P_S^* |\Delta x|}, \qquad t > t_S, \tag{38b}$$

while for the Goldstein–Stewartson singularity instead

$$y_u \sim y_{uS} - \frac{\tau_w}{UU'} + O(|\Delta x| \ln |\Delta x|), \qquad t \to \infty? \tag{38c}$$

For the downstream-slipping singularity y_u is as yet unknown [3.49].

From the results of this paper, some important comparison material may be derived as an aid in the computation of upstream-slipping separation. In view of the assumption made, the outer flow boundary $U(x, t)$ is here required to be regular. For the irregular Helmholtz-type outer-flow boundary condition the reader is referred to Sychev's paper [3.54]. First of all, the wall shear is regular (Eq. (29)) and usually negative:

$$\tau_w < 0, \qquad \tau_{w,x}, \tau_{x,t} = O(1). \tag{39a}$$

From (3.15) it follows that

$$(\delta_{,x})^{-1} \sim a|\Delta x| [1 + O(|\Delta x|^{1/2})]. \tag{39b}$$

Hence, the location of separation may be determined by plotting the inverse gradient of the boundary-layer thickness as close to separation as accuracy can be maintained and extrapolating toward zero. As a check, the square root of this quantity plotted against $|\Delta x|^{1/2}$ should pass through the origin more or less linearly:

$$(\delta_{,x})^{-1/2} \sim a^{1/2}|\Delta x|^{1/2}[1 + O(|\Delta x|^{1/2})]. \tag{39c}$$

The boundary-layer thickness δ can be the usual one, or the displacement thickness can be used instead. In the latter case a coordinate system moving with the separation in the streamwise direction should be used for the constant a to retain the same meaning. But in this moving coordinate system the momentum and energy thicknesses should not be used in (39b) and (39c), as they are finite at separation. Finally, the velocity and the viscous particle acceleration vanish singularly along the zero vorticity line:

$$(u - u_S)_{\omega=0} \sim a^2(u_{,yy})_{\omega=0} = O(|\Delta x|^{1/2}). \tag{39d}$$

Shen conjectures that the generation of the singularity is similar to wave steepening in compressible flow [3.10]. The present results support this: near the forming singularity the pressure and viscous effects are negligible; hence on the zero vorticity line

$$\frac{\partial u}{\partial t} + u\frac{\partial u}{\partial x} = 0. \tag{40}$$

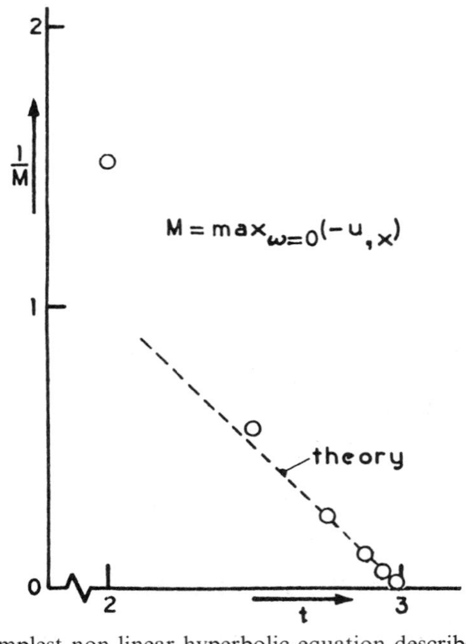

Figure 9. The simplest non-linear hyperbolic equation describes the generation of separation. (Mesh 145 × 65 × 480).

This simple hyperbolic equation does indeed describe wave steepening; compare the Burgers equation. Many results of interest follow very easily from (40): for example, that $u_{,x}$ blows up as $1/(t - t_S)$ on the zero vorticity line. This is in good agreement with the numerical data according to Fig. 9. Note that (40) describes wave steepening on the part of the zero vorticity line near zero wall shear, but wave stretching on the part near the Proudman–Johnson rear stagnation point.

Mention should further be made of the boundary-layer solutions for the impulsively rotated sphere obtained by Banks and Zaturska [3.55] and for counterrotating plates by Bodonyi and Stewartson [3.16]. In both these cases the boundary-layer solution also becomes singular in finite time, when boundary-layer material is ejected. By rewriting the results in Lagrangian coordinates [3.49], it can be seen that these cases, like the present two and that of Sychev [3.54], all correspond to an inviscid, transverse "explosion" of a particle inside the boundary layer.

5 The New Picture

If the boundary-layer problem for the impulsively started circular cylinder had had a smooth solution for all time, the physical picture would have been that the vorticity is essentially diffused out over the entire flow field, similarly to the solution of Proudman and Johnson [3.4, 3.58] for a rear stagnation point at a plane wall. Instead, it turned out that the solution becomes singular because part of the vorticity layer moves away from the wall in an inviscid manner, while essentially preserving its diffusion-determined thickness. Hence the consistent physical picture for high-Reynolds-number flow is one in which thin streaks of vorticity traverse a more or less irrotational flow field, at least for moderately long time t. This suggests to defining the separation point as the point where a vorticity streak departs from the wall. Only for a nearly stready upstream boundary layer need this be a station of zero wall shear.

Acknowledgment

The authors are indebted to Professor A. A. Szewczyk of Notre Dame University for the computer facilities used to obtain the present numerical results and for other kind assistance; we also want to acknowledge valuable discussions with Professors R. Betchov of Notre Dame University, A. D. D. Craik of the University of St. Andrews, W. Fiszdon of Warsaw University, and S. Leibovich and F. K. Moore of Cornell. In particular we also want to thank Professor K. Stewartson who, traveling from Cornell to California, took along some of our displacement-thickness plots for Dr. Cebeci to compare with his accurate Eulerian scheme.

This work was supported by the U.S. Navy under contract number N00014-77-C-0033.

CHAPTER 18

Some Characteristics of Unsteady Two- and Three-Dimensional Reversed Boundary-Layer Flows

H. A. Dwyer* and F. S. Sherman†

Introduction

The purpose of this investigation is to study some of the characteristics of unsteady two- and three-dimensional boundary-layer flows with flow reversal. The primary focus of the study will be the time development of the flow-reversal region due to adverse pressure gradients, and differences between two- and three-dimensional flows. Some other features which have been investigated are the following: (a) determination of the existence of steady solutions with flow reversal; (b) development of a generalized coordinate transformation to remove singularities due to leading edges and impulsive starting; and (c) development of a numerical method for integrating the two- and three-dimensional equations in flow-reversed regions. Since these topics cover a very large range of material, the three-dimensional investigations have been limited to flows along a line of symmetry where the influence of three-dimensionality is felt through the cross-flow derivative. However, the addition of cross flow does exert a very substantial influence on the structure of the boundary layer, as will be seen in this paper.

The study of flow reversal in steady and unsteady boundary layers has been an active area of research for many years [2.124, 3.56]. Steady boundary-layer flows have consistently exhibited a singular behavior at the flow reversal or "separation point", which has severely limited boundary-layer theory in its

* University of California, Davis, CA 95616.
† University of California, Berkeley, CA 94720.

ability to predict complicated separated flows. This behavior is still an active area of research, and little is known about the nature of the equations in three-dimensional flow. In recent years, interest in both the two- and the three-dimensional behavior of the equations has been revived for the unsteady case, due to applications in aerodynamics such as the performance of helicopter rotors [3.11, 3.31]. At the present time there are arguments pro and con on the singularity and its nature in unsteady flow, and a strong statement about the singularity has recently been made by Cebeci [3.22]. Cebeci has stated a definite *no* to the existence of a singularity in a finite time interval. However, it will be shown in this paper that a more meaningful question to ask is, "Is the solution physically meaningful for finite times?" The answer to this question is that the solution *is not* physically meaningful. Also, it will be shown in the present paper that three-dimensional flows can be very different from their two-dimensional counterparts.

Basic Equations and Transformations

The time-dependent, incompressible boundary-layer equations along a line of symmetry have the following form (these equations also contain the two-dimensional case):

$$\frac{\partial u}{\partial x} + \frac{\partial v}{\partial y} + \zeta = 0, \qquad \zeta = \frac{\partial w}{\partial z},$$

$$\frac{\partial u}{\partial t} + u\frac{\partial u}{\partial x} + v\frac{\partial u}{\partial y} = -\frac{1}{\rho}\frac{\partial p}{\partial x} + \nu\frac{\partial^2 u}{\partial y^2},$$

$$\frac{\partial \zeta}{\partial t} + u\frac{\partial \zeta}{\partial x} + v\frac{\partial \zeta}{\partial y} + \zeta^2 = -\frac{1}{\rho}\frac{\partial^2 p}{\partial z^2} + \nu\frac{\partial^2 \zeta}{\partial y^2}, \tag{1}$$

where u and v are the velocity components in the x- and y-directions; ζ the cross-flow derivative, p the pressure, ρ the density, ν the kinematic viscosity, z the cross-flow direction, and t the time. For the solution of many boundary-layer problems it is necessary to rescale the variables and remove singularities that can appear at leading edges and at the unsteady start of the flow. A transformation that will rescale the variables and remove leading-edge singularities as well as the impulsive start singularity is

$$\xi = x, \qquad \tau = t, \qquad n = y\left(\frac{1 + U_e t/x}{\nu t}\right)^{1/2}. \tag{2}$$

In terms of these new independent variables, the set of equations (1) become

$$\xi\frac{\partial f'}{\partial \xi} + \xi\frac{\partial f'}{\partial n}\frac{\partial n}{\partial x} + \frac{\partial v'}{\partial n} + \beta_x f' + \frac{\zeta\xi}{U_e} = 0,$$

$$\frac{\xi}{U_e}\frac{\partial f'}{\partial t} + \xi f'\frac{\partial f'}{\partial \xi} + \left[\frac{\xi}{U_e}\frac{\partial n}{\partial t} + f'\frac{\partial n}{\partial x} + v'\right]\frac{\partial f'}{\partial n}$$

$$= (1 - f')\beta_\tau + (1 - f'^2)\beta_x + \varepsilon\frac{\partial^2 f}{\partial n^2},$$

$$\frac{\xi}{U_e}\frac{\partial\psi}{\partial t} + \xi f'\frac{\partial\psi}{\partial\xi} + \left[\frac{\xi}{U_e}\frac{\partial n}{\partial t} + \xi f'\frac{\partial n}{\partial x} + v'\right]\frac{\partial\psi}{\partial n}$$

$$= (1-\psi)\psi_t + (1 - f'\psi)\psi_x + \xi\frac{\zeta_e}{U_e}(1-\psi^2) + \varepsilon\frac{\partial^2\psi}{\partial n^2}, \quad (3)$$

where the following definitions have been employed:

$$f' = \frac{u}{U_e}, \qquad \psi = \frac{\zeta}{\zeta_e},$$

and the subscript e denotes the boundary-layer edge condition:

$$\beta_x = \frac{x}{U_e}\frac{\partial U_e}{\partial x}, \qquad \beta_t = \frac{x}{U_e^2}\frac{\partial U_e}{\partial t},$$

$$\psi_x = \frac{x}{\zeta_e}\frac{\partial\zeta_e}{\partial x}, \qquad \psi_t = \frac{x}{U_e\zeta_e}\frac{\partial\zeta_e}{\partial t},$$

$$v' = \frac{v\xi}{U_e}\left(\frac{1 + U_e t/x}{\nu t}\right)^{1/2}, \quad (4)$$

$$\varepsilon = \frac{\xi + U_e t}{U_e t}.$$

Some interesting limiting forms of these questions will be given in the following sections when the equations are applied to specific flow situations.

Unsteady Two-Dimensional Flow

The two-dimensional flow to be studied is a variation on the investigation of Howarth [3.56]. The inviscid velocity distribution is imposed impulsively and has the following distribution in space:

$$U_e = 1 - Ax + x^2. \quad (5)$$

The attractive feature of this distribution is that the size of the reversed-flow region can be adjusted by use of the constant A. The general flow development can be seen with the use of Fig. 1, where the axes x, n, and t are shown along with typical "separation" and "reattachment" lines. At an early stage of the flow development the boundary-layer and inviscid flow are in the same direction; however, as the flow develops, a flow-reversed region appears near the wall, due to the term $-Ax$. Further downstream the x^2 term dominates and the flow-reversed region ends. The size of the flow-reversed region depends on the value of the coefficient A; values near 1 are primarily used in the present paper.

Before describing the results of the study, the limiting forms of equation set (3) and the numerical method will be given. The physical plane singularity at

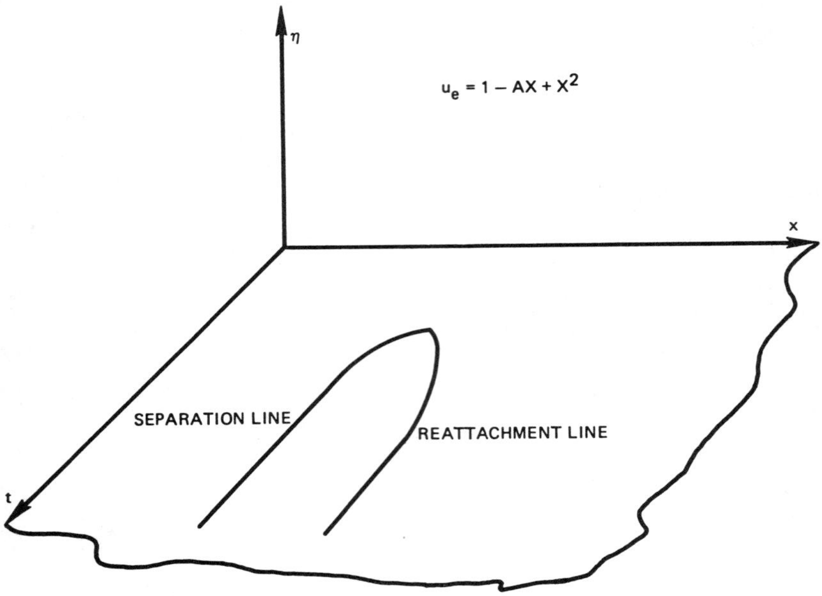

Figure 1. Qualitative development of flow wall structures.

$x = 0$ is of the Blasius type, and it is removed with the transformation given in Eq. (2). Taking the limits $\xi \to 0$ and $U_e \to 1$, we obtain

$$\varepsilon \to 1, \qquad \frac{\partial n}{\partial t} \to 0, \qquad \frac{\partial n}{\partial x} \to -\frac{n}{2x},$$

and the continuity and x-momentum equations become

$$f''' + f''f = 0 \tag{6}$$

(note: $\zeta = 0$ for two-dimensional flow) and this is the well-known Blasius equation [3.56]. Therefore, the conditions along the plane $x = 0\,(0 \le y \le \infty$ and $t > 0)$ are given by the tabulated values of Eq. (6).

The physical-plane singularity at $t = 0$ and $x > 0$ is also easily removed by the transformation, as can be seen by taking the limits $\tau \to 0$ and $\xi \neq 0$. The results are

$$\begin{gathered} \frac{\partial n}{\partial x} \to 0, \qquad \frac{\partial n}{\partial t} \to -\frac{n}{2t}, \\ v' \to 0, \qquad \beta_\tau \to 0, \qquad \beta_x \to 0, \\ \frac{n}{2}\frac{\partial f'}{\partial n} + \frac{\partial^2 f'}{\partial n^2} = 0 \end{gathered} \tag{7}$$

Equation (7) is also well known, and tabulated values can be found in Rosenhead [2.124]. Again, it is seen that the transformation developed in this paper is useful in removing difficulties in the physical plane, and it also poses the initial conditions in terms of well-known functions without scaling problems.

With the boundary and initial conditions well posed and the scaling difficulties due to the singularities removed, the numerical solution can proceed in a straightforward fashion. For the diffusion coordinate n, implicit central differences have been applied, while for the two timelike coordinates τ and ξ, a windward-type difference is used. The real time derivative, $\partial/\partial\tau$, becomes a backward difference, while for the convective derivative $\partial/\partial\xi$, a test must be applied to determine the direction of u.

Typical distributions of the wall primary flow-velocity derivative are shown in Fig. 2 for the case $A = 1$. For early times (case ①), the wall derivative is positive over the entire x-line. At twice this time (case ②), a significant region of flow reversal has developed; this region is smooth, and the numerical calculations are converged. As time develops, the flow-reversed regions increase in a regular fashion, however, a sharp change in the wall derivative begins to appear at the point of flow reversal. This sharp change can be seen in time with the help of Fig. 3. In this figure, the wall derivatives before and after the flow-reversal point are shown. The non-flow-reversed point ($x = 0.19$) is always regular and converged, while the flow-reversed point ($x = 0.20$) initially is well behaved, but quickly begins to behave erratically. The reason for this erratic behavior seems to be a very large growth of the boundary layer, as shown in Fig. 4. In this figure, the boundary-layer profiles are given at the flow-reversed point as a function of time. Even when the edge of the boundary

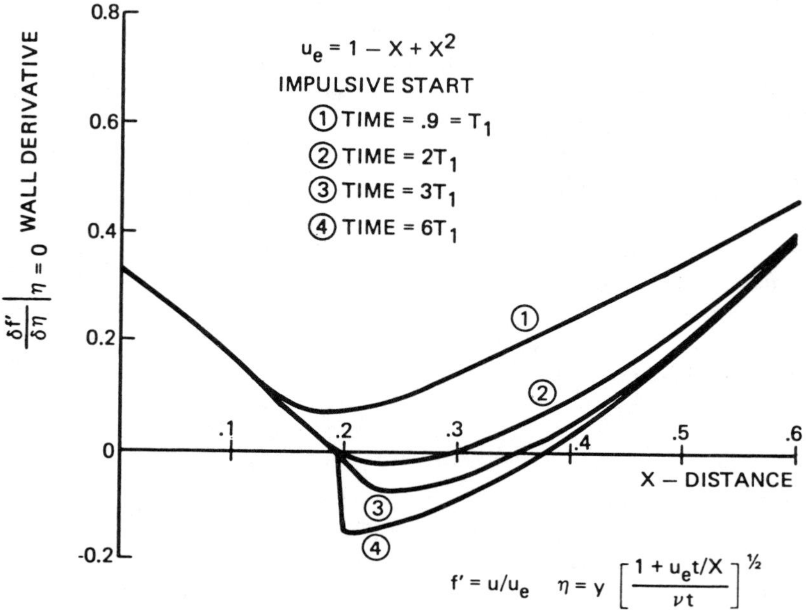

Figure 2. Wall velocity-derivative development with time.

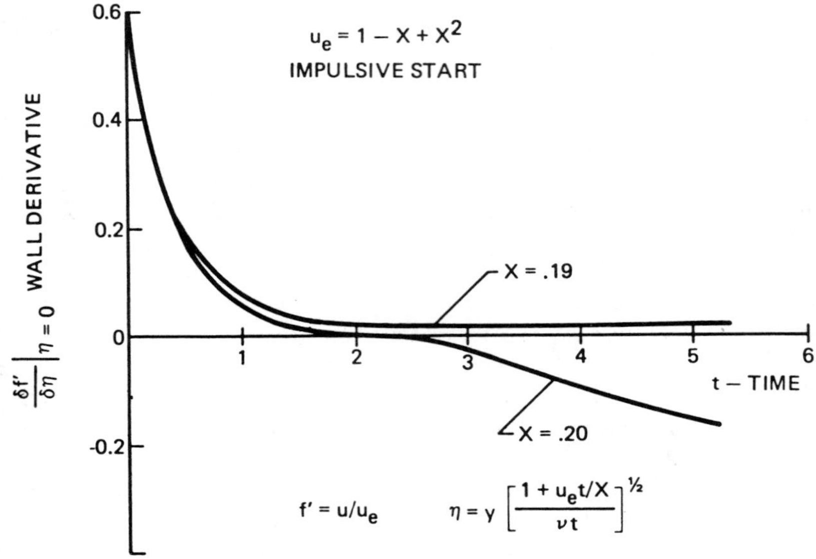

Figure 3. Wall velocity-derivative development at flow reversal.

layer was placed at $n = 100$, the same behavior was observed to occur, and when the flow-reversed region arrives at the boundary-layer edge, serious oscillations occur in the solution. As these oscillations propagate downstream the numerical solution rapidly deteriorates. It should also be mentioned that no serious oscillations seem to start at the "reattachment" point.

At this point in the study it is appropriate to compare these results with those of Cebeci [3.22]. It is seen in the present calculations that for long times the boundary-layer growth becomes excessively thick and the numerical results become meaningless. It is true, as Cebeci said, that no apparent singularity appears in the calculation as the boundary-layer thickness, but the usefulness of these calculations is not clear. Certainly, in real flows, the boundary layers will not get as thick as in these calculations, and there does not seem to be any practical way of determining how thick is too thick.

It may be that the difficulties encountered in the present investigation were not encountered by Cebeci because his calculations were not carried out for a long enough time. The dimensionless variable to consider in the time development of significant structures [3.57] in a time-dependent flow is the following:

$$T = \frac{U_e t}{x}.$$

In the present calculations the dimensionless time had a value of at least 5 at all points in the flow field. If T has values less than 1, there is not sufficient real time for a structure to develop, and if it is assumed in Cebeci's calculations that U_0 has a value 1 (not specified by Cebeci), then his maximum value of $\lambda = 0.5$ is not of sufficient magnitude for a flow structure to develop in the boundary layer.

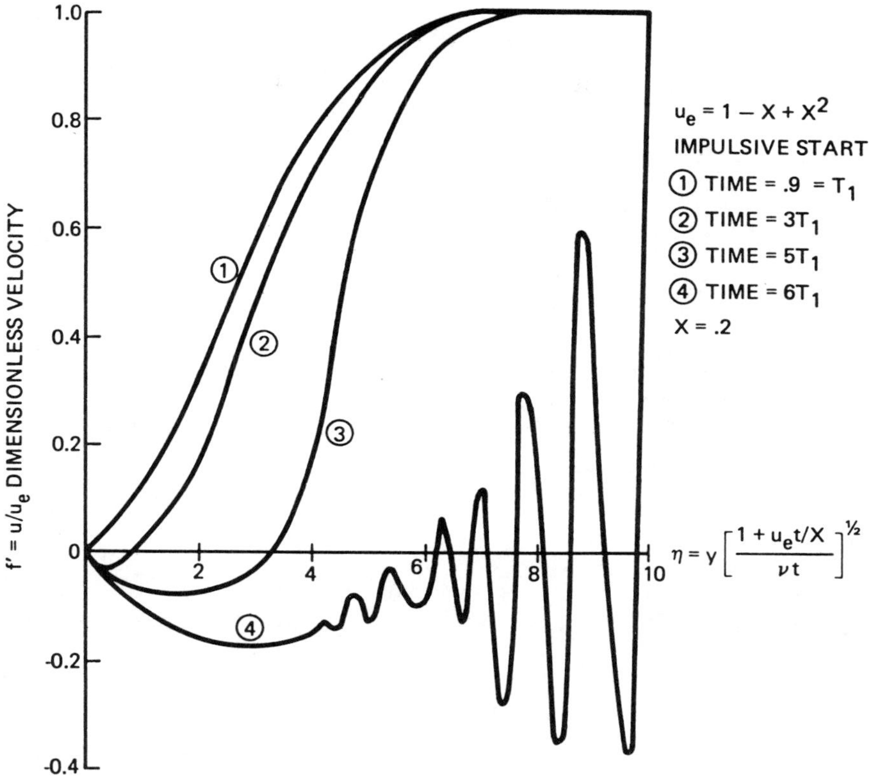

Figure 4. Velocity profiles at flow-reversal point.

At this point it is appropriate to discuss other conditions of the flow (values of A) that were studied. When the constant A in Eq. (5) is increased, the size of the flow-reversed region increases, and the excessive boundary-layer growth occurs more rapidly in time. As A is decreased in magnitude, the flow-reversed region decreases in size and the excessive growth takes a longer time to appear. For $A \approx 0.96$ and $\Delta x = 0.01$, the flow-reversed region only occurs at one station and the excessive growth appears to no occur. However, if the x-step is refined and more grid points placed in the flow-reversed region, the excessive growth occurs again, but with a slower pace in time.

Unsteady Three-Dimensional Flow

The second part of the investigation is concerned with the nature of unsteady flow reversal for a three-dimensional boundary layer. The flow studied is an unsteady version of a case that has been investigated by Cooke and Robins [3.58, 3.59]. The impulsively started edge conditions are

$$U_e = Ax(1 - x),$$

$$\frac{\partial W_e}{\partial z} = B, \qquad C = \frac{A}{B}.$$

This three-dimensional flow applies along a line of cross-flow symmetry between nodal and saddle points of attachment. The initial conditions at $t = 0$ have the same nature as two-dimensional flow and the same tabulated values can be used to start the calculation. At the stagnation point, or nodal point of attachment, Eq. (3) becomes

$$\frac{n}{2}\frac{\partial f'}{\partial n} + \frac{\partial v'}{\partial n} + f' + \frac{\zeta}{a} = 0,$$

$$\frac{1}{a}\frac{\partial f'}{\partial t} + \left[-\frac{f'n}{2} + v'\right]\frac{\partial f'}{\partial n} = (1 - f'^2) + \frac{\partial^2 f'}{\partial n^2}, \tag{8}$$

$$\frac{1}{a}\frac{\partial \psi}{\partial t} + \left[-\frac{f'n}{2} + v'\right]\frac{\partial \psi}{\partial n} = \frac{\xi_e}{a}(1 - \psi^2) + \frac{\partial^2 \psi}{\partial n^2},$$

and this system must be solved simultaneously with the basic equation set (3). At the rear stagnation point the transformation used in Eq. (2) is not valid, because

$$U_e \to 0, \qquad x \to 1,$$

and the expression for ε is singular. However, since it is known that this problem has self-similar solutions at $x = 1$, it was decided to use the condition

$$\frac{\partial f'}{\partial \xi} = \frac{\partial \psi}{\partial \xi} \to 0 \quad \text{as } x \to 1$$

to determine the downstream flow. (Note: this is a statement of self-similarity in the transformed coordinates.) Further details of this procedure and some difficulties will be given after some numerical results are first presented.

For very early times, no flow reversal occurs, and then the reversal starts near the rear stagnation point. The flow-reversed region quickly moves both forward and rearward, and the forward flow-reversal point reaches a steady state rather quickly. The structure of this flow near the onset of flow reversal is radically different than the two-dimensional one just studied, and this can be seen clearly with the help of Fig. 5.

In Fig. 5 the wall derivative immediately before and after the flow-reversal point is plotted as a function of time ($A = 1$). As can be readily seen, a steady value is quickly attained and remains fixed in time. This behavior is in marked contrast to the excessive growth shown by the two-dimensional flow. The key to this difference is contained in the cross flow, which allows for mass to flow sideways from the retarded region. The side flux eliminates the excessive increase in the y-velocity v, and the boundary layer does not grow excessively.

The distribution of the wall derivative is shown at various times in Fig. 6. There are two regions of this figure which are interesting: the flow-reversal point ($0.77 \leq x \leq 0.78$) and the rear stagnation point. At the flow-reversal point the flow is not singular, and this is opposed to the conjectures of Cooke and Robins. From the present results it seems that the singularity that Cooke and Robins suggested does not exist, and probably the only reason they could

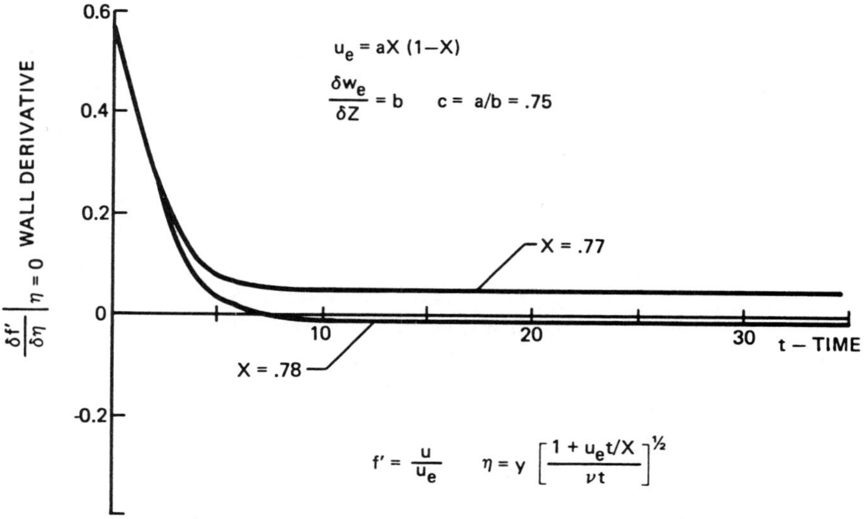

Figure 5. Wall velocity-derivative development at flow reversal.

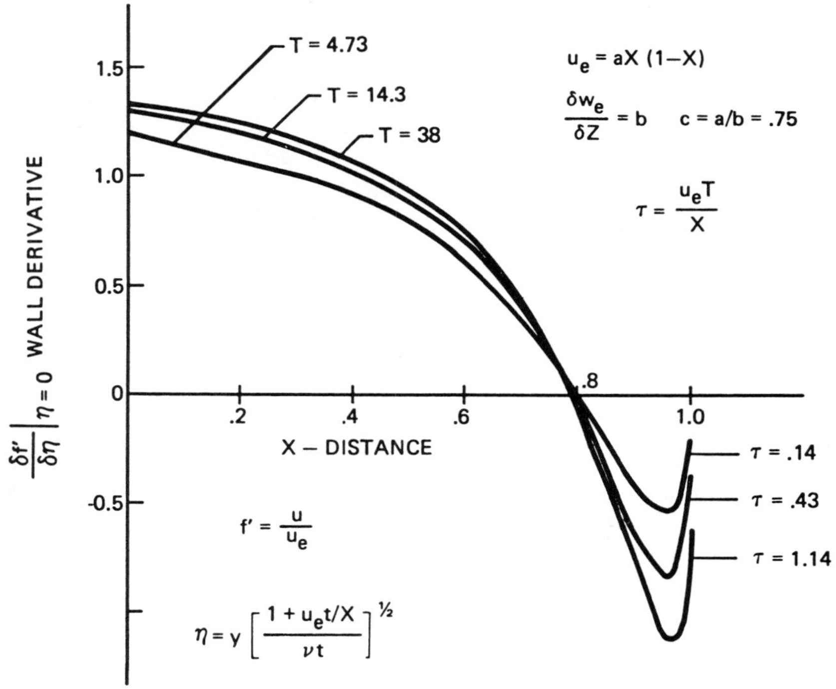

Figure 6. Wall velocity-derivative development with time.

not integrate further was their unstable numerical method. (It is extremely important when investigating regions of flow reversal to use a stable numerical method, or else it is impossible to distinguish a singularity from numerical error growth.)

The distribution of the wall derivative near the rear stagnation point is disturbing, due to the slow approach to steady state, the rapid change of wall shear near the rear stagnation point, and the influx of rotational fluid from the upstream flow. The forward stagnation region also has small values of T, but the favorable pressure gradient and the incoming inviscid fluid quickly set up a flow close to steady state. The value of T at $x = 1$ is always zero, since U_e equals zero at that location, and values of T can be made arbitrarily small by placing points near $x = 1$. This behavior is further amplified by the slow convection of rotational fluid into the region, and the net effect is that it takes a very long time for steady state to be achieved. In fact, for the present calculations, the time necessary was too great for the computer funds available, and the question of whether a steady state exists has not been fully answered. It also seems very dubious to these authors whether the rear stagnation point can be reached with boundary-layer-type equations.

It can be seen from Fig. 6 that the wall derivative is continuing to take on larger negative values; however, a study of the velocity profiles shows very little thickening of the boundary layer. It seems that the time development shown in Fig. 6 is the normal slow development of the flow at the rear stagnation point; however, the final answer will have to await further expensive calculations.

Another difficulty exhibited by Fig. 6 is the lack of time-dependent similarity shown at the rear stagnation point. Again, this should not be surprising, because of the slow convection into this part of the flow field. Unfortunately, although these results seem to agree with the physics of the problem, they cast doubt on the accuracy of the calculations, since the flow at $x = 1$ (not shown in Fig. 6) was found by specifying $\partial f'/\partial\xi$ and $\partial\psi/\partial\xi$ to be zero. Therefore, future calculations will also have to treat this part of the flow field more carefully.

Conclusion

The major conclusions of this study are the following:

1. The coordinate transformation introduced to remove both leading-edge and impulsive singularities enabled numerical solutions to be obtained easily without scaling difficulties.
2. Unsteady two-dimensional boundary-layer flow seems to be well behaved for short times with flow reversal. However, for a prolonged flow reversal the flow seems to exhibit excessive growth at the point of reversal, and it is not obvious how to determine when the solution breaks down. The excessive growth seemed to be independent of the size of the flow-reversed region for the present problem.

3. Unsteady three-dimensional flow is very different from two-dimensional flow. This difference can be attributed to the added degree of freedom for mass to flow away from the retarded boundary layer. For the problem studied there was no excessive growth of the boundary layer at the flow-reversal point, and the solution was well behaved. A difficulty with the investigation was the slow approach to steady state at the rear stagnation point.

CHAPTER 19

Large-Time Boundary-Layer Computations at a Rear Stagnation Point Using the Asymptotic Structure

P. G. Williams*

1 Introduction

A major difficulty in continuing the computation of the laminar boundary layer at a rear stagnation point to large times is the exponential growth with time of the boundary-layer thickness. The large-time asymptotic structure of the velocity profile has been elucidated by Proudman and Johnson [3.38], who found that a double structure was needed with the usual y-scale in the inner region and a scale of ye^t in the outer region, and that this structure could be matched approximately to a numerical solution. Subsequently Robins and Howarth [3.4] extended the asymptotic expansion and matched it to a more accurate numerical solution, but still carried out with a uniform step in y. Ultimately 10,000 y-steps had to be taken to reach $t = 5.5$. We therefore consider here how the asymptotic structure could be used to reduce this computational effort.

Since the double structure is applicable in the vicinity of a rear stagnation point as well as actually at the point, we include the x-variation in the development and later specialize to the case treated by Robins and Howarth, in which the x-variation is removed. We show that the implementation of the double structure can be achieved in a relatively straightforward manner by using the Keller–Cebeci [3.60] box method, but find that a certain difficulty arises which has not arisen in previous double-structure calculations [2.106, 3.61]. We propose a scheme here for overcoming this difficulty which

* University College, London.

has the feature that, in order to join the inner region, with a y-step of h, say, to the outer region, with a y-step of he^t, an intermediate region is introduced with a propitiously chosen nonuniform y-step. Results are compared with those of Robins and Howarth.

2 Transformation of the Equations

In nondimensional boundary-layer-scaled variables the time-dependent boundary-layer equation for the stream function ψ reads

$$\frac{\partial^2\psi}{\partial y\,\partial t} + \frac{\partial\psi}{\partial y}\frac{\partial^2\psi}{\partial x\,\partial y} - \frac{\partial\psi}{\partial x}\frac{\partial^2\psi}{\partial y^2} - u_e\frac{du_e}{dx} - \frac{\partial^3\psi}{\partial y^3} = 0, \tag{1}$$

where u_e is the longitudinal velocity at the edge of the boundary layer. If we scale out the mainstream variation by writing $\psi = u_e f$, the momentum equation becomes

$$\frac{\partial f'}{\partial t} + u_e\left(f'\frac{\partial f'}{\partial x} - f''\frac{\partial f}{\partial x}\right) - \frac{du_e}{dx}(1 - f'^2 + ff'') - f''' = 0, \tag{2}$$

where $'$ denotes $\partial/\partial y$. At a rear stagnation point $u_e = 0$ and $du_e/dx = -1$, so that (2) becomes

$$\frac{\partial f'}{\partial t} + 1 - f'^2 + ff'' - f''' = 0, \tag{3}$$

which is the equation solved by Robins and Howarth together with the boundary and initial conditions

$$\begin{aligned} f = f' &= 0 \quad \text{on } y = 0 \text{ for } t \neq 0,\\ f' &\to 1 \quad \text{as } y \to \infty,\\ f' &= 1 \quad \text{at } t = 0 \text{ for } y \neq 0, \end{aligned} \tag{4}$$

for the case when $u_e = -x$ and f depends only on y and t.

It is well known that the initial development can be calculated more accurately by using a transformed y-coordinate based on the small-time structure, namely $Y = y/t^{1/2}$. We wish to explore the efficacy of employing the corresponding structure for large times. This is more complicated in that it has two regions, and in this sense the situation is similar to the small-time double structure occurring in the moving-stagnation-point problem [3.61]; however, there are differences which may cause difficulty, so we concentrate here on the simpler problem where f does not depend on x.

The Proudman–Johnson theory, as extended by Robins and Howarth, suggests that in the outer region

$$f = e^tF_0(e^{-t}y) + F_1(e^{-t}y) + e^{-t}F_2(e^{-t}y) + \cdots, \tag{5}$$

while in the inner region

$$f = f_0(y) + ce^{-t}f_1(y) + (ce^{-t})^{3/2}f_2(y) + O(te^{-2t}). \tag{6}$$

We assume that the change to this double structure can be imposed at $t = 1$, and for convenience take $e^{1-t}y$ as the outer variable, which then coincides with y at the changeover point. Because of the form of the outer expansion we try first taking e^{1-t} instead of t as time variable for $t < 1$.

We now consider the general form the momentum equation takes when the various transformations are made, introducing at the same time the first-order form which is needed for the application of the box method [2.34]. To avoid introducing extra notation we use the same set of capital letters to denote the appropriate variables, whichever transformation is currently being used (this expedient is used in the computer program anyway); thus lowercase letters have a global significance, but uppercase letters have a local meaning which depends on the region being considered. We can summarize by defining for

(a) small time: $T = t,\quad Y = y/t^{1/2},\quad F = f/t^{1/2},$
(b) intermediate time: $T = t,\quad Y = y,\quad F = f,$
(c) large time (inner): $T = e^{1-t},\quad Y = y,\quad F = f,$
(d) large time (outer): $T = e^{1-t},\quad Y = e^{1-t}y,\quad F = e^{1-t}f.$

Taking $u = U$ for uniformity and introducing $V = \partial U/\partial Y$, we can write the boundary-layer equations in the fixed form

$$U = \frac{\partial F}{\partial Y}, \qquad V = \frac{\partial U}{\partial Y}, \tag{7}$$

$$c_1 \frac{\partial U}{\partial T} + c_2\left(U \frac{\partial U}{\partial X} - V \frac{\partial F}{\partial X}\right) + c_3(1 - U^2 + FV) + c_4 \frac{\partial V}{\partial Y} + c_5 V = 0 \tag{8}$$

if the coefficients c_k, which do not depend on the solution, are defined appropriately. From the specific definitions below it can be seen that c_1 and c_4 may depend on T, c_2 and c_3 may also depend on X, and c_5 depends only on Y. They can be set out in a table for the different cases as follows:

Case	c_1	c_2	c_3	c_4	c_5
(a)	T	Tu_e	$-T\,du_e/dx$	-1	$-\frac{1}{2}Y$
(b)	1	u_e	$-du_e/dx$	-1	0
(c)	$-T$	u_e	$-du_e/dx$	-1	0
(d)	$-T$	u_e	$-du_e/dx$	$-T^2$	$-Y$

(9)

3 Application of the Box Method

We introduce a rectangular grid in the X, Y, T space with steps DX, DY, DT and apply the box method [2.34] to approximate the momentum equation at the midpoint M of each elementary box. The grid spacings may be non-uniform, and in fact we will find that there are regions where it seems to be necessary to take both DY and DT as variable. This implicit method solves simultaneously for a column of solution values at each (X, T) station. Let the current station be denoted by A, and neighboring ones at previous X- and T-stations by B, C, D as in Fig. 1, so that they have coordinates $A(X, T)$, $B(X - DX, T)$, $C(X, T - DT)$, $D(X - DX, T - DT)$. It is convenient to avoid a running subscript for the X- and T-directions because only the last X- and T-stations are normally involved in the current computation at station A. We distinguish the values at $X - DX$ by a bar and those at $T - DT$ by a star, so that in particular $\bar{X} = X - DX$ and $T^* = T - DT$.

For the Y-direction a running subscript is unavoidable, and if there are n grid points from the wall to the edge of the boundary layer, we denote their positions by Y_j, $j = 1, \ldots, n$, with $Y_1 = 0$, and correspondingly use F_j, U_j, V_j for the F, U, V-values to be calculated at the current (X, T) station. If the local box is contained between Y levels $j - 1$ and j, the box method involves sums over these two levels at each relevant (X, T) station; these sums we denote by the corresponding capital subscript, so that, for example, U_A denotes the sum $U_j + U_{j-1}$ and F_B denotes $\bar{F}_j + \bar{F}_{j-1}$. Differences between values on the two Y-levels involve only V-values and will in fact be summed over the four (X, T) stations, so let V_J denote this sum for level j and V_{J-1} for level $j - 1$. Further sums and differences are needed for approximating the terms in (8) by second-order centered approximations, and these are again characterized by the corresponding capital subscripts. For example, for values at the midpoint M of the box we make the following typical replacements:

$$\frac{\partial \hat{U}}{\partial T} = \frac{U_T}{4DT}, \qquad \frac{\partial \hat{U}}{\partial X} = \frac{U_X}{4DX}, \qquad \hat{U} = \frac{U_M}{8}, \tag{10}$$

where $\hat{}$ signifies midpoint values and

$$U_T = U_A + U_B - U_C - U_D, \qquad U_X = U_A - U_B + U_C - U_D,$$
$$U_M = U_A + U_B + U_C + U_D,$$

which are the relevant combinations for representing the T-derivative, X-derivative, and midpoint value.

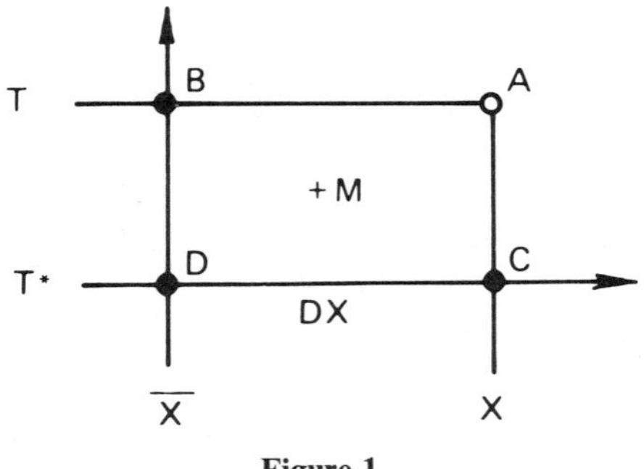

Figure 1.

Approximating (8) at the midpoint of the box then yields

$$C_1 U_T + C_2(U_M U_X - V_M F_X) + C_3(64 - U_M^2 + F_M V_M) + C_4(V_J - V_{J-1}) + C_5 V_M = 0, \quad (11)$$

where the coefficients C_k are related to the c_k by

$$C_1 = \frac{\hat{c}_1}{DT}, \quad C_2 = \frac{\hat{c}_2}{8DX}, \quad C_3 = \frac{\hat{c}_3}{16}, \quad C_4 = \frac{\hat{c}_4}{DY}, \quad C_5 = \frac{\hat{c}_5}{2}. \quad (12)$$

Finally, if we apply similar centered differencing to (7) at the current station A, we obtain the set of equations

$$(r_1)_j = 0, \quad (r_2)_j = 0, \quad (r_3)_j = 0, \qquad j = 1, \ldots, n, \quad (13)$$

where for $j = 2, \ldots, n$

$$(r_1)_j = -F_j + F_{j-1} + \frac{DY}{2} U_A, \qquad (r_2)_j = -U_j + U_{j-1} + \frac{DY}{2} V_A, \quad (14)$$

$$(r_3)_{j-1} = C_1 U_T + C_2(U_M U_X - V_M F_X) + C_3(64 - U_M^2 + F_M V_M) + C_4(V_J - V_{J-1}) + C_5 V_M. \quad (15)$$

The remaining equations come from the boundary conditions and read

$$(r_1)_1 = -F_1, \qquad (r_2)_1 = -U_1, \qquad (r_3)_n = 1 - U_n. \quad (16)$$

These are solved in the standard way by Newton iteration and block elimination. Since the quantities U_T, U_M, etc. have been chosen so that their derivatives with respect to the unknown they contain is unity, no further factors are introduced into the Newton coefficients.

If the current station A is in a region where a double structure is used, these equations have to be modified by the addition of equations representing continuity of f, u, v at the join between the two regions. This we consider in the next section.

4 Double-Structure Modifications

We assume that the change to the large-time double structure takes place at $t = 1$; otherwise we could rescale t so that it did, and this would merely introduce an extra scale factor that could be absorbed in c_1 in (8). At $t = 1$ we also have $T = 1$, where $T = e^{1-t}$, so that the inner and outer scalings are the same at $t = 1$. As t increases, T decreases from 1 to 0 and the mesh cells in the (T, Y) plane are as shown in Fig. 2 if the T-spacing and Y-spacing in each region are uniform. We take $j = \mu$ to be the last j-level of the inner region, so that the line separating the two regions in the (T, Y) plane is given by $Y = y_\mu T$. For programming reasons it is convenient to introduce extra storage locations with $j = \mu + 1$ to contain the first j-level of the outer region, which is physically at the same location in the (T, Y) plane. Then from

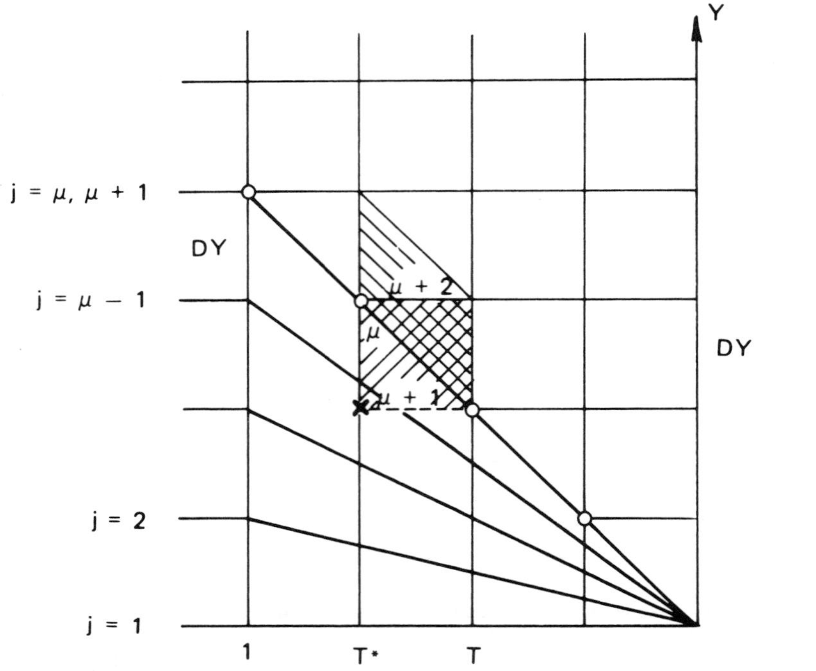

Figure 2. (T, Y) plane with uniform grids.

the interface conditions, which connect the inner variables to the outer variables by continuity, we have

$$\begin{aligned}(r_1)_{\mu+1} &\equiv TF_\mu - F_{\mu+1} = 0,\\ (r_2)_{\mu+1} &\equiv U_\mu - U_{\mu+1} = 0,\\ (r_3)_\mu &\equiv -V_\mu + TV_{\mu+1} = 0.\end{aligned} \tag{17}$$

Each time we progress from T^* to T, we introduce one more grid point into the outer region (with the small-time double structure for the moving stagnation point we removed one point from the outer region), and in order to keep values corresponding to the same Y-level also at the same j-location, we have to shift the values at T^* out one location. At $t = 1$ a preliminary shift has to be made to make room for the copy at the $j = \mu + 1$ level of the values at the $j = \mu$ level. Recalling that j indicates the level at the top of each box, we see that the box of the outer region with $j = \mu + 2$ overlaps the box of the inner region with $j = \mu$. If we are to relate the outer solution at $j = \mu + 1$ to that at $j = \mu + 2$, it seems that we must use the momentum equation for this box; however, the values at T^* with $j = \mu + 1$ are not necessarily at grid points of the inner region. One way to deal with this situation is to interpolate, say with cubic interpolation, but as T decreases to 0 this interpolation point will move further and further into the inner region if the T-step is uniform, which will presumably lead to inaccuracy.

A way to avoid the interpolation is to use a zigzag approximation by satisfying the momentum equation at the center of the diagonal cell hatched in Fig. 2. However, it seems that a more important source of inaccuracy, because of the higher-order terms in the inner expansion (6), might be the use of uniform steps in T. This raises the question of how should we choose nonuniform steps in T. From the form of the third term in (6) we might suppose that uniform steps in $T^{1/2}$ would be appropriate, but there might still be a small problem from the behavior of the fourth term. A natural way of generating nonuniform steps in T can be found if we observe that no interpolation will be required if we choose the T-step so that the interpolation point, × in Fig. 2, coincides with the inner y-position at T^* for level $j = \mu - 1$. Since $Y = Ty$, a uniform step of DY in y translates at T^* into a uniform step of T^*DY in Y; also the dividing line is given by $Y = y_\mu T$, so from the geometry we see that we must take

$$DT = -T^* \frac{DY}{y_\mu}. \tag{18}$$

It should be noted that this generates a uniform step in t, for since $t = 1 - \ln T$ we have

$$t - t^* = \ln \frac{T^*}{T} = \ln \frac{T^*}{T^* - DT} = \ln \frac{1}{1 - DY/y_\mu}. \tag{19}$$

The nonuniform steps in T automatically generate nonuniform steps in Y in the outer region from the dividing line $Y = y_\mu T$ to the line $Y = y_\mu$, as shown in Fig. 3, where it is clear that we have an expanding region in which the Y-step varies on any T-line from T^*DY to DY, the variation being geometric with common ratio $(1 - DY/y_\mu)^{-1}$. Thus the Y-step varies smoothly from the outer value DY at the line $Y = y_\mu$ down to a value that matches the inner step at the dividing line.

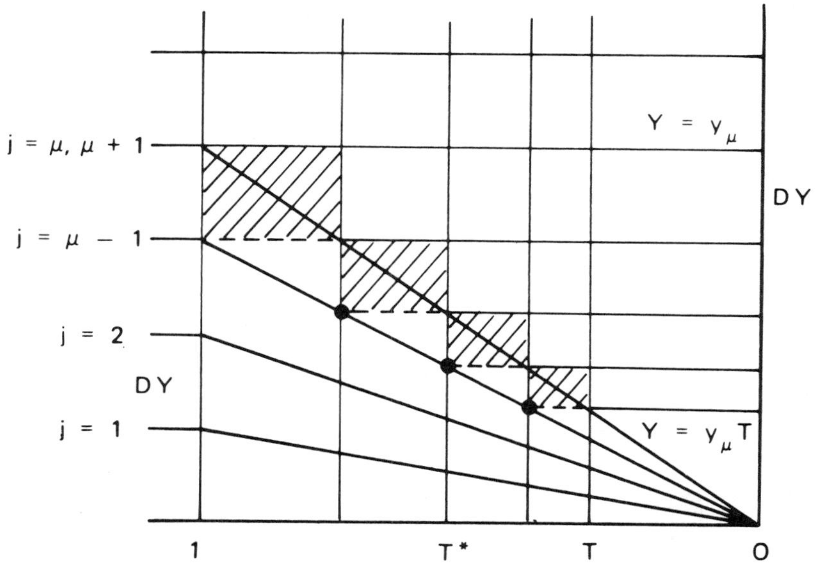

Figure 3. Varying steps in T and Y.

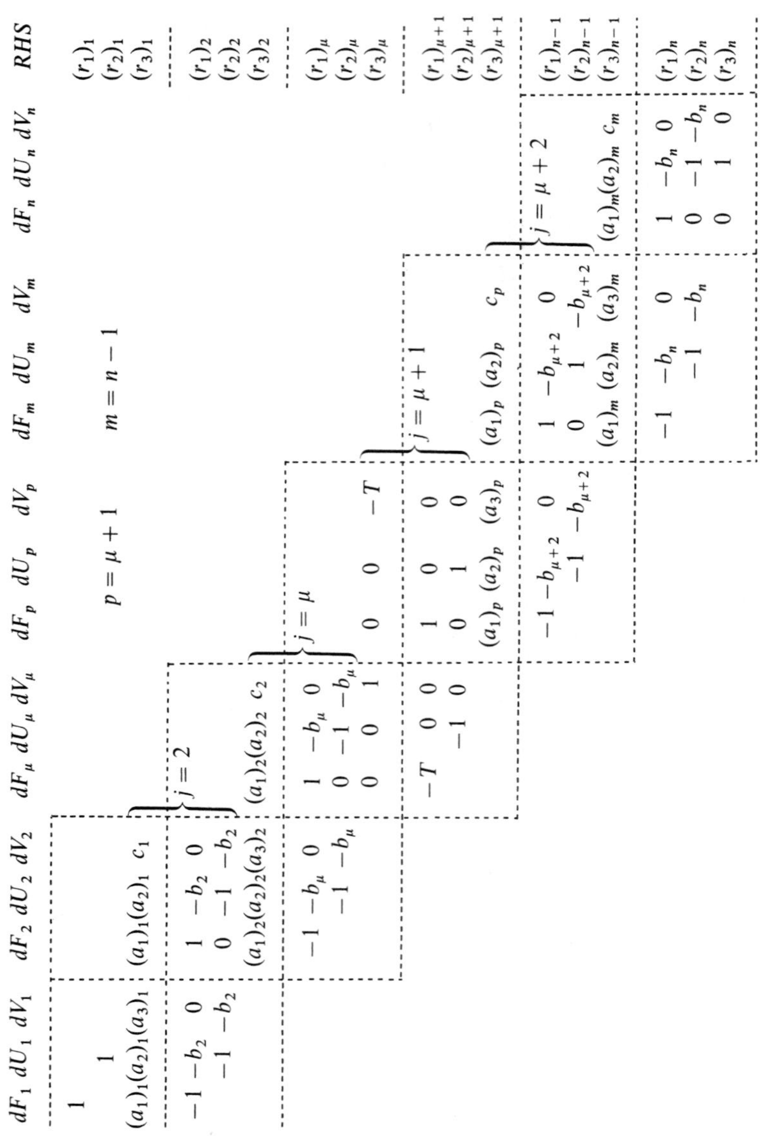

Figure 4. Structure of Newton equations illustrated for $\mu = 3$, $n = 6$.

To compute the new profiles at T we solve (13) by Newton iteration with the equations (17) inserted and the C_k in (15) computed according as $j \leq \mu$ or $j \geq \mu + 2$. For $j = \mu + 1$ the coefficients are obtained from (17). The Newton equations have coefficient matrix and right-hand sides of the form in Fig. 4, where

$$\begin{aligned} (a_1)_{j-1} &= (C_2 - C_3)V_M, \qquad & b_j &= \tfrac{1}{2}(DY)_{j-1}, \\ (a_2)_{j-1} &= -C_1 - C_2(U_X + U_M) + 2C_3 U_M, & & \\ (a_3)_{j-1} &= C_4 - C_5, & c_j &= -C_4 - C_5. \end{aligned} \tag{20}$$

The block elimination is carried out with a UL decomposition subroutine and a solve subroutine.

5 Results for Flow at a Rear Stagnation Point

The profiles at a rear stagnation point can be computed by the scheme described if the stations A and B are taken at the stagnation point, the profiles at C and D are set to zero, u_e is set to zero, du_e/dx is set to -1, and

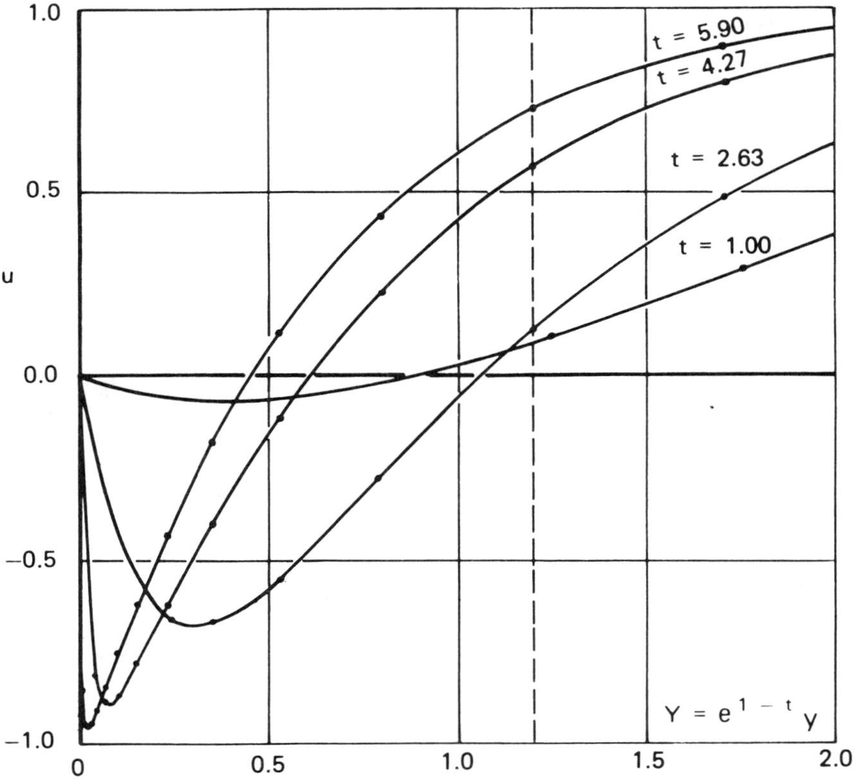

Figure 5. Velocity profiles plotted against $e^{1-t}y$.

appropriate modifications made to the coefficients C_k. The small-time variable $y/t^{1/2}$ was used right up to $t = 1$ with a time step of 0.05 and $DY = 0.05$. The initial value of n was 161, and this was allowed to increase if the expansion of the boundary layer warranted it. At $t = 1$, n had reached 231. For $t > 1$ the double structure was used with $\mu = 26$, which gave a subsequent time step of 0.0408. As points were inserted as described in Section 4, n increased gradually except where points were removed from the outer edge if u attained its limiting value of 1 to within $\frac{1}{2}10^{-6}$ too soon. In fact n ultimately dropped to 221 at $t = 6$. A tolerance of $\frac{1}{2}10^{-6}$ on the maximum change in u was used for the Newton iterations. Quadratic convergence was observed in three iterations for all values of t considered. In fact the convergence criterion would have been satisfied in two iterations if the tolerance had been set very slightly higher.

Results are compared with those of Robins and Howarth in Figs. 6 and 7. Figure 5 shows some typical u-profiles plotted against $e^{1-t}y$, which emphasizes the development of the double structure. To show the way the grid varies relative to the profiles, points have been marked which have ten grid intervals separating them, the point on each profile nearest to the wall corresponding to the outermost point of the inner region. The broken line

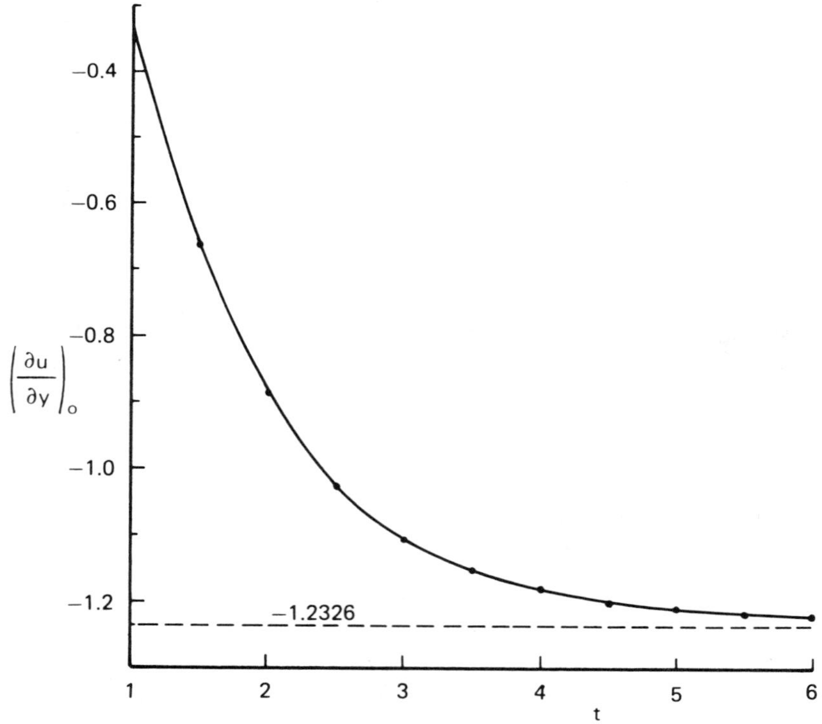

Figure 6. Skin friction tending to its limiting value.

indicates the value of Y beyond which the outer step is uniform. Since $\mu = 26$, there are twenty-five intervals between the wall and the first marked point.

In Fig. 6 sample values of the results obtained by the double-structure scheme are compare with the curve obtained by Robins and Howarth. As can be seen, the agreement is excellent; in fact the agreement was considerably better than the graph suggests.

In Fig. 7 a similar comparison is made for the displacement thickness. As can be seen, a discepancy begins to appear at about $t = 4$, and thereafter the results begin to diverge. Current tests that we have made have failed to find any significant error, but perhaps it should be pointed out that the stream function, on which the displacement thickness depends, does exhibit unstable behavior in the outer region. Clearly further tests should be made to check the sensitivity of the outer solution.

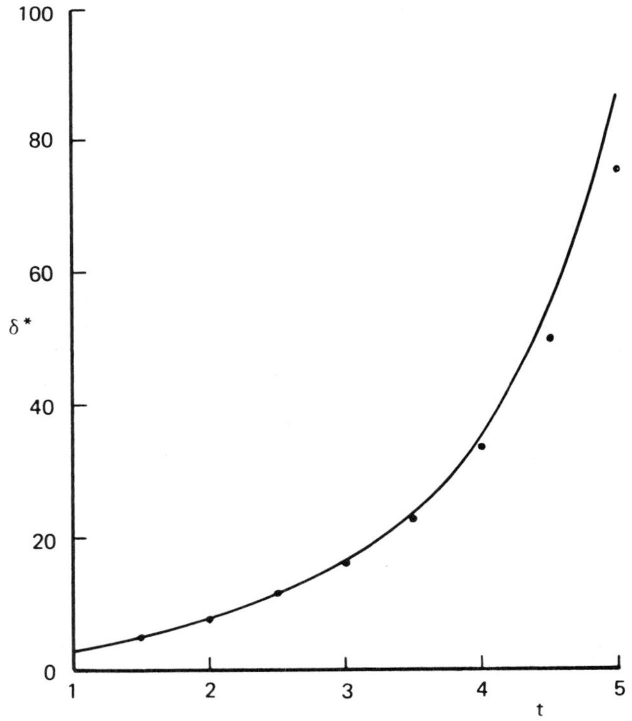

Figure 7. Displacement thickness.

CHAPTER 20

Form Factors near Separation

Claude Barbi*† and Demetri Telionis†

1 Introduction

The past decade has seen a large number of publications in the area of unsteady separation, as reflected by recent review articles (Sears and Telionis [3.35], Williams [3.62], Shen [3.10], Telionis [3.5]). In most of the contributions emphasis is concentrated on the behavior of the boundary-layer equations. Although the mathematical characteristics of these equations are quite important, no conclusion about the physics of the problem can be drawn, since the approximation hypothesis breaks down a little upstream of separation.

A familiar characteristic of the solution in the neighborhood of separation is the appearance of a singularity, which has been suggested as a criterion for the phenomenon [3.34]. Controversial results have been published since then. Unfortunately, no rigorous proof has been presented establishing the existence or nonexistence of the singularity. The work of Williams and Johnson [3.7, 3.63], Dwyer and Sherman (Chapter 18), and Van Dommelen and Shen [3.36] essentially supports the existence of a singularity: Cebeci [3.12, 3.22] presents calculations of separating flows free of singularities. It is interesting that the results of Wang [3.33] indicate singular behavior near separation; however, he claims that the opposite is true.

* On leave from Institut de Mécanique des Fluides de Marseille.

† Virginia Polytechnic Institute and State University, Blacksburg, VA 24061.

A few examples have been worked out by the present authors, and more results in favor of the existence of a singularity are presented here. However, the main thrust of this paper is the search for a more practical criterion.

In many applications engineers have been measuring or calculating integral quantities. Separation criteria have been proposed that are based on specific values of ratios of such quantities. The most popular ratio is the form factor, which is the ratio of the displacement thickness to the momentum thickness. The form factor takes approximately the value of 2.6 as laminar separation is approached. In a true boundary-layer calculation, one would not expect to find a fixed value of the form factor at separation. However, numerical results presented here indicate that characteristic values of the form factor are attained in the immediate neighborhood of separation. This behavior may serve as a signal for separation.

In this paper two sets of numerical results are presented. The first pertains to the moving-wall case, which has been recently looked at via an integral method [3.64]. The second set of results represents a truly unsteady situation, whereby a recirculating bubble is generated in a boundary layer which was attached at earlier times. In all cases two form factors are calculated—the ratio of the displacement to the momentum thickness, H, and the ratio of the energy to the momentum thickness, K:

$$H = \frac{\delta_1}{\delta_2}, \qquad K = \frac{\delta_3}{\delta_2}, \tag{1}$$

where δ_1, δ_2, and δ_3 are the displacement, momentum, and energy thicknesses, respectively.

The behavior of these quantities is studied in the neighborhood of separation and results are presented as functions of downstream distance, or, following Fansler and Danberg [3.64], on a phase plane of H and K.

2 Steady Flow—Moving Walls

The relationship between separation of steady flow over moving walls and unsteady flow over fixed walls was described in the fifties by Sears [3.3], Moore [3.1], and Rott [3.2]. Only recently, however, has this connection been established rigorously; Williams and Johnson [3.7, 3.63] introduced a transformation that maps a certain class of unsteady separating flows onto steady flows separating over moving walls.

Moore [3.1] attempted to establish the singular character of the equations at the MRS station by investigating the self-similar solutions over a moving skin. Steady boundary-layer equations for the flow over a wedge of included angle $\beta\pi$ reduce to an ordinary differential equation

$$f''' + ff'' + \beta(1 - f'^2) = 0 \tag{2}$$

with η the similarity variable and boundary conditions

$$f(0) = f'(0) = 0, \qquad f(\infty) \to 1. \tag{3}$$

The function f is proportional to the stream function, and therefore f' and f'' are proportional to the velocity and the shear respectively.

Similar calculations were carried out later by Telionis and Werle [3.65, 3.66] and Williams and Johnson [3.7, 3.63]. In Fig. 1 the skin-friction function, $f''(0)$, is plotted versus the pressure gradient parameter β. For a fixed wall, it can be seen that the vanishing of $f''(0)$ (that is, the zero-skin-friction profile) is approached with an infinite slope at $\beta = \beta_0 \equiv 1.98$. In fact it has been established that this is a square-root singularity with respect to the variable $\beta - \beta_0$. It was anticipated that for moving walls, the singular point would coincide with the MRS point. However, this is not the case, as pointed out in [3.1, 3.66]. In Fig. 1, the loci of the singular point and the MRS point are marked for comparison.

Nonsimilar calculation of the boundary-layer equations were carried out by Telionis and Werle [3.6] and Williams and Johnson [3.7] for downstream-moving walls, and by Tsahalis [3.9] for upstream-moving walls. For the first case, it was established that indeed the MRS station is approached with a square-root singularity, while this is approximately true for the second case.

Nonsimilar calculations were performed also by Fansler and Danberg [3.64] via an integral method. The self-similar profiles of Eq. (2) were employed in [3.64] to facilitate integration of integral equations. Fansler and Danberg proceeded to calculate the form factors H and K, which are plotted

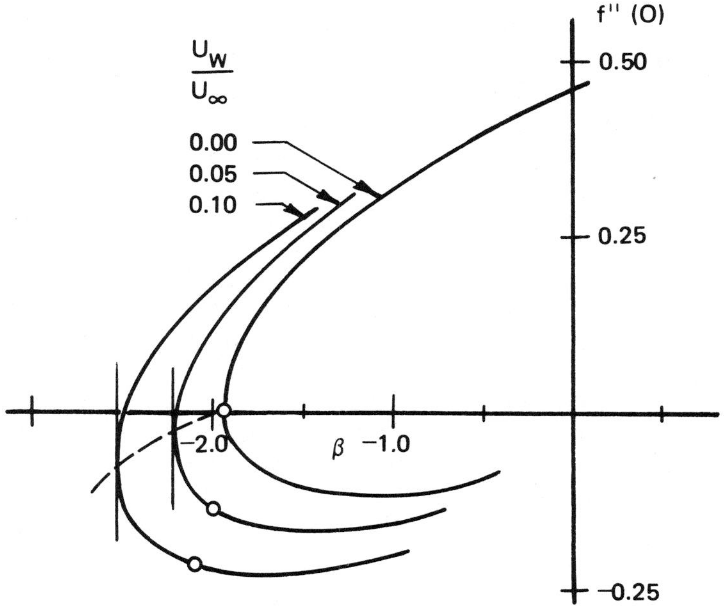

Figure 1. The skin-friction function $f''(0)$ plotted against the pressure-gradient parameter for different skin speeds (from [3.65]). ---, locus of β-θ singularity; ○, MRS profile.

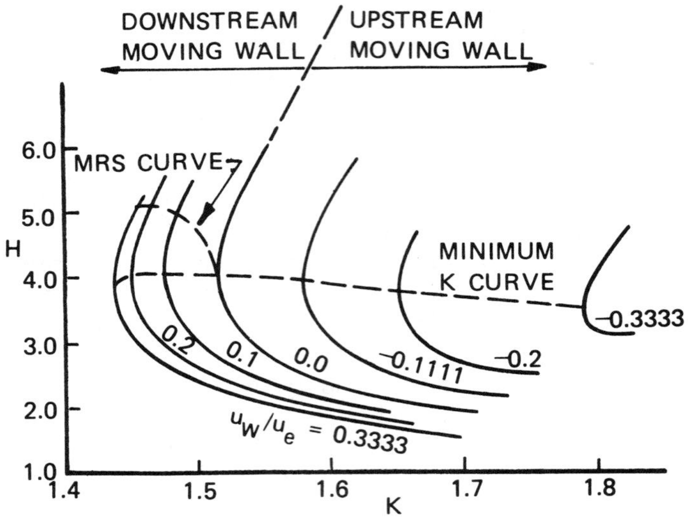

Figure 2. The form-factor phase plane showing H versus K for different skin speeds (from [3.65]).

in Fig. 2. As separation is approached, H increases and K decreases, for all values of the skin velocity. What seems to be intriguing is the fact that all curves turn sharply up and eventually change their slope from negative to positive at a point of infinite slope. It has been pointed out [3.64] that in fact the singularity in H is of the square-root type, a fact that once more stimulates the investigator to attempt to link it with the separation singularity. This is not the case, as pointed out in [3.64]. However, this should not be surprising, in view of the fact that the self-similar profiles parametrized with β are used, and in the self-similar domain a similar discrepancy has been discovered earlier [3.66], as shown in Fig. 1. Namely, the point of the singularity in the plot of Fig. 1 does not coincide with the MRS point.

Fansler and Danberg [3.64] have also calculated the MRS station, which is shown in Fig. 2. The departure of the MRS point from the point of singularity in Figs. 1 and 2 is qualitatively very similar. Note that marching towards separation, one encounters first the singularities of $f''(0)$ and H, and then the MRS point. Fansler and Danberg [3.64] go one step further to propose as a criterion of separation the singular point in the H-K curve. As evidence in favor of this proposal they derive the separation point displacement from calculations of the flow over a rotating cylinder and compare it with the experimental data of Vidal [3.67] and Brady and Ludwig [3.68] as shown in Fig. 3. One of the main goals of the present paper is to demonstrate that actually the difference between the points of the singularity and the MRS point is very small, if measured in terms of actual physical distance. The MRS criterion therefore should be an adequate criterion for steady flows over moving walls. Moreover, the present work provides

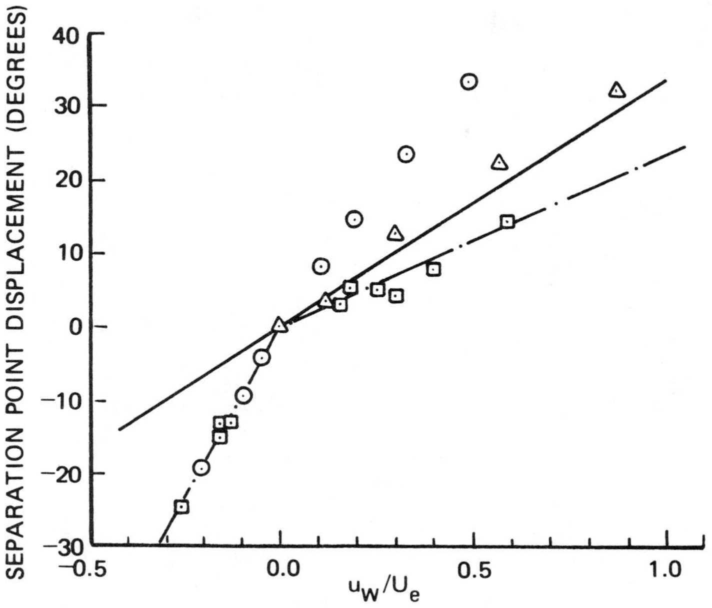

Figure 3. The displacement of separation over a rotating cylinder (Δ, Brady and Ludwig [3.68]; □, Vidal [3.67]; –·–·–, ———, Fansler and Danber [3.64]; ○, present results.

numerical results of the complete equation, to be contrasted with the approximate calculations of [3.64].

The boundary-layer equations were integrated by a finite-difference method described adequately in [2.25, 3.65]. The system of equations was recast in Görtler variables:

$$2\xi \frac{\partial F}{\partial \xi} + F + \frac{\partial V}{\partial \eta} = 0, \tag{4}$$

$$2\xi F \frac{\partial F}{\partial \xi} + V \frac{\partial F}{\partial y} + \beta(F^2 - 1) = \frac{\partial^2 F}{\partial \eta^2}, \tag{5}$$

where F and V are reduced velocity components, ξ and η are stretched coordinates, and β is the pressure-gradient parameter $\beta = (2\xi/U_e)\, dU_e/d\xi$. Numerical integration was marched downstream by an implicit method.

For the boundary-layer developing over a rotating cylinder with radius R, the boundary conditions are

$$F = u_w, \quad V = 0 \qquad \text{at } \eta = 0, \tag{6}$$

$$F \to U_e = 2 \sin \frac{x}{R} \quad \text{as } \eta \to \infty, \tag{7}$$

where x is the distance from the stagnation point, measured along the skin.

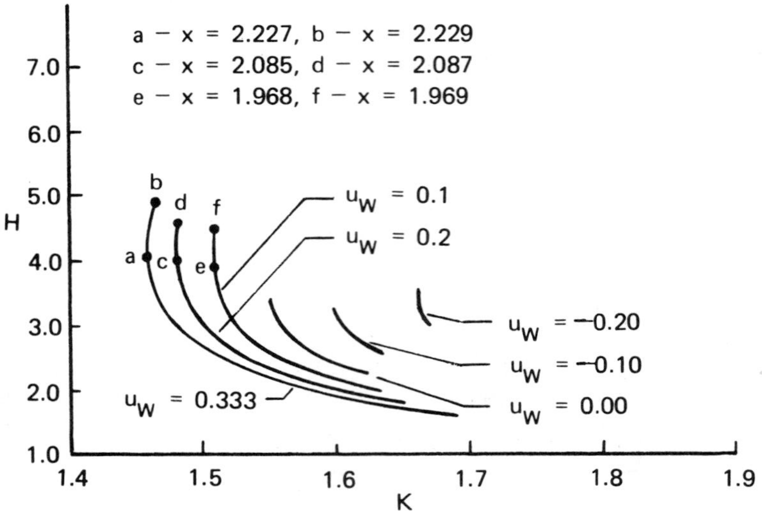

Figure 4. The *H*-*K* curves calculated by the present method.

The results of the present numerical calculations are in good agreement with those presented in Fig. 2, at least for points away from separation. Considerable deviations are encountered for values of H larger than approximately 2.5. Plotting the present results on top of the curves of Fig. 2 would have made it very difficult to differentiate between the two sets of curves. For this reason the *H*-*K* plots calculated by the present method are shown

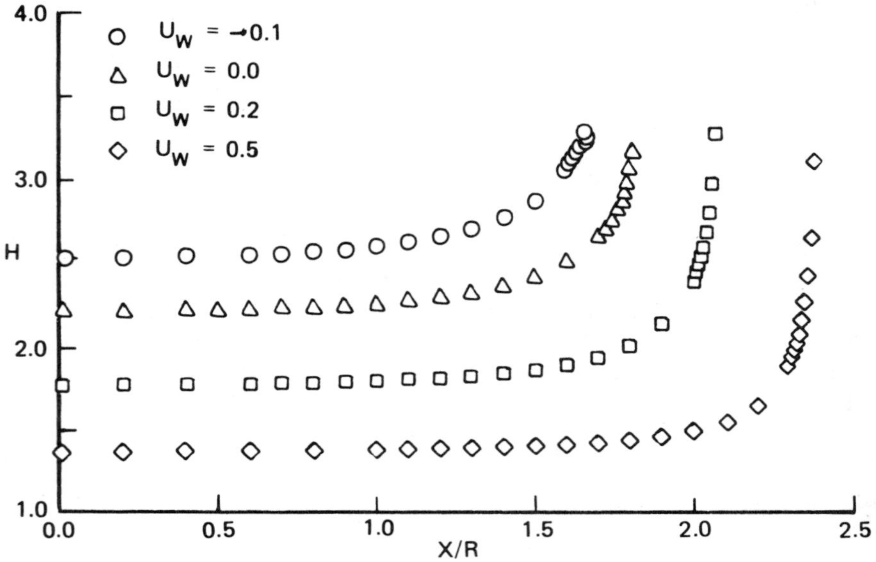

Figure 5. The behavior of the form factor H in the physical domain for different skin speeds.

separately, in Fig. 4. The curves seem indeed to pass the infinite-slope point in the H-K plot without any evidence of singular behavior in the physical domain. A very short distance downstream, the familar singularity gradually appears, and the calculations do not converge at the station of the MRS profile. The upper end of the curves corresponds to the MRS station, and the locus of these points would appear smoother if smaller integration steps were chosen.

The singular character of the solution is most clearly demonstrated if the form factor is plotted against the physical distance x/R. This is shown in Fig. 5 for a few wall-velocity ratios. A comparison of Fig. 5 and Fig. 4 is enough to convince one that the minima of K are passed without any anomalies in the H-x plot and with no influence on the impending singularity which follows. The displacement of separation as calculated by the present method is included in Fig. 3 for the flow over a rotating circular cylinder.

3 Unsteady Flow—Growing Recirculating Bubble

To test the form-factor criteria in unsteady flow, a problem was selected which has been suggested as a good test case. Starting with a uniform outer-flow velocity distribution, a decrease followed by an increase of the velocity is imposed gradually. Such outer flows were proposed by Nash and Patel [3.69], Cebeci [3.22], and Dwyer and Sherman (Chapter 18). In terms of dimensionless quantities, Nash and Patel [3.69] considered

$$U_e = U_0(1 - cxt) \qquad \text{for } t > 0, 0 < x < x_0, \tag{8}$$

$$U_e = U_0[(1 - ct(x_1 - x)] \qquad \text{for } t > 0, x_0 < x < x_1. \tag{9}$$

This distribution has a kink at $x = x_0 = 0.714$.

To avoid the discontinuity in the pressure gradient, Cebeci [3.12] proposed a similar but smoother velocity profile. Dwyer and Sherman (Chapter 18) chose a similar nonlinear distribution. The distribution given by Eqs. (8) and (9) has been suggested by the panel of the AGARD symposium on unsteady aerodynamics held in Ottawa in 1977, and has been adopted in the present paper.

All such distributions have one characteristic in common. They lead the boundary-layer flow into a region of partially reversed flow which grows in time. However, the flow downstream of the bubble is attached. In this way it is guaranteed that the Courant–Lewy–Friedrichs criterion is met for all times. Calculations which terminate with an open recirculating bubble can also be implemented [3.70], provided the domain of integration at each time instant is reduced to exclude regions that have received the upstream-propagating message of the end of the calculations.

The unsteady version of Eqs. (4) and (5) was integrated first in the $t = 0$ plane, marching in the direction of increasing downstream physical coordinate x. Time was incremented, and the marching process in the positive x-direction was repeated in the new time plane. A zigzag scheme [3.70] was used throughout to facilitate integration through regions of partially reversed flow.

A plot of the form factor H versus the axial distance x/c is shown in Fig. 6. A considerable increase is observed as the point $x/c = 0.650$ is approached. Time $t = 6.75$ is the last instant that the calculation can converge along the entire surface. Beyond this point, the calculations terminate at $x/c = 0.644$. Cutting the steps of integration indicates that larger values of H can be found in the neighborhood of $x/c = 0.644$, but it is not possible to cross this station and move further downstream. These are the typical features of the separation singularity which have been studied extensively in the past. The case at hand has the form of a separation singularity which developed in time, in a domain which was free of singularities at earlier times.

It should be mentioned here that in response to the jump of the pressure gradient at $x/c = 0.714$, the calculated skin friction shows at the same point a very sharp variation. Unfortunately, velocity profiles have not been plotted and it is not possible to estimate whether the upstream propagation of the message of the apparent discontinuity influences the region where the separation singularity appears. This may be the case in the calculations of Nash and Patel, who found for turbulent flow a singularity at the point $x/c = 0.66$.

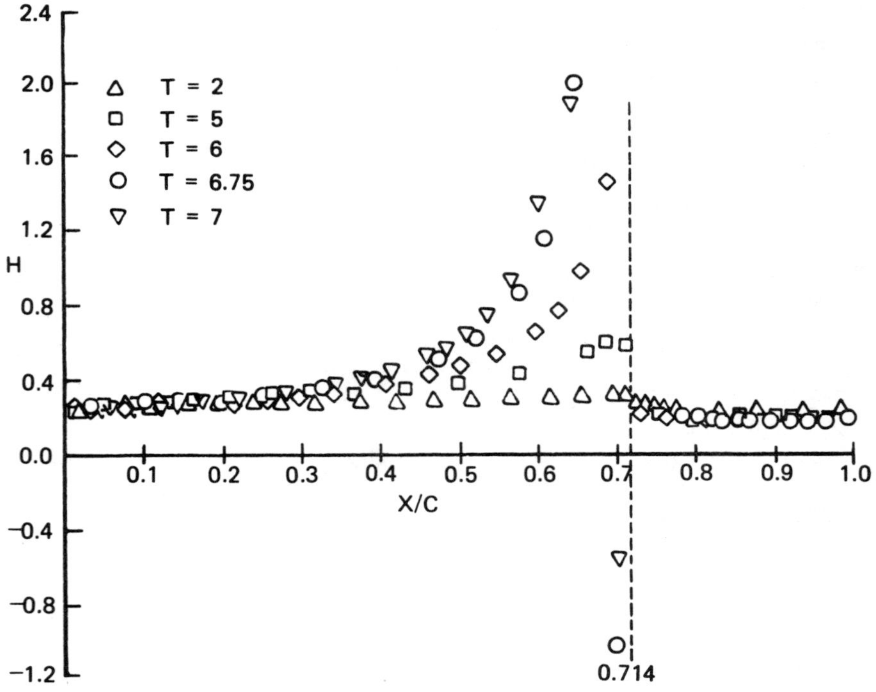

Figure 6. The form factor H along the axial physical distance x/c for the flow of Eqs. (8), (9) and different values of time.

4 Conclusions

One of the main objectives when calculating boundary layers is the estimation of the location of separation. Whether this is altogether possible for unsteady flows is still an open question. However, most investigators believe that the uninteracted boundary layer contains enough information to signal at least the neighborhood of separation. Various criteria have been proposed, but no experimental evidence is yet available to compare their relative performance. Of all the suggested criteria only a few are convenient to employ in practice. In the present paper we have discussed the behavior of form factors in the neighborhood of separation.

As separation is approached, the displacement thickness δ, and therefore the form factor H, grow sharply. The boundary-layer equations, although invalid in the vicinity of separation, simulate this behavior nicely. For the case of steady flow, comparison with experimental data indicates that analytical models predict with reasonable accuracy the location of separation. More specifically, it is demonstrated that for H-values between 2.6 and 3.0 the location of the separation is practically at hand. The present evidence indicates that beyond the singular point in the H-K curve, large changes of H correspond to extremely small changes of the axial distance x. The discrepancies between the appearance of the H-K singularity and that of the separation singularities are for all practical purposes unimportant.

An example of laminar unsteady flow considered here is in qualitative agreement with earlier work of Nash and Patel, who studied a turbulent case. The form factor grows sharply in a region well downstream of the point of zero skin friction. This example however may be subject to criticism. It may be argued that the effect of a discontinuity in the slope of the outer velocity distribution may be propagated upstream to influence the development of the singularity. Defending the present point of view, one may argue that if numerical instabilities propagate upstream, they should continue extending further and further upstream as time increases. On the contrary, the sharp changes of all singular properties seem to center at a fixed point upstream of the pressure-gradient kink and downstream of the point of zero skin friction.

CHAPTER 21

Unsteady Development of the Boundary Layer in the Vicinity of a Rear Stagnation Point

James C. Williams, III*

Introduction

The early studies of Rott [3.2, 3.71], Moore [3.1], and Sears [3.3] led to the development of a model for unsteady boundary-layer separation which has become known as the Moore–Rott–Sears model. In this model, unsteady separation is characterized by the vanishing of both the shear and the velocity at an interior point of the boundary layer, as seen in a coordinate system moving with separation. Moore [3.1] and Sears and Telionis [3.35] also argue that in the solution of the boundary-layer equations, a singularity occurs at the unsteady separation point.

In recent years the results of a number of analytical investigations have verified the Moore–Rott–Sears model, both with respect to the physical characteristics associated with separation and to the existence of a separation singularity. These investigations include the works of Telionis, Tsahalis, and Werle [3.70], Williams and Johnson [3.7, 3.72], and Telionis and Tsahalis [3.31]. In the last of these Telionis and Tsahalis investigated the boundary-layer development on a circular cylinder which is impulsively set into motion, at time zero, at a uniform speed. They found that at a non-dimensional time (tU/a) of 0.65 and "in the neighborhood of $\theta = 140°$, a singularity appears, that travels upstream following a point of zero wall shear." Two more recent investigations of the flow development past a

* Auburn University, Auburn, Alabama 36849.

circular cylinder, those of Wang [3.33] and Cebeci [3.12], find no singularity; they suggest that the flow is smooth for all finite time. The lack of a singularity in the solution of Proudman and Johnson [3.38] for the flow development in the vicinity of a rear stagnation point, impulsively set into motion, is also taken as evidence of the lack of a singularity associated with separation.

These conflicting results raise a question regarding the singularity postulated at separation by Moore and by Sears and Telionis. In the present work, the indications of a singularity associated with unsteady separation are reviewed within the framework of semisimilar solutions to the unsteady boundary-layer equations. It is shown that when the boundary-layer equations are formulated in semisimilar form there is an indication that a singularity exists in the solutions to these equations. Furthermore, solutions to these equations indicate that, in cases where the singularity occurs and in which separation is expected, it occurs at or near a point which has all the features of separation postulated in the Moore–Rott–Sears model.

In addition, new solutions are presented for the initial development, according to the boundary-layer equations, of the viscous flow in the vicinity of a rear stagnation point on a body with a sharp trailing edge. These solutions yield additional insight into the development of such flows and the existence of a singularity in such flows.

Analysis

The objective of the present investigation is to determine the physical and mathematical characteristics of boundary-layer separation in unsteady flow based on solutions of the unsteady boundary-layer equations. Let x and y be the coordinates along and normal to the body surface respectively, and u and v be the corresponding velocity components. Also, let t represent time, and ν represent the kinematic viscosity of the fluid. The equations which describe the two-dimensional motion of an incompressible fluid within the boundary layer on a body undergoing unsteady motions are

$$\frac{\partial u}{\partial x} + \frac{\partial v}{\partial y} = 0, \tag{1}$$

$$\frac{\partial u}{\partial t} + u\frac{\partial u}{\partial x} + v\frac{\partial u}{\partial y} = \frac{\partial u_\delta}{\partial t} + u_\delta\frac{\partial u_\delta}{\partial x} + \nu\frac{\partial^2 u}{\partial y^2}. \tag{2}$$

The boundary conditions applicable to the solutions of these equations are the usual no-slip conditions at the wall and the condition that the x-component of velocity, $u(x, y, t)$, matches the known inviscid flow solution, $u_\delta(x, t)$, as the distance from the wall becomes large.

A major problem encountered in the solution of Eqs. (1) and (2) is the existence of three independent variables (x, y, t) in the problem. The technique of semisimilar solutions seeks to eliminate this difficulty by reducing the number of independent variables from three (x, y, t) to two (η, ξ) by appropriate scalings. In the spirit of this transformation, we introduce two new scaled coordinates, η and ξ, defined by

$$\eta = \frac{y\sqrt{U/\nu l}}{g^*(x^*, t^*)}, \qquad \xi = \xi(x^*, t^*).$$

Here $x^* = x/l$ and $t^* = tU/l$ are normalized x and time coordinates, U is a characteristic velocity, and l is a characteristic length for the problem. The function $g^*(x^*, t^*)$ may be thought of as a scaling function for the normal, or y, coordinate, and $\xi(x, t)$ may be thought of as a new x-coordinate which has been scaled with time. In addition, a new nondimensional stream function $f(\xi, \eta)$ is introduced. This nondimensional stream function is related to the usual stream function $\psi(x, y, t)$ by

$$\psi(x, y, t) = \sqrt{\nu U l}\, g^*(x^*, t^*) u^*(x^*, t^*) f(\xi, \eta)$$

in which $u_\delta^* = u_\delta/U$. The continuity equation is satisfied identically by the introduction of the stream function, and the momentum equation becomes, in terms of the dimensionless stream function and new coordinates,

$$\frac{\partial^3 f}{\partial \eta^3} + (d + e) f \frac{\partial^2 f}{\partial \eta^2} + d\left[1 - \left(\frac{\partial f}{\partial \eta}\right)^2\right] + a\left[1 - \frac{\partial f}{\partial \eta}\right] + \frac{b}{2}\eta \frac{\partial^2 f}{\partial \eta^2} - c\frac{\partial^2 f}{\partial \eta\, \partial \xi} + h\left[\frac{\partial f}{\partial \xi}\frac{\partial^2 f}{\partial \eta^2} - \frac{\partial f}{\partial \eta}\frac{\partial^2 f}{\partial \eta\, \partial \xi}\right] = 0. \quad (3)$$

The coefficients a, b, c, d, e, and h are defined by

$$a = g^{*2}\frac{1}{u_\delta^*}\frac{\partial u_\delta^*}{\partial t^*}, \qquad b = \frac{\partial g^{*2}}{\partial t^*}, \qquad c = g^{*2}\frac{\partial \xi}{\partial t^*},$$

$$d = g^{*2}\frac{\partial u_\delta^*}{\partial x^*}, \qquad e = \tfrac{1}{2}u_\delta^*\frac{\partial g^{*2}}{\partial x^*}, \qquad h = u_\delta^* g^{*2}\frac{\partial \xi}{\partial x^*},$$

and if semisimilar solutions are to exist, these coefficients must be functions of ξ alone. Additional details of the method of semisimilar solutions may be found in Williams and Johnson [3.72]. At this point it is convenient to write Eq. (3) as a system of two equations:

$$\frac{\partial^2 W}{\partial \eta^2} + \alpha_1\frac{\partial W}{\partial \eta} + \alpha_2 W + \alpha_3 = \alpha_4\frac{\partial W}{\partial \xi}, \quad (4)$$

$$\frac{\partial f}{\partial \eta} = W, \quad (5)$$

where

$$\alpha_1 = (d+e)f + h\frac{\partial f}{\partial \xi} + \frac{b}{2}\eta, \quad \alpha_2 = -dW - a, \quad \alpha_3 = d + a, \quad \alpha_4 = c + hW.$$

This form, in which W is treated as one of the dependent variables, emphasizes the parabolic nature of Eq. (3). Furthermore, this is the form in which Eq. (3) is generally formulated for numerical solution. Stewartson [3.73] has pointed out that in an equation which has the form of Eq. (4), a singularity may occur when the coefficient of a leading term (i.e., α_4) of the differential equation vanishes.

Examples of Flows with a Singularity

Several cases of familiar steady and unsteady flows in which a singularity occurs will now be reviewed to provide some indication of the physical significance of singularities which occur and also to provide guidance in evaluating new solutions to the unsteady boundary-layer equations. In each of the cases studied below, the solution has been obtained using the same basic computer program in which equations (4) and (5) are solved using an implicit finite-difference scheme. In each case, a similar solution exists for $\xi = 0$. Solutions for subsequent values of ξ are obtained by marching the solution in the direction of increasing ξ. In each case, the approach of the singularity (of the vanishing α_4) is heralded by an increase in the number of iterations required to obtain convergence of the local velocity profile until, at one station, convergence cannot be obtained in some reasonable number of iterations. The flows reviewed, the semisimilar scalings employed, and the corresponding expressions for α_4 are presented in Table 1.

The first flow reviewed is the classical case of steady Hawarth's linearly retarded flow. Since this flow is well understood, it provides a basis for interpreting subsequent flows. Figure 1 shows the variation of the coefficient α_4 with ξ. Here α_4 is calculated at $\eta = 0.08$ (this is the value of η corresponding to the station closest to, but not on, the wall). The coefficient α_4 increases from zero at $\xi = 0$, reaches a maximum value, and then decreases. As ξ approaches 0.1198, α_4 approaches zero. This point corresponds to the classical point of separation for this well-known flow. The variation of velocity profiles with ξ (or x^* in this case) is well known and is not reproduced here. It is only necessary to point out that the condition $\alpha_4 = 0$ corresponds very closely to the point of vanishing shear, which is classically accepted as the condition of separation for steady flow. The variation of the vertical component of velocity, v^*, with ξ at fixed η is also well known and is not reproduced here. It is only necessary to remember that the reciprocal of the square of v^*, at fixed η, approaches zero linearly as separation is approached, indicating the well-known Goldstein singularity at separation.

As a second case to be reviewed, we consider the case of Falkner–Skan flows impulsively set in motion, which includes the special case of the impulsively set into motion flat plate, studied extensively by Stewartson [3.73] and others. This case is of interest because in it a singularity appears, but the singularity is not related to separation.

Table 1 Flows Reviewed

	(x^*, t^*)	$g^{*2}(x^*, t^*)$	u_δ^*	α_4
Howarth's linearly retarded flow	x^*	x^*/u_δ^*	$1 - x^*$	x^*W
Impulsively set into motion Falkner–Skan Flows [3.74] $(m = 0.5)$	$1 - \exp\left(\frac{-u_\delta^* t^*}{x^*}\right)$	$x^*\xi/u_\delta^*$	Cx^{*m}	$\xi(1 - \xi)\{1 + (1 - m)\ln(1 - \xi)\}$
Unsteady variation of Howarth's linearly retarded flow [3.63] $(\lambda = 1)$	$x^*/(1 - \lambda t^*)$	x^*/u_δ^*	$1 - \xi$	$\lambda\xi^2/(1 - \xi) + \xi W$
Unsteady variation of the Falkner–Skan flows [3.72] $(m = -0.05)$	$x^* \exp(1 + \lambda t^*)$	x^*/u_δ^*	$x^{*m} \exp(m - 1)(1 + \lambda t^*)$	$\xi(\lambda\xi^{1-m} + W)$

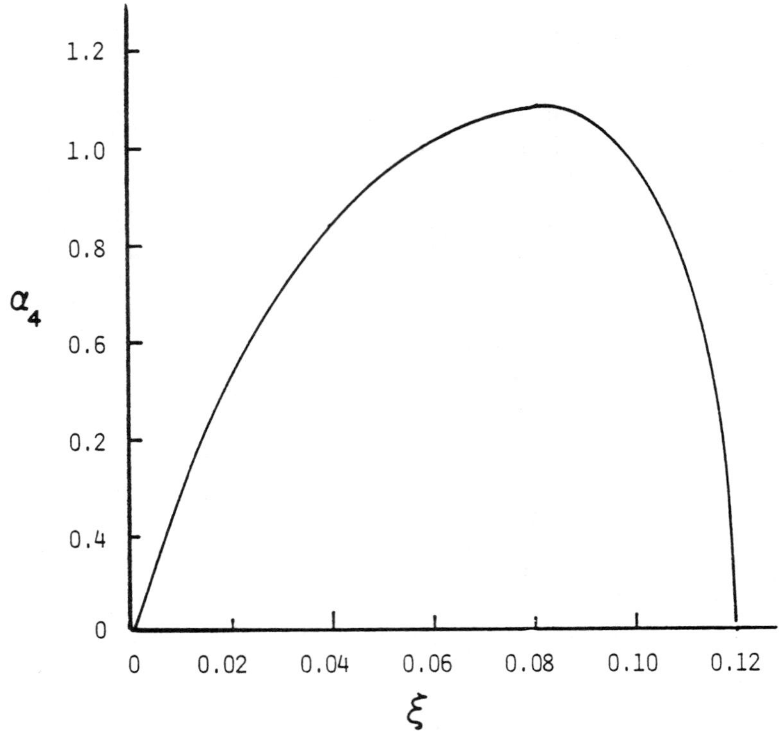

Figure 1. Variation of α_4 at $\eta = 0.08$ with ξ for steady Howarth's linearly retarded flow.

For the general Falkner–Skan flows we employ the scalings developed by Williams and Rhyne [3.74]. The coefficients in the semisimilar transformation are easily calculated. In this case the coefficient c is positive, while the coefficient h is negative, so that α_4 may vanish with values of W greater than zero. In fact, the minimum value of α_4 for this case always occurs at the upper edge of the boundary layer where $W = 1$.

Figure 2 shows the variation of the minimum value of α_4 with ξ for the case of a wedge impulsively set in motion, with $m = 0.5$. The minimum value of α_4 increases from zero at $\xi = 0$ to a maximum value and then decreases. The coefficient α_4 vanishes at $\xi = 0.865$. Figure 3 shows the velocity profiles for this flow for $\xi = 0$, 0.4, and 0.86. Clearly the velocity profiles for this flow show neither the tendency toward zero shear as α_4 approaches zero that one would expect for steady separation, nor the reverse flow that one might expect for unsteady separation. In fact, for this flow, as ξ increases, the boundary layer becomes thinner and the wall shear increases. The variation of $1/v^{*2}$ (evaluated at the upper edge of the boundary layer) with ξ is shown in Figure 4. We note that in this case, $1/v^{*2}$ does not approach zero as α_4 approaches zero. Although α_4 approaches zero, indicating

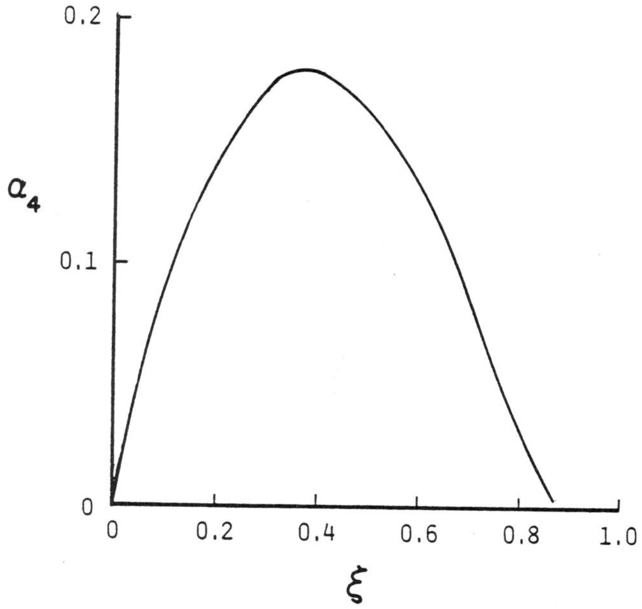

Figure 2. Variation of the minimum value of α_4 for an impulsively set into motion wedge.

a singularity in the flow, this singularity does not indicate separation. Instead, the singularity in the case is indicative of the fact that during the early stages of the flow development the local flow behaves as if it were unaware of the existence of the leading edge. The singularity occurs at the time when the signal from the leading edge first reaches the particular x-station in question [3.74]. In this case, then, a singularity in the flow and the phenomenon of

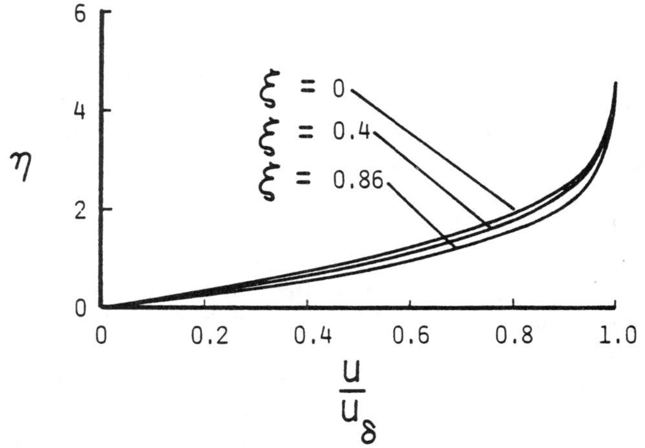

Figure 3. Velocity profiles for an impulsively set into motion wedge.

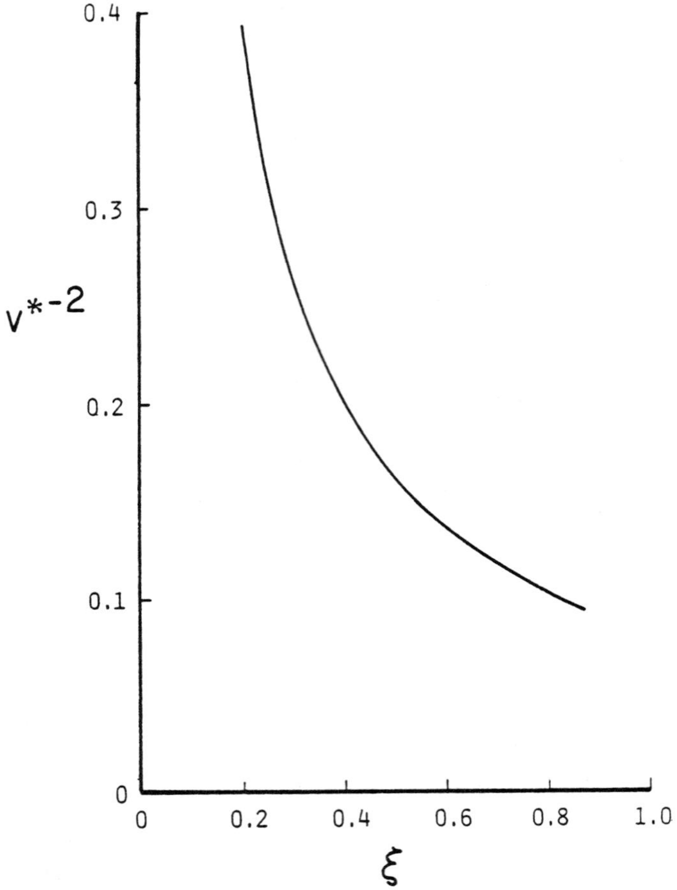

Figure 4. Variation of $1/v^{*2}$ with ξ for an impulsively set into motion wedge.

separation are not linked. The difference between this flow and the previous one is most pronounced in the variation of the vertical component of velocity.

Next we reconsider two unsteady-flow solutions in which α_4 approaches zero and hence a singularity occurs, and for which the flow characteristics in the vicinity of the singularity are exactly those postulated in the Moore–Rott–Sears model. The first of these flows is the unsteady variation of Howarth's linearly retarded flow studied by Williams and Johnson [3.7]; the second is the unsteady variation of the Falkner–Skan flows, also studied by Williams and Johnson [3.72]. The results in the two cases are very similar, so that only the detailed results for the latter case will be presented here.

The scales employed in each case are again presented in Table 1. We note that in both cases, with λ positive, the only way α_4 can become zero is for W to become locally negative, i.e., for there to be reverse flow. Integration into regions of reverse flow is possible as long as α_4 is positive. In this respect, we note the analogy between α_4 in Eq. (4) and the coefficient of thermal conductivity in the heat-conduction equation. If the coefficient α_4 is positive, information is transmitted in the direction of increasing ξ; if α_4 is negative, information is transmitted in the direction of decreasing ξ. Thus, integration into regions where α_4 is negative cannot proceed unless some additional information is provided in the form of a downstream boundary condition.

For both flows α_4, evaluated at a fixed η, increases from zero at $\xi = 0$ to a maximum value and decreases sharply toward zero, as shown in Figure 5 for the unsteady Falkner–Skan flow with $\lambda = 1.0$ and $m = -0.05$. Here α_4 is evaluated along the line $\eta = 2.4$, which corresponds to the minimum value of W at $\xi = 0.308$ (close to separation). The sharp decrease in α_4 as α_4 approaches zero occurs because in this region W is increasingly negative. By extrapolation, we estimate the singularity to occur at $\xi = 0.310$.

Since the singularity occurs at a fixed value of ξ, it is easy to determine the motion of singular point. The velocity at the separation point is given in general by

$$U_s^* = \left.\frac{dx^*}{dt^*}\right|_{\xi_{sep}}$$

For the unsteady variation of Howarth's linearly retarded flow we obtain $U_s = -\lambda\xi_{sep}$, and for the unsteady variation of the Falkner–Skan flows we obtain $U_s = -\lambda\xi_s \exp(-1 + \lambda t^*)$. Since our desire is to investigate the Moore–Rott–Sears model for unsteady separation, it is important that we

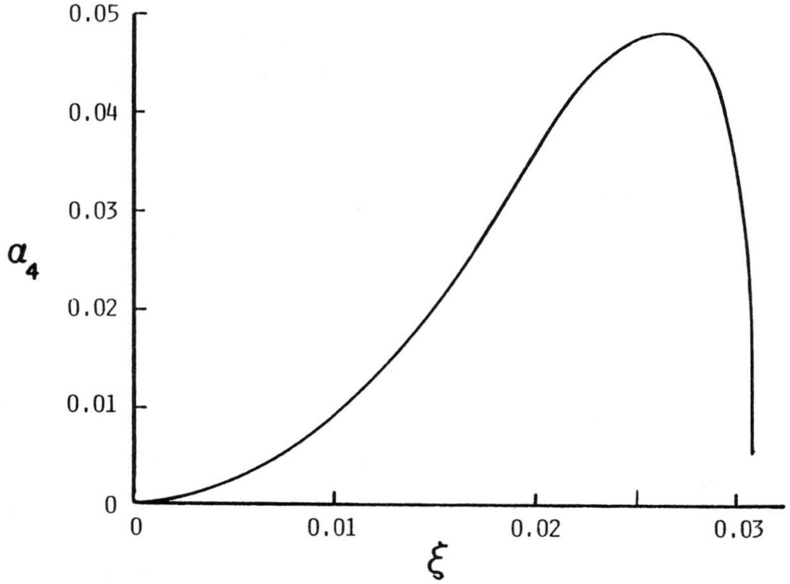

Figure 5. Variation of the value of α , for $\lambda = 1$ and $\eta = 2.4$, with ξ for an unsteady form of Howarth's linearly retarded flow.

study the flow in the coordinate system moving with the singular point. The x-component of velocity in the coordinate system moving with the singularity, u_m, is related to the x-component of velocity in the body-fixed coordinate system by

$$u_m^* = u^* - U_s^*.$$

The velocity profiles in the coordinate system moving with the singularity (with separation) are shown in Fig. 6 for $\xi = 0$, 0.25, and 0.308. Clearly, as

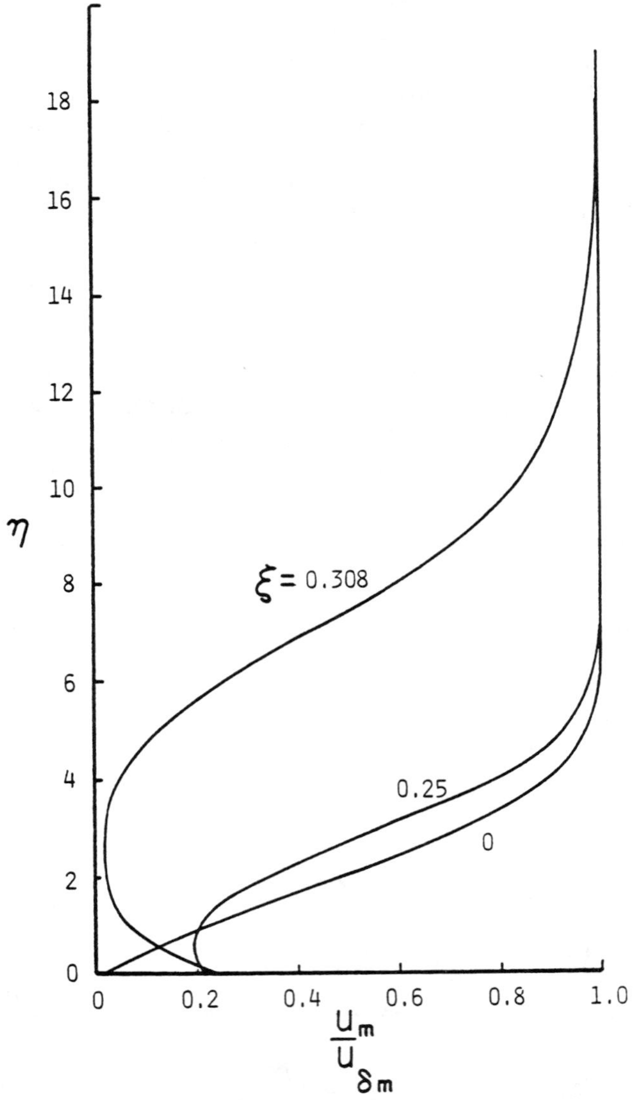

Figure 6. Velocity profiles, in the moving coordinate system, for an unsteady form of the Falkner-Skan flows.

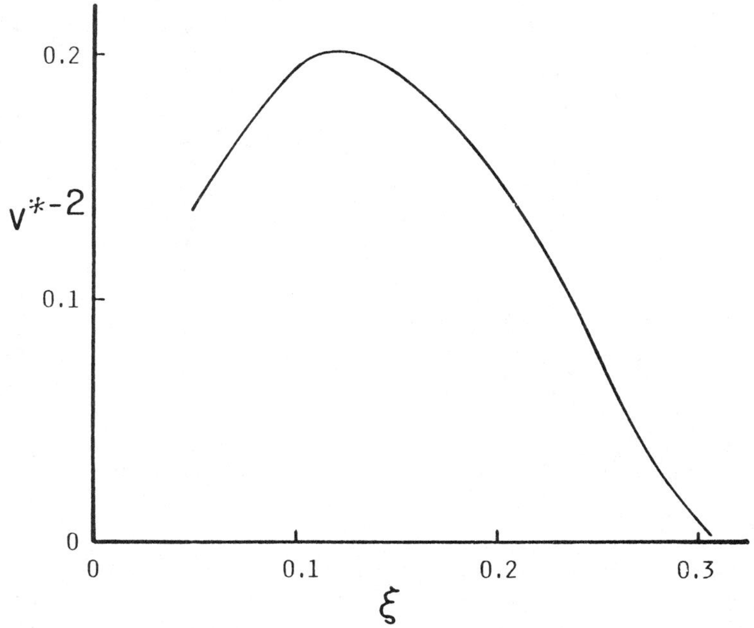

Figure 7. Variation of $1/v^{*2}$ with ξ for an unsteady form of the Falkner-Skan flows.

the singular point is approached, the velocity profile approaches one in which the velocity and shear are zero, at a point within the boundary layer, as viewed by an observer moving with the singular point (with separation). The results obtained for this family of flows, and for the unsteady Howarth flows, tend to verify the Moore–Rott–Sears model for unsteady separation. Additional verification of the singularity is presented in Fig. 7, where the inverse square of the vertical velocity, $1/v^{*2}$, along the line $\eta = 2.4$, is plotted as a function of ξ. Clearly, this quantity approaches zero as α_4 approaches zero, indicating an unbounded vertical velocity and an unbounded boundary layer thickness at the separation (singular) point. A similar result is obtained in the case of the unsteady Howarth flows.

Rear Stagnation Point

The previous examples of flows in which a singularity occurs provide an indication of what physical characteristics are to be expected in the vicinity of the singularity. These examples are all drawn from previously studied flows. We now turn our attention to a new flow. We consider the development, according to the boundary-layer equations, of the flow in the vicinity

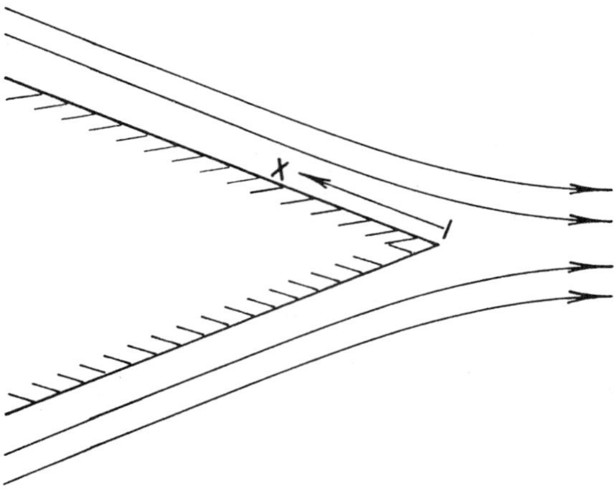

Figure 8. Geometry-flow in the vicinity of a rear stagnation point.

of the sharp trailing edge on a symmetrical body which is impulsively set into motion. An appropriate idealization for such a flow is the flow in the vicinity of a wedge, with the flow leaving the wedge (as opposed to flowing onto the wedge). The geometry for this flow is shown in Figure 8. The potential-flow solution for the idealized flow in the neighborhood of the trailing edge is simply the reverse of the potential flow for a sharp wedge.

The velocity component parallel to the surface, denoted by u_δ^*, is simply

$$u_\delta^* = -Cx^{*m}.$$

Small values of m correspond then to small trailing-edge angles, while large values of m correspond to large trailing-edge angles. The value $m = 1$ corresponds to a blunt (bluff) trailing edge.

This is the inviscid flow, which develops instantaneously when the body is impulsively set into motion. We now turn our attention to the development of the boundary layer, bounded on one side by the body and on the other side by the above potential flow.

It may seem odd that it is possible to integrate the boundary-layer equations and determine the flow development locally in the vicinity of the trailing edge, without regard to the flow upstream on the forebody. This is only possible for a certain period of time after initiation of the flow: the period of time required for the effect of the leading edge of the body to be transmitted downstream to the region of the trailing edge. Although this period cannot be made infinite, it can be lengthened arbitrarily. The time required for the leading-edge effect to be transmitted downstream depends upon the distance from the leading edge to the point in question divided by the average velocity at the edge of the boundary layer upstream of that point. This time can be extended, within practical limits, by altering the length and shape of the forebody.

This problem of the flow in the vicinity of the trailing edge on a body, impulsively set in motion, can be also handled by the method of semisimilar solutions. In this case we employ the scalings

$$g^{*2} = \xi / Cx^{*m-1}, \qquad \xi = 1 - \exp(-Cx^{*m-1}t^*).$$

With these scalings the coefficients in Eq. (3) become

$$a(\xi) = 0, \quad b(\xi) = 1 - \xi, \quad c(\xi) = \xi(1 - \xi), \quad d(\xi) = -m\varsigma,$$

$$e(\xi) = (m - 1)\frac{\xi + (1 - \xi)\ln(1 - \xi)}{2},$$

$$h(\xi) = (m - 1)(1 - \xi)\ln(1 - \xi),$$

and the coefficient α_4 becomes

$$\alpha_4 = (1 - \xi)\left\{\xi + (m - 1)\ln(1 - \xi)\frac{\partial f}{\partial \eta}\right\}.$$

It is interesting to note, at this point, that for the rear stagnation point on a bluff body (i.e., the case where $m = 1$), α_4 can never vanish in the region of integration $0 \leq \xi \leq 1$. Thus no singularity can occur in the case of a rear stagnation point on a bluff body. This is exactly the result obtained by Proudman and Johnson [3.38]. For $m \neq 1$, however, a singularity can occur. In the present case, $m > 0$ corresponds to flows with an adverse pressure gradient. Thus we might expect separation to occur in these flows.

Equations (4) and (5) have been integrated, using the same computer program as used in studying the previous flows, for values $m = 0, 0.2, 0.4, 0.6, 0.8$, and 1.0. The results of these calculations for the wall shear are presented in Fig. 9, where the normalized wall shear

$$F''_W = \sqrt{\frac{\nu x}{UCx^{*m}}}\frac{1}{u^*_\delta}\frac{\partial u^*}{\partial y^*}\bigg|_0$$

is plotted as a function of the nondimensional time

$$\tau = Cx^{*m-1}t^*/2$$

The calculations, for which results are presented in Fig. 9, are terminated in one of three ways, depending upon the manner in which the solution proceeds. For $m = 0$, the computations were terminated at a value of τ approximately 3.8. For this value of m and for $\tau < 3.8$, there is no evidence of a singularity. This case was terminated at this value of τ because continuation of the calculations would require excessive computer time. Further, in view of the above discussion, it did not seem realistic to extend the calculation indefinitely, since sooner or later the effects of the forebody would be felt.

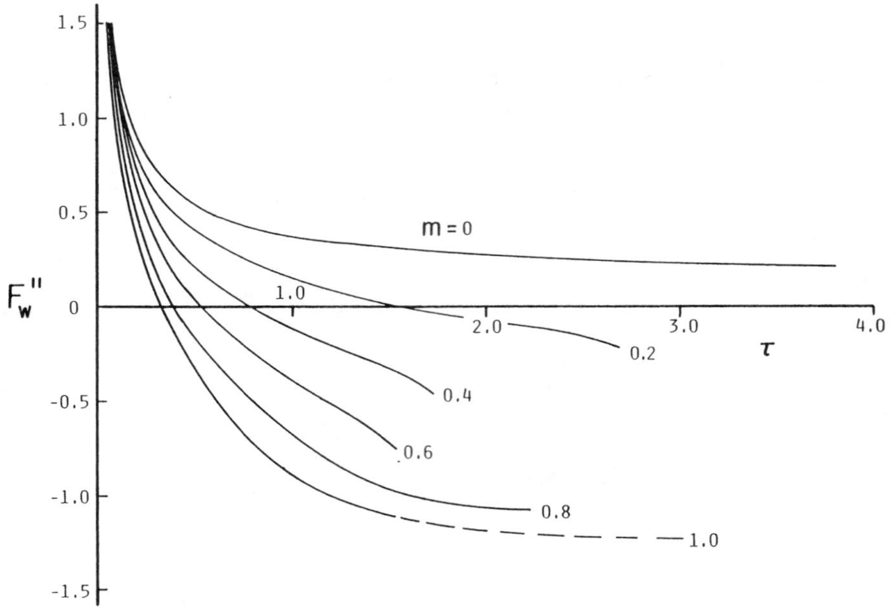

Figure 9. Variation of the normalized wall shear with τ for rear stagnation point flows.

For $m = 0.2$, $m = 0.4$, and $m = 0.6$, the computations terminated abruptly, exhibiting the same numerical characteristics associated with a singularity in the previously studied flows. The case where $m = 0.4$ will be investigated further shortly.

For $m = 0.8$ and 1.0, the computations were terminated at the values of ξ indicated because the boundary-layer thickness exceeded an arbitrary value corresponding to $\eta = 60$. For $m = 0.8$, a singularity would have eventually terminated the calculation as in the cases of $m = 0.4$ and 0.6. However, the validity of the calculations for $m = 0.8$ when the boundary layer becomes excessively thick is certainly open to question. On the other hand, the solution for $m = 1$ can be carried further (dashed curve) without regard to the boundary-layer thickness, since the solution for $m = 1$ is a solution to the full Navier–Stokes equations.

We return now to the calculations for $m = 0.4$, where the computation appeared to be terminated by a singularity. Figure 10 shows the variation with ξ of α_4 along the line $\eta = 2.7$ for this flow. Again $\eta = 2.7$ is the value of η corresponding to the minimum value of u at the last station where convergence was obtained. Clearly, α_4 first increases from $\alpha_4 = 0$ at $\xi = 0$ to a maximum value and then decreases sharply, approaching zero at ξ approximately 0.970, indicating a singularity in the vicinity of this point.

The velocity profiles in a coordinate system moving with the singularity for this flow are shown in Fig. 11 for $\xi = 0.2, 0.8, 0.969$. Here again the

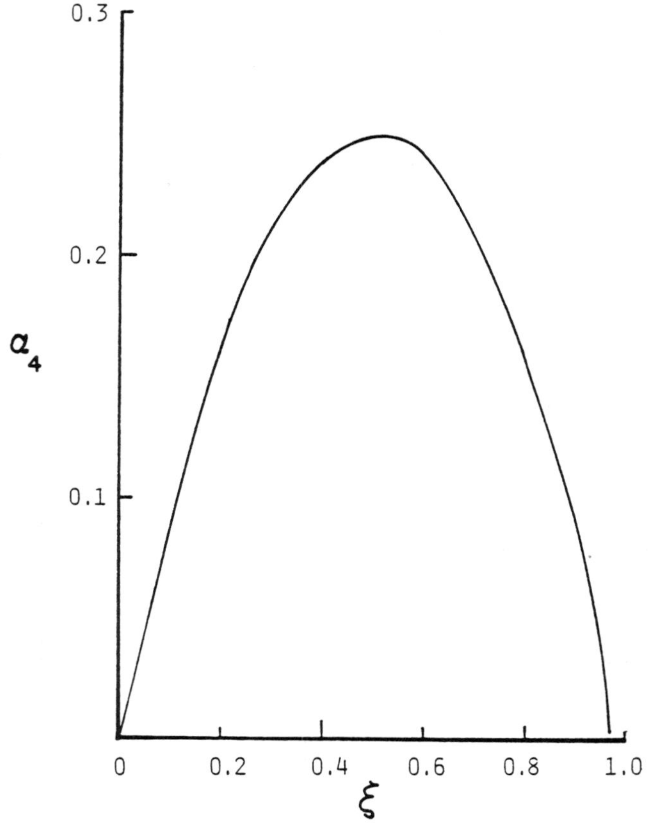

Figure 10. Variation of α_4, for $\eta = 2.7$, with ξ for rear stagnation point flows ($m = 0.4$).

velocity profiles, in the coordinate system moving with the singularity, approach a velocity profile in which the velocity and velocity gradient are zero at an interior point of the fluid. This result tends to verify the model of Moore, Rott, and Sears.

Figure 12 shows the variation of the inverse square of the vertical component of velocity (along $\eta = 2.7$) with ξ. As in the other cases where the singularity was related to separation, this quantity approaches zero as the singularity is approached.

It appears, then, that for small or moderate trailing-edge angles (small or moderate adverse pressure gradient), unsteady separation, as postulated in the Moore–Rott–Sears model, occurs early in the flow development process. For large trailing-edge angles, however, the boundary layer may become so thick that the external inviscid flow is altered before separation occurs. In any event, within the limitations of boundary-layer theory, unsteady separation never occurs at a blunt trailing edge.

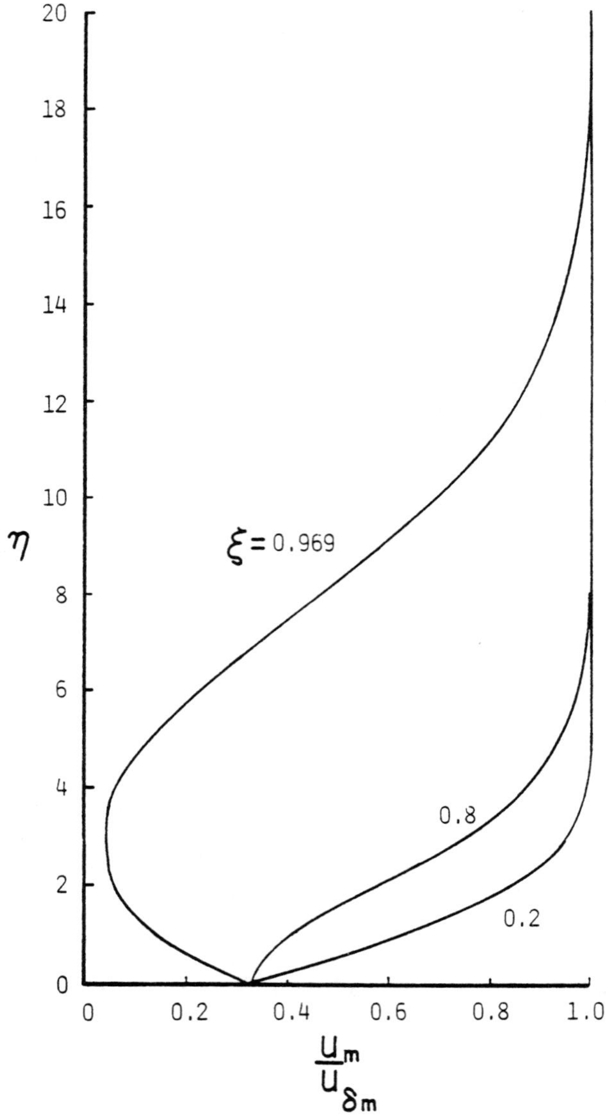

Figure 11. Velocity profiles, in the moving coordinate system, for a rear stagnation point flow ($m = 0.4$).

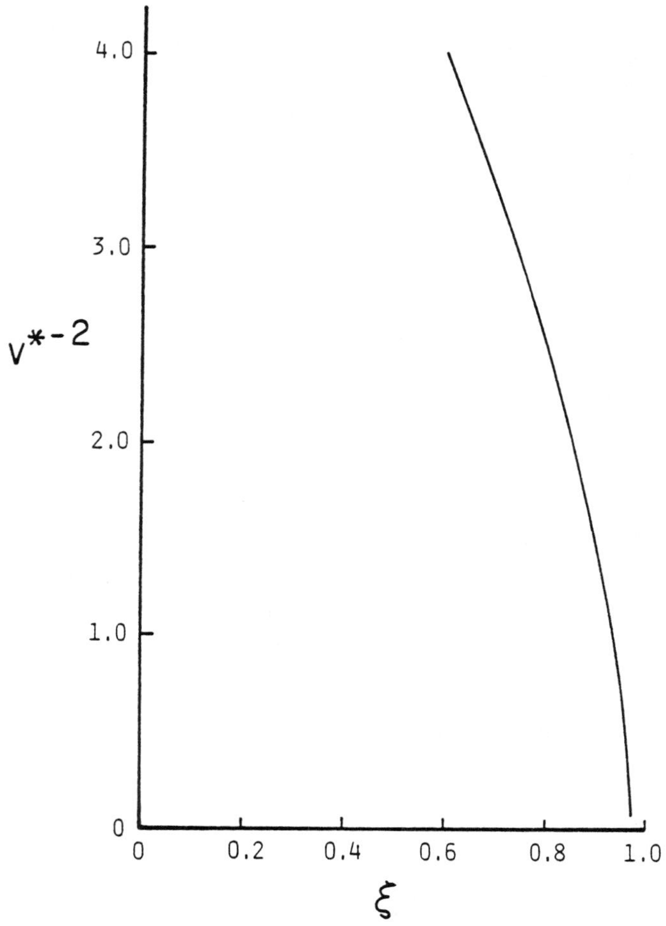

Figure 12. Variation of $1/v^{*2}$ with ξ for a rear stagnation point flow ($m = 0.4$).

Concluding Remarks

A number of existing solutions of the unsteady boundary-layer equations have been reviewed to determine the relationship between singularities encountered in these solutions and the physical behavior of the boundary layer in the vicinity of the singularity. To these solutions has been added a new set of solutions for the flow development in the vicinity of a rear stagnation point on a body which is impulsively set in motion. In general, these solutions indicate that the singularities encountered are associated with one of two possible flow phenomena.

In one case, the singularity is encountered in the solutions for flow over bodies impulsively set in motion and signifies the limiting time in which the local flow may develop without being influenced by the effect of the leading edge of the body. For this case, the velocity profiles at (or near) the singularity show no special characteristics, and the vertical component of velocity remains bounded as the singularity is approached.

In the second case, the singularity is characterized by a vertical velocity which becomes unbounded (singular) as the singularity is approached. In the steady case, as indicated by the solution for Howarth's linearly retarded flow, the unbounded increase in the vertical velocity and the phenomenon of separation are well known and intimately linked. In the unsteady case, the singularity and the unbounded increase in the vertical velocity are indicative of separation. Furthermore, the flow in the vicinity of the singularity has just the characteristics postulated in the Moore–Rott–Sears model for unsteady separation; i.e., the velocity and shear vanish at some point within the boundary layer when viewed in a coordinate system moving with separation. *All of the existing semisimilar solutions to the unsteady boundary-layer equations for flows in which separation is expected, exhibit a moving singularity and a flow in the vicinity of the singularity which tends to substantiate the Moore–Rott–Sears model.*

The new solutions for the flow development in the vicinity of a sharp trailing edge clearly show that no singularity exists in the flow when the included trailing-edge angle is π. This is consistent with the results of Proudman and Johnson [3.38]. These results also indicate that for blunt enough trailing edges the boundary layer becomes excessively thick before the singularity appears, while for moderately sharp trailing edges the singularity occurs fairly early in the flow development. These results suggest that the flow near a rear stagnation point on a bluff body may not be the best flow in which to test for a separation singularity. It is suggested that a better test for the development of the singularity might be a study of the flow development over a sharp-trailing-edge body (e.g., a Kármán–Trefftz airfoil) which is impulsively set in motion.

References for Part 3

[3.1] F. K. Moore: *Boundary-Layer Research*, ed. H. Gortler (Springer-Verlag, Berlin 1958), 296.
[3.2] N. Rott: Quart. J. Appl. Math. **13**, 444 (1956).
[3.3] W. R. Sears: J. Aero. Sci. **23**, 5, 490 (1956).
[3.4] A. J. Robins, J. A. Howarth: J. Fluid Mech. **56**, 161 (1972).
[3.5] D. P. Telionis: J. Fluids Engrg. **101**, 29 (1979).
[3.6] D. P. Telionis, M. J. Werle: J. Appl. Mech. **40**, 369 (1973).
[3.7] J. C. Williams, III, W. D. Johnson: AIAA J. **12**, 1388 (1974).
[3.8] S. N. Brown: Phil. Trans. Roy. Soc. London **A257**, 409 (1965).
[3.9] D.Th. Tsahalis: AIAA J. **15**, 561 (1977).
[3.10] S. F. Shen: Adv. Appl. Mech. **18**, 177 (1978).
[3.11] L. W. Carr, K. W. McAlister, W. J. McCroskey: NASA TN D-8382 (1977).
[3.12] T. Cebeci: J. Comp. Phys. **31**, 153 (1979).
[3.13] L. L. Van Dommelen, S. F. Shen: J. Comp. Phys. **38**, 125 (1981).
[3.14] L. L. Van Dommelen, S. F. Shen: Private communication (1980).
[3.15] T. Cebeci: Unpub. work (1980).
[3.16] R. J. Bodonyi, K. Stewartson: J. Fluid Mech. **79**, 669 (1977).
[3.17] C. Simpson: Private communication (1980).
[3.18] V. C. Patel, J. Nash: *Unsteady Aerodynamics*, *1*, ed. by R. B. Kinney (Univ. of Arizona Press, Tucson 1975).
[3.19] P. Bradshaw, D. H. Ferriss, N. P. Atwell: J. Fluid Mech. **28**, 593 (1967).
[3.20] T. Cebeci, L. W. Carr: Proc. Royal Soc., A380, 291 (1982).
[3.21] T. Cebeci, A. M. O. Smith: *Analysis of Turbulent Boundary Layers* (Academic Press, New York 1974).
[3.22] T. Cebeci: AIAA J. **16**, 1305 (1978).
[3.23] P. Bradshaw: AIAA J. **17**, 790 (1979).

[3.24] L. W. Carr: NASA TM-78-445 (1977).
[3.25] D. Catherall, K. W. Mangler: J. Fluid Mech. **26**, 163 (1966).
[3.26] P. Bradshaw, T. Cebeci, J. H. Whitelaw: *Engineering Calculation Methods for Turbulent Flow* (Academic Press, London 1981).
[3.27] A. E. P. Veldman: AIAA J. **19**, 79 (1981).
[3.28] G. E. Raetz: Northrop Aircraft Co. Rept NA1 58-73 (1957).
[3.29] K. C. Wang: J. Fluid Mech. **48**, 397 (1971).
[3.30] K. W. Wang: Phys. Fluids **18**, 951 (1975).
[3.31] D. P. Telionis, D.Th. Tsahalis: Acta Astronautica **1**, 1487 (1974).
[3.32] J. H. Phillips, R. C. Ackerberg: J. Fluid Mech. **58**, 561 (1973).
[3.33] K. C. Wang: Martin Marietta Lab. Rept. No. TR-79-16c (1979); see also Proceedings 15th Intl. Congr. Theor. Appl. Mech., Toronto (1980).
[3.34] W. R. Sears, D. P. Telionis: *Recent Research on Unsteady Boundary Layers*, ed. E. A. Eichelbrenner (Les Presses de L'universite Laval, Quebec 1972), 404.
[3.35] W. R. Sears, D. P. Telionis: SIAM J. Appl. Math. **28**, 215 (1975).
[3.36] L. L. Van Dommelen, S. F. Shen: XIII Symp. Adv. Prob. & Math. in Fluid Mech., Oltztyn-Kortowo, Poland (1977).
[3.37] K. C. Wang: Lockheed Georgia Co. Viscous Flow Symp., LG 77ER0044, Atlanta (1977); also Martin Marietta Labs. Rept. TR 76-54c (1976).
[3.38] I. Proudman, K. Johnson: J. Fluid Mech. **12**, 161 (1962).
[3.39] R. M. Terrill: Phil. Trans. Roy Soc. London **A253**, 55 (1960).
[3.40] H. Schlichting: *Boundary-Layer Theory*, 7th ed. (McGraw-Hill, New York 1979), 29, 150, 415, 427.
[3.41] H. Blasius: Z. Math. Phys. **56**, 1 (1908).
[3.42] O. Tietjens: *Stromungslehre*, **2**, 1st ed. (Springer-Verlag, Berlin 1970), 105.
[3.43] R. Bouard, M. Coutanceau: J. Fluid Mech. **101**, 583 (1980).
[3.44] O. A. Oleinik: PPM, J. Appl. Math. Mech. **30**, 505 (1966).
[3.45] O. A. Ladyzhenskaya: *The Mathematical Theory of Viscous Incompressible Flow*, 2nd Engl. ed. (Gordon & Breach, New York 1969), 141.
[3.46] M. Schwabe: Ing. Archiv. **6**, 34 (1935).
[3.47] S. F. Shen, J. P. Nenni: *Unsteady Aerodynamics*, ed. R. B. Kinney, **1** (Univ. of Arizona Press, Tucson 1975), 245.
[3.48] Sir H. Lamb: *Hydrodynamics*, 6th ed. (Dover, New York 1945), 12.
[3.49] L. L. Van Dommelen: Ph.D. Thesis, Cornell University (1981).
[3.50] R. J. Belcher, O. R. Burggrad, J. C. Cooke, A. J. Robins, K. Stewartson: *Recent Research on Unsteady Boundary Layers*, **2**, ed. E. A. Eichelbrenner (Les Presses de L'universite Laval, Quebec 1972), 1444.
[3.51] L. L. Van Dommelen, S. F. Shen: XVth Int. Congr. Theor. Appl. Mech., Toronto, Canada (1980).
[3.52] K. Stewartson: Quart. J. Mech. Appl. Math. **11**, 399 (1958).
[3.53] S. N. Brown, K. Stewartson: Ann. Rev. Fluid Mech. **1**, 56 (1969).
[3.54] V. V. Sychev: Fluid Dyn. **14**, 829 (1980).
[3.55] W. H. H. Banks, M. B. Zaturska: J. Engrg. Math. **13**, 193 (1979).
[3.56] L. Howarth: Proc. Roy. Soc. London **A164**, 547 (1938).
[3.57] H. A. Dwyer: AIAA J. **6**, 2447 (1968).
[3.58] J. C. Cooke, A. J. Robins: J. Fluid Mech. **41**, 823 (1970).
[3.59] A. Davey: J. Fluid Mech. **10**, 593 (1961).
[3.60] H. B. Keller, T. Cebeci: *Lecture Notes in Physics—8, Proc. 2nd Int. Conf. on Numerical Methods in Fluid Dynamics* (Springer-Verlag, New York 1971).
[3.61] T. Cebeci, K. Stewartson, P. G. Williams: SIAM J. Appl. Math. **36**, 190 (1979).

[3.62] J. C. Williams, III: Annual Review Fluid Mech. **9**, 113 (1977).
[3.63] J. C. Williams, III, W. D. Johnson: AIAA J. **12**, 1427 (1974).
[3.64] K. S. Fansler, J. E. Danberg: AIAA J. **15**, 274 (1977).
[3.65] D. P. Telionis, M. J. Werle: VPI Rept. E-72-13 (1972).
[3.66] D. P. Telionis: J. Fluids Engrg. **97**, 117 (1975).
[3.67] R. J. Vidal: Wright Air Development Center, TR-59-75, Wright Patterson AFB, Dayton, Ohio (1959).
[3.68] W. G. Brady, G. R. Ludwig: AFAPL-TR-65-115 (1965).
[3.69] J. F. Nash, V. C. Patel: NASA CR-2546 (1975).
[3.70] D. P. Telionis, D.Th. Tshalis, M. J. Werle: Phys. Fluids **16**, 968 (1973).
[3.71] N. Rott: *Theory of Laminar Flows*, ed. F. K. Moore (Princeton Univ. Press 1964), 431.
[3.72] J. C. Williams, III, W. D. Johnson: *Unsteady Aerodynamics*, ed. R. B. Kinney (Univ. of Arizona Press, Tucson 1975), 261.
[3.73] K. Stewartson: Quart. J. Mech. Appl. Math. **4**, 182 (1951).
[3.74] J. C. Williams, III, T. B. Rhyne: SIAM J. Appl. Math. **38**, 215 (1980).

PART 4

TRANSONIC FLOWS

CHAPTER 22

Advances and Opportunities in Transonic-Flow Computations

W. F. Ballhaus, Jr.,*† G. S. Deiwert,*‡
P. M. Goorjian,*§ T. L. Holst,*§ and P. Kutler*††

1 Introduction

Three tools serve the aerodynamicist in acquiring an understanding of the physics of transonic flows and applying this understanding to the development of engineering hardware. These tools are *analysis*, *computations*, and *experiment* or *testing*. Analysis provides an understanding of the influence of design variables and flight conditions on aerodynamic performance, thereby establishing guidelines for configuration design (e.g., the area rule). Computations provide detailed flow-field information, an efficient means for design optimization, and a data source for calibrating experimental measurements. Testing provides accurate predictions of aerodynamic performance, an understanding of complex flows, and a means to verify computational methods.

During the last decade, progress has been made in refining all three tools. Some of the improvements in analytical methods are described in the paper in this volume by Cheng. A number of evolutionary advances have occurred in experimental techniques, including nonintrusive measurement techniques, high-Reynolds-number capability, adaptive walls to minimize wind-tunnel-wall interference, and improved data acquisition methods to improve productivity and reduce the cost per data point.

* Ames Research Center, NASA, Moffett Field, CA 94035.

† Director of Astronautics.

‡ Research Scientist, Computational Fluid Dynamics Branch.

§ Research Scientist, Applied Computational Aerodynamics Branch.

†† Chief, Applied Computational Aerodynamic Branch.

While the advances in analysis and testing have for the most part been *evolutionary*, there has been a major *revolution* in computational techniques—the 1970s were truly the decade for the advent of the computational aerodynamicist. In 1970, a solution to the Euler equations for transonic flow about an airfoil required hours of computer time. Better solutions can now be obtained in minutes, and solutions to the full potential equation require only a few seconds. In 1971, small-disturbance solutions for flows about wings consumed many hours of computer time. Now, solutions for multiple-component configurations can be computed in minutes, and full potential solutions require less than an hour. The second half of the decade also saw significant advances in numerical solution procedures for solving unsteady inviscid transonic flows. Progress in the treatment of inviscid flows, both steady and unsteady, is reviewed in Sections 2 and 3. The inviscid methods have advanced to the point where they are being used effectively to develop engineering designs for cruise conditions, where viscous effects normally are not dominant.

In the latter half of the 1970s, pioneering work was initiated for the treatment of viscous-dominated flows about aircraft near their boundaries of performance. These flow fields are characterized by flow separation and can be described by solutions to the Reynolds-averaged approximation of the Navier–Stokes equations with an appropriate turbulence model for closure. Progress in this area is reviewed in Section 4.

In Section 5 we briefly describe some of the research areas in which advances are required to meet the challenge of accurately and efficiently computing solutions to transonic flows about aircraft.

In our attempt to review the discipline, we have purposely avoided dealing with work that is adequately treated in other sections of this volume. One notable omission is the coupling of inviscid flows with boundary-layer solutions. Such methods provide adequate viscous corrections for cases in which separation is nonexistent or confined to small regions of the flow.

Other chapters in this volume describe advances of interest to computational aerodynamicists. Cheng has contributed an interesting and thorough review of analytical transonic theories for both high- and low-aspect-ratio wings. These theories and related computations provide physical insight and inexpensive approximate solutions. In this paper Cheng points out and clarifies discrepancies and differences in the efforts of several investigators, discusses the lift correction to the transonic equivalence rule for low-aspect-ratio wings, treats high-aspect-ratio swept and oblique wings as lifting-line problems, describes an oblique-wing computation analogy to an unsteady airfoil computation, and presents a solution for flow about an oblique wing with an embedded shock. Hafez describes an analytical perturbation analysis for transonic flows with shock waves. He first presents a general formulation of the perturbation problem and introduces a new perturbation sequence expansion. He then applies it to the treatment of unsteady effects, three-dimensional effects, and wind-tunnel-wall corrections.

The paper by Sears provides observations with supporting evidence regarding the difficult task of defining the freestream vector (speed and direction) in wind-tunnel testing. Using a simple panel-method computation, he shows how the adaptive-wall concept might be used to accurately establish arbitrarily chosen values of Mach number and angle of attack. The success of the technique, Sears concludes, hinges primarily on success in solving the instrumentation problem.

Four papers describe advances designed to improve or extend present computational capability. The work of Caughey and Jameson presents and evaluates two improvements to their finite-volume scheme for solving the inviscid full potential equation: (a) a second-order-accurate artificial viscosity for supersonic flow regions and (b) modification of flux balances in the finite-volume approximation to permit freestream conditions to be satisfied exactly. The result presented for the first improvement tested indicates second-order-accurate convergence as the volumes are successively reduced toward zero. This provides more accurate predictions of C_l and C_d for a given grid than the commonly used first-order-accurate scheme. It also provides better agreement with exact (hodograph) solutions for the highly sensitive supercritical shock-free airfoil cases. The stated conclusion on the second improvement is that restructuring the flux balance to recover freestream conditions exactly has only a small effect and is probably not worth the effort.

Malmuth et al. present small-disturbance and full potential formulations and accompanying solution algorithms for upper-surface-blown airfoils. Upper-surface blowing can increase usable lift for enhancing V/STOL capability at low speeds and high maneuverability at transonic speeds. Inviscid computed results in their paper indicate that these enhancements can be significant. The inviscid results, compared with experiment, indicate the need for refinements incorporating wave-interaction phenomena near the jet exit as well as viscous interaction processes in the downstream portion of the wall jet. Using a finite-difference boundary-layer module, Malmuth et al. were able to quantify for the first time computationally the amount of blowing required to maintain attached flow.

Yoshihara describes the complicated interaction between an imbedded shock wave and a boundary layer. He then presents an empirically determined (but phenomenological) modification that can be simply incorporated in inviscid codes to model this interaction at little additional expense in computer time. This model has been tested on several airfoil flow fields, including one with shock-induced separation. It should prove useful for steady and unsteady flow-field computations as an inexpensive alternative to detailed viscous-flow simulations.

The work of Seebass and Fung deals with time-linearized computations for unsteady transonic flows. Such computations should prove useful in efficiently providing transonic aerodynamic input to aeroelastic analyses. Their work improves upon the indicial-approach studies by Ballhaus and Goorjian by (a) providing specific treatment of moving shock waves and (b) incorporating a new far-field boundary condition at the grid boundaries to reduce the energy of (spurious) reflected waves. Both improvements contribute significantly to reducing the computation costs. Seebass and Fung also provide analytical expressions for the dependence of the force, moment, amplitude, and phase on reduced frequency. These useful expressions are derived by introducing simple approximations to indicial responses.

2 Status—Inviscid Transonics (Steady)

In the last decade, significant progress has been made in the treatment of potential formulations for the numerical solution of steady transonic flows. These improvements have appeared for both major potential formulations: the *transonic small-disturbance* (TSD) formulation, which utilizes relatively simple stretched and/or sheared meshes with small-disturbance boundary conditions, and the *full potential* (FP) formulation with general body-oriented mapping procedures. Advances have been made in spatial discretization schemes allowing improved algorithm reliability and accuracy, in iteration schemes allowing improved computational efficiency, and in grid generation–patching techniques allowing more complete and accurate treatment of the geometrical complexities of realistic aircraft. These advances, coupled with the anticipated availability of new large-scale scientific computers, indicate significant near-term increases in the use of transonic flow-field simulations for aircraft design. Some of the more significant of these advances are reviewed in this section.

The field of computational transonics is a new one, having essentially started in 1970. Murman and Cole [4.1] were the first to achieve a stable transonic solution for the two-dimensional TSD equation by using the concept of type-dependent differencing. Soon after, this procedure was extended to three dimensions for swept-wing calculations by Ballhaus and Bailey [4.2] and wing–cylinder calculations by Bailey and Ballhaus [4.3]. At about this same time, numerical procedures for solving the transonic FP equation were being developed with suitable mapping procedures. Notable contributions tions are due to Steger and Lomax [4.4] and Garabedian and Korn [4.5], both with nonconservative FP formulations for airfoil configurations. The first three-dimensional calculation (nonconservative) was introduced by Jameson [4.6] and was used to solve the transonic flow about wings. Subsequently, a solution procedure for the conservative form of the FP equation was introduced by Jameson [4.7] and extended to three dimensions by Jameson and Caughey [4.8, 4.9].

These early pioneering calculations set the stage for the flourish of activity in computational transonics that has occurred in the last several years. Much of the recent work in steady transonic potential methods has been devoted to finding better ways of solving the full potential equation in conservative form. A number of authors have developed variations on the Jameson conservative FP spatial-differencing scheme, including Eberle [4.10, 4.11, 4.12] with a finite-element formulation, Holst and Ballhaus [4.13] with an upwind-biased density formulation, and Hafez et al. [4.14] with an artificial compressibility formulation. All of these spatial-differencing schemes are similar and are based on an upwind evaluation of the density. This "artificial density" approach is both simple and reliable, and allows for easy extension to three dimensions, even for completely general curvilinear coordinate systems.

Another area of research that has recently received much attention is the development of more computationally efficient relaxation schemes. Such schemes are required because transonic flow problems are inherently nonlinear, and no direct (noniterative) solution procedures are known to exist. The most notable advances have been made by adapting elliptic-type solution procedures [fully implicit approximate factorization (AF) and/or multigrid] to the nonlinear, mixed (elliptic–hyperbolic) transonic flow problem. The first application of an AF scheme to transonic flow calculations was by Ballhaus et al. [4.15] for the solution of the TSD equation about two-dimensional airfoil configurations. This algorithm (called AF-2) was extended to the two-dimensional full potential equation by Holst and Ballhaus [4.13, 4.16] and to the three-dimensional full potential equation by Holst [4.17]. Other AF schemes (primarily of the ADI type) have been used in a variety of different applications, including those by Deconinck and Hirsch [4.18, 4.19] for two-dimensional cascade calculations, Baker [4.20] for airfoil calculations (nonconservative form), Chattot et al. [4.21] for wing calculations, and Sankar et al. [4.22] also for wing calculations. In general, the AF formulation displays significant improvement in computational speed relative to the conventional successive-line-overrelaxation (SLOR) algorithms used in [4.1–4.9]. A convergence-history comparison ([4.23]) between SLOR and AF2 both applied to a transonic Korn airfoil calculation is shown in Fig. 1. The rms error in the airfoil surface pressures (E_{rms}) is plotted vs. computer time (CDC 7600 computer) for each algorithm. For this case the AF scheme is approximately five times faster than SLOR in achieving an acceptable level of E_{rms}.

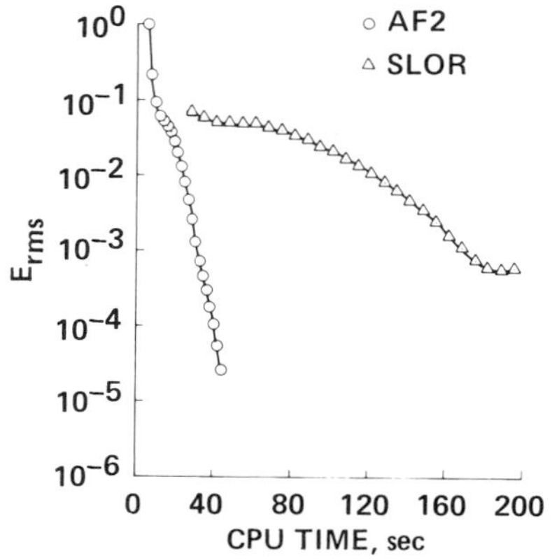

Figure 1. Convergence History Comparisons for Two Full Potential Equation Algorithms; Korn Airfoil, $M\infty = 0.74$, $\alpha = 0°$ (Holst [4.23]).

The multigrid iteration scheme is a convergence acceleration technique that requires an independent base iteration scheme, e.g., SOR, SLOR, AF, etc. Multigrid schemes have existed for quite some time, having been first introduced by Fedorenko [4.24] in 1964, but have only recently gained prominence in transonic flow calculations. The multigrid technique uses a sequence of grids ranging from very coarse to fine. Each grid is used to eliminate one small range of errors in the error frequency spectrum, namely, the highest frequency supported on each mesh. The attractive feature of this approach is that the high-frequency error on the coarsest mesh is actually the lowest-frequency error existing in the problem. Because this usually troublesome low-frequency error is efficiently dealt with on a coarse mesh, very little computational work is expended in removing it from the solution. Thus a tremendous convergence-rate enhancement is obtained relative to SLOR, which requires extremely large numbers of iterations to reduce low-frequency errors to an acceptable level.

Pioneering calculations utilizing multigrid techniques were first performed by Brandt [4.25, 4.26]. South and Brandt [4.27] were the first to present a multigrid algorithm using SLOR as the base algorithm for a transonic flow calculation. The convergence of this multigrid algorithm was five times as fast as that of the base SLOR scheme on uniform meshes, and twice as fast on stretched meshes. To date, the most successful application of a multigrid convergence acceleration scheme for a practical transonic flow problem is the work of Jameson [4.28]. In this study, the full potential equation in conservative form is solved using a multigrid scheme based on a newly devised AF base iteration scheme. Convergence histories for a typical transonic-flow-field solution, which were computed using different numbers of meshes (from one mesh—i.e., no multigrid—up to five meshes), are shown in Fig. 2. Increasing the number of meshes greatly improves the convergence rate. Some other examples in which multigrid algorithms were used to solve transonic flow problems include Arlinger [4.29, 4.30] and McCarthy and Reyhner [4.31].

One last subject needs to be examined before leaving steady transonic potential methods—grid generation-patching techniques. Some progress has been made in this area for both standard potential formulations (TSD and FP). For the TSD formulation, patching or embedded-grid techniques have been utilized to the extent that nearly complete aircraft configurations can now be numerically analyzed in the transonic regime. This type of application is best exemplified by the work of Boppe and Stern [4.32, 4.33]. An example calculation from [4.33] is provided in Fig. 3 for an aircraft configuration in which the wing, fuselage, pylon, nacelles, and winglets have all been modeled numerically. Good correlation with experiment is obtained for this calculation.

For more challenging cases that cannot be treated using TSD, complicated body-oriented mappings are used to generate finite-difference grids. The development of procedures to generate such grids has been a slow, tedious process and can be considered the pacing item in steady inviscid transonic

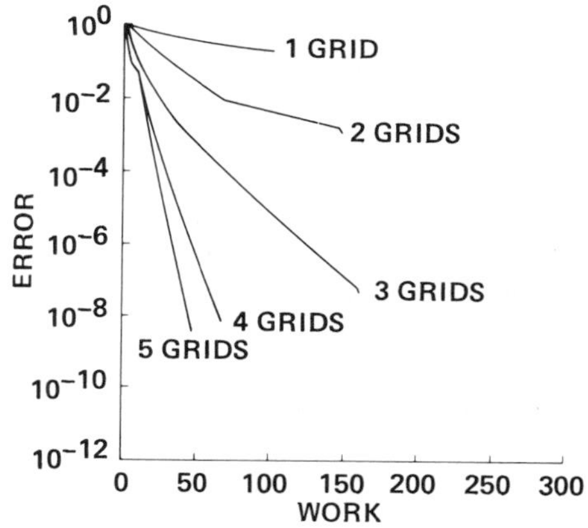

Figure 2. Multi-Grid Convergence Histories Using Different Numbers of Grids; NACA 64A410, $M \infty = 0.72$, $\alpha = 0°$ Jameson [4.28]).

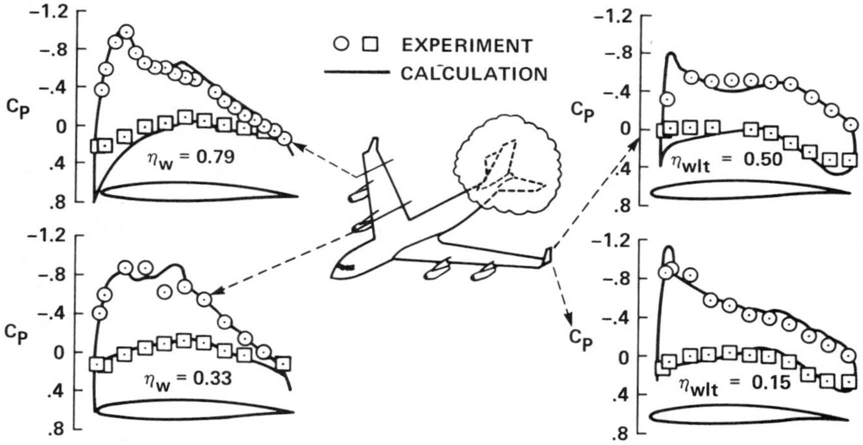

Figure 3. Present Capability for Aircraft Simulation Using TSD Formulation; KC-135 Aircraft, Wing/Fuselage/Pylon/Nacelle/Winglet Configuration (Boppe and Stern [4.33]).

flow computations. Nevertheless, good work has been initiated in several major areas, including numerical transformation procedures and algebraic mapping procedures. Specific examples include Thompson et al. [4.34, 4.35], Thomas and Middlecoff [4.36], Steger and Sorenson [4.37], and Yu [4.38] for grid generation schemes based on numerical procedures, and Eiseman [4.39–4.41] and Eriksson [4.42] for grid generation schemes based on algebraic schemes.

Another area of research associated with steady transonic flows involves the use of the Euler equations. Because these equations represent an improved approximation to the transonic flow problem, especially for the stronger shock cases and cases with significant vorticity production, interest in utilizing this approach has been increasing. The more demanding computational requirements (both storage and execution times) associated with Euler-equation algorithms are rapidly being overcome by the introduction of larger, faster scientific computers and more efficient computational algorithms. Basic research involving the Euler formulation for steady transonic applications is currently increasing and is expected to continue to do so as the potential algorithms enter a more production-oriented research phase.

3 Status—Inviscid Transonics (Unsteady)

During the last five years, research in the area of computational unsteady transonic aerodynamics has undergone a period of exciting growth. This activity was motivated by the need to provide engineers with computational tools to solve important unsteady aerodynamic and aeroelastic problems in the often troublesome transonic regime. For example, in the development of supercritical wings, flutter analysis is most crucial and most difficult in the transonic regime, where alarming unexpected dips in flutter boundaries have been observed experimentally [4.43–4.45]. Other examples of important unsteady transonic flow problems include advancing helicopter rotor blades [4.46], turbomachinery, and the analysis of buffet and aileron buzz [4.47]. The development of codes that can be applied to such engineering problems has only recently become feasible with the availability of fast computers and the development of efficient algorithms that can correctly account for the motion of shock waves [4.48, 4.49]. In the following brief review of unsteady transonic flows, we will mention some of these algorithm developments and some of the benchmark calculations made by them. (For more extensive reviews, see [4.50–4.55].)

First we will review computations that use the transonic, small-disturbance potential formulation. A large part of the effort in developing codes for routine engineering applications has been based on solving this formulation. This choice was made because of the development of very efficient methods for solving the small-distrubance equation in comparison to methods available for more exact equations. For example, as shown in Fig. 4, the calculations of Ballhaus and Goorjian [4.56] based on the low-frequency, small-disturbance equation are over 180 times faster than an explicit method used to solve the Euler equations. Even when more efficient implicit methods are used for the Euler equations, such as those by Chyu and Davis [4.57], they are still prohibitively expensive on existing computers [4.58].

Algorithm development for the small-disturbance equation in two dimensions was started by Ballhaus and Lomax [4.59] who applied semi-implicit methods to the general small-disturbance equation and its low-frequency approximation. Next a fully implicit method was developed by Ballhaus and Steger [4.60] for the low-frequency equation. This alternating-direction implicit algorithm was implemented into a computer code LTRAN2

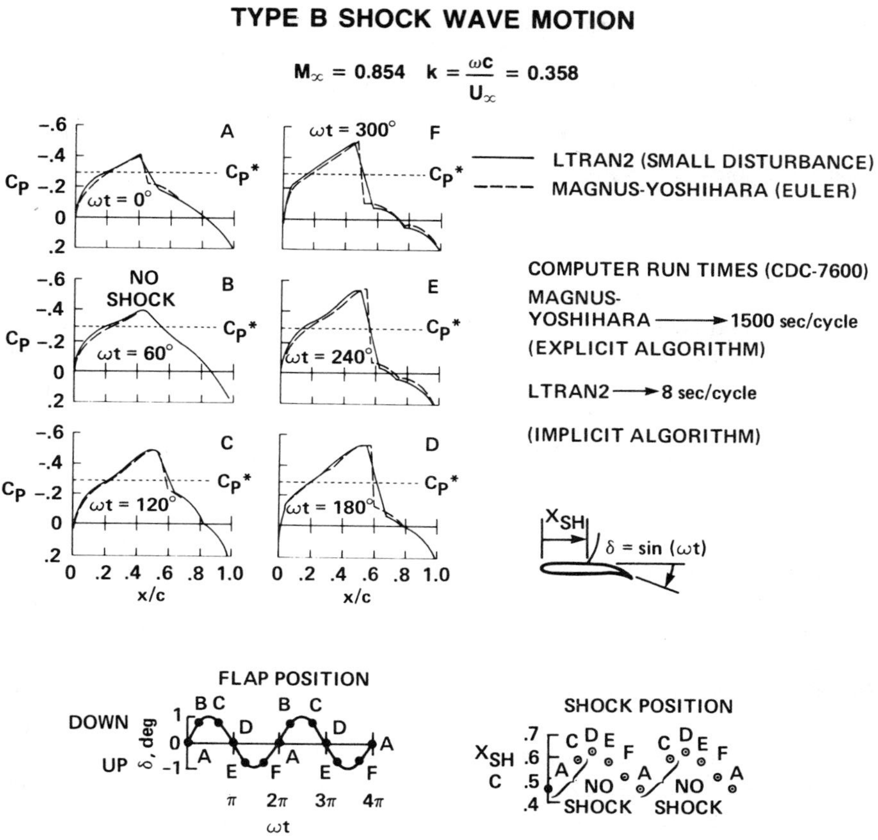

Figure 4. Unsteady upper surface pressure coefficients for a NACA 64A006 airfoil with oscillating trailing edge flap, thowing type B shock wave motion (disappearing shock), calculated using the Euler and small-disturbance equations.

by Ballhaus and Goorjian [4.56]. Figure 4 shows a demonstration calculation using the code, in which comparisons of accuracy and speed are made with calculations using the Euler equations. These calculations show a shock-wave motion that agrees qualitatively with the type-B shock-wave motion produced experimentally by Tijdeman [4.61].

Further developments of the algorithm used in LTRAN2 were made by Houwink and van der Vooren [4.62] and by Couston and Angeline [4.63]; both groups extended the frequency range of LTRAN2 by adding high-frequency terms to the wake condition and the equation for the pressure coefficient. Rizzetta and Chen [4.64] extended the algorithm further by including a high-frequency term in the governing equation. Rizzetta and Yoshirhara [4.65] made a detailed investigation of the influence of the high-frequency terms, as well as developing a viscous wedge model for incorporating viscous effects on the motion of the shock wave. Kwak [4.66] implemented nonreflecting far-field boundary conditions into LTRAN2 to increase its efficiency by reducing the computational domain. More recently, Goorjian and van Buskirk [4.67] have increased the efficiency of LTRAN2 an order of magnitude by using monotone methods in the algorithm. Finally, Borland et al. [4.68] have extended the LTRAN2 algorithm into three dimensions and incorporated it into a three-dimensional code, LTRAN3, for wings.

Now we review some of the benchmark calculations performed by LTRAN2. First LTRAN2 was used to calculate [4.56] the three types of shock-wave motion that have been produced experimentally by Tijdeman [4.61]. Next LTRAN2 was used to compute the indicial response flow field [4.69] and, by the use of Duhamel's integral, unsteady lifts and moments for various frequencies of harmonic motion. The indicial method had first been applied to transonic flows by Beam and Warming [4.70], who solved the Euler equations by an explicit method. Nixon [4.71] extended the indicial method to calculations of unsteady pressures in transonic flows.

LTRAN2 has also been applied to simple transonic aeroelastic problems in which the constrained motion of airfoils was found by simultaneously solving the structural motion and aerodynamic equations. Figure 5 shows the first nonlinear transonic aeroelastic computation [4.69] of an airfoil undergoing simple pitching motion. Rizzetta [4.72] extended this method to airfoils with three modes of motion; Yang et al [4.73] made a comprehensive study using this coupled-equation approach for transonic flutter analysis.

Wind-tunnel-wall modeling [4.74] was used in LTRAN2 to investigate the influence of the walls on the type of shock-wave motion. The calculations showed that the discrepancy between experimental observations [4.53] and computations could be due in part to wind-tunnel-wall interference. Finally, LTRAN2 was used in a study [4.75] of flutter suppression on an airfoil by the use of flaps as active controls.

Code development based on implicit-methods for solving the full potential equation is still in the initial stages. Isogai [4.76] developed a semi-implicit algorithm for this equation in nonconservation form. Next Chipman and Jameson [4.77] developed a conservative implicit algorithm that uses the velocity as a primitive variable. Then Goorjian [4.78], Steger and Caradonna [4.79], Sankar and Tassa [4.80], and Chipman and Jameson [4.81] developed implicit algorithms to solve the full potential equation in conservation form. Steger and Caradonna [4.82] extended their method to three dimensions. Figure 6 shows a sample calculation [4.78], which has served as a testbed in the development of this algorithm.

For a review of the Euler equations, see [4.48–4.53]. Here we will only note the benchmark calculations of Magnus and Yoshihara [4.83], who solved the Euler equations by an explicit method. Subsequently, more efficient implicit methods were developed, such as those of Beam and Warming [4.84]. Currently, there is an intense effort to increase still further the efficiency of algorithms for the Euler equations.

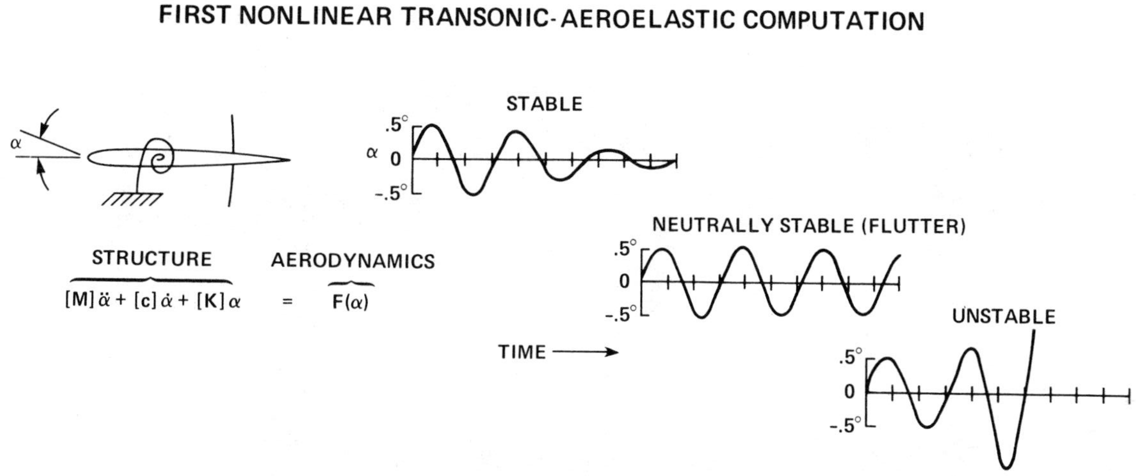

Figure 5. Aeroelastic computation, calculated using coupled aeroelastic and nonlinear, transonic, aerodynamic equations, showing evolving airfoil pitch amplitudes that result from three values of the structural damping coefficient.

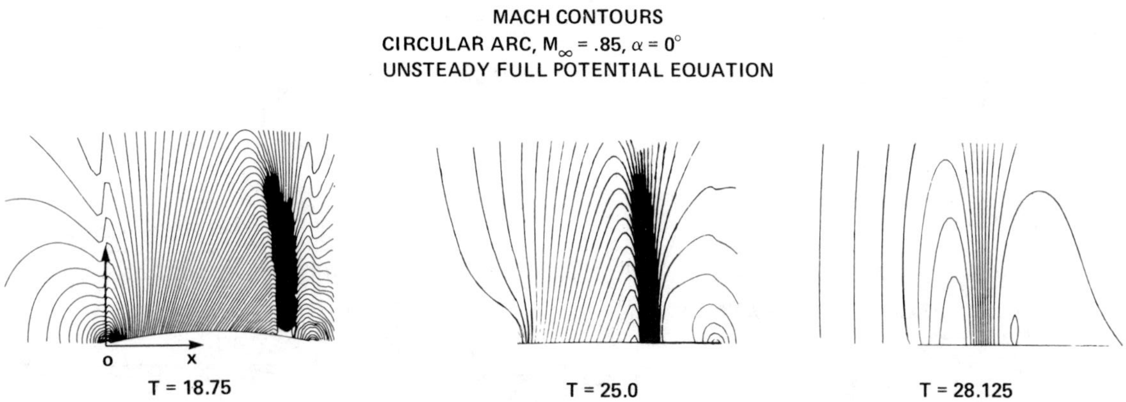

Figure 6. Mach contours, showing the upstream propagation of a shock wave resulting from the motion of a thickening-thinning airfoil, calculated by the unsteady full potential equation in conservation form. Time, T, is measured in chord lengths of travel.

4 Status—Viscous Transonics

Some transonic flows can be dominated by viscous phenomena such as the influence of boundary-layer displacement on circulation, strong viscous–inviscid interactions, viscous–viscous interactions between two different streams of fluid, flow separation, and viscosity-induced unsteady flow. Consideration of these effects is generally critical when studying aerodynamic performance near performance boundaries. To analyze such flows, it is necessary to use an equation set sufficient to describe the dominant physics. In most cases, this requires a suitable subset of the Reynolds-averaged Navier–Stokes equations. These equations describe the inviscid flow field, the boundary-layer behavior and its interaction with the inviscid field, and the interactions between two or more viscous flow fields.

The first numerical study using the Navier–Stokes equations of a transonic flow with strong viscous effects was by Deiwert [4.85, 4.86] in 1974. He extended the explicit finite-difference method of MacCormack [4.87, 4.88] to nonorthogonal, nonuniform curvilinear meshes for arbitrary body shapes, and solved the Reynolds-averaged Navier–Stokes equations for flow over a two-dimensional biconvex circular arc. This first study was done in support of the design of a companion experimental effort to document the flow over the same configuration. Freestream Mach numbers were considered such that a shock impinged on the boundary layer with strength sufficient to induce separation. Computations were used to determine the tunnel test-section design and equivalent freestream Mach numbers. Subsequently, several studies [4.89–4.92] were performed and comparisons made between computation and experiment. Typical examples of these comparisons are shown in Figure 7, where surface pressure distributions are compared. Figure 7(a) shows pressure comparisons for the design conditions at a Mach number of 0.775. The computed and experimental pressure distributions on both the airfoil surface and the test-section walls are in good agreement except near the trailing edge of the airfoil, where the flow is separated. The computed solution indicates more extensive separated flow than does the experiment. Figure 7(b) shows similar pressure comparisons for the off-design condition at a Mach number of 0.786. Here the agreement on both the test-section walls and the airfoil surface is good ahead of the shock, but diverges sharply aft of the shock. The experiment indicates the presence of a weak, oblique shock, and the computations predict a strong, nearly normal shock. This agreement is due to a mismatch of outflow boundary conditions between the experiment and computation. As pointed out recently by T. J. Coakley (private communication), the flow at off-design conditions should be treated as an internal flow problem, and downstream boundary conditions on pressure should be imposed that correspond to the experimental flow field. This treatment permits the proper downstream influence and results in the prediction of an oblique shock and hence good agreement with the experimentally determined pressure recovery at the airfoil trailing edge. An improved solution by Coakley is included in Fig. 7(b).

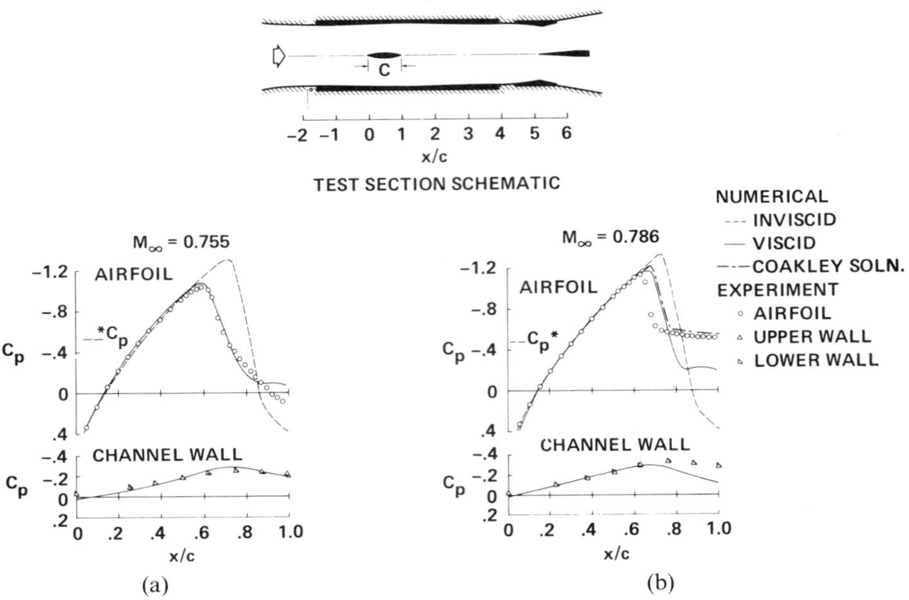

(a) (b)

Figure 7. Comparison of numerical solutions with experiment: a) $M_\infty = 0.775$, $\mathrm{Re}_{c,\infty} = 2 \times 10^6$; b) $M_\infty = 0.786$, $\mathrm{Re}_{c,\infty} = 10 \times 10^6$.

Lifting airfoil configurations were first treated in 1976 [4.93]. In this study, the viscous–viscous interaction between the upper- and lower-surface boundary layers at the trailing edge was considered. Figure 8 shows a comparison of measured and computed surface pressure distributions over the supercritical Korn 1 airfoil at slightly off-design conditions. The free-stream Mach number is 0.755, the angle of incidence is 0.12, and the Reynolds number is 21 million. Included as well are results of an inviscid computation using the Garabedian–Korn code [4.94]. The agreement between the viscous and the experimental results of Kacprzynski et al. [4.95, 4.96] is excellent. Both distributions indicate that at this slightly off-design condition the flow is still shock-free. The inviscid solution indicates the presence of a shock and predicts significantly greater lift than either the experiment or viscous computation. This comparison vividly illustrates the importance of viscous displacement effects for some transonic flows even in the absence of strong viscous–inviscid interactions.

The lifting-airfoil code was further modified [4.97] to use the improved MacCormack algorithm [4.98] and an adaptive grid in the vicinity of the shock, and was used to study other airfoil shapes [4.99, 4.100]. Levy [4.101] used the code to study experimentally observed buffeting on the 18% biconvex circular arc considered in the first study. In this effort he included the influence of the wind-tunnel wall by appropriate boundary conditions in the computations. The Mach-number range where buffet was experimentally observed was remarkably well predicted by the Navier–Stokes code,

		M	α, deg	C_D	C_L
○	EXP	0.755	0.12	0.012	0.38
♦	N-S	0.755	0.12	0.012	0.42
——	INV	0.755	0.12	0.004	0.68

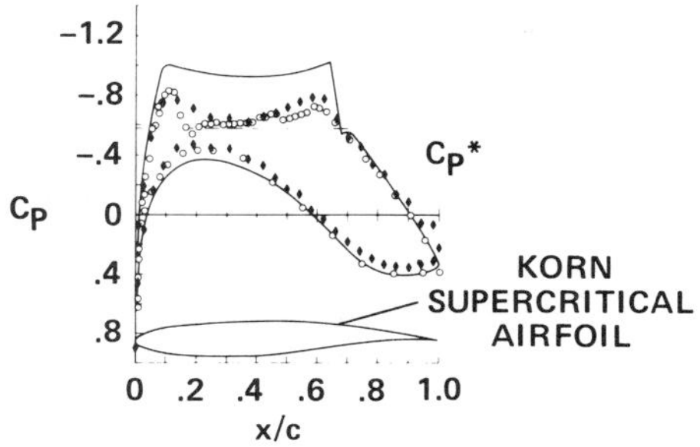

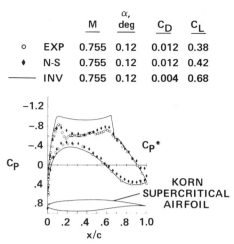

Figure 8. Surface pressure distribution over Korn 1 airfoil at near-design conditions; $Re_c = 21 \times 10^6$.

as were the frequency and character of the surface pressure variation. Figure 9, from Seegmiller et al. [4.102], shows a comparison between the computed and measured surface pressure variation with time at four positions on the biconvex circular arc. The positions correspond to the midchord point and the 77.5% chord point on both the upper and lower surfaces. The pressure variation on the upper surface is a half period out of phase with the variation on the lower surface. The details of the pressure rise and decay agree well with experiment. Additional comparisons of the computed transient behavior of the shock–boundary-layer interaction and the separated flow pattern show good agreement with experimental high-speed shadowgraph movies.

Another unsteady-flow application of this code was the determination of buffet boundaries for the Korn 1 airfoil [4.103]. Figure 10 shows a lift-drag polar and a lift curve for the Korn 1 supercritical airfoil for a nominal Mach number of 0.75. The computations are compared with the experimental data of Kacprzynski et al. [4.95, 4.96]. The computed drag polar indicates the onset of buffet somewhat after maximum lift has been realized. The lift curve indicates that the onset of buffet occurs at an angle of incidence near 3°. The computations were performed assuming free boundaries at the nominal wind-tunnel test conditions, and no adjustments were made to account for Mach-number or flow-angularity corrections due to wall interference. Neglecting Mach-number corrections, comparisons with the lift-curve data suggest equivalent angle-of-attack corrections of roughly $-0.3°$ and $-1.3°$ for the 6%- and 20.5%-wall-porosity experiments, respectively.

More recently, the code has been modified [4.104] to include adaptive meshing in the wake region and used to look at various turbulent transport models. Additional studies of turbulent models have been performed with similar codes for transonic internal flows in [4.105] and [4.106].

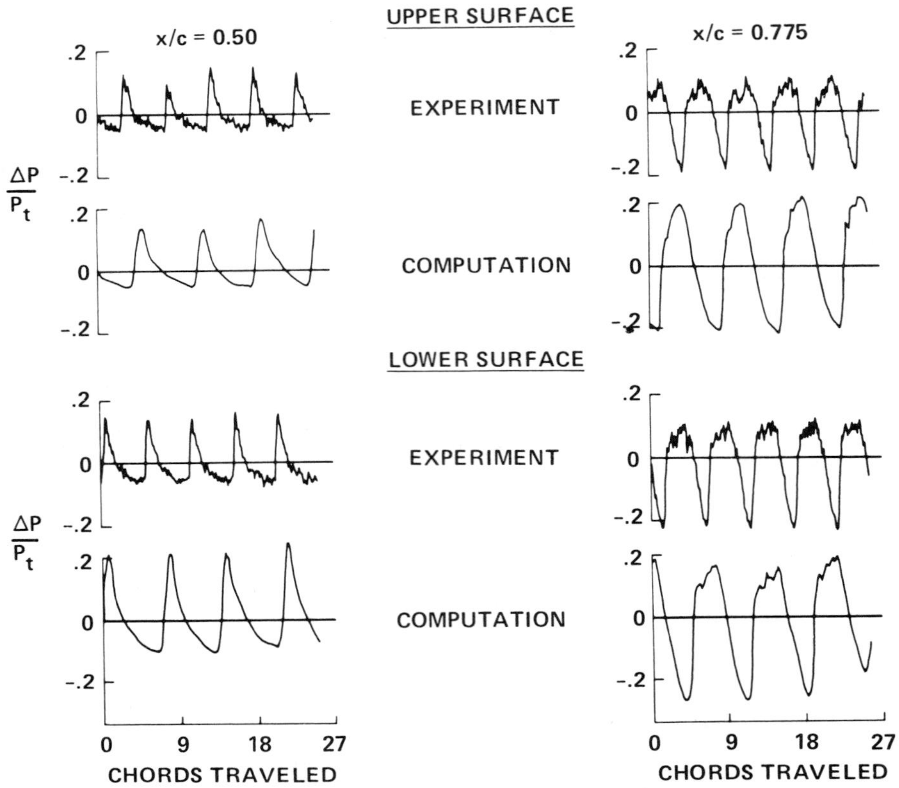

Figure 9. Surface pressure time histories on the 18% circular arc airfoil with unsteady flow; $M_\infty = 0.76$, $\mathrm{Re}_c = 11 \times 10^6$, $\alpha = 0°$.

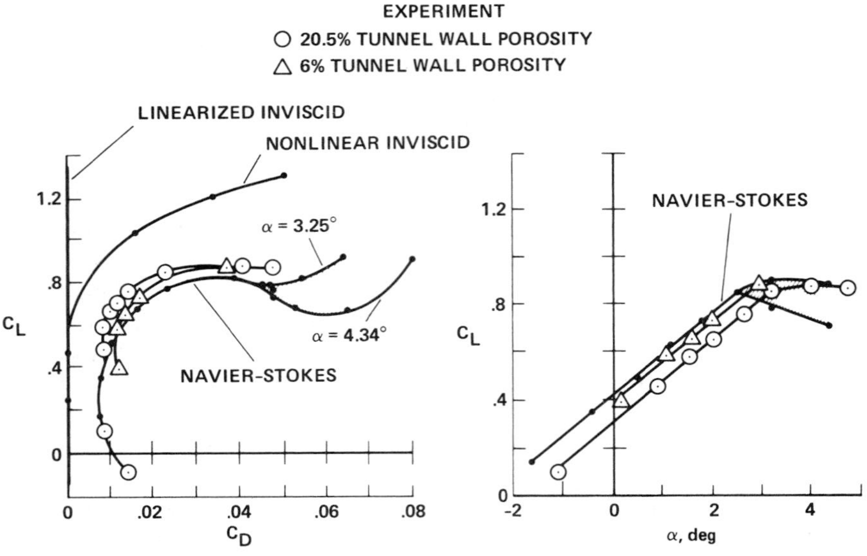

Figure 10. Drag polar and lift curve for Korn 1 airfoil at nominal Mach number of 0.75; $\mathrm{Re}_c = 21 \times 10^6$.

In an independent study, Peery and Forester [4.107] used MacCormack's explicit method to simulate multistream nozzle flows. In this work there were strong viscous–inviscid interactions as two supersonic coflowing jets exhaust into a subsonic coflowing stream. The influence of the external flow on the nozzle flow field was found to be significant.

Steger [4.108] applied the implicit algorithm of Beam and Warming [4.84, 4.109] to transonic flows, and the code was used to calculate transonic aileron buzz by Steger and Bailey [4.47] and flow through cascades by Steger et al. [4.110]. Nietubicz et al. [4.111] applied the method to axisymmetric flows, and Pulliam and Steger [4.112] applied it to three-dimensional flows over bodies of revolution. This code is presently being used by Deiwert to study transonic afterbody flows [4.113]. As an example of this work, Fig. 11 shows the computed surface pressure distribution and computed shear patterns in the conical afterbody region of a cylindrical forebody with a cylindrical plume simulator for a freestream Mach number of 0.9 and an angle of incidence of 6°. The topological features of the pressure map and

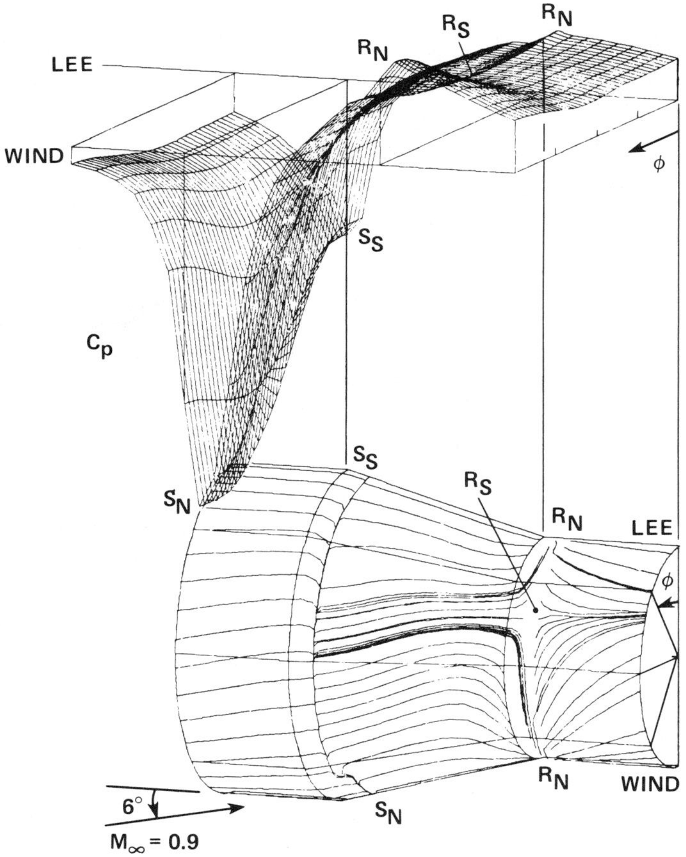

Figure 11. Surface flow pattern and pressure map for Shrewsbury afterbody, $M_\infty = 0.9$, $\mathrm{Re}_d = 2.9 \times 10^6$, $\alpha = 6°$.

the surface shear pattern are identical. Here the singular points S (for separation) and R (for reattachment) are subscripted by S and N for the saddle points and nodes, respectively. The separation bubble is closed, with the line of separation S_S–S_N occurring just upstream of the shock impingement location. Downstream of the reattachment line R_N–R_S–R_N the flow again separates, this time in the crossflow direction, as a free shear layer in the form of a vortex sheet. This is evidenced by the convergence of surface streamlines downstream of R_S.

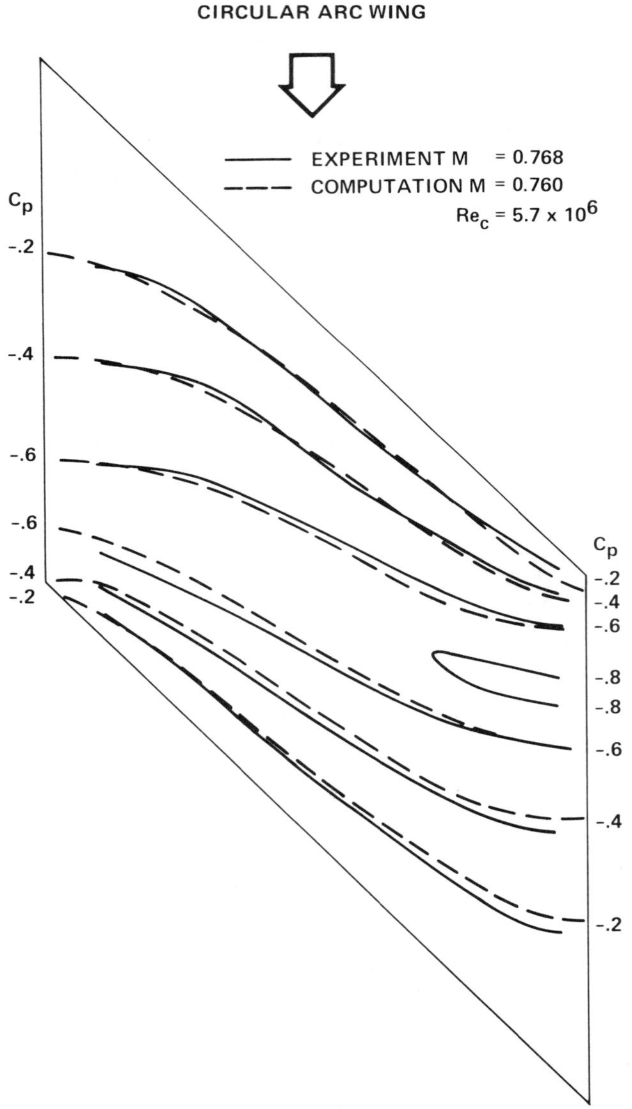

Figure 12. Comparison of Computational and Experimental Isobars.

Another example of the treatment of three-dimensional transonic flow is given by Bertelrud et al. [4.114] in their study of transonic flow about swept wings. The computational algorithm is similar to the MacCormack hybrid method [4.115] and uses a two-equation turbulence model. Figure 12 shows a comparison of computed and experimental surface isobars on a swept, 10%-thick biconvex circular-arc wing. The argreement is quite good. In this case the inclusion of viscous effects is of minor importance; the solution, however, represents a first attempt to simulate viscous flows over three-dimensional wings with a Navier–Stokes code.

All the examples discussed to this point have been computed with either the MacCormack explicit or hybrid-type method or the Beam–Warming implicit method. Recently, Cosner [4.116] reported on transonic simulations about axisymmetric afterbodies using a velocity-splitting method based on the work of Dodge [4.117, 4.118]. This method shows promise of improved computational efficiency for many transonic flow problems. Several investigators are pursuing this approach, and Cosner has extended his work to include three-dimensional flows. This work is unpublished.

Two transonic-flow reviews by Wu and Moulden [4.119] and Spreiter and Stahara [4.120] discuss methods, other than Navier–Stokes, used to consider viscous effects when the interactions are weak. As mentioned earlier, these methods will not be reviewed here. More recent examples of some of these techniques are described by Liou [4.121], Messiter [4.122], and Melnik [2.6].

5 Research Opportunities

The long-range goal of the computational fluid dynamicist is to compute the "real" flow about a complete flight vehicle (e.g., commercial transport, fighter aircraft, reentry vehicle, etc.) in a matter of minutes on the computer. To date only real-flow computational simulations have been obtained for either simplified configurations or components of complicated configurations. To accomplish the goal will require a quantum jump in the state of the art of computation for treating three-dimensional configurations. This includes improvements in three-dimensional body-definition and mesh-generation capabilities, the development of faster and more accurate algorithms for solving the governing equations, and enhancements of turbulence models which describe the true physical flow behavior in the viscous-dominated regions of the flow. In addition, future computations such as these will require considerably larger and faster computers than those currently available or envisioned for the near future.

In this section some of the research areas of opportunity are discussed in which advances will be required to accomplish the stated objective. In addition, some potential applications of computational transonic fluid dynamics are considered, along with some requirements in computer-technology development for exercising the developed computer programs.

In the solution methodology employed by the computational transonic fluid dynamicist, there are basically three areas which will require continued research and development: (a) geometry, which includes the body-definition and mesh-generation processes, (b) algorithm efficiency, and (c) turbulence modeling. These three items were presented by Chapman [4.123] as being principal pacing items for stages of approximation to the Navier–Stokes equations (see Fig. 13).

APPROXIMATION LEVEL TO EQUATIONS OF MOTION	PACING ITEMS
LINEARIZED INVISCID	NONE
NONLINEAR INVISCID WITH COUPLED BOUNDARY LAYER	• GRID GENERATION FOR COMPLEX GEOMETRIES
REYNOLD-AVERAGED NAVIER-STOKES	• ADVANCED COMPUTERS • IMPROVED EFFICIENCY OF NUMERICAL METHODS • IMPROVED TURBULENCE MODELS – LARGE SEPARATED REGIONS – TRANSITIONAL TYPE SEPARATIONS – HYPERSONIC FLOW
TURBULENT EDDY-SIMULATION	• ADVANCED COMPUTERS { • TIME-DEPENDENT THREE-DIMENSIONAL LAW OF THE WALL (HIGH Re APPLICATIONS) • CODE DEVELOPMENT (LOW Re APPLICATIONS) } • IMPROVED EFFICIENCY OF NUMERICAL METHODS • IMPROVED SUBGRID SCALE TURBULENCE MODELS

Figure 13. Pacing items for approximations of full Navier-Stokes Equations.

Before the researcher can generate a finite-difference mesh about a given configuration, he must have an accurate representation of the body geometry. In general this information is normally available to him in the form of blueprints containing longitudinal cross sections and various views, or in the form of digitized data. The process of constructing the required analytically defined vehicle from this type of information is quite tedious and relies heavily on man–machine interaction. Obviously, research in this area which could yield a more automated user-oriented process is highly desirable.

Once the body has been defined analytically, the process of discretizing the surrounding flow field is the next step in the solution process. There exist a scenario of possibilities for discretizing the three-dimensional flow field about a complicated configuration. Schemes such as the "block approach" from Boeing, the "composite subregion approach" from Lockheed (Palo Alto), or the "component adaptive approach" being developed by Lockheed (Georgia) are all viable candidates. Which approach will dominate is moot: all have advantages and deficiences. However, some ideal characteristics which any discretization method should strive for include (a) different surface orientations (i.e., coordinate alignment with the body and/or the peripheral bow shock), (b) a minimum number of coordinate singularities, (c) a minimum number of total grid points with a maximum concentration of grid points near the body, (d) a minimum number of interpolations in going from one grid to the other, for example, in the case of component adaptive overlapping approach, and (e) a minimal adverse effect on the solution procedure with regard to convergence speed and reliability.

Once the method has been selected, the actual process of determining coordinate-point locations can be performed. To accomplish this there are basically three candidate approaches: (a) conformal mappings, (b) algebraic, and (c) differential processes, each with its own advantages and disadvantages. In generating grids with these procedures, properties in the resulting mesh such as coordinate-line skewness, transformation-Jacobian variation, and grid-point clustering should be observed to minimize drastic variations in these properties, which could generate solution-dominating truncation errors.

As the size and complexity of flow-field codes increases to the extent of saturating the computer, the computational fluid dynamicist will begin to devise schemes for making better use of his available number of grid points. One approach for doing this is termed "solution-adaptive" and has already received considerable attention. It should become even more popular in the future. The ultimate objective in such an approach is to distribute the available grid points as the solution develops throughout the three-dimensional flow field so that the solution truncation error is uniform everywhere. It should be realized that this is much easier said than done.

If one is permitted to extrapolate recent progress into the future in the area of numerical-algorithm development, one can expect one or possibly two orders of magnitude of improvement in speed and accuracy over the next decade. Most of this effort will be concentrated on developing algorithms to solve the Reynolds-averaged Navier–Stokes equations. Novel ideas which employ multigrid procedures or zonal methods will probably be used to attain the speed required to solve the complicated flows about complete configurations. This algorithm development process will be closely linked with the computer and its particular architecture, in order to enhance the global efficiency of procedures to be developed.

In the area of turbulence modeling, the building-block experiments as well as turbulent eddy simulations will continue to provide the modelers with information to improve the numerical simulation of turbulence, for flows with small and even large regions of separation. In addition, the modeling will become more physically realistic for transitional separated flows and for hypersonic flows. In the meantime, viscous-flow calculations will continue, and as the improved turbulence models are developed, the computed flows will become more accurate.

Computational transonic fluid dynamics is a new discipline, but over the last decade it has proven to be a potential tool for enhancing the aeronautical design process. The demonstration of this potential has relied heavily on the most powerful computers in existence, and its exploitation in the future will demand even larger and faster computers. To attain the goal of computing the real flow field about complex three-dimensional configurations will require more computational power than is currently available or planned for the near future.

The computational requirements for the flow field about a complete aircraft require at least a machine with speed of 1000 Mflops (million floating point operations per second) and a memory of 240 million words. The largest currently available (Class VI) machines, such as the ILLIAC-IV (16 million words, 25 Mflops), CYBER 205 (4 million words, 80 Mflops), and CRAY-1s (4 million words, 30 Mflops), do not qualify. A summary of the computer speed and memory requirements for existing or planned computers is displayed graphically in Fig. 14 for a typical Navier–Stokes calculation.

In the past NASA has participated in the development and operation of a number of major computational facilities involving the most advanced machines such as the ILLIAC-IV and CYBER 203. The ILLIAC-IV, located at NASA Ames Research Center, provided substantial benefits in the areas of computer technology, algorithm development, and computational fluid dynamics. Its availability as a Class VI machine has permitted the research frontier of computational fluid dynamics to be advanced by as much as 4 to 5 years.

The ILLIAC-IV pioneered new computer technology that is in common use today. It was the first computer to use (a) an interconnection network between multiple processors and multiple memories, (b) emitter-coupled logic for random-access memories, rather than thin-film memories, and (c) the barrel switch for high-speed arithmetic and logic.

The machine architecture of the ILLIAC-IV offered the first radical departure from conventional sequential computers. As a result, new algorithms were developed to take advantage of the machine's parallel features. These newly formulated codes often ran faster in their parallel form than in the original sequential version when they were processed on more conventional scientific computers such as the CDC 7600.

The ILLIAC-IV has provided the means for studying the physics of fluid flows numerically at a level of detail heretofore not practical. For example, the complete Navier–Stokes equations, without approximation, have been solved in investigations of the sources of aerodynamic noise, in the transition to turbulence in the flow of a jet, and in the formation and structure of turbulence spots on flat plates. The ILLIAC-IV has also been used to investigate practical aerodynamic problems that could not be treated effectively in the past because of speed and memory limitations of conventional computers. Examples of phenomena not previously simulated numerically are aileron buzz, airfoil buffet, and reacting-gas nonequilibrium flow about the Shuttle Orbiter during high-speed entry into the atmosphere.

Ames Research Center is proposing to develop a Numerical Aerodynamic Simulator (NAS) which should provide the required computational power in a user-oriented mode to support the objective stated above. The major elements of the NAS system are shown schematically in

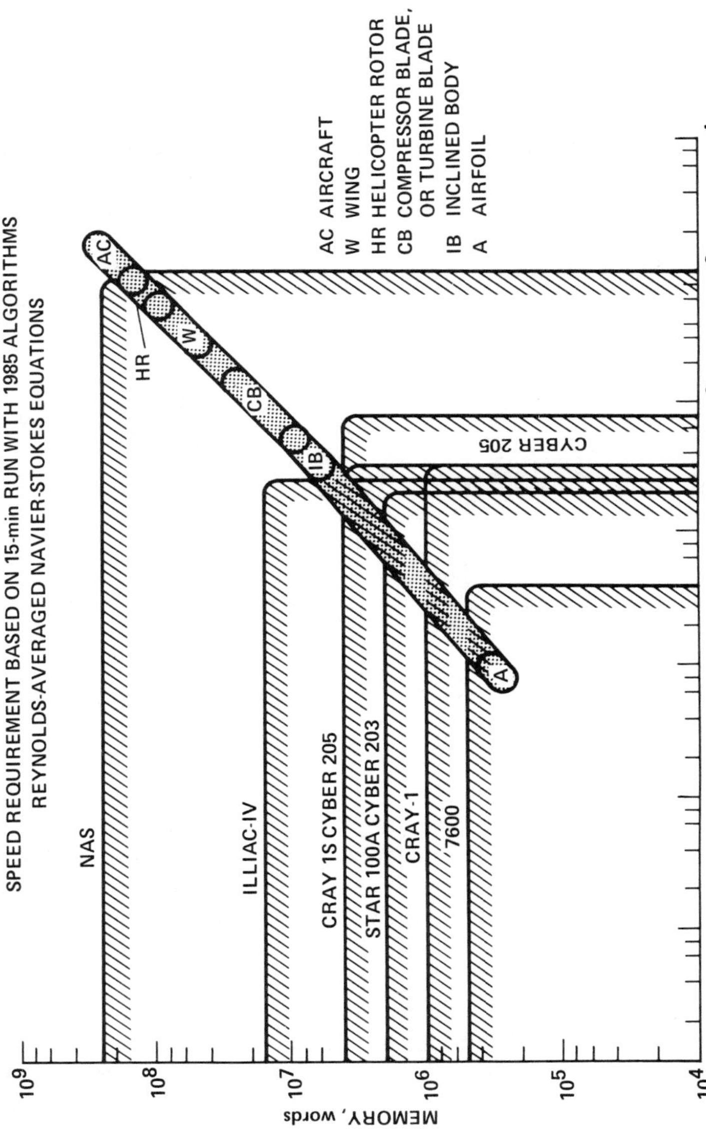

Figure 14. Computer speed and memory requirements compared with computer capabilities.

Fig. 15. The two major elements of the system are the Flow Model Processor and the Support Processor System. The Flow Model Processor consists of a very high-speed computing engine capable of a sustained computational rate of 1 billion floating-point operations per second (1000 Mflops), a 40-million-word random-access main memory, a 200-million-word secondary memory, and a controller. The Support Processing System will provide the general-purpose processing, data storage and manipulation, input–output, user interface, and operational management capabilities required to support the Flow Model Processor. It will also provide 2 billion words of on-line file storage and 100 billion words of off-line storage. A collection of intelligent terminals and associated storage devices as well as a number of graphical display devices will be provided. Data-communication interfaces will be provided to serve local as well as remote users.

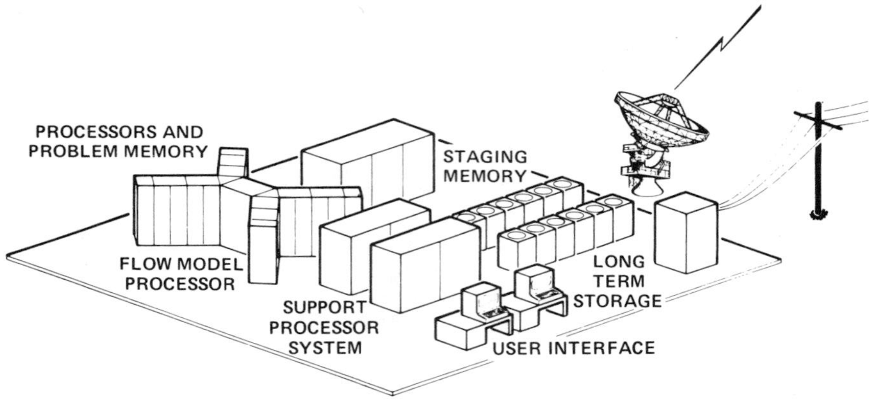

Figure 15. Numerical aerodynamic simulator—System Configuration.

With the advent of the "supercomputers" the computational fluid dynamicist will have to (a) be aware of the machine architecture and use it effectively in the design of his software, and (b) be prepared to reduce the enormous amount of data produced by such a machine. The form of an efficient solution algorithm depends not only on the equations to be solved but also the computer architecture. The vectoral architecture of the new machines and compiler characteristics will also lead to the design of new programming languages and coding philosophies.

The solution of three-dimensional problems on supercomputers is going to require the researcher to efficiently process input, intermediate, and final output data. This pre-processing, on-line processing, and postprocessing of bulk data can only be done effectively using computer graphics. Thus the effective utilization of on-site supercomputers will necessitate networking them with peripheral mini-computers linked with interactive graphics, control, and display devices.

The sophistication of the complicated computer codes to be developed in the future will require the implementation of software development standards and the management of the software development process. Today's transonic computer codes are generally developed by a single researcher, are very difficult to modify, and are often not reliable. In general the computer codes that will be designed and utilized for supercomputers in the future will be constructed not by a single researcher but by a team of researchers and programmers. Constraints such as those depicted in Fig. 16 will dominate

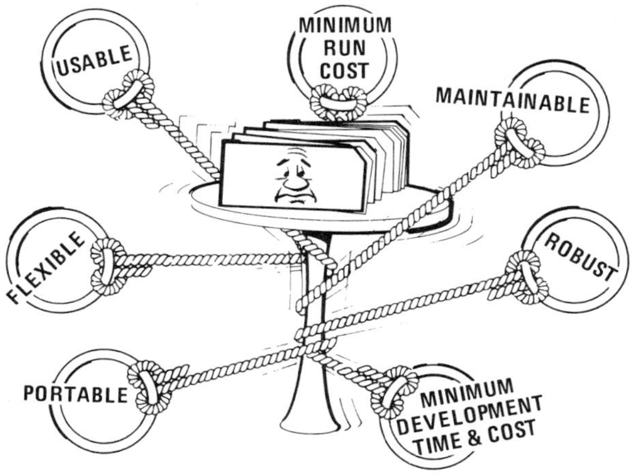

Figure 16. Software development considerations.

the construction process. It is thus mandatory that before these codes are exported to the user community, they be carefully tested, certified, and well documented. That is, the true capabilities and limitations of the codes must be known, and their actual performance and cost must be specified.

With improvements in computational techniques and advances in computer capability, aerodynamic researchers must and will find more effective ways of applying computational tools in design and analysis. One area that should receive considerable attention is the prediction of aeroelastic performance in the transonic regime. No clearly optimum manner of using nonlinear unsteady transonic codes for prediction of the flutter boundary of an aircraft wing presently exists. However, substantial progress is anticipated in this area during the 1980s as a result of continuing improvements in unsteady transonic codes and a tendency in design toward the use of configurations and materials that require improved aeroelastic predictions. With the growing popularity of design concepts such as aeroelastic tailoring and swept-forward wings, computational aeroelastic simulations will begin to play a dominant role in the design process.

6 Concluding Remarks

In attempting to predict advances in research, there is a tendency to overestimate what can be achieved in the short term and underestimate what can be achieved in the long term. With this in mind, let us consider in general terms how the discipline might progress in the 1980s.

There will continue to be evolutionary advances in steady and unsteady inviscid transonic flows. The emphasis will be on applications. With these tools now in the hands of industry, we can expect extensions to the treatment of realistic, complex aerodynamic configurations and the adaptation of unsteady codes for use in aeroelastic predictions.

We expect revolutionary advances in the treatment of viscous transonic flows. These advances will be fueled by (a) advances in numerical solution techniques, (b) larger, faster computers, and (c) improved turbulence models resulting from detailed experiments and computational large-eddy simulations. This capability will be in the hands of industry and will be an integral part of the design process, especially for fighter aircraft, by the end of the decade.

CHAPTER 23

The Transonic-Flow Theories of High- and Low-Aspect-Ratio Wings

H. K. Cheng*

1 Introduction

This paper reviews two theoretical developments in the three-dimensional (3-D) transonic flow studies, which represents an extension of the lifting-line and the slender-wing concepts in the classical theory [4.124, 4.125, 4.126] to the analyses of the lift-dominated, nonlinear mixed flows.

The lifting line in Prandtl's original model for a high-aspect-ratio wing was assumed to be *straight* and *perpendicular* to the main flow direction [4.124], as it is in Van Dyke's asymptotic theory [2.127]. Cook and Cole [4.127] formulated the corresponding problem in the framework of the transonic small-disturbance theory. This extension as such does not reflect the full scope of the lifting-line concept. From the viewpoint of transonic aerodynamics, the influence of wing sweep [4.128] is crucial and occupies a central position in the present paper. The various features associated with the high- and low-aspect-ratio wings to be discussed below may well be regarded as consequences brought about by the different magnitudes of the leading-edge sweep angle.

As in most asymptotic theory, the approaches represented by the two opposite limits of the aspect ratio share the goal of reducing the 3-D problem locally to a 2-D one. Thereby, one may gain better physical insight, in addition to greater simplicity in the problem analysis. This may in turn facilitate design analyses and data correlation for both experiment and computation.

* Department of Aerospace Engineering, University of Southern California, Los Angeles, CA 90007.

The material used in the exposition is taken mostly from work by the author and his associates, and is believed to be more complete than comparable results from other sources. Remarks on existing work in the areas of interest will be given. Specific comments on technical details have been omitted, along with certain portions of the original manuscript, to conserve space. Fuller discussions may be found in the cited references.

The review has focused mainly on the nonlinear aspects of the steady transonic flows. Nonlinear regimes can be identified, in which unsteady effects may alter the 3-D problems analyzed. For studies of unsteady transonic flow in the linear slender-wing regime, readers are referred to Liu, Platzer, and Ruo's paper [4.129].

A serious aerodynamic study cannot ignore the important influence of the boundary layers, but a successful treatment of the inviscid outer flow is a prerequisite for developing a sound approach to the problem of inviscid-viscous interaction. The possibility for adopting the present approach to the study of boundary layers on a swept wing is indicated in Section 4.

2 Transonic Slender Wings: The Equivalence Rule Involving Lift

2.1 The Area Rule

The 3-D flow field far above an obstacle determines the drag rise associated with the shock loss in the outer flow. The most important aerodynamic concept which may be used to control this outer flow is the equivalence rule of Oswatitsch, Keune, and Whitcomb [4.130–4.132]. It is well established from the classical linear theory [4.133–4.135] that the flow far from a slender body at small incidence may be identified with that produced by an (equivalent) *axisymmetric* body having the same (axial distribution of) cross-sectional area $S_c(x)$. The contributions by Whitcomb [4.130] and Oswatitsch [4.131, 4.132], as well as Cole and Messiter [4.136], on the transonic equivalence rule lie essentially in their demonstrations, through experimental verification and theoretical formalism, that the same rule applies also for a *nonlinear* transonic *outer* flow. In the cited works, the typical apex angle λ of the slender wing is assumed to be as small as the wing thickness ratio τ. There is also a restriction on the degree of asymmetry in the cross flow, namely, $\alpha/\tau \ll 1$, where α denotes an angle of attack measured from zero lift.

The theoretical basis for this version of the equivalence rule is brought out most clearly, perhaps, by Ashley and Landahl [4.137], who relax the unnecessary restriction on α/τ to $\alpha/\tau = O(1)$.

The theory should apply just as well to wings with a unit-order apex angle. The formulation for this case was brought out formally by Spreiter and Strahara [4.138]; its validity has been examined on the basis of experimental field measurement and computer analyses by the same authors quite recently [4.139].

Underlying the cited works and related studies by Oswatitsch, Berndt, and Drougge [4.140–4.144] is the concept that the body affects the nonlinear transonic outer flow in the same manner as does a *line source* with a strength proportional to dS_c/dx (x being the distance along the wind direction). In this sense, it has been referred to as the *area* rule [4.140].

2.2 Asymmetry and the Nonlinear Lift Contribution

Domains of α, τ, and λ exist in which the line source does not dominate the far field. For $\lambda = O(1)$, it is quite easily shown that the line doublet associated with the lift may contribute equally to the nonlinear outer flow as soon as α reaches an order of $\sqrt{\tau}$. Cheng and Hafez pointed out that most transonic transport designs fall within this domain [4.145, 4.146]. Owing to the line doublet, the outer flow is no longer axisymmetric, and the correct version of the equivalence rule involving lift was not obtained until subsequent analyses brought out a number of unsuspected lift corrections.

This begins with Barnwell's [4.147] discovery of a sourcelike correction to the outer flow overlooked by Cheng and Hafez [4.145, 4.146], which arises from matching the nonlinear terms of the inner and outer solutions. Subsequent formulations by Cheng and Hafez [4.148–4.150] have uncovered several additional sourcelike corrections of comparable orders. These result from the second-order nonlinear corrections to the lift-dominated inner (Jones) solution, not completely accounted for by the standard transonic small-disturbance (TSD) theory [4.137, 4.151, 4.152], and producing effects of first-order importance in the outer flow.

Although Barnwell's original contribution [4.147] does not lead to a correctly formulated reduced problem (even apart from the omission of several nonlinear lift corrections mentioned), his work is significant for bringing out a key element in the development. A subsequent report by Barnwell [4.153] contains results similar to those of Cheng and Hafez [4.148–4.150], using body-oriented coordinates and adopting some of the mathematical results in the Appendix of [4.149]. An error appears in Barnwell's report but can be easily corrected; it is traceable to Eq. (40) of [4.153], where a correction to the transferred tangency condition was overlooked.

The importance of the nonlinear lift correction to the equivalence rule has been studied by Cheng [4.154], using a transport design as a concrete example (see Section 2.6 below).

Recently, Cramer [4.155] repeated the analysis for a limiting case in Cheng and Hafez's work [4.148–4.150], which pertains to a lifting wing without thickness, i.e. $\tau \equiv 0$. His analysis reproduces most results of [4.148–4.150] for this case; however, he also claimed to have simplified and improved on existing work, in view of the absence of certain second-order terms in his results. This discrepancy may be traced to the incompleteness in Cramer's matching analysis. Completing his analysis to the level in question brings agreement. A note by Dr. Cramer to this effect is forthcoming [4.156].

2.3 The Asymptotic Theory: Three Sources of Nonlinear Corrections

The theory determining the lift contribution in the equivalence rule [4.148–4.150, 4.153] considers solutions in two distinct flow regions. The linear dimension of the inner region in the cross-flow plane is taken to be comparable to the half span b, as in [4.131, 4.132, 4.136–4.144], but the transverse length scale of the outer region is $b' = b/\varepsilon$, with the small parameter ε chosen to allow the influence of τ and λ, as well as α, on the structure of the outer flow. For the present purpose, it suffices to take

$$\varepsilon \equiv \sqrt{(\gamma + 1)M_\infty^2 \tau\lambda} \tag{2.1}$$

which reduces to $\varepsilon = \sqrt{\gamma + 1}\, M_\infty \tau$ for a slender wing or body with $\lambda = \tau$. The outer flow, to which the transonic equivalence rule will apply is governed by three parameters [4.148–4.150],

$$K \equiv \frac{(1 - M_\infty^2)\lambda^2}{\varepsilon^2}, \qquad \sigma_*^2 \equiv \frac{\sqrt{(\gamma + 1)|\ln \varepsilon|\lambda^3}\, M_\infty \alpha}{\sqrt{\tau}}, \tag{2.2a}$$

$$\Gamma_* \equiv \frac{8}{(\gamma + 1)|\ln \varepsilon|\lambda^2}, \tag{2.2b}$$

which controls three distinctly different nonlinear effects. The K in Eq. (2.2a) may be identified with the transonic similarity parameter (reflecting the importance of $\phi_x \phi_{xx}$ in the TSD equation [4.151, 4.152]; the σ_* controls the linear and nonlinear lift contributions to the outer flow, and finally the Γ_* indicates the relative importance of $(\partial/\partial x)(\phi_y^2 + \phi_z^2)$, absent from the standard TSD equation.

The asymptotic theory yielding the equivalence rule involving lift [4.148–4.150] is first developed for the limit $\varepsilon \to 0$, keeping σ_*, Γ_*, and K fixed. The formulation remains valid also for σ_* etc. vanishing with ε. For an *unbounded* σ_* corresponding to $\tau \to 0$, rescaling of the outer flow based on α and λ can be made, but the reduced equation systems for the bounded and unbounded cases prove to be completely equivalent [4.148–4.150] therefore, the use of the set of equations (2.1), (2.2) suffices.

2.4 Nonlinear Corrections to Jones's Solution and Their Significance

An important modification of the transonic area rule results from corrections to Jones's slender-wing solution for the inner region. This can be quite readily seen from the PDE for the perturbation potential,

$$\phi_{yy} + \phi_{zz} = (M_\infty^2 - 1)\phi_{xx} + M_\infty^2 U_\infty^{-1} \frac{\partial}{\partial x}\left[\left(1 + \frac{\gamma - 1}{2} M_\infty^2\right)\phi_x^2 + \phi_y^2 + \phi_z^2\right] + \cdots. \tag{2.3}$$

The three nonlinear terms on the right are the second-order compressibility corrections to the divergence-free equation in the slender-body approximation, and may be interpreted as a *source distribution* in the cross-flow plane. Far enough from the x-axis, the integrable part of this distribution produces an effect equivalent to a *line source*. The strength of the latter does not vanish with the thickness, as long as the ϕ gradient is not identically zero. The ratio of this source strength to that which arises from the thickness, i.e. to dS_c/dx, must be proportional to α^2/τ, which is essentially $\sigma_*^2/|\ln \varepsilon|$. The nonintegrable part of the right-hand member of Eq. (2.3) leads to corrections which are singular. Among these is one proportional to $\alpha^2(\ln r)^2$, where $r = \sqrt{y^2 + z^2}$; this gives rise to a correction for ϕ in the far field proportional to $(\alpha^2|\ln \varepsilon|\ln r)/\tau$, and hence a correction to the line source proportional to σ_*^2.

The last two square terms in Eq. (2.3) signify the kinetic energy of the cross flow and are absent from the TSD equation [4.151, 4.152]. These terms have been treated by Van Dyke [4.157] and Lighthill [4.158] for slender bodies in supersonic flow, although their effect in that context is far less significant. Interestingly, the line source contributed by these terms give rise to a non-vanishing total volume flux, and hence an equivalent afterbody with a cross-section area proportional to the vortex drag. There is also a second-order nonlinear lift correction to the wing boundary condition which contributes to a line sink, resulting in a reduction of the cross-section area of the equivalent afterbody by one-half [4.149, 4.150].

The last feature is not producible by the standard TSD theory. The cross-flow kinetic-energy terms were omitted in Hayes's study [4.159, 4.160] on the grounds that waves tend to approach planar ones, rendering their effect noncumulative. This argument overlooks the nonlinear contribution to the equivalent source effect which provides the initial data for the waves.

2.5 The Equivalence Rule

An application of Cheng and Hafez's [4.148–4.150] asymptotic theory is the deduction of a transonic equivalence rule allowing for the lift contribution. The formulation shows that the outer nonlinear perturbation field is completely determined to the leading order by a line doublet and a line source; the analysis succeeds in identifying the strengths of the pair with the axial distributions of the lift $F(x)$ and of the *cross-section area* of an *equivalent body*, $S_e(x)$. The latter can be explicitly calculated from the geometrical cross-section area $S_c(x)$ and the lift distribution over the wing surface. The rule may then be stated as: the outer flow at a given transonic parameter K is the same as long as $F(x)\sigma_*|\ln \varepsilon|^{-1}$ and $S_e(x)$ remain unchanged.

Essential in the theory is the matching of the inner and the outer solutions, which identifies the line doublet and the line source in the outer solution with its counterparts in the inner solution. Apart from the doublet and the source, the parts matched contains the nonlinear corrections noted in Section 2.4, which are determined once the axial lift distribution $F(x)$ is specified. The outer solution has parts which are regular near the axial and also have to be matched; their matching determines the pressure in the inner region and other feedback effects from the far field. In short, the line-source and line-doublet pair suffices to determine ϕ in the outer region (to the leading order).

The strength of the line source for the outer flow, $q(x)$, determined from the matching is found to be expressible as that representing the perturbed flow over a slender body with a cross-section area $S_e(x)$. More precisely [4.148–4.150],

$$q(x) \propto \frac{d}{dx} S_e = \frac{d}{dx} \{S_c(x) + \sigma_*^2[(8\pi)^{-1}(1 + |2 \ln \varepsilon|^{-1})F_x^2 + |2 \ln \varepsilon|^{-1} T(x) + 8^{-1}\Gamma_* E(x)]\}, \tag{2.4}$$

where $E(x)$ is a normalized kinetic energy in the Treffz plane, which gives rise to a nonvanishing afterbody, and, interestingly, $T(x)$ assumes a form identical to that of $E(x)$ with the differential pressure (across the vortex sheet) replacing the differential side wash. Both E and T are nonnegative; thus, the theory yields an inequality

$$S_e(x) > S_c(x), \tag{2.5}$$

showing that the equivalent body has an *increased* cross-section area depending on the lift.

In passing, we remark that the asymptotic theory of Cheng and Hafez [4.148–4.150] is developed up to the (relative) order $\varepsilon^2(\ln \varepsilon)^2$. However, for wings lacking bilateral symmetry, the version of the equivalence rule described above is subject to a larger error of order $\varepsilon \ln \varepsilon$.

The formulation of the outer problem in the reduced variables also furnishes a transonic similitude; accordingly, local pressure and Mach number can be correlated as functions x of $\eta \equiv r/b'$, $\omega \equiv \tan^{-1}(z/y)$, and K; the shock boundary as $x = x^D(\eta, \omega, K)$; and the drag rise due to shock loss in the outer flow as $D_W/\rho_\infty U^2 b^2 \tau^2 M_\infty^2 = f(K)$, or

$$c_{D_w} = \tfrac{1}{2}\tau^2 M_\infty^2 \mathrm{AR}\, f(K), \tag{2.6}$$

where AR is the aspect ratio $4b^2$/(wing area). If a shock is *fully developed* in the outer flow, $f(K)$ is a unit-order function of K, except at large σ_*, for which $f = O(\sigma_*^2)$.

Equation (2.6) represents that part of the drag rise that *cannot* be controlled by the use of supercritical wing sections; whether the effect in question is important may be inferred from the magnitude of $\frac{1}{2}\tau^2 M_\infty^2 \mathrm{AR}$ gauging the order of c_{D_w} in Eq. (2.6). A rise of 0.002 in c_{D_w} may be considered a substantial increase and has been used by Goodmanson [4.161] to define the drag-rise Mach number. For the example studied in [4.161] ($M_\infty = 0.98$, $\tau = 0.0263$, and AR $= 6.3$), the gauge factor in Eq. (2.6) is 0.0023. The equivalence rule is thus seen to be relevant. (The τ above includes the body contribution [4.149, 4.150].)

2.6 Examples

The importance of the nonlinear effect in question is best illustrated by a comparison of lift corrections to the equivalent cross-section area $S_e(x)$ of an appropriate transonic airplane model, for several alternative lifting-surface arrangements. Figure 1 presents the result of such a study taken from [4.154], in which the general specifications for the Mach 0.98 design of Goodmanson [4.161] is adopted. With gross weight 287,300 lb, altitude 39,000 ft, $b = 60$ ft, $c_0 = 80$ ft (axial length of the lifting surface), $l = 171$ ft, and $S_{c\,\max} = 270\ \text{ft}^2$, we have $\varepsilon = 0.0512$ and Eq. (5) becomes

$$S_e - S_c = 0.0499 F_x^2 + 0.180T + 0.267E. \tag{2.7}$$

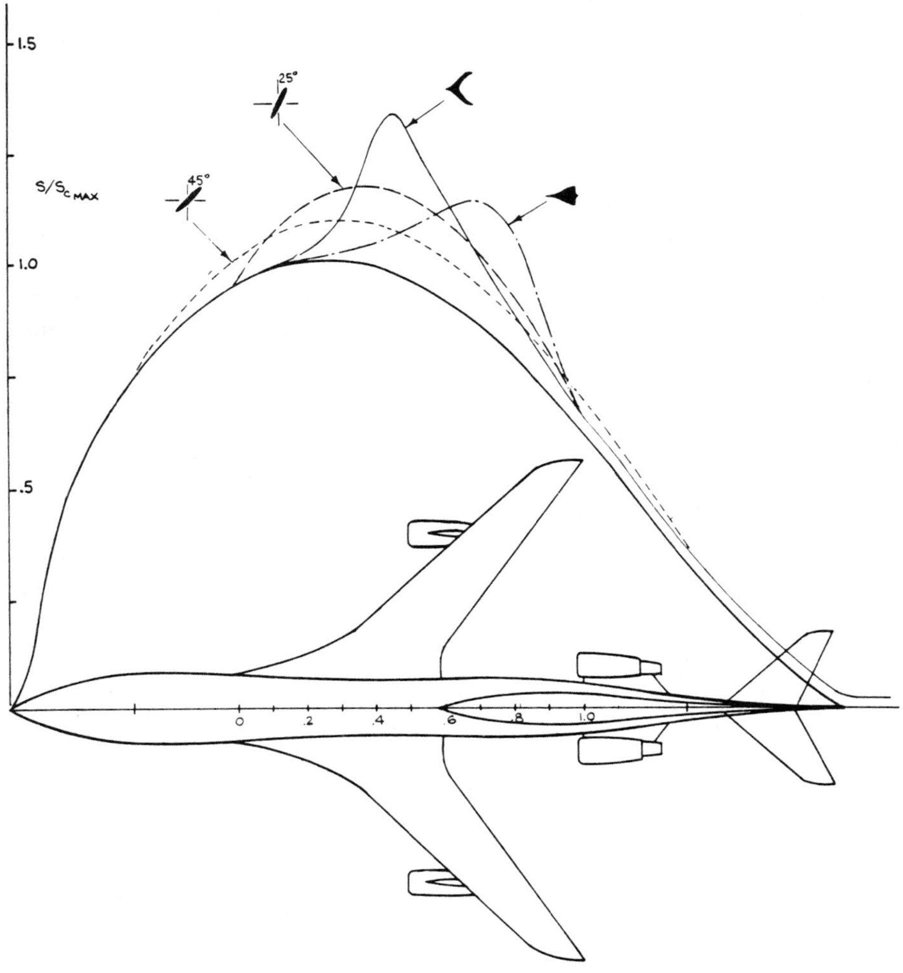

Figure 1. Equivalent-body cross-section area of a Mach 0.98 transport design with four alternative lifting-surface arrangement.

The functions F_x, T, and E have been calculated for various lift distributions; results computed for four representative planforms (two oblique [4.162, 4.163], one symmetric swept, and one Concorde) are shown. The very smooth full solid line in the figure represents the geometrical cross-section distribution $S_c(x)$ taken from the original Goodmanson design study [4.161].

Except for one of the oblique arrangements, all planforms have the same leading-edge swept angle. The increase in the cross-section area of the equivalent body in all cases and their (somewhat surprising) differences are clearly evident. For a conventional design, a rise in S_e by 35% over the zero-lift maximum occurs over a distance of a root chord. This study clearly indicates the advantage of an oblique-wing design [4.162, 4.163] for reducing the far-field shock loss in the transonic range. (A smoother S_e distribution for a symmetric swept wing is, of course, possible through modification of the lift distributions near the wing roots in conjunction with a change in $S_c(x)$.)

2.7 Untreated Aspects

Inasmuch as the area rule remains an essential consideration in transonic aircraft design, the significant lift modifications to the outer flow shown above must be of equal relevance. Nevertheless, its importance has yet to be confirmed by experiments and flow-field computations. The lask of computational assessment is due in part to inadequate code and computer capability as well as cost considerations. Computational studies focusing attention on the nonlinear outer-flow region far from the wing should prove helpful.

Except for limited efforts in Hafez's thesis work reported in [4.149] (and a Flow-Research Company Memorandum by Hafez), which assumes a small lift correction, meaningful examples illustrating the asymmetric, nonlinear transonic outer flow consistent with the theory of [4.148–4.150] do not exist. Of particular interest would be the identification of those (normalized) S_e and F distributions which will provoke shock loss in the outer flow.

Among aspects untreated in the theory is the problem in which a shock in the outer flow reaches and terminates at the axis (wing's vicinity). The theory is not strictly valid when the wing aspect ratio becomes unbounded; the main flow structure may still be retained in this case, which represents an interesting overlapping domain for the slender-wing and the lifting-line theories (cf. Section 4).

3 The Transonic Lifting-Line Problem

3.1 Supercritical Airfoils and Lifting-Line Theories

The use of wing sweep to control the compressibility effect has been practiced by aircraft engineers since Busemann's early study [4.164–4.167]. Current interest in swept-wing design has focused on the potentiality for adopting 2-D supercritical airfoil data, abundant through extensive research during the last decade [4.168–4.172, 4.95]. Implicit in this design idea is, thus, the lifting-line concept, but the flow problem in question must deal with the crucial elements of the wing sweep and the nonlinear, mixed component flow.

For a straight unswept wing, the 3-D effect would appear mainly as a local upwash correction. Cook and Cole [4.127] have elucidated the asymptotic analysis of this problem; Cook [4.173] also carried out a formal study of the uniqueness of the solution. A local wing sweep brings forth two elements not found in the classical work. One is a pronounced upwash induced by the trailing vorticity next to the trailing edge; the other is an additional compressibility correction in the continuity equation, arising from the spanwise density variation. The second element represents a fundamental departure from Prandtl's concept in that, on account of the sweep, the component flow around each wing section can no longer be treated as a genuine *plane* flow.

Comparable contributions to the upwash and to the continuity equation arise also from the curvature of the planform centerline, wherever its magnitude becomes suitably large; their relevance to the studies of transonic oblique wings as well as aquatic-animal swimming propulsion were noted in [4.174]. Among the earlier works involving sweep and centerline curvature are the linear analyses by Weissinger [4.175], Krienes [4.176], Dorodnitsyn [4.177], and Thurber [4.178]. A more relevant study in this category is provided by the relatively recent analysis of Cheng for a high-aspect-ratio oblique wing [4.179, 4.180], where comparison with the panel method is also made (see Section 3.5).

The theory and results to be discussed below are based primarily on the works of Cheng and Meng [4.181–4.183], and also Cheng, Meng, Chow, and Smith [4.184, 4.185]. An important theoretical development in [4.181–4.183] is the uncovering of a local similarity in the 3-D flow structure for a certain class of swept wings, permitting the reduced 2-D equations to be solved *once* for *all* span stations (Section 3.4). Equally important, perhaps, is their comparison with full-potential solutions (Section 3.6). As noted, Cheng and Meng's theory treats both centerline sweep and curvature effects for an oblique wing, but is applicable also to the more conventional swept wing (barring the vicinity of the wing apex; see Section 3.6).

A formulation for the transonic problem of a straight oblique wing has been made by Cook [4.186], using oblique coordinates. The upwash expression presented therein appears to be in accord with that given by Cheng and Meng [4.181–4.183]. Some discrepancies will be noted below.

The following discussion will be confined mainly to analyses of oblique wings and swept wings with a centerline made up of a finite number of *straight segments*, so that the centerline curvature will not enter directly into the local, inner-problem formulation. It suffices in this case to employ Cartesian coordinates x', y', z', with z' increasing in the lift direction and y' running along the centerline.

3.2 The Three Parameters in the Asymptotic Theory

Through Section 3, we let α characterize the thickness ratio or the incidence (measured from the zero lift) of the wing sections. Denote the local sweep angle by Λ; the local component Mach number is then $M_n = M_\infty \cos \Lambda$. There are three parameters controlling the reduced inner problem:

$$K_n \equiv \frac{1 - M_n^2}{\alpha^{2/3} M_\infty}, \qquad \Theta \equiv \frac{\Lambda}{\alpha^{1/3}}, \qquad \varepsilon \equiv \frac{1}{\alpha^{1/3}\, \mathrm{AR}_1}, \tag{3.1}$$

where K_n is basically the transonic similarity parameter for the component flow [4.151, 4.152, 4.187, 4.1], Θ is a rescaled sweep angle, and ε is the reciprocal of a reduced aspect ratio for the transonic range. (The α and ε defined here are not to be confused with those in Section 2.) The sweep range of interest is that which will keep K_n at order unity; this amounts to requiring the sweep angle to be sufficiently small, i.e., Θ to remain also at unit order. The asymptotic theory of [4.181–4.183] pertains to the limit $\varepsilon \to 0$, holding K_n, Θ, and $\alpha^{1/3}\varepsilon^{-1}$ *fixed*. The requirement on $\alpha^{1/3}\varepsilon^{-1}$ is made so that the remainder in the equations can be estimated to be of order ε^2. Restricting to a high subsonic and a linear sonic outer flow, the requirement on K_n and Θ can be put as

$$\Theta^2 \le K_n = O(1). \tag{3.2}$$

3.3 The Reduced Inner Problem and an Unsteady Analogy

The inner variables for the inner problem are readily suggested by the 2-D TSD theory as

$$\hat{x} \equiv \frac{2x'}{c_0}, \qquad \hat{z} \equiv \frac{2M_n^{2/3}\alpha^{1/3}z'}{c_0},$$
$$\hat{y} \equiv \frac{y'}{b}, \qquad \hat{\phi} \equiv \frac{2\phi}{\alpha^{2/3}M_n^{-2/3}U_n c_0} \tag{3.3}$$

The PDE governing the inner problem, specialized to the case of a straight centerline ($d\Lambda/dy' = 0$), is

$$\frac{\partial}{\partial \hat{x}}\left(K_n \hat{\phi}_x - \frac{\gamma + 1}{2}\hat{\phi}_{\hat{x}}^2\right) + \hat{\phi}_{\hat{z}\hat{z}} = 2\varepsilon\Theta\hat{\phi}_{\hat{x}\hat{y}} + \cdots, \tag{3.4}$$

omitting terms of order ε^2. The left-hand member is that in the von Kármánn equation [4.187, 4.1] and the right-hand member is the additional 3-D compressibility correction mentioned earlier. Interestingly, the latter leads to a line-source effect in the far field, and can be explicitly determined in some cases [4.181–4.183]. To be sure, the source effect in question is that for the *volume* flux. Though unsuspected, similar effect should rise also in a linear problem, unless the wing has no thickness. This feature was absent in Cook's work [4.186].

Unlike the inner problem for swept wings in the linear regime, the perturbation velocity in this inner region may be considered continuous across the trailing vortex sheet; this follows from the relatively small sweep angle $\Lambda = O(\alpha^{1/3})$ stipulated in the analysis.

If $\hat{y}$ is taken as a normalized time, Eq. (3.4) can be interpreted as one governing a 2-D TSD flow *near* the quasisteady limit, in which the right-hand member represents an unsteady correction of order ε. An example of the solution via the unsteady analogy will be shown later.

3.4 The Perturbation Problem and 3-D Similarity Structure

The inner problem defined by Eq. (3.4) and its boundary conditions is amendable to a perturbation analysis for small ε, in which $\hat{\phi}$, as well as the interior shock boundary $\hat{x} = \hat{x}^D(z, y)$, if it exists, can be developed as

$$\begin{aligned} \hat{\phi} &= \hat{\phi}_0(\hat{x}, \hat{z}; \hat{y}) + \varepsilon\hat{\phi}_1(\hat{x}, \hat{z}; \hat{y}) + \cdots, \\ \hat{x}^D &= \hat{x}_0^D(\hat{z}; \hat{y}) + \varepsilon\hat{x}_1^D(\hat{z}; \hat{y}) + \cdots, \end{aligned} \tag{3.5}$$

anticipating a weak (logarithmic) dependence of $\hat{\phi}_1$ and x_1^D on ε. The equations governing $\hat{\phi}_1$ and $\hat{x}_1^D$ may accordingly be *linearized*. As in most perturbation analysis involving shock, the jump condition will be analytically transferred to the unperturbed position. Thus, the reduced problem is essentially a perturbation of the basic 2-D component flow, with the right-hand member of Eq. (3.4), and similar terms in other equations, appearing as a *nonhomogeneous* term determined by $\hat{\phi}_0$.

One may see that the problem of uniqueness for $\hat{\phi}_1$ is the same as that for the perturbation in a purely 2-D problem for $\hat{\phi}_0$ due to a *small variation* in the wing geometry. This was precisely the problem addressed by Cook [4.173] in connection with the lifting-line theory for straight and unyawed wings (my comments on [4.173] appear in the *Mathematical Review*, vol. **58**, No. 3916, 1979).

An interesting property of the equation system for the reduced inner problem is its admission of a similarity structure which greatly facilitates computation work. This becomes possible owing to the linearity in $\varepsilon\hat{\phi}_1$, permitting its decomposition into *separate* parts, each of which has similarity solution independent of $\hat{y}$, as does the basic solution $\hat{\phi}_0$. The geometrical requirement for this similarity structure is that the wing section at each span station be generated from the *same* airfoil profile, keeping the same thickness ratio and incidence. More precisely, the ordinates for the two ($\pm$) surfaces have to be

$$z' = \frac{c_0}{2}[\alpha\hat{Z}^{\pm}(\hat{x}) + \alpha^{1/3}\hat{Z}_B(\hat{y}) + \alpha\varepsilon\hat{I}(\hat{y})(\hat{x} - \hat{x}_0)], \tag{3.6}$$

where $Z^{\pm}$, $\hat{Z}_B$, and I are of unit order. The function $\hat{Z}_B$ represents an upward wing bend, and $\hat{I}$ wing twist, adequate for the control of the 3-D effects. With the similarity variables defined as

$$\tilde{x} \equiv \frac{\hat{x}}{\hat{c}}, \qquad \tilde{z} \equiv \frac{\hat{z}}{\hat{c}}, \qquad \tilde{y} \approx \hat{y}, \qquad\qquad \tilde{\phi} \equiv \frac{\hat{\phi}}{\hat{c}}, \tag{3.7}$$

the similarity flow structure admissible to Eq. (3.4) for a fixed K_n and Θ is given by

$$\tilde{\phi} = \tilde{\phi}_0(\tilde{x}, \tilde{z}) + \varepsilon\Theta \frac{d\hat{c}}{d\hat{y}} \cdot \tilde{\phi}_1(\tilde{x}, \tilde{z}) - \varepsilon\sqrt{K_n}\tilde{\mathbb{C}}_1^i(\tilde{y})\tilde{z} + \varepsilon\left[\sqrt{K_n}\tilde{\mathbb{C}}_1^i(\tilde{y}) + \hat{I}(\tilde{y}) + \Theta\frac{d}{d\tilde{y}}\hat{Z}_B(\tilde{y})\right]\tilde{\phi}_2(\tilde{x}, \tilde{z}), \tag{3.8}$$

where $\tilde{\mathbb{C}}_1^i(y)$ is a normalized upwash correction to be discussed shortly. Through Eq. (3.8), the 3-D compressibility correction in Eq. (3.4) and the upwash correction (including wing twist and bend) can be independently accounted for by $\tilde{\phi}_1$ and $\tilde{\phi}_2$. The resulting system of equations for $\tilde{\phi}_0$, $\tilde{\phi}_1$, and $\tilde{\phi}_2$ is solved *only once* for *all* span stations [4.181, 4.182].

With $\tilde{\mathbb{C}}_1^i(y)$ determined by the outer solution (Section 3.5), the upwash influence can be controlled through Eq. (3.8). Of interest is the possibility of eliminating the wake-induced velocity completely (to within order $\varepsilon^2 \ln \varepsilon$) by setting

$$\sqrt{K_n}\tilde{\mathbb{C}}_1^i(\tilde{y}) + \hat{I}(\tilde{y}) + \Theta\frac{d}{d\tilde{y}}\hat{Z}_B(\tilde{y}) = 0.$$

3.5 The Near-Wake and Far-Wake Influence

The upwash at a straight, unswept lifting line in the classical theory may be considered a *far-wake* influence, because the vorticity shed locally, $d\Gamma/dy$, does not contribute to the upwash directly. In approaching a lifting line at sweep, however, the "induced velocity" (obtained after subtracting out the local singularity of the bound vortex) increases with decreasing distance logarithmically. This logarithmic singularity in the velocity, therefore, will raise the level of the induced upwash on the wing by a factor of $\ln \varepsilon$, as is known from earlier *ad hoc* analyses [4.175–4.178].

The existence of the logarithmic singularity may be explained by a decomposition of the *locally shed* vorticity into two components, one parallel to the centerline and the other normal to it. The first component gives a jump in $\partial\phi/\partial x'$ proportional to $\sin \Lambda \, d\Gamma/dy$ across the trailing vortex sheet; this jump in $\partial\phi/\partial x'$ will give a logarithmic singularity in the upwash, as is well known in the classical 2-D thin-airfoil theory. The resulting incidence correction determined by the matching is then expected to be proportional to $\varepsilon \ln \varepsilon \cdot \sin \Lambda \cdot d\Gamma/dy$. Its relative importance is generally amplified in the outer portion of wing, where $|d\Gamma/dy|$ is larger. This and other corrections proportional to a local $d\Gamma/dy$ represent the *near-wake* influence.

To provide a more concrete idea of the velocity induced by trailing vorticities as well as the self-induced velocity *at* the lifting line, we examine below, for some typical cases, the finite part of the upwash obtained in the inner limit of the outer lifting-line solution. This quantity, being a functional of the span loading $\bar{\Gamma}_0(\bar{y})$, can be written in two parts [4.182–4.185]:

$$\Sigma(\bar{y}) + \Sigma^c(\bar{y}).$$

The first function Σ depends on the local swept angle Λ and contains also the logarithmically large near-wake contribution

$$\frac{1}{2\pi} \sin \bar{\Lambda} \cdot \ln(\sqrt{K_n}\varepsilon^{-1}) \cdot \frac{d}{d\bar{y}} \bar{\Gamma},$$

where $\sin \bar{\Lambda} \approx \Theta K_n^{-1/2}$ and $K_n^{1/2}\varepsilon^{-1} \approx (1 - M_\infty^2)^{1/2}\, \mathrm{AR}_1 \sec \bar{\Lambda}$. The second function Σ^c is a functional of both $\bar{\Gamma}_0(y)$ and the centerline shape; Σ^c vanishes for a completely straight centerline (but not so for centerlines made up of straight segments with different sweeps). This upwash function determines the upwash correction $\tilde{\mathbb{C}}_1^i$ for the inner solution through matching as

$$\begin{aligned}\Sigma + \Sigma^c &= -\sqrt{K_n}\hat{\mathbb{C}}_1^i + (2\pi)^{-1}(\ln 2)\Theta\, K_n^{-1/2} \frac{d\bar{\Gamma}_0}{d\bar{y}} \\ &= -\sqrt{K_n}\tilde{\mathbb{C}}_1^i + (2\pi)^{-1}\Theta K_n^{-1/2}\tilde{\Gamma}_0 \frac{d\hat{c}}{d\hat{y}} \ln \frac{2}{\hat{c}},\end{aligned} \tag{3.9}$$

where $\tilde{\Gamma}_0 \equiv \bar{\Gamma}_0/\hat{c}$.

Figure 2, taken from [4.185], shows the spanwise distribution of this upwash function for a symmetric swept wing at five different degrees of sweep, including two with forward sweeps (in dashes). The curves are illustrated for $\sqrt{1 - M_\infty^2}\, \mathrm{AR}_1 = 9.18$ and for an *elliptic* span loading $\bar{\Gamma}_0(\bar{y})$. The value $-\frac{1}{4}$ for $\bar{\Lambda} = 0$ corresponds to Prandtl's uniform upwash for an unswept straight wing.

The need for a washout on a swept-back wing ($\bar{\Lambda} > 0$) is quite evident from the significant reduction of the downwash in the outer wing panel shown by the full solid curves. Near the apex ($\hat{y} = 0$) the upwash function shows the expected trend of a downwash increase,

$$\Sigma^c \sim -\frac{\bar{\Gamma}_0(0)}{4\pi} \frac{\sin \bar{\Lambda}}{\bar{y}} \tag{3.10}$$

due to the dominating influence of the bound vortices on the opposite wing panel.

Except for the absence of the apex singularity, the upwash behavior for a straight oblique wing is similar to that shown for the symmetric swept wing

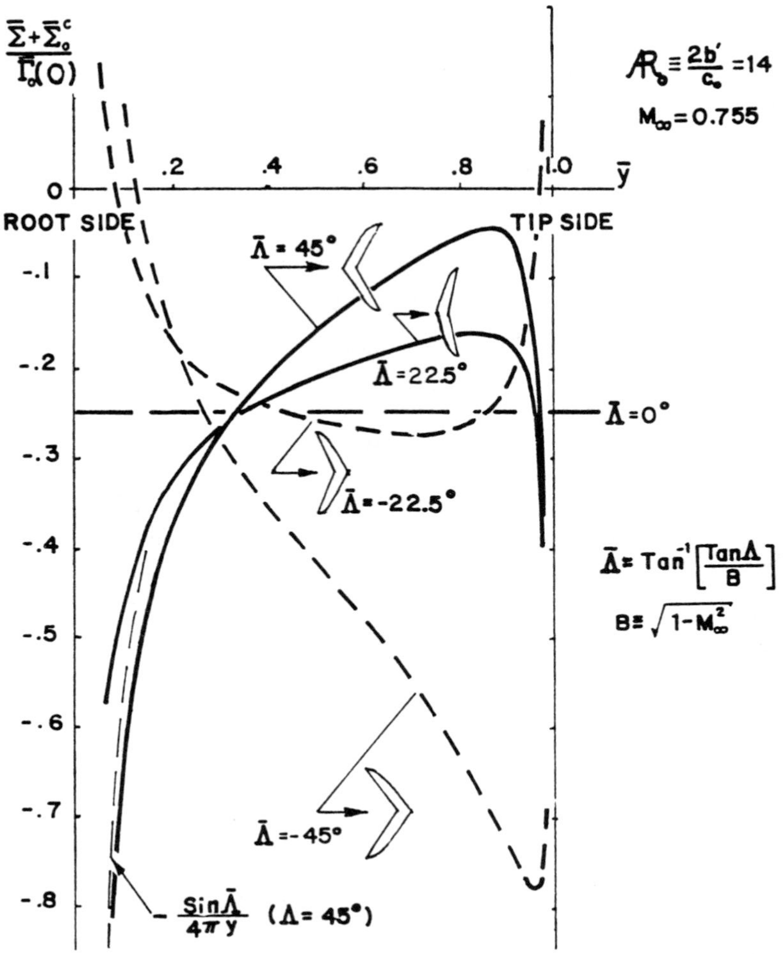

Figure 2. The upwash function of elliptically loaded, V-shaped lifting lines illustrated for $\sqrt{1 - M_\infty^2}\ AR_1 = 9.18$.

in that there is a significant reduction of the downwash in the aft wing panel and an increase in the fore wing panel [4.180, 4.184]. This asymmetry is responsible for the pronounced unbalanced rolling moment on an oblique wing [4.162, 4.163].

It is not clear whether the lifting-line approach may enjoy the same degree of usefulness as Prandtl's original theory. Figure 3 presents the result of such a test for an oblique wing in the linear regime, taken from [4.179], in which the theory is applied to an elliptic flat plate at incidence employing the upwash function Σ calculated in [4.180]. The figure shows the span loading at $M_\infty = 0$ for a 16.78:1 elliptic flat plate, of which the straight axis is located at the 40% chord line and has a sweep angle of 45°. As shown, the difference between the theory and the panel method (in open circles) is small at most stations, in contrast to that from the uncorrected local 2-D (strip) theory (in dashes). Similarly encouraging comparison is found with other planforms as well as in a lower range of aspect ratio [4.180].

3.6 Shock-Free Solutions: Comparison With Full-Potential Results

Apart from the uncertainty of its adequacy in treating the far-wake influence, the scarcity of the span stations available in the current 3-D transonic-flow programs [4.189–4.191, 4.56] could constitute a source of discrepancies. In applications to oblique wings, only ten span stations are available on each wing panel. The wing tips are generally represented by a trapezoid. Discernable differences exist among solutions for a given geometry, depending on the manner in which the wing-geometry data and grid spacing are managed [4.184, 4.185]. This naturally delimits the degree of certainty in our conclusion on the comparison. For the same reason, generation of full-potential results from two independent users is considered helpful in discerning the relatively weak 3-D effects of interest.

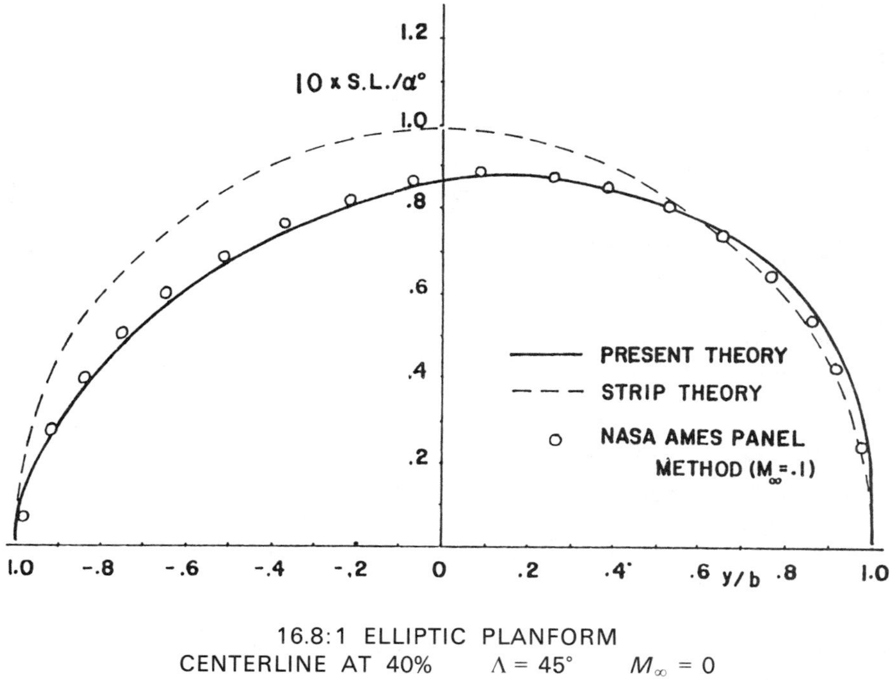

Figure 3. Span loading of a 16.78 : 1 elliptic flat plate with a straight 40% chord line pivoted at 45°.

The oblique-wing examples studied by Cheng and Meng [4.182, 4.183] have elliptic planforms whose major axes coinciding with the mid (50%) chord; the wing sections are generated from a *single* profile so that the similarity solutions are applicable. Figure 4 presents one of the comparisons in surface pressure coefficient taken from [4.184] and [4.185] for the case of a 14:1 elliptic wing with the NASA 3612-02, 40 airfoil, pivoted at $\Lambda = 30°$ at $M_\infty = 0.768$. This gives $M_n = 0.665$, $K_n = 3.45$, $\Theta = 1.062$, and $\varepsilon = 0.145$; the C_p-value for the local component Mach number to reach unity is -0.689.

The NASA Ames FLO 22 data (in the thinner solid and dashed curves) are seen to agree reasonably well with the similarity solution, except near the leading edge, where the small-disturbance assumption for the present theory breaks down. Noticeable discrepancy can also be seen near the trailing edge, where great accuracy of both the theory and the FLO 22 code is suspect. The degree of agreement shown is encouraging indeed, observing that neither the thickness ratio, nor the sweep angle, nor $1 - M_n^2$ appears to be as small as the small-disturbance assumption would require. Note that the relative error in the asymptotic theory is comparable to ε^2 or $\alpha^{2/3}$, and with the 12% thickness ratio, $\alpha^{2/3}$ is 0.243 in this case.

For a V-shape wing, there is a sweep-angle discontinuity at the apex, and it is not *a priori* clear if the comparison can be similarly encouraging. Figure 5 presents such a comparison. Except for the discontinuity at the apex, the wing retains the same geometry, and the same Θ and ε, as the oblique one in the preceding figure. The freestream Mach number is, however, slightly lower, being $M_\infty = 0.755$; thus $K_n = 3.60$. The component flow may be considered as being high-subcritical (note the local component Mach number would reach 1 at $C_p = -0.727$ in this case). Except in the vicinities of the wing tip and the symmetry plane, the agreement of the theory with the two sets of FLO 22 data is seen to be about as good as in the preceding comparison. We note that the FLO 22 data generated at the Grumman Aerospace Corporation (in filled circles) gives noticeable higher peaks for $-C_p$ than the corresponding data from the NASA Ames Research Center (in open circles) at most inboard stations. (The discrepancy in the lower surface pressure between theory and the FLO 22 may also be noted at most stations; it is believed to result from inadequate convergence in the similarity solutions generated originally in [4.181].

The most encouraging conclusion from this comparison is the surprisingly limited extent of the region affected by the apex breakdown. The agreement of the theory and the full potential computation remains reasonable at stations as close to the symmetry plane as one-tenth of the semispan. The discrepancy there is seen to be even less than that shown in the tip region, where inadequate representation in the planform geometry by the code could very well be responsible.

Numerical solution to the reduced lifting-line problem via the unsteady analogy has been studied by T. Evans, using an ADI algorithm quite similar to Ballhaus and Goorjian's [4.156]. A number of nonlifting examples have been obtained, and the results compared well with solutions generated from

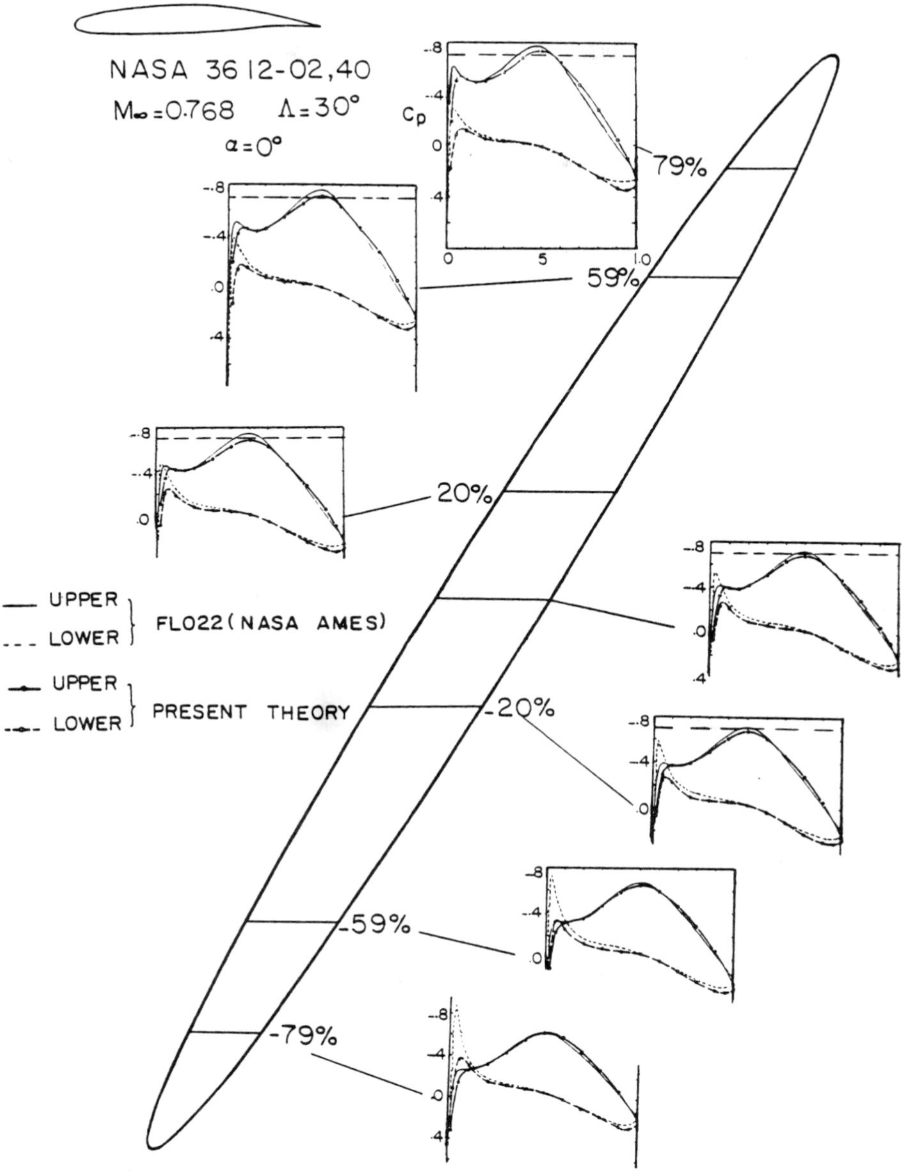

Figure 4. A consolidated plot showing surface C_p on seven span stations of an oblique wing with sweep angle 30° at Mach 0.767.

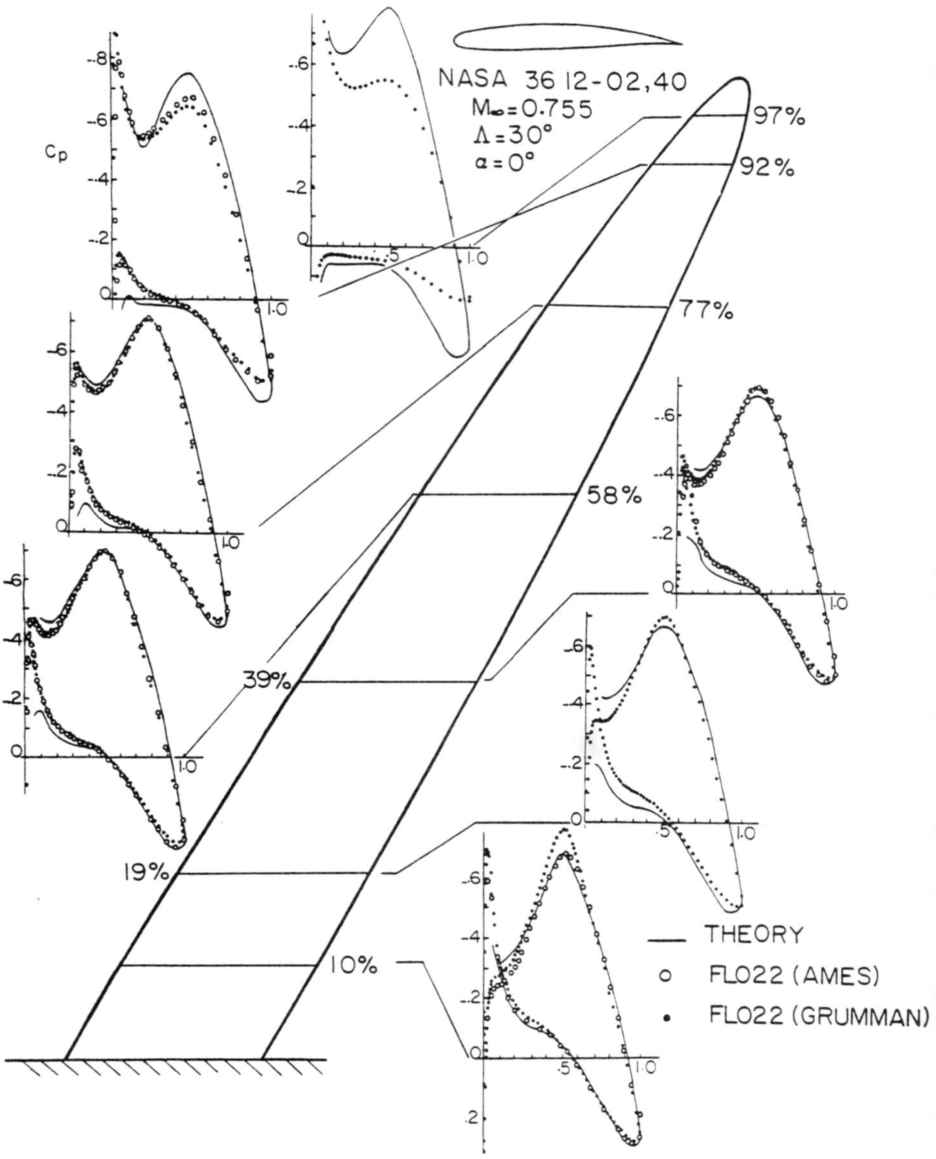

Figure 5. A consolidated plot showing surface C_p on seven span stations of a symmetric V-shaped wing with a 30° sweep-back at Mach 0.755. The airfoil is NASA 3612–12, 40.

the similarity structure [4.183]. The case involving lift requires special program implementations on the far-field and wing boundary conditions; these, together with the leading-edge singularity, appear to cause instability, which is brought under control by using very fine mesh. Figure 6 shows a typical pressure distribution from one example analyzed by Evans, which is made for an elliptic oblique wing with the same sweep angle and geometry as in Fig. 4 except $M_\infty = 0.755$. Evans's result (in dashes) is seen to compare reasonably well with both the similarity solution and the FLO 22 data for the same span station ($\hat{y} = 0.69$).

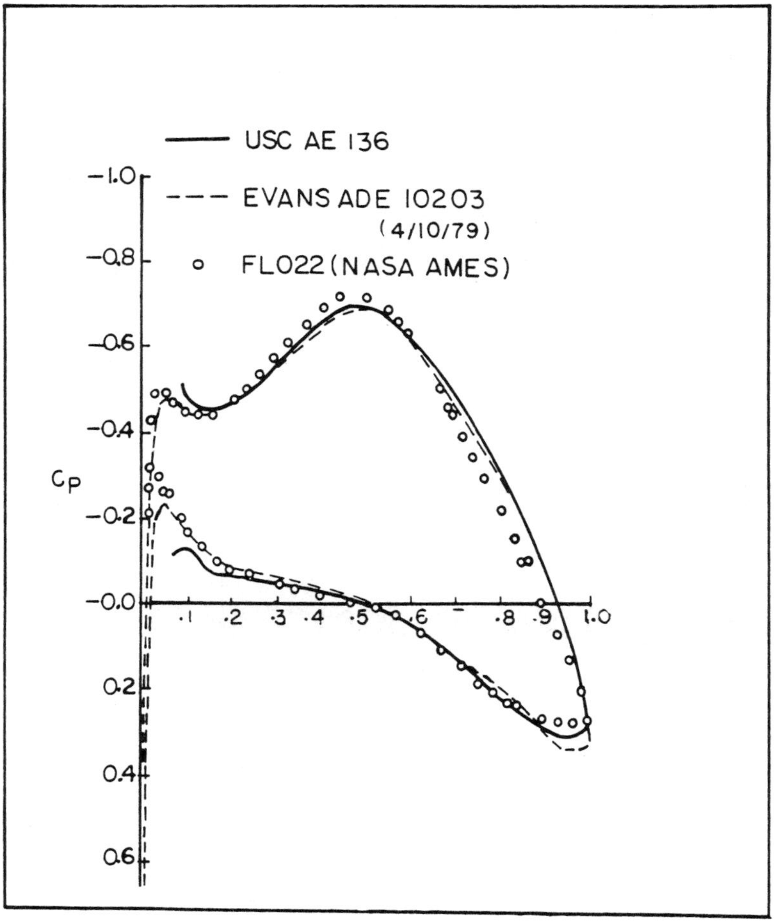

Figure 6. Comparison of the solution via an unsteady analogy (in dashes) with those of the similarity structure (in solid curves) and the full-potential solution (FLO 22, in open circles). The surface pressure is computed for the span station at 69% semispan of an elliptic planform pivoted at 30° at Mach 0.755. The airfoil is NASA 3612–02, 40.

In passing, we mention that comparison with results from 3-D small-disturbance codes (e.g. [4.190]) has not been made, owing to the inapplicability of the existing codes to the oblique wings, which was the focus of the original study in [4.181–4.183].

3.7 The Problem Involving a Shock

One untested aspect of the present theory is the treatment of a shock wave. The development along the line of unsteady analogy is promising, inasmuch as the shock-capturing capability of the ADI scheme has been quite effectively demonstrated by Ballhaus and Goorjian [4.56].

An alternative is to analyze the shock perturbation along with other 3-D corrections, which has the merit of *preserving* the 3-D similarity structure shown in Section 3.4. But it may fail for a very weak shock, and in the vicinity of a shock–sonic-boundary intersection, since the latter could lead to a nonuniformity in the approximation. With this method, one would be left hopeless in attempting to study shock formation from perturbations in an otherwise shock-free flow. With these reservations, we proceed to discuss the shock perturbation problem and some unpublished results of S. Y. Meng.

An unambiguous procedure for computing numerically 3-D corrections involving shock must begin with shock fitting, which gives the shock boundary as a discontinuity surface. The part of the (difference) algorithm treating the perturbation in the shock jump and in the shock boundary must be consistent with the shock-fitting procedure originally applied to the basic (2-D) component flow. For the problem of $\tilde{\phi}_0$, it is convenient to follow the procedure worked out earlier by Hafez and Cheng [4.191] for 2-D TSD flows, which has a built-in capability to distinguish a forward-inclined shock from a backward-inclined one. For the perturbation analysis, the ability to make these distinctions is no longer necessary, since the interior shock boundary of the reduced problem has been fixed by $\tilde{x} = \tilde{x}_0^D(z)$ in the basic 2-D solution; but it is essential to split the calculation of the $\partial\tilde{\phi}_1/\partial\tilde{z}$ jumps into two parts along different neighboring vertical lines across the shock. This is done to ensure the shock as a surface of slope continuity for $\tilde{\phi}$. Similar to the shock fitting for $\tilde{\phi}_0$, the difference equation at a "shock point" downstream of $\tilde{x} = \tilde{x}_0^D$ in the original relaxation procedure is replaced by those based on

$$2\tilde{x}_0 - (\gamma + 1)\langle\tilde{\phi}_{1\tilde{x}} + \tilde{x}_1^D\tilde{\phi}_{0\tilde{x}\tilde{x}}\rangle = -2\frac{d}{d\tilde{z}}\tilde{x}_0^D \cdot \frac{d}{d\tilde{z}}\tilde{x}_1^D, \tag{3.11a}$$

$$[\![\tilde{\phi}_1 + \tilde{x}_1\tilde{\phi}_{0\tilde{x}}]\!] = 0, \tag{3.11b}$$

where $\langle\ \rangle$ and $[\![\ ]\!]$ denote the arithmetical mean and the jump across the boundary $\tilde{x} = \tilde{x}_0^D(\tilde{z})$, respectively. Implementations via analytic continuation are needed at neighboring points, wherever a grid point in the difference operator is lost to the other side of the shock [4.191]. Throughout, the rule of forbidden signals is observed in forming the difference equations in the supercritical region (see Moretti's work [4.192, 4.193]).

The mesh distribution, the algebraic system, and the iterative procedure are basically the same as those used for the shock-free examples [4.181–4.183]; the form of the difference equations used in treating the shock in the $\tilde{\phi}_0$ system differs from that in Hafez and Cheng [4.191], and is believed to represent a slight improvement in accuracy.

There is a reexpansion singularity at the shock root over a curved solid surface, as is well known [4.194]. The latter manifests itself in a logarithmic singularity in $\tilde{\phi}_{1\tilde{x}}$ as may be inferred from Eq. (3.11a). Without special treatment, the perturbation analysis will fail to yield the proper correction to the shock jump. Cheng and Meng [4.182] and [4.183] have shown how the shock jump and the correct reexpansion flow behavior can be inferred from an accurate numerical solution for $\partial\tilde{\phi}_1/\partial\tilde{x}$. An alternative and simpler method for deducing the correct value of $\hat{\phi}_{\hat{x}}$ behind the shock root (sr) is to make use directly of the relation

$$\langle \phi_{\hat{x}} \rangle_{\text{sr}} = \frac{K_n}{\gamma + 1} + 2\varepsilon\Theta \frac{\partial}{\partial \hat{y}} \hat{x}_0^D. \tag{3.12}$$

As an example of treating 3-D perturbation of an imbedded shock, Meng considers a wing generated from NACA 0012 at 2° incidence, assuming $K_n = 2.375$, which gives a rather strong shock for the $\tilde{\phi}_0$ solution ($[\![\tilde{\phi}_{0\tilde{x}}]\!] \approx 2.5$ on the top surface). The similarity solution for $\tilde{\phi}_0$, including the shock boundary and the jumps in $\tilde{\phi}_{0\tilde{x}}$ and $\tilde{\phi}_{0\tilde{y}}$, is first determined. Using the revised difference equations near the shock mentioned, the shock polar is found to be adequately reproduced by the solution which employs a mesh of 81×65 distributed over the entire field $|\tilde{x}| \lesssim 5$, $|\tilde{z}| \lesssim 6$, without further grid refinement. The linear system for $\tilde{\phi}_1$, with $\partial\tilde{\phi}_1/\partial\tilde{z} = 0$ on the wing, is then solved with the same mesh.

The numerical solution for $\tilde{\phi}_2$ corresponding to a *unit* value of $\partial\tilde{\phi}_2/\partial\tilde{z}$ on the wing appears to have a convergence problem for the same mesh and the particular relaxation parameters used (1.8 in the elliptic and 0.8 in the hyperbolic region). Iterative solutions for $\tilde{\phi}_2$ employing grid refinement, modification of boundary conditions, or other strategies has not been thoroughly explored up to the time of this writing. The result for $\tilde{\phi}_2$ obtained by Meng in this case was determined from the difference of $\tilde{\phi}_0$ at two slightly different incidences, since $\tilde{\phi}_2$ accounts only for the local upwash and incidence changes [4.181–4.183]. The incidence change $\Delta\alpha_i$ used is 0.5°, corresponding to $\Delta\alpha_i/\tau = 0.073$. The accuracy in $\tilde{\phi}_2$ so obtained is believed to be no worse than that of the finite-difference solution from an 81×65 mesh. The surface distributions of $\tilde{\phi}_{0\tilde{x}}$, $\tilde{\phi}_{1\tilde{x}}$, and $\tilde{\phi}_{2\tilde{x}}$ determined are used to compute the surface pressure coefficient from $C_p = -2\cos^2\Lambda\,(\alpha/M_n)^{2/3}\hat{\phi}_{\hat{x}}$ with $\hat{\phi}$ from Eq. (3.8).

As an example, this set of similarity solutions is applied to a pivoted 14:1 elliptic planform with $\Lambda = 30°$ at $M_\infty = 0.8683$. Taking α to be the airfoil thickness ratio, 0.12, the result for K_n is very close to that of the similarity solutions, $K_n = 2.375$. Figure 7 illustrates the upper and lower surface pressure at three span stations $\tilde{y} = 0, \pm 0.8$. The shock on the upper wing surface is seen to locate slightly forward of the midchord position at $\hat{y} = 0$ and -0.8; it moves downstream away from the midchord at $\hat{y} = +0.8$ on the aft wing panel. The C_p-values immediately behind the shock are determined from Eq. (3.12), using the $\hat{\phi}_{\hat{x}}$ immediately upstream of the shock

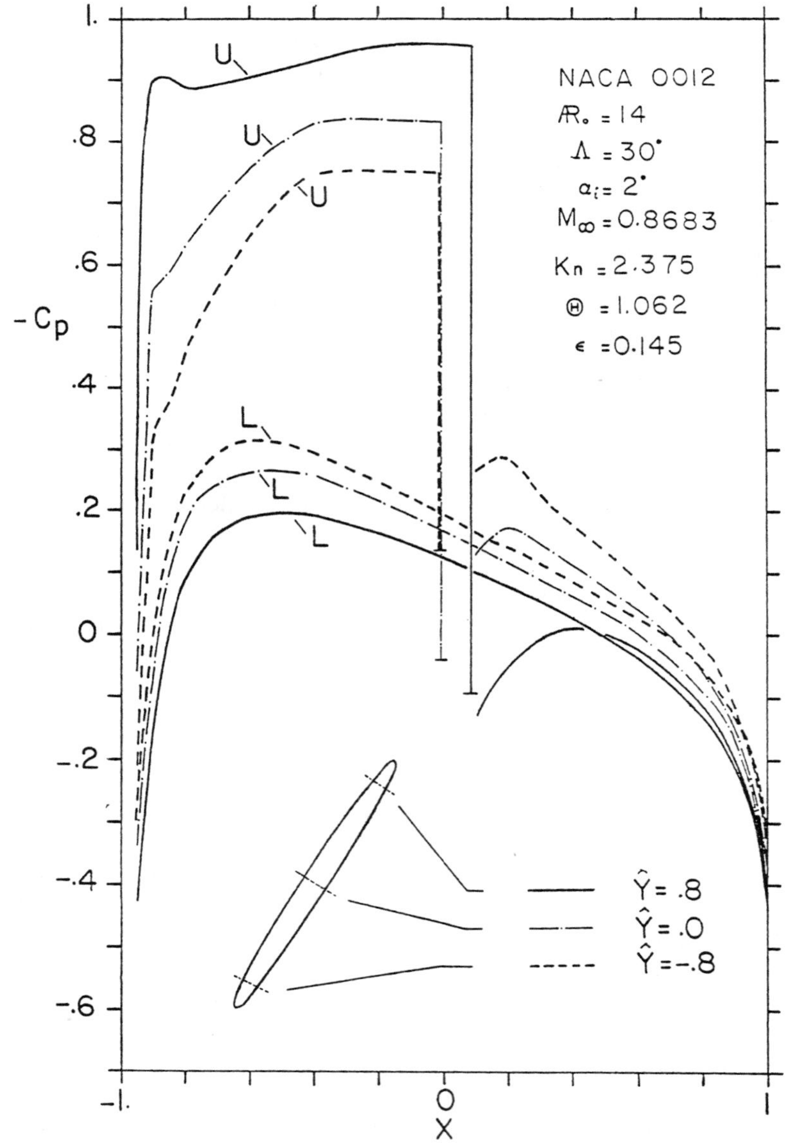

Figure 7. Surface pressure at three span stations on an elliptic oblique wing with 30° sweep at Mach 0.8683. The airfoil is NACA0012 at 2° incidence.

and the $\partial \hat{x}_0^D/\partial \hat{y}$ determined from the similarity solution. The stronger shock jump observed on the aft panel is anticipated from the upwash and compressibility corrections discussed earlier. In cases where the difference between C_p minimum and C_p at the shock root is small, a local description of the reexpansion flow behavior can be quite easily computed. This step of the calculations was not performed for Fig. 7.

A meaningful comparison of results in Fig. 7 with corresponding data from a full potential code is not available, since the FLO 22 code is not successful in capturing shocks, as is well known. Results from another code based on a PDE of a full-conservation form (FLO 27) has been made available to us through the NASA Ames Research Center, but the code does not apply to an oblique wing. Comparison of the theory with FLO 27 has recently been made for a V-shaped wing; this, and a more thorough discussion of the procedures and results, are presented in a subsequent paper [4.195].

4 Concluding Remarks

The author has taken the opportunity to review two complementary theoretical developments in transonic aerodynamics for high- and low-aspect-ratio wings. Using singular-perturbation techniques, the formulations have extended the classical slender-body and lifting-line concepts to the study of transonic nonlinear mixed flows, bringing out several features unsuspected, or not clearly identified, in earlier work. As indicated, the approach aims at gaining better physical insight, in addition to greater simplicity in the problem analysis. In particular, the studies should facilitate works of calculation and data correlation in aerodynamic design studies; this promise is substantiated in part by the results shown in Figs. 1, 4, and 5. Key results and highlights of the developments have been summarized in the Introduction and in the text, and will not be repeated here. It may be added that the similarity structure, when applicable, may reduce the computation work sufficiently to allow the use of a PDP-11 class minicomputer for the entire system of the reduced 2-D problems.

The swept-wing analyses discussed in Section 3 have stipulated a high-subsonic or sonic-free stream, for which the outer flow can be described by the linear Prandtl–Glauert equation in the leading order. Regimes exist in which the flow field far from the wing section is not linearizable, e.g. when $K_n < 0$, corresponding to a sonic or supersonic component flow under a supersonic freestream. Still untouched is the original problem of a supersonic swept (oblique or V-shaped) wing ($M_\infty > 1$) with a high subcritical freestream component ($0 < K_n = O(1)$). The parameter controlling the 3-D effect in this case is $1/\alpha^{2/3}\,\mathrm{AR}_1$, signifying a correction much larger than $\varepsilon \equiv 1/\alpha^{1/3}\,\mathrm{AR}_1$.

Interestingly, a domain for the transonic swept wing also exists, where the outer flow removed from the wing section is governed by the linear Jones theory (to leading order) but has a nonlinear mixed flow in the far field. The latter corresponds to the outer region in the theory for the transonic equivalence rule (Section 2). Thus, the theory for a slender wing in Section 2 and that for a high-aspect-ratio wing in Section 3 do overlap. This domain has yet to be treated, since the derivation of the lift correction given in Eq. (2.4) has stipulated a unit-order wing aspect ratio, not strictly compatible with the assumption $\varepsilon \ll 1$ in Section 3.

It may be pointed out that an asymptotic analysis for the boundary layer over a high-aspect-ratio wing can be developed, in which the leading approximation is that over an infinitely extended, yawed cylinder. With the similarity structure in the outer flow, the boundary-layer problem can be reduced to a 2-D analysis. In this regard, the lifting-line development in [4.196] based on the full potential theory may be useful in treating 3-D influence in the inviscid–viscous interaction problem [2.7].

Acknowledgment

It is quite fitting to dedicate this work to my friend, Morton Cooper, upon his retirement from the O.N.R., from whom my work has derived much encouragement and support, and whose contribution in his earlier years at NASA Langley Research Center also contains some novel results of swept-wing aerodynamics [4.128].

This work was done with the support of the Office of Naval Research, Contract No. N00014-75-C-0520. The author would like to express his appreciation for the valuable advice and assistance given by D. A. Caughey, T. Evans, M. M. Hafez, S. Y. Meng, R. M. Hicks, and M. D. Van Dyke. He want to thank G. X. Jia and G. Karpouzian for their help in uncovering several errors and furnishing the corrected data in Fig. 7.

Note Added after Proof. The dash curves in Fig. 2 are not correctly presented. The correct curves for the swept-forward wings can be obtained from data presented for the swept-back wings (in solid curves) by turning the set of (solid) curves to the opposite side of the horizontal line corresponding to the zero sweep. That is, the nonuniform part of the upwash in this case has a skew symmetry with respect to the sweep angle.

CHAPTER 24

Perturbation of Transonic Flow with Shocks[1]

Mohamed Hafez*

Introduction

Perturbation analysis is used to study transonic flows including nonlinear inviscid compressibility effects as in transonic small-disturbance theory, as well as perturbations of boundary conditions and flow parameters (e.g. Mach number M_∞, angle of attack α, thickness ratio τ, wing aspect ratio AR, sweep angle Λ, reduced frequency ω, tunnel height H, etc.).

Perturbation analysis is also used to study the numerical stability and convergence of a discrete system approximating a continuous problem, as well as the stability and convergence of the iterative solution of the large algebraic system of equations resulting from the discretization process.

Transonic flow is inherently nonlinear; the nonlinearity is responsible for the mixed type of the transonic equation as well as the formation of a discontinuous solution—namely shock waves. Recalling that the linearized theory breaks down in the transonic regime and both direct expansions in τ and M_∞ fail to present compressibility effects, the linearization of transonic flows resulting from the usual perturbation analysis is questioned. Indeed, such equations are, in general, inadequate, and extra conditions have to be imposed explicitly in order to determine the solution uniquely.

In the following, a general formulation of the perturbation problem is studied. A new approach, perturbation sequence expansion, is introduced

[1] Dedicated to Morton Cooper on the occasion of his retirement.

* The George Washington University/NASA-Langley Research Center, Hampton, VA 23665.

and applied to three examples: unsteady effects, three-dimensional effects, and wind-tunnel wall corrections.

General Formulation

The inviscid potential transonic equation is a kinematical relation representing conservation of mass. Strictly speaking, the partial differential equation is not valid across a shock. Nevertheless, if the nonlinear equation is written in a conservation form, it admits a weak solution with a jump condition consistent with the mass conservation law. More precisely, let the nonlinear equation be

$$N(\phi) = 0, \tag{1}$$

and let the jump condition admitted by the weak solution of (1) be

$$[Q]_{x_s} = 0, \tag{1'}$$

where $[Q]$ is the jump of Q across the shock at x_s. In the numerical solution of (1), the shock may be fitted according to (1′), or conservative finite differences (finite elements) may be used where the shock is captured and (1′) is automatically satisfied.

In the usual perturbation analysis, ϕ is expanded in a perturbation series:

$$\phi = \sum_i \varepsilon_i \phi_i, \tag{2}$$

or simply

$$\phi = \phi_0 + \varepsilon\phi_1 + \varepsilon^2\phi_2 + \cdots. \tag{2'}$$

ϕ_0 is governed by a nonlinear equation, and $\phi_1, \phi_2, \ldots$ are governed by linear equations. Writing these equations in a conservation form, the admissible jump conditions are

$$\begin{aligned} N_0(\phi_0) &= 0, \qquad [Q_0]_{x0} = 0, \\ L_1(\phi_1) &= 0, \qquad \left[\frac{\partial Q}{\partial \varepsilon}\right]_{x0} = 0. \end{aligned} \tag{3}$$

In solving the ϕ_0-problem by conservative differences, the shock is captured at x_0. The ϕ_1-equation is linear with discontinuous coefficients; its solution is discontinuous only at the same place the coefficients are, namely x_0. A solution may be obtained by "conservative" finite differences, but it does not conserve mass.

To take into account the effect of shock movement due to perturbations, Cheng and Hafez [4.146] used analytical continuation of the correct jump conditions at x_s to obtain the proper relation at x_0, which has to be explicitly imposed on the ϕ_1-equation to determine uniquely a solution which conserves mass.

Consistent with (2′), the expansion for the shock position and the jump conditions are

$$x_s = x_0 + \varepsilon x_1 + \varepsilon^2 x_2 + \cdots \tag{4}$$

and

$$\left[Q_0 + \varepsilon Q_1 + \cdots + \frac{\partial Q_0}{\partial x} \varepsilon x_1 + \cdots \right]_{x_0} = 0. \tag{5}$$

The resulting first-order jump condition at x_0 is

$$\left[Q_1 + \frac{\partial Q_0}{\partial x} x_1 \right]_{x_0} = 0. \tag{6}$$

Equation (6) is, in general, different from Eq. (3). A shock-fitting procedure must be used to impose Eq. (6) on the ϕ_1-equation; otherwise, mass is not conserved.

Recently, Nixon [4.197] suggested an alternative approach using strained coordinates. A transformation is used such that the shock does not move, i.e., the perturbed shock x_s and the unperturbed shock x_0 are the same in the transformed domain. Conservative finite differences are applicable, since it can be shown that the weak solution of such a perturbed equation conserves mass. The transformation is, however, in terms of the unknown shock position x_s. Not only is the equation more complicated due to the transformation, but also the boundary condition. The analytical continuation method, in the physical domain, is indeed more practical and straightforward.

Later, Nixon [4.198] used strained coordinates to linearly interpolate between two discontinuous solutions of the full system $N(\phi) = 0$, at two different values of a certain parameter assuming that the shock is linearly perturbed. The interpolation in the physical domain is of the same accuracy, assuming linear perturbation of the shock position, the jump conditions, and the strength of any singularities. For the cases of interest here, the full system $N(\phi) = 0$ is replaced by simpler problems $N_0(\phi_0) = 0$, $L_1(\phi_1) = 0, \ldots$.

Present Method

The perturbation series expansions, whether using analytical continuation in the physical domain or strained coordinates, are not valid when a shock appears or disappears in the flow due to perturbations. In order to analyze the formation and diminution of shocks, the following perturbation sequence expansion is proposed:

$$\phi = \lim_{i \to \infty} \phi^i, \tag{7}$$

$$x_s = \lim_{i \to \infty} x^i, \tag{7'}$$

such that

$$\phi^0 = \phi_0 \qquad\qquad x^0 = x_0$$
$$\phi^1 = \phi_0 + \varepsilon\phi_1 \qquad\qquad x^1 = x_0 + \varepsilon x_1$$
$$\phi^2 = \phi_0 + \varepsilon\phi_1 + \varepsilon^2\phi_2 \qquad x^2 = x_0 + \varepsilon x_1 + \varepsilon^2 x_2$$

and

$$\begin{aligned} N_0(\phi^0) &= 0, \qquad [Q^0]_{x^0} = 0, \\ N_1(\phi^1) &= 0, \qquad [Q^1]_{x^1} = 0. \end{aligned} \tag{8}$$

Conservative finite differences are used to solve the equations (8); the shock moves from x^0 to x^1 to x^2, and the solution is conservative at each level. From a numerical point of view, the solution of a linear problem with variable discontinuous coefficients is almost as difficult as the solution of a nonlinear problem. In the following, three different examples are discussed.

Example 1: Unsteady Effects

Consider the following one-dimensional problem:

$$\begin{aligned} -\tfrac{1}{2}(\phi_x^2)_x &= 2\phi_{xt}, \\ \phi(0, 0) = \phi_x(0, 0) = 0, &\qquad \phi(2, t) = \varepsilon \operatorname{Re} e^{i\omega t}. \end{aligned} \tag{9}$$

The jump conditions admitted by (9) are

$$\langle\phi_x\rangle_{x_s} = 2\frac{dx_s}{dt}, \qquad [\phi]_{x_s} = 0, \tag{9'}$$

where $\langle a\rangle_{x_s}$ denotes the average of a across the shock at x_s.

Perturbation Series

Let

$$\begin{aligned} \phi(x, t) = \phi_0(x) &+ \varepsilon \operatorname{Re} e^{i\omega t}\phi_1(x) \\ &+ \varepsilon^2(\phi_{2,0} + \operatorname{Re} e^{2i\omega t}\phi_{2,2}) + \cdots. \end{aligned} \tag{10}$$

The equation for ϕ_0 and the associated jump conditions are

$$-\tfrac{1}{2}(\phi_{0x}^2)_x = 0$$

and

$$\langle\phi_{0x}\rangle_{x_0} = 0, \qquad [\phi_0]_{x_0} = 0. \tag{11}$$

The equation for ϕ_1 is

$$-(\phi_{0x}\phi_{1x})_x = 2i\omega\phi_{1x}, \tag{12}$$

and the jump condition of the linear equation (12) at x_0 is

$$[\phi_{0x}]_{x_0}\langle\phi_{1x}\rangle_{x_0} + 2i\omega[\phi_1]_{x_0} = 0. \tag{12'}$$

The jump conditions which conserve mass at x_0 are given by Hafez, Rizk, and Murman [4.199]:

$$\begin{aligned}[\phi_{0x}]_{x_0}\langle\phi_{1x}\rangle_{x_0} + \langle 2i\omega - \phi_{0xx}\rangle_{x_0}[\phi_1]_{x_0} &= 0,\\ [\phi_1]_{x_0} &= -[\phi_{0x}]_{x_0}x_1.\end{aligned} \tag{13}$$

The geometrical interpretation of the second equation of (13) is sketched in Fig. 1.

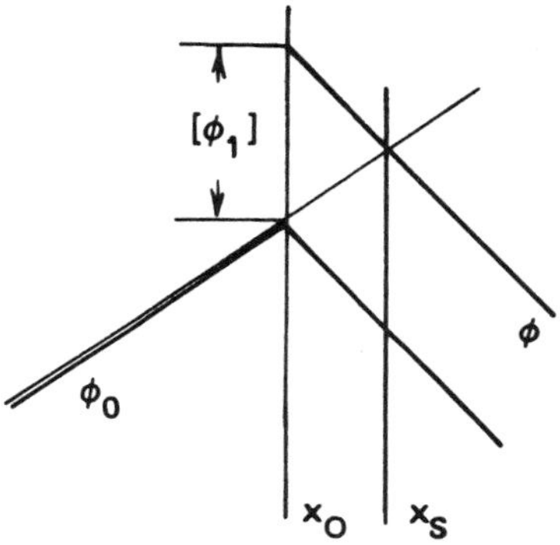

Figure 1. A sketch of ϕ_0 and ϕ in the vicinity of the unperturbed shock.

In this particular example, Eq. (12′) and the first equation of (13) are identical, since $\langle\phi_{0xx}\rangle_{x_0}$ is zero. Equations (12) and (12′) have a unique solution of the form $Ae^{2i\omega x}$, where A is determined by the size of the perturbation.

Perturbation Sequence

The equations for ϕ^0 is the same as (11), and the equation for ϕ^1 is

$$-\tfrac{1}{2}(\phi_x^{1\,2})_x = 2i\omega(\phi^1 - \phi^0)_x. \tag{14}$$

The weak solution of (14) admits a jump at x^1 such that

$$\langle\phi_x^1\rangle_{x^1}[\phi_x^1]_{x^1} - 2i\omega[\phi^0]_{x^1} = 0 \tag{14′}$$

and the analog of Eq. (13) becomes

$$[\phi^0]_{x^1} = \varepsilon|x^1|[\phi_x^0]_{x^0}, \tag{15}$$

where $|x^1|$ is the magnitude of x^1.

The geometrical interpretation of Eq. (15) is sketched in Fig. 2. Substituting Eq. (15) in Eq. (14′) yields

$$\langle\phi_x^1\rangle_{x^1}[\phi_x^1]_{x^1} = 2i\omega\varepsilon|x^1|[\phi_x^0]_{x^0}. \tag{16}$$

Equation (16) is consistent with (9′).

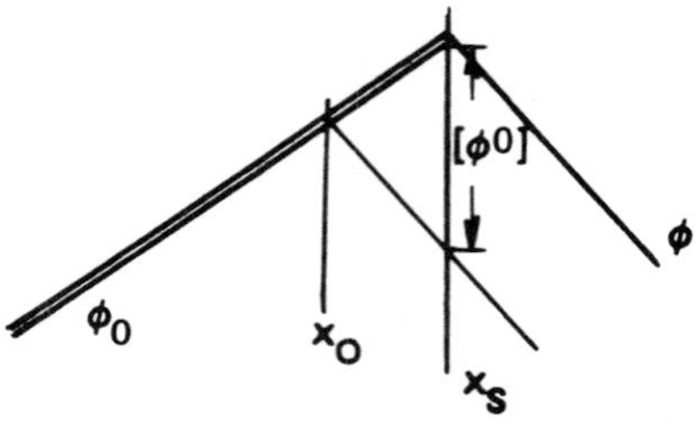

Figure 2. A sketch of ϕ_0 and ϕ in the vicinity of perturbed shock.

Two-dimensional Flows

Consider the airfoil problem where the governing equation is

$$(K\phi_x - \tfrac{1}{2}\phi_x^2)_x + (\phi_y)_y = 2\phi_{xt} \tag{17}$$

and the boundary condition is

$$\phi_y(x, 0) = f_0'(x) + \varepsilon \operatorname{Re} e^{i\omega t} f_1(x). \tag{17'}$$

The perturbation-series analysis reduces to

$$\begin{gathered}(K\phi_{0x} - \tfrac{1}{2}\phi_{0x}^2)_x + (\phi_{0y})_y = 0,\\ \phi_{0y}(x, 0) = f_0'(x)\end{gathered} \tag{18}$$

and

$$\begin{gathered}(K\phi_{1x} - \phi_{0x}\phi_{1x})_x + (\phi_{1y})_y = 2i\omega\phi_{1x}\\ \phi_{1y}(x, 0) = f_1(x)\end{gathered} \tag{18'}$$

The weak solution of (17) admits the following jump relations:

$$\begin{gathered}\left(\frac{\partial x_s}{\partial y}\right)^2 + 2\frac{\partial x_s}{\partial t} = -\langle K - \phi_x \rangle.\\ [\phi] = 0\end{gathered} \tag{19}$$

Expansion in terms of ε gives

$$\begin{gathered}\langle K - \phi_{0x} \rangle_{x0} = -\left(\frac{dx_0}{dy}\right)^2,\\ [\phi_0]_{x_0} = 0\end{gathered} \tag{20}$$

and

$$2\frac{dx_0}{dy}\frac{dx_1}{dy} + 2i\omega x_1 = \langle \phi_{1x} + \phi_{0xx}x_1 \rangle_{x0},$$

$$[\phi_1]_{x_0} = -[\phi_{0x}]_{x_0} x_1,$$

where

$$\frac{dx_0}{dy} = -\frac{[\phi_{0y}]}{[\phi_{0x}]},$$
$$\frac{dx_1}{dy} = -\frac{1}{[\phi_{0x}]}\left([\phi_{1y} + x_1\phi_{0yx}] + \frac{dx_0}{dy}[\phi_{1x} + x_1\phi_{0xx}]\right) \tag{21}$$

assuming shocks are locally normal, dx_0/dy and dx_1/dy are negligible, and the jump conditions derived for the one-dimensional problem are applicable except $\langle\phi_{0xx}\rangle \neq 0$. In fact, at the root of the shock, $\phi_{0x} \sim x_1 \log x_1$, and hence $\langle\phi_{0xx}\rangle$ is logarithmically large. A comparison between the solutions of the perturbation equations with and without the shock jump conditions is given in Fig. 3 after Ehler [4.200].

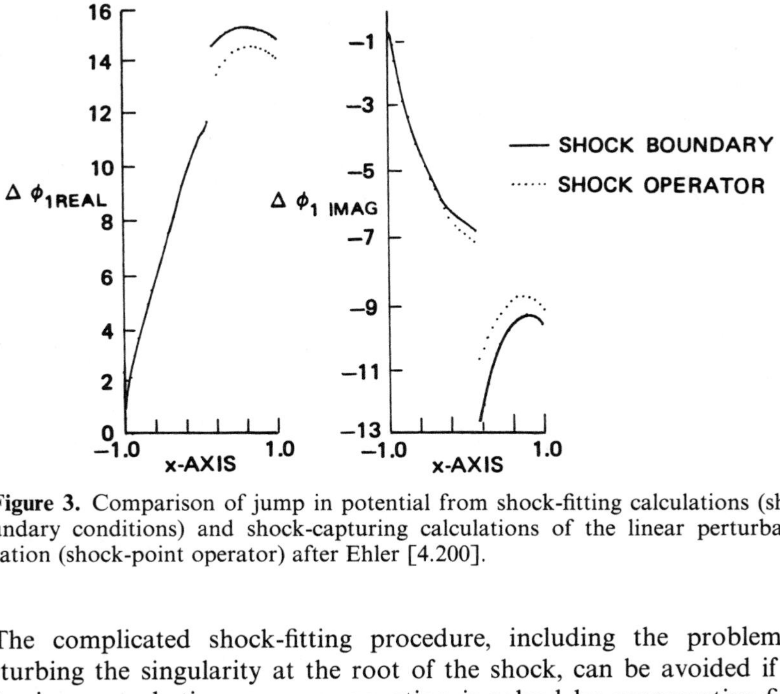

Figure 3. Comparison of jump in potential from shock-fitting calculations (shock boundary conditions) and shock-capturing calculations of the linear perturbation equation (shock-point operator) after Ehler [4.200].

The complicated shock-fitting procedure, including the problem of perturbing the singularity at the root of the shock, can be avoided if the following perturbation sequence equation is solved by conservative finite differences:

$$(K\phi_x^1 - \tfrac{1}{2}\phi_x^{12})_x + (\phi_y^1)_y = 2i\omega(\phi_x^1 - \phi_x^0)$$
$$\phi_y^1(x, 0) = f'_0(x) + \varepsilon \operatorname{Re} e^{i\omega t} f_1(x). \tag{22}$$

Notice that the boundary condition is a function of ωt and Eq. (22) has to be solved for different values of ωt to construct the periodic response. The shock, however, may disappear in a part of the period, as in Ballhaus and Goorjian's work [4.56]. The next example deals with such a shock motion.

Example 2: Three-dimensional Effects

In the extension of classical slender-body and lifting-line theories to transonic flows, the correction equations for 3-D effects are linear with variable discontinuous coefficients, and a shock-fitting procedure has to be implemented to satisfy the jump conditions representing conservation of mass. Alternatively, perturbation sequence equations are solved with shock-capturing methods.

Equivalence Rule

In Cheng and Hafez's [4.150] analysis, as well as in Barnwell's [4.153], the inner problem is mainly governed by a cross-flow Laplace equation, while the outer problem is governed by the classical 3-D transonic small-disturbance equation with an equivalent line source and a line doublet at the axis. The doublet strength is proportional to the total lift distribution, which can be obtained by linear theory (i.e., the total lift at any station x is proportional to the square of half the span). The source strength consists of two parts: the first is related to the geometrical cross-sectional area, and the second is the nonlinear lift contribution. The outer solution has to be obtained numerically by solving

$$\begin{gathered}(K - \phi_x)\phi_{xx} + \frac{1}{r}(r\phi_r)_r + \frac{1}{r^2}\phi_{\theta\theta} = 0, \\ \lim_{r\to 0} \phi \sim S_e'(x)\log r + \sigma\frac{D(x)}{r}\sin\theta + \cdots.\end{gathered} \tag{23}$$

Lift perturbations can be analyzed using series or sequence expansions as follows:

Perturbation series. Assuming

$$\phi(x, r, \theta) = \phi_0(x_1 r) + \sigma\phi_1(x_1 r) + \sin\theta + \cdots, \tag{24}$$

we have

$$\begin{gathered}(K\phi_{0x} - \tfrac{1}{2}\phi_{0x}^2)_x + \frac{1}{r}(r\phi_{0r})_r = 0, \\ \phi_0 \sim S_e'(x)\log r\end{gathered} \tag{25}$$

and

$$\begin{gathered}(K\phi_{1x} - \phi_{0x}\phi_{1x})_x + \frac{1}{r}(r\phi_{1r})_r - \frac{\phi_1}{r^2} = 0, \\ \phi_1 \sim \frac{D(x)}{r},\end{gathered} \tag{25'}$$

together with the proper jump conditions.

Perturbation sequence. Assuming

$$\phi^0 = \phi_0, \qquad \phi^1 = \phi_0 + \sigma\phi_1 \sin\theta, \tag{26}$$

we have

$$\begin{aligned}(K\phi_x^1 - \tfrac{1}{2}\phi_{1x}^2)_x + \frac{1}{r}(r\phi_r^1) - \frac{\phi^1 - \phi^0}{r^2} &= 0,\\ \phi^1 \sim S_e'(x)\log r + \sigma\frac{D(x)}{r}\sin\theta.&\end{aligned} \tag{27}$$

A comparison between the pressure distributions (in the vertical plane) of the solutions of Eqs. (25′) and (23) is given in Fig. 4. The agreement is good except in the shock region. The shock positions are shown in Fig. 5. In this calculation, the axisymmetric shock is smeared out as shown in Fig. 4, and consequently the shock position of the lift perturbation problem is not accurate. The sonic lines in different θ-planes of the solutions of Eqs. (23) and (27) for another example are shown in Fig. 6. Due to the proper account taken of the nonlinearity in the perturbation-sequence procedure, the shock and the sonic line are in good agreement with the three-dimensional calculations, and assuming that ϕ is sinusoidal in the θ-direction does not lead to a sinusoidal shock as in the perturbation-series expansion.

Lifting Line

Following Van Dyke's [2.127] singular perturbation analysis of Prandtl lifting-line theory, Cheng and Meng [4.182], Cook and Cole [4.127], and Cook [4.186] extended the classical theory to transonic flows including oblique wings. Here, the outer solution is mainly governed by the linear 3-D Prandtl–Glauert equation and the wing shrinks to a line of singularities (vortices). Further corrections due to compressibility effects are obtained in the case of subsonic far field by iterating on the nonlinear terms (of the transonic small-disturbance equation) as usual. The inner problem is basically nonlinear. It is obvious that as the aspect ratio becomes infinite the flow at any spanwise station becomes two-dimensional. For an oblique wing, the governing equation, in rotated coordinates (perpendicular to the leading edge) is

$$(K_n\phi_{x'} - \tfrac{1}{2}\phi_{x'}^2)_{x'} + (\phi_{z'})_{z'} = \varepsilon\left(2\theta\phi_{x'y'} + \frac{d\theta}{dy'}\phi_{x'}\right) \tag{28}$$

together with surface and wake boundary conditions, where x', y', z' are the rotated coordinates, K_n is the transonic similarity parameter in the plane normal to the leading edge, ε is proportional to 1/AR, and θ is proportional to the sweep angle Λ. Matching with the outer solution provides a boundary condition (induced velocity) for the inner problem.

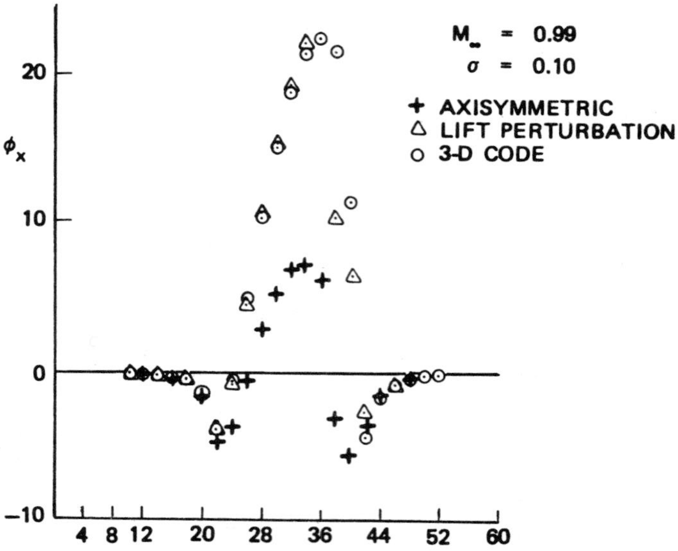

Figure 4. Pressure distributions along the axis in the vertical plane.

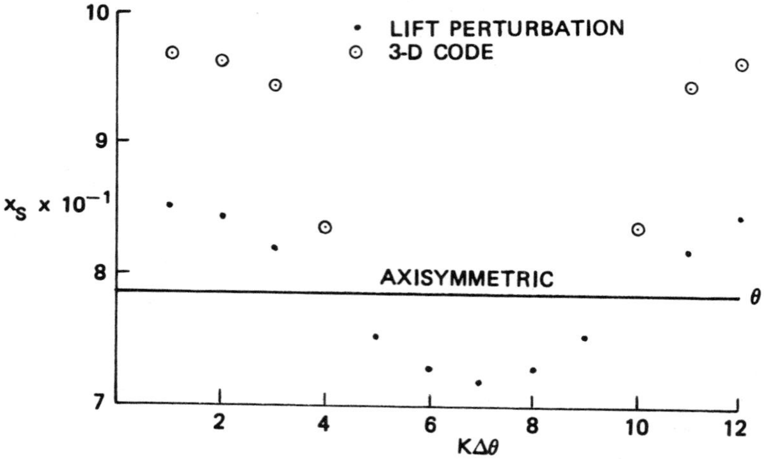

Figure 5. Shock position from perturbation-series calculation.

Neglecting the curvature ($d\theta/dy' = 0$), Eq. (28) is analogous to the unsteady transonic flutter equation, and Tom Evans successfully used an ADI algorithm to solve the corresponding initial–boundary-value problem (see [4.182]).

Meanwhile, perturbation-series analysis for small ε yields

$$\begin{aligned} \phi &= \phi_0(x', z', y') + \varepsilon\phi_1(x', z'; y') + \cdots, \\ x_s &= x_0(z'; y') + \varepsilon x_1(z'; y') + \cdots, \end{aligned} \tag{29}$$

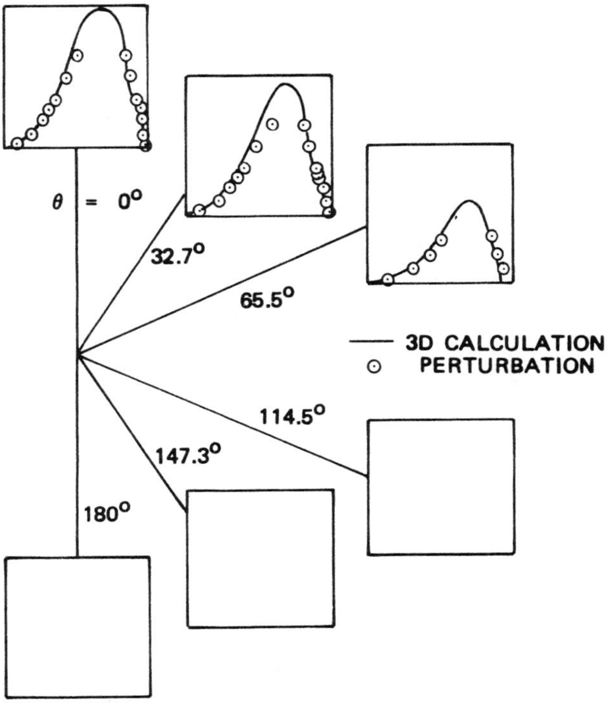

Figure 6. Sonic line and shock position of perturbation-sequence results in different θ-planes.

where the equations for ϕ_0 and ϕ_1 are

$$(K_n\phi_{0x'} - \tfrac{1}{2}\phi_{0x'}^2)_{x'} + (\phi_{0z'})_{z'} = 0 \tag{30}$$

and

$$(K_n\phi_{1x'} - \phi_{0x'}\phi_{1x'})_{x'} + (\phi_{1z'})_{z'} = 2\theta\phi_{0x'y'} \tag{30'}$$

together with the proper jump and boundary conditions. ϕ_0 is a 2-D solution to the component flow, and ϕ_1 is the first-order three-dimensional correction.

Cheng and Meng [4.182] solved the equations (30) numerically, and their results compare favorably with the full potential three-dimensional calculations of Jameson. The special case of a straight unyawed wing ($\theta = 0$) has been solved by Small [4.201, 4.202]. A uniqueness proof of the corresponding linear ϕ_1-problem is given by Cook [4.173] for shock-free flows.

On the other hand, perturbation-sequence analysis yields:

$$(K_n\phi_{x'}^1 - \tfrac{1}{2}(\phi_{x'}^1)^2)_{x'} + (\phi_{z'}^1)_{z'} = 2\theta\phi_{0x'y'} \tag{31}$$

with proper boundary conditions. Equation (31) is nonlinear and admits the proper jump conditions conserving mass. It is much easier to solve Eq. (31) than the initial–boundary-value problem of Eq. (28), and the difference is of higher order.

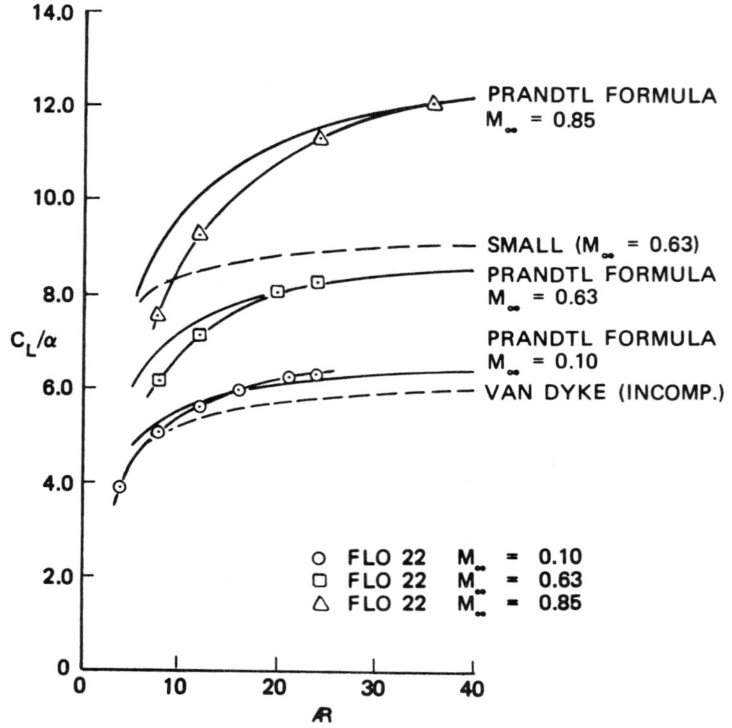

Figure 7. Comparison between Prandtl formula for C_L/α and results from three dimensional calculations.

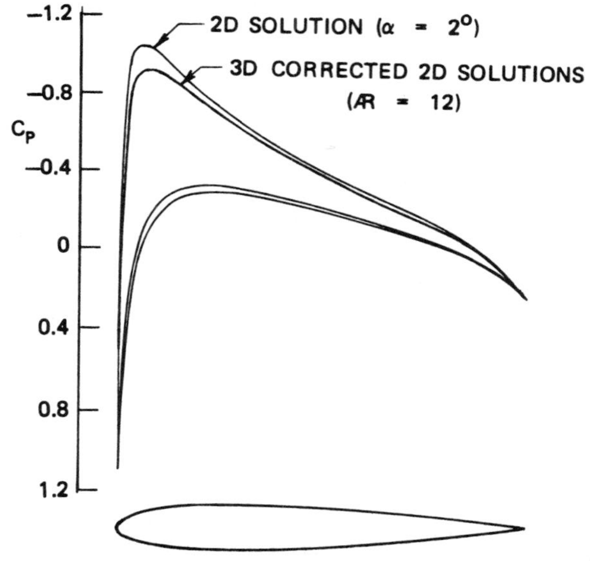

Figure 8. Surface pressure distribution at mid span, $M_\infty = 0.7$.

Finally, the small-disturbance assumptions can be removed and the theory can be based on the full potential equation and exact surface and wake boundary conditions [4.196].

For the special case of a straight unyawed elliptic wing, the induced velocity is constant along the span and the three-dimensional correction according to Prandtl theory is simply to modify the angle of attack. This simplification carries over to the transonic regime, namely,

$$\alpha_{\text{eff}} = \alpha_{2\text{-D}} \frac{\beta \, \text{AR}}{\beta \, \text{AR} + 2}, \tag{32}$$

where $\beta = \sqrt{1 - M_\infty^2}$. The corresponding formulas for $C_{L\alpha}$ are:

$$C_{L\alpha} = \begin{cases} 2\pi \dfrac{\text{AR}}{\text{AR} + 2} & \text{(incompressible)} \\ \dfrac{2\pi}{\beta} \dfrac{\beta \, \text{AR}}{\beta \, \text{AR} + 2} & \text{(subsonic)} \\ C_{L\alpha \, 2\text{-D}} \dfrac{\beta \, \text{AR}}{\beta \, \text{AR} + 2} & \text{(transonic).} \end{cases} \tag{33}$$

Figure 7 shows the validity of the above formulas. Figures 8 and 9 show the 2-D, 3-D, and corrected 2-D pressure distribution at mid span for $M_\infty = 0.63$ and 0.85 respectively (AR = 12). In Fig. 10 the shock locus is plotted. The general case of nonelliptic swept wings is studied by Melson and Hafez [4.203].

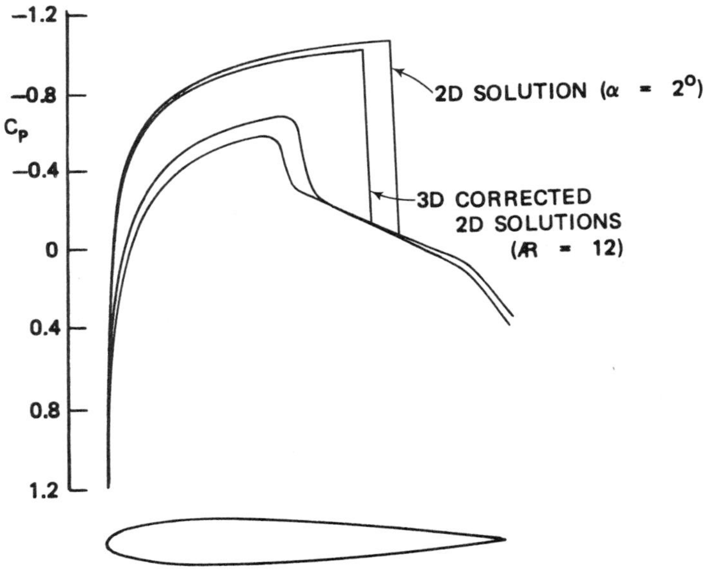

Figure 9. Surface pressure distributions at mid-span, $M_\infty = 0.85$.

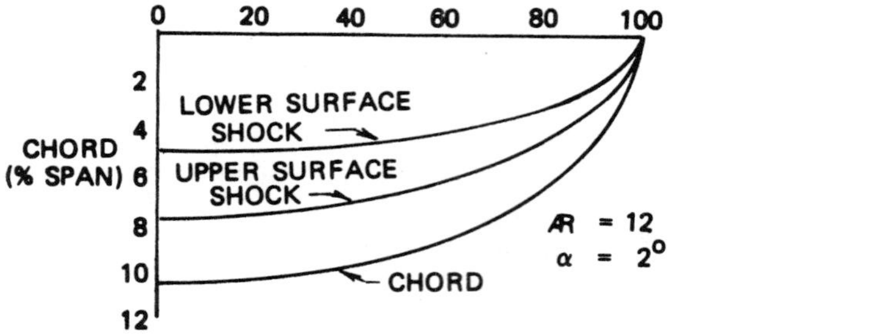

Figure 10. The shock locus on elliptic wing at $M_\infty = 0.85$ and $\alpha = 2°$.

Example 3: Wind-Tunnel Wall Correction

Recently, Chan [4.204, 4.205] analyzed two-dimensional wind-tunnel inferences as a singular-perturbation problem. In the outer region, near the wall, the perturbation potential ϕ can be developed as

$$\phi\left(x_1 y; \frac{1}{H}\right) = \phi_0(x_1 y) + \frac{1}{H}\phi_1(x_1 y) + \frac{\log H}{H}\phi_2(x_1 y) + \cdots. \tag{34}$$

The transonic small-disturbance equation can then be written for each order as

$$\begin{aligned} K\phi_{0xx} + \phi_{0yy} &= 0, \\ K\phi_{1xx} + \phi_{1yy} &= (\gamma + 1)\phi_{0x}\phi_{0xx}, \\ K\phi_{2xx} + \phi_{2yy} &= 0, \end{aligned}$$

with the wall boundary conditions

$$\phi_{nx} \pm \frac{1}{p}\phi_{ny} = 0, \quad n = 0, 1, 2, \quad \text{at } y = \pm H, \tag{35}$$

where p is the wall porosity factor. The inner expansion*, near the airfoil, is

$$\phi(x_1 y) = \phi_0(x_1 y) + \frac{1}{H}\phi_1(x_1 y) + \cdots, \tag{36}$$

where

$$\left(K\phi_{0x} - \frac{\gamma + 1}{2}\phi_{0x}^2\right)_x + \phi_{0yy} = 0,$$

$$\phi_{0y} = f'(x) - \alpha$$

* This asymptotic development is questioned. See Cole *et al.* AIAA Paper 82-0933 (1980).

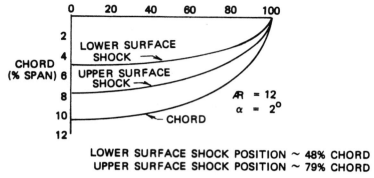

and

$$(K\phi_{1x} - (\gamma + 1)\phi_{0x}\phi_{1x})_x + \phi_{1yy} = 0,$$
$$\phi_{1y} = 0 \tag{37}$$

together with proper jump conditions. ϕ_0 is the free-air solution, and ϕ_1 is the interference-flow solution. By proper matching, the inner expansion of the outer solution can be written as

$$\begin{aligned}\phi \sim & -\frac{\Gamma_0}{2\pi}(\theta + a_1 y) \\ & + \frac{\log H}{H}\left[\frac{\gamma+1}{4\pi^2 K}\left(\frac{\Gamma_0}{2}\right)^2\left(\frac{\cos\theta}{r} + \frac{b_2 x}{K}\right)\right] \\ & + \frac{1}{H}\left[\frac{d_1}{2\pi\sqrt{k}}\left(\frac{\cos\theta}{r} + \frac{b_2 x}{K}\right) + \frac{\gamma+1}{4\pi^2 K}\left(\frac{\Gamma_0}{2}\right)^2 \frac{\log r}{r}\cos\theta \right. \\ & - \frac{\gamma+1}{16\pi^2 K}\left(\frac{\Gamma_0}{2}\right)^2 \frac{\cos 3\theta}{r} \\ & \left. + \text{complementary functions}\right]. \end{aligned} \tag{38}$$

The interference solutions are presented in a_1, b_2 and the complementary functions (to satisfy the wall boundary condition). Γ_0 and d_1 are the circulation and the doublet strength of the airfoil respectively. Solution of (35) yields the apparent angle of attack $\Delta\alpha$ in the outer field due to the wall restriction. The lift correction is of order $1/H$ and can be evaluated by solving (37) with the outer boundary condition provided from matching. The shock condition for the interference flow is transferred to the zero-order shock location (free-air solution) by analytic continuation [4.206].

Notice that an effective thickness (doublet) is induced by the nonlinear compressibility effects, in terms of the lift Γ_0. Using the approximations

$$\frac{\log r}{r} \approx \frac{1}{r}, \quad r \geq 1 \quad \text{and} \quad \frac{\cos 3\theta}{r} \approx -3\frac{\cos\theta}{r}, \quad \theta \simeq \frac{\pi}{2},$$

the strength of the equivalent doublet is given by

$$d_e = \frac{\gamma+1}{2\pi\sqrt{k}}\left(\frac{\Gamma_0}{2}\right)^2 [\tfrac{7}{4} + \log H]. \tag{39}$$

(The equivalent doublet strength due to lift in free air is smaller; the difference is the $\log H$ term.) The angle-of-attack correction [4.206] is found to be close to the classical results (linear theory), but the Mach-number correction is different.

On the other hand, Kemp [4.207] and Murman [4.208] analyzed the wall effect as the difference between a free-air calculation ϕ^0 and a tunnel calculation ϕ^1 using the wall boundary condition or a measured pressure (ϕ_x^1) near the wall. (They used an equivalent body accounting for viscous effects. For the sake of simplicity, viscous effects are neglected here.) The angle of attack and the Mach number are varied in the free-air calculations until the calculated surface pressure matches the surface pressure in the tunnel calculations, thus yielding ΔM and $\Delta\alpha$ corrections.

To calculate such corrections for a slender wing, a line-source and a line-doublet distribution are used in free-air and in tunnel calculations. Typical shock locations from such calculations are plotted in Fig. 11. The angle of attack and the Mach number could have been varied so that the two shock loci were matched (say in a least-squares sense). Instead, a Mach-number correction is calculated based on an axisymmetric calculation of an equivalent body of revolution. The shock movement, in free-air calculations, around such a body of revolution is found to be linearly dependent on the freestream Mach number; hence no optimization technique, as in [4.208], is needed. Also, the shock location varies linearly with small angle-of-attack perturbations, as shown from three-dimensional as well as lift-perturbation calculations (see Fig. 12). Hence, assuming the validity of the equivalence rule, the perturbation-sequence analysis is adequate to determine the angle-of-attack and the Mach-number corrections due to wall interferences.

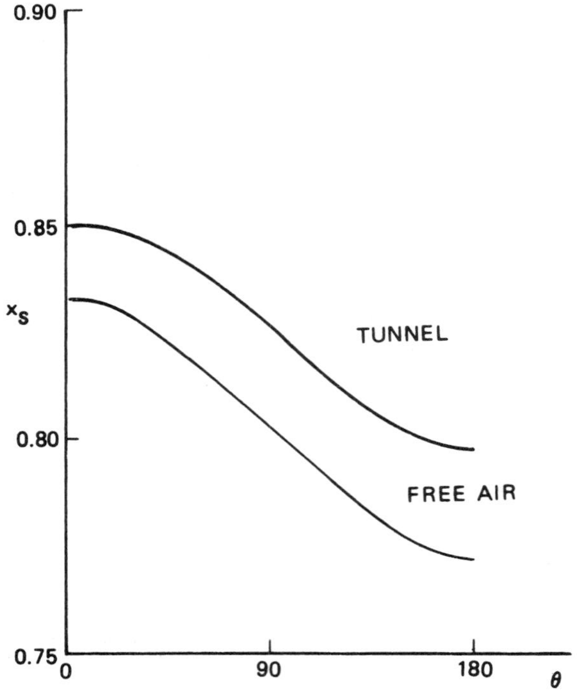

Figure 11. Shock position from a line source and a line doublet: 3-D calculations in free air and in a solid-wall tunnel.

(a)

(b)

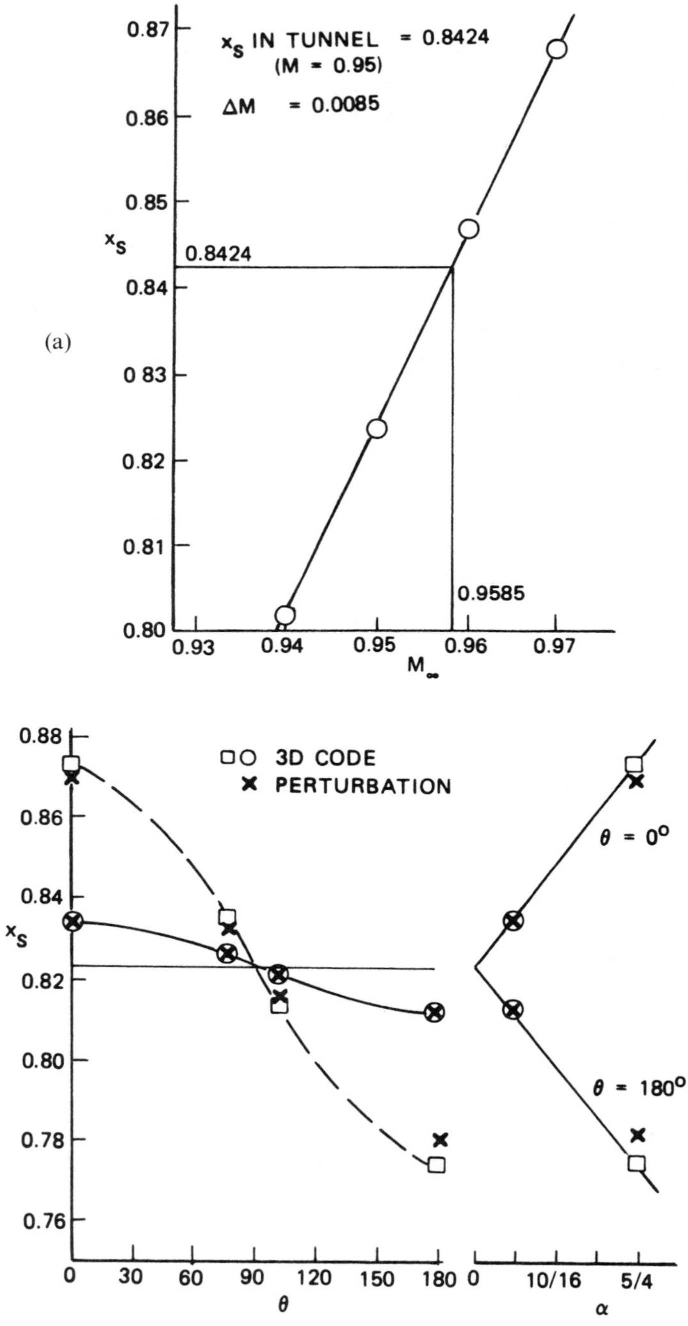

Figure 12. (a) Variation of shock position with freestream Mach number. (b) Variation of shock position with freestream angle of attack.

Concluding Remarks

Perturbation-series analysis of nonlinear transonic problems yields a set of linear perturbation equations which are in general sufficient to obtain a smooth solution. Perturbation of jump conditions admitted by the weak solution of the proper conservation form provides the necessary conditions to impose on the linear perturbation equations in order to obtain a conservative weak solution. In general, such conditions are not contained in the linear perturbation equations, and a shock-fitting procedure must be used to satisfy the proper jump conditions. The perturbed shock strength and position are part of the solution. Such a linearization is limited to a small perturbation where shocks are not generated or diminished by perturbation.

In this paper, an alternative approach, namely, perturbation sequences, is studied. The complications of shock fitting are avoided. The perturbation equations are nonlinear and written in a proper conservation form; they can be solved by shock-capturing methods. Moreover, the formation and disappearance of shocks can be easily handled. Three examples are given: unsteady effects, three-dimensional corrections to axisymmetric and two-dimensional flows as in slender-body and lifting-line theories, and finally wind-tunnel wall corrections.

Acknowledgment

The work reported here was supported in part by NASA-Langley Research Center. The author wishes to thank Donald Lovell, Duane Melson, and Larry Green of Theoretical Aerodynamic Branch for their help with the computations.

CHAPTER 25

On the Definition of Freestream Conditions in Wind-Tunnel Testing

W. R. Sears*

Introduction

The concept of adaptive wall control in wind tunnels was advanced [4.209, 4.210] as a solution to the problem of boundary interference, especially in testing models in the transonic regime. It had become apparent that wall effects could be significant, their character and magnitudes unknown, and reliable theory for the correction of measured data nonexistent. Adaptive wall control is based on the observation that the existence of unconfined-flow conditions in a wind tunnel—and, in fact, a quantitative assessment of departure from such conditions—can be ascertained by calculation from data measured in the flow field. Changes can then be made, systematically, in the tunnel boundary configuration, leading in an iterative way to unconfined flow, for any arbitrary model configuration.

Considerable work has now been done in efforts to realize this concept in practical hardware. (See, for example, [4.211].) In briefest summary, it might be said that (a) the concept has been proved correct, (b) the calculation mentioned above has not been found to be particularly difficult, even in three-dimensional transonic flow, but (c) measurement of the required flow-field data, to the required accuracy, poses some challenging technical problems.

The present paper is not a progress report on this developmental program, but presents instead some observations and some evidence concerning the utility of the adaptive-wall scheme in solving a more basic problem, slightly

* University of Arizona, Tucson, AZ 85721.

different from the classical wall-interference problem that was the original goal. This is the problem of the definition of freestream speed and direction—in other words, the determination of the freestream vector.

It is customary, in wind-tunnel testing, to determine the tunnel speed from the readings of instruments that provide a calibration of the empty tunnel; the assumption is that this instrumentation is unaffected by the presence of the model. It is usually recognized that this is an approximation, and limits are therefore placed upon the size of models tested, especially as regards "blockage" (e.g., the ratio of the maximum model cross-sectional area to the area of the working section). It is usually necessary to assume that the free-stream direction is parallel to the tunnel's axis, as it is when the model is absent. Clearly, the accuracy of this assumption depends on such details as the distance between model and nozzle and the extent of the model's disturbance field.

Unfortunately, the errors introduced by these approximations are difficult to assess and therefore are unknown in many experiments. At the same time, accurate knowledge of both the speed (Mach number) and the stream direction is important in aeronautical engineering. There is a certain anomaly here: we are not usually interested in knowing the angle of attack to great accuracy, but we do require the stream direction precisely. For example, an error of only $\frac{1}{10}^\circ$ in flow angle introduces an error of $0.0017C_L$ in drag coefficient, where C_L is the lift coefficient. Considerably better accuracy than this is desired, but is hard to achieve in any tunnel.

Stream Definition in an Adaptable-Wall Tunnel

The principle of the adaptable-wall scheme is to define an interface S between a calculated exterior flow field and an experimentally produced interior field wherein the boundary conditions are provided by the model. The interface is defined by instrumentation, i.e., S is the locus of points where the flow variables are measured. The far-field boundary condition of the external field consists of flow at the freestream speed and direction. When matching of interior and exterior flows is achieved at the interface, by iterative adjustment of the tunnel-wall controls, the flow about the model is the desired subregion of the flow field determined by the boundary conditions at the model and at infinity.

The iterative process is one of successive modification of both inner and outer flows by means of the wall controls while the boundary conditions remain unchanged. Assuming that these controls are powerful enough—i.e., that the tunnel operator can achieve the increments of flow variables at S called for in the process of iteration—the tunnel flow will always be driven to the desired stream speed and direction. Thus, at least in principle, correct simulation of unconfined flow, at whatever stream speed and direction are selected by the tunnel operator for his external-flow calculation, is always arrived at, regardless of initial conditions in the tunnel, provided only that the tunnel stream is steady and homenthalpic. The initial flow (before iteration) may have the wrong speed and flow inclination, and any kind of boundary interference; when the iteration converges—i.e., matching is obtained at the interface—all the boundary conditions are satisfied. The converged flow may or may not involve inclination relative to the tunnel's geometrical axis. (We are led to the idea of changing the angle of attack without moving the model.) If the converged, unconfined flow does involve such inclination, clearly this must be recognized in decomposing the resultant force on the model into "lift" and "drag."

The adaptable-wall wind tunnel therefore provides, by definition, a "self-calibration" whose accuracy is unlimited in principle. In practice, its accuracy will depend on that of the instrumentation provided at the interface and on the precision to which the matching at the interface can be carried out; this, in turn, will depend upon the power, precision, and sophistication of the wall-control organs. The former—the instrument accuracy—is the easier to discuss. To establish the stream direction very accurately obviously means to measure accurately the local flow inclination at points of S, relative to the axes of the balance system; this requirement suggests that optical instrumentation may be desired.

A remark about the power and precision of wall controls may be in order. One of the most powerful and potentially precise appears to be the conventional tunnel-speed control—e.g., the r.p.m. of the main drive system. This is hardly a "wall-control organ," but may be so treated in the present discussion. Some tunnel operators have also observed that differential plenum pressure—just above and below the working section, for example—exerts a powerful and sensitive control over flow inclination in ventilated tunnels.

Some Numerical Examples

Studies are currently being carried out, under O.N.R. sponsorship, at the University of Arizona concerning the exploitation of the adaptable-wall idea for low-speed testing of V/STOL configurations. These studies are not entirely pertinent to the subject of this chapter; nevertheless, they do involve some numerical simulations of the adaptable-wall process, including iterations to unconfined flow, and these can be used to illustrate some of the points made in this paper.

Simulation of the adaptive-wall process involves essentially two parts:

(i) Calculation of external flow from data given at the interface S. This is the same procedure that would be involved in actual use of an existing tunnel.

(ii) Calculation of internal flow, including the model, from data given on S. This is a simulation of the experiment.

In our present studies, the interface consists of 5 sides of a rectangular box (Fig. 1) inside which the model is mounted. Calculation (i) is carried out by panel methods, assuming incompressible, irrotational flow. Four rectangular

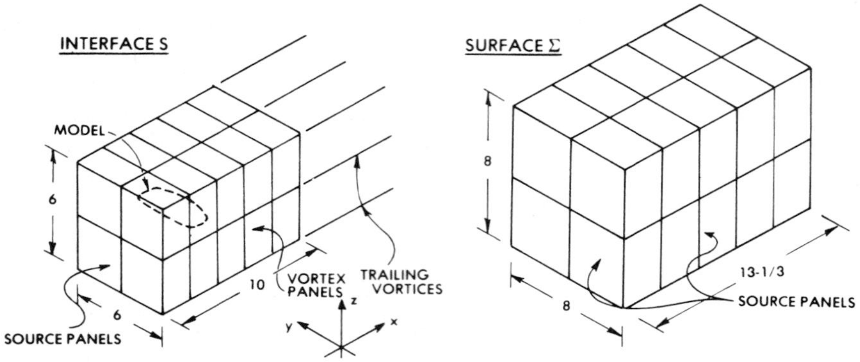

Figure 1. Sketch showing interface S (left) and surface Σ (right).

panels of uniform source strength are used on the upstream (front) face of S, and 40 rectangular distributed-vortex panels (uniform distributions of elemental horseshoe vortices) are used on the sides, top, and bottom.

For the simulation (ii) of internal flow, 5 sides of a larger box, Σ, are used (Fig. 1). Rectangular uniform-source panels, 44 in number, are arranged on the 5 sides of Σ.

For the purposes of this study, a very simple wing model has been chosen, namely, a plane horseshoe vortex of span 4, centrally located with respect to S and Σ, its circulation determined, linearly, by the x- and z-components of fluid velocity at the midpoint of the lifting line.

The demonstration therefore begins with some selected initial flow involving deviations of the stream from the preselected vector $\mathbf{V}$ and/or wall effects represented by image singularities or the like: there is an initial perturbation field due to these features and due to the wing, whose initial circulation is determined from this perturbation field.

The process of iteration is that described in earlier papers on adaptive wind tunnels [4.209, 4.210]. The external field, represented by [ ], uses boundary values of tangential velocity v_t on S, satisfies the far-field boundary condition, and leads to a distribution of normal velocity $v_n[v_t]$. The inner field is then simulated, including the responsive wing, whose circulation is determined by the flow, using as boundary data updated values of v_n, namely,

$$v_n^{(p+1)} = v_n^{(p)} + k\delta^{(p)}v_n, \tag{1}$$

where

$$\delta^{(p)}v_n = v_n^{(p)} - v_n[v_t^{(p)}], \tag{2}$$

where p denotes the number of the iteration, $p = 1, 2, \ldots$. Here k is a "relaxation factor," chosen here, on the basis of previous experience, to be 0.25. Convergence to unconfined flow is represented by $\delta^{(p)}v_n \to 0$, for then Eq. (2) states that the inner and outer flows match at S and both inner (at model) and outer (far-field) boundary conditions are satisfied. The combined inner–outer field is singularity-free except at the model.

Results are presented graphically in Fig. 2, where the variations of some interesting quantities, as the iteration proceeds, are plotted against p. Table 1 shows what cases are presented and gives the key to Fig. 2.

Table 1 Cases iterated.

	Initial			Final (unconfined)	
Case	V	Flow inclination	Images	V	Flow inclination
(a)	0	0	None	1	0.25 rad
(b)	0.8	0	None	1	0.25 rad
(c)	0.8	0	4 image horshoe vortices	1	0.25 rad

In all cases, the unconfined-flow stream speed V was chosen to be 1.0, and the constants of the horseshoe vortex were selected to give a circulation, in unconfined flow, equal to 2.30. (Since the wingspan is 4 and an approximation to Γ is, in customary notation,

$$\Gamma = \tfrac{1}{2}VcC_l \approx \pi Vc\alpha,$$

it will be recognized that this represents a case of large lift and appreciable flow perturbation.)

The quantities plotted in Fig. 2 are:

1. The wing circulation $\Gamma^{(p)}$.
2. The "error signal" $|\delta v_n^{(p)}|$. The value plotted is the maximum on S for each p. As mentioned, $\delta v_n^{(p)} \to 0$ denotes convergence to unconfined flow.
3. The average absolute error. This is the average value of the absolute difference between $v_n^{(p)}$ and the exact value for a plane horseshoe vortex, divided by the maximum value of v_n on S and averaged over all 44 field points of S—i.e., center points of the panels—after p iterations.

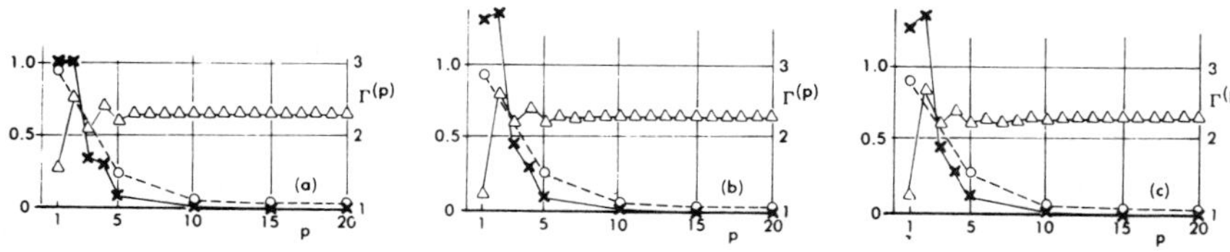

Figure 2. Results of iterations of cases outlined in Table 1. p denotes the iteration number. $k = 0.25$. —Δ—Δ—Δ—, wing circulation $\Gamma^{(p)}$ (right-hand scale); —×—×—×—, "error signal" $\delta^{(p)}v_n$ (left-hand scale); —O—O—O—, dimensionless average absolute error in $v_n^{(p)}$ (left-hand scale). (The points are connected only for clarity.)

Conclusion

Figure 2 shows that satisfactory convergence was obtained with $k = 0.25$ in all cases. There is measurable error in the final, converged, unconfined flow field, as measured by the average absolute error on S. In all cases this value is less than one percent of the stream speed; this is a measure of the ability of our simple panel methods to model the exact flow field. It is clear that the iteration procedure corrects for errors in the stream vector—magnitude and/or direction—as well as for the extraneous flow field of "image" vortices.

The author concludes that adaptive-wall technology may provide a technique for setting the stream vector accurately, in magnitude and direction, in wind-tunnel testing. Success of this technique hinges primarily upon success in solving the instrumentation problem.

Acknowledgment

The research reported here is supported by the U.S. Office of Naval Research under Contract No. N00014-79-C-0010P00001. This work has been monitored by Mr. Morton Cooper, to whom this paper is dedicated. Ever since the inception of the adaptable-wall wind-tunnel program, Mr. Cooper has been involved in it and has, in characteristic style, made innumerable valuable technical contributions to it. His intelligent and perceptive insight and encouragement has been deeply appreciated and will be sorely missed.

CHAPTER 26

Basic Advances in the Finite–Volume Method for Transonic Potential-Flow Calculations

D. A. Caughey* and Antony Jameson†

1 Introduction

The finite-volume methods of Jameson and Caughey [4.8, 4.9, 4.212] provide a general framework within which it is fairly easy to calculate the transonic potential flow past essentially arbitrary geometrical configurations. Like finite-element methods, these methods use only local properties of the transformations which generate the difference grid. This feature essentially decouples the solution of the transonic flow equations from the grid-generating step, so that minor modifications of a universal algorithm can be applied in any boundary-conforming coordinate system. Although the initial variants of these methods used line relaxation to solve the difference equations, the multigrid alternating-direction-implicit (MAD) scheme of Jameson [4.213] has also been applied to provide high rates of convergence to very small residuals for two-dimensional calculations [4.214].

Two particular features of the formulation of the finite-volume methods will be addressed in the present paper, with the aim of improving the accuracy and consistency of the method. The first is an improved artificial viscosity, which allows retention of formal second-order accuracy in supersonic zones; the second is a modification of the scheme which allows the freestream conditions to be satisfied identically by the difference equations. In the

* Associate Professor, Sibley School of Mechanical and Aerospace Engineering, Cornell University, Ithaca, NY 14853.

† Professor, Department of Mechanical and Aerospace Sciences, Princeton University, Princeton, NJ 08540.

following sections, the fully conservative finite-volume method will first be briefly reviewed, including the changes necessary to retain second-order accuracy in supersonic zones. The problem of consistency with the freestream solution will then be discussed and a remedy proposed. Finally, results of calculations incorporating these changes will be presented and discussed.

2 Analysis

2.1 Finite-Volume Scheme

For convenience, here and throughout the paper the analysis will be described for a two-dimensional problem, and only distinguishing features of the extension to three-dimensional problems will be discussed. The equations of steady, inviscid, isentropic flow can be represented as follows. Let x, y be Cartesian coordinates, and u, v be the corresponding components of the velocity vector $\mathbf{q}$. Then the continuity equation can be written as

$$(\rho u)_x + (\rho v)_y = 0, \tag{1}$$

where ρ is the local density. This is given by the isentropic law

$$\rho = \left(1 + \frac{k-1}{2} M_\infty^2 (1 - q^2)\right)^{1/(k-1)}, \tag{2}$$

where k is the ratio of specific heats, and M_∞ is the freestream Mach number. The pressure p and the speed of sound a follow from the relations

$$p = \frac{\rho^k}{k M_\infty^2} \tag{3}$$

and

$$a^2 = \frac{\rho^{k-1}}{M_\infty^2}. \tag{4}$$

Consider now a transformation to a new set of coordinates X, Y. Let the Jacobian matrix of the transformation be defined by

$$H = \begin{Bmatrix} x_X & x_Y \\ y_X & y_Y \end{Bmatrix}, \tag{5}$$

and let h denote the determinant of H. The metric tensor of the new coordinate system is given by the matrix $H^T H$, and the contravariant components of the velocity vector U, V are given by

$$\begin{Bmatrix} U \\ V \end{Bmatrix} = H^{-1} \begin{Bmatrix} u \\ v \end{Bmatrix} = (H^T H)^{-1} \begin{Bmatrix} \phi_X \\ \phi_Y \end{Bmatrix} \tag{6}$$

where ϕ is the velocity potential. Equation (1), upon multiplication by h, can then be written

$$(\rho h U)_X + (\rho h V)_Y = 0. \tag{7}$$

The fully conservative finite-volume approximation corresponding to Eq. (7) is constructed by assuming separate bilinear variations for the independent and dependent variables within each mesh cell. Numbering the cell vertices as illustrated in Fig. 1, and assuming that the local coordinates $X_i = \pm\frac{1}{2}$, $Y_i = \pm\frac{1}{2}$ at the vertices, the local mapping can be written

$$x = 4\sum_{i=1}^{4} x_i(\tfrac{1}{4} + X_i X)(\tfrac{1}{4} + Y_i Y), \tag{8}$$

where x_i is the x-coordinate of the ith vertex. Similar formulas are assumed to hold for y and ϕ. At a cell center, this transformation yields formulas for derivatives such as

$$x_X = \tfrac{1}{2}(x_2 - x_1 + x_3 - x_4). \tag{9}$$

If we introduce the averaging and differencing operators

$$\begin{aligned} \mu_X f_{i,j} &= \tfrac{1}{2}(f_{i+1/2,j} + f_{i-1/2,j}), \\ \delta_X f_{i,j} &= (f_{1+1/2,j} - f_{i-1/2,j}), \end{aligned} \tag{10}$$

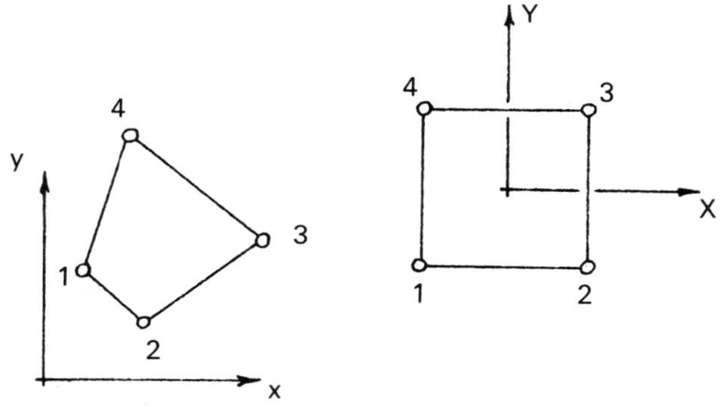

Figure 1. Mesh cell in physical and computational domains.

then the transformation derivatives, evaluated at the cell centers, can be expressed by formulas such as

$$\begin{aligned} x_X &= \mu_Y \delta_X x, \\ x_Y &= \mu_X \delta_Y x, \end{aligned} \tag{11}$$

with similar expressions for the derivatives of y and the potential. Such formulas can be used to determine ρ, h, U, and V at the center of each cell using Eqs. (2), (5), and (6). Equation (7) is represented by conserving fluxes across the boundaries of auxiliary cells whose faces are chosen to be midway between the faces of the primary mesh cells. This can be represented as

$$\mu_Y \delta_X(\rho h U) + \mu_X \delta_Y(\rho h V) = 0. \tag{12}$$

This formula can also be obtained by applying the Bateman variational principle that the integral of the pressure

$$I = \int p \, dx \, dy \tag{13}$$

is stationary, and approximating I by a simple one-point integration scheme in which the pressure at the center of each grid cell is multiplied by the cell area. For subsonic flow the finite-volume method can equally well be regarded as a finite element method with isoparametric bilinear elements.

The extension to treat transonic flows is accomplished by adding an artificial viscosity to introduce an upwind bias. The use of the one-point integration scheme leading to Eq. (12) has the advantage of requiring only one density evaluation per mesh point, but also has the undesirable effect of tending to decouple the solution at odd- and even-numbered points of the grid; suitable recoupling terms can be added to improve the stability of the solution. If we represent the influence coefficients of the terms containing ϕ_{XX} and ϕ_{YY} in the expanded form of Eq. (12) as

$$\begin{aligned} A_X &= \rho h\left(g^{11} - \frac{U^2}{a^2}\right), \\ A_Y &= \rho h\left(g^{22} - \frac{V^2}{a^2}\right), \end{aligned} \tag{14}$$

where g^{ij} are the elements of $(H^T H)^{-1}$, then the compensated equation can be written

$$\delta_X \mu_Y(\rho h U) + \delta_Y \mu_X(\rho h V) - \frac{\varepsilon}{2} \delta_{XY}(A_X + A_Y)\delta_{XY}\phi = 0, \tag{15}$$

where $0 \leq \varepsilon \leq \frac{1}{2}$. In practice $\varepsilon = \frac{1}{2}$ is generally used. An alternative method of obtaining the recoupling terms is to use a higher-order integration scheme which takes account not only of the pressure at the center of each cell, but also of its x- and y-derivatives.

Second-Order Viscosity. The original scheme was stabilized in supersonic regions by the explicit addition of an artificial viscosity, chosen to simulate the directional bias introduced by the rotated difference scheme of Jameson [4.6]. We defined

$$\begin{aligned} \hat{P} &= \frac{\rho h \sigma}{a^2}(U^2 \delta_{XX} + UV\delta_{XY})\phi \\ \hat{Q} &= \frac{\rho h \sigma}{a^2}(UV\delta_{XY} + V^2\delta_{YY})\phi, \end{aligned} \tag{16}$$

where the switching function

$$\sigma = \max(0, 1 - (M_c/M)^2) \tag{17}$$

is nonzero only for values of the local Mach number $M = q/a$ greater than some critical Mach number M_c. Then, after defining

$$P_{i+1/2,j} = \begin{cases} \hat{P}_{i,j} & \text{if } U \geq 0, \\ -\hat{P}_{i+1,j} & \text{if } U < 0, \end{cases} \tag{18}$$

with a similar shift for Q, we represented Eq. (15) as

$$\delta_X(\mu_Y(\rho h U) + P) + \delta_Y(\mu_X(\rho h V) + Q) - \frac{\varepsilon}{2}\delta_{XY}(A_X + A_Y)\delta_{XY}\phi = 0. \tag{19}$$

The difference equations (15) approximate the original differential equation (7) to within a formal truncation error of second order in the mesh spacing in the physical plane when the mesh is smooth. Since, however, the additional fluxes P and Q added in supercritical regions are of the order of the physical mesh spacing, the equations (19) approximate Eq. (7) to within a truncation error of only first order in the mesh spacing. The error resulting from the introduction of the artificial viscosity can be reduced to second order at nearly all points in the flow field if we define

$$P_{i+1/2,j} = \begin{cases} \hat{P}_{i,j} - (1 - \kappa\delta_X\rho)\hat{P}_{i-1,j} & \text{if } U \geq 0, \\ -\hat{P}_{i+1,j} + (1 - \kappa\delta_X\rho)\hat{P}_{i+2,j} & \text{if } U < 0, \end{cases} \tag{20}$$

where κ is a constant of order unity. In regions where the solution is smooth, the term $\kappa\delta_X\rho$ is of first order in the mesh spacing, and the viscosity is formally a second-order quantity. Near a shock, the quantity $1 - \kappa\delta_X\rho$ becomes small, and the equations (20) approximate the equations (18)—i.e., the viscosity reverts to a first-order quantity. This hybridization of the second-order scheme has been found necessary to stabilize computations for solutions containing strong shocks.

Freestream Consistency. The implementation of the algorithm is simplified by introducing the reduced potential G describing perturbations from a uniform free stream, inclined at an angle α to the x-axis:

$$G = \phi - x\cos\alpha - y\sin\alpha. \tag{21}$$

The contravariant velocities are then calculated in two steps, first determining

$$\begin{Bmatrix} u \\ v \end{Bmatrix} = (H^T)^{-1}\begin{Bmatrix} G_X \\ G_Y \end{Bmatrix} + \begin{Bmatrix} \cos\alpha \\ \sin\alpha \end{Bmatrix}, \tag{22}$$

and then using the first equation of (6) to determine U and V. For two-dimensional problems, this procedure has the attractive feature that the freestream conditions (i.e., $G = 0$) identically satsify the difference equations. This is easily verified, since at each cell center in this case

$$h\begin{Bmatrix} U \\ V \end{Bmatrix} = \begin{Bmatrix} \mu_X\delta_Y y\cos\alpha - \mu_X\delta_Y x\sin\alpha \\ -\mu_Y\delta_X y\cos\alpha + \mu_Y\delta_X x\sin\alpha \end{Bmatrix}. \tag{23}$$

The density is calculated using the Cartesian velocities from (22) in Eq. (2). For the freestream condition, this gives $\rho = 1$, and Eq. (12) becomes

$$\mu_Y \delta_X(\mu_X \delta_Y y \cos \alpha - \mu_X \delta_Y x \sin \alpha) + \mu_X \delta_Y(-\mu_Y \delta_X y \cos \alpha + \mu_Y \delta_X x \sin \alpha) = 0, \quad (24)$$

since the averaging and differencing operators commute.

In three-dimensional problems, the difference equations do not admit $G = 0$ as a solution, because the elements of H^{-1} consist of nonlinear products of the mapping derivatives, and the averaging and differencing operators do not cancel identically. For example, the x-component of the freestream is given by

$$h\begin{Bmatrix} U \\ V \\ W \end{Bmatrix} = \begin{Bmatrix} y_Y z_Z - y_Z z_Y \\ y_Z z_X - y_X z_Z \\ y_X z_Y - y_Y z_X \end{Bmatrix} \cos \alpha. \quad (25)$$

At the cell centers, the fluxes are represented by

$$h\begin{Bmatrix} \overline{U} \\ \overline{V} \\ \overline{W} \end{Bmatrix} = \begin{Bmatrix} \mu_{XZ}\delta_Y y \mu_{XY}\delta_Z z - \mu_{XY}\delta_Z y \mu_{XZ}\delta_Y z \\ \mu_{XY}\delta_Z y \mu_{YZ}\delta_X z - \mu_{YZ}\delta_X y \mu_{XY}\delta_Z z \\ \mu_{YZ}\delta_X y \mu_{XZ}\delta_Y z - \mu_{XZ}\delta_Y y \mu_{YZ}\delta_X z \end{Bmatrix} \cos \alpha, \quad (26)$$

and the (three-dimensional equivalent of) Eq. (12) becomes

$$\mu_{YZ}\delta_X(h\overline{U}) + \mu_{XZ}\delta_Y(h\overline{U}) + \mu_{XY}\delta_Z(h\overline{W}) = R, \quad (27)$$

which is not identically zero for an arbitrary smooth mesh. It can be verified, in fact, that Eq. (27) results in

$$\begin{aligned} R = \tfrac{1}{2}(&\mu_X\delta_X((\mu_Y\delta_Y y)(\mu_Z\delta_Z z) - (\mu_Z\delta_Z y)(\mu_Y\delta_Y z)) \\ &+ \mu_Y\delta_Y((\mu_Z\delta_Z y)(\mu_X\delta_X z) - (\mu_X\delta_X y)(\mu_Z\delta_Z z)) \\ &+ \mu_Z\delta_Z((\mu_X\delta_X y)(\mu_Y\delta_Y z) - (\mu_Y\delta_Y y)(\mu_X\delta_X z))) \cos \alpha. \end{aligned} \quad (28)$$

If the Cartesian coordinates are expanded in Taylor series about their values at the central point shared by the eight neighboring mesh cells contributing to the residual in Eq. (28), the residual can be verified to be a fourth-order quantity in the local mesh spacing. Since in our formulation the residuals of Eq. (12) are of second order in the mesh spacing, this effect corresponds to a second-order error in the residual. This is consistent with the overall second-order accuracy of the scheme in subsonic regions, but since the mesh cells are necessarily quite large far from the body in three-dimensional calculations, it was thought that the error introduced by this discretization of the freestream contribution might be important.

In order to assess the significance of this error, calculations were performed with this source of error removed from the residual by rewriting Eq. (12) as

$$\mu_{YZ}\delta_X(\rho h U - h\overline{U}) + \mu_{XZ}\delta_Y(\rho h V - h\overline{V}) + \mu_{XY}\delta_Z(\rho h W - h\overline{W}) = 0, \quad (29)$$

where $h\overline{U}$, $h\overline{V}$, $h\overline{W}$ were calculated using formulas similar to (26), but which also included the y-component of the freestream velocity. Since these formulas can be considered approximations to (25), it is easily verified that the added terms do cancel identically when evaluated analytically. When evaluated numerically, however, they cancel exactly the error introduced by the contribution of the freestream to the total fluxes. At boundary points it is necessary to devise reflection rules for these artificial fluxes. Our procedure was to continue the fluxes in the plane parallel to the boundary across the boundary unchanged; the flux normal to the boundary was corrected by retaining the first term in a Taylor series expansion of the component of Eq. (25) normal to the boundary, evaluating all differences at the centers of the cells immediately adjacent to the boundary. This scheme, when applied to the two-dimensional form of the difference equations, maintains its self-canceling property; in the three-dimensional case, it has the effect of replacing the differences in Eq. (28) taken normal to the boundary by one-sided formulas.

2.2 Boundary Conditions and Grid Generation

An important advantage of the finite-volume method is its decoupling of the solution procedure from the grid-generation step. This permits the grid to be generated in any convenient manner, and allows application of an essentially universal algorithm to any problem for which a boundary-conforming coordinate system can be generated.

The airfoil calculations to be described were computed on a mesh generated by weakly shearing the conformal mapping to a circle of the Joukowsky airfoil most closely approximating the actual airfoil in the leading-edge region. Details of this coordinate system are contained in [4.214]. The three-dimensional calculations were computed on a mesh generated by the cylindrical–wind-tunnel mapping sequence described in [4.9] and [4.212].

Two types of boundary conditions must be specified to determine solutions for the potential-flow problems considered herein. The no-flux condition must be enforced across any solid boundaries (such as the airfoil, wing, and fuselage surfaces); and appropriate far-field boundary conditions must be specified at the necessarily finite limits of the computational domain. In addition, for the airfoil calculations, a discontinuity in potential across some branch cut must be incorporated if the airfoil has lift. For the three-dimensional calculations, a linearized treatment of the vortex sheet is used. It assumes that constant-strength vortex filaments trail downstream of the wing trailing edge along the nearly streamwise computational surfaces. The values of the reduced potential on the sheet are determined by requiring the velocity normal to the assumed location of the sheet to be continuous across the sheet.

The solid-surface boundary conditions are quite easy to enforce in boundary-conforming coordinate systems, because the difference scheme is formulated in terms of the contravariant components of the velocity. The appropriate condition is that the out-of-plane component is zero. This is incorporated by reflection of the normal-flux contributions for the cells immediately adjacent to the boundary.

A disadvantage of the finite-volume schemes is the need to truncate the usually infinite domains of aerodynamic interest to finite computational regions. This is in contrast to methods in which the equation can be analytically transformed with suitable stretching functions so that the difference mesh extends to infinity. (See, e.g., [4.6], [4.188], and [4.215].) In the analyses treated here, a reduced potential is introduced to describe the perturbations upon an otherwise uniform stream. This potential is set equal to values appropriate for a compressible vortex of circulation Γ:

$$G = \frac{\Gamma}{2\pi} \arctan\left(\frac{y\sqrt{1 - M_\infty^2}}{x}\right) \tag{30}$$

on the far-field boundaries of the computational domain for the airfoil calculations. The reduced potential is set equal to zero on the upstream and lateral far-field boundaries for the three-dimensional calculations; the first derivative in the streamwise direction of the reduced potential is set equal to zero on the downstream boundary. This is consistent with the assumption that the flow properties have become invariant in the streamwise direction. If the freestream is in the x-direction, irrotationality then implies that $\partial u/\partial y = \partial v/\partial x = 0$ and $\partial u/\partial z = \partial w/\partial x = 0$, and consequently that the streamwise velocity component has its freestream value.

2.3 Solution of Difference Equations

The difference equations resulting from this formulation are solved iteratively. A fairly conventional successive-line overrelaxation (SLOR) scheme is used in the computer codes which solve the three-dimensional problems, with care taken in the formulation of the algorithm to model the correct domains of dependence. The solutions to be presented were calculated on a sequence of three grids, each containing eight times the number of mesh cells of the preceeding one. On each of the coarser grids, 200 relaxation sweeps were performed, and the solution was used as an initial estimate for the solution on the next grid. Only 100 iterations were performed on the finest grid. This is sufficient to remove nearly all of the high-wave-number error, but the lift may not be completely converged. Since we will only compare results obtained in a similar manner, however, this lack of convergence should not introduce a serious systematic error. Details of the relaxation scheme can be found in [4.8] and [4.9]. The two-dimensional results to be presented were calculated using a finite-volume generalization of Jameson's multigrid alternating-direction (MAD) algorithm [4.213, 4.214].

3 Results

Results will now be presented to demonstrate the effects of the improved accuracy of the second-order viscosity and the removal of the truncation error of the freestream. The results for the improved viscosity will be presented for two-dimensional calculations; the results with the freestream contribution to the residuals subtracted out will necessarily be presented for a three-dimensional geometry.

Figure 2(a) and (b) show the surface pressure distributions calculated for the flow at a freestream Mach number of 0.75 past an NACA 0012 airfoil at 2° angle of attack. Both results were calculated using the same grid, consisting of 128×32 mesh cells in the circumferential and radial directions, respectively. Figure 2(a) shows the result using the first-order accurate viscosity; Fig. 2(b) shows the result using the second-order formulation. The result using the first-order viscosity clearly underpredicts the size of the supersonic pocket above the airfoil and consequently underpredicts the lift. That the second-order accurate result is, in fact, nearer the exact answer is verified by the convergence study plotted in Fig. 3 for this case, calculated using sequences of grids for each scheme. The calculated lift and drag coefficients are plotted against mesh spacing for the first-order scheme, while the results of the second-order scheme are plotted against the square of the mesh spacing. (NX is the number of mesh cells in the circumferential direction.) Both schemes clearly converge in the limit of zero mesh spacing to the same lift and drag coefficients, but on the 128×32 grid, the error in lift coefficient calculated using the first-order scheme is still almost 10%. For the second-order scheme the error is less than 3%.

Results of the first- and second-order schemes for a shock-free solution are shown in Fig. 4(a) and (b). The surface pressure distributions are plotted for the Korn airfoil (Catalog No. 75-06-12) [4.168] at a freestream Mach number of 0.75 and zero angle of attack. These conditions correspond to the shock-free design point for this airfoil, so the exact solution should have a smooth recompression back through the sonic velocity at the downstream boundary of the supersonic pocket above the airfoil. Results are shown for the first-order scheme in Fig. 4(a) and for the second-order scheme in Fig. 4(b). Both solutions were obtained on identical grids containing 128×32 mesh cells. The second-order-accurate scheme produces an almost shock-free result, whose pressure distribution is much nearer the hodograph solution. The calculated drag coefficient still differs appreciably from zero, but the solution calculated on a 256×64 mesh results in a drag coefficient of only 0.0005.

The three-dimensional calculations were performed for the ONERA wing M-6, mid-mounted on a circular cylinder. The wing geometry is described in

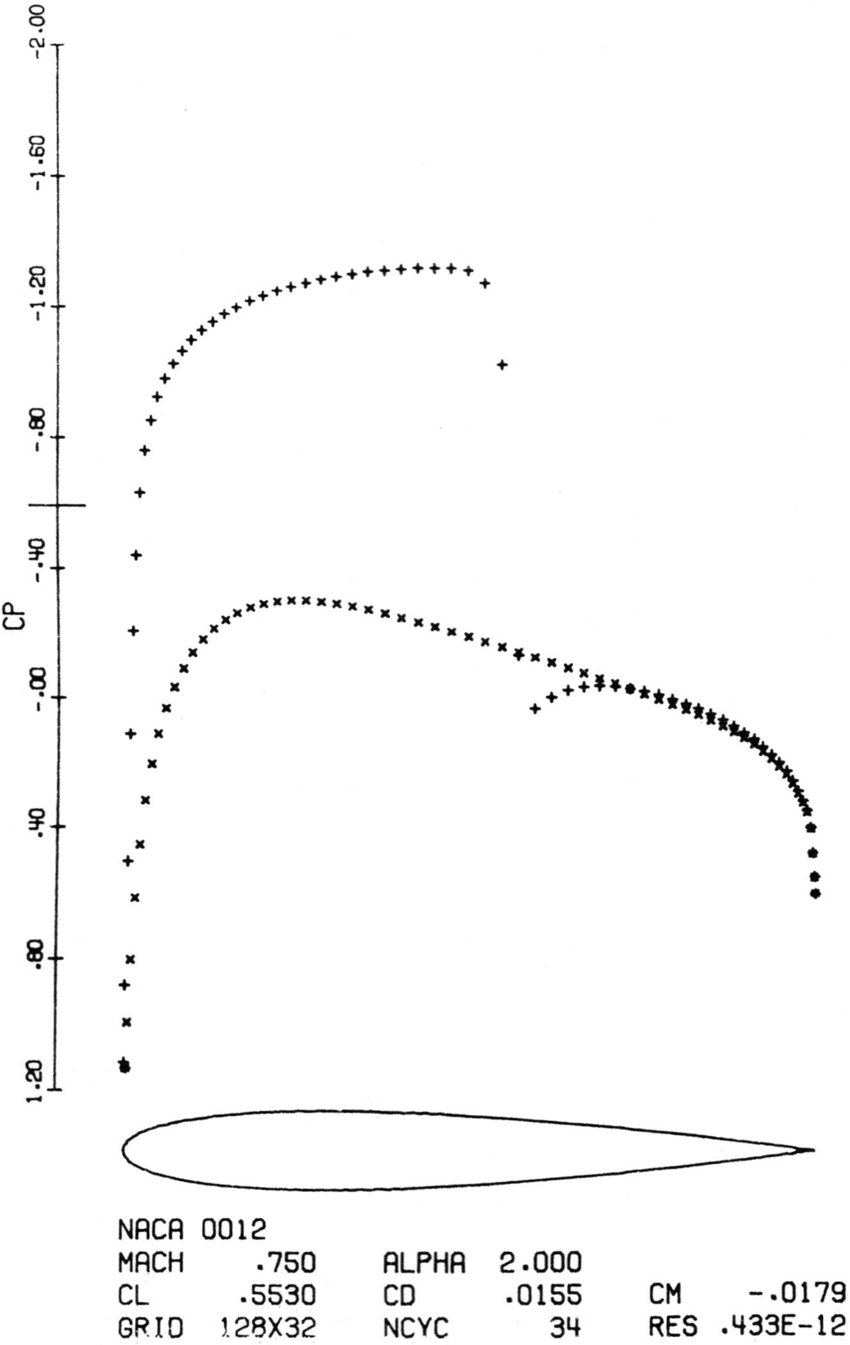

Figure 2a. Surface pressure distributions for transonic flow past NACA 0012 airfoil at 0.75 Mach number and 2 degrees angle of attack: first-order scheme.

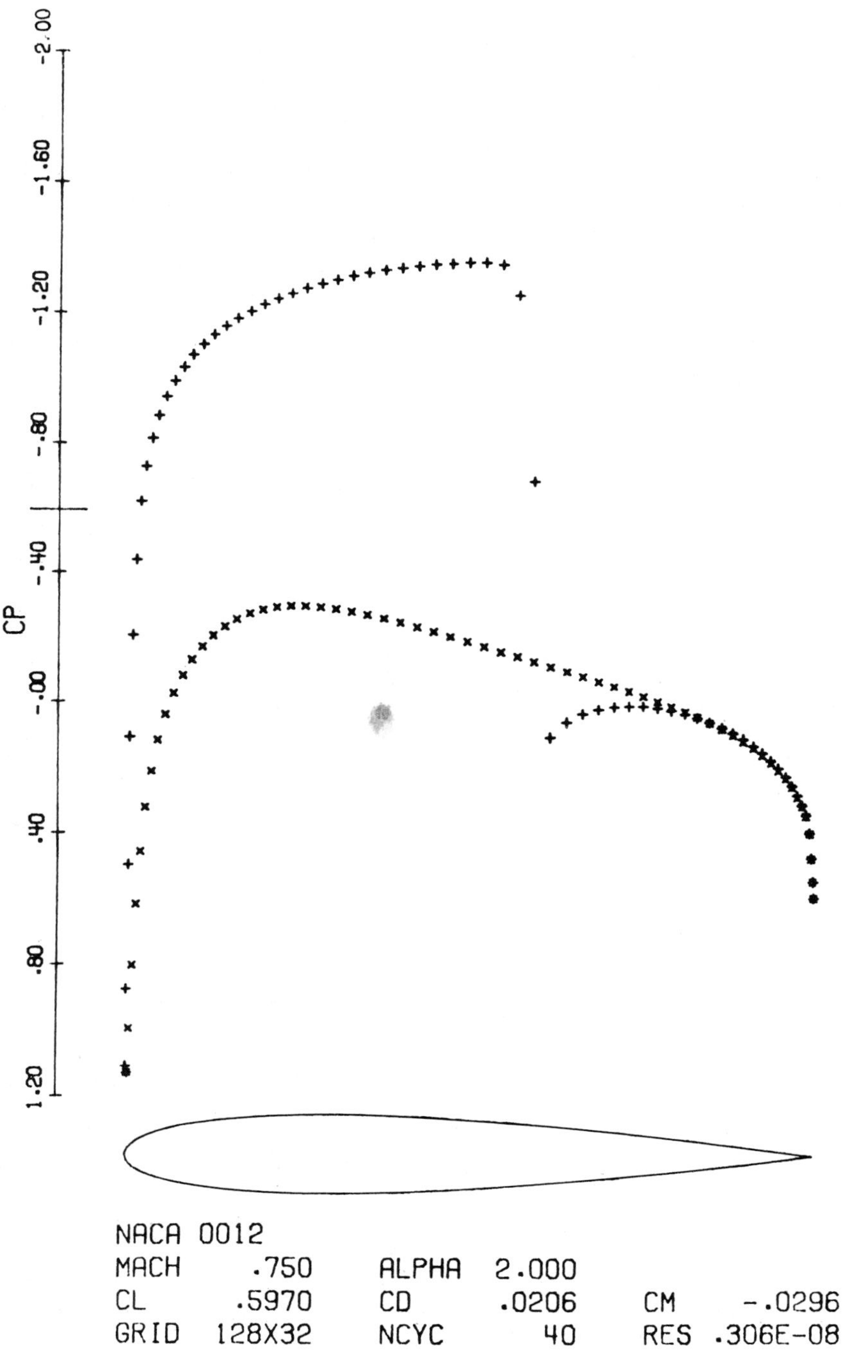

Figure 2b. Surface pressure distributions for transonic flow past NACA 0012 airfoil at 0.75 Mach number and 2 degrees angle of attack: second-order scheme.

[4.216], and other calculations for this wing–cylinder combination are presented in [4.8], [4.9], and [4.212]. A perspective view of the wing–fuselage grid (corresponding to the coarsest of the three grids used) is pictured in Fig. 5. Calculations were performed on an identical grid containing 160 × 24 × 32 mesh cells, using the basic algorithm of [4.8] and [4.212] and the modification described herein to remove the freestream contribution to the residual. The wing lift coefficients were 0.3016 and 0.3120 for the original and modified schemes, respectively. The discrepancy in lift coefficient was therefore less than $\frac{1}{2}\%$. At inboard stations, the wing surface pressure distributions are virtually indistinguishable. The streamwise surface pressure coefficients are compared at two outboard span stations in Fig. 6(a) and (b). As can be seen, even here the details of the pressure distributions are in excellent agreement, even at the 72%-semispan station where the leading- and trailing-edge shocks are beginning to merge. The excellent agreement between the two sets of results indicates that the truncation error introduced by the inconsistency between the freestream conditions and the difference equations introduces no serious

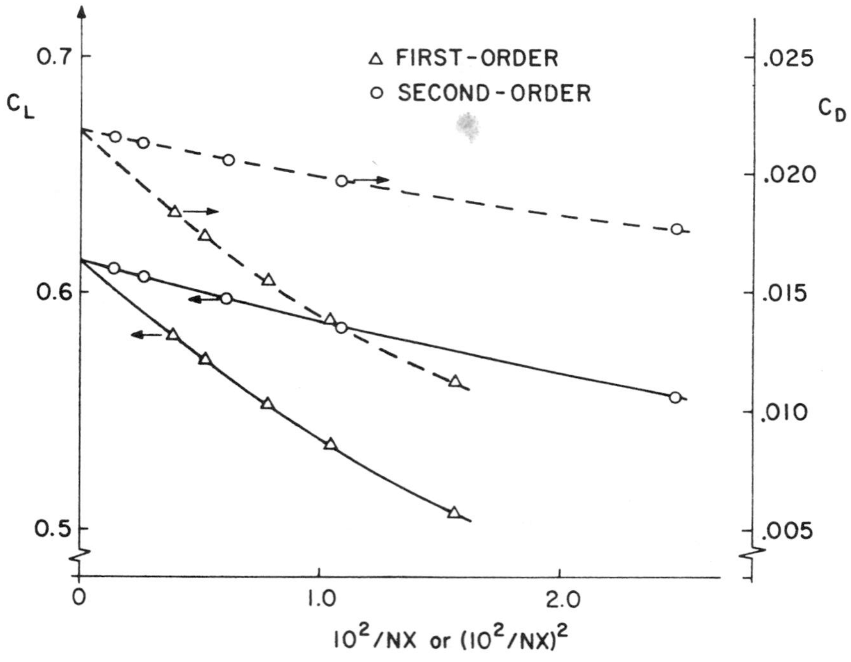

Figure 3. Convergence of lift and drag coefficients with mesh spacing for first- and second-order schemes. NACA 0012 at 0.75 Mach number and 2 degrees angle of attack. Triangles represent first-order scheme; circles represent second-order scheme. Solid line connects results for lift coefficient; dashed line connects results for drag coefficient.

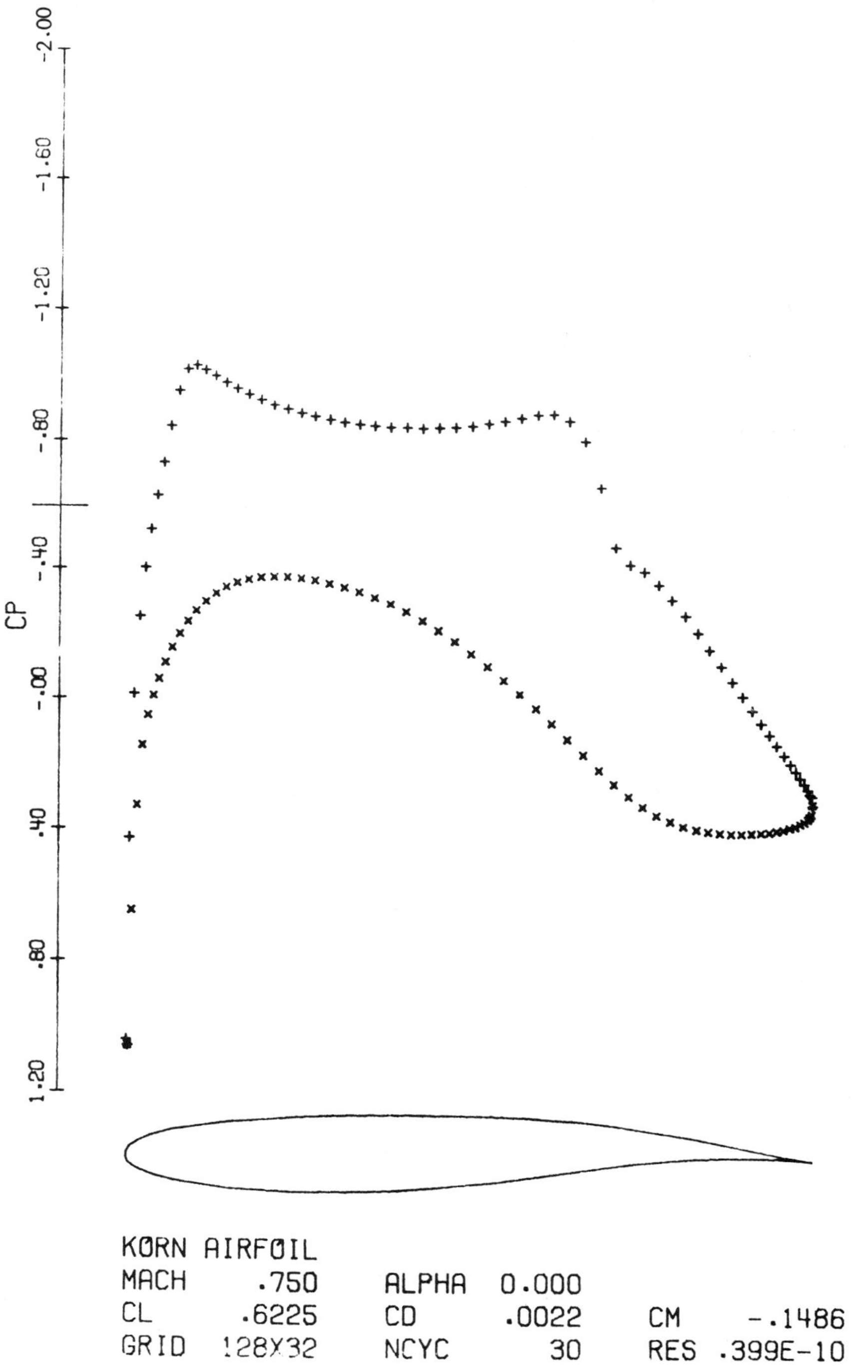

Figure 4a. Surface pressure distributions for transonic flow past Korn airfoil at shock-free design point: first-order scheme.

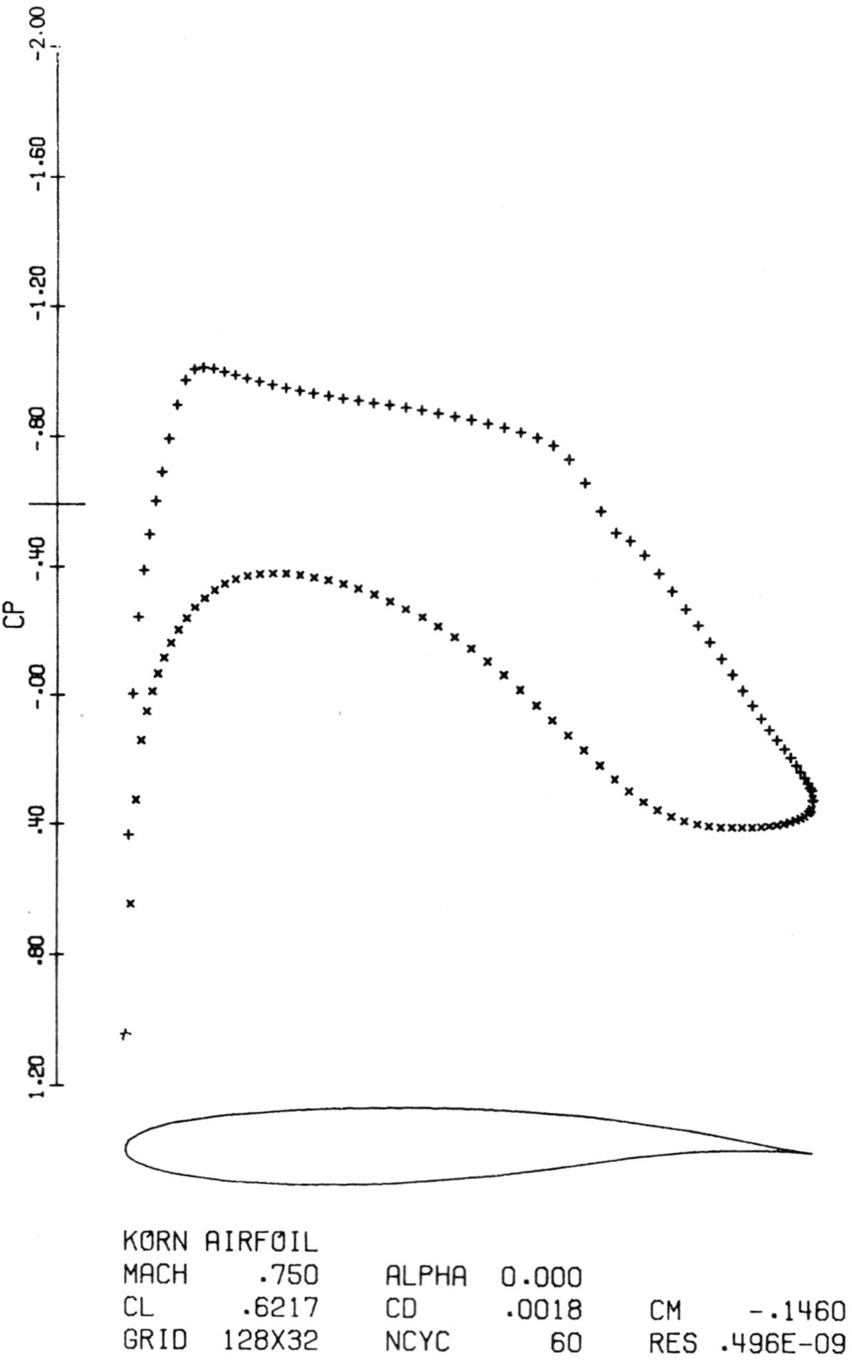

Figure 4b. Surface pressure distributions for transonic flow past Korn airfoil at shock-free design point: second-order scheme.

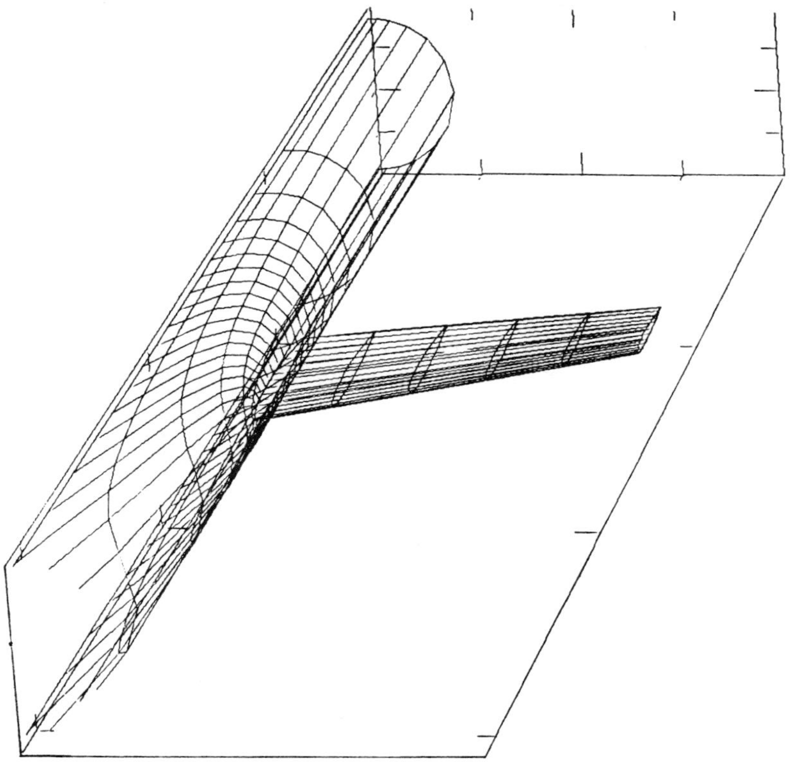

Figure 5. Wing and fuselage grid system for ONERA wing/cylinder configuration.

error, at least in the vicinity of the body. A similar inconsistency in a finite-volume scheme for the Euler equations developed by Pulliam and Steger [4.112] has also been reported to cause no serious error. In view of the additional labor involved in calculating the freestream fluxes, particularly at the boundaries in the present scheme, the original scheme is probably to be preferred.

4 Conclusions

Several fundamental improvements to the finite-volume method for the calculation of transonic potential flows have been presented. The incorporation of an improved artificial viscosity which retains the second-order accuracy of the basic scheme except in the immediate vicinity of shocks is shown to be necessary for the proper prediction of lift when calculations are performed on reasonably coarse meshes. The lack of consistency of the difference equations with a uniform freestream is shown, by comparison with a modified scheme which subtracts this contribution, to produce small errors.

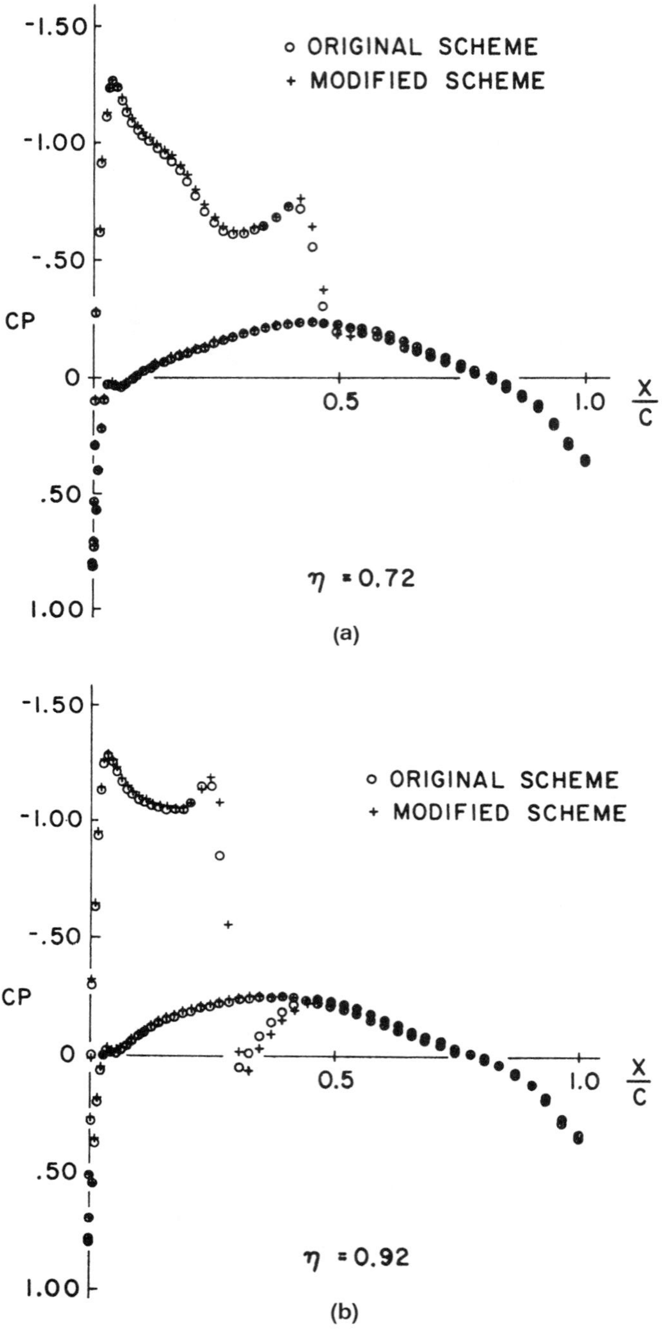

Figure 6. Streamwise wing surface pressure distributions for ONERA wing/cylinder combination. Circles represent original scheme; crosses represent modified scheme with freestream fluxes removed from residuals. a) 72 percent semi-span. b) 92 percent semi-span.

Acknowledgment

This work was supported in part by the Office of Naval Research under Contracts N00014-77-C-0032 and N00014-77-C-0033, and by NASA under Grants NSG-1579 and NGR-33-016-201. It is a particular pleasure on this occasion to thank Morton Cooper for his continued interest in and support of this work.

CHAPTER 27

Flow Structures Associated with Upper-Surface-Blown Airfoils

N. D. Malmuth,*† W. D. Murphy,*‡ V. Shankar,*‡ J. D. Cole,§ and E. Cumberbatch¶

1 Introduction

Upper-surface blowing (USB) has been proposed as a means of increasing usable lift and thereby enhancing V/STOL capability at low speeds in landing configurations. At transonic Mach numbers, it has the further application of achieving low turn radii in dogfight scenarios. The attendant high accelerations are accomplished through elimination of separation by suppression of adverse pressure gradients in the viscous boundary layer, and also movement of shocks downstream of the trailing edge, thereby discouraging shock-induced separation and buffet at high maneuver incidences. Further applications of laminar-flow control through stabilization using tangential blowing to achieve favorable pressure gradients is of strong interest currently. In the application of this concept, the engine bleedoff, thrust, and structural penalties required to achieve the foregoing aerodynamic advantages is of importance to the designer. To obtain this relationship, a knowledge of the associated flow fields is required. Although attention has been given to the jet flap in theoretical investigations, relatively little analysis has been performed on upper-surface-blown configurations. For incompressible speeds, the work of Spence [4.217] represents the classical thin-airfoil treatment of the jet-flap

* Rockwell International Science Center, Thousand Oaks, CA 91360.

† Manager, Fluid Dynamics.

‡ Mathematical Sciences Group.

§ Professor, Structures and Mechanics Department, University of California at Los Angeles, Los Angeles, CA 90024.

¶ Professor, Department of Mathematics, Purdue University, West Lafayette, IN 47906.

problem. At transonic Mach numbers, a computational jet-flap solution based on small-disturbance theory was developed for airfoils, and generalized for three-dimensional wings by Malmuth and Murphy [4.218, 4.219]. In these analyses, the classical Karman–Guderley model was applied with a generalized version of the Murman–Cole successive-line overrelaxation scheme [4.1] to treat the free-jet boundaries. The jet was assumed to be thin, and it was assumed on a heuristic basis that the Spence boundary conditions were applicable across it. These conditions involve equilibration between the normal pressure gradient and the centrifugal force associated with the momentum in the jet.

In this paper, the applicability of the conditions will be analyzed for a compressible rotational jet in the context of blowing upstream of the trailing edge on the upper surface, i.e., upper-surface blowing, in contrast to the jet-flap configuration, in which the jet emanates from the trailing edge. The aspect of the paper involving fine structure of the jet layer represents an extension of the earlier work of Malmuth and Murphy [4.220] on transonic wall jets. From these analyses, the paper will describe the numerical approach to the USB problem, and various results showing possibilities for lift augmentation will be presented.

Pressure distributions arising from these solutions will be presented and compared with experimental data. Sources of error will be identified, particularly those associated with wave-interaction phenomena at the jet exit. Others involving viscous modifications of the wall jet will be considered. In particular, the impact of the tangential blowing on the boundary layers will be analyzed as the first step of a viscous-interaction procedure, to be presented elsewhere, for such flows. Results from the computational methods will be presented that provide inexpensive quantifications for the first time of shock-induced separation delay due to tangential blowing.

2 Formulation and Analyses

2.1 Full Potential Theory

In Fig. 1, a USB configuration is shown. For purposes of providing a general framework for the subsequent sections which deal with a small-disturbance mode, a subsumptive full potential formulation is indicated. Two separate potentials are introduced, Φ_I and Φ_E for the jet and external flow respectively. Appropriate to the solution of the full potential equation indicated, boundary conditions on the airfoil surface and slip lines such as AB are shown. Additional conditions of pressure continuity across the latter jet boundaries also hold. These are discussed, in addition to far-field and trailing-edge aspects, in [4.221].

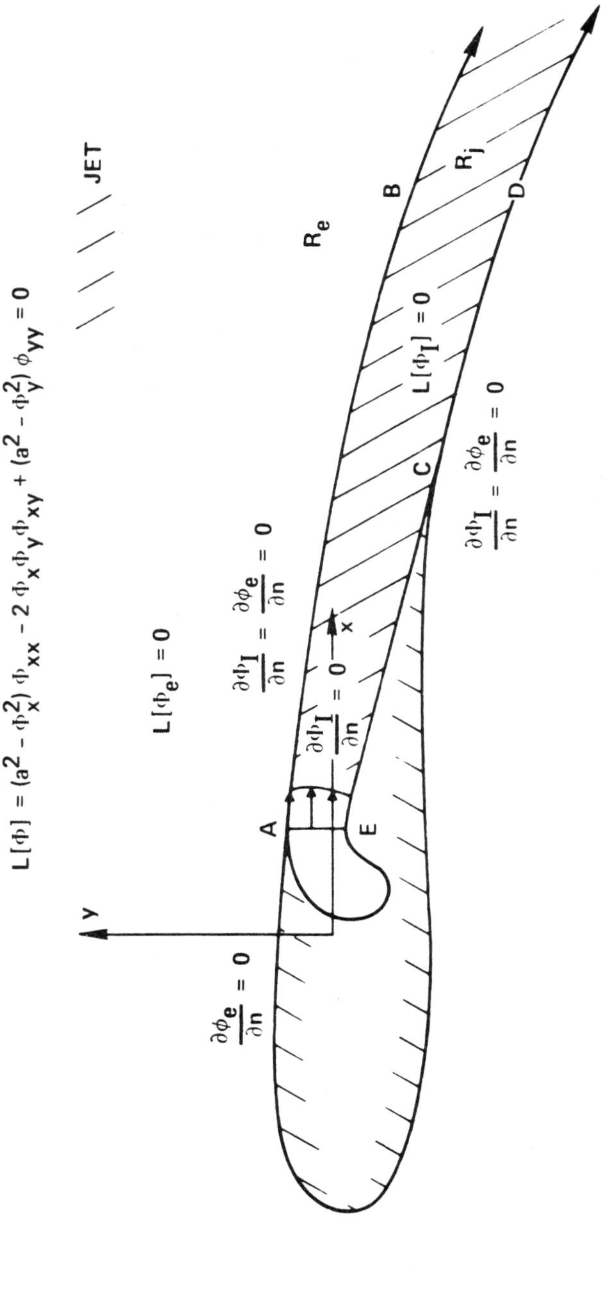

Figure 1. Transonic upper surface blowing—full potential equation formulation.

2.2 Thin-Jet Theory

As an essential ingredient of a small-disturbance formulation, the jet structure is developed in this section for purposes of specification of the boundary conditions. In particular, it will be shown how the Spence theory of [4.217] and the heuristic framework of [4.218] (which is, crudely speaking, a small-deflection approximation of the formulation of the preceding section) can be derived from a systematic approximation procedure. For this purpose, we relax the irrotational assumptions implicitly embodied in the previous section.

Referring to Fig. 2, a section of the jet of Fig. 1 is detailed. A curvilinear coordinate system is embedded in the jet as indicated. The lines η = constant are parallel to a reference line (the ξ-axis), which only under special circumstances coincides with the centerline of the jet. Otherwise, the ξ-axis is a reference line which is the centerline of an approximate parallel flow to be discussed subsequently. In this coordinate system, the lines ξ = constant are normals to ξ-axis. In what follows, the incompressible case will be discussed. The generalizations to compressible flow are straightforward.

To obtain an approximate incompressible set of equations prototypic of the compressible case, the thin-jet limit is considered. The characteristic jet thickness is shown in Fig. 2, where the jet boundary is denoted as $\eta = \tau b(\xi) = \tau b_0 + \tau^2 b_1 + \cdots$.

We now define a thin jet limit

$$\tau \to 0, \quad \xi, \eta^* = \eta/\tau \quad \text{fixed}, \tag{1}$$

where the boundary-layer coordinate η^* is introduced to keep the jet slip lines in view in the limit process. In (1), the appropriate representation of the velocity components q_ξ, q_η shown in Fig. 2 and the pressure p to yield a nontrivial structure are

$$\frac{q_\xi(\xi, \eta; \tau)}{U} = \frac{1}{\sqrt{\tau}} u_0(\xi, \eta^*) + \sqrt{\tau} u_1(\xi, \eta^*) + \cdots, \tag{2a}$$

$$\frac{q_\eta}{U} = \sqrt{\tau} v_0 + \tau^{3/2} v_1 + \cdots, \tag{2b}$$

$$\frac{p}{\rho U^2} = p_0 + \tau p_1 + \cdots, \tag{2c}$$

where U is some typical freestream velocity and all coefficients of the gauge functions involving τ are $O(1)$ as $\tau \to 0$. These orders are consistent with the massless-momentum-source model of Spence [4.217].

Substitutions of these expansions into the exact equations and boundary conditions gives a hierachy of problems for the various terms in these developments. Details of the solutions for the first- and second-order quantities are

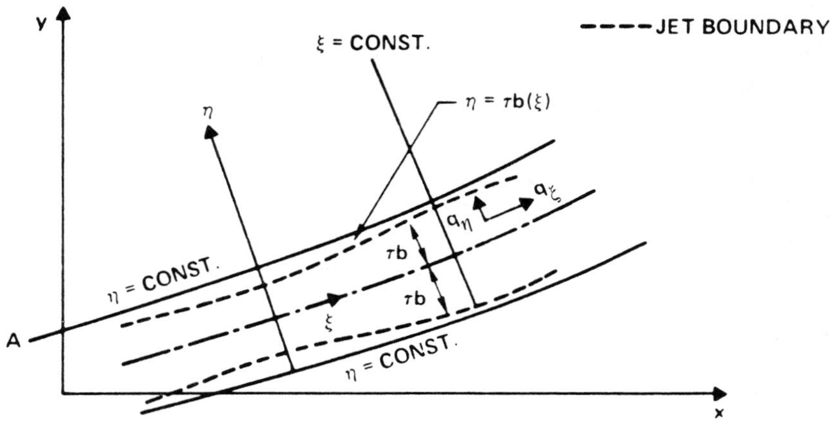

Figure 2. Section of jet and curvilinear coordination system.

given in [4.221]. For a constant-velocity jet exit—i.e., $q_\xi(0, \eta^*) = C$, $q_\eta(0, \eta^*) = 0$—we obtain the following solution:

$$u_0 = u_0(\psi) = C, \tag{3a}$$

$$v_0 = 0, \tag{3b}$$

$$p_0(\xi, \eta^*) = (1 - \eta^*)\frac{C^2}{R} + q_u(\xi), \tag{3c}$$

$$b_0 = 1, \tag{3d}$$

$$\psi = C\eta^*, \tag{3e}$$

where ψ is the zeroth-order stream function, and q_u is the pressure on the upper slip line in units of twice the freestream dynamic pressure.

Discussion. The equations (3) describe a parallel flow jet. The total jump in pressure across the jet from (3c) is

$$[p_0] = p_0(\xi, 1) - p_0(\xi, -1) = 2C^2/R, \tag{4}$$

where R is the radius of curvature of the jet centerline. The equations (3) are consistent with the Spence model. It should be noted that in contrast to the latter, no assumption regarding irrotationality is required to obtain (4), in contrast to the results of previous workers. The radius of curvature of the jet is approximately R upstream of the trailing edge, which in turn is approximately given by that of the blown upper surface. Downstream of the trailing edge, R is determined from applying (4) to the treatment of the flow outside of the jet. Upstream of the trailing edge, the wall pressure is evaluated by (4), since R is known and is given by

$$p(\xi, -1) = \frac{2C^2}{R} + q_u(\xi). \tag{5}$$

External Flow. At distances large compared to the jet width, the fine structure of the jet is important only insofar as it provides matching conditions to the irrotational flow field outside itself. In incompressible flow, this external "outer" flow can be determined by thin-airfoil theory. At transonic speeds, small-disturbance theory is appropriate for this region. Details of the asymptotic matching procedure have been discussed in [4.221]. Based on these developments and the earlier ones for arbitrary-deflection thin jets in Section 2.2, the boundary conditions for the outer flow in the incompressible and transonic cases for the jet flap and upper surface blowing are now indicated.

Jet Flap. Let us represent the equation of the jet as

$$y = \delta g(x),$$

where δ is the thickness ratio of the airfoil, and in a small disturbance approximation consider the "outer" expansion pressure coefficient to be given as

$$\frac{p - p_\infty}{\rho U^2} = \delta P(x, y) + \cdots.$$

Then by virtue of a generalization of (4) we obtain

$$[P(x, 0)] = -C_j g''(x) = -[\phi_x], \tag{6}$$

where

$$C_j \equiv \frac{\rho \int_{-\tau}^{\tau} q_\xi^2 \, d\eta}{\rho U^2} = O(1) \tag{7a}$$

and ϕ is a perturbation potential.

Equation (6) is used in conjunction with the jet tangency relation

$$\phi_y(x, 0) = g'(x) \tag{7b}$$

and the airfoil boundary conditions to determine the external flow field. These relations coincide with those derived by Spence. They can be generalized for transonic flow by placing the ρ inside the integrand in (7a) and replacing δ in the pressure coefficient formula by $\delta^{2/3}$.

Upper-Surface Blowing. To treat conditions on the blown part of the airfoil, Eq. (6) can be applied by approximating the radius R by $(f'')^{-1}$ to obtain the wall pressures, and using the airfoil and jet boundary conditions to determine the upper slip line jet pressures.

From the arbitrary-deflection thin-jet theory derived in Section 2.2, it can be seen that rotational flow produces the same pressure jumps across the jet in the dominant approximation as the irrotational Spence models. Correspondingly, it can be shown that to within factors involving the density, qualitatively similar results are obtained for transonic flow. Another important aspect of the asymptotic representations derived here is that they lead to higher approximations for the structure of the jet and external flow which can be systematically obtained. Finally, the analytical solutions described above and in [4.221] allow the systematic assessment of the effects of initial vorticity and skewness which are inaccessible to other theories.

3 Results and Discussion for Transonic Upper-Surface Blowing

A successive-line overrelaxation (SLOR) scheme within a Karman–Guderley framework [4.218] has been used to compute the flow field over an upper-surface-blown airfoil. On the blown portion, the jump conditions across the jet are determined by the asymptotic results given in previous sections, i.e., Eqs. (6) and (7b). Providing that the region is not too close to the jet exit or trailing edge, the streamwise gradients can be neglected in the entropy and velocity component parallel to the wall. Away from these regions, the pressure gradient perpendicular to the streamlines is balanced by centrifugal force. For the region near the jet exit, these assumptions become invalid. Here, the scale of the gradients in the streamwise direction becomes important, principally due to the influence of wave interactions with the slip line. Similar fine structures occur near the trailing edge, where the flow can stagnate on the unblown side, depending on the ratio of the stagnation pressure of the jet to the ambient stagnation value. For incompressible flow, [4.221] discusses the tristable equilibrium at the trailing edge corresponding to the value of the stagnation-pressure ratio; the dividing streamline leaves tangent to the upper surface if this is greater than unity. Consistent with the previous discussion, the appropriate generalization to transonic flow was assumed also to be this arrangement for a single-valued pressure without a shock in that location. This assumption has been altered to assess the sensitivity of the flow to the dividing-streamline angle. In this connection, surface pressures for the dividing streamline bisecting the trailing-edge angle (as it would in incompressible flow) were compared with those for the upper-surface tangent configuration. Based on these studies, significant differences are anticipated only for large incidences and trailing-edge angles.

Typical results obtained from the computational model are shown in Fig. 3, in which the flow over a thick airfoil designed at Rockwell's Columbus Aircraft Division (CAD) was analyzed with the SLOR code. Here, the pressures for various values of the blowing coefficient C_j are compared against those for the unblown case at a freestream Mach number $M_\infty = 0.703$, and angle of attack $\alpha = 0°$. Substantial lift augmentation is evident for blowing. Also evident is the associated rearward motion of the shock with increased blowing and sectional loading, as if the incidence were increased.

In Fig. 4 the corresponding increase in lift coefficient C_L with slot downstream movement is also shown, as well as the increase in the size of the supersonic region.

Similar increases of lift with blowing coefficient, as well as the size of the supersonic region, have been illustrated in [4.221].

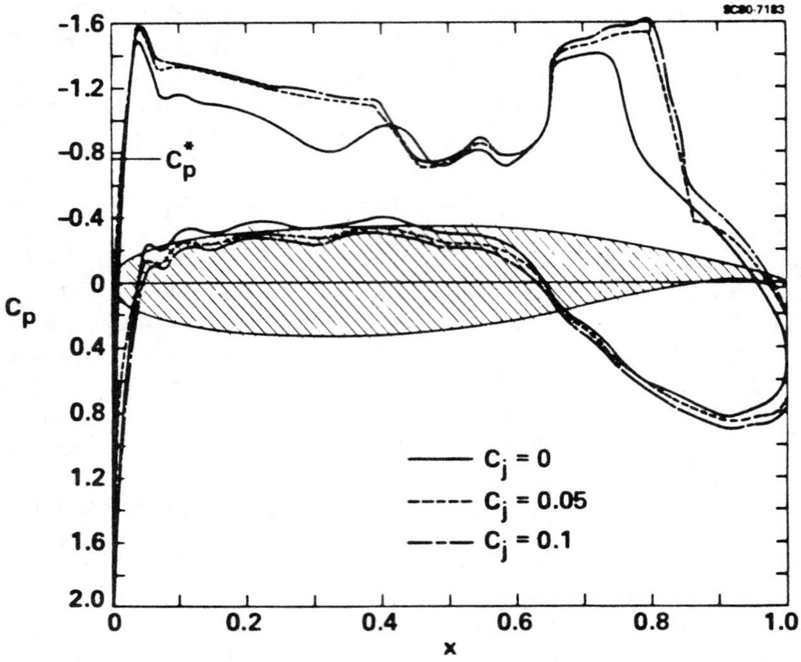

Figure 3. Effect of blowing coefficient, (C_j), variations on chordwise pressures for CAD USB supercritical airfoil, $M_\infty = 0.703$, $\alpha = 0°$, (slot location at 65% chord).

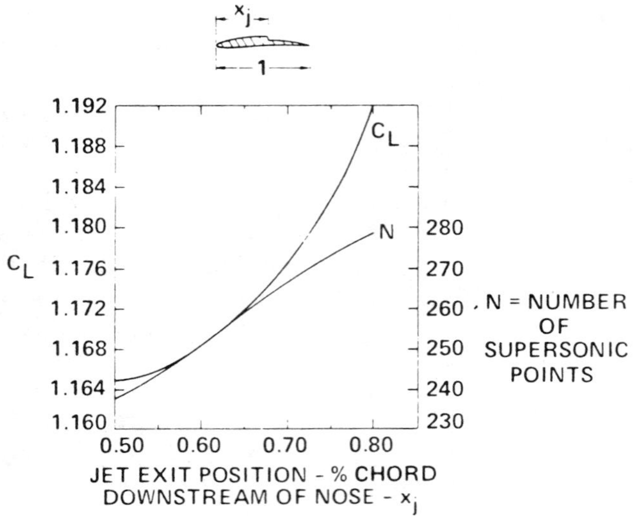

Figure 4. Behavior of C_L and criticality as a function of extent of blowing, CAD USB airfoil, $C_j = 0.1$, $M_\infty = 0.703$, $\alpha = 0°$.

Tests of the adequacy of the foregoing model to simulate realistic transonic USB airfoil flows have been inhibited by the lack of suitable experimental data. Information exists only for highly three-dimensional configurations, large thickness, or incidence in ranges beyond the validity of the assumptions of small-disturbance theory. Another restriction is the unavailability of the associated geometric data and flow diagnostics accompanying the tests. The results of Yoshihara [4.222] and his coworkers were useful in this connection and allowed us to make comparisons with the jet-flap specialized version of the USB theory in [4.218]. For the simulations described in this paper, tests performed by N. C. Freeman at NPL on a USB modified 6% thick RAE 102 airfoil, and described in [4.223], appear to be the most suitable results for comparison at present. Unfortunately, the angle of attack associated with the NPL data is 6°, which is marginal for the application of a small-disturbance model.

Figure 5 indicates comparisons of chordwise pressures for various values of C_j. Also shown are schlierens indicating the associated flow-field structure. Turning to the $C_j = 0$ results (part (a)), massive shock induced separation is indicated and is apparently initiated at the downstream limb of the lambda shock on the upper surface. This is reflected in the classical erosion of the suction plateau and is responsible for the indicated disagreement between the inviscid computational results and the data. For these tests, nominal tangential blowing with a slot height of 0.07% of the chord was used. The slot location is 15% downstream of the nose. The Mach number M_∞ immediately above the slip line at the slot (point A in Fig. 1) is approximately 1.29 for both C_j's indicated. For $C_j = 0.017$, the slot Mach number M_e has been estimated as 1.79 for $C_j = 0.048$, $M_e \approx 2.36$.

Comparison between theory and experiment in part (b) of Fig. 5 indicates reduced discrepancies on the upper surface associated with the limited separation. In part (c), the agreement is correspondingly further improved.

To achieve adequate realism, it is important to discuss factors responsible for the disagreements. One feature not captured by the USB simulation is the pressure spike at the slot location. Based on the slot size, the streamwise scale for this phenomenon is at least an order of magnitude greater than the characteristic wavelength of a Mach diamond pattern in the wall jet. These fluctuations may not be resolvable with conventional pressure-tap arrangements for the thin slot employed in the tests. If a rough model of a coflowing inviscid supersonic wall jet over a flat plate is used to describe the flow near the slot, the approach to a final steady state may be damped oscillatory or monotone, according as the reflection coefficient

$$R = \frac{\lambda - 1}{\lambda + 1},$$

where

$$\lambda = \frac{M_e^2 \beta}{M^2 \beta_e}, \qquad \beta = \sqrt{M^2 - 1}, \quad \beta_e = \sqrt{M_e^2 - 1},$$

is positive or negative.

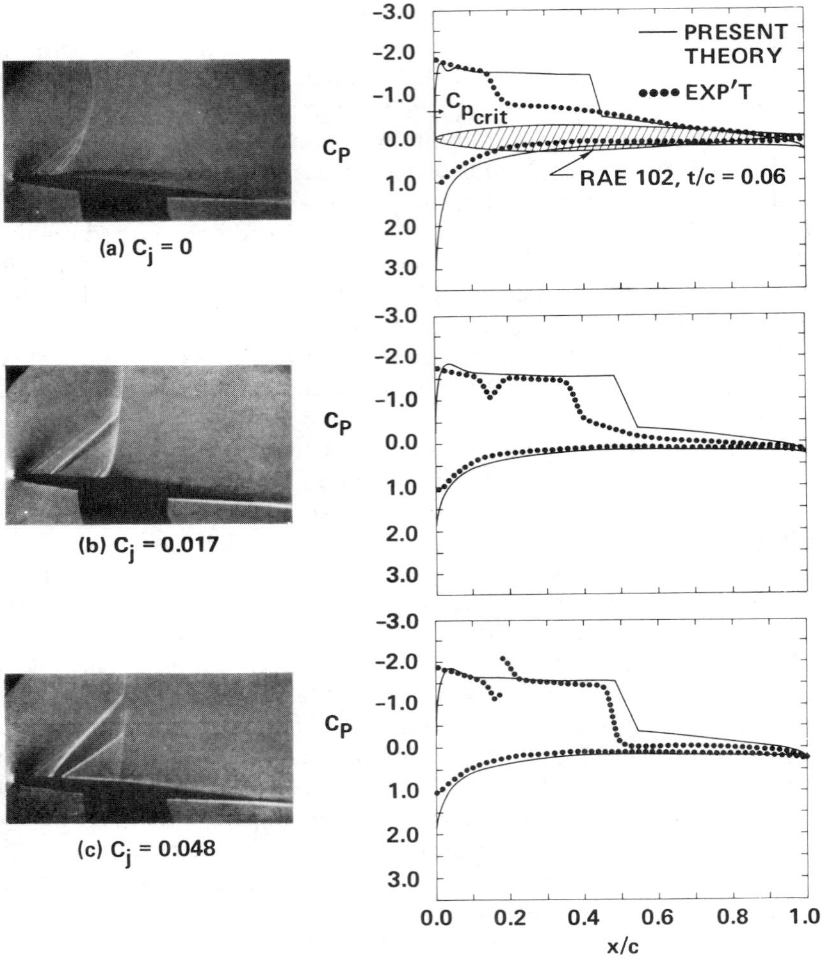

Figure 5. Comparison of USB theory of this paper with NPL tests of Freeman [4.223], $M_\infty = 0.75$, $\alpha = 6°$, c = chord, t = maximum thickness.

The relaxation length L to achieve the downstream pressure in units of the exit height is of the order of ln R, which can be approximately 5 to 50 in the present case, depending on the accuracy of the estimate for M_e. Note in this connection that

$$R < 0 \quad \text{for } 1 \leq M \leq \frac{M_e^2}{\beta_e} \text{ and } M_e \leq M \leq \infty,$$

$$R \geq 0 \quad \text{for } \frac{M_e^2}{\beta_e} \leq M \leq M_e.$$

For the submerged case, $R \to 1$ ($M_e \gg M$) and the Prandtl periodic pattern is obtained, with no radiation of energy to the external flow.

These facts suggest that one factor that may be responsible for the observed spike is the internal decay process in the jet. If transonic effects and wall curvature are accounted for, the presence of "ballooning" and throats in the jet may also be contributory. These aspects are further discussed in [4.221].

Turning now to the discrepancy of the values shown on the rear surface (downstream of 0.5c) in Fig. 5(c), we note that in spite of the obvious elimination of separation, a thick viscous wall jet is present. Downstream diffusion will affect the application of the Spence relation on the blown portion as well as the shock jump. In view of the wall-jet thickness shown on the schlierens, this factor appears to be more significant than obliqueness at its foot. A near-term refinement is being implemented employing a coupled inviscid–viscous model using second-order boundary-layer corrections to the Spence boundary conditions accounting for axial gradients of the displacement and momentum thickness. The first phase of this model involves a blown boundary-layer module. Preliminary results from this component illustrating the effectiveness of blowing in delaying separation is provided in the next section.

4 Viscous Effects

A laminar boundary-layer module has been constructed based on a generalization of the box method of [4.224] to tangential blowing. Results have been computed by coupling this element to the inviscid framework previously discussed to provide a blown-boundary-layer algorithm, formulational details of which will be reported elsewhere. Using a NACA 0012 airfoil shown in Fig. 6 with a smoothed small-disturbance SLOR chordwise pressure distribution

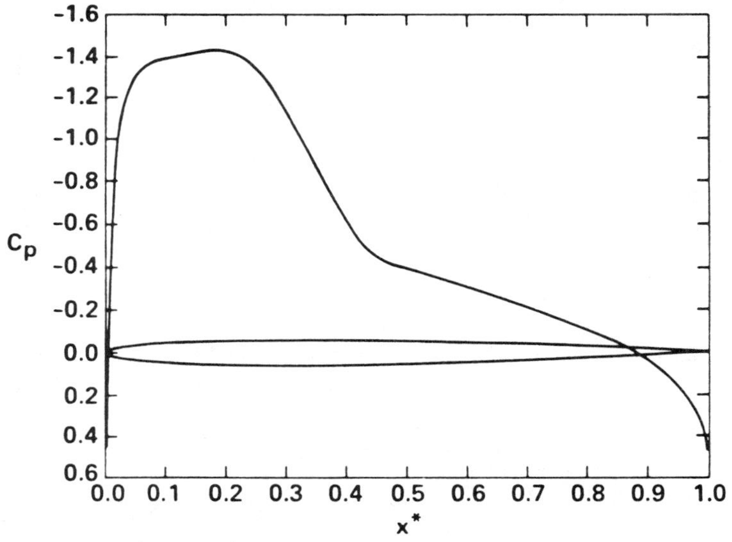

Figure 6. Upper surface pressure distribution on NACA 0012 airfoil $M_\infty = 0.7$, $\alpha = 3°$.

on its upper surface for $M_\infty = 0.7$ and $\alpha = 3°$ as a basis of illustration, typical results are shown in Fig. 7, which indicates the streamwise evolution of velocity profiles upstream of the slot and uncorrected for viscous-interaction effects on the external flow field. The normal coordinate ζ is a Blasius reduced version of the normal curvilinear physical coordinate η. Despite the rather severe adverse pressure gradients, particularly those associated with the nose singularity, and their significant influence on the source term in the momentum equations as the coefficients of the reduced form in the Blasius variables, the box scheme is robust enough to treat such variations. For the case at hand, the loss in fullness in the profile resulting from the adverse pressure gradient is evident. Associated decreasing wall shear stress is also apparent.

Regarding the influence of slot blowing, Fig. 8 shows the evolution for the same airfoil of the profiles downstream of a slot located at $x^* = 0.2$; hereinafter (x^*, y^*) refer to Cartesian coordinates erected at the mean camber line in the usual way. For the examples selected, an initial parabolic slot profile was utilized in which the velocity function at the slot station $x^* = x_s^*$ is given by

$$\frac{u}{u_e} = f'(x_s^*, \zeta^*) = -A(\zeta^* - \zeta_d)\zeta^*, \qquad 0 < \zeta^* < \zeta_d, \tag{8a}$$

$$A = \frac{4\left(\dfrac{u}{u_e}\right)_{\max}}{\zeta_d^2}, \qquad \zeta^* = \zeta + \zeta_d,$$

and for the range of ζ^* above the slot lip,

$$f'(x_s^*+, \zeta^*) = f'(x_s^*-, \zeta^*), \qquad \zeta_d < \zeta^* < \infty, \tag{8b}$$

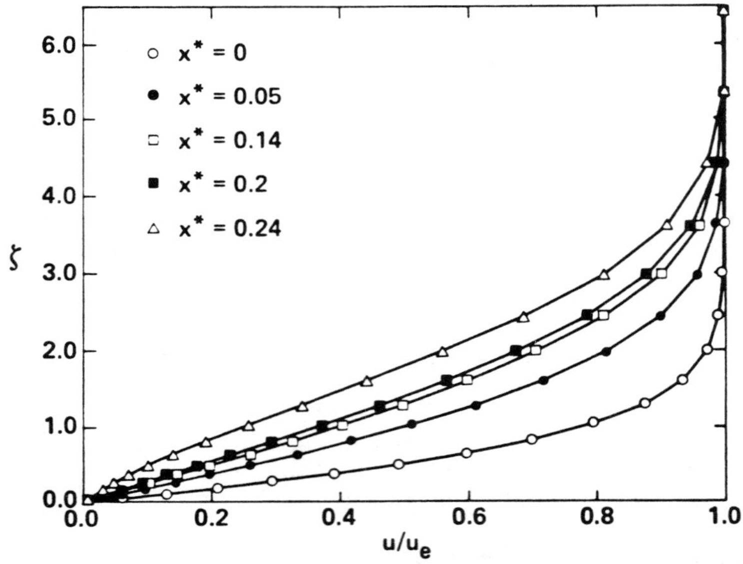

Figure 7. Unblown streamwise profile development for case of Figure 6.

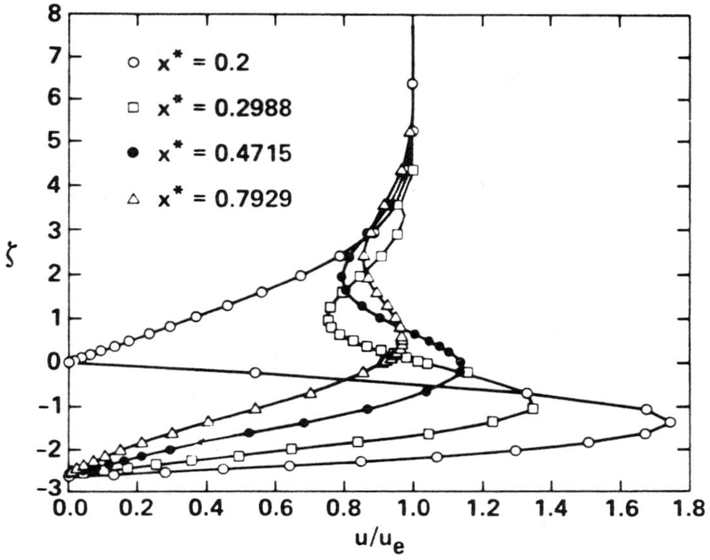

Figure 8. Blown profiles, $A = 1$, $\zeta_d = 2.65$.

where the solution which had been marched from upstream to the slot location is assumed to have a continuous velocity component at the slot (excluding the pathological case of shocks at the slot location).

From (8a),

$$f = -\frac{A\zeta^{*3}}{3} + A\zeta_d\frac{\zeta^{*2}}{2}, \qquad 0 < \zeta^* < \zeta_d, \tag{9a}$$

$$f(\zeta_d) = \frac{A\zeta_d^3}{6}. \tag{9b}$$

For more general slot profiles in which

$$f'(x_s^*, \zeta^*) = F'(\zeta^*), \qquad 0 < \zeta^* < \zeta_d,$$

the blowing coefficient is $C_\mu = J\delta^{1/3}M_\infty^{3/4}/\rho_\infty U^2 c$, where J is the momentum flux per unit span and c is the chord, is given by

$$C_\mu = M^{3/4}\delta^{1/3}\left(\frac{x_s^*}{\text{Re}_\infty}\right)^{1/2}\int_0^{\zeta_d} F'^2(\zeta^*)\,d\zeta^*, \tag{10}$$

where Re_∞ is the freestream Reynolds number based on the chord, and x_s^* is the streamwise position of the slot in units of the chord. Actually, (10) is an approximation to within terms of $O(\delta^{2/3})$. In (10), the quantity ζ_d is given by

$$\zeta_d = \frac{d}{c}\frac{\bar{\rho}}{\rho_e}\left(\frac{\text{Re}_\infty}{x_s^*}\right)^{1/2}, \tag{11}$$

where d is the dimensional slot height and $\bar{\rho}$ is the mean density across the slot.

For the example indicated in Fig. 8, the slot was located at 20% chord, very close to the onset of separation, which occurred at 27.6% chord for this case. It should also be noted that the location of the shock, which had been slightly smoothed for purposes of initial checkout of the algorithm (which can handle nonsmooth cases), is at about 23% chord. The characteristic diffusion of the profile associated with mixing of the jet is evident in the figure. More significant and not clearly indicated (but shown later), is the fact that the separation point has now been moved downstream to 79% chord. Other calculations show that with modest further increases in ζ_d, separation can be completely eliminated. (Note in this context that the peak velocity for (8a) is $A\zeta_d^2/4$.)

In [4.221] the corresponding profile development for an "underblown" case where the peak is less than the freestream value is discussed. Relaxation to a conventional profile occurs as previously. However, this blowing configuration actually results in premature separation as compared to the case of no blowing. This is indicated in Fig. 9, where the effect of blowing on the separation-point location is shown. Despite the initially higher shear stress at the slot in the underblown case $\zeta_d = 1$, $A = 2$, the higher vorticity diffusion and lower overall momentum in the layer leads to earlier separation, which can be seen as the leftmost solid circle in Fig. 9. This level should be related to the unblown result shown in dashed lines in the same plot. Moreover, the case $\zeta_d = 2.65$, $A = 1$, moves the separation point substantially downstream to almost 80% chord. Other cases are shown in Fig. 9, such as the one corresponding to the slot position at approximately 11% chord. This demonstrates the dramatic role of the slot location and upstream boundary-layer

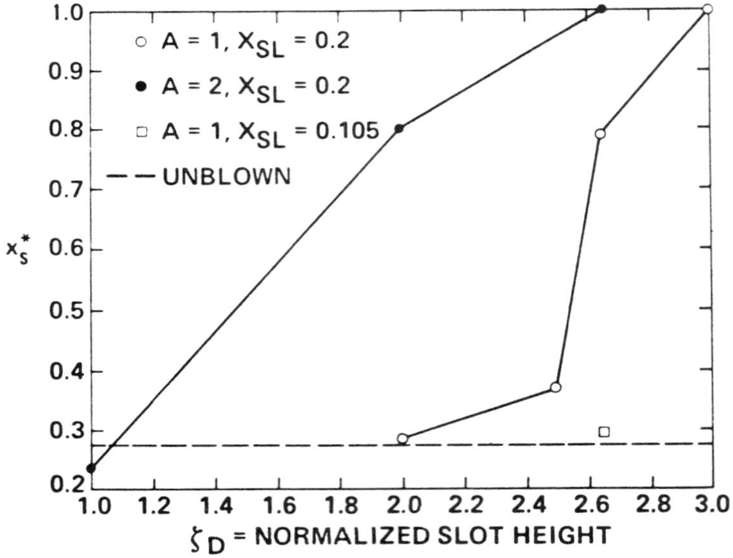

Figure 9. Position of separation point, x_S^*, with blowing—NACA 0012, $M_\infty = 0.7$, $\alpha = 3°$.

thickness in delaying separation: for the same slot height, the upstream slot location gives an almost trivial downstream movement of the separation location, in contrast to the potent effect of the downstream slot position.

The results of Fig. 9 can be replotted as in Fig. 10 to show the trends as a function of a momentum-flux parameter which is proportional to C_μ. By Eq. (10), $C_\mu \propto A^2\zeta_d^5$ for a parabolic profile. For the limited number of cases run, there is a suggestion in Fig. 10 that the curves of Fig. 9 collapse to a single universal band of results. In view of the roles of the pressure gradient and the multiple extrema in the velocity profile, this assertion must be regarded as tentative at best. What is significant, however, is that for the first time, the delay effect of tangential blowing on natural and shock-induced separation over transonic airfoils has been inexpensively quantified.

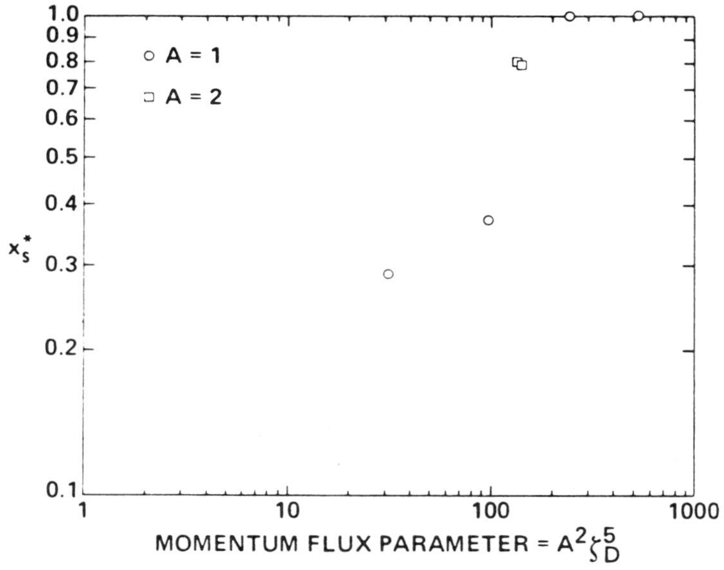

Figure 10. Reduced plot up separation location x_S^* as a function of momentum flux parameter $A^2\zeta_d^5$.

5 Conclusions

Asymptotic and computational models have been used to obtain the flow over upper-surface-blown (USB) airfoils at incompressible and transonic speeds. The treatment involves a detailed analysis of the flow in the jet. The analytical and computational results indicate that:

1. In the thin-jet small-deflection approximation, the pressure jumps associated with the Spence theory prevail even if the flow is rotational and compressible.

2. The asymptotic developments provided allow further systematic refinements.
3. Effects associated with initial skewness and vorticity inaccessible to other theories can be assessed.
4. Computational results obtained for transonic USB configurations indicate significant enhancement of lifting pressures associated with blowing.
5. Comparisons with experiment indicate the need for refinements incorporating wave-interaction phenomena near the jet exit as well as viscous-interaction processes in the downstream portion of the wall jet.
6. Substantial downstream movements in the shock-induced separation point are achievable with application of tangential blowing. By computational schemes, these delays can be quantified less expensively than with experimental methods.

With the viscous module and the appropriate iterative algorithm for coupling to the external flow (to be implemented in the near future), optimization between separation suppression, wave-drag minimization, and supercirculation control will be possible. It is envisioned that the design techniques contained in [4.225–4.227] will augment this capability by providing methods to modulate shock formation in concert with the blowing effects.

Acknowledgment

A major portion of this effort was sponsored by the Office of Naval Research under Contract N00014-76-C-0350. The support of Morton Cooper for this effort is gratefully acknowledged.

CHAPTER 28

Transonic-Shock–Boundary-Layer Interaction: A Case for Phenomenology

J. Wai and H. Yoshihara*

Introduction

Viscous interactions play a significant role in transonic flows over airfoils, greatly affecting the forces and moments on the airfoil. Of the interactions, that between the shock wave and the boundary layer plays the greatest role. In Fig. 1 the complexities of such interactions are sketched. The resulting flow is unsteady, due to the interaction of the coherent large-scale turbulence with the shock. Schlieren pictures, for example, show the foot of the shock fluctuating in reaction to the oncoming large-scale eddies. The shock wave penetrates into the boundary layer, weakens, and vanishes as it approaches and reaches the sonic line. The subsonic layer beneath the shock further serves to cushion the airfoil from the shock, attenuating the shock pressure rise.

For applications, the primary interest is not in the complex flow sketched above, but in the time-averaged flow yielding an airfoil pressure distribution as measured in the wind tunnel with pressure taps. Here the shock acquires a profile due to the time averaging of the fluctuating shock and the subsonic cushioning.

With such a complex flow on hand, we must understandably turn to a simplified phenomenological model for the interaction.

* Boeing Company, Seattle, WA 98124.

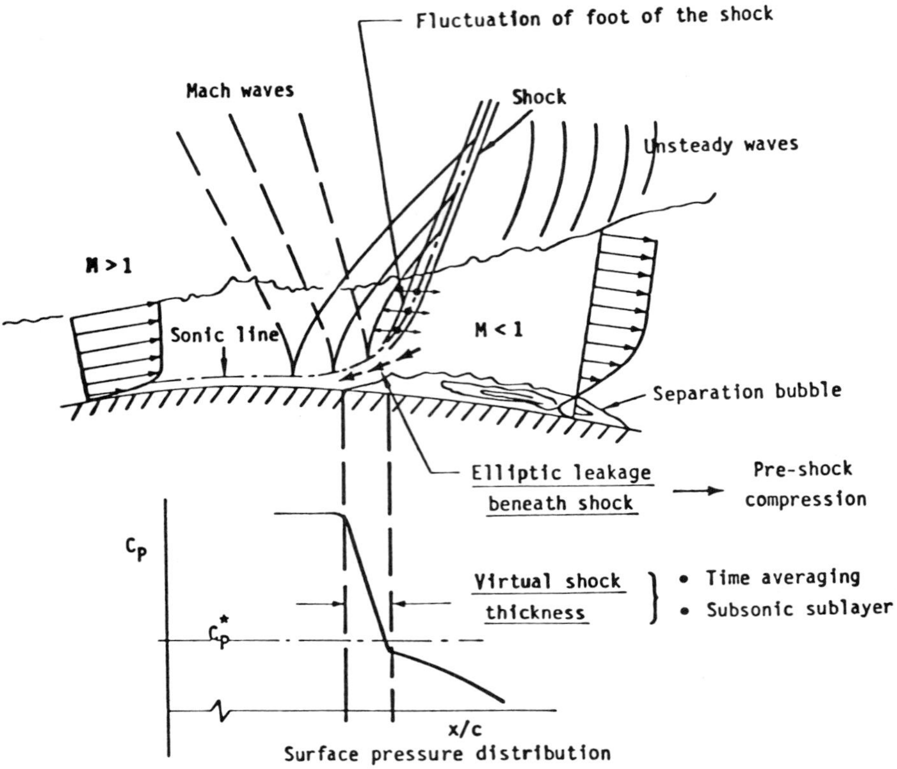

Figure 1. Complexities of the shock-boundary layer interaction.

Shock–Boundary-Layer Interaction Model

The basis of the modeling is the viscous-ramp method proposed earlier and used more recently in [4.228] and [4.229]. Here a wedge-nosed ramp is placed at the base of the shock to convert the normal shock in an inviscid flow to an oblique shock yielding the experimentally observed reduced shock pressure rise. Such a ramp simulates the abrupt thickening of the boundary-layer displacement aft of the shock. The postshock pressure prescribed is modeled from wind-tunnel measurements. In Fig. 2 a plot of such pressures is given for a variety of airfoils over a wide range of Reynolds numbers. To a reasonable approximation, the mean of the plotted points is represented by a constant postshock pressure $p_2/H = 0.54$, H being the total pressure. From the oblique shock conditions (shock polar in Fig. 2) a wedge angle can be defined, as a function of the preshock Mach number M_1, corresponding to the prescribed postshock pressure. The viscous wedge tied to the base of the shock then adjusts itself so that its nose angle corresponds to this value as a function of the instantaneous M_1. The overall shape of the

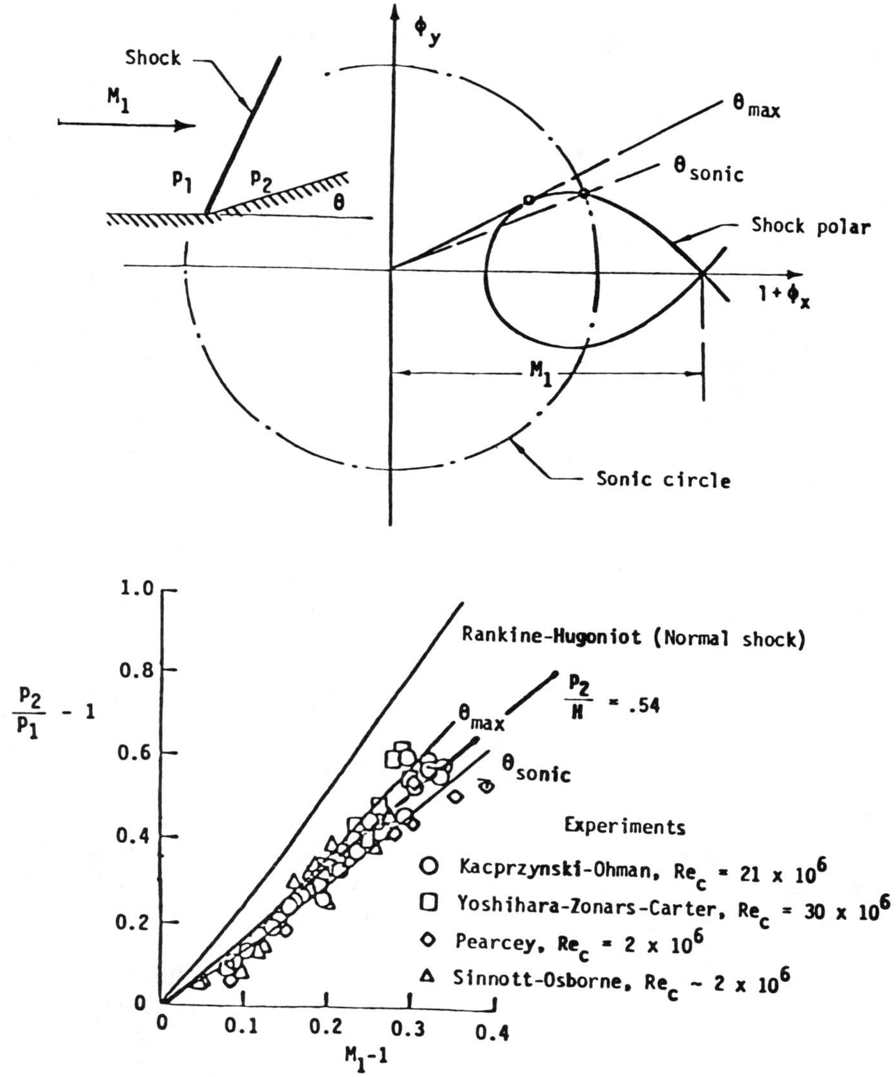

Figure 2. Modeling of the post-shock pressure and the shock polar.

viscous ramp must be empirically developed, and a typical example is given later (Fig. 7).

In summary, in the viscous ramp method a prescribed ramp is placed at the base of the shock in an inviscid calculation to yield the essential feature of the shock–boundary-layer interaction so far as the inviscid flow is concerned. Here no special refinement of the mesh is required. Obtaining the proper reduced shock pressure rise largely ensures a reasonable positioning of the shock.

In most airfoil flows the viscous displacement effects over the aft portions of the airfoil and the effects of the near wake also must be taken into account. Thus to complete the modeling of the shock–boundary-layer interaction, we must also provide the postshock values of the boundary-layer variables which serve as initial values for the downstream boundary layer and wake. For this purpose we shall use Green's lag-entrainment-integral boundary-layer equations [4.230] based on the boundary-layer approximations. From the sketch of the interaction flow in Fig. 1, one must question the validity of the latter approximations, and hence the applicability of Green's equations. However, calculations of East, Smith, and Merryman [4.231] showed that reasonable results are obtained with Green's equations for shock–boundary-layer interactions even with separation present. This has motivated the present approach.

Green's equations comprise three first order ordinary differential equations. Two of the equations are derived by integrating the continuity and streamwise momentum equations across the layer, while the third is derived from the Bradshaw–Ferriss turbulent energy equation. These boundary-layer equations are nonsingular for cases of interest, but they possess stiffness features which must be respected. This can be seen most simply using an alternate form of the equations derived by East, Smith, and Merryman [4.231], given by

$$\delta_x^* = F_1 + F_2\phi_{xx} \qquad \text{(direct form),} \tag{1}$$

or alternatively, rearranging terms,

$$\phi_{xx} = F_2^{-1}(\delta_x^* - F_1) \qquad \text{(indirect form).} \tag{2}$$

Here δ^* is the displacement thickness, ϕ is the perturbation potential, F_1 and F_2 are functions of the boundary-layer variables, and the subscripts denote differentiations, with x the streamwise coordinate. Equation (1) or (2) must be supplemented by a suitable number of the original boundary-layer equations to form a complete set.

For unseparated flows, East, Smith, and Merryman pointed out that $|F_1| \gg |F_2|$, so that the direct equation (1) represents a stiff equation for δ_x^*. That is, the displacement slopes are relatively insensitive (stiff) to the boundary-layer input pressure gradient ϕ_{xx}. On the other hand, for fully separated flow $|F_2| \gg |F_1|$, so that the indirect equation (2) now represents the desired stiff equation for the pressure gradient. In this case the proper input is the displacement slope.

To complete the calculation of the shock–boundary-layer interaction, the postshock values of the boundary-layer variables are now calculated using Green's equations in the proper stiff form as follows. The first step is the calculation of the inviscid flow, using the viscous ramp, to the postshock point (the first postshock subsonic point). The resulting pressure distribution is then inputted into the direct equation (1), and the latter integrated starting at a point just upstream of the shock where initial values of the boundary-layer variables are given. If the boundary layer is not separated, the integration is carried out to the postshock point, yielding directly the desired postshock values of the boundary-layer variables.

If, however, the boundary layer separates prior to reaching the postshock point, then Green's equations are switched to the indirect form given in Eq. (2), and the equations are integrated from the point of separation to the postshock point, inputting the displacement slopes from the viscous ramp. In this way the use of the proper stiff form of the equations permits a reasonable determination of the postshock boundary-layer variables with approximate input data.

There is, however, one question that must be resolved in the above procedure. In the calculation of the inviscid flow with a difference algorithm, shock waves are captured with a profile and a thickness that are mesh-size-dependent, being generated by truncation-error viscosity. That is, halving the streamwise mesh size will lead to a doubling of the shock pressure-rise gradient. Use of such an unphysical pressure-gradient distribution would suggest an unphysical shock–boundary-layer interaction. However, calculations with Green's equations showed that if the mesh size is not excessive (less than approximately 3% chord), the postshock values of the boundary-layer variables are nearly independent of the mesh size. The above unphysical mesh-dependent shock pressure-rise profile would therefore be of less concern, provided of course the invariant values are acceptable.

In the above calculations a shock pressure profile was first postulated as an arc-tangent curve defined with a fixed postshock pressure $p_2/H = 0.54$, a variable preshock pressure p_1 (with Mach number M_1), and a variable thickness T/C, C being the airfoil chord. In the case that the boundary layer remained unseparated through the shock profile, the above shock pressure-gradient profile was inputted into the direct form of Green's equations, varying M_1 and T/C. If however the boundary layer separated prior to reaching the postshock pressure, Green's equations were switched to the indirect form at the point where the displacement slope attained the viscous-wedge angle. The latter occurred just upstream of the separation point. The viscous ramp slopes were then inputted here, and the integration carried out until the postulated postshock pressure was obtained.

In Fig. 3 the resulting postshock displacement and momentum thicknesses are plotted as a function of the shock thickness and M_1. It is seen here that the above quantities approach a plateau for T/C less than approximately 0.0625. Since the shock thickness is typically two meshes wide, the above thickness would correspond to a mesh of approximately 3% chord. (This corresponds to the mesh in the TSFOIL code used later.) In Fig. 3 the separated cases are identified by the filled circle symbols.

In Fig. 4 we next show the pressure distributions through the shock for a shock thickness of 0.0625 and various M_1. The point at which the equations are switched from the direct to the indirect form is indicated, as well as the point of separation. It is further seen here that after separation the pressure-rise gradient is significantly reduced from that in the unseparated region. This then suggests that the proper postshock pressure should be taken,

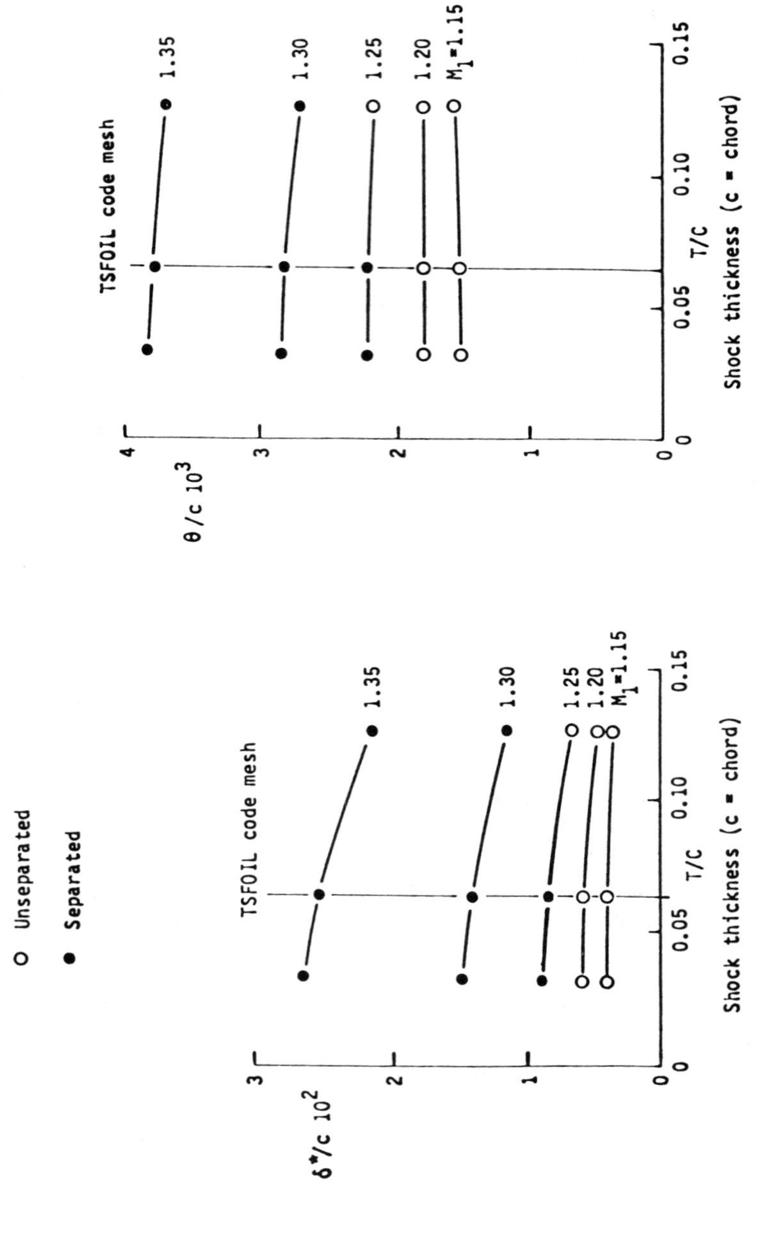

Figure 3. Post-shock δ^* and θ^* as a function of shock thickness.

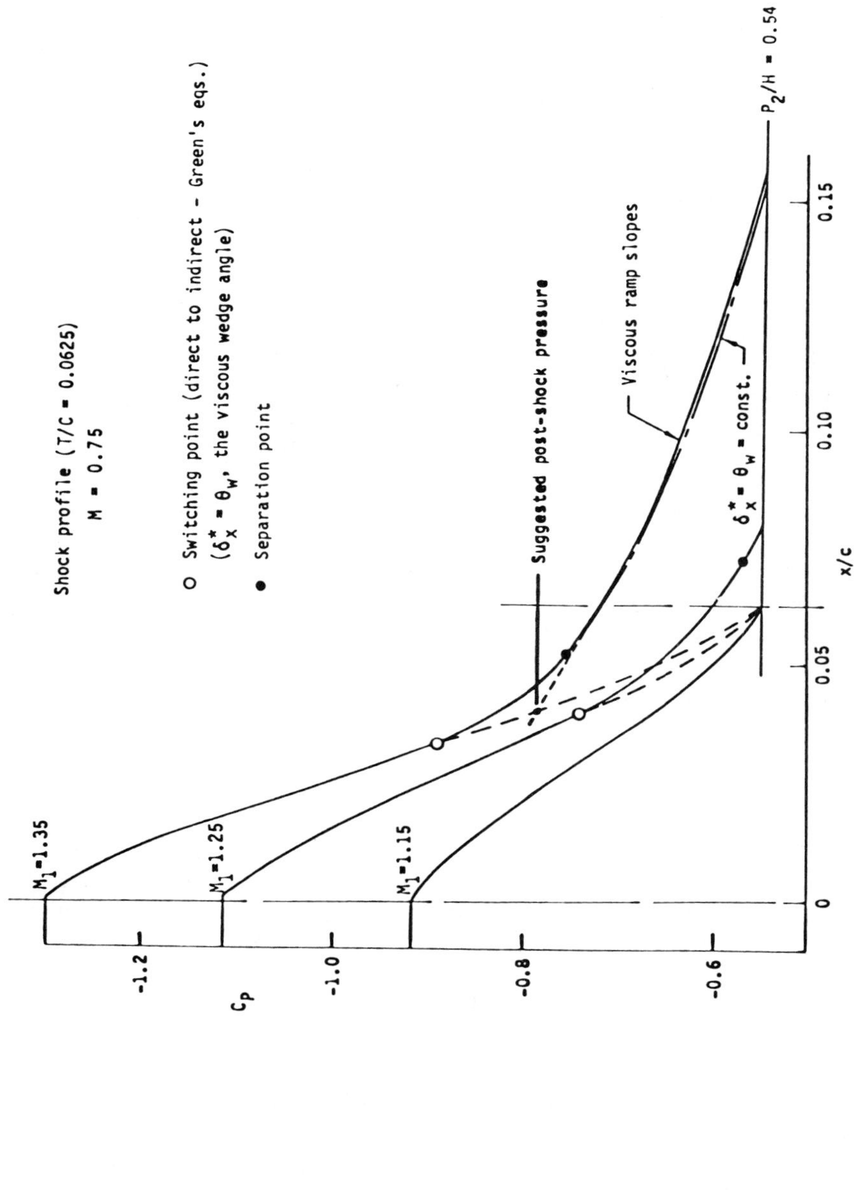

Figure 4. Shock pressure profile.

not as $p_2/H = 0.54$, but at the supersonic value indicated in Fig. 4 near the separation point. That is, the primary shock should be an oblique shock, to be then followed by a weaker terminating shock. Figure 2 also suggests that $p_2/H = 0.54$ may be questionable for the strong case $M_1 = 1.35$, particularly for wind-tunnel Reynolds numbers. In the calculation for $M_1 = 1.35$, to check the stiffness of the pressure distribution relative to the displacement slope input, a constant displacement slope equal to the viscous-wedge angle was inputted instead of the previous viscous-ramp slopes. The results indicated negligible effect of this displacement-slope change, verifying the desired stiffness of the equations.

In summary, in the above procedure the inviscid portion of the flow within the shock profile was determined using the viscous ramp at the base of the shock. The use of the latter ensured the desired postshock pressure and hence a realistic shock location. The required postshock values of the boundary-layer variables were then obtained by inputting the resulting pressure distribution or the viscous-ramp slopes into Green's equations.

As a result of this procedure there will arise an inconsistency of the boundary-layer and inviscid flows within the shock profile. That is, for example, there will be a difference between the displacement slopes determined in the unseparated portion of the boundary layer and the viscous-ramp slopes used in the inviscid calculations. This inconsistency will not be of concern so long as the postulated postshock pressure in the inviscid flow and the correct postshock state of the boundary layer are obtained. Here the former will be assured in the inviscid region by the use of the viscous ramp, while latter is achieved by the use of the proper stiff form of Green's equations. With respect to the latter, it may be recalled that in the unseparated case, the postshock boundary-layer variables depended, not on the input pressure-gradient distribution, but only on the pressure change across the shock. In the separated case, after the switch to the indirect form, the pressure distribution was relatively insensitive to the input displacement thickness.

Examples

Two examples are presented to demonstrate the use of the proposed shock–boundary-layer interaction model. These examples contain in addition viscous interactions over the aft portion of the airfoil as well as in the wake. Details of the treatment of these interactions and the inviscid flow are omitted here, but they may be found in the cited references.

The first example is the case of the RAE 2822 airfoil at $M = 0.75$, angle of attack 3.19°, and a chord Reynolds number of 6.2 million [4.232]. The experimental results for the pressure distribution and the boundary-layer variables on the upper surface as well as for the wake are available for this case for comparison with the calculated results [4.233]. The experiments indicate that shock-induced separation is present with reattachment just upstream of the trailing edge.

The inviscid flow here was calculated with the transonic small disturbance code TSFOIL. This procedure employs a successive-line overrelaxation algorithm.

In Figs. 5 and 6 the calculated and experimental results are compared. Reasonably good agreement in the pressure distribution is seen here, but the displacement thickness is overpredicted. The calculations furthermore do not predict reattachment, undoubtedly due to the overprediction of the displacement thickness. In general the test–theory comparison shown here is good, but further refinement of the interaction procedure as well as the boundary-layer equations is needed.

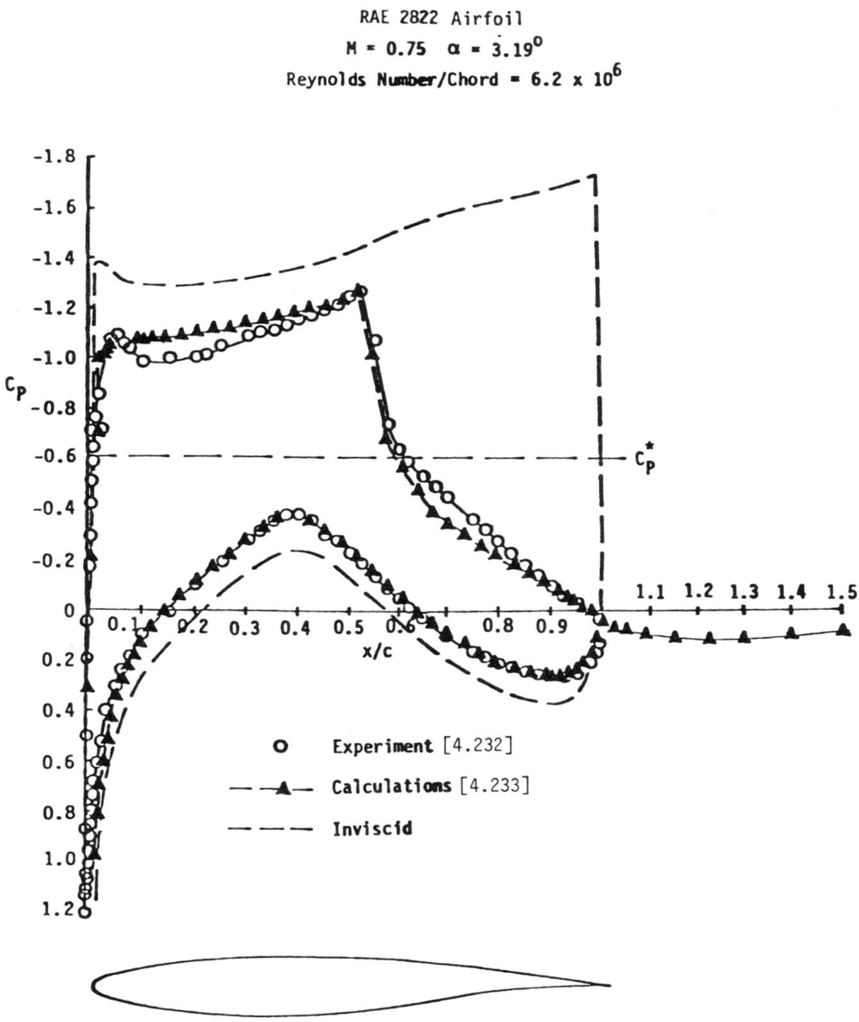

Figure 5. Test-theory comparison RAE 2822 airfoil (pressure distribution).

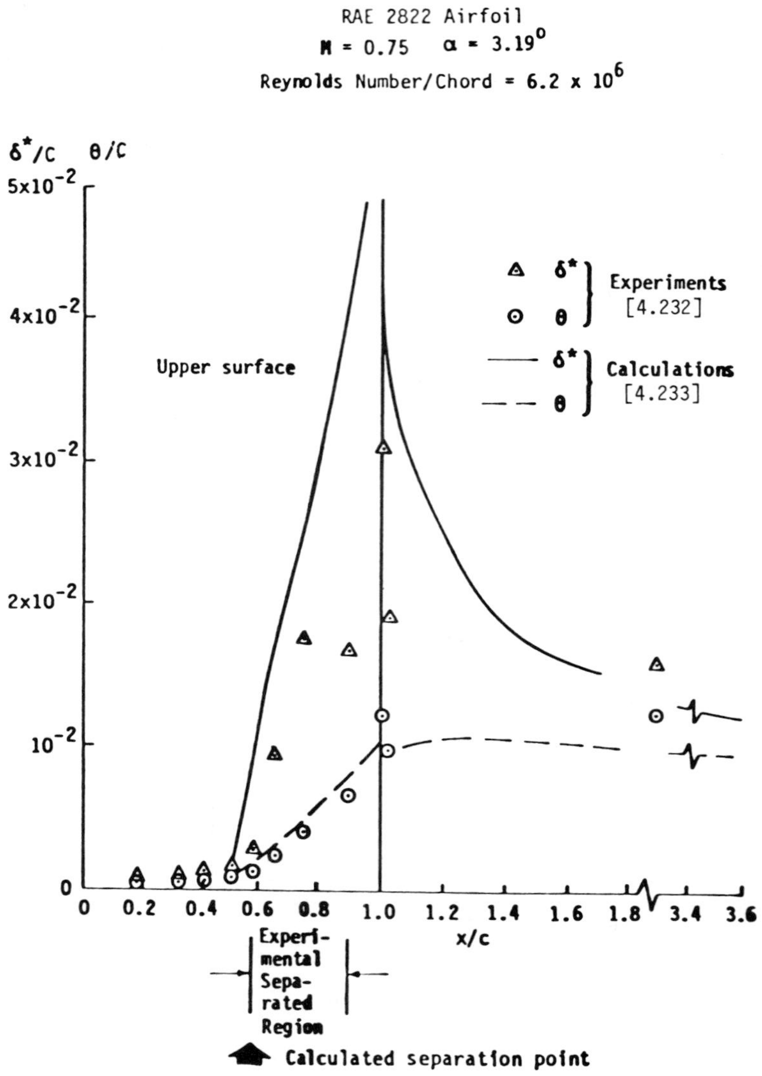

Figure 6. Test-theory comparison RAE 2822 airfoil (boundary layer and wake characteristics.

In this example, the standard TSFOIL mesh is used, which in the region of the shock is 3% of chord. The addition of the viscous equations required little additional computing time, amounting to 20% relative to the inviscid flow computation time, which was approximately 50 seconds CPU time on the CYBER 175 computer.

The second example [4.228] is for the 16.8%-thick NLR 7301 airfoil in harmonic pitching oscillation at $M = 0.75$. The pitching amplitude is 0.5°

about a mean incidence of 0.37°, with a reduced frequency of 0.6. Measurement of the unsteady pressures on the upper surface, and the steady pressures on both surfaces at the mean incidence, are available to check the computations [4.234]. The calculations were carried out using an unsteady transonic small-disturbance code employing an alternating-direction implicit algorithm. The shock–boundary-layer interaction was modeled with the viscous ramp shown in the top part of Fig. 7. The sizing of the ramp was assumed

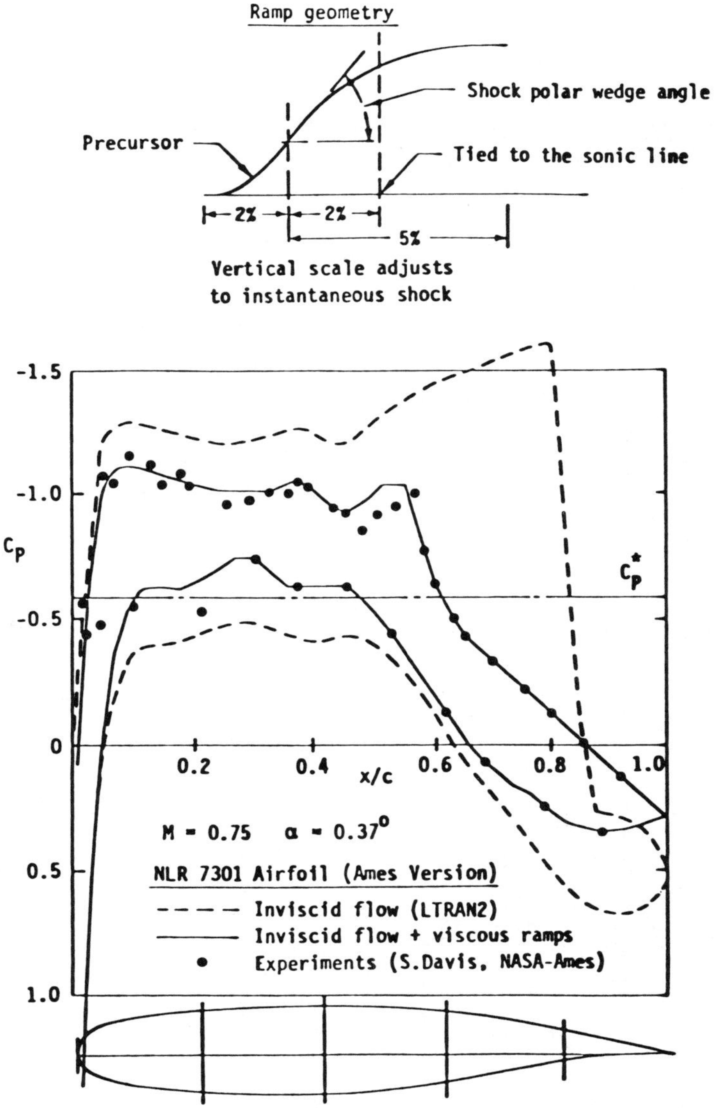

Figure 7. Test-theory comparison NLR 7301 airfoil: steady mean flow.

to be in quasisteady equilibrium with the instantaneous shock. In addition aft displacement ramps were determined following the procedure given by Yoshihara [4.235] using the steady pressure distribution at the mean incidence, which is assumed to be given as input data. The following rationale is used for the aft displacement ramps. First, it is assumed that these displacement layers are a function primarily of the aft pressure gradients. In aft-cambered supercritical airfoils, such as the 7301 airfoil, the steady aft pressures are nearly independent of the angle of attack. Thus the aft pressures and hence the aft ramps can be expected to be essentially invariant during the pitching cycle, permitting the steady aft ramps determined above to be frozen onto the airfoil.

In Fig. 7 the calculated and measured steady pressure distributions at the mean incidence are compared. Also plotted here is the inviscid distribution, showing the large effects of the viscous interactions. In Fig. 8 the unsteady

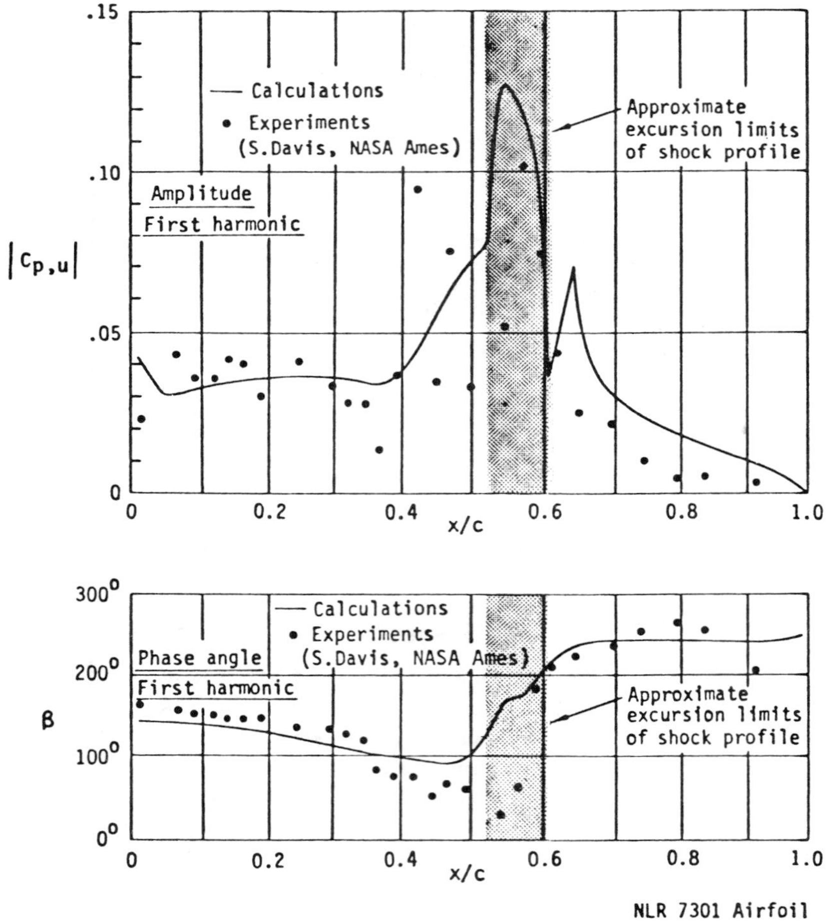

Figure 8. Amplitude and phase of the first harmonic.

pressures are compared in terms of the amplitude and phase of the first harmonic of local pressure history. The agreement seen here is good.

Concluding Remarks

The complexities of the shock–boundary-layer interaction, particularly with separation present, indicate the need of phenomenological modeling. The results shown here suggest that the use of the viscous ramp, together with Green's lag-entrainment equations in the proper stiff form, constitutes a reasonably good model which does not add significantly to the computer time. Most importantly, it holds promise for the separated case as well as for both steady and moderately unsteady flows.

CHAPTER 29

Unsteady Transonic Flows: Time-Linearized Calculations

A. Richard Seebass and K.–Y. Fung*

Introduction

A combination of technical advances should improve the fuel efficiency of transport aircraft by fifty percent in the next decade. Analogous improvements in the transonic performance of military aircraft should also be realized. These large gains will come from a combination of improvements in engine, structural, and aerodynamic efficiency. More than half will come from improvements in the aerodynamic efficiency, including active control, and the use of composite materials in the primary structure. Part of the improvement in aerodynamic efficiency will result from flight at supercritical Mach numbers with subcritical levels of lift-to-drag ratio.

This improved transonic performance mandates an accurate prediction of aeroelastic behavior at transonic Mach numbers. Of special concern are flutter boundaries. In 1976 Farmer and Hanson [4.45] reported that the flutter boundaries of two dynamically identical wings were markedly different at transonic Mach numbers due to very minor differences in wing profile thickness. The results of their measurements are shown in Fig. 1, indicating a more severe flutter boundary for the wing with a "supercritical" profile.

Today we understand well the qualitative behavior of inviscid steady and unsteady transonic flows, and we have rudimentary understanding of viscous effects. For flight regimes that involve unseparated flows the main ingredient in the calculation of flutter boundaries is an accurate determination of the

* Aerospace and Mechanical Engineering, University of Arizona, Tucson, AR 85718.

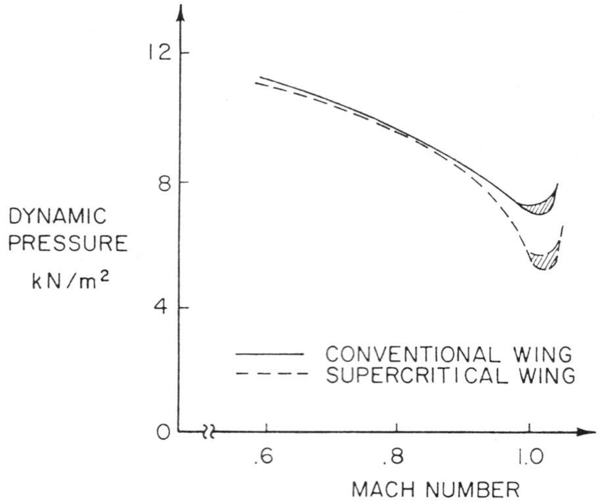

Figure 1. Flutter boundary for two dynamically identical TF-8A wings [4.45].

steady pressure field. This may be determined either by experiment or by calculation. But once it is known, the response of the wing to pitching, plunging, or aileron motion may be found by numerical means. The most essential ingredient in predicting this response is an accurate prediction of the motion of any shock waves present in the flow (see, e.g., [4.51]). Numerical algorithms that capture shock waves must use relatively fine grid spacing near the shock wave if they are to predict its motion, due to the small changes of interest in flutter studies. But this shock motion may be predicted accurately by a time-linearized algorithm using a relatively coarse grid if the calculations are done correctly. This has not generally been the case; other investigators have ignored this essential effect. This point is also noted by Hafez (Chap. 24).

We report here on two-dimensional, time-linearized computations which not only properly account for shock-wave motion, but also resolve it even though the grid used is coarse. To avoid unnecessarily large computational domains, the linearized far field for an unsteady vortex with a circulation determined by the airfoil's lift is used to evaluate the potential there [4.238]. This avoids the reflection of the unsteady disturbances from the grid system. The airfoil's response to a given mode of motion is determined by superposition from that for an indicial motion. In many cases this indicial response. can be modeled in a simple way, providing an instructive analytic result for the dependence of the lift and moment-coefficient amplitudes and phase lags on reduced frequency [4.239].

The computational efficiency of the time-linearized calculation of the amplitudes and phase lags for a range of reduced frequencies is compared with nonlinear and frequency-domain computations. The time-linearized calculation of an indicial response can result in a factor of ten or more reduction in computational effort; this is especially significant in three dimensions.

Governing Equations

As noted above, time linearization about a known steady state is an effective mechanism for determining the unsteady response to a given mode of motion. For flows that are unseparated and not at incipient separation, an inviscid treatment of the unsteady flow should be adequate for flutter studies. It is important, however, that the nonlinear and viscous aspects of the underlying steady state be determined accurately. This may be done either by experiment, or by a reliable computational algorithm such as Grumfoil [2.6]. In any case, we assume that an accurate steady-state pressure distribution has been used to provide the input for the inverse calculation of the airfoil shape that will provide this pressure distribution when the steady state flow is computed using the small-perturbation approximation.

As Lin et al. [4.240] observed more than thirty years ago, within the context of the small-perturbation approximation, the basic equation governing the unsteady motions is linear unless the reduced frequency, k, is $O(\delta_0)$. Here δ_0 is the measure of the perturbation potential and $k = \omega c/U$, where ω is the frequency, c the chord, and U the freestream speed. Thus, if we write the velocity potential, Φ, as

$$\Phi = Uc\{x + \delta_0 \phi\},$$

where ϕ is the perturbation potential and δ_0 is some measure of the disturbance, the governing equation is

$$-\frac{2k}{\delta_0} M_\infty^2 \phi_{xt} + \left\{\frac{1 - M_\infty^2}{\delta_0} - (\gamma + 1)M_\infty^2 \phi_x\right\}\phi_{xx} + \phi_{yy} = 0, \qquad (1)$$

where terms $O(k^2\phi_{tt}/\delta_0)$ and $O(k\phi_t \phi_{xx})$ have been neglected because $k = O(\delta_0)$. (Here the time has been nondimensionalized by the circular frequency, the spatial coordinates by the airfoil chord, and the y-coordinate by $\delta_0^{1/2}$.) The $k\phi_t \phi_{xx}$ term is of little consequence. Neglecting the ϕ_{tt} term is equivalent to disregarding one of the characteristics and assuming disturbances propagate downstream at infinite speed; this has important computational advantages.

The boundary condition at the body is

$$\delta_0^{3/2}\phi_y(x, 0, t) = \delta Y'(x) + \tilde{\delta}\tilde{Y}_x(x, t) + k\tilde{\delta}\tilde{Y}_t(x, t), \qquad (2)$$

where the body is given by

$$y = \delta Y(x) + \tilde{\delta}\tilde{Y}(x, t).$$

Across the airfoil wake the jump in the pressure coefficient must vanish. This implies that in the wake

$$\Delta\phi(x, 0, t) = \Gamma(t - kx). \qquad (3)$$

In both these boundary conditions we have retained terms of $O(k)$, which is not consistent with the approximation made in Eq. (1). But as [4.62] demonstrates, this gives good agreement with the results of linear theory for values of

k up to, and even above, 1.0. This term is, of course, also retained in the evaluation of the pressure coefficients.

As noted in [4.51], the appropriate measure of δ_0 is $\max[\delta^{3/2}, \tilde{\delta}^{3/2}, (k\tilde{\delta})^{3/2}]$ and normally $\delta^{3/2}$. Far from the airfoil (see [4.238]),

$$\phi(x, y', t') = \frac{1}{2\pi} \int_0^{t'} f(x, y', t' - t'_0) \frac{d\Gamma(t'_0)}{dt'_0} \, dt'_0 \tag{4}$$

where

$$\begin{aligned} f(x, y', t') = {} & H(t' + x - \sqrt{x^2 + y'^2}) \\ & \times \left\{ \tan^{-1} \frac{\sqrt{t'^2 + 2xt' - y'^2} + t'}{y'} \right. \\ & \left. - \tan^{-1} \frac{\sqrt{t'^2 + 2xt - y'^2} - t'}{y'} \right\}, \end{aligned}$$

and

$$t' = t \frac{1 - M_\infty^2}{kM_\infty^2}, \qquad y' = y \sqrt{\frac{1 - M_\infty^2}{\delta_0}}.$$

Here H is the Heaviside unit step function.

In addition to Eq. (1) and the boundary conditions (2)–(4), a shock jump condition needs to be imposed if the shock wave is to be treated as a discontinuity rather than "captured" by the numerical calculations. Because the former is the intent here, we need to note that

$$\begin{aligned} & - \frac{2kM_\infty^2}{\delta_0} [\![\phi_x]\!]^2 \left(\frac{dx}{dt}\right)_s \\ & - \left[\frac{1 - M_\infty^2}{\delta_0} - (\gamma + 1) M_\infty^2 \overline{\phi_x} \right] [\![\phi_x]\!]^2 + [\![\phi_y]\!]^2 = 0 \end{aligned} \tag{5a}$$

on

$$\left(\frac{dy}{dx}\right)_s = - \frac{[\![\phi_x]\!]}{[\![\phi_y]\!]}, \tag{5b}$$

where $[\![(\cdots)]\!]$ and $\overline{(\cdots)}$ indicate the jump in and average of $(\cdots)$ across the shock wave. Equation (5a) ensures the conservation of mass. The conservation of momentum is replaced by the irrotationality condition (5b) or its equivalent, $[\![\phi]\!] = 0$.

Time Linearization

We use the ADI technique introduced by Ballhaus and Steger [4.60] to compute the steady-state solution of Eq. (1), $\phi_0(x, y)$, subject to the steady boundary conditions implied by Eqs. (1)–(4), using the coordinate stretching of [4.242]. Aside from the far-field condition (4) and the inclusion of terms of $O(k)$ relative to $O(1)$ in (2) and (3), this is equivalent to the NASA Ames computer code LTRAN2 of Ballhaus and Goorjian [4.56]. We next linearize about this steady state by assuming that

$$\phi(x, y, t) = \phi_0(x, y) + \frac{\tilde{\delta}}{\delta}\tilde{\phi}(x, y, t) + o\left(\frac{\tilde{\delta}}{\delta}\right),$$

where $\tilde{\delta}/\delta = o(1)$. This gives, with $\delta_0 = \delta^{3/2}$,

$$-\frac{2kM_\infty^2}{\delta_0}\tilde{\phi}_{xt} + \left\{\left[\frac{1 - M_\infty^2}{\delta_0} - (\gamma + 1)M_\infty^2\phi_{0x}\right]\tilde{\phi}_x\right\}_x + \tilde{\phi}_{yy} = 0, \tag{6a}$$

subject to

$$\tilde{\phi}_y(x, 0, t) = \tilde{Y}_x(x, t) + k\tilde{Y}_t(x, t), \tag{6b}$$

and, with

$$\Gamma(x, t) = \Gamma_0(x) + \frac{\tilde{\delta}}{\delta}\tilde{\Gamma}(x, t),$$

we have

$$\Delta\tilde{\phi}(x, 0, t) = \tilde{\Gamma}(t - kx). \tag{6c}$$

With this linearization of the solution about a steady state at time $t = 0$, we use (4) for the potential far from the airfoil as the circulation departs from its steady-state value.

As noted earlier, taking proper account of shock motions is of prime importance in unsteady transonic flow. We use the procedure of [4.242] to account for shock motions. Because the shock waves are nearly normal to the freestream, we assume that this is the case and impose

$$[\![\phi]\!] = 0$$

on the normal-shock approximation to Eq. (5), viz.,

$$\frac{dx_s}{dt} = \frac{\gamma + 1}{2k}\left\{\frac{M_\infty^2 - 1}{(\gamma + 1)M_\infty^2} + \delta_0\bar{\phi}_x\right\}.$$

Again we linearize about the steady state, writing

$$x_s(t) = x_{0s} + \frac{\delta_0\tilde{\delta}}{\delta}\chi(t)$$

which gives

$$\frac{d\chi(t)}{dt} = \frac{\gamma + 1}{2k}\bar{\tilde{\phi}}_x(x_{0s}, 0, t) \tag{7}$$

as the equation that keeps track of the shock-wave position. Straightforward linearization of $[\![\phi_0 + (\tilde{\delta}/\delta)\tilde{\phi}]\!] = 0$ gives the expression that determines $\tilde{\phi}$ behind the shock from its value ahead of the shock [4.242]:

$$[\![\tilde{\phi}(x_{0s}, t, y)]\!] = -\frac{\gamma + 1}{2k}[\![\phi_{0x}(x_{0s}, y)]\!]\int_0^t \tilde{\phi}_x(x_{0s}, 0, \hat{t})\,d\hat{t}. \tag{8}$$

This must be integrated in conjunction with (6a). The ADI procedure of [4.60] is adopted, as outlined in [4.242], to effect a solution of Eq. (6) in conjunction with Eq. (8), subject to the time-linearized boundary conditions in Eqs. (6a) and (6b).

Indicial Response

One of the major advantages of time linearization is that, for a given mode of motion, the amplitude and phase lag of the lift or moment coefficient for a given reduced frequency may be computed by a linear superposition of the results obtained for a step change. For example, if the change in lift coefficient as a function of time for a step change in angle of attack, $C_{l\alpha}$, is that sketched in Fig. 2, then the lift coefficient for an angle-of-attack variation $\alpha(t)$ is

$$C_l(t) = C_{l\alpha}(t)\alpha(0) - \int_0^t C_{l\alpha}(\tau)\frac{d\alpha(t-\tau)}{d\tau}\,d\tau. \tag{9a}$$

Thus, for a periodic motion $\alpha(t) = \alpha_0 + \tilde{\alpha}e^{i\omega t}$,

$$C_l(t) = \tilde{\alpha}e^{i\omega t}\{C_{l\alpha}(\infty) - i\omega\int_0^\infty [C_{l\alpha}(\infty) - C_{l\alpha}(\tau)]e^{-i\omega\tau}\,d\tau\} + C_{l\alpha}(\infty)\alpha_0. \tag{9b}$$

The low-frequency approximation made in Eq. (1), viz., that $k^2\phi_{tt}$ was negligible, is, of course, not valid for the high-frequency components of the indicial response calculation. It is, however, perfectly satisfactory for the computation of the indicial response, provided this response is only used to compute motions for which $k = o(1)$.

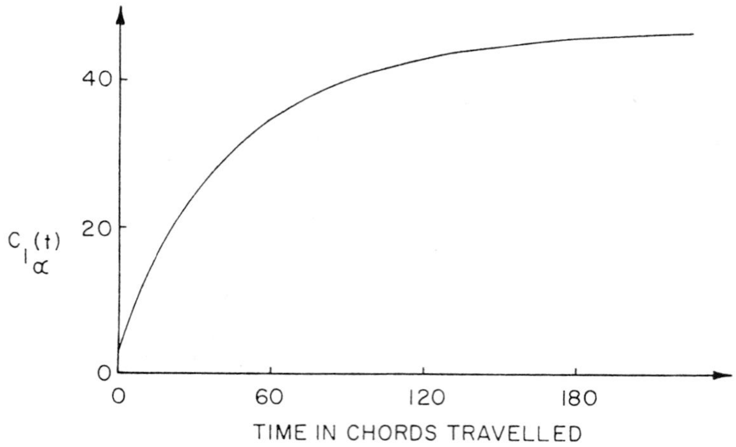

Figure 2a. Lift coefficient as a function of time for a step change in angle of attack; NACA 64A006, $M = 0.88$.

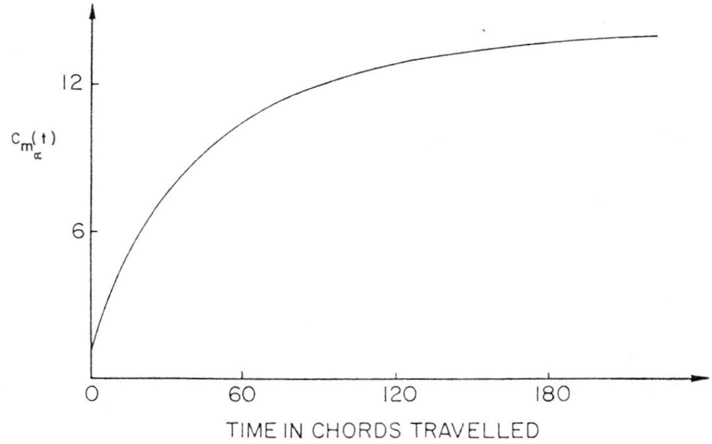

Figure 2b. Moment coefficient as a function of time for a step change in angle of attack; NACA 64A006 $M = 0.88$.

Far Field

In order to calculate the response for low reduced frequencies we must accurately resolve the indicial response as the motion approaches its asymptotic state. Typically, this requires the computation of the indicial response for 300 chord lengths of airfoil motion. In this time the unsteady perturbations have also traveled a little more than 300 chord lengths. As a consequence, any boundary condition imposed on a $|y|$ = constant boundary that is less than 150 chord lengths away can contaminate the indicial response through a reflection from a boundary. Our experience has been that an erroneous boundary condition such as $\phi = 0$ has to be imposed at $|y|$ greater than 80 chord lengths in order to avoid serious errors in the phase lag determined from an indicial response. The same must be true for the computation of a harmonic motion, although it would be more difficult to determine that the phase lag was in error in such a computation. This same observation should also be noted for unsteady wind-tunnel tests. If there are significant acoustic reflections from the wind-tunnel walls, the observed phase lags may be in error. This experimental difficulty warrants further investigation, especially in two-dimensional studies.

On the other hand, we know that with the appropriate steady-state value of the potential applied at about 20 chord lengths, the steady-state solution is perfectly adequate. With the imposition of the unsteady boundary condition (4), or its time-linearized analog, we find that once again 20 chord lengths will suffice. For low to moderate reduced frequencies, viz., $k = 0.1$ to 1.0, the acoustic wavelengths associated with the motion are about 10 to 1.0 chords. The grid spacing employed may be stretched, but grid spacings comparable to or larger than the acoustic wavelength will result in acoustic reflections from the grid itself. Thus, while a grid stretching is employed in the calculations, the largest grid spacing used remains a fraction of a chord length.

Harmonic-Oscillator Modeling

Typically, the indicial response of the lift coefficient to a step change in angle of attack or flap angle, or the imposition of plunging velocity, is like that shown in Fig. 2(a). The same is also approximately true for the moment coefficient taken about the airfoil's leading edge (Fig. 2(b)). We see that, to a first approximation, the response is nearly exponential and governed by a simple first-order differential equation, and further, that to a second approximation, the difference between this response and an exponential function can be modeled by a damped harmonic oscillator. Thus, if we let $u(t)$ be a normalized indicial response such that $u(0) = 0$, $u(\infty) = 1$, we should write

$$u(t) = 1 + u_0(t) + u_1(t) + \cdots . \tag{10}$$

where

$$L_0 u_0 \equiv \dot{u}_0 + \lambda u_0 = 0, \tag{11}$$

and, in general,

$$L_i u_i \equiv \ddot{u}_i + 2p_i\dot{u} + q_i u_i = 0.$$

The constants λ, p_i, q_i are determined to best model the indicial response. That is, the solutions to these equations, viz.,

$$u_0(t) = -e^{-\lambda t} \tag{12a}$$

and

$$u_i(t) = u_i(0)e^{-p_i t}\frac{\sin \Omega t}{\Omega}, \tag{12b}$$

where

$$\Omega = \sqrt{q_i - p_i^2},$$

are combined to best approximate the indicial response. To be specific, if we let $\varepsilon_0 = u(t) - 1 - u_0(t)$; then we choose λ such that

$$I_0(\lambda) = \int_0^\infty [L_0 \varepsilon_0]^2 \, dt \tag{13}$$

is minimum. Setting $\partial I_0/\partial\lambda = 0$, we find

$$\lambda^{-1} = 2\int_0^\infty \{1 - u(t)\}^2 \, dt. \tag{14}$$

In an analogous manner we let $\varepsilon_1 = u(t) - 1 - u_0(t) - u_1(t)$ and choose p_1 and q_1 so that

$$I_1(p_1, q_1) = \int_0^\infty [L_1\varepsilon_1]^2 \, dt \tag{15}$$

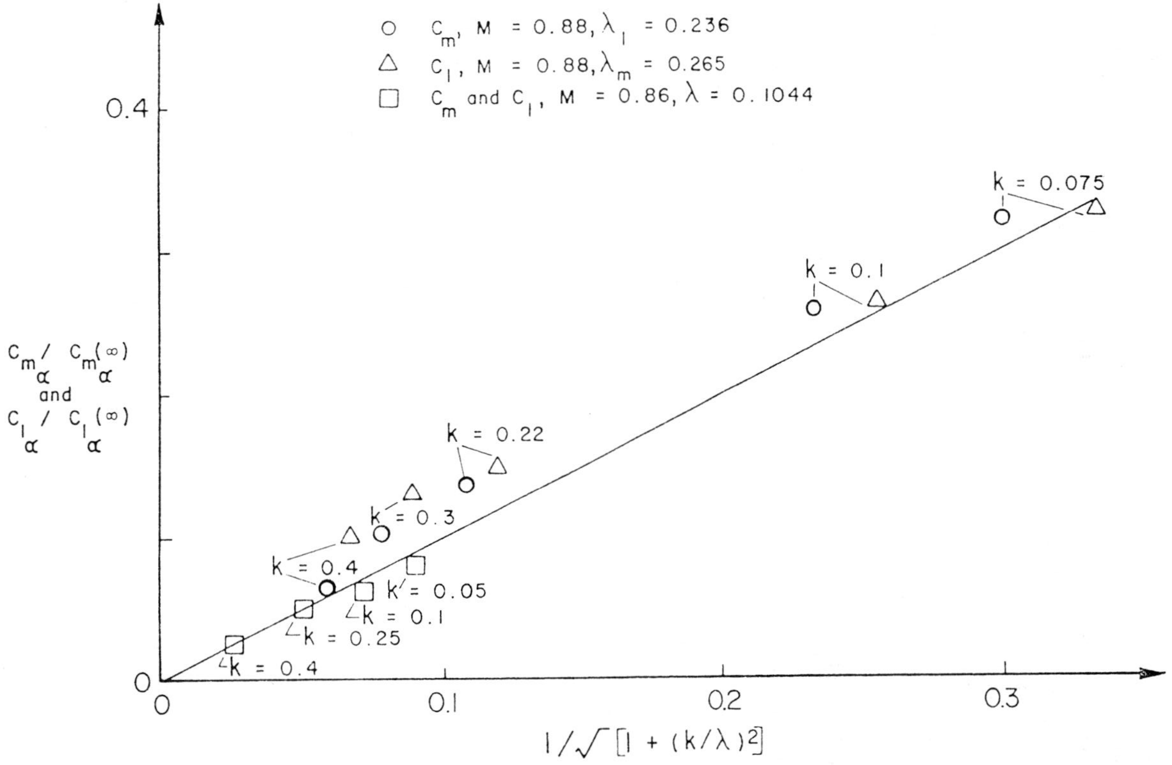

Figure 3. Lift and moment amplitudes normalized by their quasi-steady value as a function reduced frequency for an NACA 64A006 airfoil.

is minimized. This gives

$$p_1 = \frac{(\dot{u}(0) - \lambda)^2}{2 \int_0^\infty (\dot{u} - \dot{u}_0)^2 \, dt} \quad \text{and} \quad q_1 = \frac{\int_0^\infty (\dot{u} - \dot{u}_0)^2 dt}{\int_0^\infty (u - u_0) \, dt}. \tag{16}$$

The extent to which the simple first approximation is justified for selected examples is shown in Figs. 3 and 4. In many instances an acceptable determination of the phase lag requires the second approximation. This will be discussed more fully in [4.239]. The constant λ and the $I_0(\lambda)$ of the first approximation can be determined immediately from Eqs. (13) and (14) as the indicial response is being calculated. If I_0 is not sufficiently small, then the constants p_1 and q_1 can be calculated from (16). If the second approximation is not judged sufficiently accurate because I_1 is not acceptably small, the modeling is abandoned, as the computational expense of computing p_1 and q_1 is comparable to that required for three reduced frequencies.

Some time ago it was noted by Tijdeman [4.53] and the authors [4.242] that the amplitude of a harmonic response decays like k^{-1} with increasing k. A somewhat more general result is implied by (12). If $F(t)$ is a harmonic response, e.g., $C_{l\alpha}(t)$, then with

$$F(t) = F_0(t) + F_1(t) + \cdots, \tag{17}$$

we find that the first approximation gives

$$F_0/A = (1 + k'^2)^{-1/2} \sin(kt - \theta_0) \tag{18a}$$

where

$$\sin \theta_0 = k'/(1 + k'^2)^{1/2}. \tag{18b}$$

Here A is the amplitude of the indicial response to a unit change and $k' = k/\lambda$. The second approximation gives

$$\frac{F_1}{A} = \frac{\dot{u}_1(0) k' \Omega' \sin(kt - \theta_1)}{[\{1 + (k' + \Omega')^2\}\{1 + (k' - \Omega')^2\}]^{1/2}} \tag{19a}$$

where

$$\sin \theta_1 = \frac{(k'^2 - \Omega'^2 - 1)}{\{[1 + (k' + \Omega')^2][1 + (k' - \Omega')^2]\}^{1/2}} \tag{19b}$$

and $\Omega' = \Omega/\lambda$.

We see immediately from Eq. (18) that in the first approximation the amplitude of the harmonic response is

$$[1 + (k/\lambda)^2]^{-1/2},$$

which behaves like k^{-1}, for large k, and that the phase lag is

$$\sin^{-1} \frac{k/\lambda}{\{1 + (k/\lambda)^2\}^{1/2}},$$

which grows linearly with k/λ for small k/λ, and thereafter is nearly independent of k/λ.

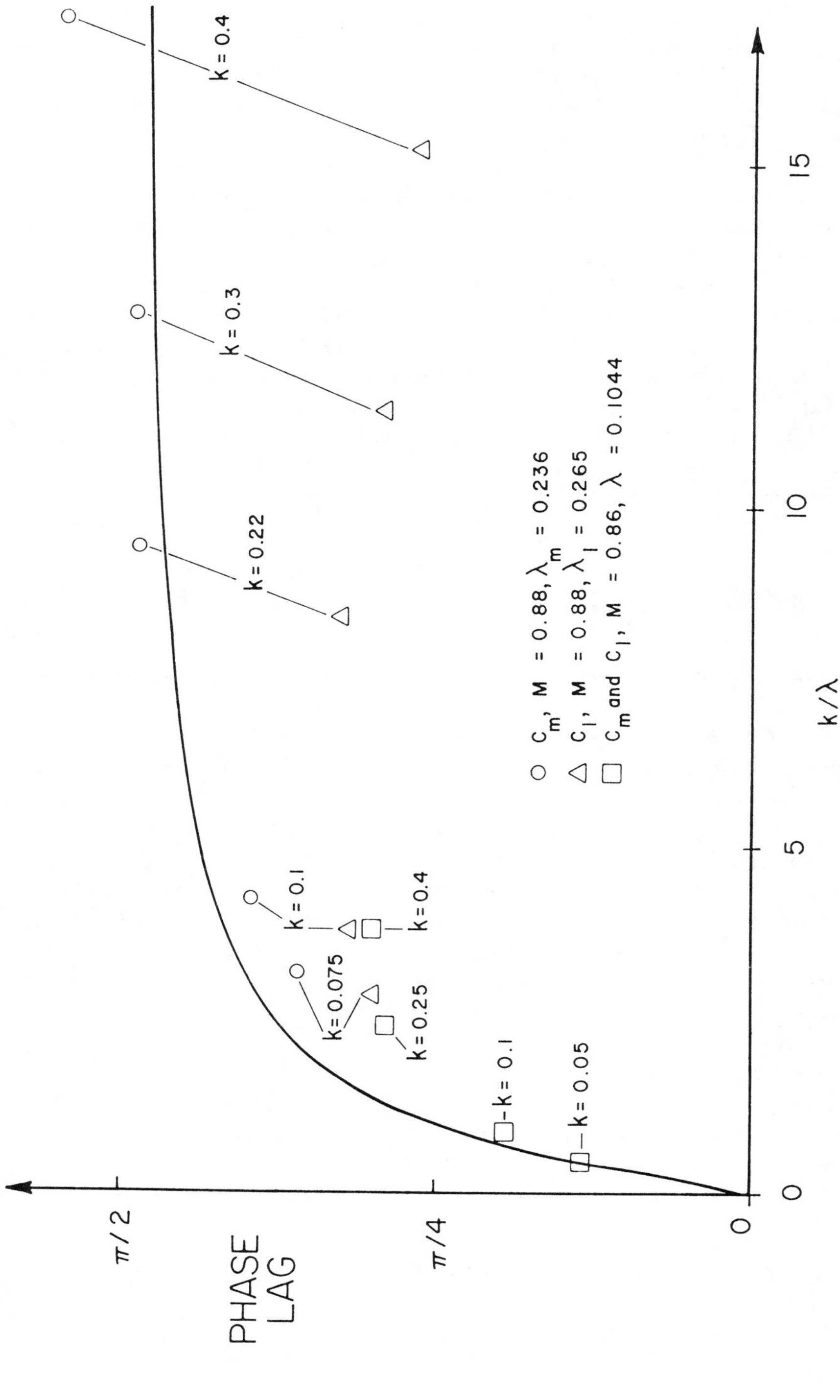

Figure 4. Phase lag for the lift and moment as a function of reduced frequency for an NACA 64A006 airfoil.

We limit our discussion to the simple variation of the amplitude and phase lag of the lift and moment coefficients for an NACA 64A006 in pitch with reduced frequency. Earlier more detailed results depicting the shock motion etc. are to be found in [4.242] and [4.243]. Because we have linearized about a steady, small perturbation solution, we draw no practical conclusions from our study. Tijdeman (personal communication) reports that the application of LTRAN2 to determine the response about an experimental steady state for the F29 airfoil at varying incidence, in conjunction with strip theory and the results of panel methods for subcritical flow to account for three-dimensional effects, was used to predict the flutter boundary of the F29 wing.

Figures 3 and 4 give the amplitudes and phase lags of the lift and moment coefficients for the NACA 64A006 airfoil oscillating in pitch at selected reduced frequencies with $M_\infty = 0.86$ and 0.88. Individual results are shown by symbols with the reduced frequency noted below them. Generally, they are well described by the first approximation of the harmonic-oscillater model (solid line). For $M_\infty = 0.86$ the lift and midchord moment results are indistinguishable, but their phase lags are not correctly captured by the modeling of the first approximation.

Computational Effort

As noted earlier, a time-linearized calculation requires substantially less computational effort than a nonlinear one. We delineate those differences here, noting that the larger grid spacing that may be used with our shock-fitting procedure implies a computational saving in addition to the considerations discussed here.

If we let T represent the number of time steps required to calculate a time-periodic solution in an $L \times M \times N$ spacial domain, then the total computational effort for a flutter study using either a nonlinear or a time-linearized algorithm is proportional to the product $TLMN$. If we do not time-linearize, this must be done at K reduced frequencies, to give a total computational effort that is proportional to $KTLMN$. If a time-linearized procedure is used to compute a single indicial response, the effort is $TLMN$. This can be used to generate the mode response for any reduced frequency in about T additional steps, giving a total computational effort proportional to $TLMN + \text{const} \cdot T$. Typically, T is 500 and L, M, N are about 50, 25, 25, respectively. Thus, a nonlinear analysis at ten reduced frequencies has a computational effort of about

$$10 \times 500 \times 50 \times 25 \times 25 \approx 10^8.$$

On the other hand, a time-linearized computation requires a computational effort of

$$500 \times 50 \times 25 \times 25 + \text{const} \times 500 \approx 10^7,$$

providing a factor-of-K reduction from the nonlinear analysis. The less refined L grid spacing required also favors the time-linearized algorithm, contributing roughly another factor-of-ten reduction from the nonlinear computation.

Conclusion

The amplitudes and phase lags of the lift and moment coefficients, at selected reduced frequencies, can be computed accurately and efficiently by time linearization about a measured or computed pressure field. For three-dimensional studies the linear superposition of the results of a single indicial response substantially reduces the computational effort. Further reductions are achieved by using the linearized far field for the unsteady flow, and by treating the shock wave as a discontinuity in the computations. In many cases, the indicial response can be modeled by a simple harmonic oscillator, and this provides an analytical result for the dependence of the lift and moment coefficients on reduced frequency, further reducing the computational effort.

Acknowledgment

This research was supported by ONR Contract No. N00014-76-C-0182 and by AFOSR Grant No. AFOSR-76-2954I. The authors are greatly indebted to Mr. Morton Cooper for his support of their efforts to develop a program in analytical and computational aerodynamics at the University of Arizona.

References for Part 4

[4.1] E. M. Murman, J. D. Cole: AIAA J. **9**, 114 (1971).
[4.2] W. F. Ballhaus, F. R. Bailey: AIAA Paper 72-677 (1972).
[4.3] F. R. Bailey, W. F. Ballhaus: *Lecture Notes in Physics* **19**, 2 (Springer-Verlag 1972).
[4.4] J. L. Steger, H. Lomax: AIAA J. **10**, 49 (1972).
[4.5] P. R. Garabedian, D. Korn: Comm. Pure Appl. Math. **24**, 841 (1972).
[4.6] A. Jameson: Comm. Pure Appl. Math. **27**, 283 (1974).
[4.7] A. Jameson: 2nd AIAA CFD Conf. Proc., 184 (1975).
[4.8] A. Jameson, D. A. Caughey: 3rd CFD Conf. Proc., 35 (1977).
[4.9] D. A. Caughey, A. Jameson: AIAA J. **17**, 175 (1979).
[4.10] A. Eberle: MBB UFE 1325(0) (1977).
[4.11] A. Eberle: NASA TM-75324 (trans. from MBB UFE 1407(0) (1978)).
[4.12] A. Eberle: DGLR/GARTEUR 6 Symp., *Transonic Configurations*, Bad Harzburg (1978).
[4.13] T. L. Holst, W. F. Ballhaus: NASA TM 78469 (1978); also AIAA J. **17**, 2038 (1979).
[4.14] M. Hafez, J. South, E. Murman: AIAA J. **17**, 838 (1979).
[4.15] W. F. Ballhaus, A. Jameson, J. Albert: AIAA J. **16**, 573 (1978).
[4.16] T. L. Holst: AIAA J. **17**, 1038 (1979).
[4.17] T. L. Holst: AIAA J. **18**, 1431 (1980).
[4.18] H. Deconinck, C. Hirsch: 7th Intl. Conf. on *Numerical Methods in Fluid Dynamics*, Stanford (1980).
[4.19] H. Deconinck, C. Hirsch: Proc. 3rd GAMM Conf. on *Numerical Methods in Fluid Mechanics*, DFVLR, Cologne (1979).
[4.20] T. J. Baker: J. Comp. Phys. **42**, 1 (1981).
[4.21] J. Chattot, C. Coulombeix, C. Silvia Tome: Le Recherche Aerospatiale **4**, 143 (1978).

[4.22] N. L. Sankar, J. Malone, Y. Tassa: AIAA Paper 81-385 (1981).
[4.23] T. L. Holst: In: *Comput. Methods in Appl. Sci. and Eng.*, R. Glowinski and J. L. Lions (eds.) (North-Holland Publishing Co. 1980).
[4.24] R. P. Fedorenko: USSR Comp. Math. and Math. Phys. **4**, 227 (1964).
[4.25] A. Brandt: *Lecture Notes in Physics* **18**, 82 (Springer-Verlag, Berlin/New York 1973).
[4.26] A. Brandt: Math. Comp. **31**, 333 (1977).
[4.27] J. C. South, A. Brandt: Rept. 76-8, ICASE (1976).
[4.28] A. Jameson: AIAA Paper 79-1458 (1979).
[4.29] B. Arlinger: Rept. L-0-1 B439, SAAB (1978).
[4.30] B. Arlinger: 7th Internat. Conf. on *Numerical Methods in Fluid Dynamics*, Stanford (1980).
[4.31] D. R. McCarthy, T. A. Reyhner: AIAA Paper 80-1365 (1980).
[4.32] C. W. Boppe: AIAA Paper 78-104 (1978).
[4.33] C. W. Boppe, M. A. Stern: AIAA Paper 80-130 (1980).
[4.34] J. F. Thompson, F. C. Thames, C. W. Mastin: J. Comp. Phys. **15**, 299 (1974).
[4.35] J. F. Thompson, F. C. Thames, C. W. Mastin: J. Comp. Phys. **24**, 274 (1977).
[4.36] P. D. Thomas, J. F. Middlecoff: AIAA J. **18**, 652 (1980).
[4.37] J. L. Steger, R. L. Sorenson: J. Comp. Phys. **33**, 405 (1979).
[4.38] J. J. Yu: AIAA Paper 80-1391 (1980).
[4.39] P. R. Eiseman: J. Comp. Phys. **33**, 118 (1979).
[4.40] P. R. Eiseman: 4th AIAA CFD Conf. Proc., 166 (1979).
[4.41] P. R. Eiseman: Rept. 80-11, ICASE (1980).
[4.42] L. Eriksson: Workshop on *Num. Grid Generation Techniques for Partial Differential Equations*, Hampton (1980).
[4.43] S. S. Davis, G. N. Malcolm: AIAA Paper 79-0769 (1979).
[4.44] H. Ashley: J. Aircraft **27** (1980).
[4.45] G. M. Farmer, W. P. Hanson: Proc. AIAA/ASME/SAE 17th *Structure, Structural Dynamics and Materials Conf.*, King of Prussia (1976); also NASA TM X-72837 (1976).
[4.46] F. X. Caradonna, J. J. Philippe: Vertica **2**, 43 (1978).
[4.47] J. L. Steger, H. E. Bailey: AIAA J. **18** (1980).
[4.48] D. R. Chapman: AIAA J. **17** (1979).
[4.49] W. F. Ballhaus, F. R. Bailey: J. of Comp. & Fluids **8**, 133 (1980).
[4.50] W. F. Ballhaus, J. O. Bridgeman: AGARD *Special Course on Unsteady Aerodynamics*, von Karman Inst., Rhode–St. Genese (1980).
[4.51] H. Tijdeman, R. Seebass: Ann. Rev. Fluid Mech. **12**, 181 (1980).
[4.52] W. F. Ballhaus: In: *Numerical Methods in Fluid Dynamics*, H. J. Wirz and J. J. Smoldren (eds.) (1976).
[4.53] H. Tijdeman: Doctoral Thesis, Technische Hogeschool Delft (1977).
[4.54] W. J. McCroskey: Trans. ASME, J. Fluids Eng. **99**, 8 (1977).
[4.55] M. T. Landahl: Symp. *Transsonicum II*, K. Oswatitsch, D. Rues (eds.) (Springer, Berlin, 1976).
[4.56] W. F. Ballhaus, P. M. Goorjian: AIAA J. **15**, 1728 (1977).
[4.57] W. J. Chyu, S. S. Davis: AIAA Paper 79-1554 (1979); also AIAA J. (to appear).
[4.58] W. F. Ballhaus, T. L. Holst, J. L. Steger: *Lecture Notes in Physics* **90** (Springer-Verlag 1979).
[4.59] W. F. Ballhaus, H. Lomax: *Lecture Notes in Physics* **35**, 57 (Springer-Verlag 1975).
[4.60] W. F. Ballhaus, J. L. Steger: NASA TM X-73, 082 (1975).

[4.61] H. Tijdeman: Symposium *Transsonicum II*, 49 (Springer-Verlag 1976).
[4.62] R. Houwink, J. van der Vooren: AIAA Paper 79-1533 (1979); also AIAA J. **18** (1980).
[4.63] M. Couston, J. J. Angeline: J. Fluids Eng. **101**, 341 (1979).
[4.64] D. P. Rizzetta, W. C. Chen: AIAA J. **17**, 779 (1979).
[4.65] D. P. Rizzetta, H. Yoshihara: AIAA Paper 80-0128 (1980).
[4.66] D. Kwak: AIAA Paper 80-1393 (1980).
[4.67] P. M. Goorjian, R. van Buskirk: AIAA Paper 81-0331 (1981).
[4.68] C. Borland, D. Rizzetta, H. Yoshihara: AIAA Paper 80-1369 (1980).
[4.69] W. F. Ballhaus, P. M. Goorjian: AIAA J. **16** (1978).
[4.70] R. M. Beam, R. F. Warming: NASA TN D-7605 (1974).
[4.71] D. Nixon: AIAA J. **16**, 613 (1978).
[4.72] D. P. Rizzetta: AIAA J. **17** (1979).
[4.73] T. J. Yang, P. Guruswamy, A. G. Striz: AFFDL TR-79-3077 (1979).
[4.74] W. F. Ballhaus, P. M. Goorjian: AGARD CP-226 (1977).
[4.75] W. F. Ballhaus, P. M. Goorjian, H. Yoshihara: AGARD CP-227 (1977).
[4.76] K. Isogai: AIAA Paper 77-448 (1977).
[4.77] R. Chipman, A. Jameson: AIAA Paper 79-1555 (1979).
[4.78] P. M. Goorjian: AIAA Paper 80-0150 (1980); also NASA CR-152274 (1979).
[4.79] J. L. Steger, F. X. Caradonna: NASA TM-81211 (1980).
[4.80] N. L. Sankar, Y. Tassa: 7th Internat. Conf. on *Numerical Methods in Fluid Dynamics*, Stanford (1980).
[4.81] R. Chipman, A. Jameson: AIAA Paper 81-0329 (1981).
[4.82] J. L. Steger, F. X. Caradonna: AIAA Paper 80-1368 (1980).
[4.83] R. Magnus, H. Yoshihara: AIAA J. **8**, 2157 (1970).
[4.84] R. Beam, R. F. Warming: AIAA J. **16** (1978); also AIAA Paper 77-645 (1977).
[4.85] G. S. Deiwert: *Lecture Notes in Physics* **35**, D. Richtmyer (ed.), 132 (Springer-Verlag, New York (1975).
[4.86] G. S. Deiwert: AIAA J. **13**, 1354 (1975).
[4.87] R. W. MacCormack: *Lecture Notes in Physics* **8**, M. Holt (ed.), 151 (Springer-Verlag, New York 1971).
[4.88] R. W. MacCormack, A. J. Paullay: AIAA Paper 72-154 (1972).
[4.89] G. S. Deiwert, J. B. McDevitt, L. L. Levy, Jr.: NASA CP-347 (1975).
[4.90] B. S. Baldwin, R. W. MacCormack, G. S. Deiwert: AGARD-LS-73 (1975).
[4.91] G. S. Deiwert: AIAA J. **14**, 735 (1976).
[4.92] J. B. McDevitt, L. L. Levy, Jr., G. S. Deiwert: AIAA J. **14**, 606 (1976).
[4.93] G. S. Deiwert: *Transonic Flow Problems in Turbo-machinery*, T. C. Adamson and M. F. Platzer (eds.), 371 (Hemisphere Publ. Corp., Washington 1977).
[4.94] P. R. Garabedian, D. G. Korn: *Numerical Solution of Partial Differential Equations—II*, 253 (Academic Press, New York 1971).
[4.95] J. J. Kacprzynski, L. H. Ohman, P. R. Garabedian, D. G. Korn: Rept. LR-554, NRC (1971).
[4.96] J. J. Kacprzynski: Wind Tunnel Project Rept 5 × 5/0062, NRC/NAE (1972).
[4.97] G. S. Deiwert: *Lecture Notes in Physics* **59**, 159, A. I. van de Vooren and P. J. Zandberge (eds.) (Springer-Verlag, New York 1976).
[4.98] R. W. MacCormack: *Lecture Notes in Physics* **59**, 307, A. I. van de Vooren and P. J. Zandberge (eds.) (Springer-Verlag, New York 1976).
[4.99] A. Seginer, W. C. Rose: AIAA Paper 76-330 (1976).
[4.100] A. Seginer, W. C. Rose: AIAA Paper 77-210 (1977).
[4.101] L. L. Levy, Jr.: AIAA J. **26**, 564 (1978).

[4.102] H. L. Seegmiller, J. G. Marvin, L. L. Levy, Jr.: AIAA J. **16**, 1262 (1978).
[4.103] G. S. Deiwert, H. E. Bailey: NASA CP-2045 (1978).
[4.104] G. S. Deiwert: Proc. 4th U.S. Air Force and the Federal Republic of Germany Data Exchange Agreement Mtg. (1979); also NASA TM-78581 (1979).
[4.105] J. R. Viegas, T. J. Coakley: AIAA Paper 77-44 (1977).
[4.106] T. J. Coakley, J. R. Viegas: Symp. on *Turbulent Shear Flows*, University Park (1977).
[4.107] K. M. Peery, C. K. Forester: AIAA Paper 79-1549 (1979).
[4.108] J. L. Steger: AIAA Paper 77-665 (1977).
[4.109] R. Beam, R. F. Warming: J. Comp. Phys. **22**, 87 (1976).
[4.110] J. L. Steger, T. H. Pulliam, R. V. Chima: AIAA Paper 80-1427 (1980).
[4.111] C. J. Nietubicz, T. H. Pulliam, J. L. Steger: AIAA Paper 79-0010 (1979).
[4.112] T. H. Pulliam, J. L. Steger: AIAA J. **18**, 159 (1980).
[4.113] G. S. Deiwert: AIAA Paper 80-1347 (1980).
[4.114] A. Bertelrud, M. Y. Bergmann, T. J. Coakley: AIAA Paper 80-0005 (1980).
[4.115] R. W. McCormack: SIAM–AMS Proc. **11**, 130 (1978).
[4.116] R. R. Cosner: AIAA Paper 80-0193 (1980).
[4.117] P. R. Dodge: AIAA Paper 76-425 (1976).
[4.118] P. R. Dodge: AFAPL-TR-77-64 (1977).
[4.119] J. M. Wu, T. H. Moulden: AIAA Paper 76-326 (1976).
[4.120] J. R. Spreiter, S. S. Stahara: ICAS Paper 76-06 (1976).
[4.121] M. S. Liou: AIAA Paper 81-0004 (1981).
[4.122] A. F. Messiter: Z. Angew. Math. Phys. **31**, 204 (1980).
[4.123] D. R. Chapman: 7th Internat. Conf. on *Numerical Methods in Fluid Dynamics*, Stanford (1980).
[4.124] L. Prandtl: Nachrichten d.k. Gesellschaft d. Wiss **24**, 451 (1918).
[4.125] M. M. Munk: NACA TR 184 (1924).
[4.126] R. T. Jones: NACA TN 835 (1946).
[4.127] L. P. Cook, J. D.Cole: SIAM J. Appl. Math. **35**, 209 (1978).
[4.128] M. Cooper, F. C. Grant: NACA TN 3183 (1954).
[4.129] D. D. Liu, M. F. Platzer, S. V. Ruo: AIAA J. **15**, 966 (1977).
[4.130] R. T. Whitcomb: NACA TR 1273 (1956).
[4.131] K. Oswatitsch, F. Keune: Z. Flugwiss. **1** (1953).
[4.132] K. Oswatitsch, F. Keune: Z. Flugwiss. **3**, 29 (1954).
[4.133] G. N. Ward: Quart: J. Mech. & Appl. Math. **2**, 75 (1949).
[4.134] M. C. Adams, W. R. Sears: J. Aero. Sci. **20**, 85 (1953).
[4.135] W. D. Hayes: Rept. AL-222, North American Aviation, Inc. (1947); also AMS Rept. 852, Princeton Univ. (1968).
[4.136] J. D. Cole, A. F. Messiter: ZAMP **8**, 1 (1957).
[4.137] H. Ashley, M. Landahl: *Aerodynamics of Wings and Bodies* (Addison-Wesley, Reading 1965).
[4.138] J. R. Spreiter, S. S. Stahara: AIAA J. **9**, 1784 (1971).
[4.139] S. S. Stahara, J. R. Spreiter: AIAA J. **17**, 245 (1979).
[4.140] K. Oswatitsch: Appl. Mech. Review **10**, 543 (1957).
[4.141] S. B. Berndt: ZAMP **9** (1958).
[4.142] S. B. Berndt: ZAMP **35** (1955).
[4.143] S. B. Berndt: FFA Rept. 70, Aero Res. Inst., Stockholm (1956)
[4.144] G. Drougge: FFA Rept. 83, Aero Res. Inst., Stockholm.
[4.145] H. K. Cheng, M. M. Hafez: AIAA J. **10**, 1115 (1972).

[4.146] H. K. Cheng, M. M. Hafez: Rept. USCAE 121, Univ. So. Calif., Los Angeles (1972).
[4.147] R. W. Barnwell: AIAA J. **11**, 764 (1973).
[4.148] H. K. Cheng, M. M. Hafez: AIAA J. **11**, 1210 (1973).
[4.149] H. K. Cheng, M. M. Hafez: Rept. USCAE 124, Univ. So. Calif., Los Angeles (1973).
[4.150] H. K. Cheng, M. M. Hafez: J. Fluid Mech. **72**, 161 (1975).
[4.151] K. G. Guderley: *The Theory of Transonic Flow*, Pergamon Press, New York (1962).
[4.152] J. D. Cole: SIAM J. Appl. Math. **29**, 763 (1975).
[4.153] R. W. Barnwell: NASA TR R-440 (1975).
[4.154] H. K. Cheng: AIAA J. **15**, 366 (1977).
[4.155] M. S. Cramer: J. Fluid Mech. **95**, 223 (1979).
[4.156] M. S. Cramer: J. Fluid Mech. (to appear) (1981).
[4.157] M. D. Van Dyke: J. Aero. Sci. **18**, 161 (1951).
[4.158] M. J. Lighthill: In: *General Theory of High Speed Aerodynamics*, W. R. Sears (ed.), 345 (Princeton Univ. Press 1954).
[4.159] W. D. Hayes: J. Mecan. **5**, 163 (1966).
[4.160] W. D. Hayes: J. Aero. Sci. **21**, 721 (1954).
[4.161] L. T. Goodmanson: Aero & Astro., 46 (1971).
[4.162] R. T. Jones: AIAA J. **10**, 171 (1972).
[4.163] R. T. Jones: Acta Aeronautica **4**, 99 (1977).
[4.164] A. Busemann: Luftfahrtforschung **12** , 210 (1975).
[4.165] R. T. Jones: NACA Rept. 863 (1945).
[4.166] R. T. Jones, D. Cohen: In: *Aerodynamic Components of Aircraft at High Speed*, D. F. Donovan, H. R. Lawrence (eds.) (Princeton Univ. Press 1957).
[4.167] D. Kuchemann: Aero. J. **73**, 101 (1963).
[4.168] F. Bauer, P. R. Garabedian, D. G. Korn, A. Jameson: *Lecture Notes in Econ. and Math. Syst.* **108** (Springer-Verlag, Berlin 1974).
[4.169] G. Y. Niewland, B. M. Spee: Ann. Review of Fluid Mech. **5**, 119 (1973).
[4.170] J. W. Boerstoel: Rept. NLRMP 74024 II, Nat. Aero. Lab., Amsterdam (1974).
[4.171] R. T. Whitcomb: Proc. 9th Internat. Congr. on Aero. Sci. (1974).
[4.172] H. Sobieczky, N. J. Yu, K. Y. Fung, A. R. Seebass: AIAA J. **17**, 722 (1979).
[4.173] L. P. Cook: Indiana Univ. Math. J. **27** (1978).
[4.174] H. K. Cheng: Rept. USCAE 133, Univ. So. Calif., Los Angeles (1976).
[4.175] J. Weissinger: FB 1553, Berlin-Adlershof (1942); transl. as NACA TM 1120 (1947).
[4.176] K. Kreines: ZAMM **20**, 65 (1939); transl. as NACA TM 97 (1970).
[4.177] A. A. Dorodnitsyn: PMM **8**, 33 (1944).
[4.178] J. Thurber: Comm. Pure Appl. Math. **18**, 733 (1965).
[4.179] H. K. Cheng: AIAA J. **16**, 1211 (1978).
[4.180] H. K. Cheng: Rept. USCAE 135, Univ. So. Calif., Los Angeles, 1211 (1978).
[4.181] H. K. Cheng, S. Y. Meng: AIAA J. **17**, 121 (1979).
[4.182] K. K. Cheng, S. Y. Meng: J. Fluid Mech. **97**, 531 (1980).
[4.183] H. K. Cheng, S. Y. Meng: Rept. USCAE 136, Univ. So. Calif., Los Angeles (1979).
[4.184] H. K. Cheng, S. Y. Meng, R. Chow, R. C. Smith: Rept. USCAE 138, Univ. So. Calif., Los Angeles (1980).
[4.185] H. K. Cheng, S. Y. Meng, R. Chow, R. C. Smith: AIAA J. **19**, 961 (1981).

[4.186] L. P. Cook: Quart. Appl. Math., 178 (1979).
[4.187] T. von Kármánn: J. Math. Phys. **26**, 182 (1947).
[4.188] A. Jameson, D. A. Caughey: Rept. C00-3077-140, ERDA Math. Comput. Lab., Courant Inst. Math. Sci. (1977).
[4.189] P. A. Henne, R. M. Hicks: Paper 6647, Douglas Aircraft Co., Long Beach (1979).
[4.190] F. R. Bailey, W. F. Ballhaus: NASA SP-347, 1213 (1975).
[4.191] M. M. Hafez, H. K. Cheng: AIAA J. **15**, 786 (1977).
[4.192] G. Moretti: *Lecture Notes in Physics*, R. D. Ritchmeyer (ed.) (Springer-Verlag 1975).
[4.193] G. Moretti: Rept. 74-15, Brooklyn Polytech, (1974).
[4.194] K. Oswatitsch, J. Zierep: ZAMM **40**, 143 (1960).
[4.195] H. K. Cheng: *Proc. Symp. on Transonic Shock and Multidimensional Flows*, Univ. of Wisconsin, Madison (1981) (to appear).
[4.196] H. K. Cheng, R. Chow, R. E. Melnik: submitted for publication.
[4.197] D. Nixon: AIAA J. **16**, 47 (1978).
[4.198] D. Nixon: AIAA J. **16**, 699 (1978).
[4.199] M. Hafez, M. Rizk, E. Murman: Proc. AGARD Symp. Unsteady Transonic and Separated Flows (1977).
[4.200] E. Ehler: NASA Contract report (to appear).
[4.201] R. D. Small: AIAA J. **16**, 632 (1978).
[4.202] R. D. Small: Rept. ENG-7836, Univ. of California, Los Angeles (1978).
[4.203] D. Melson, M. Hafez: (to appear).
[4.204] Y. Y. Chan: ZAMP **31**, 605 (1980).
[4.205] Y. Y. Chan: J. Aircraft **17** (1980).
[4.206] Y. Y. Chan: J. Aircraft **17** (1980).
[4.207] W. Kemp: NASA CP 2045, 473 (1978).
[4.208] E. Murman: AIAA Paper 79-1533 (1979).
[4.209] A. Ferri, P. Baronti: AIAA J. **11**, 63 (1973).
[4.210] W. R. Sears: Aeronautical J. **78**, 80 (1974).
[4.211] W. R. Sears, R. J. Vidal, J. C. Erickson, Jr., A. Ritter: J. Aircraft **14**, 1042 (1977).
[4.212] D. A. Caughey, A. Jameson: AIAA J. **18**, 1281 (1980).
[4.213] A. Jameson: In: *Proc. of AIAA 4th Computational Fluid Dynamics Conf.*, 122 (1979).
[4.214] A. Jameson, D. A. Caughey, J. Steinhoff, W. H. Jou, R. Pelz: In: *Proc. of GAMM Specialist Workshop for Numerical Methods in Fluid Dynamics*, Stockholm (1979).
[4.215] D. A. Caughey, A. Jameson: AIAA J. **15**, 1474 (1977).
[4.216] B. Monnerie, F. Charpin: In: *10ᵉ Colloque d'Aerodynamique Applique* (1973).
[4.217] D. A. Spence: Proc. Roy. Soc. **A238**, 46 (1956).
[4.218] N. D. Malmuth, W. D. Murphy: AIAA J. **14**, 1250 (1976).
[4.219] W. D. Murphy, N. D. Malmuth: AIAA J. **15**, 46 (1977).
[4.220] N. D. Malmuth, W. D. Murphy: AIAA Paper 77-174 (1977).
[4.221] N. D. Malmuth, W. Murphy, V. Shankar, J. Cole, E. Cumberbatch: AIAA Paper 80-0270 (1980).
[4.222] H. Yoshihara, W. V. Carter, J. G. Fatta, R. G. Magnus: Rept. L-112173, Convair (1972).
[4.223] H. H. Pearcy: In: *Boundary Layer and Flow Control*, G. V. Lachmann (ed.), 1166 (Pergamon, Oxford/London 1961).

[4.224] H. B. Keller, T. Cebeci: AIAA J. **10**, 1197 (1972).
[4.225] V. Shankar, N. D. Malmuth, J. D. Cole: AIAA Paper 78-103 (1978).
[4.226] V. Shankar, N. D. Malmuth, J. D. Cole: In: *Proc. NASA–Langley Conf. on Advanced Technology Research* (1978).
[4.227] V. Shankar, N. D. Malmuth, J. D. Cole: AIAA Paper 79-0344 (1979).
[4.228] D. Rizzetta, H. Yoshihara: In: *AGARD Specialist Mtg. on Boundary Layer Effects on Unsteady Airloads* (1980).
[4.229] W. H. Jou, E. M. Murman: In: *AGARD Symp. on Computation of Viscous–Inviscid Interactions* (1980).
[4.230] J. Green, D. Weeks, J. Brooman: RAE R&M 3791 (1973).
[4.231] L. East, P. Smith, P. Merryman: RAE Rept. 77046 (1977).
[4.232] P. H. Cook, M. A. McDonald, M. C. P. Firmin: AGARD-AR-138 (1979).
[4.233] J. Wai, H. Yoshihara: In: AGARD Symp., Colorado Springs (1980).
[4.234] S. Davis, G. Malcolm: NASA TM 81221 (1980).
[4.235] H. Yoshihara: Rept. CASD-ERR-75-012, Convair (1975).
[4.236] K. Y. Fung: AIAA J. **19**, 180 (1981).
[4.237] K. Y. Fung: AIAA J. (to appear).
[4.238] C. C. Lin, E. Reissner, H. S. Tsien: J. Math. Phys. **3**, 220 (1948).
[4.239] R. M. Traci, J. L. Farr, E.Albano: AIAA Paper 75-877 (1977).
[4.240] K. Y. Fung, N. J. Yu, R. Seebass: AIAA J. **16**, 815 (1978).
[4.241] A. R. Seebass, N. J. Yu, K. Y. Fung: AGARD-CP-227 (1978).

PART 5

EXPERIMENTAL FLUID DYNAMICS

Introduction

A. Roshko*

Before reviewing the chapters in this part it may be interesting to provide some perspective by considering what contemporary "experimental fluid dynamics" implies. It seems possible to classify work in this area into two categories: (a) data-base experiments and (b) concept-base experiments. Of course, they are not mutually exclusive, and perhaps they are not all-embracing. In the first category the objective is to document particular flows or classes of flows, either to provide a catalog for engineering design (e.g., an airfoil series) or, more usually these days, to produce detailed information for evaluating and improving engineering methods of flow calculation. Good examples of the products of such experiments are the standard flows collected for the two Stanford conferences ([5.1] and [5.2]). Here we have in mind mainly the engineering scene, but the definition could include a more physics-oriented objective and approach, differing perhaps only in the degree of abstractness of the experiments.

A similar juxtaposition between engineering and physics can also be made for the second category, which we call "concept-base experiments". Here the approach is necessarily more explorative; the emphasis is on looking for correlations, elucidating mechanisms, noticing new effects, and in general developing or discovering concepts. The approach may be guided by theoretical ideas or by experience (intuition); it may be systematic (possibly a facet of the first category), or it may just be intelligent "fooling around". Fertile territory is provided by the turbulent shear flows, separated flows, and

* Graduate Aeronautical Lab., California Institute of Technology, Pasadena, CA 91125.

nonsteady flows which account for much of the subject matter of this volume. It is interesting that in a field as old as fluid mechanics there are still opportunities for exploration and discovery.

It is also appropriate to note, especially in view of the title of this book, that most of the above discussion might be applied not only to physical experiments, such as those described in the present part, but also to numerical experiments like some of those discussed in other parts of this volume.

The experiments by D. Johnson described in Chapter 31 were designed to provide experimental information about Reynolds stresses in the shock-wave–boundary-layer interaction region of transonic flows. Of particular interest is the use of laser anemometry with frequency offset, necessary in the region of strong interactions where high turbulence levels and possible flow reversals may occur. As shown by comparing with the experimental pressure distributions and shear-stress profiles, solutions from contemporary engineering models of the flow, based on the solutions of the *Reynolds-averaged, time-dependent, compressible Navier–Stokes equations* (called "Navier–Stokes solutions" for short) are not adequate for describing these transonic shock-wave–boundary-layer interactions.

A present trend that has been developing in experimental fluid mechanics during the past few years or so, a departure from the long-time devotion to the Reynolds-averaged view of turbulent shear flows, is exemplified in Chapter 30, by C. M. Ho. Here the nonstationary aspects of the turbulent flow are not immediately eliminated by time averaging, but rather an effort is made to understand the role of the predominant, large-scale nonstationary features. It would certainly appear to be more difficult to devise engineering models incorporating such features, but proponents of the approach are encouraged by the possibilities for "putting more physics" into models and by hopes for unforseen breakthroughs. The possibilities are illustrated in the paper by Ho, who proposes a calculation of the spreading rate of a turbulent mixing layer by a model based on the spanwise organized vortices which have been observed in such layers. Making use of the further observation that growth is accompanied by vortex merging and that these mergings can be related to saturation amplification of instability waves in an ordinary (laminar) vorticity layer, Ho ingeniously obtains a formula for the growth rate which appears to agree well with experimentally observed growth rates of *turbulent* mixing layers. As discovered by M. Lessen, who for many years has been propounding the idea that some aspects of turbulent development can be inferred from a simple instability model, there may be much skepticism as to how far such models can go in describing "real turbulence". Nevertheless these first steps in making a connection between the mean growth rate, the observed nonstationary structure, and a simple model are encouraging.

Chapter 32 presents measurements on another one of the "complex flows" being systematically studied at Imperial College under the direction of P. Bradshaw. In this experiment by Mehta, Shabaka, and Bradshaw, three examples of single streamwise vortices or vortex pairs imbedded in turbulent boundary layers are explored and measured in some detail. Their particular configurations, the corner vortices generated by a strut in a boundary layer, and the vortices produced by single or paired vortex generators are examples of what they call "skew-induced" vorticity, such as occurs in many practical cases, e.g., wing–body junctures, bodies at incidence, and other cross-flow-inducing configurations. The detailed measurements of velocity, vorticity, Reynolds-stress components, and even some triple products should be useful for developing and evaluating calculation methods for the practical cases.

Chapter 33, by Collins, Platzer, Lai, and Simmons, provides an example of the exploration of an interesting concept, namely the idea that the naturally occurring entrainment of a turbulent jet can be augmented by oscillating the jet. In the experiment, a vane placed in the centerplane of a two-dimensional jet, near its origin, could be oscillated at controlled frequencies and amplitudes. The downstream development of mean velocity profiles, as measured by pitot-static, hot-wire, and laser-Doppler anemometers, exhibits the considerable increase in jet spreading rates that can be achieved, with implied increase in local entrainment rate as high as 175 %.

Chapter 34, by Taylor, Whitelaw, and Yianneskis, uses two configurations, the flow in a square-sectioned bend and in a square-to-round transition section, to illustrate how "benchmark" measurements can be achieved in benchtop experiments. The particular point is made that use of water as the working fluid often allows a simple and convenient setup for use with laser-Doppler anemometry. This is a trend that is much in evidence, not only because of its convenience for LDA but, it might be added, because of many advantages that water has for flow visualization, which is much in fashion again.

What can clearly be described as exploratory experiments are described in the account by W. H. Young, Jr. (Chapter 35) of various experiments designed to gain insights into the difficult flow problems associated with dynamic stall on helicopter blades. The phenomena, including the flow in the boundary layer, are unsteady; separation-bubble development and bursting and vortex shedding play prominent roles. In the transonic regime, shock-wave–boundary-layer interaction and acoustic-wave coupling between various regions of the blade introduce additional phenomena. The array of interacting flow mechanisms outlined in this paper presents a challenge to experimenters and modelers alike.

The experiment described in the last chapter, by Parikh, Reynolds, and Jayaraman, has broad goals: to develop fundamental understanding of unsteady boundary-layer flows and to provide a data base for a particular carefully designed example. With an eye to the needs of the computer, an ingenious new facility, designed specifically for the purpose, allows a standard, steady flat-plate turbulent boundary layer to enter a region of linearly decreasing freestream velocity, whose gradient can be varied sinusoidally (in time). Amplitude and frequency can be controlled. Again, it was found advantageous to use water as the working medium. With the help of modern computer-aided data-processing equipment, boundary-layer profiles of the velocity and turbulent correlations were obtained at various phases of the oscillation cycle. For the range of frequencies and amplitudes in this initial phase of the work, it was found that mean velocity profiles and time-averaged Reynolds stresses were unaffected by the oscillations, i.e., their histories were quasisteady.

In summary, the seven chapters in this part provide good examples of the aspects of experimental research outlined at the beginning of this review. They also illustrate some of the contemporary techniques, the most obvious exception being flow visualization.

CHAPTER 30

Local and Global Dynamics of Free Shear Layers

Chih–Ming Ho*

Introduction

Since Brown and Roshko [5.3] pointed out that coherent structures are intrinsic features of a free shear layer, their investigation has become the mainstream of shear-turbulence research. The pairing of these structures [5.4, 5.5] was proved to be responsible for the spreading and the momentum transfer in a mixing layer. Therefore, if the evolution of these coherent structures can be well described, the dynamics of the shear layer can be understood.

Theoretical [5.6] and numerical [5.7, 5.8] studies have shown the importance of subharmonics in the vortex merging. Recently, the quantitative relationship between the subharmonic and the vortex merging was experimentally verified by Ho and Huang [5.9]. They also demonstrated that the vortex merges at the point where the vortex passage frequency becomes neutrally stable according to the linear stability analysis. Wygnanski and Oster [5.10] studied a mixing layer under very low-frequency forcing; the passage frequency scaled with the local thickness, and the average velocity in the "resonance region" was found to be equal to the neutrally stable frequency. Fiedler [5.11] further showed that the amplification rates measured in a turbulent mixing layer agree with the theoretical calculation [5.12]. These three experiments on mixing layers indicate that some of the properties of the

* Department of Aerospace Engineering, University of Southern California, Los Angeles, CA 90007.

large coherent structures can be predicted by stability analyses. It has been recognized for some time that the passage frequency at the end of a jet potential core [5.13, 5.14] and noise production [5.15, 5.16] can be predicted from the linear stability theory. This is a very attractive result, because it is then possible to understand a turbulent flow through stability analyses based upon *local* velocity profiles. However, it is also an astonishing result. How are the linear theories able to describe the highly nonlinear flows?

Several experiments that indicate that the perturbations generated by vortex mergings downstream will also affect the evolution of the coherent structures. Dimotakis and Brown [5.17] pointed out that the influence of the downstream vortices will not diminish with distance. A long time constant observed in their experiment can be explained from the feedback concept. In an impinging jet, Ho and Nosseir [5.18] showed that the coalescence of vortices near the nozzle can be greatly modified by acoustic feedback from the downstream-impinging coherent structures. Laufer and Monkewitz [5.19] found that the instability waves near the nozzle exit are modulated by a low frequency which corresponds to the passage frequency of the vortices at the end of the potential core. They also found that the forced-jet experiment by Kibens [5.20] satisfies the feedback concept. In a free jet, Gutmark and Ho [5.21] show that the locations of merging satisfy the feedback equation [5.18]. Hence, both the *local instability process* and the *global feedback phenomenon* should play crucial roles in the dynamics of a free shear layer.

In this paper, a new concept, *subharmonic evolution*, is introduced to model a free shear layer. The local instability and the global feedback effect are built into this model. The paradox of the linear theory being able to describe the highly nonlinear flow can be explained by the subharmonic evolution. Furthermore, many existing experimental observations can be interpreted through this concept.

Subharmonics and Vortex Merging

Vortex merging has been recognized as the mechanism which causes the shear layer to grow and produce momentum transfer [5.4, 5.5]. However, most of the vortex-merging studies have been limited to visualization results. The first quantitative model was proposed by Petersen [5.22]. He modeled the vortex merging through the wave dispersion relation. Ho and Huang [5.9] suggested that the subharmonics can be viewed as the catalyst of vortex merging. A conceptual picture is sketched in Fig. 1. The subharmonic amplifies and displaces the vortices laterally. Due to the velocity gradient, the lower vortices move faster than the upper ones. Finally, the two vortices merge into a single coherent structure. Ho and Huang further demonstrated this idea by applying the Nth subharmonic to the shear layer and observing the simultaneous merging of the N vortices. In a forced axisymmetric jet, the spreading rate can also be substantially changed by manipulating the vortex merging [5.23].

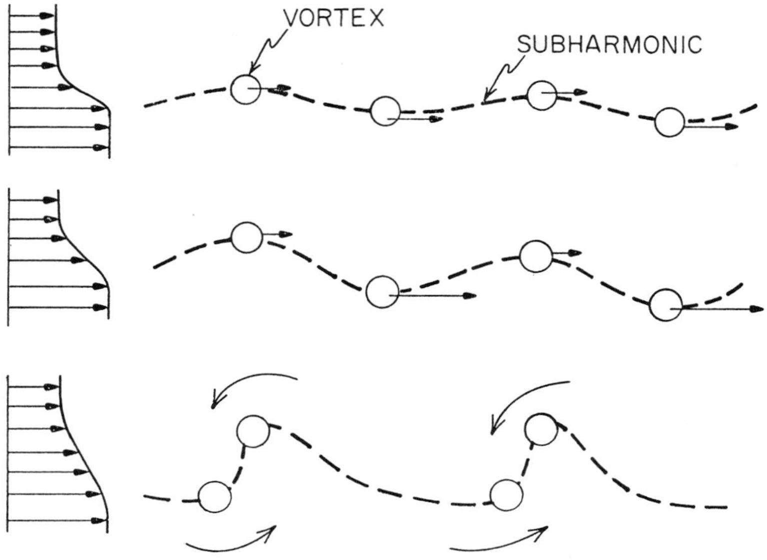

Figure 1. Subharmonic and vortex merging.

Due to the importance of the subharmonic in vortex merging, the amplification of the streamwise energy contents, $E(f)$, at the fundamental and at the subharmonic has been extensively studied [5.9]. An example is shown in Fig. 2, in which the shear layer is forced at its first subharmonic of the initial amplified frequency,

$$E(f) = \int_{-\infty}^{\infty} \frac{u'^2(f)}{2\delta U_c^2}\, dy, \tag{1}$$

and $\lambda = U_c/f_0$ is the wavelength of the fundamental waves. U_c is the mean speed of the high-speed (U_1) and the low-speed (U_2) sides of the mixing layer. f_0 is the initial instability frequency. $u'(f)$ is the rms level of the velocity fluctuations at frequency f, and δ is the maximum slope thickness. The fundamental waves ($f = 4.30$ Hz) exponentially amplify and reach a peak at $x/\lambda = 4$. The subharmonic reaches a peak at $x/\lambda = 8$.

The close relationship between the subharmonic and the vortex merging is revealed in Fig. 2 and additionally by visualization and spreading-rate measurements. From the visualization, the two merging vortices become vertically aligned at the point where the subharmonic saturates. When the spreading of the shear layer was measured, the thickness at $x/\lambda = 8$ was found to be about twice as thick as that before the vortex merging takes place. The thickness does not increase after $x/\lambda = 8$ until the next vortex merging. It is hard to pinpoint the position of vortex merging, because it is accomplished over a distance greater than several wavelengths of the initial instability waves. However, the abovementioned evidence certainly shows that the location where the subharmonic saturates can serve as a reference location for vortex

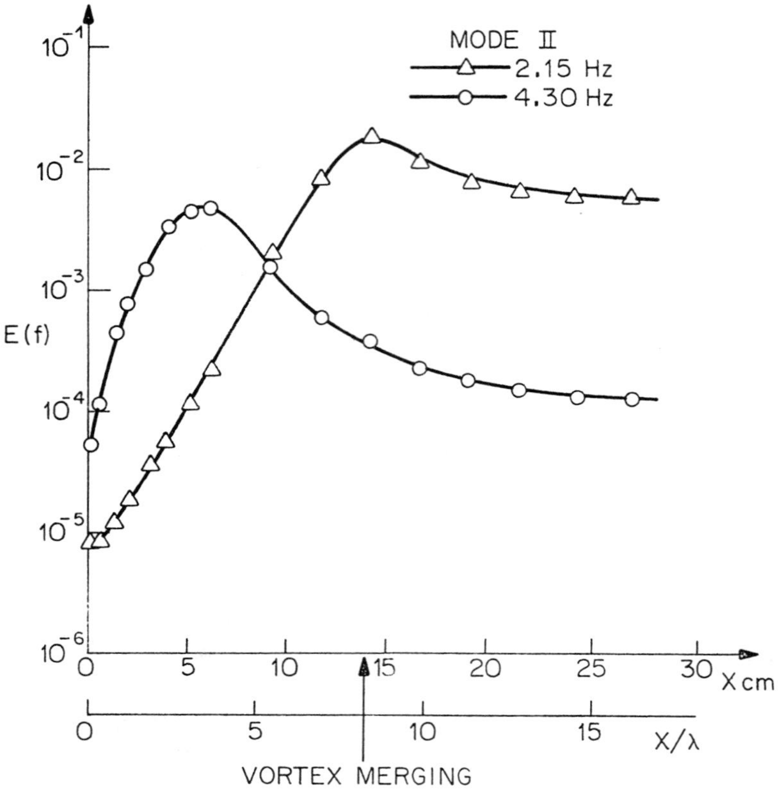

Figure 2. Amplification of fundamental and subharmonic.

merging. It is important to note that the subharmonic frequency, normalized to the thickness at its saturation position and to the mean speed U_c, is equal to the neutral stability frequency predicted by linear stability theory [5.9].

The Feedback Mechanism

Though the vortices will merge where the subharmonic wave becomes neutrally stable, the local instability alone is not enough to determine the merging. As was discussed in the introduction, the perturbation produced by vortex merging downstream will propagate upstream to force the thin shear layer near the trailing edge. New coherent structures will develop from the feedback perturbation and then merge. Hence, a feedback loop forms from the trailing edge to the merging location. The downstream- and upstream-propagating waves should satisfy the feedback equation [5.18]

$$\frac{x_M}{U_c/f_i} + \frac{x_M}{a/f_i} = N, \tag{2}$$

where x_M is the distance from the trailing edge to the ith vortex-merging position. Let x_{Mi} denote the distance between ith and $(i + 1)$th mergings:

$$x_M = x_{M0} + x_{M1} + \cdots + x_{M(i-1)}, \tag{3}$$

$$f_i = f_0/2^i. \tag{4}$$

In (2), N is an integer and is the same for all i, but it is a function of the velocity ratio, $R = (U_1 - U_2)/(U_1 + U_2)$. The quantity a is the acoustic speed. For low subsonic flow, $a \gg U_c$. Equation (2) can be simplified to

$$\frac{f_i x_M}{U_c} = \frac{f_0 x_M}{2^i U_c} = N. \tag{5}$$

From Eq. (5) we can predict the location of merging as

$$x_{M0} = x_{M1} = 2N \frac{U_c}{f_0} = 2N\lambda,$$
$$X_{Mi} = 2x_{M(i-1)} = 2^i N\lambda \quad \text{for } i \geq 2. \tag{6}$$

Kibens [5.20] forced an axisymmetric jet at very low perturbation level and could localize the vortex merging. Laufer and Monkewitz [5.19] have shown that the locations of merging satisfy the feedback equation. It is interesting to note that x_{Mi} has to be an even number of initial instability wavelengths. Furthermore, Eq. (6) requires the distance between successive mergings to double for each except the second one. The doubling in merging distance has been confirmed by experiments [5.3, 5.20].

The Subharmonic-Evolution Model

According to the discussion in the previous two sections, the model of subharmonic evolution is proposed (Fig. 3). Near the origin, the vortex develops from the initial instability wave at the local fundamental frequency; its subharmonic wave is selected by the nonlinear subharmonic resonance mechanism [5.6] and amplifies. The vortices are laterally displaced by the amplifying subharmonic. The vortex merging occurs where the first subharmonic reaches saturation. The merging location is determined by the global feedback mechanism. An important point that should be noted here is that the thickness of the shear layer doubles while the first subharmonic becomes neutrally stable. At the same location the second subharmonic becomes the locally most amplified frequency, because the ratio between the neutrally stable frequency and the most amplified frequency is about two. Hence, the *change of length scale* makes the "old" subharmonic evolve from amplifying local subharmonic to decaying local fundamental, and the "new" subharmonic starts another new cycle.

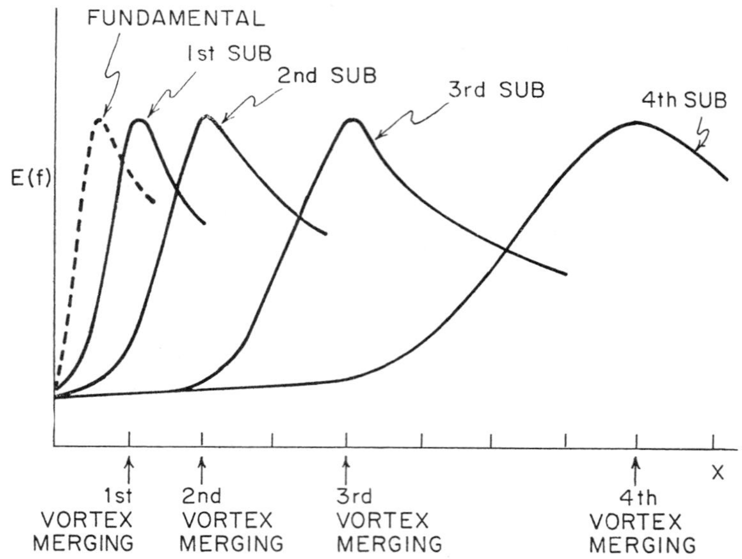

Figure 3. The subharmonic evolution model.

Discussions of the Subharmonic Evolution Model

Linear Stability Analyses and Shear Layers

The experiments used for comparison with the stability analyses can be grouped into two catagories. In one type, measurements were made close to the trailing edge and the instability waves were in the linear region. The theoretical predictions should hold. In the other type, the data were taken in the region where vortices and even vortex merging had already taken place and the flow was highly nonlinear. Agreement between the theories and these experiments not expected, but seems to be observed. A partial list of the experiments is given in Tables 1 and 2.

In Table 1, the theoretical values used for comparison are based on a tanh profile. Freymuth [5.24] examined the instability of shear flow at the exit of a round and a plane nozzle. He observed that the amplification rate followed the theoretical curve very well until the nondimensional amplification rate, $-\alpha\theta_0$, reached 0.08; then it starts to have about 15% deviation from the spatial-instability theory. Browand [5.25] studied the instability of a two-dimensional free shear layer and observed strong growth of subharmonics. Miksad [5.26] documented the amplification of all the harmonics and subharmonics in a mixing layer. Ho and Huang [5.9] studied the mixing layer at a low velocity ratio, $R = 0.31$, and found that the most amplified frequency is well predicted. Gutmark and Ho [5.21] varied the mean speed of a free jet over

Table 1 Measurements in the Linear Region.

Ref.	Velocity ratio	Strouhal No. $St = \frac{f\theta}{U_c}$ Theoretical values	Measured values	Amplification rate $-\alpha\theta$ Theoretical values	Measured values
Freymuth	1.0	0.033	0.034	0.114	0.09
Browand	1.0	0.033	0.021	0.114	0.131
Miksad	0.7	0.034	0.035	0.07	0.197
Ho & Huang	0.31	0.035	0.031	0.03	0.029
Gutmark & Ho	1.0	0.033	0.034~ 0.024	0.114	—

one order of magnitude in fine increments. The initial instability frequency varied in a wide range.

In Table 2, Crow and Champagne [5.13] forced a jet at high levels and found a preferred frequency which is much lower than the initial instability frequency. The preferred frequency, scaled with the diameter of the nozzle and the jet speed, was about 0.3; if the momentum thickness of the jet at the end of the potential is used as the length scale instead of the nozzle diameter, the nondimensional frequency is 0.1. Moore [5.27] repeated Crow and Champagne's [5.13] experiment at a much lower forcing level. However, his measurements of the stability properties were made at $x/D = 2$, where vortices have gone through several mergings already. Wygnanski and Oster [5.10] applied low-frequency forcing to a plane mixing layer; the flow went through a collective interaction [5.18], and many small vortices coalesced into a large coherent structure in a short distance. They found the passage frequency, when scaled with the local momentum thickness and the mean velocity, was close to the neutral instability frequency. Fiedler [5.11] compared the

Table 2 Measurements in the Non-Linear Region

Ref	Velocity ratio	Strouhal No. $St = \frac{f\theta}{U_c}$ Theoretical values	Measured values	Amplification rate $-\alpha\theta$ Theoretical values	Measured values
Crow & Champagne	1.0	0.033	0.1	0.114	0.092
Moore	1.0	0.033	0.07	0.114	0.08
Wygnanski & Oster	0.3	0.035	0.075	0.03	—
Fiedler	1.0	0.033	0.07	0.114	0.11
Ho & Huang	0.31	0.035	0.068~ 0.054	0.03	0.029

streamwise amplification of instability waves with the spatial instability theory. Good agreement was achieved.

It is not surprising that both the amplification rate and the passage frequency are well predicted in the linear region (Table 1), However, we need an explanation why the amplification rate is close to the linear wave amplification rate and the passage frequency is about twice the most amplified frequency in the nonlinear region. This is *not* a question of how the linear theory can be extended into the nonlinear region, because most of the experiments (Table 2) were made where vortex merging had occurred several times already. The initial instability waves and its subharmonics are highly nonlinear. It is clear that the direct application of linear theory is *incorrect*. The paradox can be explained with the subharmonic-evolution model. At downstream locations, while the passage frequency is being measured, the waves with high amplitude will contribute to most of the signal. Hence, the passage frequency is a measure of the local fundamental frequency, which equals the neutral stability frequency, and is about twice the most amplified frequency. When the amplification rate is measured, the local fundamental is either close to saturation or close to decaying. The amplification rate of the local subharmonic equals the most amplified value and represents the measured amplification rate. Therefore, the linear analyses really should not be directly applied to the nonlinear flow. The infinite new cycles of the evolving subharmonics make the linear theory become useful in the downstream region of the free shear layer.

The Phased Speed

Both the local fundamental and the local subharmonic are higher than or equal to the local most amplified frequency. The phase speeds are nondispersive and about the same as the average speed of the two streams [5.29]. Hence, the measured convection speed of the vortices is approximately equal to the mean speed.

The Distance between Vortex Mergings

The feedback mechanism requires the distance between successive mergings to double. However, the integer N in Eq. (2) has yet to be evaluated.

In many experiments with large differences in Reynolds numbers [5.9, 5.24, 5.25, 5.26, 5.28], it is always found that the amplification rates of both the fundamental and the subharmonic keep almost constant, until very close to saturation. Since vortices merge where the subharmonic wave saturates,

$$E_{si}(f) = E_{0i}(f) \exp\left[- \int_{x_{M(i-2)}}^{x_{M(i-2)} + x_{M(i-1)}} 2\alpha_i \, dx \right]. \tag{7}$$

Equation (7) can be approximated by assuming α_i to be a constant:

$$E_{si}(f) = E_{0i}(f)\exp[-2\alpha_i x_{M(i-1)}], \tag{8}$$

where $E_{si}(f)$ is the energy at saturation of the ith subharmonic. $E_{0i}(f)$ is energy level where the ith subharmonic becomes the most amplified frequency. The distance between mergings is

$$x_{M(i-1)} = -\frac{1}{2\alpha_i}\ln\frac{E_{si}(f)}{E_{0i}(f)}. \tag{9}$$

From Eq. (6) and Eq. (9),

$$\frac{\alpha_i}{\alpha_{i+1}} = 2 = \frac{x_{M(i+1)}}{x_{Mi}} \quad \text{for } i > 2 \tag{10}$$

and

$$\frac{\alpha_i}{\alpha_0} = \frac{1}{2^{i-1}} \quad \text{for } i > 2; \tag{11}$$

thus

$$x_{M(i-1)} = \frac{-1}{2^{2-i}\alpha_0}\ln\frac{E_{si}(f)}{E_{0i}(f)} \quad \text{for } i > 2. \tag{12}$$

The amplification rate at the most amplified frequency has been shown to be an almost linear function of R [5.29]. Therefore, the distances between mergings in a high-velocity-ratio flow is shorter than in a low-velocity-ratio flow. In other words, the value of N should be a function of the velocity ratio. In a mixing layer with $R = 0.31$ [5.9], the value of x_{M0} is almost equal to eight initial instability wavelengths, i.e. $N = 4$. In a jet, $R = 1$ [5.19, 5.21] and x_{M0} was found to be about equal to four initial instability wavelengths, i.e. $N = 2$.

The Effect of the Initial Condition

Near the origin of the shear layer, the instability wave amplifies and forms vortices at the initial instability frequency. After this frequency is selected, the only mechanism that can change the frequency is vortex merging. However, the thickness of the shear layer grows before the first vortex merging. Hence, the Strouhal number, $\mathrm{St} = f\theta/U_c$ (Fig. 4) increases from the origin of the shear layer to the location of the first merging. From this point on, the frequency will drop whenever vortex merging occurs. At the same time, the thickness of the shear layer doubles. Hence

$$f\theta = f_1\theta_1 \tag{13}$$

and

$$f_1 = f_0. \tag{14}$$

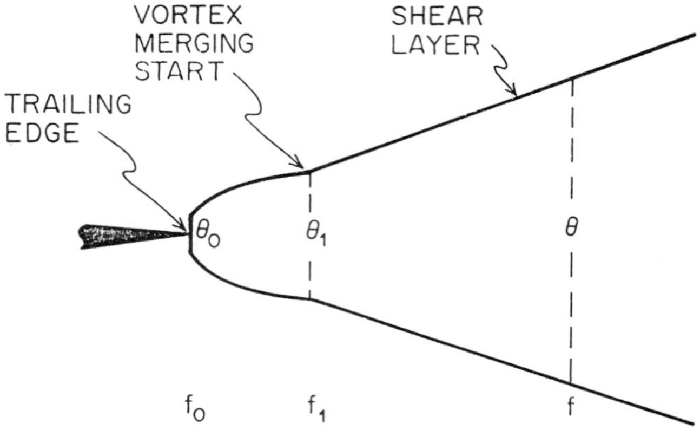

Figure 4. The initial development of a free shear layer.

The Strouhal number at any downstream location is related to the initial Strouhal number as

$$\begin{aligned}\mathrm{St} &= \frac{f\theta}{U_c} = \frac{f_0\theta_0}{U_c}\frac{f\theta}{f_0\theta_0} \\ &= \mathrm{St}_0\frac{\theta_1}{\theta_0}\frac{f\theta}{f_1\theta_1}\frac{f_1}{f_0} \\ &= \mathrm{St}_0\frac{\theta_1}{\theta_0}.\end{aligned} \tag{15}$$

The experimental values of θ_1/θ_0 are about two to three [5.9]. So the Strouhal number downstream equals the neutral stability frequency and is consistent both with the experiments (Table 2) and the subharmonic-evolution model. Furthermore, the influence of the initial condition can extend all the way to the downstream stations through the relationship of Strouhal numbers (Eq. (15)).

The Spreading Rate

The growth of the free shear layer is the result of vortex merging. Although more than two vortices involved in one merging have been observed in unforced mixing layers, the majority of vortex merging only involves two vortices. It is a reasonably good assumption to introduce a spreading model based upon pairing of two vortices. The thickness of the shear layer before the first pairing is assumed to be twice as thick as the initial thickness, i.e., $\theta_1/\theta_0 = 2$. The locations of merging follow the subharmonic evolution in model I (Fig. 5). The asymptotic spreading rate then can be determined as

$$\frac{d\theta}{dx} = \frac{4\theta_0}{x_{M1}} = \frac{8\theta_0}{x_{M2}}, \tag{16}$$

$$\frac{d\theta}{dx} = \frac{-4\alpha_0\theta_0}{\ln[E_{si}(f)/E_{0i}(f)]}. \tag{17}$$

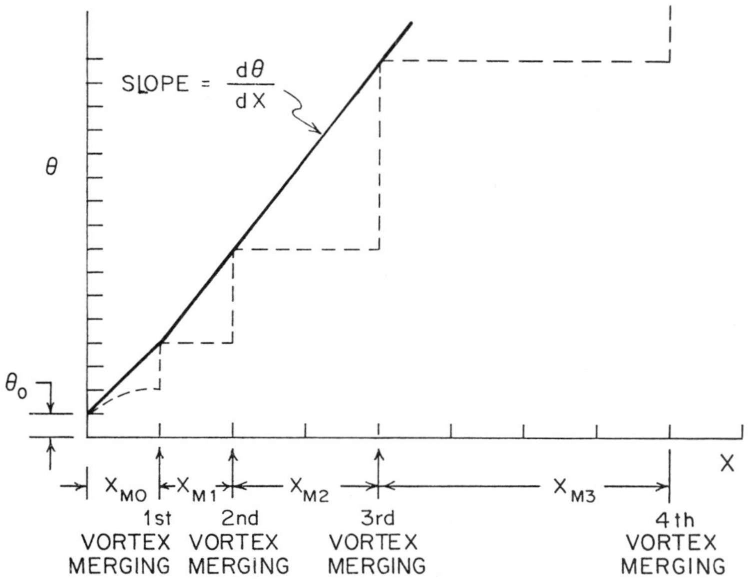

Figure 5. The spreading of a free shear layer.

Therefore, the spreading rate is a linear function of the amplification rate, which increases almost linearly with R [5.29]. Hence the rate increases with the velocity ratio. The spreading rate can also be obtained from Eq. (6) and Eq. (16):

$$\frac{d\theta}{dx} = \frac{2\theta_0}{NU_c/f_0} = \frac{2}{N}\frac{f_0\,\theta_0}{U_c} = \frac{2}{N}\,\mathrm{St}_0. \tag{18}$$

According to Monkewitz and Huerre [5.29], the most amplified frequency is not sensitive to the velocity ratios, and $f_0\,\theta_0/U_c \approx 0.034$. The initial instability frequency is always equal to or lower than the most amplified frequency [5.9, 5.21]. Hence, the upper limit of the spreading rate of a free shear layer has a very simple expression:

$$\frac{d\theta}{dx} = \frac{0.068}{N}. \tag{19}$$

Since N decreases with increasing R, the spreading rate also increases with R and is consistent with the experiments [5.3, 5.10]. Assuming the mean velocity has a tanh profile, the upper limits of spreading rates for a jet and for a mixing layer with $R = 0.31$ were calculated and are shown in Fig. 6. The results agreed well with the measured values [5.3, 5.10].

Brown and Roshko [5.3] noticed that the scatter of the spreading rate near $R = 1$ is much larger than at other velocity ratios. One possible cause of this is the amplification rate. The amplification rate is sensitive to the mean velocity for large velocity ratios, but not for small velocity ratios [5.29]. According to Eq. (17), the large scatter of the spreading rate at high velocity ratios could be due to slight differences in mean velocity profiles. Wygnanski and Oster [5.10]

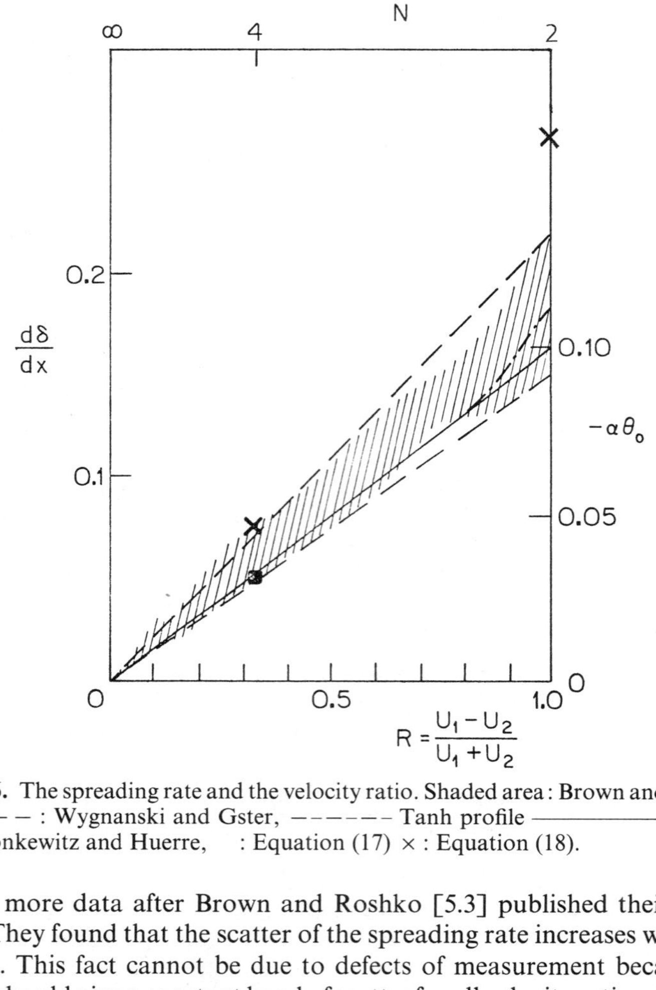

Figure 6. The spreading rate and the velocity ratio. Shaded area: Brown and Roshko, – – – – – – – : Wygnanski and Gster, ------ Tanh profile ————, Blasius profile Monkewitz and Huerre, : Equation (17) × : Equation (18).

collected more data after Brown and Roshko [5.3] published their results (Fig. 6). They found that the scatter of the spreading rate increases with velocity ratio. This fact cannot be due to defects of measurement because this problem should give a constant band of scatter for all velocity ratios. Actually, Eq. (18) indicates that the large scatter in spreading rate is a physical phenomenon. In a wide velocity range, but at constant R, the initial instability frequency will not always be the most amplified frequency [5.21]. The values of St_0 have about 40% variation from the most amplified frequency. In a forced mixing layer ($R = 0.31$) about the same amount of variation was found [5.9]. According to Eq. (18), the same amount of scatter will appear at all velocity ratios, due to the variation of St_0.

The Flight Effect on Jet Noise

The jet-noise pattern of an aircraft in flight could not be predicted from a static test until Michalke and Michel [5.30] introduced the stretching concept. The stretching constant is equal to $1/R$ and is consistent with the variation of the maximum amplification rate with R [5.29]. In a shear layer, the vortex merging can be related to the velocity ratio (Eqs. (17) and (18)). The distance between vortex mergings is proportional to $1/R$. If we further assume that vortex merging is the main noise source [5.31], the subharmonic-evolution model can also be used to interpret the flight effect on noise generation.

Conclusion

Much experimental evidence indicates that the dynamics of a free shear layer is governed by both the *local instability process* and the *global feedback mechanism.* A new concept, *subharmonic evolution,* is based upon the two mechanisms to model the coherent structures and can provide clues to many observed phenomena in free shear layers.

The idea of regeneration of subharmonics provides a rational ground to explain the paradox that linear stability analyses can describe the nonlinear free shear flow. The thickness of the shear layer doubles and the frequency halves where vortex merging occurs. The product of the frequency and the thickness is always a constant after the first merging. This fact indicates that the initial condition has long-lasting effects downstream. The changes of the local length scale modify the local amplification rate, which requires that the distance between successive mergings double. Simple formulas derived from the subharmonic-evolution model can not only provide the absolute magnitude of the spreading rate at different velocity ratios, but also offer physical explanations for the scatter of the measured spreading rate.

Acknowledgment

The author would like to express his gratitude to Professor P. Huerre for helpful discussions and to Professor J. Laufer for pointing out an error in the original version.

This work is supported by the Office of Naval Research, Contract No. N00014-77-C-0315.

CHAPTER 31

Laser Velocimetry Applied to Transonic Flow Past Airfoils

Dennis A. Johnson*

Introduction

The performance of lifting surfaces at transonic conditions is of practical importance to both fixed-wing aircraft and rotorcraft technology. In view of this, considerable effort has been expended in developing numerical prediction techniques applicable to airfoils and wings operating at transonic conditions. As a result, accurate numerical solutions are now possible for transonic lifting surfaces, provided the coupling between the inviscid and viscous flows is weak. The accuracies of the predictions, however, deteriorate as the shock-wave–turbulent-boundary-layer interaction on the suction side of the lifting surface strengths with either increasing Mach number or increasing angle of attack. For cases where this interaction is of such strength that the behavior of the turbulent boundary drastically affects the resultant surface pressure distribution, the numerical predictions have been in strong disagreement with experiment.

The measured upper-surface pressure distributions presented in Fig. 1 for an NACA 64A010 section at $M_\infty = 0.8$ demonstrate the significant role viscous effects can have at transonic conditions. The surface static pressure is plotted in terms of the pressure coefficient C_p versus chordwise position x/c. At small angles of attack α, the change in shock position with increasing angle of attack follows the trend that would be predicted by inviscid theory, that is, the shock moves aft. However, at some angle of attack greater than

* Ames Research Center, NASA, Moffett Field, CA 94035.

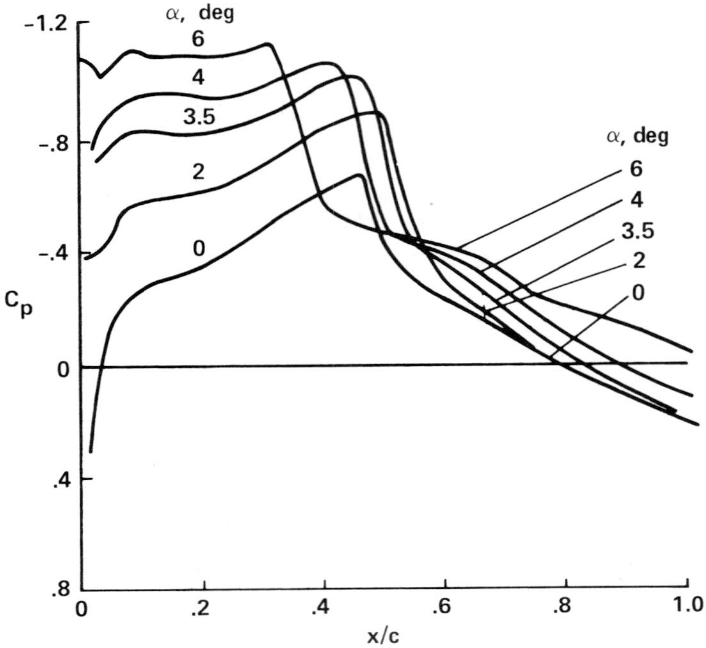

Figure 1. Measured upper surface pressure distributions for NACA 64A010 airfoil section for $M_\infty = 0.8$.

2°, the shock starts to move upstream with increasing α due to the strengthening of the shock-wave–turbulent-boundary-layer interaction. At angles of attack larger than 3.5°, the behavior of the turbulent boundary layer not only affects the shock position as seen in Fig. 1, but determines the surface static pressure levels from downstream of the shock wave to the trailing edge. The increase in trailing-edge C_p for $\alpha > 3.5°$ is a direct result of the inability of the turbulent boundary layer, once it has been severely perturbed by the shock, to support even a mild adverse pressure gradient downstream of the shock. For $\alpha \geq 3.5°$, the shock wave was sufficiently strong to produce separation at the foot of the shock wave.

Surface pressure distributions on advanced supercritical sections can be even more sensitive to the behavior of the turbulent boundary layer. Due to their relatively flat upper surfaces, the shock location is not nearly as constrained by surface geometry as for conventional sections, and the pronounced aft cambering results in a strong adverse pressure gradient near the trailing edge. Thus, there are two regions of strong inviscid–viscous interaction (at the shock and at the trailing edge) which are strongly coupled to each other. Consequently, large perturbations in the surface pressure distribution can occur due to viscous effects, even when the shock wave is not sufficiently strong to cause separation.

For these flow situations, the accurate prediction of the turbulent boundary-layer behavior is imperative. However, the available models for the turbulent Reynolds stresses appear inadequate in describing the turbulent boundary-layer behavior for strong inviscid–viscous interactions. This even includes the more sophisticated two-equation turbulence model formulations that have recently become popular. Improvements in the prediction capabilities for lifting surfaces at transonic conditions will require input from experiment. However, this input must include information on the turbulent Reynolds stresses, since the modeling of these stresses is fundamental to the problem.

Until recently, turbulent Reynolds-stress data at transonic conditions were nonexistent because an appropriate measurement technique did not exist for the transonic regime. This dilemma no longer exists with the advent of the laser velocimeter, which is capable of providing accurate measurements even in separated-flow regions. Over the past several years, for instance, this method of measurement has been successfully applied to several transonic flows of practical importance in the Aerodynamic Research Branch at Ames Research Center. These studies, and those by other researchers using similar laser-velocimeter techniques, are beginning to provide the type of data base needed to develop turbulent-flow models capable of treating the type of inviscid–viscous interactions that occur on lifting surfaces at transonic conditions.

The applications to date in the Aerodynamic Research Branch include an NACA 64A010 airfoil section [5.32, 5.33], an axisymmetric "stovepipe" model with a circular-arc section [5.34], and, most recently, the DSMA 671 supercritical airfoil section (to be published). Related to the first two applications, numerical calculations have also been performed [5.35, 5.36]. Most of the experimental results obtained to date for the NACA 64A010 airfoil section and the axisymmetric bump are for cases of extensive separation and represent an extreme test for any numerical prediction method. In the supercritical airfoil study, emphasis was placed on near-design conditions. In the present paper, some of the experimental and numerical results from these studies are examined as they relate to the goal of improved turbulent-flow modeling. First, a brief description of the laser-velocimeter technique employed in these studies is appropriate.

Laser-Velocimeter Technique

A laser-velocimeter system developed for testing in the Ames 2- by 2-Foot Transonic Wind Tunnel was used in the previously mentioned experiments. It is a dual-color system utilizing the 4880- and 5145-Å lines of an argon-ion laser. One spectral line is used to measure the streamwise velocity component u; the other to measure the vertical velocity component v. The "fringe" or "dual-scattering" principle is used wherein two parallel laser beams of equal intensity are brought to a common focus with a lens to produce a set of interference fringes. These planar fringes are equally spaced at a distance

$X_f = \lambda/(2 \sin \frac{1}{2}\theta)$, where λ is the laser wavelength and θ is the angle between the two beams. As a particle, small with respect to the fringe spacing, passes through the beam crossover point, light is scattered by the particle in proportion to the light incident upon it. A part of this scattered light is collected with a lens and imaged onto a photodetector. The output of the detector has a period τ which is related to the velocity component perpendicular to the fringes, U, simply by the expression $\tau = X_f/U$.

In the actual implementation of the "fringe" principle, a shift in frequency in one of the parallel laser beams is introduced with a Bragg cell to produce traveling fringes. The fringes travel at a speed V_f given by $V_f = X_f f_0$, where f_0 is the frequency shift. With this frequency offset, forward- and backward-traveling particles can be distinguished. The relationship between signal period and velocity becomes $\tau = (U/X_f + f_0)^{-1}$. Obviously, this capability is absolutely necessary in regions of turbulent separation. However, where counter-type signal processors are used to measure τ, moving fringes are necessary even for regions of only moderately high turbulence because these signal processors generally require at least 10 fringe crossings by the particle to affect a period measurement. With stationary fringes there is a high likelihood that this will not be achieved if the turbulence level is high, in which case the measurement system would collect a biased sample. The moving fringes produced by the frequency offset ensure a sufficient number of fringe crossings, independent of particle trajectory.

For the particle-size range of interest ($0.3\ \mu\text{m} \leq d \leq 1\ \mu\text{m}$), a significantly greater portion of the light is scattered at or near the direction of propagation of the incident laser beams than in other directions. Thus, to maximize the amount of particle-scattered light collected, the focusing lens and the light-collection lens are mounted on opposite sides of the test section. In this configuration, the system is sufficiently sensitive to operate in the Ames 2- by 2-Foot Wind Tunnel without aritificially introducing particles for light scattering. Lubrication oil within the drive system vaporizes and later condenses in the tunnel circuit to provide a generous supply of scattering centers. Measurements across a normal shock have shown that these particles are small enough in size (estimated to be $< 1\ \mu\text{m}$) to give very good response to a step change in velocity at sonic speeds. The effective sensing volume is approximately a cylinder 300 μm in diameter and 3 mm long, whose axis is aligned with the cross-stream direction.

Signal processing is accomplished with single-particle burst counters, and the individual realizations from the two channels are simultaneously recorded with a digital computer. Simultaneity in the two velocity measurements allows the velocity correlation, $\overline{u'v'}$, to be obtained in a straightforward manner by multiplying and averaging. In addition, it makes the correction of the results for velocity biasing possible. This biasing is due to particle arrival-rate dependence on the instantaneous velocity [5.37]. Each individual velocity realization is weighted by the value of $\sqrt{u^2 + v^2}$ at the time of the measurement to account for this effect. The cross-stream velocity component is not needed in this correction, due to the cylindrical shape of the probe volume. Further details of the laser-velocimeter system can be found in [5.32].

Transonic Applications

The study of [5.32] was the first attempt to measure turbulent shear stresses with the laser-velocimeter system previously described. The model was a 6-in.-chord NACA 64A010 airfoil section which spanned the entire test section. The tests were conducted at a freestream Mach number of 0.8 and a Reynolds number based on chord of 2×10^6. To ensure that the boundary layer was fully turbulent, transition strips were affixed to the airfoil section at the 17%-chord location on the upper and lower surfaces.

A high angle of attack ($\alpha = 6.2°$) was selected as the test condition of primary interest on the basis of: (a) the thick viscous layer present, which facilitated detailed probing, and (b) the inability of current numerical methods to predict even roughly a flow with such a strong inviscid–viscous interaction. An infinite-fringe interferogram obtained at this test condition and shown in Fig. 2 dramatically illustrates the thick viscous layer developed downstream of the shock. Also shown in Fig. 2 is an infinite-fringe interferogram taken at $\alpha = 0°$, which represents a case where the turbulent boundary layer remains thin and has a negligible influence on the resultant surface pressure distribution. The mean streamwise velocities $\bar{u}$ and the velocity correlations $\overline{u'v'}$, measured with the laser velocimeter at the $\alpha = 6.2°$ condition, are shown in Figs. 3 and 4. The quantity $\overline{u'v'}/u_\infty^2$ plotted in Fig. 4 closely approximates the dimensionless shear stress $\bar{\rho}\,\overline{u'v'}/\rho_\infty u_\infty^2$, since $\bar{\rho}/\rho_\infty$ does not vary significantly from unity at transonic conditions. As confirmed from oil-flow visualizations, the turbulent boundary layer at this test condition separates at the foot of the shock ($x/c = 0.37$) and remains separated to the trailing edge. The data set also included the turbulence intensities in the streamwise and vertical directions and the mean velocity in the vertical direction. These data and data obtained for the other test conditions are presented in [5.32] and [5.33].

Noteworthy from the mean velocity profiles are: (a) the relatively small reversed velocities observed in the separated-flow region, (b) the free-shear-layer shape of the profiles at the stations along the wing's surface, (c) the rapid wake closure at the trailing edge, and (d) the upward displacement of the minimum velocity location in the near wake relative to the airfoil's trailing edge. Both the mean-velocity and the turbulence-transport-property measurements indicate a fully developed free-shear-layer-type flow from $x/c = 0.67$ to the trailing edge. In this region, the mean velocity profiles can be fitted quite well with Coles's [5.38] theoretical profile, assuming the skin-friction coefficient C_f is equal to zero or very nearly so. Scaled to the shear-layer-edge velocity, the maximum turbulent shear stress is nearly constant for the three measurement stations above the airfoil's surface. In

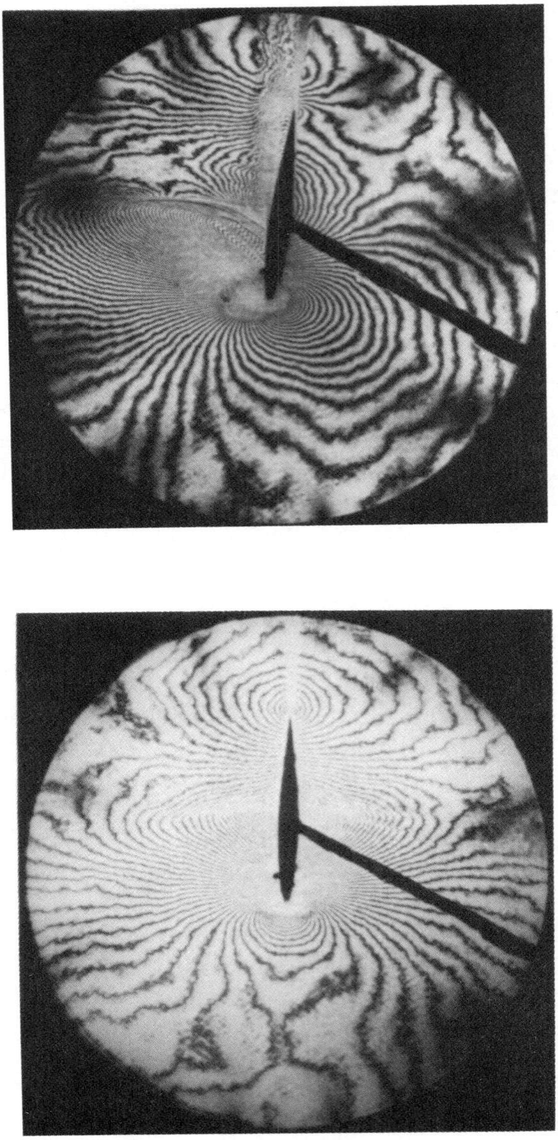

$\alpha = 6.2^\circ$ $\alpha = 0^\circ$

Figure 2. Infinite-fringe interferograms for NACA 64A010 airfoil section for $M_\infty = 0.8$.

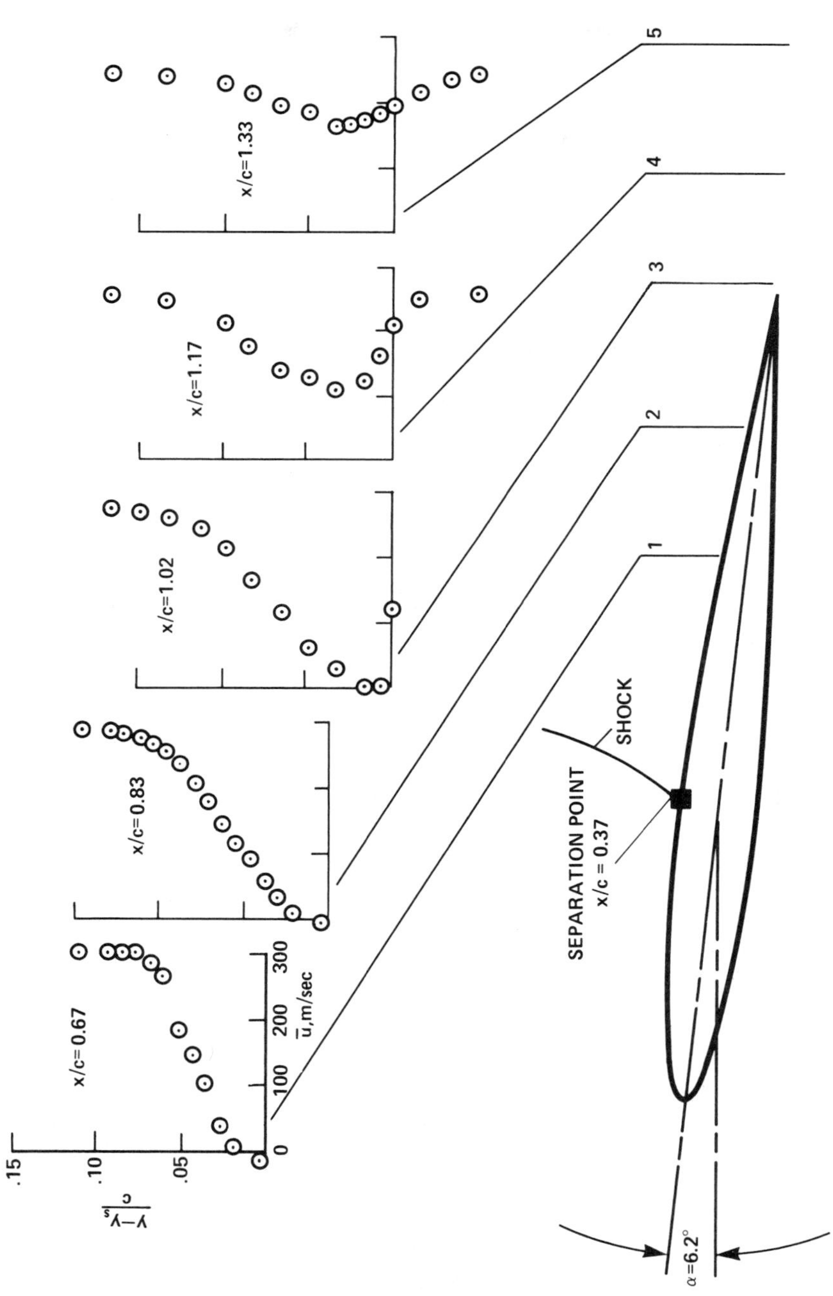

Figure 3. Mean velocity profiles for NACA 64A010 airfoil section: $M_\infty = 0.8$; $\alpha = 6.2°$.

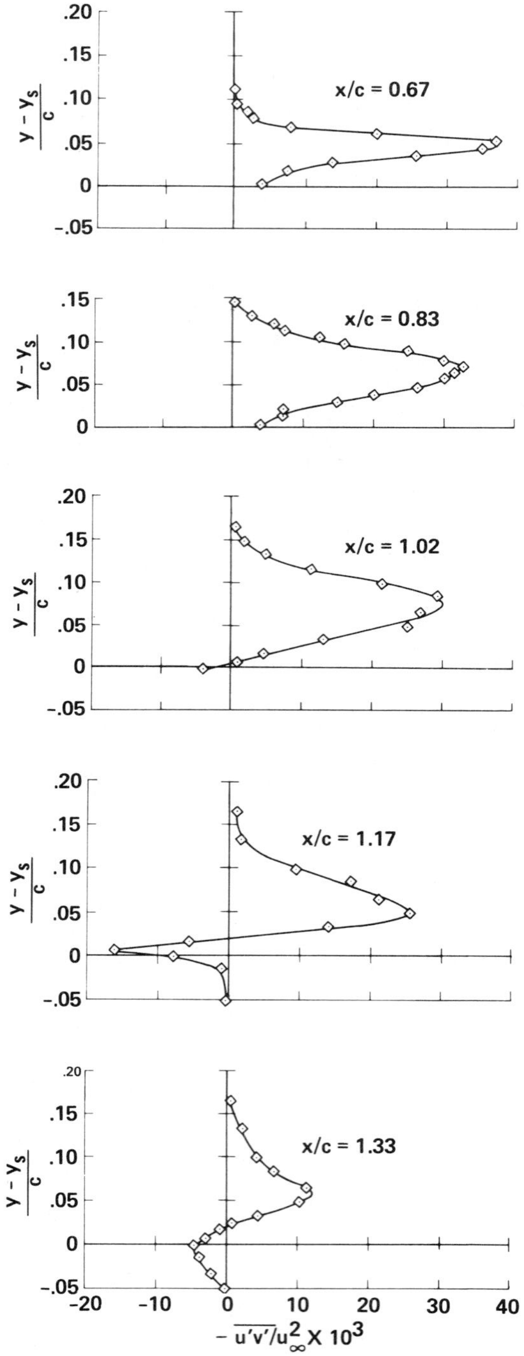

Figure 4. Velocity correlation distributions for NACA 64A010 airfoil section: $M_\infty = 0.8$; $\alpha = 6.2°$.

the central portion of the shear layer, Prandtl's mixing length l, scaled to δ, is approximately 0.1. The value of $l/\delta = 0.1$ similarly has been observed by Spencer and Jones [5.39] in a low-speed shear-layer experiment. In [5.33], measurements for $\alpha = 8°$ and for a leading-edge separation showed similar free-shear-layer characteristics. In the wake, the character of the flow quickly changes, with the turbulence fluctuations becoming more isotropic; also, Prandtl's mixing length increases rapidly.

These data do not imply that a simple algebraic turbulence model is adequate for computing this flow field. Together with the poor numerical results obtained with mixing-length models, they do imply that the inadequacies in these models lie mainly in the region near the shock, where the boundary layer is rapidly changing in character. Evidently, downstream of separation a fully developed shear layer is formed whose character can be described using local properties only. Note that the boundary layer is extremely thin just upstream of the shock ($\delta \approx 0.005c$), so the measurements in Figs. 3 and 4 are many initial boundary-layer thicknesses downstream of separation.

A major problem in experimentally documenting the turbulent boundary-layer flow on an airfoil is the extreme thinness of the boundary layer, particularly for the size airfoils that can be tested in small research facilities such as the Ames 2- by 2-Foot Transonic Wind Tunnel. For example, in the case just discussed the boundary layer is estimated to be only about 0.7 mm thick just upstream of the shock. Assuming that measurements to within at least $y/\delta = 0.1$ of the wall are needed, the velocimeter would have to be capable of measurements within 0.07 mm of the wall. This capability has not yet been attained. With this problem in mind, a flow configuration was developed [5.34] which would have a relatively thick turbulent boundary layer and yet which would simulate the type of shock-wave–turbulent-boundary-layer interactions that occur on airfoil sections. This model consists of a hollow cylinder ($\approx$15 cm in diameter) aligned with the flow direction. To attain an interaction similar to that observed on an airfoil, an axisymmetric "bump" is installed on this cylinder (Fig. 5). In the study of [5.34], this bump was a circular arc approximately 20 cm long and 19 mm thick. The bump is removable, so that different shapes can be investigated by simply replacing the existing bump with one of the desired shape. The cylinder extends well ahead of the bump to allow natural transition to a relatively thick turbulent boundary layer, $\delta \approx 1$ cm approaching the bump.

This model has several advantages over two-dimensional full-span models, both from a fluid-mechanics and from an instrumentation standpoint. Unlike the case of a plane two-dimensional model that spans the test section and thus interacts with the tunnel boundary layers, the shock on the present model terminates before reaching the tunnel wall. Furthermore, the shock strength decreases as the inverse square of the distance from the model, rather than inversely with the distance as for a plane two-dimensional model. Thus, a greater range of freestream Mach numbers (to $M_\infty = 0.9$) can be used without noticeable tunnel blockage effects. Surface flow visualization

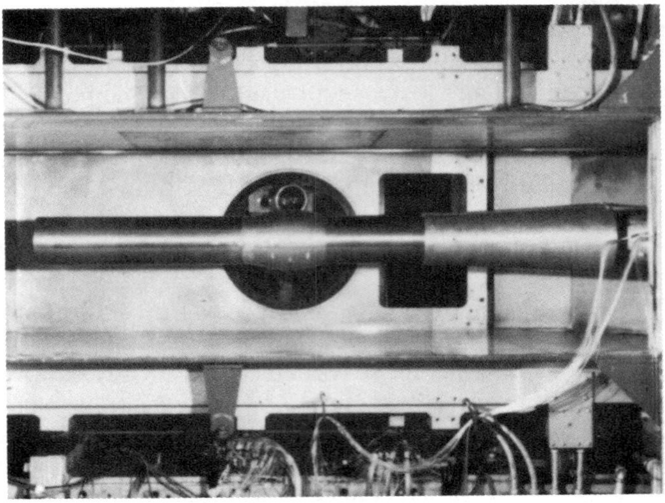

Figure 5. Photograph of axisymmetric "bump" model in Ames 2- by 2-Foot Transonic Wind Tunnel.

studies carried out for several conditions revealed that good two-dimensional flow could be realized even when extensive separation was present. For the implementation of the laser velocimeter, the model had further advantages. With the curved surfaces, diffuse reflection of the laser beams from the model surface was reduced, allowing measurements to be made very close to the wall (to within 0.2 mm).

Extensive laser-velocimeter measurements [5.34] have been acquired on this model for a freestream Mach number of $M_\infty = 0.875$ and a unit Reynolds number of $\mathrm{Re}/m = 13.6 \times 10^6/m$. At these conditions, the shock that occurred at a downstream location of $x/c = 0.66$ was not, in itself, strong enough to produce separation. However, the combination of the perturbation introduced into the boundary layer by the shock and the trailing-edge gradient caused the flow to separate downstream of the shock at $x/c \approx 0.70$. Reattachment occurred downstream of the bump at $x/c \approx 1.1$. The measured surface pressure distribution and the displacement-thickness distribution as determined from the laser-velocimeter measurements for this test condition are shown in Fig. 6. Included in Fig. 6 are comparisons with solutions of the Reynolds-averaged, time-dependent, compressible Navier–Stokes equations (discussed later in the paper).

Note from Fig. 7 that the viscous layer at the trailing edge is nearly as thick as that for the NACA 64A010 at $\alpha = 6.2°$. Also, the turbulent-shear-stress levels and the shape of the mean velocity profiles are similar at this station, as seen in Fig. 7. However, larger reversed-flow velocities were observed for the bump model. For the axisymmetric bump model, a value of $l/\delta \approx 0.1$ across the central portion of the boundary layer was only observed upstream of the bump at $x/c = -0.25$, where the boundary layer was in

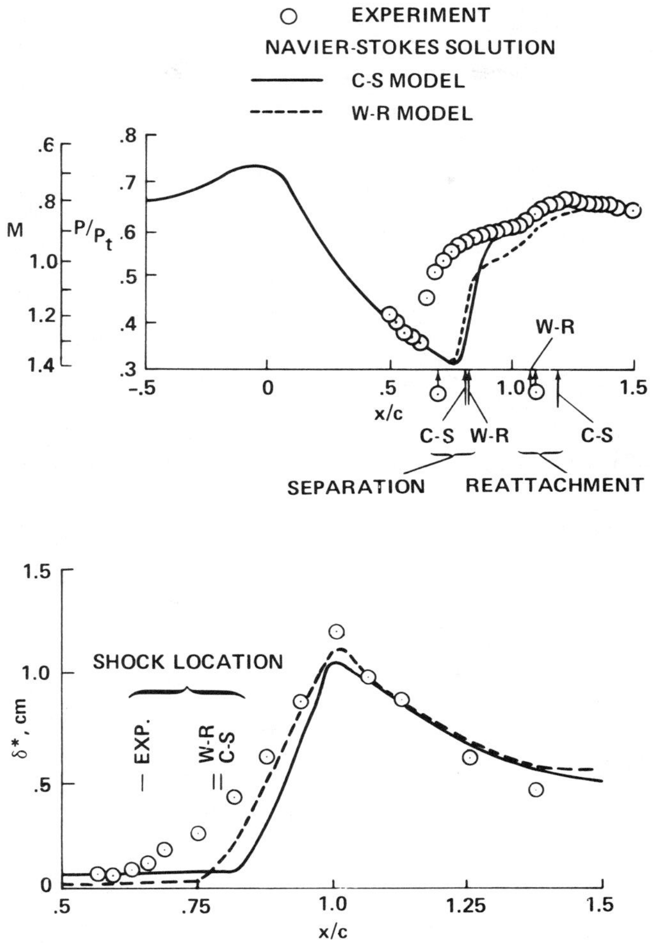

Figure 6. Surface pressure and displacement thickness distributions for axisymmetric "bump" model: $M_\infty = 0.875$; $Re/m = 13 \times 10^6$.

near-equilibrium, and at the trailing edge, where the turbulent shear stress attained its maximum value. Between the shock and the trailing edge there was a trend toward smaller mixing lengths due to the strain rate increasing more rapidly than the increase in turbulent shear stress. Downstream of the trailing edge, where the maximum turbulent shear stress was monotonically decreasing, the mixing lengths grew as the strain rate decayed more rapidly. The situation was not the same as observed for the airfoil section, where l/δ remained nearly constant between measurement stations in the region of the shear layer where the turbulent shear stress was a maximum. Recall that separation occurs much later for the bump model than for the 64A010 airfoil section ($x/c = 0.70$ vs. $x/c = 0.37$) and that the boundary-layer thickness upstream of the shock is much thicker in the bump experiment.

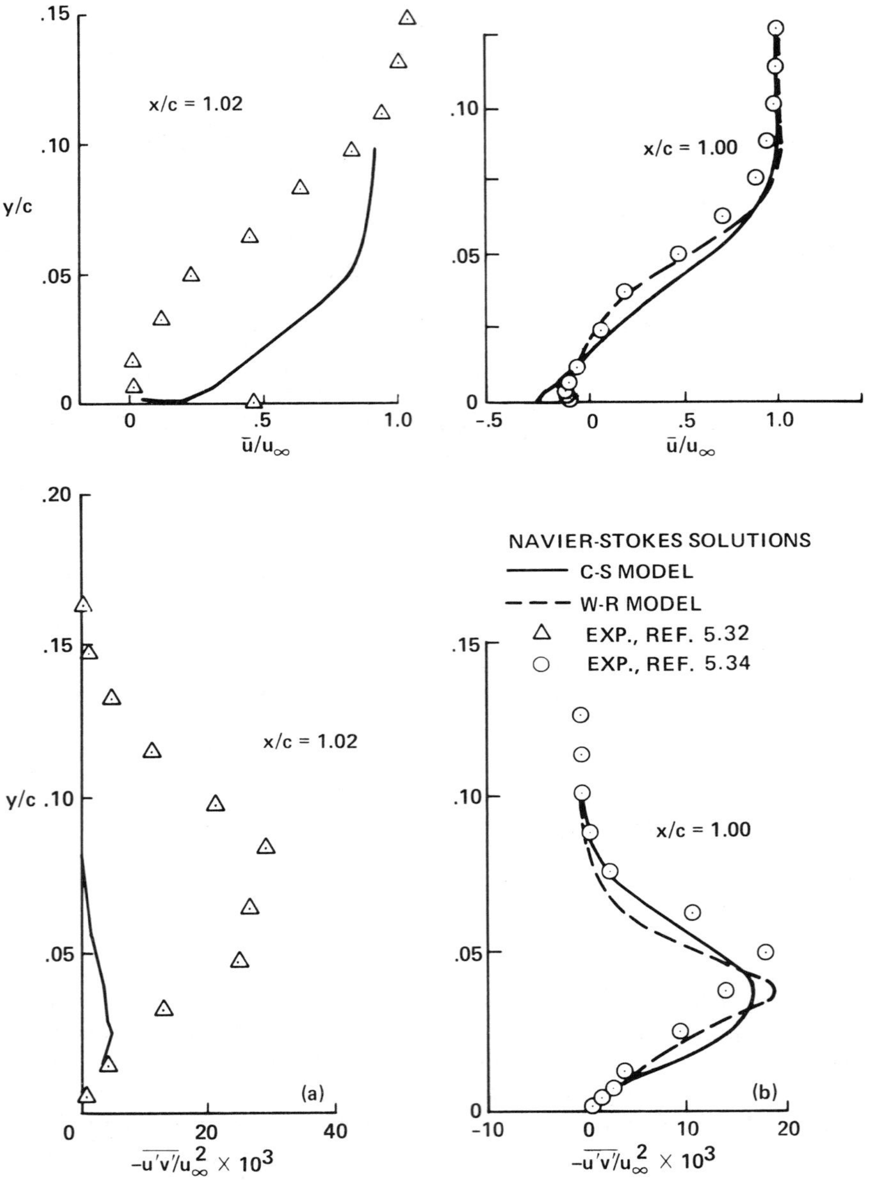

Figure 7. Boundary layer properties at trailing edges: (a) NACA 64A010, $M_\infty = 0.8$ and $\alpha = 6.2°$; (b) axisymmetric "bump" model, $M_\infty = 0.875$.

Thus, it is not surprising that the development of a fully developed shear layer was not observed for the bump model.

In [5.35] solutions to the Reynolds-averaged, time-dependent, compressible Navier–Stokes equations were obtained for this flow case. Numerical solutions were realized for two turbulence models: the Cebeci–Smith (C–S) eddy-viscosity model and the Wilcox–Rusesin (W–R) two-equation kinetic-energy model. As anticipated, measurements and predictions agreed quite well upstream of the shock curve. However, both turbulence models predicted the shock location approximately 10% chord too far downstream (Fig. 6). Somewhat fortuitously, both models did predict the measured boundary-layer properties reasonably well near the trailing edge of the model, as seen in Figs. 6 and 7.

Similar solutions [5.36] for the flow field with the NACA 64A010 section at $\alpha = 6.2°$ and $M_\infty = 0.8$ disagree even more with experiment. Results from these calculations for the Cebeci–Smith algebraic model are included in Figs. 7 and 8. Effects of the tunnel walls had to be taken into account in these calculations. This was done using the measured static pressure field one chord above and below the airfoil as boundary conditions for the calculations. The qualitative differences in the calculations for the axisymmetric "bump"

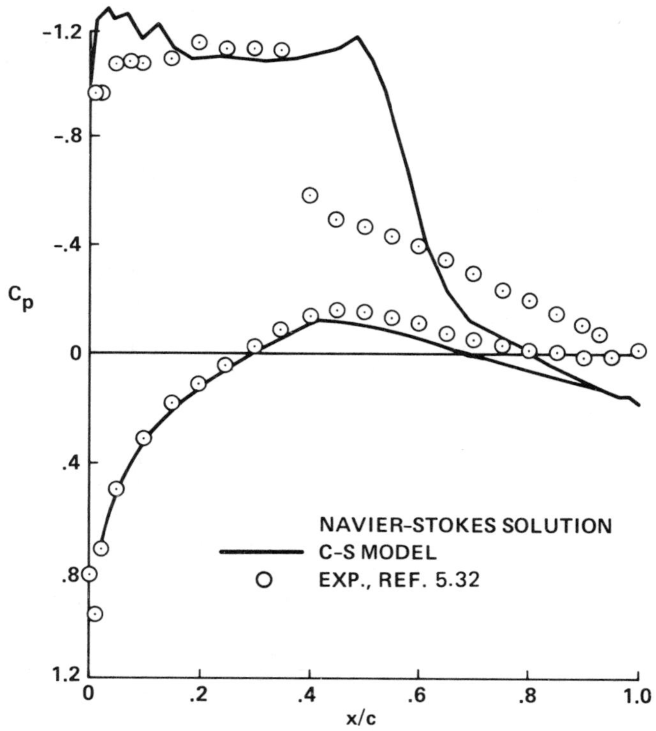

Figure 8. Surface-pressure comparison between experiment and Navier–Stokes solution with pressure boundary condition: NACA 64A010, $M_\infty = 0.8$ and $\alpha = 6.2°$.

model and the NACA 64A010 may be due to differences in the inviscid flows for these two models. For the circular-arc model the shock becomes considerably stronger as it moves aft on the model, whereas for the NACA 64A010 section at $\alpha = 6.2°$, the shock does not. Apparently, due to the stronger shock calculated for the circular-arc section, an extensive region of separated flow is predicted, whereas, for the 64A010 section, only a small bubble is predicted.

One of the difficulties in analyzing the differences in the predicted and measured turbulent flow behavior is that the experimental and calculated boundary layers are experiencing different inviscid flows due to the misprediction of the shock location. In this regard, there is an advantage in using the inverse boundary-layer method for evaluating turbulent-flow models. With this method, the shock location and strength can be set to agree with the experiment by using the experimentally determined surface static pressures as input. Moreover, the required computational time for a solution can be reduced by orders of magnitude, which makes the evaluation of a large number of turbulence models more practical. The applicability of the inverse boundary-layer method to a separated flow as extensive as that for the axisymmetric "bump" experiment was verified in [5.35]. For all practical purposes, this method was able to reproduce the same results obtained from the solution of the Reynolds-averaged, time-dependent Navier–Stokes equations with the N–S-predicted skin-friction distribution as input (Fig. 9).

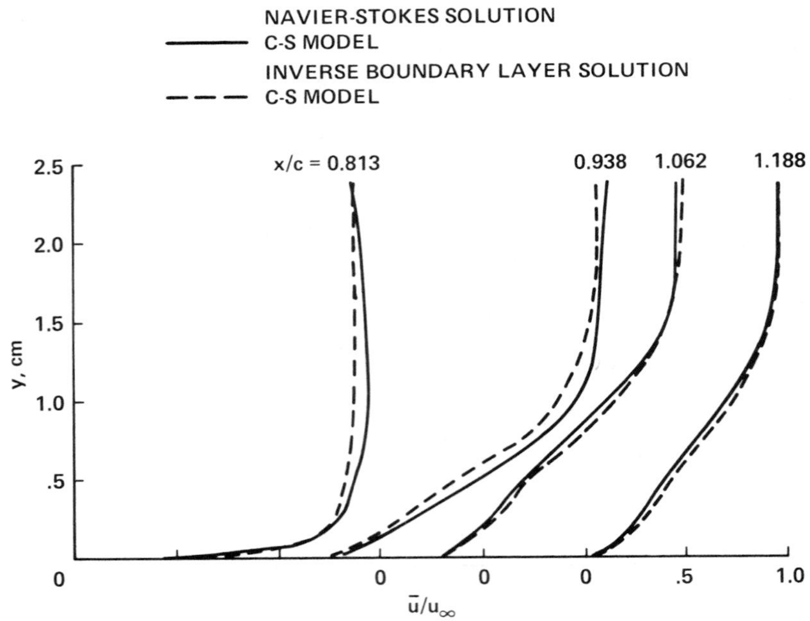

Figure 9. Comparison of inverse boundary-layer solution with Navier–Stokes solution for axisymmetric "bump" model with $M_\infty = 0.875$.

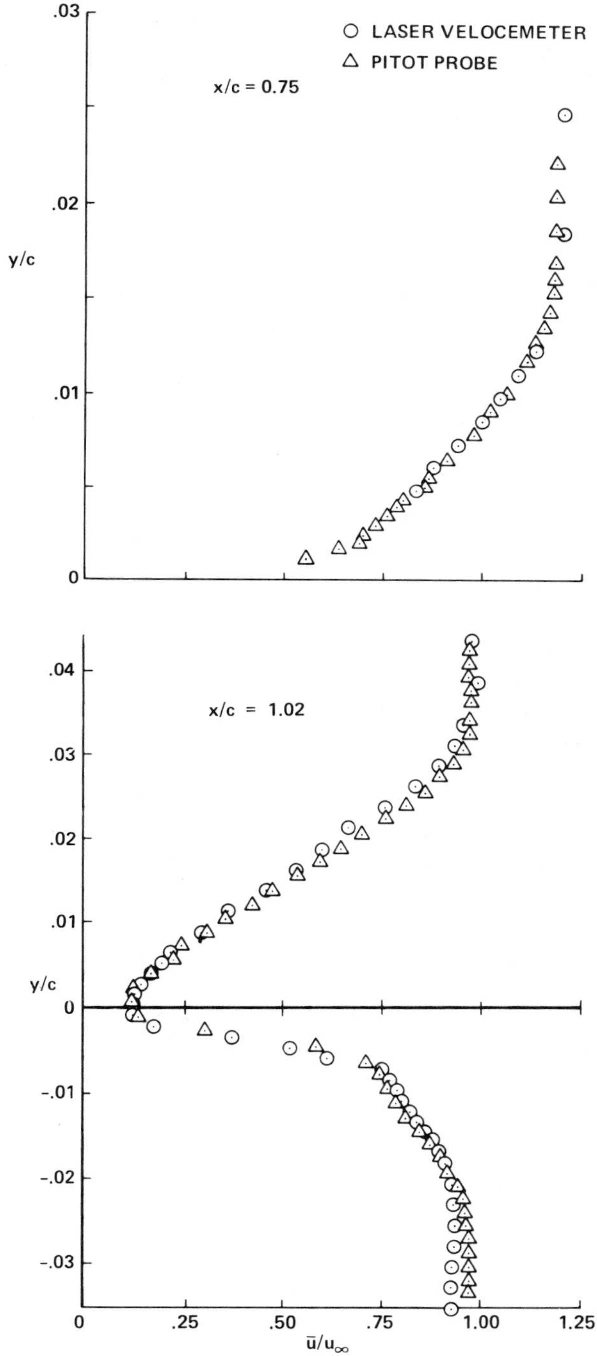

Figure 10. Example of mean velocity measurements for a supercritical section (DSMA 671) at cruise conditions.

The two cases previously described represent extreme cases in aerodynamics. In a very recent study (yet to be published), the same laser-velocimeter system was applied to a supercritical airfoil section (DSMA 671) in the NASA–Ames 2- by 2-Foot Transonic Wind Tunnel for a cruise condition. Measurements were obtained upstream of the shock (where the boundary layer was too thin to probe with the laser velocimeter), along the aft portion of the airfoil, and in the near-wake by pitot-tube methods similar to those described in [5.40]. At downstream stations, laser-velocimeter measurements were also obtained. Since the boundary layer remained attached, the pitot-tube results are believed to be reliable. Except for the stations upstream of the shock, data were only taken where the probe mechanism had a negligible influence on the shock location and surface pressure distribution. With the combination of these two measurement techniques, the boundary-layer and near-wake flows were well documented. Shown in Fig. 10 are some sample results. Future comparisons of the results obtained in this study with numerical prediction methods should prove very enlightening.

Concluding Remarks

With the advent of the laser velocimeter technique, the complicated inviscid-viscous interactions that occur in transonic flows can begin to be examined in detail experimentally. This technique has the capability of providing measurements of the turbulent Reynolds stresses as well as mean velocities. It is also capable of accurately probing regions of turbulent separation—a capability sorely needed in experimental fluid dynamics. In this paper, measurements obtained by this technique for cases of transonic shock-induced separation have been presented. Also, these results have been compared with solutions of the Reynolds-averaged, time-dependent, compressible Navier–Stokes equations. Overall, the predictions are not in good agreement with the experiments, due to deficiencies in the models for the turbulence Reynolds stresses.

The development of adequate turbulence model formulations for these types of flows is indeed a formidable task. It will require our best efforts in experimentally documenting the turbulent flow behavior for a wide range of flow configurations and test conditions. However, with the advances that have been made in both experimental and computational fluid dynamics, future progress in this area seems possible.

CHAPTER 32

Imbedded Longitudinal Vortices in Turbulent Boundary Layers

R. D. Mehta, I. M. M. A. Shabaka, and P. Bradshaw*

1 Introduction

Sufficiently strong lateral deflection ("skewing") of the mean streamlines in a shear layer can lead to the generation of discrete longitudinal vortices. The mechanism is basically inviscid, given the presence of an initial shear: this "skew-induced" vorticity should be distinguished from the generation of streamwise vorticity by Reynolds stresses, which is important only in very long, straight streamwise corners and is confined in practice to noncircular ducts. In most cases of practical interest the boundary layer and the imbedded vortex will be turbulent but the decay of the vortex under the action of Reynolds stresses is slow, because its circulation is reduced only by the effect of the *spanwise* component of skin friction. That is, the skew-induced vortices can influence the flow for a very long distance downstream. The present paper is a description of measurements of the mean and turbulent properties of decaying vortices and vortex pairs; the process of formation (being "inviscid") does not require such detailed study as the slow decay. The configurations used are idealizations of those found in practice, but the data should be useful for developing and testing calculation methods intended for real-life cases.

The most obvious case in which lateral deflection of a shear layer leads to the generation of discrete vortices is divergence of the boundary layer flowing over a surface around a tall obstacle protruding from the surface. The horseshoe vortex wrapped round the junction between a wing and a body,

* Department of Aeronautics, Imperial College, London, SW7 2BY.

between an axial flow turbomachine blade and a hub, or between a hull and a fin is formed in this way. A single vortex trails downstream in each wing–body corner, while downstream of the trailing edge the two vortices merge to form a pair with the "common flow" between them directed towards the surface. Vortices can also be generated on nearly flat surfaces, by sufficiently strong cross flow. The vortex pair on the leeward side of a body of revolution at incidence, and the bilge vortices generated near the bow of a ship with a full hull form, are two examples. A vortex *pair* can be formed in a sufficiently strong convergent cross flow, as in the bottom of a "S-bend" dorsal intake or the wall of a wind-tunnel contraction: in this case the "common flow" is away from the surface and leads to rapid thickening of the boundary layer. The Taylor–Gortler vortices that form in laminar or turbulent boundary-layer flow over concave surfaces have some points of similarity with the decaying vortices discussed above, but will not be discussed further in the present paper: detailed investigations are being carried out by the present authors' colleagues.

The experiments to be described in the present paper relate to:

1. one leg of a horseshoe vortex in a 90° streamwise corner;
2. an isolated vortex in the boundary layer on a flat surface, the vortex being artificially generated far upstream;
3. vortex pairs on flat surfaces: again, the vortices are artificially generated far upstream, and the "common flow" can have either sign.

2 Test Rigs

All the experiments were done in a 30-in. × 5-in. (762-mm × 127-mm) open-circuit blower wind tunnel (Fig. 1). Working-section lengths up to about 3 m could be used. The nominal tunnel speed for all experiments was 100 ft/sec (30 m/s). Standard Pitot tubes, three-hole Conrad yawmeters, and conventional cross hot-wire probes were used for the measurements. The fluctuating signals from hot wires were recorded on analog magnetic tape, and later transcribed to digital magnetic tape for computer analysis, including linearization. Statistics involving both v- and w-components were deduced from measurements with the cross wire probe rotated through various angles about its axis.

The corner vortex was generated by an idealized wing, spanning the 5-in. height of the tunnel with its leading edge 57 in. from the working-section entrance (Fig. 2). The model had a half-elliptical nose 6 in. long, followed by a slab of 2-in. constant thickness. Thus, pressure gradients were negligible except in the leading-edge region, and there was no tendency for the vortex to drift away from the corner under the influence of the pressure gradient. The boundary layer on the "body" (actually the floor of the wind tunnel) was about 1 in. thick at the leading edge.

The isolated vortices and vortex pairs on flat surfaces are generated by half-delta wings mounted in the settling chamber of the wind tunnel, as

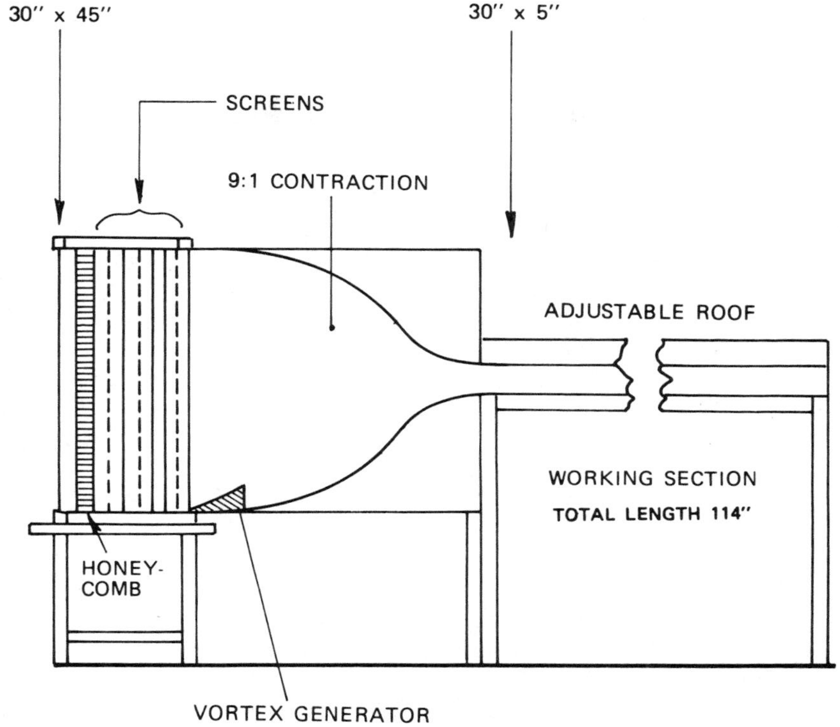

Figure 1. Test rig (blower not shown).

shown in Fig. 1. As the flow passes through the contraction, the circulation around the vortex is conserved, but the percentage velocity decrement in the wake of the delta wing is very much reduced, so that at exit from the 9-to-1 two-dimensional contraction we have a concentrated vortex with nearly uniform axial velocity. (Although the contraction is two-dimensional, the vortex rapidly recovers its circular shape.) For the vortex-pair experiments, two delta-wing vortex generators, set close together, are used: the vortex-generator configurations were developed by flow-visualization experiments in another wind tunnel before quantitative measurements began.

3 Results

Some features are common to the results of all the experiments. The circulation around the isolated vortices, and even the circulation around one vortex of a pair, decreases only very slowly with increasing downstream distance. Even vortices that are initially very small grow quite rapidly to fill the boundary layer, the growth rate being comparable with that of plumes of passive contaminants. The diameter of the corner vortex remains roughly

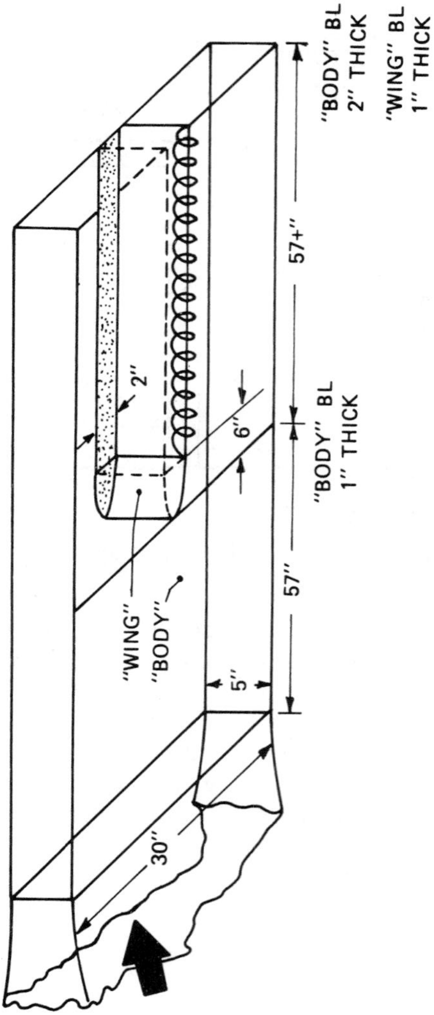

Figure 2. Wing–body-junction test section.

equal to the local thickness of the body boundary layer (which is thicker than the wing boundary layer), but the region of velocity defect in the isolated vortices and vortex pair projects well outside the undisturbed boundary layer. Another feature common to all the configurations is the complicated behavior of the various components of eddy viscosity, which become negative in significant regions of the flow: this implies that the use of a simply behaved eddy viscosity to predict the spreading rate of the vortex is likely to be unsatisfactory unless only gross properties are required. Unfortunately, the behavior of the triple velocity products which appear in the Reynolds-stress transport equations is also complicated, so that although calculation methods based on transport equations are needed in principle for the calculation of vortex flows, they may be difficult to develop in practice. Although this paper is not directly concerned with calculation methods, it should be pointed out that a calculation method which has been optimized for predicting secondary flows of the second kind (the "stress-induced" secondary flows found in long noncircular ducts) will not necessarily be very good at predicting secondary flows of the first kind (the skew-induced flows discussed in the present paper).

3.1 The Corner Vortex

This work is fully described by Shabaka [5.41], and a synopsis is given by Shabaka and Bradshaw [5.42]. The results are archived on magnetic tape in the Thermosciences Division, Stanford University, as part of the data set for the 1980/81 Conferences on Complex Turbulent Flows [5.2]. The "wing" used is shown in Fig. 2, which also gives details of the initial boundary layers. Measurements were not made in the leading-edge region in which the horseshoe vortex is generated, and the measurements made at the start of the parallel section of the wing give the initial conditions from which the vortex decays.

Mean-velocity contours at three streamwise positions are shown in Fig. 3. The position of the vortex center, deduced from results presented below, is shown approximately as a guide to understanding. The distortion of the mean-velocity contours by the vortex is mild, but, as shown in Fig. 4, cross-flow angles of several degress are found near the wing and body surfaces. Figure 4 hints, and Fig. 5 shows plainly, that the vorticity is not a maximum at the position which Fig. 4 indicates to be the center of the vortex, but is largest near the surface of the body, where of course it consists mainly of the secondary shear $\partial W/\partial y$, $\partial V/\partial z$ being comparatively small. Surface cross-flow angles are large enough to be visible in surface oil-flow pictures [5.41]. The effect of transport of x-component momentum by the vortex on the skin-friction coefficient can be deduced qualitatively from the velocity contours near the surface in Fig. 3: the velocity profiles follow the logarithmic law over significant distances, except very near the corner.

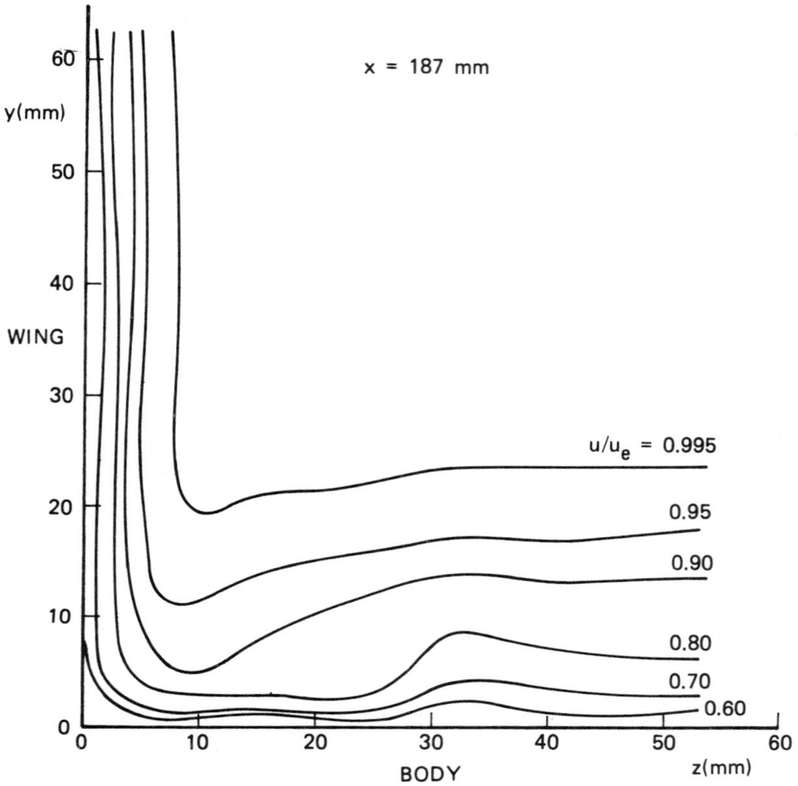

Figure 3. Streamwise velocity contours in wing–body-junction flow: (a) $x = 187.1$ mm.

Contours of turbulence intensity, of which examples are given in [5.42], look qualitatively like the axial-velocity contours. Figure 6 shows isometric views of the profiles of the Reynolds shear stresses. Far from the corner, the profiles of $\overline{uv}$ (the body shear stress) and of $\overline{uw}$ (the wing shear stress) revert to two-dimensional behavior, but $\overline{uv}$ in particular behaves very oddly near the corner, in a way that could not be accounted for by convection of the basic boundary-layer stress pattern by the secondary flow. The shear stress in the y-z plane, $\overline{vw}$, is fairly small except in the vicinity of the vortex center, where it changes sign rapidly. The eddy viscosities, defined as $-\overline{uv}/(\partial U/\partial y)$ and $-\overline{uw}/(\partial U/\partial z)$, have negative regions near the wing and near the body, respectively: the shear stresses in the region of negative viscosity are significant, so that the transfer of axial momentum away from the corner would be seriously miscalculated in the region near the solid surfaces if an eddy viscosity based on two-dimensional boundary-layer practice were used.

It is not possible, in a short paper, to discuss the rather complicated behavior of the rather large number of relevant triple products. Figure 7 shows profiles of $\overline{uv^2}$, the rate of turbulent transport of $\overline{uv}$ (the body shear

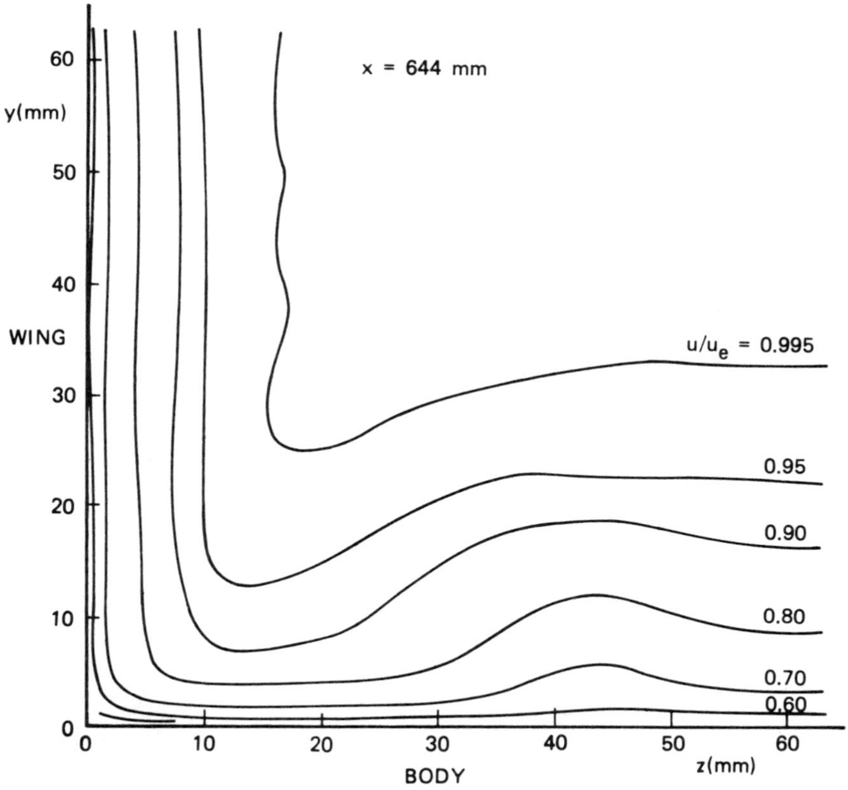

Figure 3. Streamwise velocity contours in wing–body-junction flow. (b) $x = 644.3$ mm.

stress) in the y-direction (away from the body). For large values of z, results are typical of isolated boundary layers, but the profiles become greatly distorted nearer the corner, and $\overline{uv^2}$ appears to asymptote to nonzero values at y within the wing boundary layer, although values at *very* large y (i.e. in the two-dimensional wing boundary layer far from the body) must tend to zero.

3.2 Single Isolated Vortex

Figure 8 shows the axial velocity contours just downstream of the start of the working section, where the floor boundary layer is about 4 mm thick. The height of the vortex center is expected to be at about $y = 10$ mm, and its z-coordinate is obviously about 25 mm (the z-origin being effectively arbitrary). Figure 9 shows axial-velocity contours, and secondary-flow vectors, about 1.3 m downstream of the start of the working section. For larger negative values of z than those shown here, the velocity contours asymptote to the approximately two-dimensional state shown for large

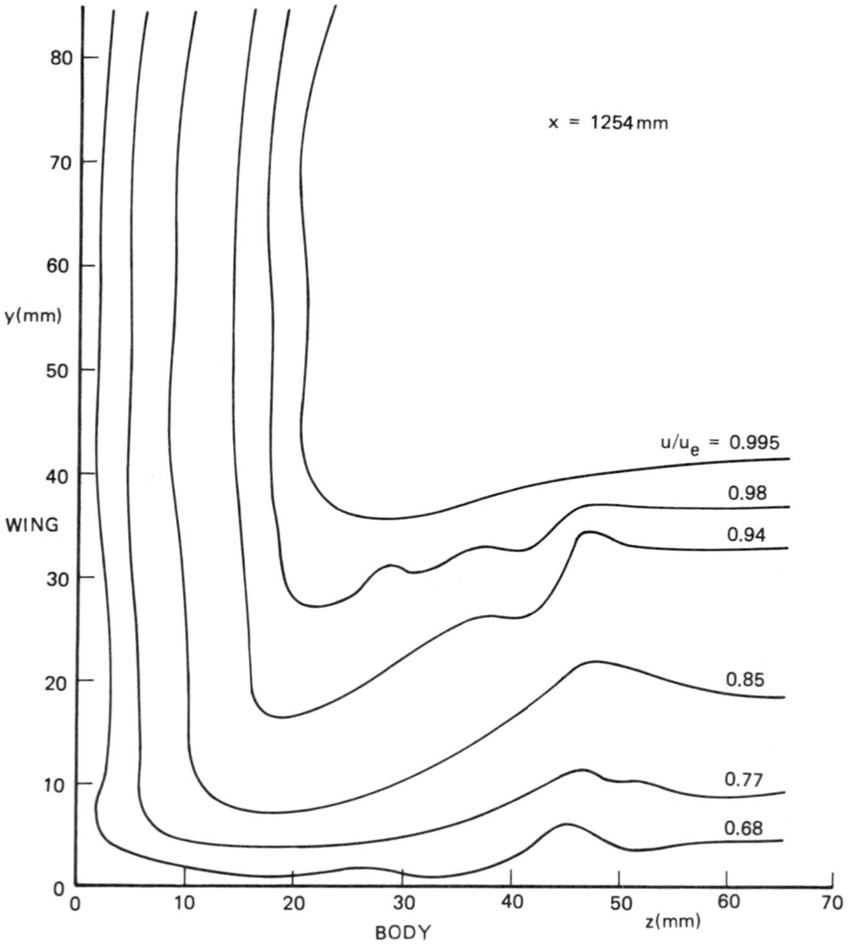

Figure 3. Streamwise velocity contours in wing–body-junction flow: (c) $x = 1254$ mm.

positive z. The vortex center is quite clearly defined, at roughly the same z-position as in Fig. 8, but at about 20 mm from the surface, indicating that the vortex moves away from the surface as the boundary layer thickens. (Contours at $x = 1.94$ m are nearly geometrically similar to those shown here.)

As in the case of a corner vortex, maximum cross-flow angles and maximum streamwise vorticity occur near the surface. However, the region of total pressure deficit in the vortex extends out to about twice the thickness of the undisturbed boundary layer: since the initial total-pressure deficit in the vortex (Fig. 8) is quite small, this low-total-pressure fluid must have been transported from near the surface by the secondary flow, assisted by turbulent diffusion.

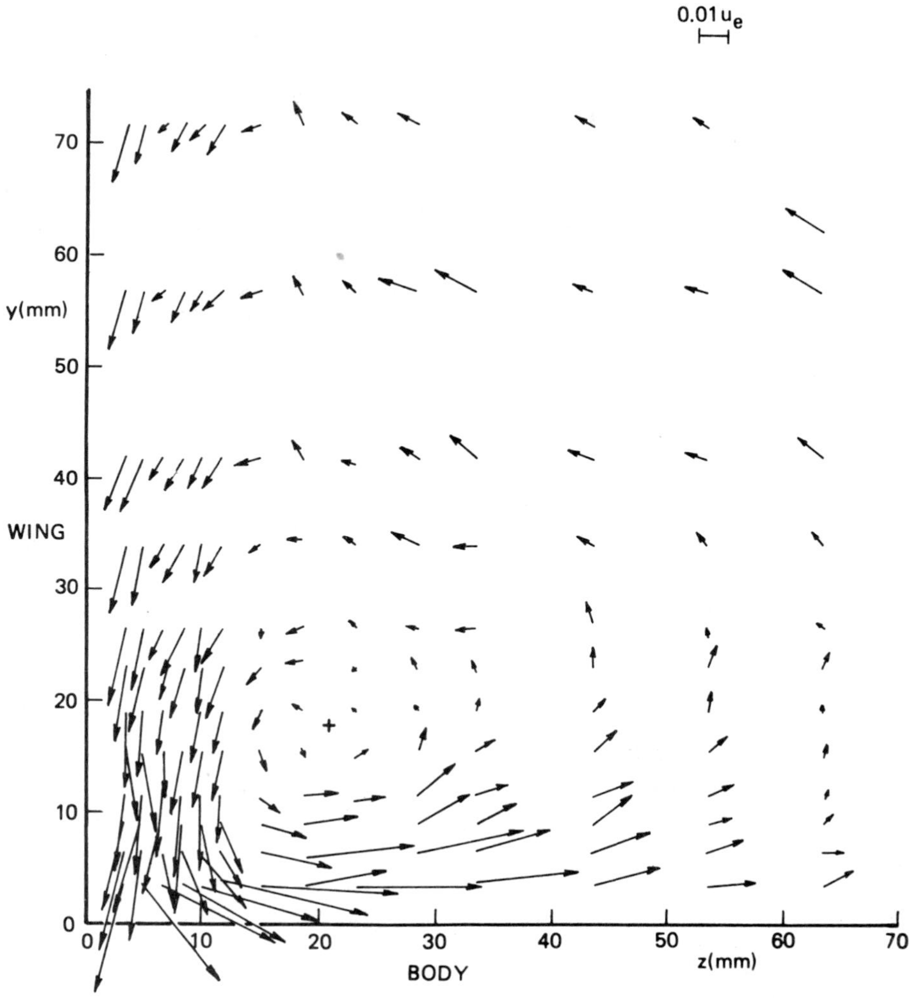

Figure 4. Secondary flow vectors at $x = 614$ mm.

The effect of z-wise convection of fluid by the vortex, in an action broadly similar to the rolling of sheet metal, is simply demonstrated by the behavior of the skin-friction coefficient shown in Fig. 10: velocity profiles near the surface obey the logarithmic law at all values of z, so that the skin friction is roughly proportional to the square of the velocity at, say, $y = 0.1\delta$. The skin-friction profiles are qualitatively explicable by the transfer of high-momentum fluid towards the surface in the region of $z = -40$, and transfer away from the surface in the region of $z = 0$: the secondary peak in c_f at $z = 20$, corresponding to the dip in the velocity contours near this position, is more difficult to explain. The secondary-velocity contours at station 9

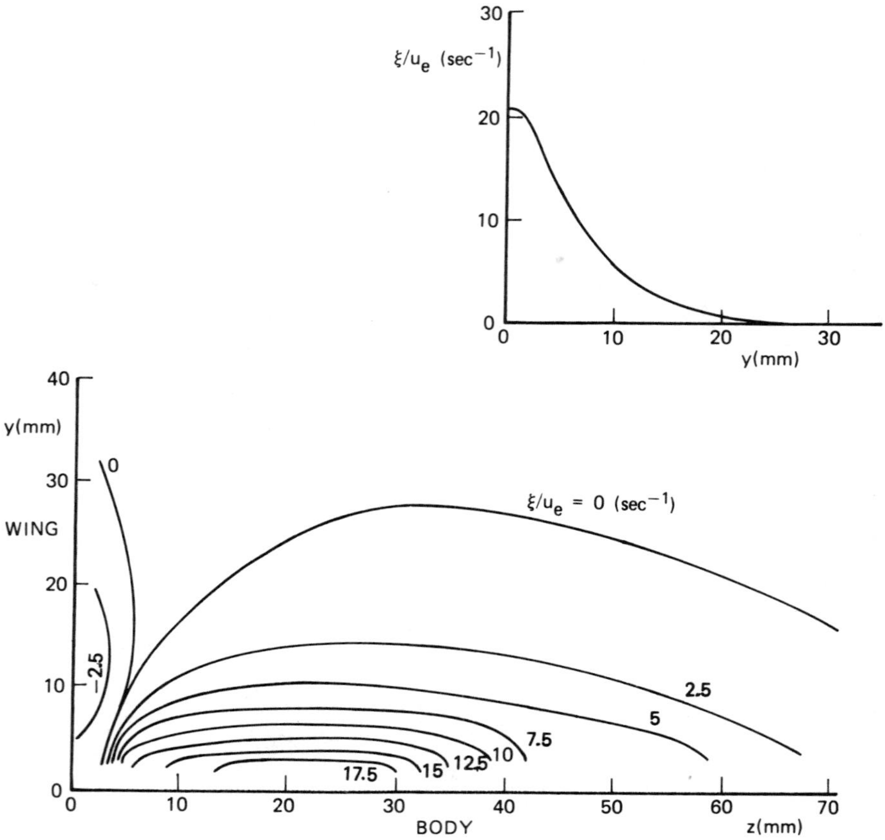

Figure 5. Streamwise vorticity contours at $x = 614$ mm.

show some hints of a second vortex, with its axis at about $z = 20$, but it is not certain exactly how this second vortex is produced.

Figure 11 shows contours of longitudinal turbulence intensity. As in the case of the corner vortex, the intensity contours look roughly similar to the velocity contours, and this similarity extends to the v- and w-component intensities as well.

The $\overline{uv}$ contours shown in Fig. 12 are chosen so that the outermost approximately coincides with the edge of the undisturbed boundary layer, but it is clear that the region of significant $\overline{uv}$ within the vortex is very much smaller than that outlined by the outermost velocity contour. Bodily rotation of the fluid in the vortex, so that eddies with a contribution to $\overline{uv}$ are rotated so as to contribute to $\overline{uw}$ instead, must be responsible. Figure 13 shows that $\overline{uw}$ (zero in a two-dimensional flow) is indeed complicated. The very rapid z-wise gradient in the region of $z = 10$ mm lines up with the position of the "tongue" in $\overline{uv}$ shown in the previous figure. Contours of "0+" and

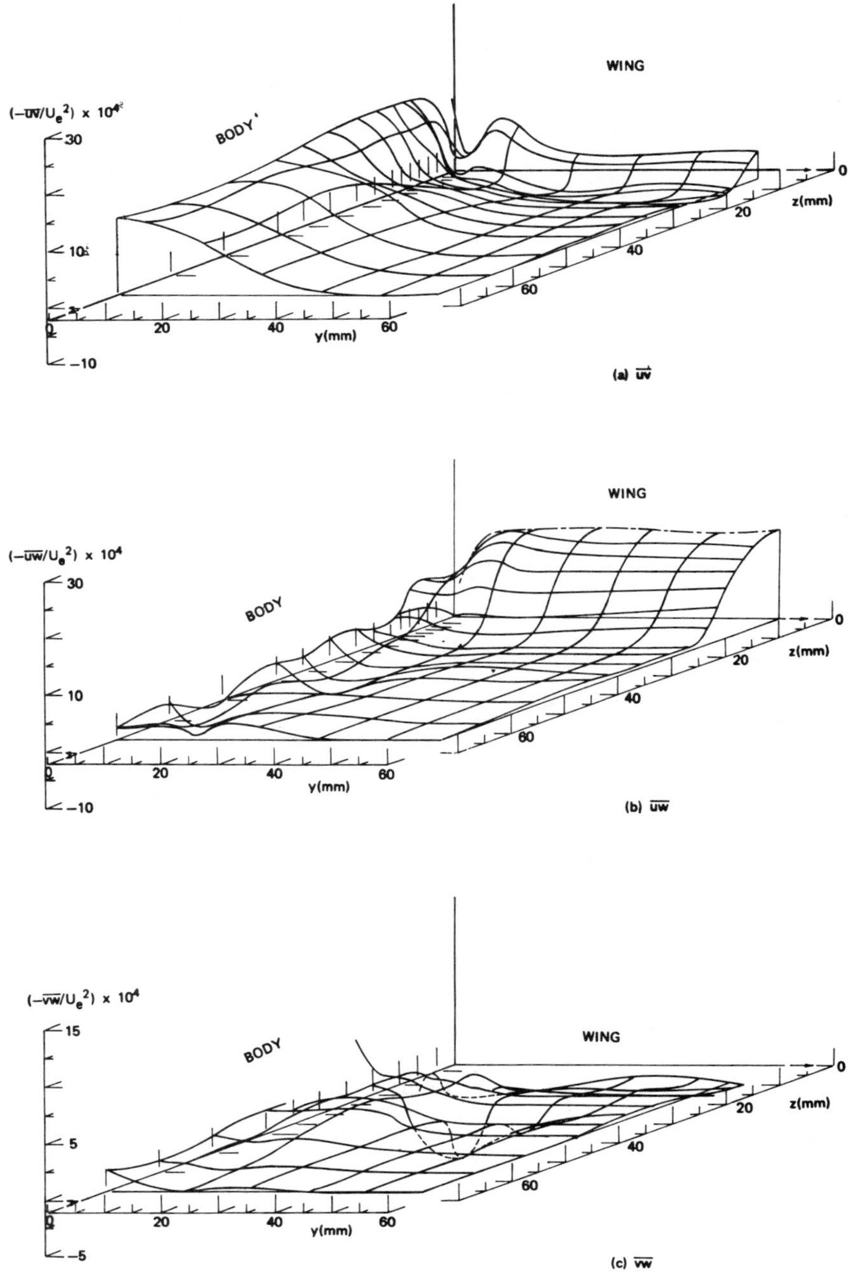

Figure 6. Reynolds-stress contours at $x = 614$ mm: (a) $\overline{uv}$, (b) $\overline{uw}$, (c) $\overline{vw}$.

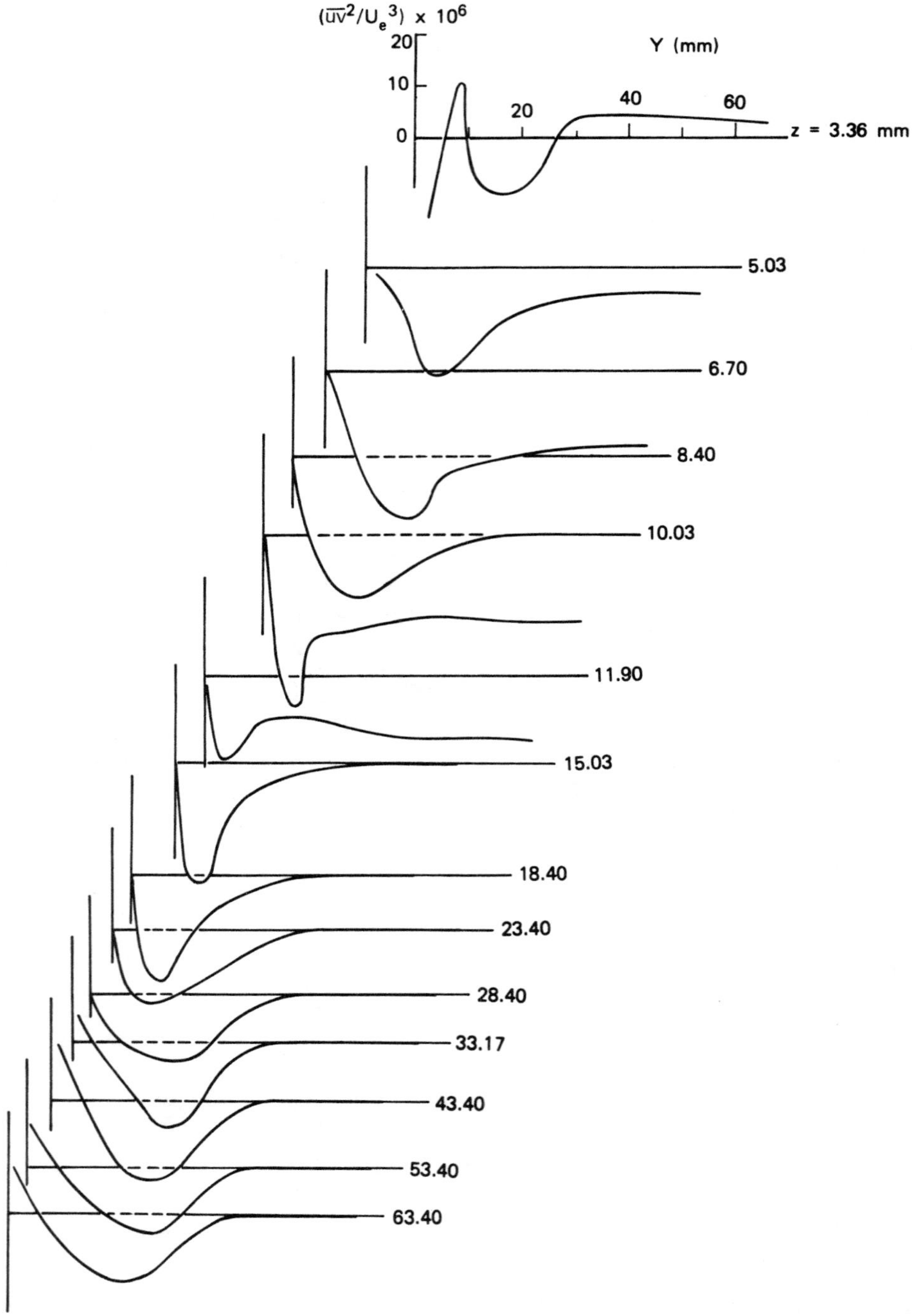

Figure 7. Profiles of triple product $\overline{uv^2}$ at $x = 614$ mm.

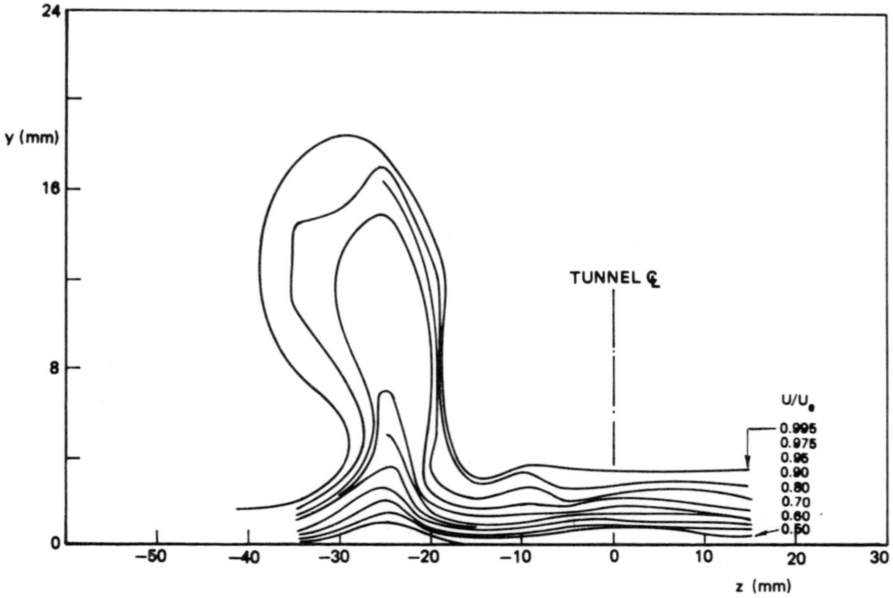

Figure 8. Mean-velocity contours in single isolated vortex at $x = 112$ mm.

"0−" are shown separately for clarity: in fact the zero contour has several cusps. The region of large positive $\overline{uw}$ at $z = -25$, where $\overline{uv}$ falls rather rapidly to zero, is in the right sense to be explained as a rotation, by the main vortex, of eddies originally contributing negatively to $\overline{uv}$. The concentrated region of negative $\overline{uw}$ near $z = 15$ is more difficult to explain, as is that near $z = -85$. Figure 14 shows $\overline{vw}$, again with "0+" and "0−" contours indicated. The general shape of the contours is believed to be trustworthy, but the large negative peak at $y = 8$, $z = 15$ appears implausible. Eddy viscosities corresponding to these Reynolds stresses have not yet been evaluated quantitatively, but even the casual observation that $\overline{uv}$ is everywhere of the same sign, whereas the velocity gradient $\partial U/\partial y$ reverses, clearly indicates that the eddy viscosity will be ill behaved.

Figure 15 shows profiles of the triple product $\overline{uv^2}$. As in Fig. 7, large departures from the two-dimensional behavior exemplified by values at large z are seen to occur.

Analysis of the results of this experiment is still in progress. It extends the mean-flow measurements of Tanaka and Suzuki [5.43] and the old NPL results quoted by Pearcey (see [5.44]).

3.3 Vortex Pair on Flat Surface ("Common Flow" Outwards)

Figures 16 and 17 show mean-velocity contours for this case. Figure 16 shows measurements near the front of the working section, the large scatter in the points for the $U/U_e = 0.995$ contour being mainly a consequence of

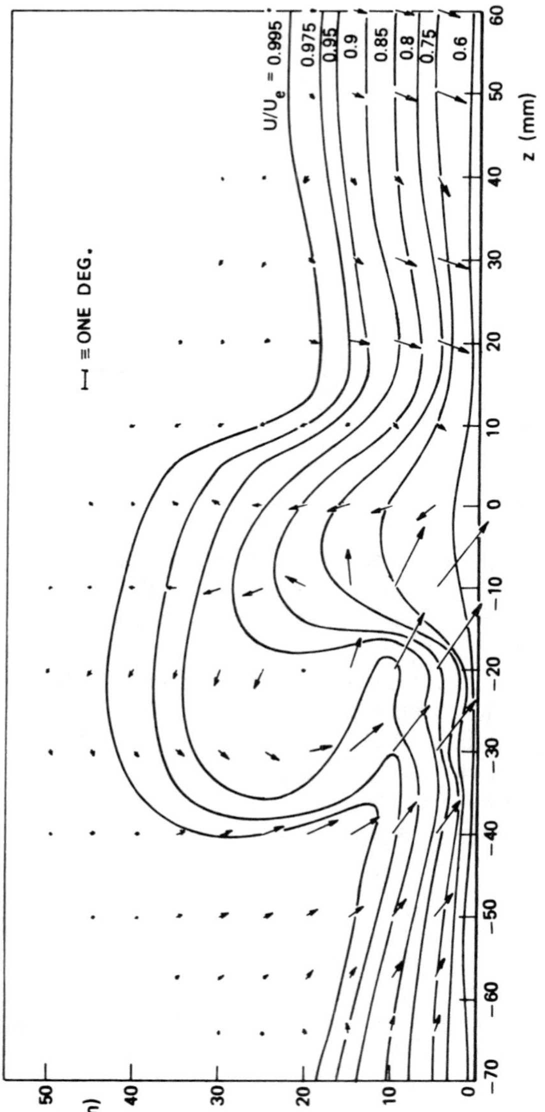

Figure 9. Mean-velocity contours and secondary-flow vectors at $x = 1332$ mm.

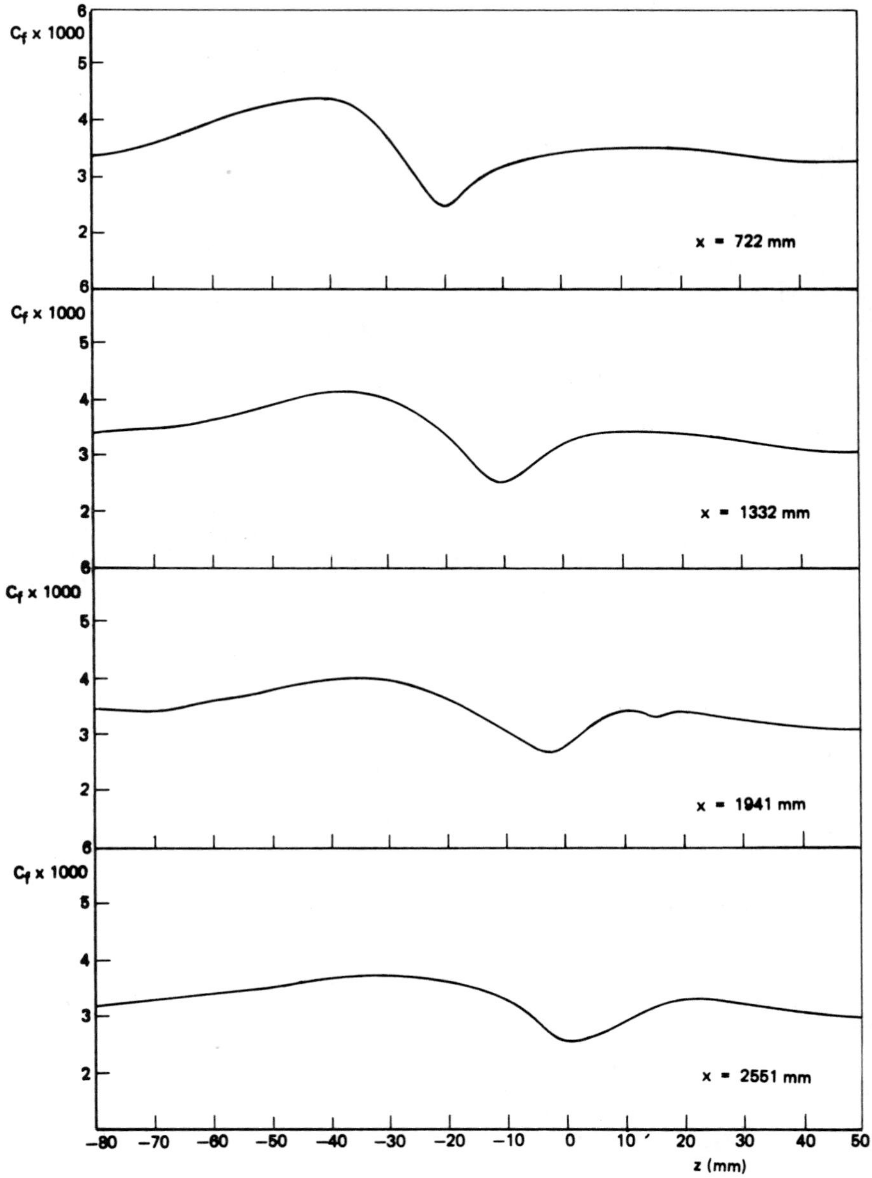

Figure 10. Skin friction in single isolated vortex.

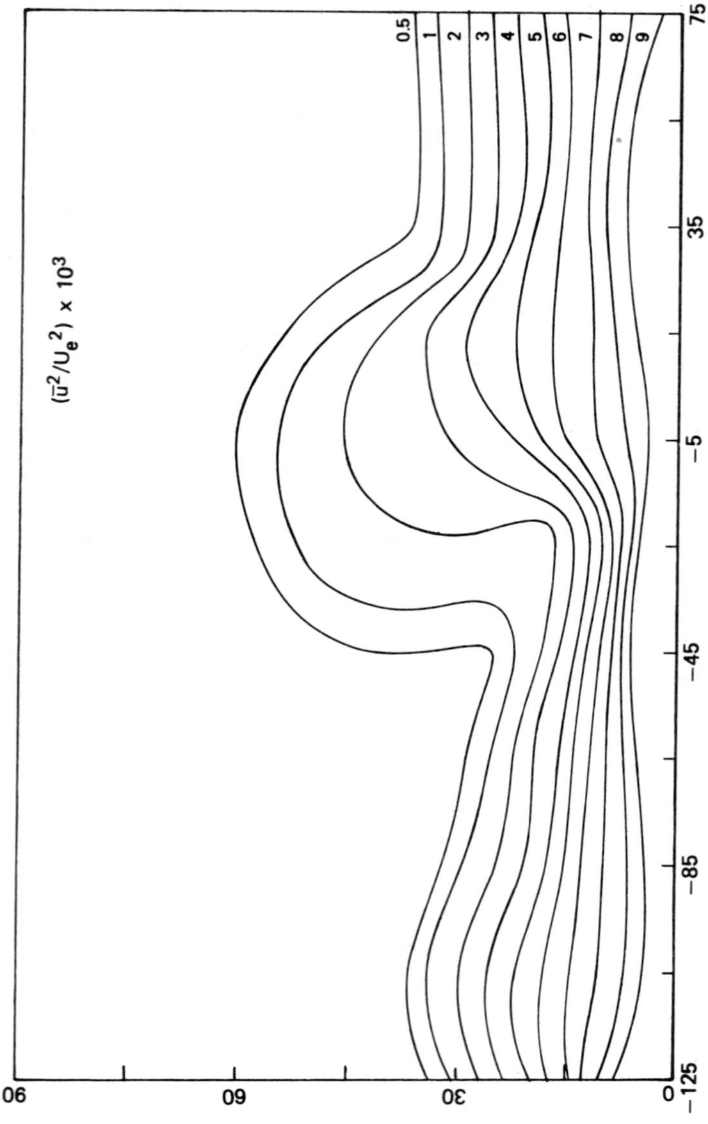

Figure 11. Longitudinal-component intensity contours.

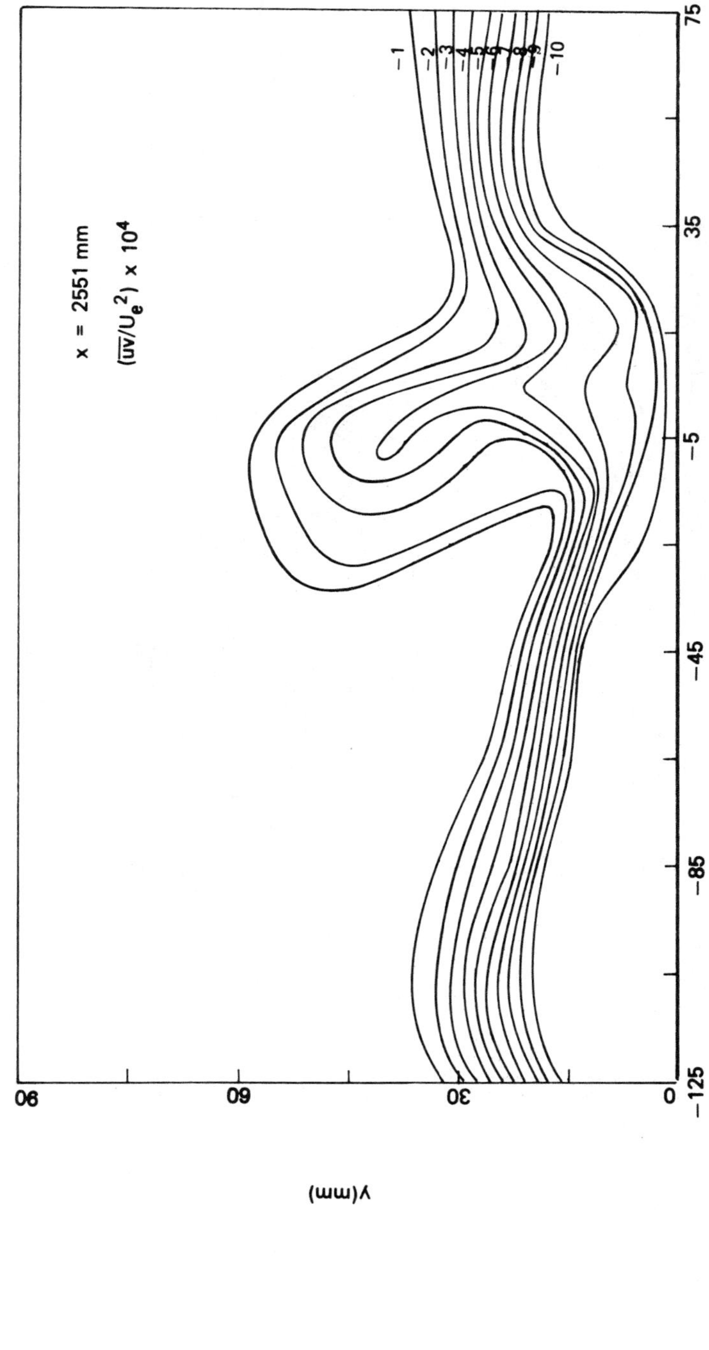

Figure 12. Primary-shear-stress contours.

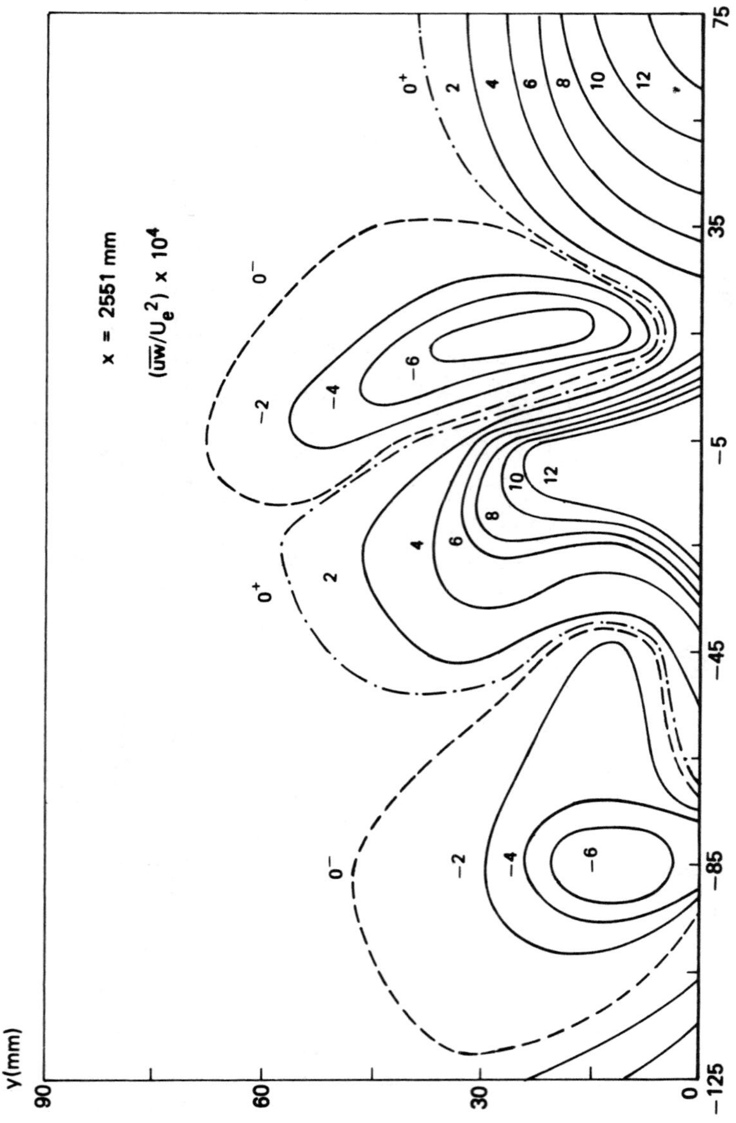

Figure 13. Secondary-shear-stress contours.

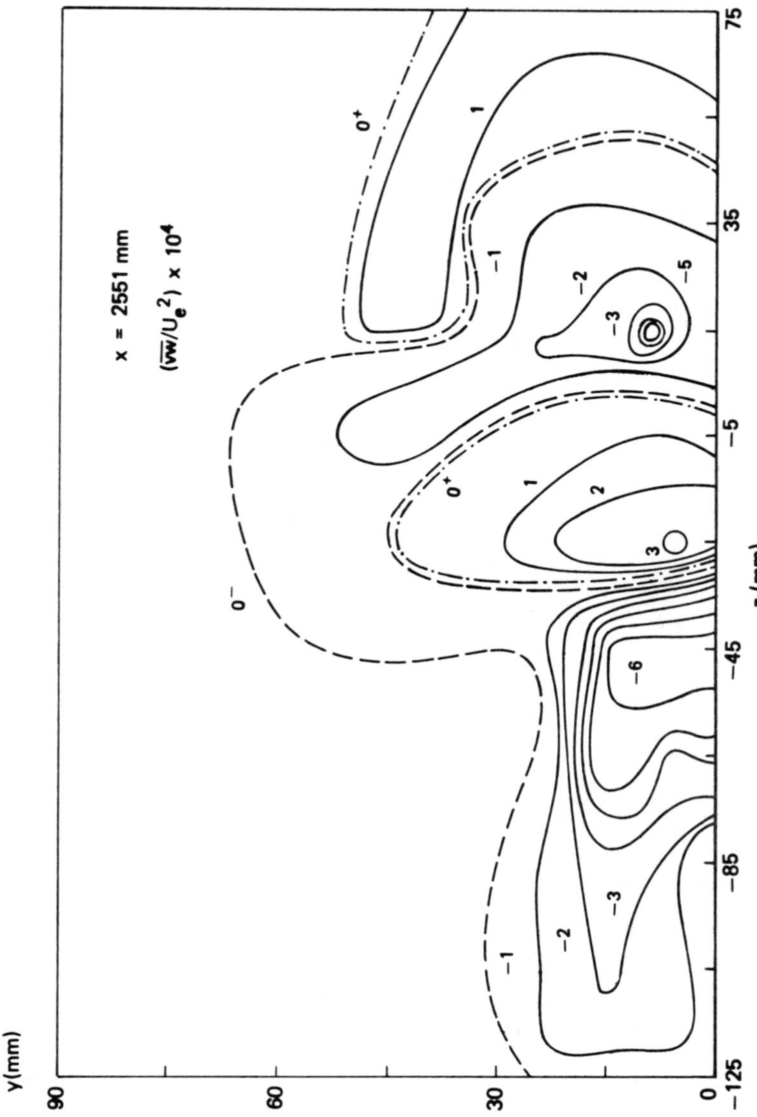

Figure 14. Cross-sectional shear-stress contours.

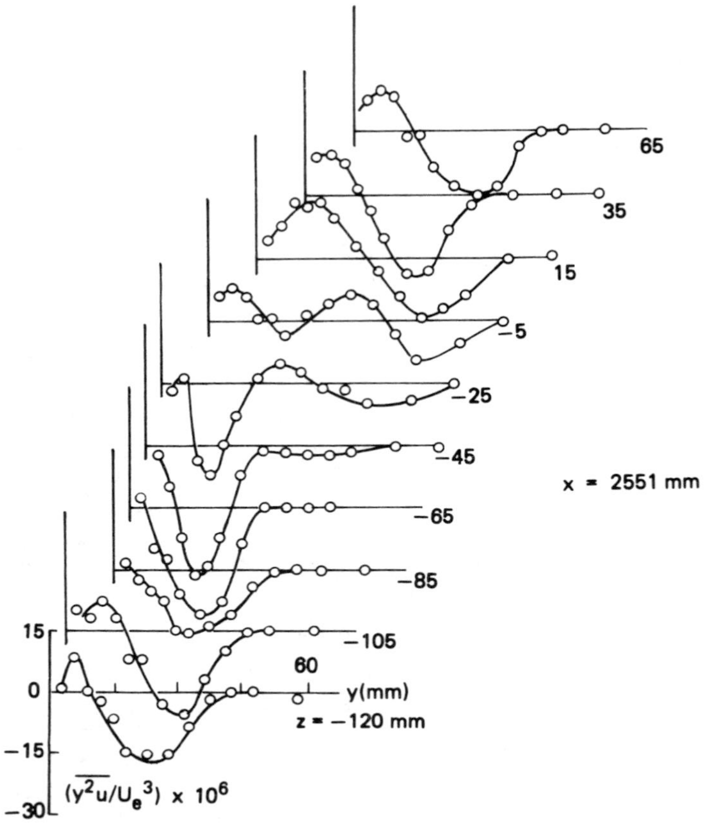

Figure 15. Profiles of triple product $\overline{uv^2}$.

the complicated total-pressure pattern in the residue of the wake of the twin-delta vortex generators in the settling chamber. In Fig. 17, the outermost contour is smooth, but the contour for $U/U_e = 0.975$ is irregular, with some suggestion that the irregularities are real rather than just scatter in a region of small gradients. In general, the contours in these figures are plausible and accurately symmetrical, though about a value of z slightly different from zero. Figure 18 shows skin-friction values at different streamwise stations. Again, the curves are accurately symmetrical, the dip at the centerline being the obvious result of inflow of low-momentum fluid from positive and negative z. The double minimum at the smallest value of x, corresponding to the dip in the velocity contour nearest the wall (Fig. 16), is evidently a result of the disturbances caused by the vortex generators, because only a single minimum is found at stations further downstream. Hot-wire measurements for this flow are in progress, and will be followed by measurements in a vortex pair with the opposite sign of rotation, so that the "common flow" is towards the surface.

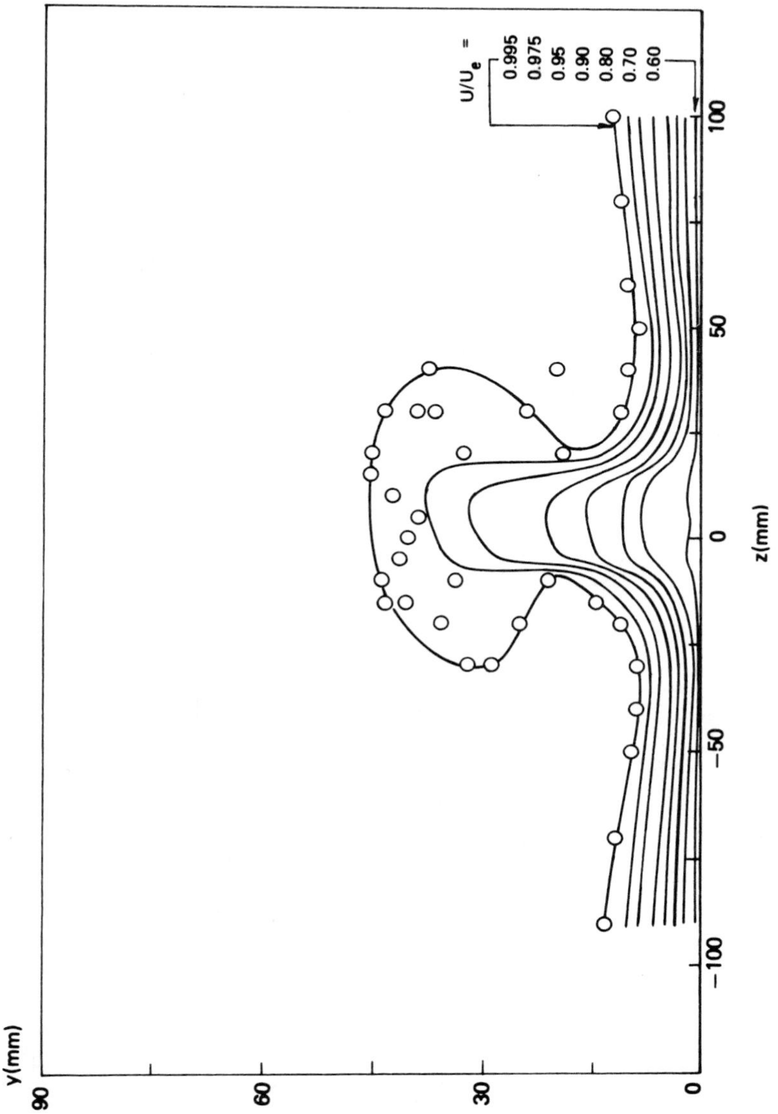

Figure 16. Mean-velocity contours in vortex pair at $x = 600$ mm.

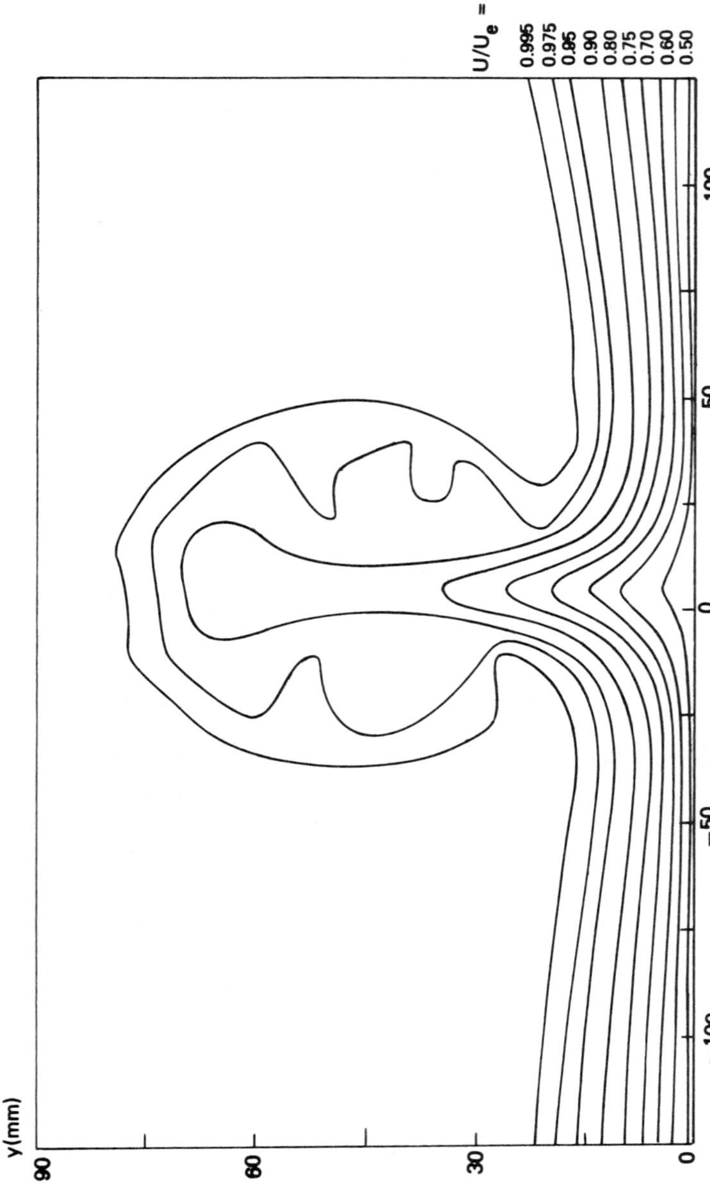

Figure 17. Mean-velocity contours in vortex pair at $x = 1350$ mm.

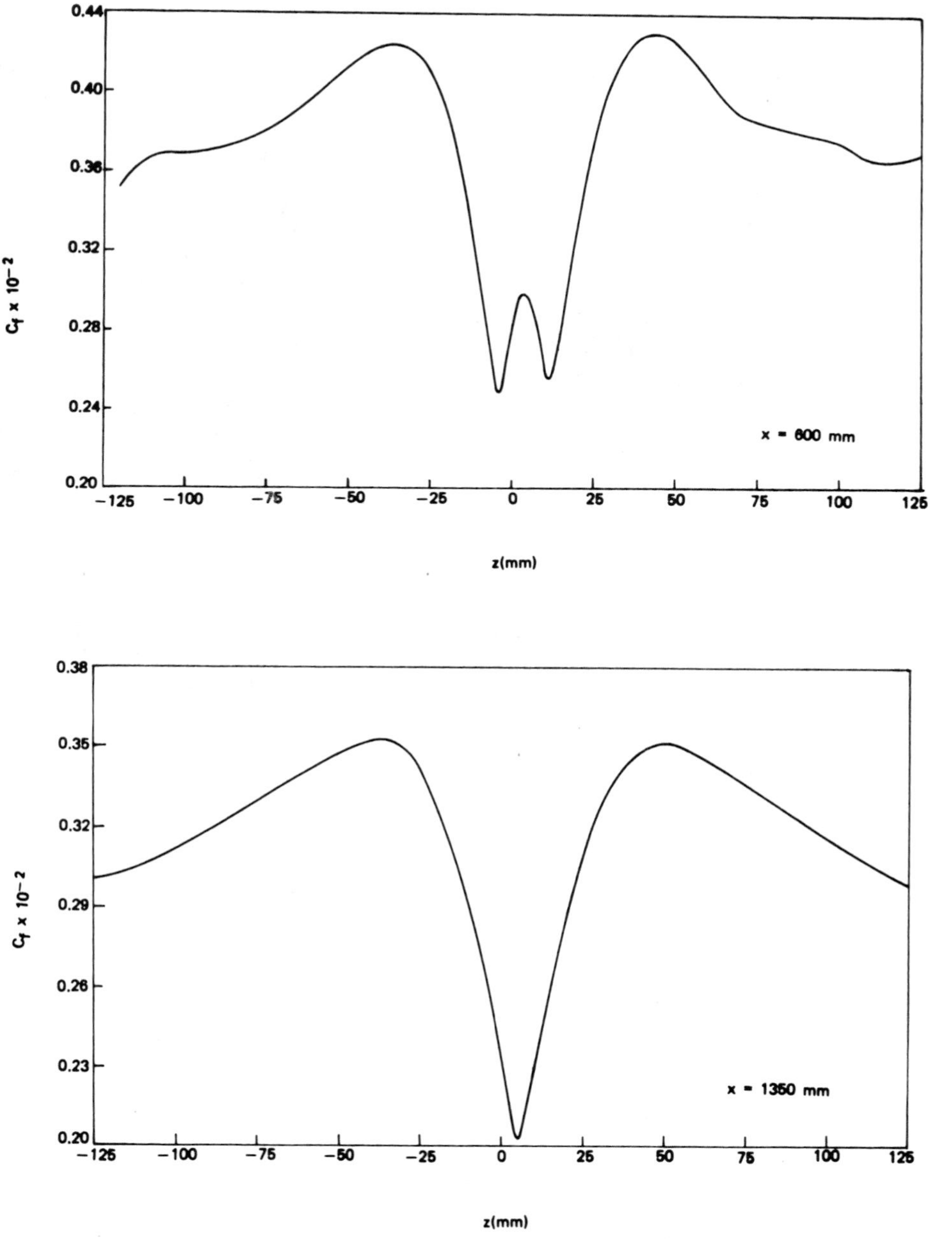

Figure 18. Skin-friction coefficient below vortex pair.

4 Conclusions

Even in their present incomplete state, the results of these experiments on longitudinal vortices imbedded in turbulent boundary layers call attention to many interesting features of the flows, in which the qualitative obviousness of the mean-velocity contours conceals a great deal of nonobvious turbulence behavior, which will have to be better understood if the mean-velocity contours are to be predicted *quantitatively*. Longitudinal imbedded vortices are such a common feature of three-dimensional turbulent flows that serious attempts to develop a calculation method for them are long overdue. The results indicate that methods using eddy viscosity or eddy diffusivity are unlikely to be satisfactory, but even transport-equation methods cannot be expected to reproduce all the details of the complicated profiles of Reynolds stresses shown above. The simple configurations tested in the present experiments correspond to a rather wide variety of real-life flows, of importance in such diverse fields as turbomachinery, aeronautical aerodynamics, ship hydrodynamics, and internal flows.

Acknowledgment

On this occasion it is a particular pleasure to acknowledge support by the United States Office of Naval Research under Contract NR/061-256 monitored by Mr. Morton Cooper. Many of the papers in this volume pay tribute to the scientific broadmindedness of ONR: this paper from overseas is a tribute to its geographical broadmindedness as well.

CHAPTER 33

Experimental Investigation of Oscillating Subsonic Jets[1]

D. J. Collins,* M. F. Platzer,* J. C. S. Lai,† and J. M. Simmons†

Introduction

Steady two-dimensional jet flows have attracted the interest of many investigators because of their fundamental and practical importance. Much of the currently available information has been summarized by Harsha [5.45], Rajaratnam [5.46], and Everitt and Robins [5.47]. Significant interest in *unsteady*-jet-flow effects first was sparked by the development of the pulse-jet engine, especially when Bertin [5.48] and Lockwood [5.49] noted the favorable effect of pulsating jet flow on secondary-flow entrainment. Lockwood [5.49] also identified the generation of ring vortices in pulsing flow, a phenomenon later verified more clearly by Curtet and Girard [5.50]. Further investigations of pulsating jet flows were performed by Johnson and Yang [5.51], Didelle et al. [5.52], Binder and Favre-Marinet [5.53], Crow and Champagne [5.13] and, most recently, Bremhorst and Harch [5.54] and Bremhorst and Watson [5.55]. A different type of unsteady jet flow is produced by time-varying jet deflection, either by mechanical oscillation of the jet nozzle or by fluidic jet actuation. The flow patterns produced by a mechanically oscillated jet exhausting into a secondary flow recently were measured in some detail by Simmons et al. [5.56], whereas fluidic jet nozzles

[1] This investigation was supported by the Naval Air Systems Command, Code AIR-310, the Australian Research Grant Committee, and the Department of Mechanical Engineering, University of Queensland.
* Naval Postgraduate School, Monterey, CA 93940.
† University of Queensland, Brisbane, Australia.

were developed and investigated by Viets [5.57, 5.58]. A comparison of the entrainment rates produced by pulsating or oscillating jets, presented by Platzer et al. [5.59], shows remarkably different rates, depending on the type of jet unsteadiness.

The present investigation was motivated by the quest for a simple, yet efficient method to increase the jet entrainment for potential use in thrust-augmenting ejectors. As pointed out by Schum [5.60], the augmenter performance is critically dependent on achieving high entrainment rates while maintaining high nozzle efficiencies if volume is to be minimized for a desired augmentation ratio. Pulsating or fluidic jet nozzles have the disadvantage of a more complicated nozzle design and of decreased efficiency; hence a new jet excitation scheme was adopted in the hope of achieving a good compromise between enhanced jet entrainment and decreased nozzle efficiency. For this purpose a small vane situated in the potential core was excited into small pitch oscillations such that both frequency and amplitude of oscillation could be varied over a significant range. The investigation of this type of jet excitation was stimulated by the encouraging results reported by Fiedler and Korschelt [5.61], who used a freely vibrating vane for jet excitation.

In the following sections the experimental setup, the measuring techniques, and the results obtained to date are presented. This paper constitutes the second phase of an investigation begun at the University of Queensland and reported earlier by Simmons et al. [5.62].

Experiments

Mean-velocity measurements were made in a vane-excited turbulent jet of air which issued into stationary air from a plenum chamber through a rectangular nozzle of length $L = 300$ mm and width $h = 6$ mm (Fig. 1). Wire gauges and honeycombs were installed upstream of the nozzle to reduce the turbulence at the jet exit. No side plates were used to contain the jet. The chamber pressure and temperature were continuously monitored during the experiment through a manometer and a thermocouple respectively.

A vane which had a symmetric airfoil section with a thickness of 1.3 mm, a span of 360 mm, and a chord of 10 mm was located symmetrically in the potentical core at $1.42h$ from the nozzle. The vane was oscillated by a reciprocating rod with an eccentricity e, driven by a motor. Various frequencies and amplitudes of oscillation of the vane about a mean position set at zero angle of attack could be attained by varying respectively the power input to the motor and the eccentricity e. Since the vane fluttered at the high nozzle pressure ratio of operation, it was supported with two bearings 4.5 mm thick each and 123 mm apart, reducing the nozzle's aspect ratio to an effective value of 1:20. This aspect ratio is still considered acceptable for two-dimensional jets (see e.g. Forthmann [5.63], van der Hegge Zijnen [5.64], and Goldschmidt and Eskinazi [5.65].

Typical jet temperatures on the centerline, at the exit and 20, 40, and 60 nozzle widths downstream, were 7, 3, 1, and 0.5°C above ambient, respec-

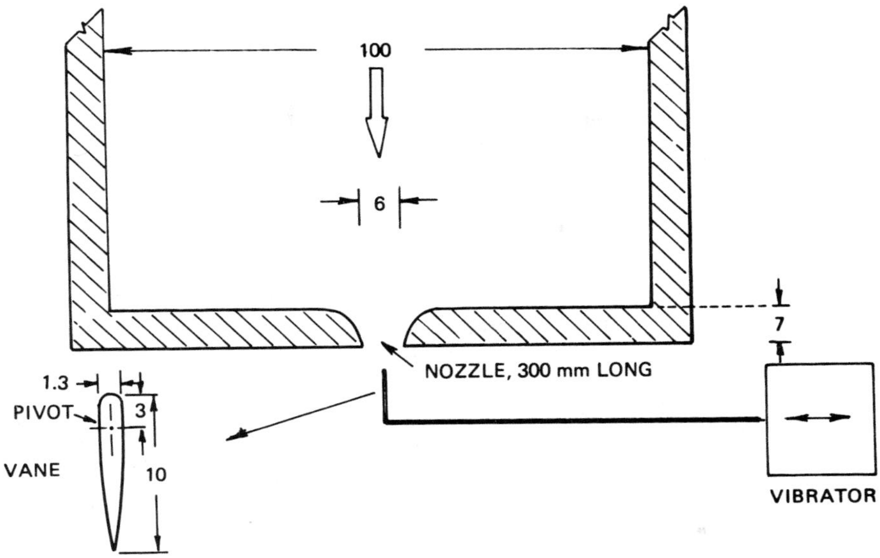

Figure 1. Configuration of nozzle and oscillating value (all dimensions in millimeters).

tively. However, these will not have any significant effect on the jet measurements.

For the convenient operation of the laser Doppler velocimeter, the facility was designed so that a velocity traverse across the jet was obtained by fixing the probe (pitot-tube or LDV) in space and moving the jet with a hydraulic lifting device.

The mean-velocity measurements were made across the width of the jet at its midspan and at distances of 20, 40, and 60 nozzle widths downstream of the nozzle for the following range of parameters:

Nozzle pressure ratio P_c/P_A: 1.008, 1.137, 1.268.
Jet exit Reynolds number: 1.47×10^4, 5.85×10^4, 8.19×10^4.
Vane amplitude of oscillation (zero-peak) ε: 2.6, 4.6, 6.9°.
Vane frequency of oscillation, f: 0, 20, 30, 40 Hz.

Here P_c is the plenum chamber pressure and P_A is the atmospheric pressure.

Instrumentation

Pitot-Static Tube

A pitot-static tube of hole internal diameter 0.74 mm was used by aligning it with the mean flow direction. The tube had four static holes located symmetrically around the periphery at 17 tube diameters. The mean pressure obtained with a pitot-static tube in a fluctuating flow is generally given by

$$P_t - P_s = \tfrac{1}{2}\rho \cdot [U^2 + K_1\overline{u^2} + K_2(\overline{v^2} + \overline{w^2})],$$

where P_t and P_s are the total and static pressures, U the mean velocity, and u, v, and w the velocity fluctuations in the streamwise and the two transverse directions. K_1 and K_2 are of order unity. Bradshaw and Goodman [5.66] concluded that in highly turbulent jets, for distances less than 150 nozzle widths, the measured static pressure is higher than the actual value. Harsha [5.45, p. 69] noted that the deviation of the total-head tube reading caused by changes in turbulence intensities is commonly ignored. Alexander et al. [5.67] observed that the total pressure decreased markedly with increasing relative turbulence level. On the other hand, Krause, Dudzinski, and Johnson [5.68] observed that in pulsating flows total-pressure tubes indicate values higher than the true average pressure. In the present measurements no attempt was made to correct for errors due to fluctuating-flow effects. However, it was noted that for both the steady and the vane-excited jet it was necessary to measure the static pressure simultaneously with the total head pressure because of a significant static pressure variation throughout the jet. Such variations were reported earlier by Miller and Comings [5.69].

Hot-Wire Anemometer[2]

A constant-temperature hot-wire anemometer was used, the wire being a platinum alloy 10 μm in diameter with its 4-mm length aligned parallel to the length of the nozzle. The anemometer was operated at a constant resistance ratio of 1.3. Typical jet temperatures on the centerline at 20, 40, and 60 nozzle widths downstream were 4, 3, and 2°C respectively above ambient, so that the cold resistance of the wire varied by at most 1%. An analysis showed that a temperature change of 8°C gives a change of only 2% in the temperature sensitivity of the wire. Hence a first-order temperature correction to the anemometer calibration was adequately achieved by operating the wire at a constant resistance ratio with its cold resistance determined by local jet conditions.

Laser Doppler Anemometer

A single-component, dual-beam laser Doppler velocimeter was used in the forward-scattering mode with an on-axis photomultiplier for detecting the signals. The laser was a Spectra-Physics Model 164 argon-ion laser operated at a wavelength of 514.5 nm with 200 to 500 mW of power. The optics system was that of a DISA two-color system with a beam separation of 27.6 and 9.2 mm. The two beams of the laser were focused on the measurement point by means of a 600-mm lens. Frequency shifting was used to separate the pedestal from the Doppler signal. Particles, generated from olive oil by a TSI Model 3075 constant output atomizer, were seeded at a rate of

[2] The hot-wire measurements were performed by the last three authors on a setup at the University of Queensland quite similar to the one shown in Fig. 1. The pitot-tube and LDV measurements were performed by the first three authors at the Naval Postgraduate School.

0.3 cc/min in the plenum chamber upstream of the nozzle. Also, before actual measurements were taken, the surrounding environment of the jet was heavily seeded to ensure a uniform distribution of particles in both the fluid originating from the nozzle and the surroundings, so that bias due to non-uniform seeding could be managed. Particle sizes, though variable, never exceeded 1 μm. The particles therefore had a size of the order of the fringe spacing in the operating mode of beam separation used and thus yielded signals with full modulation. Dynamic calculations indicate that the seed particles will track the flow with a relaxation time of the order of 10^{-6} sec. The Doppler signals received by the photomultiplier were transmitted to a DISA 55L90 LDA counter for further processing. Counters (also known as burst-signal processors) which measure the velocity of individual particles are by far the most versatile LDV signal processors as compared with trackers, frequency analyzers, and photon correlators, and they are especially suitable for low particle concentrations, provided that the signal-to-noise ratio is adequate. The signals were first bandpass-filtered to remove the high-frequency noise and the low-frequency pedestal. The filters were used in conjunction with threshold window adjustments, which set the upper amplitude limit to reject signals from large particles, and an amplifier gain control, which attenuated the input signal to reduce the noise signal to below the circuit trigger level. The signals were then validated on the basis of a $\frac{5}{8}$ comparison mode with a tolerance accuracy of $\pm 1.5\%$ or $\pm 3\%$. Details of the validation circuit and the operation of the counter are available in the DISA instruction manual (1976). The validated signals were further reduced on-line by an HP9825A microcomputer to extract the mean axial velocity component U and the aggregate axial turbulence $\bar{u}^2$. Histograms were taken of selected points during the experiment. On-line graphs of the experimental results were also obtained.

Using computer simulation, Hoesel and Rodi [5.70] showed that the error in the LDV mean velocity is inversely proportional to the bandwidth of the bandpass filter. Throughout the experiments, the signal-to-noise ratio was continuously monitored to give a value of 20, which is adequate for LDV data processed by counters. On-line histogram plots of the particle velocity distribution were used to help select the most appropriate bandpass-filter values and the value of the shift frequency. Validated signals were received at a rate of about 200–600/sec. Assuming that the particle velocity distribution is Gaussian, 95% of the data should fall within $\pm 2\sigma$ limits, where σ is the standard deviation. These limits were used to ensure that the bandpass filters were wide enough to accept at least 95% of the data.

Results

Figure 2 shows the mean-velocity profiles obtained by hot-wire anemometry at 60 nozzle widths from the jet exit. The vane amplitude was 5.2° zero-to-peak, the vane chord was 10 mm, and the vane leading edge was located at 3.3 mm

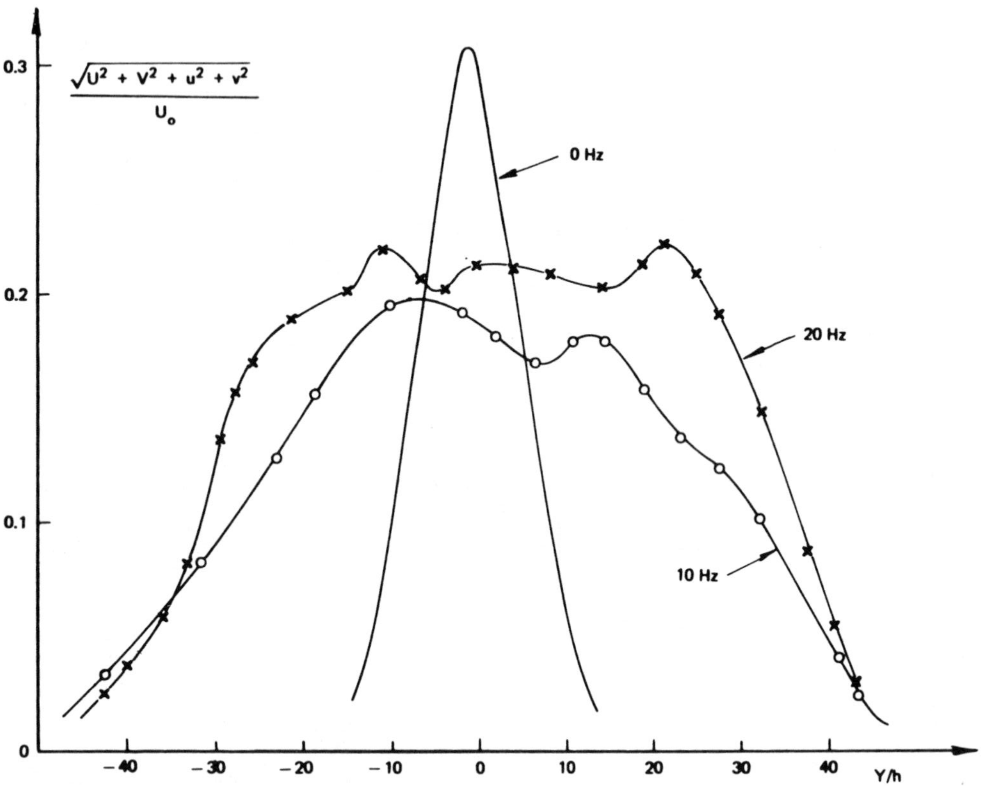

Figure 2. Profiles of mean jet velocity magnitude measured with hot wire at 60 nozzle widths downstream of nozzle. Vane is located 3.3 mm from nozzle and amplitude of oscillation is 5.2 degrees zero-peak. Jet exit velocity $u_0 = 36.4$ m/sec.

from the jet exit. The jet exit velocity was 36.4 m/sec. The vane oscillation produces a substantial spreading of the jet which is accompanied by a much faster center-velocity decay than is obtained for the steady jet. However, it should be noted that the hot wire measures the velocity magnitude rather than the *u* velocity component. Hence, the hot-wire measurements overestimate the entrainment, as is indeed confirmed by the pitot-tube and LDV measurements.

Figures 3 and 4 depict typical centerline velocity decays and jet-spreading trends as a function of downstream distance, measured by the pitot tube. These results were obtained for a pressure ratio of 1.137 and a vane amplitude of 2.6° zero-to-peak. Jet spreading is seen to increase with increasing frequency, which is accompanied by a decrease in centerline velocity with increasing frequency. Figure 5 shows a comparison of the velocity profiles measured by the pitot-static tube and by the laser Doppler velocimeter for one parameter combination.

The LDV consistently measured low compared to the pitot tube, even in steady jet flow and especially in the high-turbulence regions of the jet. Further measurements are clearly needed to evaluate the differences between the two measuring techniques in these regions.

Table 1 lists the entrainment results obtained to date for three pressure ratios, three vane amplitudes, and three frequencies (20, 30, 40 Hz) at three

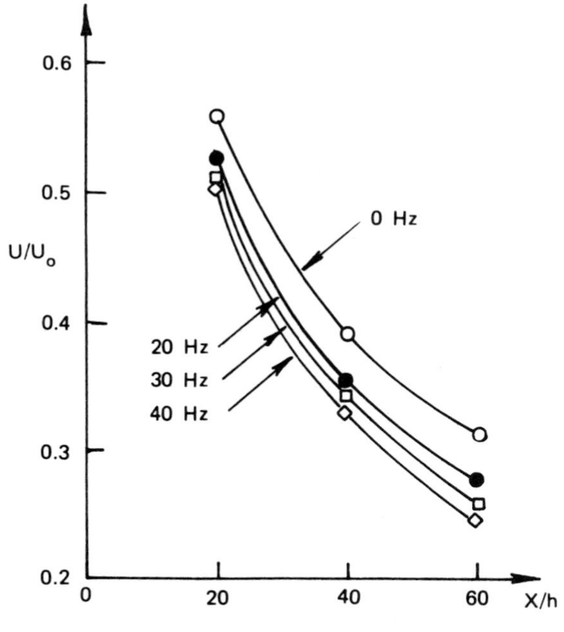

Figure 3. Mean centerline velocity decay as a function of downstream distance, measured with pitot tube at a vane frequency of 0, 20, 30, 40 Hz, amplitude of 2.6 degrees, pressure ratio 1.137.

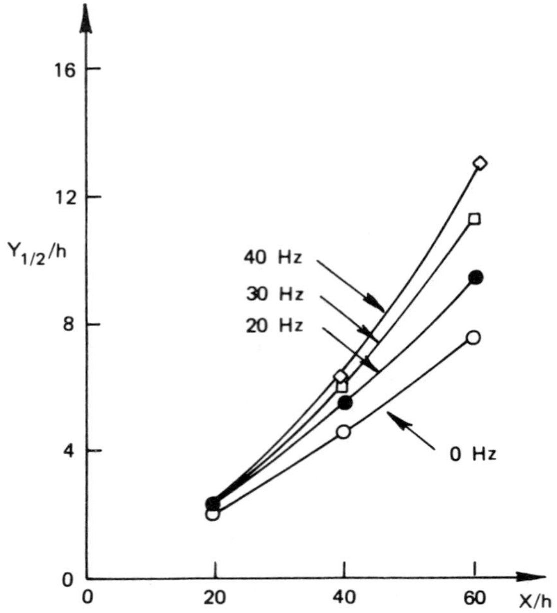

Figure 4. Mean jet-spreading (jet half-width) as a function of downstream distance, measured with pitot tube at a vane frequency of 0, 20, 30, 40 Hz, amplitude of 2.6 degrees, pressure ratio 1.137.

downstream stations (20, 40, 60x/h). Both the LDV and the pitot-tube measurements are shown. The following trends can be recognized:

1. Increasing vane amplitude and frequency substantially increase the entrainment.
2. Entrainment decreases with increasing pressure ratio.

Typical entrainment increases range from 10% to 175%, depending on the frequency, amplitude, and pressure ratio. For example, at the highest pressure ratio measured (1.268), the entrainment increase at station 60 is still between 30% and 100% for an amplitude of 6.9° at frequencies ranging from 20 to 40 Hz.

Considerable differences can be noted in some cases between the pitot-tube and LDV results, especially for large frequencies and amplitudes. This is to be expected in view of the measuring uncertainties of pitot tubes in oscillatory flow. However, significant differences also are obtained in steady jet flows which are caused mainly by the high turbulence regions of the velocity profiles. Salter [5.71] gives the following formula for the entrainment of steady rectangular jets:

$$\frac{Q(x)}{Q_E} = \left(0.6 - 0.04\,\frac{P_c}{P_A}\right)\left(\frac{x}{h}\right)^{1/2},$$

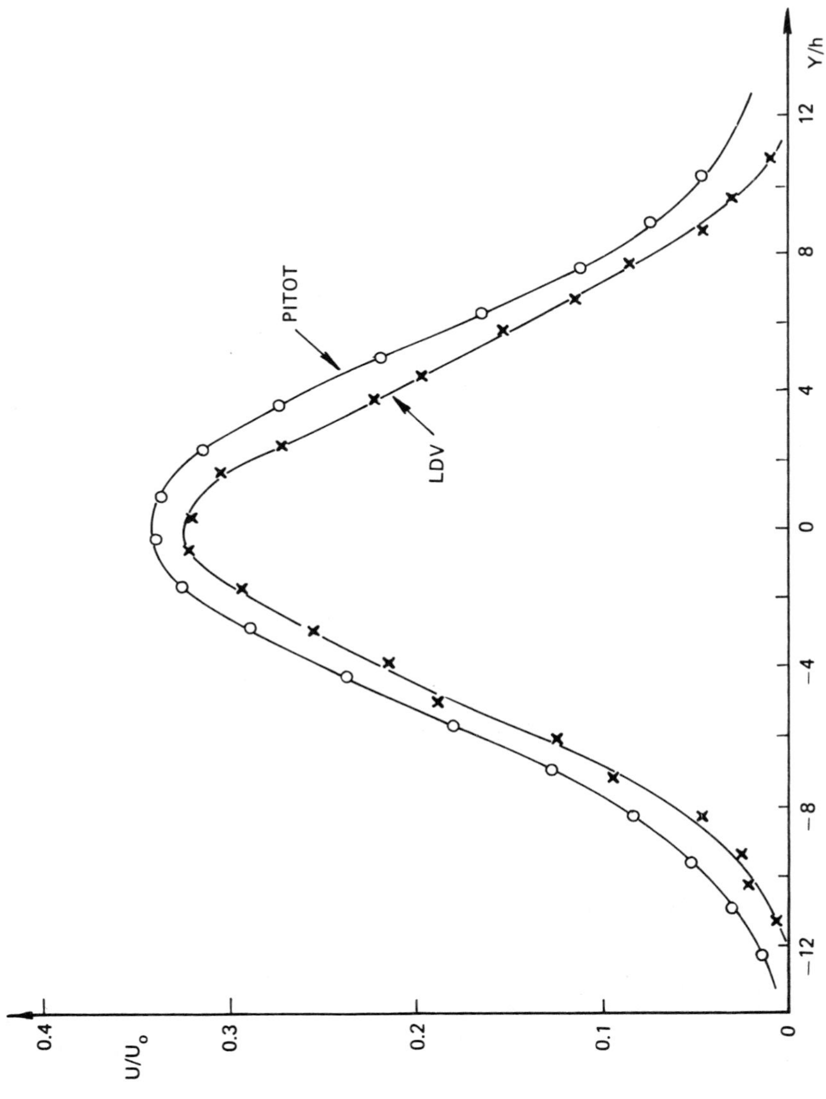

Figure 5. Comparison between measured velocity profiles at 40 nozzle widths, pressure ratio 1.137, vane amplitude 2.6 degrees zero-to-peak, frequency 30 Hz.

Table 1 Entrainment $Q(x)/Q_E$

Vane		Station					
		$20x/h$		$40x/h$		$60x/h$	
Frequency	Amplitude	LDV	Pitot	LDV	Pitot	LDV	Pitot
		(a) *Pressure ratio 1.008*					
0	0	1.45	1.45	2.52	2.7		
20	2.6	1.79	1.6	4.11	3.75		
30	2.6	1.89	1.8	4.2	4.1		
40	2.6	2.09	2.0	4.31	4.4		
20	4.6	2.57	2.1	5.18	4.6		
30	4.6	2.92	2.4	5.53	5.1		
40	4.6	3.24	2.7	6.02	5.7		
20	6.9	3.16	2.5	5.7	4.9		
30	6.9	3.19	2.9	6.19	5.4		
40	6.9	3.26	3.1	6.98	5.9		
		(b) *Pressure ratio 1.137*					
0	0	1.18	1.5	2.2	2.8	3.6	4.0
20	2.6	1.39	1.6	2.4	3.1	3.83	4.7
30	2.6	1.29	1.7	2.56	3.3	4.3	5.2
40	2.6	1.28	1.7	2.58	3.5	4.42	5.6
20	4.6	1.41	1.9	2.18	3.6	4.55	5.2
30	4.6	1.44	2.0	2.47	3.9	4.96	6.2
40	4.6	1.39	2.0	2.8	4.2	5.34	6.7
20	6.9	1.48	3.1	2.32	3.9	5.49	5.8
30	6.9	1.65	3.1	2.72	4.3	6.4	6.9
40	6.9	1.71	3.3	3.31	4.9	6.69	7.8
		(c) *Pressure ratio 1.268*					
0	0	1.08	1.5	2.2	2.7	3.16	3.8
20	2.6	1.11	1.6	2.2	2.9	3.17	4.2
30	2.6	1.15	1.6	2.3	3.1	3.6	4.5
40	2.6	1.00	1.7	2.43	3.2	4.0	4.8
20	4.6	1.01	1.8	2.24	3.3	3.67	4.8
30	4.6	1.30	1.9	2.24	3.6	4.22	5.5
40	4.6	1.33	2.0	2.36	3.9	5.87	6.0
20	6.9	1.20	2.0	2.30	3.7	4.14	5.5
30	6.9	1.21	2.1	2.63	4.0	5.7	6.1
40	6.9	1.17	2.3	2.79	4.5	6.4	7.1

which produces the following values for the three pressure ratios and stations measured:

$$\frac{Q(x)}{Q_E} - 1 = \begin{cases} \underline{20x/h \quad 40x/h \quad 60x/h} \\ 1.5, \quad 2.54, \quad 3.33 & \text{at } P_c/P_A = 1.008, \\ 1.48, \quad 2.51, \quad 3.3 & \text{at } P_c/P_A = 1.137, \\ 1.46, \quad 2.47 \quad 3.26 & \text{at } P_c/P_A = 1.268. \end{cases}$$

Analysis

Tennekes and Lumley [5.72] show that for plane turbulent jets the cross-stream momentum equation

$$U\frac{\partial V}{\partial x} + V\frac{\partial V}{\partial y} + \frac{\partial}{\partial x}(\overline{uv}) + \frac{\partial}{\partial y}(\overline{v^2}) = -\frac{1}{\rho}\frac{\partial P}{\partial y} + \nu\left(\frac{\partial^2 V}{\partial x^2} + \frac{\partial^2 V}{\partial y^2}\right)$$

can be approximated by retaining only the two terms

$$\frac{\partial}{\partial y}(v^2) = -\frac{1}{\rho}\frac{\partial P}{\partial y}. \tag{1}$$

Here, U, V are the mean velocities in the x- and y-directions, u, v the velocity fluctuations, P the static pressure, and ν the kinematic viscosity. This approximation assumes that the turbulence intensities are about half an order of magnitude smaller than the jet velocity, i.e.,

$$\frac{u}{U_s} = O\left(\frac{l}{L}\right)^{1/2},$$

where U_s is the maximum jet velocity, l is the cross-stream length scale, and L the streamwise length scale. The velocity scale for the turbulence is indicated by u, so that

$$-uv = O(u^2), \qquad u^2 = O(u^2), \qquad v^2 = O(u^2).$$

Integrating Eq. (1) and assuming no imposed external pressure gradient leads to

$$\frac{1}{\rho}\frac{\partial P}{\partial x} + \frac{\partial}{\partial x}(v^2) = 0 \tag{2}$$

Hence, the streamwise momentum equation

$$U\frac{\partial U}{\partial x} + V\frac{\partial U}{\partial x} + \frac{\partial}{\partial x}(\overline{u^2} - \overline{v^2}) + \frac{\partial}{\partial y}(\overline{uv}) = \left(\frac{\partial^2 U}{\partial x^2} + \frac{\partial^2 U}{\partial y^2}\right)$$

becomes

$$U\frac{\partial U}{\partial x} + V\frac{\partial U}{\partial y} + \frac{\partial}{\partial y}(\overline{uv}) = 0, \tag{3}$$

where only the first, second, and fourth terms have been retained, and where it has been assumed that

$$\frac{U_s}{u}\cdot\frac{1}{\mathrm{Re}}\left(\frac{l}{L}\right) \to 0,$$

which is always satisfied for sufficiently large Reynolds numbers Re. Integration of Eq. (3) with respect to y gives the well-known condition of momentum conservation

$$\rho\int_{-\infty}^{+\infty} U^2\, dy = M_0, \tag{4}$$

where M_0 is the total amount of momentum put into the jet at the origin per unit time.

The third and fourth terms in the streamwise momentum equation differ by a factor of order $O(l/L)$. For the vane-excited jet the cross-stream length scale l will be significantly larger than for the nonexcited jet. This suggests retaining these terms, thus leading to the following more accurate condition for momentum conservation:

$$\int_{-\infty}^{+\infty} (U^2 + \overline{u^2} - \overline{v^2})\, dy = \text{const.} \tag{5}$$

In Table 2,

$$M_1 = \int_{-\infty}^{+\infty} U^2\, dy \quad \text{and} \quad M_2 = \int_{-\infty}^{+\infty} (U^2 + \overline{u^2})\, dy$$

Table 2 Momentum values M_1 and M_2.

Momentum M_2					Momentum M_1				
		Station					Station		
Frequency	Amplitude	$20x/h$	$40x/h$	$60x/h$	Frequency	Amplitude	$20x/h$	$40x/h$	$60x/h$
(a) *Pressure ratio 1.008*									
0	0	1.13	1.1		0	0	1.03	0.99	
20	2.6	1.34	1.58		20	2.6	1.15	1.36	
30	2.6	1.33	1.56		30	2.6	1.11	1.36	
40	2.6	1.45	1.52		40	2.6	1.17	1.32	
20	4.6	1.66	1.85		20	4.6	1.38	1.58	
30	4.6	1.81	1.74		30	4.6	1.5	1.5	
40	4.6	1.94	1.82		40	4.6	1.64	1.57	
20	6.9	1.96	1.83		20	6.9	1.59	1.54	
30	6.9	1.91	1.8		30	6.9	1.49	1.52	
40	6.9	1.79	1.96		40	6.9	1.43	1.68	
(b) *Pressure ratio 1.137*									
0	0	0.97	0.97	1.03	0	0	0.85	0.85	0.92
20	2.6	1.09	1.02	1.09	20	2.6	0.90	0.84	0.84
30	2.6	1.05	1.05	1.12	30	2.6	0.84	0.83	0.86
40	2.6	1.04	1.01	1.06	40	2.6	0.83	0.77	0.85
20	4.6	1.13	0.97	1.18	20	4.6	0.83	0.83	0.85
30	4.6	1.13	1.0	1.12	30	4.6	0.79	0.62	0.82
40	4.6	1.08	0.99	1.14	40	4.6	0.76	0.62	0.9
20	6.9	1.07	1.0	1.2	20	6.9		0.54	0.84
30	6.9	1.13	1.02	1.2	30	6.9		0.53	0.92
40	6.9	1.13	1.08	1.15	40	6.9		0.63	0.94
(c) *Pressure ratio 1.268*									
0	0	0.94	0.98	1.0	0	0	0.85	0.86	0.88
20	2.6	0.97	0.98	1.01	20	2.6	0.79	0.81	0.79
30	2.6	0.99	0.98	1.13	30	2.6	0.80	0.72	0.82
40	2.6	0.90	1.0	1.17	40	2.6	0.70	0.79	0.85
20	4.6	0.97	0.94	1.1	20	4.6	0.74	0.66	0.77
30	4.6	1.08	0.89	1.13	30	4.6	0.62	0.59	0.74
40	4.6	1.12	0.89	1.17	40	4.6	0.77	0.57	0.79
20	6.9	0.97	0.95	1.19	20	6.9	0.60	0.54	0.7
30	6.9	0.94	0.92	1.37	30	6.9	0.58	0.54	0.9
40	6.9	0.88	0.89	1.4	40	6.9	0.51	0.53	0.97

are given from the available LDV data. For the lowest pressure ratio, 1.008, a significant increase in both jet momentum values is observed in the excited jet. The additional momentum imparted to the flow by the oscillating vane in this case is likely to be a substantial percentage of the initial jet momentum. Since this imparted momentum remains essentially constant at the higher pressure ratios, its contribution becomes less significant in comparison with the available initial jet momentum. Indeed, the momentum values at the two higher pressures ratios exhibit no significant differences between the steady and the excited jet momentum values. However, it can be noted that the M_2-values increase with increasing downstream distance and with increasing vane amplitude, in apparent violation of momentum conservation. Here, it must be remembered that the condition for momentum conservation, Eq. (5), contains at least one additional term, i.e., the fluctuating cross-flow momentum. Therefore, any definite statements about momentum conservation must await the measurement of the cross-flow fluctuations. Such measurements are planned in the next phase of this investigation. Also, a more detailed analysis will be attempted, based on the unsteady-shear-layer equations for turbulent flow, similar to the analysis of pulsating jets by Lai and Simmons [5.73] and by Carrion [5.74]. Finally, it is interesting to note that the M_1 and M_2 values differ only by about 10% for the steady jet, thus indicating the small contribution of the turbulent fluctuations. In contrast, the excited jet exhibits quite substantial variations between these two momentum integrals, again indicating the need for more detailed measurements and analyses of this type of jet flow.

Summary

Velocity measurements using pitot-tube, hot-wire and laser Doppler anemometry have been performed in jets which exhausted into still air and which were excited by a small oscillating vane located in the jet's potential core. Entrainment results obtained to date for three pressure ratios (1.008, 1.137, 1.268), three vane amplitudes (2.6, 4.6, 6.9° zero-to-peak), and three frequencies (20, 30, 40 Hz) at three downstream stations (20, 40, 60 nozzle widths) show significant entrainment increases over the steady jet. In particular, the following trends were identified: entrainment increases with increasing vane amplitude and frequency, but decreases with increasing pressure ratio.

CHAPTER 34

Measurements in Ducted Flows By Laser Doppler Anemometry

A. M. K. P. Taylor, J. H. Whitelaw, and M. Yianneskis*

Introduction

An important purpose of the present contribution is to indicate advantages of water-flow and laser Doppler anemometry for the purpose of obtaining information about ducted flows. Thus, the previous results of [1.121, 5.75–5.77] are discussed and some examples presented. In addition, new measurements are reported that were obtained in a transition duct with a square cross-section at inlet and a round cross section at outlet.

The familiar merit of water flow is that the same Reynolds number can be obtained with a smaller scale than with air. This is often outweighed by the inconvenience and difficulties with instrumentation. Laser Doppler anemometry can, however, be used to advantage, provided the containing walls are transparent, and allows an accuracy of measurement which can more than compensate for any inconvenience associated with a liquid, while retaining the advantage of scale.

For the above reasons, the measurements of [1.121, 5.75–5.77] made use of water as working fluid. The ducts, which comprised combinations of square, straight sections, and a square-sectioned 90° bend of 92-mm mean radius, were manufactured to tolerances better than those usually found in practice and ensured that imprecisions of manufacture did not influence the results. Thus, and since an important purpose was to aid the evaluation and improvement of related calculation methods, the geometric

* Department of Mechanical Engineering, Imperial College, London SW7 2BX.

boundary conditions were known precisely (see [5.79]). This is also true of the present results, where the cross-sections of the transition can be closely described by superellipses of unity shape factor.

The following section describes the flow configurations, the instrumentation, and the related advantages. As described, the use of water can allow the corresponding use of a simple form of laser Doppler anemometer which, for this application, is also the most precise. The measurement precision is discussed in raeltion to previous results, and the new results are presented in the third section. The fourth and last section presents a brief concluding discussion.

Flow Configuration, Instrumentation, and Precision

Square-sectioned bends are found in many practical applications, including, for example, the cooling passages of alternators. They are manufactured in a variety of ways, and the investigations of [1.121, 5.75–5.77] sought to provide results corresponding to a prescribed geometric shape. To achieve a flow Reynolds number of 40,000, a square cross-section of 40-mm side was chosen with a bulk velocity of 1.00 m/s. This arrangement was manufactured from 10-mm-thick Plexiglas and allowed complete optical access to a laser Doppler anemometer which made use of a low-power laser and a frequency-tracking demodulator. The use of air in the same duct with a bulk velocity of around 10 m/s would, however, require seeding of the flow with appropriate scattering centers. Alternatively, the duct could be increased in size to allow the use of hot-wire anemometry, though problems of access and uncertainties due to probe interference in regions of secondary flow would remain.

Figure 1 shows, in line-diagram form, the optical and signal-processing arrangements used in [1.121] and evolved from those of [5.75] to [5.77]. It comprises a 5-mW helium–neon laser and a transmission optics based on a diffraction grating (TNO-TH model H), although frequency shifting was not applied in this flow. The transfer function of the anemometer, corresponding to a beam-intersection half angle of 9.3°, used for the turbulent-flow measurements of [1.121], was 0.51 MHz/ms^{-1}. The light scattered in the forward direction was collected and focussed to the 0.5-mm-diameter pinhole located in front of the photomultiplier (EMI 9658B). The scattering centers, occurring naturally in water, were supplemented by minute quantities of milk to achieve a mean data rate of about 500 Hz. The output of a frequency-tracking demodulator (Cambridge Consultants model CC01) was time-averaged to give the mean and the rms of velocity components and thus avoided "velocity biasing"; errors associated with Doppler-broadening effects can be readily corrected (see [5.78]).

The instrumentation shown in Fig. 1 represents a particular simple form of laser Doppler anemometer and allows measurement precision, shown in Table 1, which is of similar order to that which would be achieved by hot-wire

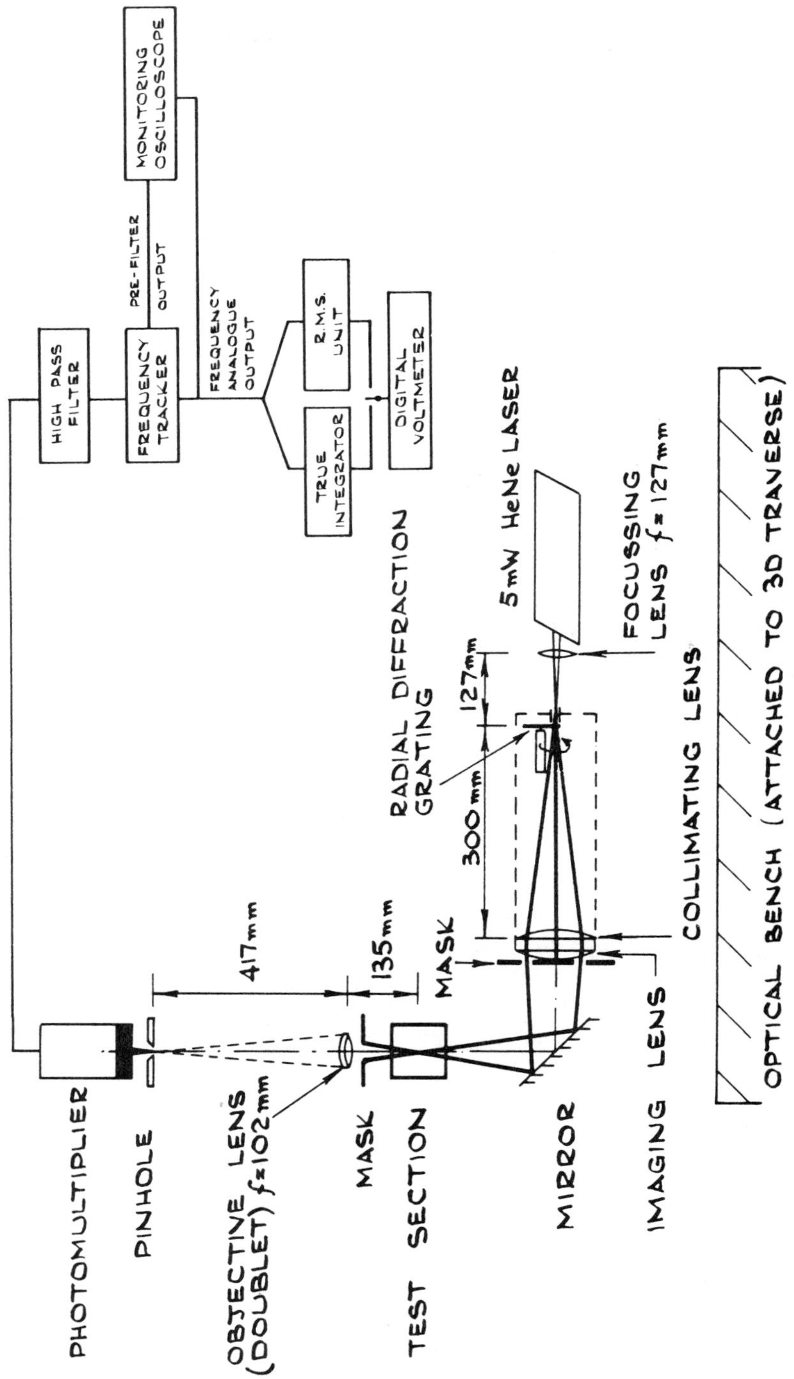

Figure 1. Laser Doppler velocimeter.

Table 1 Measurement uncertainties.[a]

Quantity	Systematic error	Random error
U/V_c	$<2\frac{1}{2}\%$	$\pm 1\frac{1}{2}\%$
V/V_c	$<3\%$	$\pm 1\frac{1}{2}\%$
W/V_c	$<3\%$	$\pm 1\frac{1}{2}\%$
$\tilde{u}/V_c$	$<3\%$	$\simeq \pm 1$ to $\pm 3\%$
$\tilde{v}/V_c$	$<3\%$	$\simeq \pm 1$ to $\pm 5\%$
$\tilde{w}/V_c$	$<3\%$	$< \pm 1$ to $\pm 5\%$
$\overline{uv}/V_c^2$	$< \pm 1\frac{1}{2}\%$	$\simeq \pm 2.5$ to $\pm 8\frac{1}{2}\%$
$\overline{uw}/V_c^2$	$< \pm 1\frac{1}{2}\%$	$\simeq \pm 2.5$ to $\pm 8\frac{1}{2}\%$

[a] Note that the systematic error in mean quantities is due mainly to gradient effects, which occur only close to walls and for which corrections can readily be made.

anemometry, in the absence of probe interference. The systematic errors are mostly attributable to velocity-gradient broadening, which is negligible except close to walls and for which precise corrections can readily be made, and to imprecision in the alignment of the sensitivity vector of the anemometer. It should be noted that the optical arrangement allows the measurement of two velocity components by rotation of the diffraction grating, and of the third by movement of the entire optical system. Simultaneous measurement of two velocity components is possible, but the simple and more easily monitored arrangement of measuring individual components is preferred.

As an example of mean-velocity results, Fig. 2 presents a small sample of those reported in [1.121] in the form of contours of the longitudinal velocity component normalized by the bulk velocity V_c. The contours were obtained by linear interpolation of a 5×10 net of individual measurements. Comparison of the laminar- and turbulent-flow results, corresponding to Reynolds numbers of 790 and 40,000, shows the greater upstream influence of the bend in the turbulent-flow case, and movement of high-velocity fluid closer to the outer radius of the bend at its exit in the laminar-flow case. The corresponding secondary flows were also measured and showed the counter-rotating nature of the secondary flows, with values up to $0.6V_c$ and $0.4V_c$ in the laminar and turbulent flows respectively. These values may be contrasted with the $0.3V_c$ measured with the more fully developed inlet conditions of [5.77] and with the normal-stress-driven secondary flows of $0.015V_c$ measured in the straight duct of [5.76].

Reference [1.121] reports on the difference in the two flows which have been revealed by the detailed measurements and shows that, with the measurement uncertainties of Table 1, the results are appropriate to the evaluation of calculation methods.

Such evaluations are reported in [5.76] and [5.77] and, in general, indicate that the laminar flow can be calculated with reasonable precision, but that the turbulent flow cannot. For the laminar flow, the grid dependence of the calculations is certainly larger than the measurement errors. For the

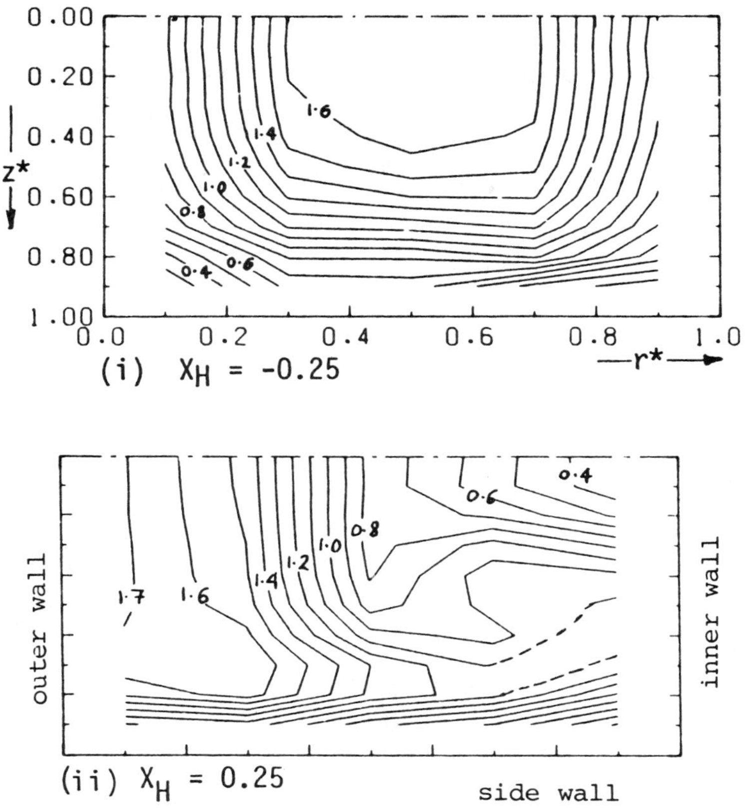

Figure 2. Laminar 90 degree bend flow, streamwise components isotachs.

turbulent flow, as indicated in [5.77], the uncertainties in the calculation are related more to numerical errors than to turbulence model assumptions.

Square-to-Round Transition

Square-to-round transition ducts are used in a wide range of engineering configurations and are of particular relevance to the intake ducts of some gas-turbine engines. The measurements presented here are intended to quantify the flow characteristics for a duct of this type and so provide better understanding and a basis for the testing of design methods. They were obtained in a duct of 460-mm length, 40-mm square cross section at inlet, and 40-mm diameter at exit, and include values of the mean axial and secondary velocities; the corresponding normal stresses and two shear stresses were also measured and are presented in [5.79].

The measurements were obtained with a laser Doppler anemometer, similar to that described in [1.121] and in the previous section, and with water flowing at a Reynolds number, based on the upstream hydraulic diameter and bulk velocity 0.89 m/s, of 35,400. The use of water allowed accurate measurements in those regions accessible to the light beams, but the surface curvature of the duct and the refractive-index gradients at the water–Plexiglas interface limited measurements in the curved parts of the duct. Matching the refractive index of the work fluid to that of Plexiglass as used in [5.80] to allow extensive measurements in a round bend, could not be used at the present turbulent Reynolds numbers, which represents a limitation of the technique.

Figure 3 shows the development of the streamwise velocity on the center-line of the duct. The velocity increases from $1.12 V_c$ upstream of the transition to $1.43 V_c$ downstream, as a consequence of the decrease in cross-section. Figures 4(a)–4(e) show profiles of the streamwise mean velocity U and the corresponding root-mean-square velocity $\tilde{u}$ at a number of stations up-stream of, within, and downstream of the transition piece. The boundary-layer thickness on the symmetry plane, defined at 95% of the maximum velocity, is about 13% of the hydraulic diameter at the inlet to the transition piece (Fig. 4(a)), but rises to about 20% at the exit (Fig. 4(d)). The change in the profiles is comparatively small, as would be expected from the centerline change in Fig. 3. The streamwise velocity distributions are in qualitative agreement with the predictions of [5.81]. Careful examination of the profiles near the sidewalls shows that the fluid near the planes x_2/D and $x_3/D = 0$ accelerates relative to that near to the fillets which form the transition, and that the boundary layers are thinner away from the fillets. The profiles of $\tilde{u}$ (and $\tilde{v}$, Fig. 4(f)) are those expected for a turbulent wall boundary layer.

Figure 5 presents profiles of the cross-stream velocity for three stations within the transition piece. The measurements quantify the symmetry about planes at 45° to the x_2 or x_3 planes, which is within the limits of the systematic error discussed in Section 2 as are the magnitudes of the cross-stream velocities on the symmetry planes (x_2/D or $x_3/D = 0$). The measurements show that large values of the cross-stream velocity are found close to the walls and to the fillets which form the transition from the square to the circular cross-section. The largest values shown are of the order of $0.07 V_c$, although the shape of the profile suggests that the maximum lies closer to the wall than $x_3/D \approx 0.9$. The magnitude and direction of these velocities indicate that their source is pressure gradients rather than Reynolds-stress gradients. The direction of the cross-stream velocities is always away from the channel diagonal (line AA' in Fig. 6) and thus cannot satisfy continuity in the cross-stream plane. This implies an acceleration of the streamwise velocity component near the symmetry planes (lines BB' in Fig. 6) which is indeed observed, particularly in Fig. 4(c).

The origin of the lateral pressure gradient, which drives the cross-stream flow, is due to the increase in pressure (P_2) near the fillet relative to that in the region of the symmetry planes (P_1). This increase is brought about by the streamline curvature occurring in planes containing the channel diagonal and the duct center-line upon entry to the transition piece, as shown in Fig. 6(a).

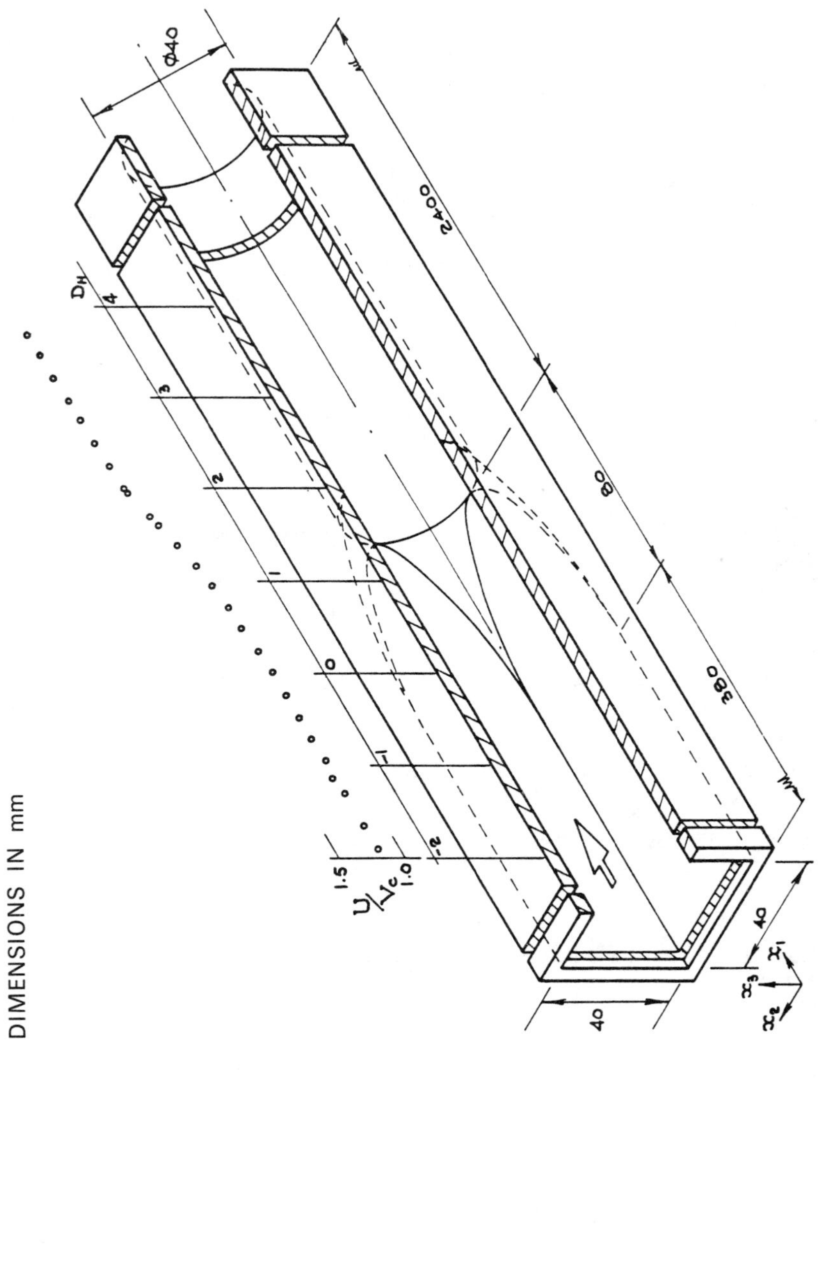

Figure 3. Flow configuration, coordinate system and centerline streamwise velocity variation.

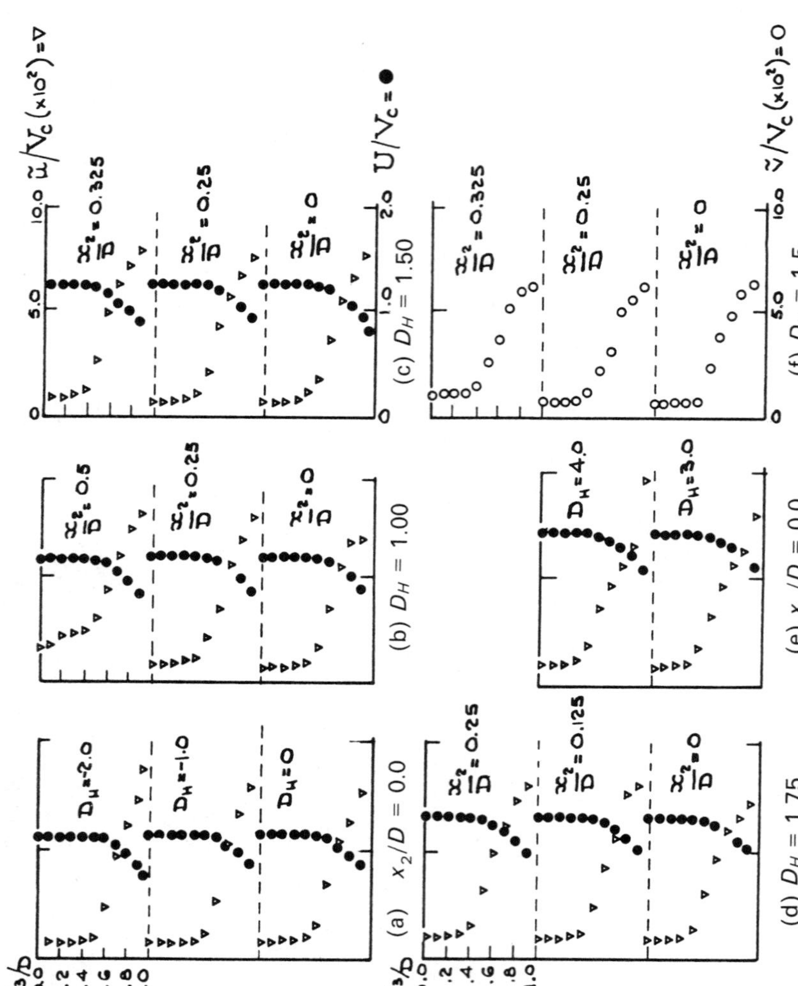

Figure 4. (a)–(e): streamwise mean velocity profiles and turbulence levels (f)—cross-stream turbulence levels.

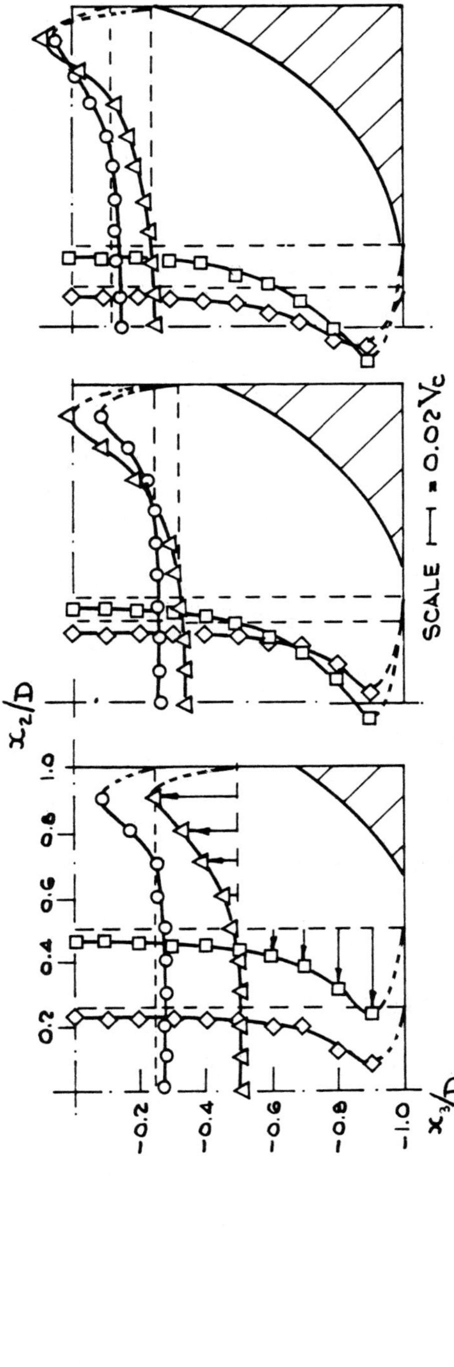

(a) $D_H = 1.0$ (b) $D_H = 1.5$ (c) $D_H = 1.75$

Figure 5. Secondary velocity profiles.

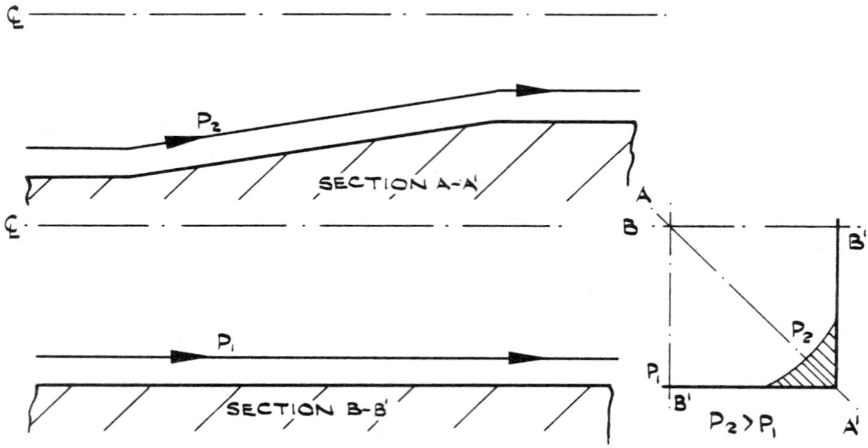

Figure 6. Effect of longitudinal streamline curvature on cross-stream pressure variation.

In contrast, no such curvature occurs with streamlines contained in planes such as that depicted in Fig. 6(b). It is noted that this mechanism should be reversed on exit from the transition piece and can be expected to cause the rapid decay of the cross-stream components.

These lateral pressure variations were also observed by Mayer [5.82]; the qualitative changes in pressure and the related secondary-flow directions are in accordance with the above discussion. However, the simultaneous occurrence of longitudinal convergence on one wall and divergence on the other wall of the transition duct results in cross-stream flows not directly comparable to those presented in this report.

Measurements of the $\overline{uw}$ and $\overline{uv}$ cross correlations were also obtained. As expected, they were very low ($<0.2 \times 10^{-3} V_c^2$) over the central region of the flow, rising to about $0.5 \times 10^{-3} V_c^2$ near the sidewalls. These values are close to the limits of accuracy of the velocimeter and are thus not presented. The results from which the figures have been plotted are tabulated in [1.121].

Concluding Remarks

The following more important conclusions may be drawn from the preceding text.

1. Precise measurements of mean-velocity components, and corresponding Reynolds stresses, have been obtained in practically relevant geometries with water as the fluid. The use of water allows these measurements to be obtained with a simple and convenient laser Doppler anemometer. The combination of fluid and instrumentation is appropriate for "benchmark" experiments in many flow configurations.

2. The flow in a square-to-round transition section, whose cross-section can be represented by superellipses, is characterized by secondary flows which have a maximum magnitude of $0.07V_c$ in the arrangement examined here and are caused by pressure forces.

Nomenclature

D — $D_H/2$
D_H — hydraulic diameter (40 mm)
r^* — normalized radial coordinate
U — mean velocity in x_1-direction
$\tilde{u}$ — rms fluctuating velocity in x_1-direction
$\overline{uv}$ — cross-correlation between u and v
$\overline{uw}$ — cross-correlation between u and w
V — mean velocity in x_2-direction
$\tilde{v}$ — rms fluctuating velocity in x_2-direction
V_c — bulk mean velocity
W — mean velocity in x_3-direction
$\tilde{w}$ — rms fluctuating velocity in x_3-direction
x_1, x_2, x_3 — coordinate directions (Fig. 3)
X_H — Axial distance upstream of inlet or downstream of exit of 90° bend in hydraulic diameters
z^* — normalized spanwise coordinate

Acknowledgment

The measurements of Section 3 were carried out under NASA Contract number NASW-3258. The authors are grateful to Mr. B. Anderson and Dr. L. Povinelli for useful discussion on this aspect of the paper.

CHAPTER 35

Fluid-Mechanics Mechanisms in the Stall Process of Airfoils for Helicopters

Warren H. Young, Jr.*

Introduction

The stalling of helicopter rotor blades occurs on the retreating side of the rotor disk. The lower velocity on the retreating side (typically 0.3 to 0.5 Mach number) is compensated for by higher lift coefficients, which require higher angles of attack. When stall occurs, the blade dynamic and elastic properties become important in determining the subsequent changes in angle of attack. The aerodynamics of dynamic stall govern the aeroelastic response and can lead to a dynamic response known as stall flutter. In describing the fluid mechanics involved in dynamic stall, the blade motions must be taken into account. The three basic structural modes that are excited by dynamic stall are the blade torsional mode, the bending (normal to the chord) mode, and the flapping mode. The excitation of these three modes is illustrated in Fig. 1 from Crimi's aeroelastic analysis of a two-dimensional section of a helicopter blade in [5.83] and [5.84]. The three basic structural modes of motion, at three different frequencies, contribute to the angle of attack. The mixture of modes causes the sequence of stall and unstall occurrences to be at irregular time intervals. The structural response most important to the angle of attack is the pitching-angle displacement. The sharp spikes in the aerodynamic pitching moment act as a series of impulses to cause pitching oscillations. The unsteady lift causes low-frequency motions due to blade

* Structures Laboratory, U.S. Army Research and Technology Laboratories (AVRADCOM), NASA Langley Research Center, Hampton, VA 23665.

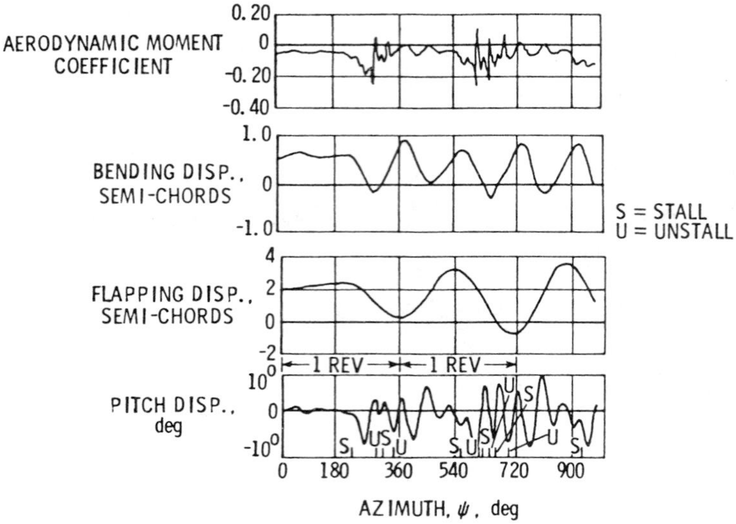

Figure 1. The interaction of dynamic stall and helicopter blade motions (S denotes stall, U denotes unstall). [5.83].

bending and flapping. These two plunge motions are of secondary importance in determining the angle of attack and oscillatory structural response. These oscillations cause vibrations and blade loads which often limit the flight envelope.

Unlike classical (unstalled) flutter, the oscillations due to stall flutter result from a series of large impulsive aerodynamic forces. The aerodynamic damping (negative or positive) of a dynamic stall cycle does not have enough time to significantly amplify or diminish the helicopter-blade oscillations. After only a few torsional oscillations the blade has rotated to the advancing side of the rotor disk, and the mean angle of attack is too small to sustain stall flutter. Even in a steady freestream, however, the amplitude of flutter oscillations is usually limited to 8–20°. Stall flutter also differs from classical flutter in that the torsional and bending frequencies are not close together even though both modes contribute to the stall flutter.

Most of the dynamic-stall experimental work has been for a two-dimensional airfoil section oscillating sinusoidally in pitch about the quarter-chord point. Much of this research has been done at the Aeromechanics Laboratory of USARTL (AVRADCOM) by W. J. McCroskey and his associated. This work has clarified the features of dynamic-stall aerodynamics up to a Mach number of 0.3. Reference [4.54] is an overview of this research, and the most recent results are described in [5.85]. The present paper will examine several recent experimental and analytical studies against the background of this research.

The purpose of this paper is to identify the many aerodynamic physical flow mechanics which must be modeled by a numerical solution of dynamic stall. The phenomena that control the flow during the stall portion of a dynamic-stall cycle are described, and their influence on blade motion is outlined. Four mechanisms by which dynamic stall may be initiated are identified. The interaction of the flow and the helicopter structural dynamics is considered in order to assess the relative importance of the various flow phenomena to a dynamic-stall calculation. The fluid mechanics that contribute to the identified flow phenomena are summarized, and the usefulness of a model that incorporates the required fluid mechanics mechanisms is outlined.

Airfoil Upper-Surface Flow

Vortex Shedding

Vortex shedding is responsible for the large excursions in the pitching moment that occur after stall. A large vortex (or series of several vortices) shed from near the leading edge of the airfoil controls the flow over the upper surface of the airfoil. A two-dimensional solution of the Navier–Stokes equations by Mehta [5.86] has demonstrated that the speed of downstream movement and size of this vortex (Fig. 2) can be calculated for laminar flows. Definition of the movement of the vortex is particularly important in calculating dynamic, aerodynamic, and structural interactions. The shed vortex contains a region of low pressure and generates a large unsteady lift force on the airfoil. As this low-pressure region moves from the leading edge to the trailing edge, large nose-down pitching moments are generated on the airfoil. Because of the torsional flexibility of helicopter blades, these pitching moments cause deflections which reduce the angle of attack. When the vortex is convected past the trailing edge and the nose-down pitching moment is diminished, the structure will tend to oscillate at the torsional natural frequency. If the time for the pitching moment change is equal to one-half the torsional natural period of the blade, a powerful aeroelastic amplification of the blade pitch oscillation will occur. The changes in total lift will simultaneously interact with bending of the blade. These torsion and bending displacements must be included in a determination of the aerodynamic loads on the airfoil.

The time history of the movement of the pressure pulses has been defined by Carta [5.87]. Pressures measured on a pitching airfoil were presented as pseudo-three-dimensional plots of pressure time histories (Fig. 3). The ridges define regions of low pressure, and the valleys define regions of higher pressure. The time history at each chordwise location has a peak. Connecting the peaks from leading to trailing edge defines a ridge. The slope of the ridge is related to the speed of a pressure pulse that proceeds from the leading to the trailing edge. The speed depends strongly on the reduced frequency and

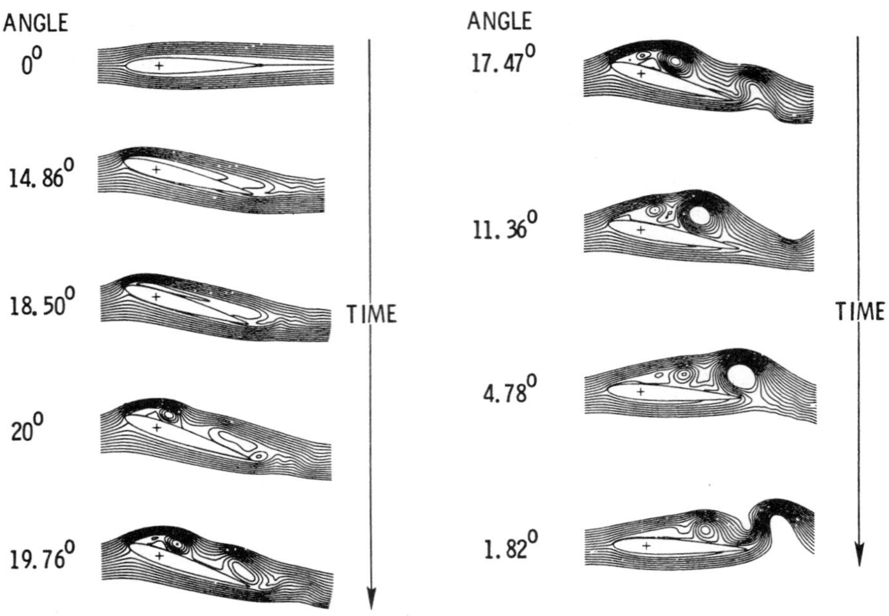

Figure 2. Calculations of streamlines for one cycle of dynamic stall in laminar flow. [5.86].

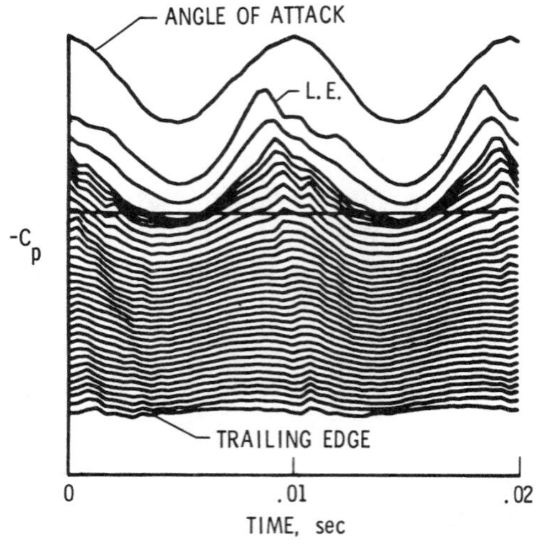

Figure 3. Time histories of airfoil pressures in a pseudo-three-dimensional format. [5.87].

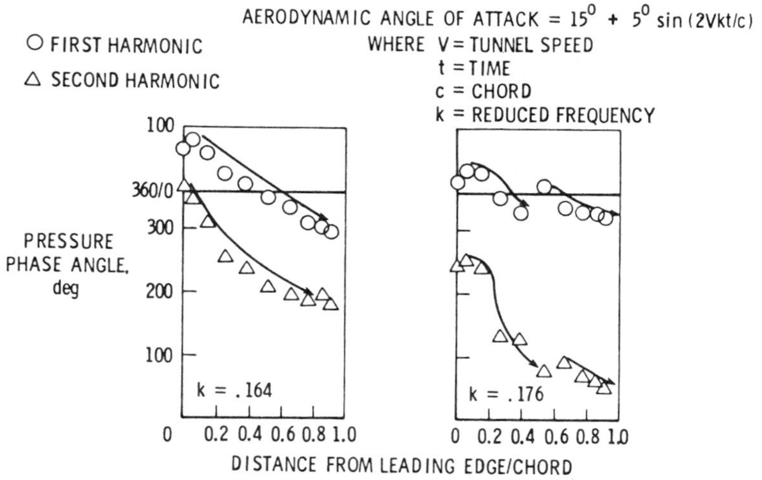

Figure 4. Phase angles of pressure harmonics for airfoil dynamic stall. [5.88].

to a lesser degree on the mean angle of attack. There is also some evidence that the speed depends on the amplitude of oscillation of the angle of attack.

The time history of the pressure pulses becomes even more complex when plunging motion is considered. In a separate investigation [5.88], Carta found that plunge motion caused patterns in the pressures that were not observed in pitching motion. Figure 4 compares the phases of the first two harmonics of the pressures on the airfoil upper surface. Arrows have been added to Carta's figure to indicate that decreasing phase angle represents rearward movement of the pressure associated with the harmonic. Although this test was at a fairly low speed (Mach number 0.18 and Reynolds number 0.6×10^6), it is clear that empirical approximations to vortex motion based on pitch alone are inadequate for the mixed pitch and plunge motions that occur on helicopter rotor blades. The downstream motion of vortices over pitching airfoils, although dependent on frequency and amplitude, is at least a smooth function of time. However, plunge motion apparently disrupts the orderly downstream propagation of vortices.

Recovery from Stall

The part of the dynamic stall cycle that is dominated by shed vortices is terminated by the recovery from stall. Once the airfoil aerodynamic angle of attack has decreased sufficiently to allow the flow over the airfoil upper surface to reattach, initially the process of vortex shedding will cease, then previously shed vortices will be convected downstream, and eventually the airfoil will become completely unstalled. This is the least investigated and most analytically intractable part of the dynamic stall cycle. An early investigation [5.89] showed that the recovery process was nonrepeatable.

Figure 5 shows both lift and pitching-moment variations for successive cycles of an airfoil oscillating in pitch. The cause of the nonrepeatability is unknown.

An attempt to gain insight into the flow field after the shed vortices have passed downstream was summarized in [5.90]. The experiments measured the velocities about the center span of an aspect-ratio-8 wing at about 19.5° angle of attack. These static-stall results were obtained in two experiments, which are identified as the high-Mach-number ($M = 0.49$) case and the low-Mach-number ($M = 0.148$) case. The velocities above the airfoil in the low-Mach-number case (Fig. 6) form a classical turbulent free shear layer. The velocity profiles (Fig. 7) are matched with Gortler profiles using a spreading parameter taken from shear-layer experiments with free jets [5.91]. The solid line in Fig. 6 was calculated by the inviscid analysis of [5.92]. The solid line is the track of the vortex sheet that the inviscid analysis used to approximate the free shear layer. The experimental center of the free shear layer (the dashed line) agrees with the analysis amazingly well. The high-Mach-number airfoil was found to shed strong vortices from near the airfoil crest. As these vortices were convected downstream, the paths that successive vortices followed grew increasingly random as they approached the trailing edge. The convection speed increased from about 30% of the freestream speed near the crest to near the freestream speed at the trailing edge.

These experiments suggest that as the shed vortices in dynamic stall are convected downstream, a free shear layer is established behind them. This

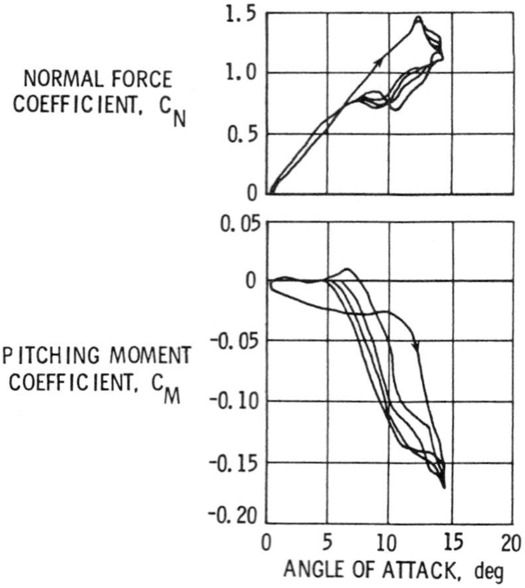

Figure 5. Successive cycle-to-cycle variation in C_N and C_M during pitch oscillations. [5.89].

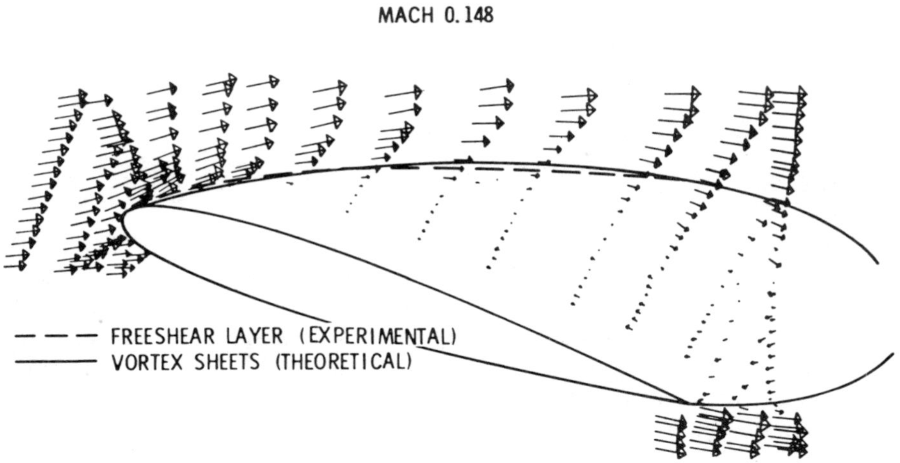

Figure 6. Measured velocities and correlation of free shear layer locations for low-mach-number case. [5.90].

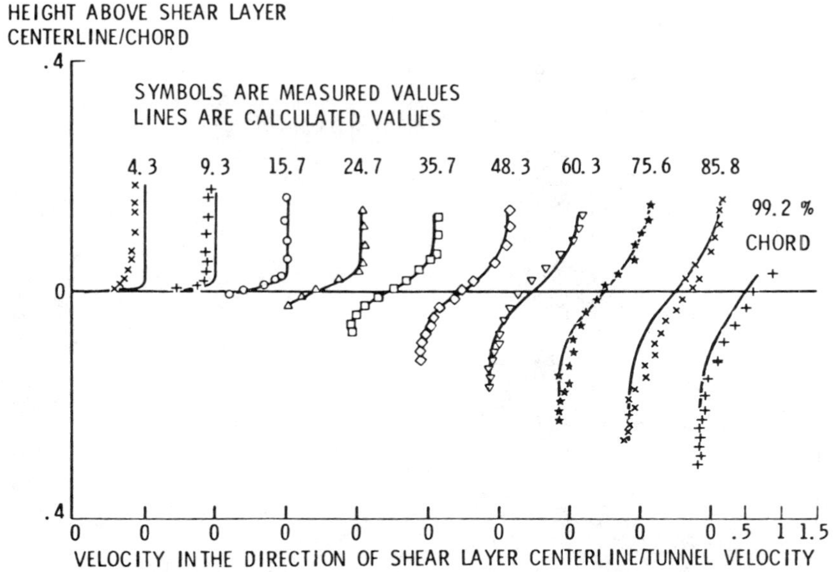

Figure 7. Measured velocities above a stalled airfoil compared to turbulent free shear layer profiles. [5.90].

free shear layer is amenable to analysis because it tends in the steady state toward simple jetlike velocity profiles. However, the number and timing of the vortices shed from near the crest of the airfoil are not known.

Wake Vorticity

As the shed vortices move downstream, a large wake region is formed behind the airfoil. Information about the wake comes from low-Reynolds-number flow visualizations and from measurements in static stall wakes. The inviscid static stall model of [5.92] does not require the vortex sheets shed from near the airfoil crest and from the trailing edge to meet downstream. Figure 6 shows a gap between the two vortex sheets calculated by Maskew and Dvorak. Through this gap fluid is convected forward into the separated region. This has been observed in the flow fluctuations made visible by the vapor-screen technique illustrated in Fig. 8. Fluid is lost from the separated region by entrainment. Figure 9 is a detail of Fig. 6 showing the entrainment caused by the shear layer at the trailing edge. Because of this entrainment there can be no wake closure in static stall.

In dynamic stall the velocity shed into the wake during one cycle can be as much as one-half the bound vortex strength required to give the airfoil its maximum lift. Figure 5 shows the lift coefficient decreasing from 1.4 to 0.7 during the stall process. The change in circulation on the airfoil is of course equal to the circulation contained in the shed vortices. The velocity induced by a shed vortex will be a significant influence on the airfoil angle of attack

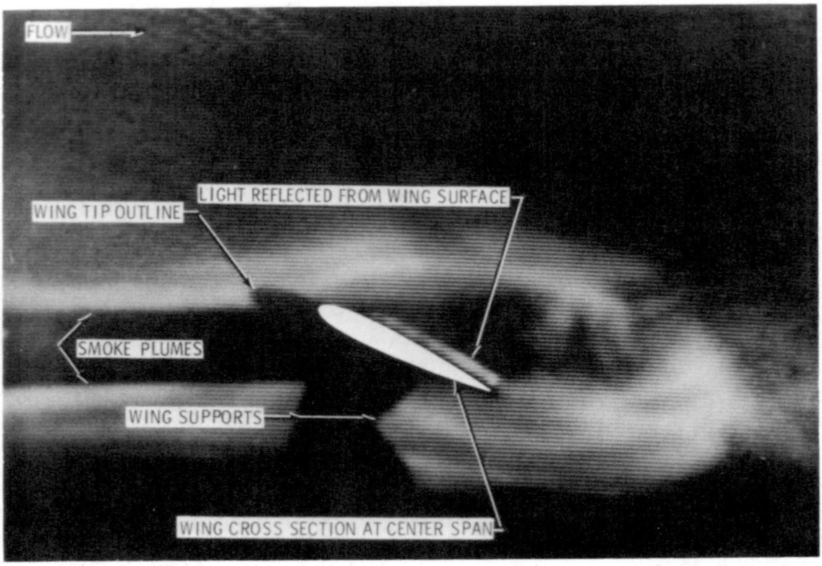

Figure 8. Smoke flow visualization at the wing center span in static stall. [5.80].

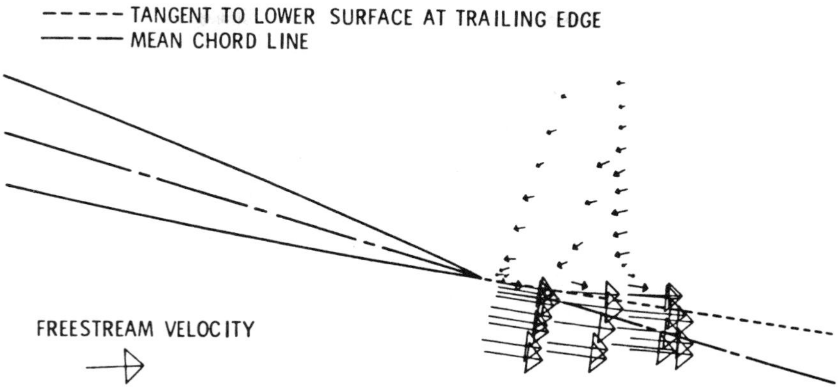

Figure 9. Resultant mean velocity vectors behind the trailing edge of a stalled airfoil. [5.90].

even after it has been convected 20 or 30 chord lengths downstream. The very detailed flow visualizations of McAlister and Carr [5.93] show a thick wake even before stall occurs. During stall recovery (Fig. 10) the shed vortices in the wake are convected far above the plane of the airfoil. Therefore, not only the distance downstream of the trailing edge but also the height of the vortex is significant. The position of the shed vortices will be a strong influence on the airfoil upper-surface pressure distribution.

Stall Triggers

Attached Flow

The many schemes available to calculate the flow about unstalled airfoils remain valid for dynamic stall only until the angle of attack becomes sufficiently large to cause disturbances in the boundary layer. The boundary

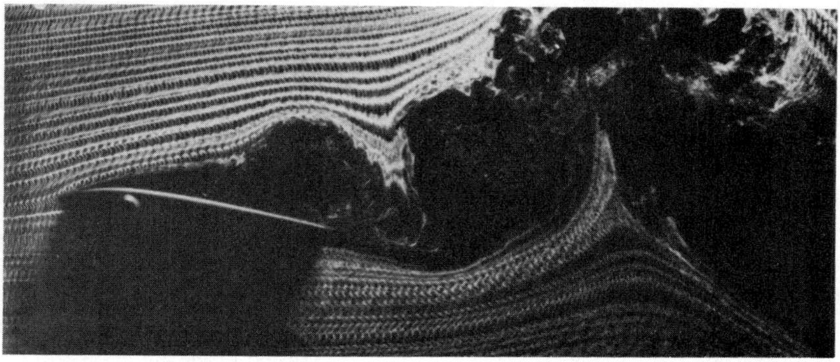

Figure 10. Water tunnel flow visualization of recovery from stall of an oscillating airfoil. [5.93].

layer is unsteady, and, at the Reynolds numbers of interest (about 1 to 8 million based on airfoil chord), it is laminar about the nose and turbulent at the trailing edge. Four boundary-layer phenomena have been identified as potential triggering mechanisms for the shedding of vortices:

1. Bursting of a bubble of separated flow that forms at moderate angles of attack on the airfoil upper surface at the boundary-layer transition point.
2. The arrival at the leading-edge region of a turbulent-boundary-layer flow-reversal point that is moving forward from the trailing edge.
3. Shock-wave–boundary-layer interaction behind the airfoil crest.
4. Acoustic waves underneath the airfoil propagating from the trailing edge upstream near the airfoil lower surface and perturbing the stagnation and flow-separation points.

The initiation of the vortex-shedding process may involve more than one of these mechanisms. The mechanism that first succeeds in triggering vortex shedding will depend on airfoil shape, Mach number, and Reynolds number.

Flow Reversal and Separation

The first stall trigger to be considered is the bursting of the separation bubble. For many airfoils the transition from laminar to turbulent boundary-layer flow is carried out in a separation bubble. The bubble begins at the separation point of the laminar boundary layer. At low angles of attack, the zone of flow separation remains small, and the flow above the separation zone changes from laminar to turbulent. The onset of turbulence will, in moderately adverse pressure gradients, allow the boundary layer to reattach and close the separation bubble. If the pressure gradient is so adverse that the flow cannot reattach, or if the displacement thickness grows so rapidly that the transition to turbulence is too far from the airfoil surface to reattach the boundary layer, then the bubble is said to have burst. The bursting is characterized by massive separation and vortex shedding. The displacement thickness of the separation bubble tends to alleviate the pressure gradient, making it less adverse. Figure 11 is taken from an early study by Shamroth and Kreskovsky [5.94] that ignored the pressure-gradient alleviation in calculating the separation bubble. The bubble grew far beyond any reasonable value without any sign of bursting. This anomaly was explained by experimental measurements using hot films on the airfoil upper surface by Carr, McAlister, and McCroskey [5.95] and by an analytical study by Scruggs, Nash, and Singleton [5.96]. These two works identified the second stall trigger.

The second stall trigger is the flow-reversal movement in the turbulent boundary layer on the airfoil upper surface. The progressive separation of the turbulent boundary layer from trailing edge to leading edge as the angle of attack increases is a common cause of stall in quasisteady experiments.

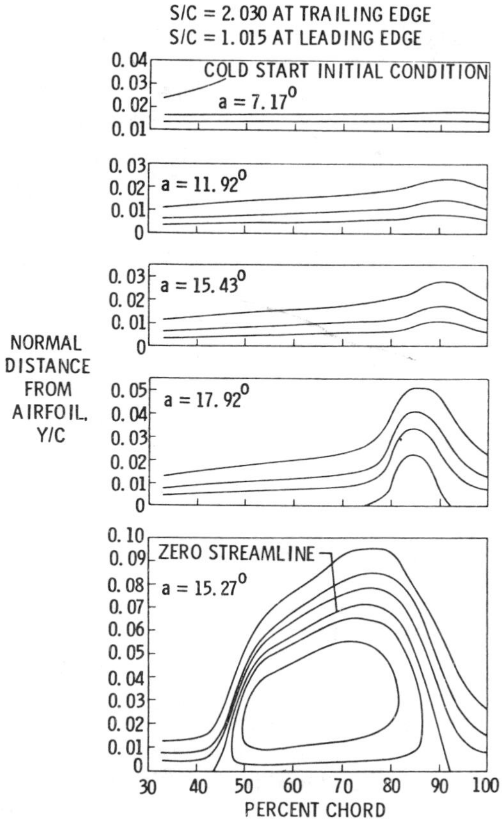

Figure 11. Calculated streamlines about a leading edge bubble for an oscillating airfoil. [5.94].

This behavior is illustrated in Fig. 12. For the dynamic-stall case the pitch rate affects both the potential flow and the turbulent boundary layer. Unsteady effects retard the forward movement of the separation point. However, once forward movement began, the speed calculated by this incompressible analysis could exceed the sonic speed for practical helicopter applications. Furthermore, the analysis showed that even though the flow near the wall was in the upstream direction, the viscous region remained thin and caused little disturbance in the pressure distribution. This flow reversal without massive separation is short-lived. Experiments showed that vortex shedding was soon initiated. The means of interaction between the flow reversal in the turbulent boundary layer and the separation bubble is not known, but flow reversal has been found to precede vortex shedding for most of the airfoils tested. Thus the separation of the compressible, turbulent, unsteady boundary layer and the transition in the separation bubble are important mechanisms in the initiation of dynamic stall.

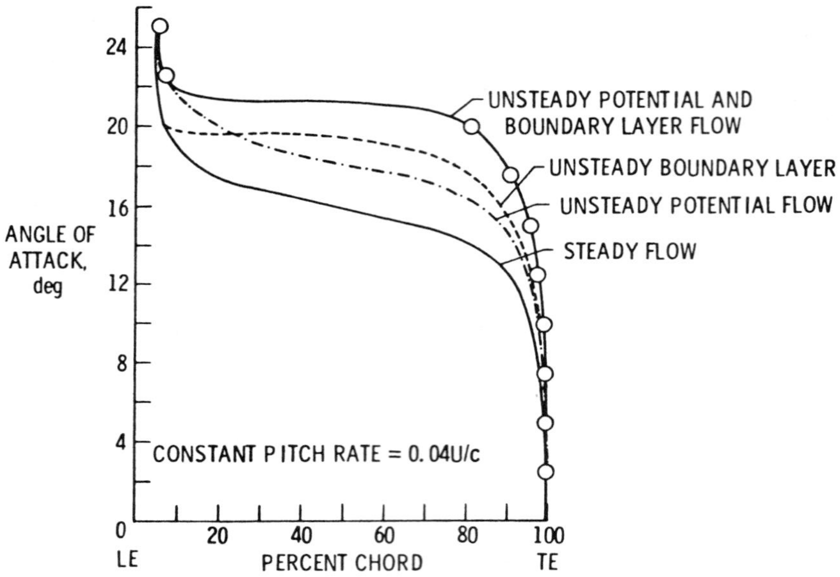

Figure 12. Movement of boundary layer flow reversal point on an airfoil pitching from 0° to 25° angle of attack. [5.96].

Mach-Dependent Stall Initiation

The last two stall triggers are deduced from static stall experiments. Schlieren flow visualizations, such as Fig. 13 (from [5.97]), have not yet been obtained for dynamic stall. Even though supersonic flow speeds over the airfoil crest have been measured for a freestream Mach number near 0.3, the results [5.85] show no shock-wave formation. The influence of the Mach number on the performance of an airfoil undergoing dynamic stall has been well documented [5.98], but the role of a shock in stall initiation has not.

Another possible, but unproven, stall trigger mechanism was identified in static stall by St. Hilaire, Carta, Fink, and Jepson [5.99]. A meticulous scrutiny of the airfoil surface pressures revealed that identifiable pulses in the pressure time histories were moving upstream on the airfoil lower surface. The speed of movement roughly coincided with the sonic speed minus the local flow velocity. The pulses were observed to move through the stagnation point with a 180°-phase shift and to go around the nose of the airfoil. Vortex shedding was usually preceded by the arrival of a pressure pulse, although not all pulses triggered vortex shedding. This phenomenon must depend strongly on the freestream Mach number. Thus the compressibility of the flow both inside and external to the viscous regions is an important fluid-flow mechanism, because both shock waves and acoustic waves have the potential of triggering vortex shedding.

α=0°;M=0.70

α=4°;M=0.70

α=8°;M=0.70

Figure 13. Schlieren visualization of shock induced separation in steady flow. [5.97].

Summary of Flow Phenomena

The pressure distribution on airfoils in static stall has been satisfactorily predicted by a combination of attached-boundary-layer calculations and an inviscid model with free shear layers. All static-stall models are explicitly steady-state in that the spontaneous shedding of vortices, such as those shed from a circular cylinder, are excluded from consideration. For dynamic stall the calculation of the shedding mechanism is essential, but the shedding of at least the initial vortex should be much easier to predict than spontaneous shedding in static stall. The list of possible triggering mechanisms will probably be reduced by additional experiments, but at least three fluid-flow mechanisms must be considered: (a) a separation bubble near the leading edge that begins at the laminar-boundary-layer separation point and ends in a turbulent reattached boundary layer, (b) a compressible, unsteady, turbulent boundary layer on both the upper and lower airfoil surfaces, (c) shock- or sonic-wave interaction with the boundary layer. Immediately after stall onset, a large shed vortex is convected over the airfoil upper surface. Laminar calculations for pitch motion show that the vortex shape and speed are strongly influenced by its close proximity to the airfoil upper surface. The process by which secondary vortices are shed, although calculable for laminar flow, may have a different trigger mechanism from the initial shed vortex. Certainly the pressure patterns on the upper surface of a plunging airfoil suggests a more complex pattern of vortex shedding and movement than in pitching motion. The reattachment of the flow and the formation of attached boundary layers is nonrepeatable from cycle to cycle. Speculative extrapolation from static stall suggests that free shear layers may form above the airfoil upper surface and behind the trailing edge before the flow reattaches.

Flow Modeling

The use to which a flow model might be put is one of the factors that determine the degree of completeness of the model needed. The lack of experimental measurements in the flow field puts severe demands on any analytical model of dynamic stall. The model must include the flow phenomena that can reasonably be expected to occur, without the usual guidance of quantitative experiments. In fact, one of the major needs for analytical calculations is to guide experiments. Although a model that is sufficiently complete to use with confidence at this time will not be efficient enough for rotor aeroelastic calculations, three continuing needs for the best possible analysis are foreseen. New airfoil designs or boundary-layer control devices which may change the stall trigger mechanism should first be analyzed by a theory which contains all the possible trigger devices. Extensions of the theory beyond two-dimensional flow is most safely done with a complete model. Finally, the increasing expense of wind-tunnel experiments justifies the most complete model for guidance of the tests. Careful experiments, and not just a complete analytical model, should be available to measure the accuracy of the simpler and cheaper analyses that are needed for rotor performance and aeroelastic calculations.

Based on these requirements and the present knowledge of flow phenomena, a complete model of airfoil dynamic stall must account for the following flow phenomena:

1. A compressible laminar boundary layer with a moving separation point or transition point.
2. A separation bubble with possible transition to turbulence.
3. A turbulent, compressible, unsteady boundary layer with moving separation and with flow reversal in the thin viscous layer.
4. Shock-wave interaction with the boundary layer.
5. The shedding of vorticity from the boundary-layer separation point or from the shock wave.
6. The movement of shed vorticity due to convection, diffusion, and shear forces.
7. Free shear layers above the airfoil upper surface and behind the trailing edge.
8. Induced velocities from previously shed vortices in a compressible freestream flow.
9. Acoustic-wave propagation below the airfoil.
10. Arbitrary airfoil motion in pitch and plunge.

Concluding Remarks

The fluid flow mechanisms which have been identified include both proven and potential factors to be included in the analysis of helicopter-blade dynamic stall. The uncertainties in the list of mechanisms are primarily due to the lack of flow-field measurements in the pertinent ranges of Mach and Reynolds numbers. Until further experimental results are available, the most promising means of testing the importance of an item on the list is a theoretical calculation that includes all the identified phenomena.

CHAPTER 36

On the Behavior of an Unsteady Turbulent Boundary Layer

P. G. Parikh, W. C. Reynolds, and R. Jayaraman *

1 Introduction

The term "unsteady turbulent flow" is applied to a turbulent flow field containing a well-defined, time-dependent, organized velocity component. Unsteady turbulent boundary layers occur in a variety of engineering applications. Boundary layers on the surface of a gas-turbine blade rotating in the wakes of stator vanes, on the surface of a helicopter rotor blade performing pitch oscillations, on the walls of a reciprocating-engine cylinder, and in nuclear-reactor components during transients are but a few examples where unsteadiness is imposed on a turbulent boundary layer.

Despite the obvious practical importance of unsteady turbulent boundary layers, very few basic experiments aimed at the development of a fundamental understanding of such flows were conducted until very recently. This was primarily due to the difficulties associated with analog processing of a velocity signal from such flows. The availability of digital signal-processing instrumentation in recent times has prompted several new experiments in the area of unsteady turbulent boundary layers.

In parallel with the development of advanced measurement and data-processing techniques, great strides have been made in the art of turbulent-boundary-layer prediction methods. Any computational scheme must, however, rely on a particular turbulence model which is based on available data for a given flow situation. Furthermore, the validity of any prediction

* Department of Mechanical Engineering, Stanford University, Stanford, CA 94305.

scheme can be checked only by comparison with available experimental data.

The objectives of the Stanford Unsteady Turbulent Boundary Layer Program are: to develop a fundamental understanding of such flows, to provide a definitive data base which can be used to guide turbulence-model development, and to provide a test case which can be used by computors for comparison with their predictions.

The earliest systematic investigation of unsteady turbulent boundary layers was by Karlsson [5.100], who subjected a flat-plate turbulent boundary layer to oscillation amplitudes up to 34% of the mean freestream velocity at frequencies ranging from 0 to 48 Hz. Karlsson found that imposed oscillations had practically no effect on the mean velocity profile in the boundary layer. The measurements of the amplitude and phase of the periodic component as well as of the turbulence intensity were rather uncertain, due to the limited accuracy of the signal-processing instruments employed.

More recently, several investigators have studied unsteady turbulent boundary layers experimentally. Patel [5.101] studied the behavior of a flat-plate turbulent boundary layer subjected to traveling waves in the freestream, and reported measurements of mean, periodic, and random parts of the streamwise velocity components. Kenison [5.102] continued the work of Patel [5.101] with adverse mean pressure gradient. Schachenmann and Rockwell [5.103] reported measurements of the boundary layer on the wall of a conical diffuser with unsteady through flow. The group of Cousteix et al. at Toulouse conducted experiments in both zero and adverse-mean-pressure-gradient boundary layers [5.104, 5.105]. In order to generate large amplitudes of imposed freestream oscillations, the latter experiments were run at a fixed frequency, which was the resonant frequency of the tunnel. Simpson [5.106] conducted experiments on separating adverse-pressure-gradient turbulent boundary layers. Unsteady-boundary-layer experiments with large imposed oscillation frequencies are currently under way at IIHR under the direction of Ramaprian and Patel [5.107]. The ranges of parameters employed in these experiments will be compared with those for the present experiment in Section 2.

All the experiments summarized above are characterized by unsteady flow at the inlet to the unsteady region. The distinctive feature of the present experiments is that the boundary layer at the inlet to the unsteady region is a standard, *steady*, flat-plate turbulent boundary layer. It is then subjected to well-defined and controlled periodic oscillations of the freestream. This feature is especially important from the point of view of a computer, which needs a rather precise specification of the initial conditions at the beginning of an unsteady computation.

2 Freestream Boundary Condition of the Present Experiment

A new water tunnel has been designed and built for the present experiments. The desired freestream velocity distribution in the tunnel is shown in Fig. 1. The freestream velocity remains *steady* and uniform for the first 2 meters of

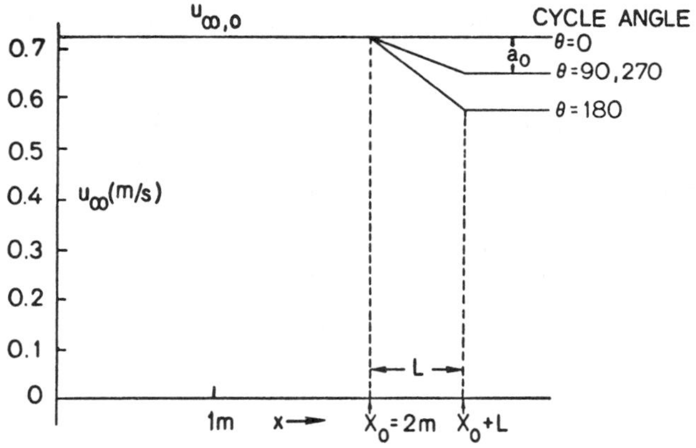

Figure 1. Desired free-stream velocity distribution.

boundary-layer development. It then decreases linearly in the test section so that the magnitude of the velocity *gradient* varies sinusoidally from zero to a maximum value during the oscillation cycle. The mean freestream velocity distribution in the test section is thus linearly decreasing and corresponds to the distribution at the cycle phase angle of 90°, while the amplitude of imposed freestream oscillations grows linearly in the streamwise direction, starting at zero at the entrance, to a maximum value of a_0 at the exit. Both the amplitude and frequency of the imposed freestream oscillations may be varied in the present experiment. The freestream velocity distribution is thus governed by the equation

$$u_\infty(x, t) = \begin{cases} u_{\infty,0} & \text{if } x < x_0, \\ u_{\infty,0} - \dfrac{a_0(x - x_0)}{L}[1 - \cos \omega t] & \text{if } x_0 \le x < x_0 + L. \end{cases}$$

The important parameters of this problem are:

1. an amplitude parameter

$$\alpha = \frac{a_0}{u_{\infty,0}},$$

and

2. a frequency parameter

$$\beta_\delta = \frac{f\delta_0}{u_{\infty,0}},$$

where $f = \omega/2\pi$ and δ_0 is the thickness of the boundary layer at the inlet to the unsteady region. The ranges of parameters in the present experiments are as follows:

$$u_{\infty,0} = 0.73 \text{ m/sec}, \quad x_0 = 2 \text{ m}, \quad L = 0.6 \text{ m}, \quad \delta_0 = 0.05 \text{ m},$$

$$0 < f < 2 \text{ Hz}, \qquad 0 < \alpha < 0.2, \qquad 0 < \beta_\delta < 0.14.$$

Table 1 Ranges of experimental parameters in various investigations.

Investigator	$\tilde{u}$ (m/sec)	f (Hz)	$\beta_\delta = \frac{f\theta\delta}{\bar{u}_\infty}$
Karlsson (1959)	5.0	0.33–48	0.0055–0.686
M. H. Patel (1977)	19.8	4–12	0.004–0.012
Kenison (1978)	22.0	0–6	0–0.005
Schachenmann & Rockwell (1976)	3.0	5–35	0.0072–0.058
Cousteix et al. (1977)	33.6	43	0.002
Cousteix et al. (1979)	30.0	38	0.002
Simpson (1977)	22.0	0.6	?
Ramaprian and V. C. Patel (1980)	1.0	0.1–8.0	0.005–0.38
Present	0.73	0.1–2.0	0.007–0.14

A comparison of the parameter range of this experiment with that of previous unsteady-boundary-layer experiments is shown in Table 1.

It should be mentioned that the value of the frequency parameter β_δ at the so-called "bursting frequency" in turbulent boundary layers is about 0.2 [5.108]. Thus the imposed oscillation frequencies used in the present experiments cover the range from quasisteady ($f \approx 0$) to values approaching the bursting frequency. The results reported here are for a nondimensional amplitude $\alpha = 0.05$. An amplitude of $\alpha = 0.15$ is sufficient to cause separation of the turbulent boundary layer in the test section during a part of the oscillation cycle. Results for large-amplitude experiments including periodic separation will be presented in a future publication.

3 Experimental Facility

A schematic of the water tunnel built for these experiments is shown in Fig. 2. The system operates at a constant head and a constant flow resistance, so that the total flow remains constant. The tunnel consists of a 16 : 1 nozzle contraction followed by a 2-m-long development section. The test boundary layer is grown on the top wall of the tunnel. The freestream velocity in the development section is maintained uniform along x by bleeding a small amount of flow from the bottom wall.

The linear decrease in freestream velocity in the test section is accomplished by uniformly bleeding off a fraction of the total flow through a perforated plate which forms the bottom of the test section. The remainder of the flow exits through a second perforated plate farther downstream. Flow through each perforated plate then passes on to the exit header, which contains two compartments separated by a partition plate. The flow from the two compartments exits the tunnel through a sliding-gate valve which contains a series of longitudinal slots which run perpendicular to the partition plate. This design ensures that, regardless of the position of the sliding-gate valve,

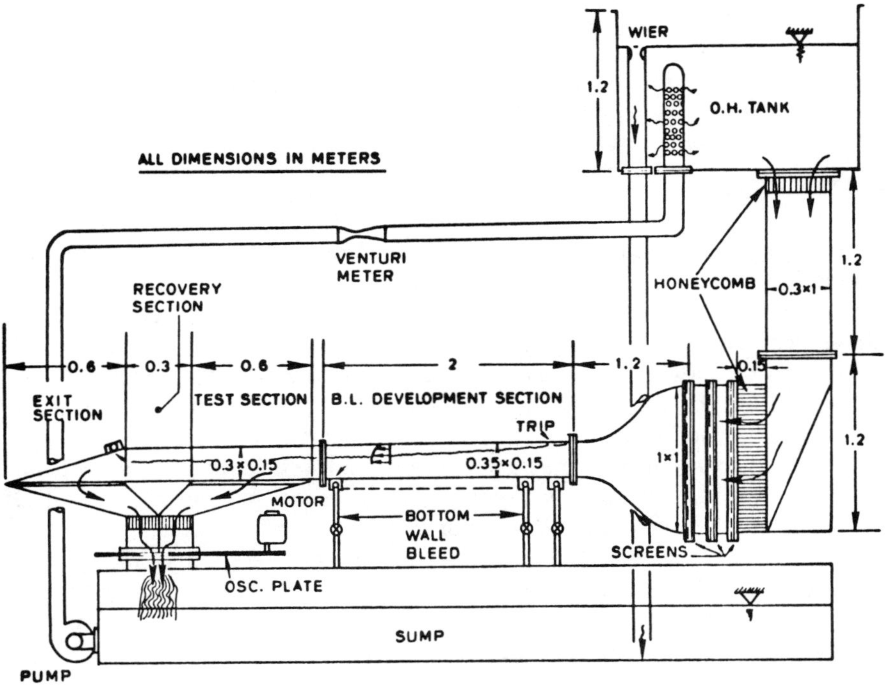

Figure 2. Schematic of the unsteady boundary layer water tunnel.

the total flow area of the slots remains the same. The slots are sized so that their total area constitutes the controlling resistance of the entire fluid circuit. As the sliding-gate valve is oscillated by means of a DC-motor-driven scotch yoke mechanism, the fraction of the total flow bled from the bottom wall of the test section varies in a harmonic fashion while the total flow through the tunnel remains constant. By this arrangement, a linearly decreasing periodic freestream distribution is established in the test section while the upstream flow (in the development section) remains steady.

4 Measurement and Data-Processing Techniques

The techniques of pitot, hot-film, and laser anemometry are employed for the measurement of velocity in the present experiments. The primary technique applied to date for unsteady measurements has been a single-channel, forward-scattering DISA laser anemometer with a frequency-shift capability.

Following Hussain and Reynolds [5.109], the instantaneous velocity signal from an unsteady turbulent flow may be decomposed into three parts:

$$u = \bar{u} + \tilde{u} + u', \tag{1}$$

where $\bar{u}$ is the mean, $\tilde{u}$ is the time-dependent, organized (deterministic) component, and u' is the contribution of random fluctuations. $\bar{u}$ may be determined by a simple long-time averaging of u. In the present experiments, $\tilde{u}$ is periodic, and it may be determined by first phase-averaging the instantaneous velocity signal and then subtracting out the mean. Thus,

$$\tilde{u} = \langle \mathrm{u} \rangle - \bar{u}, \tag{2}$$

where $\langle u \rangle$ is the phase average, which is determined by constructing an ensemble of samples taken at a fixed phase angle of the imposed oscillation cycle and then averaging each ensemble after a sufficient number of samples have been collected. In the present experiments, it was found that, for an imposed harmonic oscillation of the freestream (at $\alpha = 0.05$), the response at points within the boundary layer was almost sinusoidal, with a contribution of higher harmonics less than 5% of the fundamental. In view of this fact, the periodic component $\tilde{u}$ may also be extracted sufficiently accurately from the instantaneous signal u by a cross-correlation with the pulser signal. The latter technique was used in obtaining the results reported here.

A digital correlator (HP 3721A) was used for the determination of cross-correlations. More recently, a DEC MINC-11 laboratory minicomputer system has been in use for automatic data acquisition and processing, which includes the determination of the phase average $\langle u \rangle$.

5 Apparatus Qualification Tests

A series of qualification tests were conducted to ascertain satisfactory operation of the tunnel as designed, and modifications were made to improve its performance. Of particular importance was the requirement of a steady upstream flow while the flow in the test section was oscillatory. Trimming adjustments to the slot area of the pulser were required to ensure that the total flow remained constant and that the upstream propagation of the disturbances was acceptably small. An rms organized disturbance level of 0.3% was achieved, and this was comparable to the background freestream turbulence level in the development section.

Next, a series of boundary-layer mean-velocity profiles were measured in the development section. The local momentum-thickness Reynolds numbers (Re_{δ_2}) were computed from these profiles. The local skin-friction coefficient (C_f) was determined from the mean profiles by the Clauser plot technique, and the data were found to follow the correlation

$$C_f = 0.0128\ \mathrm{Re}_{\delta_2}^{-1/4},$$

which is accepted as a standard for a steady flat-plate turbulent boundary layer [5.110]. The mean-velocity profile at the entrance to the unsteady region (i.e., at $x = x_0$) had $\mathrm{Re}_{\delta_2} = 3000$ and, when plotted in velocity-defect coordinates, followed Coles's law of the wake.

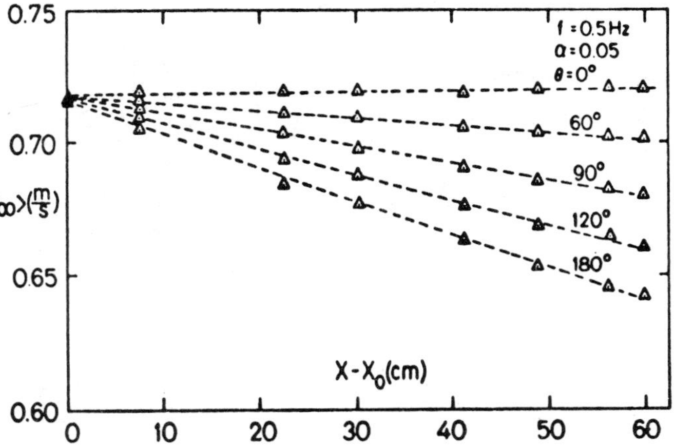

Figure 3. Phase-averaged free-stream velocity distribution.

The measured phase-averaged freestream velocity distribution is plotted in Fig. 3 with the cycle phase angle as the parameter. Notice that the distribution is quite linear and that the organized disturbance at $x = x_0$ is quite small (rms $\approx 0.003u_{\infty,0}$).

6 Experimental Results

The measurements reported here were taken at a fixed streamwise location near the end of the test section at $x - x_0 = 0.568$ m.

The measured mean-velocity profiles at three stationary-pulser positions $\theta = 0$, $90°$, $180°$ are shown as dashed curves in Fig. 4(a). These represent phase-averaged profiles at zero frequency, i.e., quasisteady profiles. They would indeed be the phase-averaged profiles at very low frequencies of imposed oscillations. At this value ($\alpha = 0.05$) of the amplitude of imposed oscillations, the response of the boundary layer is almost linear, so that the profile corresponding to $\theta = 90°$ lies nearly midway between the $\theta = 0$ and $180°$ profiles, and further that the $\theta = 90°$ profile represents the mean profile for quasisteady oscillations. Also, the difference between the $\theta = 0$ and $90°$ profiles at a fixed y-location represents the amplitude of quasisteady (very low frequency) oscillations at that location in the boundary layer. It may be seen that the quasisteady amplitudes in the boundary layer are larger than the freestream amplitude. We shall return to this point later on in the discussion.

The mean-velocity profiles measured under oscillatory conditions at different frequencies are also shown in Fig. 4(a). It may be seen that the profiles at various frequencies are identical with the one measured under stationary condition with pulser angle set at $\theta = 90°$. The latter profile, of course, represents the mean-velocity profile for quasisteady oscillations. It

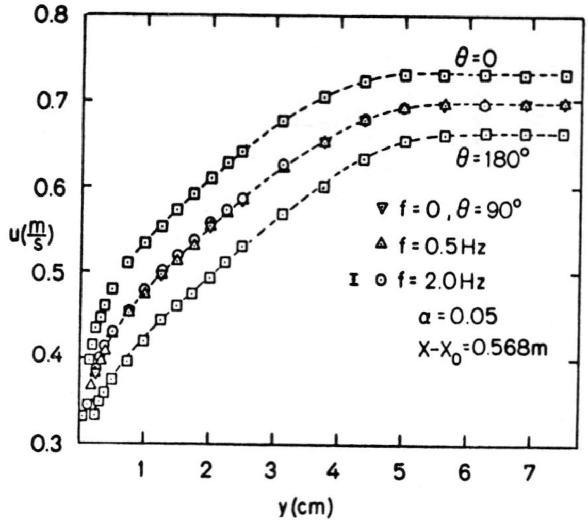

Figure 4. a) Mean velocity profiles (outer region).

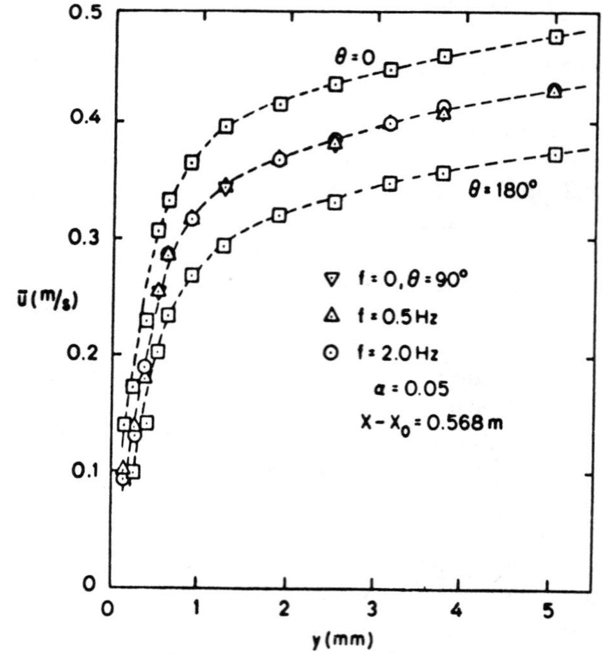

Figure 4. b) Mean velocity profiles (near-wall region).

may be concluded that the mean-velocity profile (at a fixed amplitude $\alpha = 0.05$) is independent of the imposed oscillation frequency in the entire range $0 \leq f \leq 2$ Hz. The same behavior persists all the way up to the wall, as shown in Fig. 4(b) for the near-wall region.

This behavior of the mean velocity profile may be explained by an examination of the governing equations. Insertion of the three-part decomposition of the velocity field, as shown in Eq. (1), into the momentum equation of the boundary layer and subsequent time-averaging yields the following equation for the mean field:

$$\bar{u}\frac{\partial \bar{u}}{\partial x} + \bar{v}\frac{\partial \bar{u}}{\partial y} = -\frac{1}{\rho}\frac{d\bar{p}}{dx} + \nu\frac{\partial^2 \bar{u}}{\partial y^2} - \frac{1}{\rho}\frac{\partial}{\partial y}[\overline{u'v'} + \overline{\tilde{u}\tilde{v}}]. \tag{3}$$

Equation (3) may be recognized as the same one governing an ordinary turbulent boundary layer, except for the addition of the term $\overline{\tilde{u}\tilde{v}}$, which represents Reynolds stresses arising from the organized fluctuations in the boundary layer.

Now, as long as boundary-layer approximations remain valid (i.e., before separation occurs), it may be assumed that the freestream pressure is impressed upon the boundary layer. The time-mean pressure gradient $d\bar{p}/dx$ may be shown to be independent of the imposed oscillation frequency and is indeed the same as that obtained with the pulser stationary at the $\theta = 90°$ position. Therefore, the mean velocity field will be frequency-dependent if and only if one or both of the following happen:

1. The distribution of Reynolds stress $\overline{u'v'}$ is altered under oscillatory conditions and is dependent on the frequency of imposed oscillations.
2. The Reynolds stress $\overline{\tilde{u}\tilde{v}}$ arising from organized fluctuations becomes significant compared with $\overline{u'v'}$.

We shall next show that neither one of the above two is the case.

Figure 5 shows the measured distribution of u'_{rms} under stationary conditions with the pulser at a mean position of $\theta = 90°$, as well as those measured

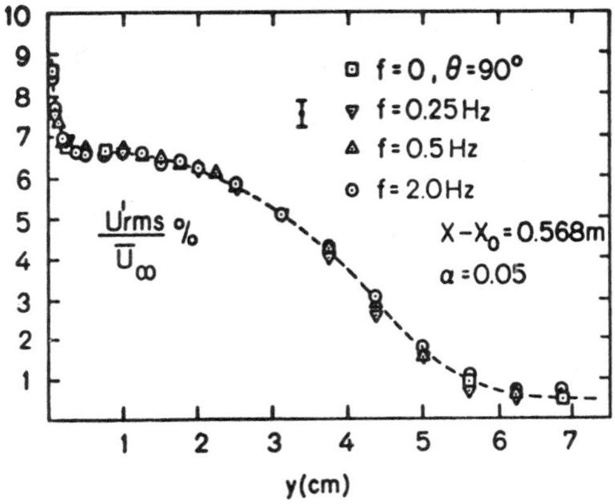

Figure 5. Profiles of longitudinal turbulence intensity.

under oscillatory conditions at frequencies up to 2 Hz. During measurements of u'_{rms} under oscillatory conditions, the contribution of the organized fluctuations was subtracted out, so that only the random part of the fluctuations is represented in Fig. 5. It may be seen that the distribution of u'_{rms} is independent of the imposed oscillation frequency, and further, that it is the same as that measured under stationary conditions with the pulser fixed at the mean position of $\theta = 90°$. As our present measurement system does not have the capability of measuring two components of velocity simultaneously, data on $\overline{u'v'}$ are not reported here. However, it is reasonable to assume that the $\overline{u'v'}$ distribution would also remain independent of the oscillation frequency and would be the same as that measured under stationary conditions at $\theta = 90°$.

In Fig. 6, a comparison is shown between measured values of $\overline{\tilde{u}\tilde{v}}$ at 2 Hz and data on $\overline{u'v'}$ obtained by Andersen et al. [5.111] in a steady adverse pressure gradient boundary layer at comparable conditions. The present data on $\overline{\tilde{u}\tilde{v}}$ were obtained by separate LDA measurements of $\tilde{u}$ and $\tilde{v}$ and their respective phases. It may be seen that the contribution of $\overline{\tilde{u}\tilde{v}}$ to the total Reynolds stress is insignificant over almost the entire boundary layer.

In view of the data presented in Figs. 5 and 6, it is not surprising that the mean-velocity profile remains independent of the imposed oscillation frequency and is indeed the same as the one measured under stationary conditions with $\theta = 90°$.

The behavior of the periodic component will next be examined. The profiles of amplitudes a_1 measured in the boundary layer and normalized by the freestream amplitude $a_{1,\infty}$ are shown in Fig. 7. The profile for quasi-steady ($f \approx 0$) oscillations was determined, as explained earlier, from the mean-velocity profiles measured under stationary conditions with the pulser

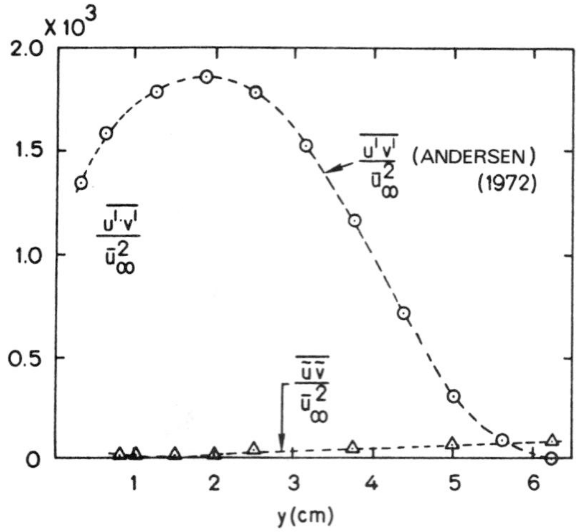

Figure 6. Comparison of Reynolds stress of organized fluctuations with that due to random fluctuations.

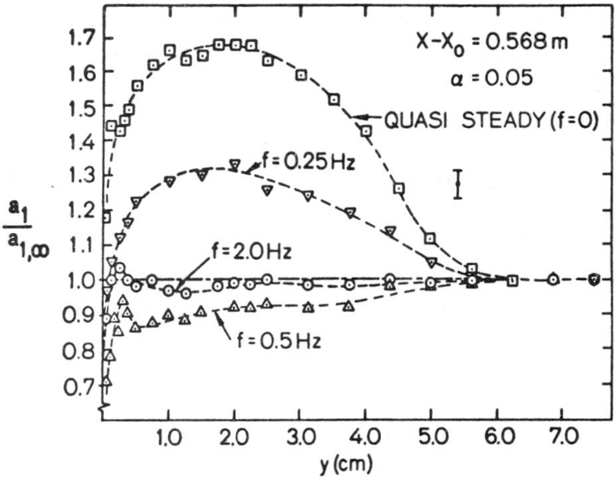

Figure 7. Amplitude distribution across the boundary layer.

set at $\theta = 0$, 90°, and 180° (see Fig. 4(a), (b)). Note that, during quasisteady oscillations, the amplitude in the boundary layer overshoots the freestream amplitude by as much as 70%. It may be mentioned that the data for $f = 0.1$ Hz, although not shown in Fig. 7, do indeed come very close to the quasisteady behavior.

The effect of dynamics is seen as the attenuation of the amplitudes within the boundary layer. Surprisingly, however, the present data show that the measured amplitudes for $f = 0.5$ are slightly below those for the $f = 2.0$-Hz case. At high frequencies of imposed oscillations, the behavior of the periodic component profile is clear: the amplitudes in most of the boundary layer are the same as the imposed freestream amplitudes. Near the wall the amplitude of the periodic component rapidly drops to zero.

The phase differences between the boundary-layer oscillations and freestream oscillations are shown in Fig. 8. For quasisteady oscillations,

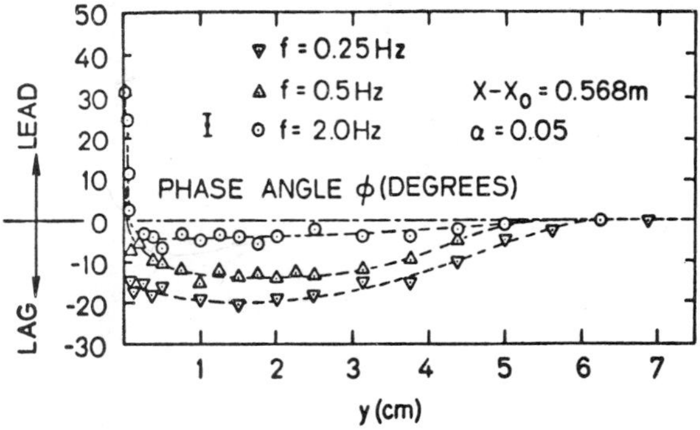

Figure 8. Phase variation across the boundary layer.

obviously, there is no phase difference. The largest phase lags (of the four cases) in the outer region of the boundary layer were observed for $f = 0.25$ Hz The effect of increasing the frequency is to reduce the phase lag in the outer region, but to introduce large phase *leads* in the region very close to the wall. Clearly, the asymptotic behavior of the *outer region* for large imposed frequencies of oscillation is once again a zero phase lag with respect to freestream oscillations, as in the quasisteady case.

The combination of the asymptotic behaviors of $a_1/a_{1,\infty}$ and ϕ in the outer region for large frequencies of imposed oscillations, together with the fact that the mean-velocity profile is unaffected by imposed oscillations, has the effect of freezing the boundary-layer thickness at large frequencies. This effect is shown in Fig. 9, where the instantaneous boundary-layer thickness $\langle\delta_{0.99}\rangle$ is plotted as a function of the cycle phase angle for several frequencies of imposed oscillations. The quasisteady behavior of $\langle\delta_{0.99}\rangle$ is quite obvious: at $\theta = 0$, the boundary layer in the test section continues to develop under a zero pressure gradient and is the thinnest at this point in the entire cycle. As the quasisteady phase angle is increased, pressure gradients of increasing adversity are imposed on the boundary layer, causing it to thicken. The maximum thickness is attained at $\theta = 180°$ when the magnitude of the adverse pressure gradient in the test section is the maximum. The quasisteady boundary-layer thickness oscillates 180° out of phase with the freestream oscillations.

Under oscillatory conditions at $f = 0.25$, 0.5, and 2.0 Hz, two things happen: a significant phase lag develops from quasisteady behavior, and the amplitude attenuates with increasing frequency. For the $f = 2.0$-Hz case, the variation over the complete cycle is less than 1 % and the boundary-layer thickness is practically frozen during the oscillation cycle.

It may be shown by a simple argument based on a mixing-length model of boundary-layer turbulence that the freezing of the boundary-layer thickness

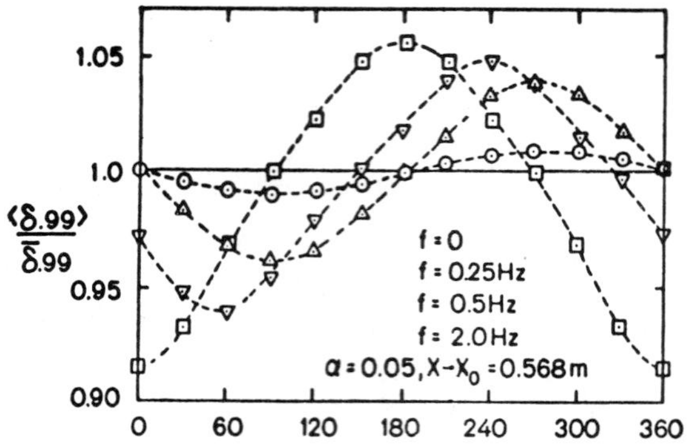

Figure 9. Variation of instantaneous boundary layer thickness over the oscillation cycle.

at large frequencies is also accompanied by freezing of the Reynolds stress over the oscillation cycle. To prove this, we hypothesize that the phase-averaged Reynolds-stress distribution may be related to the phase-averaged velocity profile in the same manner as for a steady boundary layer, i.e.,

$$\langle u'v' \rangle = \varepsilon_m \frac{\partial \langle u \rangle}{\partial y}, \tag{4}$$

where

$$\varepsilon_m = l^2 \left| \frac{\partial \langle u \rangle}{\partial y} \right|. \tag{5}$$

Now, in the *outer region* of the boundary layer, the mixing length l may be modeled as

$$l = \lambda \langle \delta_{0.99} \rangle \tag{6}$$

where λ is nearly a constant. Now,

$$\langle u \rangle = \bar{u} + \tilde{u} = \bar{u} + a_1(y) \cos[\omega t + \phi(y)]. \tag{7}$$

However, in the high-frequency limit,

$$\begin{aligned} a_1(y) &= a_{1,\infty} = \text{const}, \\ \phi(y) &= 0 \end{aligned} \tag{8}$$

and

$$\langle \delta_{0.99} \rangle = \delta_{0.99} = \text{const}.$$

Therefore

$$\frac{\partial \langle u \rangle}{\partial y} = \frac{\partial \bar{u}}{\partial y}. \tag{9}$$

Combining the above results,

$$\langle u'v' \rangle = \lambda^2 \delta_{0.99}^2 \left[\frac{\partial \bar{u}}{\partial y} \right]^2 = \overline{u'v'}, \tag{10}$$

i.e., the phase-averaged Reynolds stress in the outer region also becomes frozen at a mean value $\overline{u'v'}$.

Finally, the variation of the instantaneous skin-friction coefficient $\langle C_f \rangle$ is shown as a function of the cycle phase angle in Fig. 10. The instantaneous value of $\langle C_f \rangle$ was determined by the Clauser plot technique from the phase-averaged velocity profiles $\langle u \rangle$, assuming a law-of-the-wall equation for the log region:

$$u^+ = \frac{1}{0.41} \ln y^+ + 5.0. \tag{11}$$

The quasisteady variation of $\langle C_f \rangle$ is seen to be in phase with the freestream velocity oscillations, and the quasisteady amplitude is about 12% of the

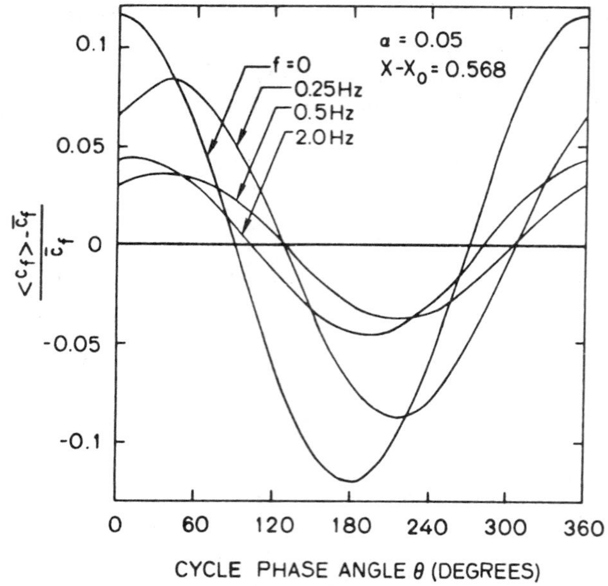

Figure 10. Variation of skin-friction as determined by Clauser plot technique over the oscillation cycle.

mean. The effect of dynamics is to introduce a phase shift and attenuate the amplitude of skin-friction variation over a cycle.

The asymptotic behavior of the skin-friction variation appears to be close to the variation shown for $f = 2.0$ Hz. It should be mentioned, however, that the applicability of the Clauser plot technique may be questionable in view of the large phase variations that are present in the near-wall region at high frequencies (see Fig. 8). Careful measurements in the region very close to the wall (i.e., the region between the wall and the log region) are currently under way to determine the dependence of $\bar{u}$, a_1, θ, and u' in this region on the frequency of imposed oscillations.

7 Conclusions

The conclusions from the first phase of the present experiments (at $\alpha = 0.05$) may be summarized as follows:

1. The mean velocity profile in the boundary layer is unaffected by imposed freestream oscillations in the range of frequencies employed, and it is in fact the same as the one measured with a freestream velocity distribution held steady at the mean value.
2. This behavior of the mean velocity field is a consequence of two observations: (a) the time-averaged Reynolds-stress distribution across the boundary layer is unaffected by the imposed oscillations and is indeed

the same as the one measured with the freestream velocity distribution held steady at the mean value; and (b) the Reynolds stresses arising from the organized velocity fluctuations under imposed oscillatory conditions are negligible compared to the Reynolds stresses due to the random fluctuations.

3. The amplitude of the periodic component in the boundary layer under quasisteady oscillations ($f \rightarrow 0$) is as much as 70% larger than the imposed freestream amplitude. However, at higher frequencies of imposed oscillations, the peak amplitude in the boundary layer is rapidly attenuated toward an asymptotic behavior where amplitudes in the outer region of the boundary layer become the same as the freestream amplitude, dropping off to zero in the near-wall region.
4. The quasisteady boundary-layer response is in phase with the imposed oscillations. As the frequency of imposed oscillations is increased, phase lags begin to develop in the outer region of the boundary layer. The magnitude of this phase lag reaches a maximum and then decreases with increasing frequency until an asymptotic limit is reached where the outer region once again responds in phase with the free stream. Near the wall, however, large lead angles are present at higher oscillation frequencies.
5. A consequence of conclusions 3 and 4 above is that the boundary-layer thickness becomes nearly frozen over the oscillation cycle at higher frequencies.
6. A consequence of conclusions 3, 4, and 5 above is that the Reynolds-stress distribution in the outer region of the boundary layer also becomes frozen over the oscillation cycle at higher frequencies.
7. The skin-friction coefficient determined by the Clauser plot technique from the phase-averaged profiles under oscillatory conditions shows that the amplitude of the skin-friction oscillations at the wall appears to reach an asymptotic value at high frequencies and that this asymptotic amplitude is considerably attenuated from the quasisteady amplitude.

Acknowledgment

This research is sponsored by the Army Aeromechanics Laboratory, NASA–Ames Research Center, and the Army Research Office. The authors wish to express their gratitude to Drs. Larry Carr and James McCroskey (AML), Mr. Leroy Presley (NASA–Ames), and Dr. Robert Singleton (ARO) for their continued support and cooperation.

References for Part 5

[5.1] S. J. Kline, M. V. Morkovin, S. Sovran, D. S. Cockrell (eds.): *Computation of Turbulent Boundary Layers—1968 AFOSR-IFP*-Stanford Conf. **I** (Stanford Press, Stanford 1969).

[5.2] S. J. Kline et al. (eds.): *Proceedings, 1980/81 AFOSR/HTTM*—Stanford Conf. on Complex Turbulent Flows: Comparison of Computation and Experiment (Stanford Press, Stanford 1980).

[5.3] G. L. Brown, A. Roshko: J. Fluid Mech. **64**, 775 (1974).

[5.4] C. D. Winant, F. K. Browand: J. Fluid Mech. **63**, 237 (1974).

[5.5] F. K. Browand, P. D. Weidman: J. Fluid Mech. **76**, 127 (1976).

[5.6] R. E. Kelly: J. Fluid Mech. **22**, 547 (1967).

[5.7] G. M. Corcos, F. S. Sherman: J. Fluid Mech. **73**, 241 (1976).

[5.8] J. J. Riley, R. W. Metcalfe: AIAA Paper 80-0274 (1980).

[5.9] C. M. Ho, L. S. Huang: J. Fluid Mech. **119**, 443 (1982).

[5.10] I. Wygnanski, D. Oster: J. Fluid Mech. (to appear).

[5.11] H. E. Fiedler: *Lecture Notes in Physics* (Springer-Verlag) (to appear).

[5.12] A. Michalke: J. Fluid Mech. **23**, 521 (1965).

[5.13] S. C. Crow, F. H. Champagne: J. Fluid Mech. **48**, 567 (1971).

[5.14] D. G. Crighton, M. Gaster: J. Fluid Mech. **77**, 397 (1976).

[5.15] C. K. W. Tam: J. Fluid Mech. **46**, 747 (1971).

[5.16] L. Merkine, J. T. C. Liu: J. Fluid Mech. **70**, 353 (1975).

[5.17] P. Dimotakis, G. L. Brown: J. Fluid Mech. **78**, 535 (1976).

[5.18] C. M. Ho, N. S. Nossier: J. Fluid Mech. **105**, 119 (1981).

[5.19] J. Laufer, P. Monkewitz: AIAA Paper 80-0962 (1980).

[5.20] V. Kibens: AIAA J. **18**, 434 (1980).

[5.21] E. Gutmark, C. M. Ho: Bull. Amer. Phys. Soc. **25**, 1102 (1980).

[5.22] R. A. Petersen: J. Fluid Mech. **89**, 469 (1978).

[5.23] G. E. Bouchard, W. C. Reynolds: Bull. Amer. Phys. Soc. **25**, 1086 (1980).
[5.24] P. Freymuth: J. Fluid Mech. **25**, 683 (1966).
[5.25] F. K. Browand: J. Fluid Mech. **26**, 281 (1966).
[5.26] R. W. Miksad: J. Fluid Mech. **56**, 695 (1972).
[5.27] C. J. Moore: J. Fluid Mech. **80**, 321 (1977).
[5.28] D. Bechert, E. Pfizenmaier: J. Fluid Mech. **72**, 341 (1975).
[5.29] P. Monkewitz, P. Huerre: Phys. Fluids (to appear).
[5.30] A. Michalke, U. Michel: J. Sound Vib. **67**, 341 (1979).
[5.31] L. Laufer: In: *Ommaggio a Carlo Ferrari*, 451 (1974).
[5.32] D. A. Johnson, W. D. Bachalo: AIAA J. **18**, 16 (1980).
[5.33] D. A. Johnson, W. D. Bachalo, F. K. Owen: AIAA Paper 79-1500 (1979).
[5.34] W. D. Bachalo, D. A. Johnson: AIAA Paper 79-1479 (1979).
[5.35] D. A. Johnson, C. C. Horstmann, W. D. Bachalo: AIAA Paper 80-1047 (1980).
[5.36] L. A. King, D. A. Johnson: AIAA Paper 80-1366 (1980).
[5.37] D. K. McLaughlin, W. G. Teiderman: Phys. Fluids **16**, 2082 (1973).
[5.38] D. Coles: J. Fluid Mech. **1**, 191 (1956).
[5.39] B. W. Spencer, B. G. Jones: AIAA Paper 71-613 (1971).
[5.40] F. W. Spaid, L. S. Stivers, Jr.: AIAA Paper 79-1501 (1979).
[5.41] I. M. M. A. Shabaka: Ph.D. Thesis, Imperial College, London (1978).
[5.42] I. M. M. A. Shabaka, P. Bradshaw: AIAA J. **19**, 131 (1981).
[5.43] I. Tanaka, T. Suzuki: In: *Proc. Internat. Symp. on Ship Resistance, SSPA*, Goteberg (1978).
[5.44] G. V. Lachman (ed.): *Boundary Layer and Flow Control*, 1277 (Pergamon Press, Oxford 1961).
[5.45] P. T. Harsha: Rept. AEDC-TR-71-36, Arnold Engineering Dev. Center (1971).
[5.46] N. Rajaratnam: *Turbulent Jets* (Elsevier Sci. Publ. Co., New York 1976).
[5.47] K. W. Everitt, A. G. Robins: J. Fluid Mech. **88**, 563 (1978).
[5.48] J. Bertin: Academie Sci. **240**, 1855 (1955).
[5.49] R. M. Lockwood: Rept. ARD-308, Hiller Aircraft Co. (1963).
[5.50] R. M. Curtet, J. P. Girard: In: *Proc. of ASME Symp. on Fluid Mechanics of Mixing*, 173 (1973).
[5.51] W. S. Johnson, T. Yang: ASME Paper 68-WA/FE-33 (1968).
[5.52] H. Didelle, G. Binder, A. Craya, R. Laty: Rept., Laboratories de Mecanique des Fluids, Univ. Grenoble (1972).
[5.53] G. Binder, M. Favre-Marinet: In: *Proc. of ASME Symp. on Fluid Mechanics of Mixing*, 67 (1973).
[5.54] K. Bremhorst, W. H. Harch: In: *Turbulent Shear Flows*, 37, F. Durst, B. E. Launder, F. W. Schmidt, J. H. Whitelaw (eds.) (Springer-Verlag, Berlin/Heidelberg 1979).
[5.55] K. Bremhorst, R. D. Watson: J. Fluids Eng. (to appear).
[5.56] J. M. Simmons, M. F. Platzer, T. C. Smith: J. Fluid Mech. **84**, 33 (1978).
[5.57] H. Viets: AIAA J. **13**, 1375 (1975).
[5.58] H. Viets: In: *AGARD Conf. Proc. on Turbulent Boundary Layers* (1979).
[5.59] M. F. Platzer, J. M. Simmons, K. Bremhorst: AIAA J. **16**, 282 (1978).
[5.60] E. F. Schum: In: *Proc. of the Workshop on Prediction Methods for Jet V/STOL Propulsion Aerodynamics, Naval Air Systems Command*, 639 (1975).
[5.61] H. Fiedler, D. Korschelt: In: *2nd Symp. on Turbulent Shear Flows*, Imperial College, London (1979).

[5.62] J. M. Simmons, J. C. S. Lai, M. F. Platzer: Rept. 10/79, Univ. of Queensland, Brisbane (1979).
[5.63] E. Forthmann: Ing. Archiv **5**, 42 (1934); also NACA TM 789 (1936).
[5.64] B. G. van der Hegge Zijnen: Appl. Sci. Res. **7**, 256 (1958).
[5.65] V. W. Goldschmidt, S. Eskinazi: J. Appl. Mech. **33**, 735 (1966).
[5.66] P. Bradshaw, D. G. Goodman: R&M 3527, Aero. Research Council (1968).
[5.67] L. G. Alexander, T. Baron, E. W. Comings: Bull. 413, Univ. of Illinois (1953).
[5.68] L. N. Krause, T. J. Dudzinski, R. C. Johnson: Prog. in Astro. & Aero. **34**, 193 (1974).
[5.69] D. R. Miller, E. W. Comings: J. Fluid Mech. **3**, 1 (1957).
[5.70] W. Hoesel, W. Rodi: In: *Proc. of the LDA-Symposium, Copenhagen*, 251 (1975).
[5.71] G. R. Salter: ARL-TR-75-0132, A. F. Aerospace Research Labs (1975).
[5.72] H. Tennekes, J. L. Lumley: *A First Course in Turbulence* (MIT Press, Cambridge 1972).
[5.73] J. C. S. Lai, J. M. Simmons: Rept. NP S67-80-015, Naval Postgraduate School (1980).
[5.74] S. C. Carrion: M. S. Thesis, Calif. State Univ., Long Beach (1981).
[5.75] A. Melling, J. H. Whitelaw: J. Fluid Mech. **78**, 289 (1976).
[5.76] J. A. C. Humphrey, A. M. K. P. Taylor, J. H. Whitelaw: J. Fluid Mech. **83**, 509 (1977).
[5.77] J. A. C. Humphrey, J. H. Whitelaw, G. Yee: J. Fluid Mech. **103**, 443 (1981).
[5.78] F. Durst, A. Melling, J. H. Whitelaw: *Principles and Practice of Laser-Doppler Anemometry*, 2nd ed. (Academic Press, 1981).
[5.79] A. M. K. P. Taylor, J. H. Whitelaw, M. Yianneskis: Rept. FS/80/30, Imperial College, London (1980): also NASA Rept. 3367.
[5.80] Y. Agrawal, L. Talbot, K. Gong: J. Fluid Mech. **85**, 497 (1978).
[5.81] P. R. Eisman, R. Levy, H. McDonald, W. R. Briley: NASA CR 3029 (1978).
[5.82] E. Mayer: VDI-Forschungsheft 389 (1938).
[5.83] P. Crimi: NASA CR-2573 (1975).
[5.84] P. Crimi: NASA CR-2322 (1973).
[5.85] W. J. McCroskey, K. W. McAlister, L. W. Carr, S. L. Pucci, O. Lambert, R. F. Indergand: AHS 80-1 (1980).
[5.86] U. B. Mehta: AGARD CP-227 (1977).
[5.87] F. O. Carta: NASA CR-2394 (1974).
[5.88] F. O. Carta: NASA CR-3172 (1979).
[5.89] L. Gray, J. Liiva, F. J. Davenport: Rept. 68-89A, USAAVLABS (1969).
[5.90] W. H. Young, Jr., D. R. Hoad: AIAA Paper 79-0147 (1979).
[5.91] S. F. Birch, J. M. Eggers: NASA SP-321, 11 (1972).
[5.92] B. Maskew, F. A. Dvorak: J. Amer. Helicopter Soc. **23**, 2 (1978).
[5.93] K. W. McAlister, L. W. Carr: NASA TM-78446 (1978).
[5.94] S. J. Shamroth, J. P. Kreskovsky: NASA CR-132425 (1974).
[5.95] L. W. Carr, K. W. McAlister, W. J. McCroskey: NASA TN D-8382 (1974).
[5.96] R. M. Scruggs, J. F. Nash, R. E. Singleton: NASA CR-2462 (1974).
[5.97] C. L. Ladson: NASA TN D-7182 (1973).
[5.98] L. Gray, J. Liiva: Rept. 68-13B, USAAVLABS (1968).
[5.99] A. O. St. Hilaire, F. O. Carta, M. R. Fink, W. D. Jepson: NASA CR-3092 (1979).
[5.100] S. K. F. Karlsson: J. Fluid Mech. **5**, 622 (1959).
[5.101] M. H. Patel: Proc. Royal Soc. London A **353**, 121 (1977).

[5.102] R. C. Kenison: AGARD Conf. Proc. No. 227 (1978).
[5.103] A. A. Schachenmann, D. O. Rockwell: J. Fluid Eng., **98**, 695 (1976).
[5.104] J. Cousteix, A. Desopper, R. Houdeville: In: *Turbulent Shear Flows—I*, F. Durst et al. (eds.) (Springer-Verlag, 1979).
[5.105] J. Cousteix, R. Houdeville, M. Raynaud: 2nd Symp. on Turbulent Shear Flows, Imperial College, London (1979).
[5.106] R. L. Simpson: AGARD Conf. Proc. No. 227 (1978).
[5.107] B. R. Ramaprian, V. C. Patel: Private communication, Univ. of Iowa (1980).
[5.108] K. N. Rao et al.: J. Fluid Mech. **48**, 339 (1971).
[5.109] A. K. M. F. Hussain, W. C. Reynolds: J. Fluid Mech. **41**, 241 (1970): **54**, 241 (1972).
[5.110] W. M. Kays, R. J. Moffat: Rept. HMT-20, Stanford University (1975).
[5.111] P. S. Andersen, W. M. Kays, R. J. Moffat: Rept. HMT-15, Stanford University (1972).